International Series in Operations Research & Management Science

Volume 331

The book series **International Series in Operations Research and Management Science** encompasses the various areas of operations research and management science. Both theoretical and applied books are included. It describes current advances anywhere in the world that are at the cutting edge of the field. The series is aimed especially at researchers, advanced graduate students, and sophisticated practitioners.

The series features three types of books:

• Advanced expository books that extend and unify our understanding of particular areas.

• Research monographs that make substantial contributions to knowledge.

• Handbooks that define the new state of the art in particular areas. Each handbook will be edited by a leading authority in the area who will organize a team of experts on various aspects of the topic to write individual chapters. A handbook may emphasize expository surveys or completely new advances (either research or applications) or a combination of both.

The series emphasizes the following four areas:

Mathematical Programming : Including linear programming, integer programming, nonlinear programming, interior point methods, game theory, network optimization models, combinatorics, equilibrium programming, complementarity theory, multiobjective optimization, dynamic programming, stochastic programming, complexity theory, etc.

Applied Probability: Including queuing theory, simulation, renewal theory, Brownian motion and diffusion processes, decision analysis, Markov decision processes, reliability theory, forecasting, other stochastic processes motivated by applications, etc.

Production and Operations Management: Including inventory theory, production scheduling, capacity planning, facility location, supply chain management, distribution systems, materials requirements planning, just-in-time systems, flexible manufacturing systems, design of production lines, logistical planning, strategic issues, etc.

Applications of Operations Research and Management Science: Including telecommunications, health care, capital budgeting and finance, economics, marketing, public policy, military operations research, humanitarian relief and disaster mitigation, service operations, transportation systems, etc.

This book series is indexed in Scopus.

Chandrasekar Vuppalapati

Artificial Intelligence and Heuristics for Enhanced Food Security

Springer

Chandrasekar Vuppalapati
San Jose State University
San Jose, CA, USA

ISSN 0884-8289 ISSN 2214-7934 (electronic)
International Series in Operations Research & Management Science
ISBN 978-3-031-08742-4 ISBN 978-3-031-08743-1 (eBook)
https://doi.org/10.1007/978-3-031-08743-1

This Springer imprint is published by the registered company Springer Nature Switzerland AG
The registered company address is: Gewerbestrasse 11, 6330 Cham, Switzerland

The author dedicates his works, book, and efforts of this book to his late father Murahari Rao Vuppalapati, who had inspired him to write the book and guided the author's pursuits of life, and to his loving mother Hanumayamma Vuppalapati, who was the inspiration to develop company and dairy products for the betterment of humanity and future generations.

Preface

The world is at a critical juncture. In 2020, nearly one in three people did not have access to adequate food—between 720 and 811 million people faced hunger. Compared with 2019, 46 million more people in Africa, almost 57 million more in Asia, and about 14 million more in Latin America and the Caribbean were affected by hunger. Based on forecasts of global population growth, current deficit to feed people around the world, and increased demand for greener fuel and biodiesel, food security will remain an important economic development issue over the next several decades. As the food-versus-fuel tension becomes more intense, the day will come when more agricultural products will be used for energy than food. Adding to the conundrum, the COVID-19 pandemic has changed the face of the earth in terms of supply chain, resource availability, and human labor and has exposed our vulnerabilities in food security to an even greater extent. In essence, humanity is at a critical juncture, and what this unprecedented movement in our lives has thrusted upon us—the practitioners of the agriculture and technologists of the world—is to innovate and become more productive to address the multipronged food security challenges.

Agricultural innovation is key to overcoming concerns of food security; the infusion of data science, artificial intelligence, sensor technologies, and heuristics modeling with traditional agricultural practices such as soil engineering, fertilizers, and agronomy is one of the best ways to achieve this. Data science helps farmers to unravel patterns in equipment usage, transportation and storage costs, yield per hectare, and weather trends to better plan and spend resources. Artificial intelligence (AI) enables farmers to learn from fellow worldwide farmers to apply best techniques that are transferred learning from the AI. The sensor technologies play an important role in getting real-time farm field data and provide feedback loops to improve overall agricultural practices and can yield huge productivity gains. Heuristic modeling is an essential software technique that codifies farmers' tacit knowledge such as better seed per soil, better feed for dairy cattle breed, or production practices to match weather pattern that was acquired over years of their hard work to share with worldwide farmers to improve overall production efficiencies, the best antidote to the food security issue. In addition to the paradigm shift, economic sustainability of small farms is a major enabler of food security. As part of the book, I have proposed macroeconomic pricing models that data-mine macroeconomic signals and the influence of global economic trends on small farm sustainability to provide actionable insights to farmers to avert any financial disasters due to recurrent economic crises.

The mission of the book, *Artificial Intelligence and Heuristics for Enhanced Food Security*, is to prepare current and future software engineering teams with the skills and tools to fully utilize advanced data science, artificial intelligence, heuristics, mathematical optimization, linear programming (LP), constraint programming (CP), mixed-integer programming (MIP), mathematical solvers, and economic models to develop software capabilities that help achieve sustained food security for years to come and ensure enough food for the future of the planet and generations to come! Ensuring food security is my personal and professional obligation! The sole theme of the works of the book is to help achieve food security, making the world a better place than what I have inherited. The book also covers:

- Advanced data science techniques for signal mining agricultural practices and policies
- Linear programming, mixed-integer programming, constraint programming, and AI/ML linkage models
- Heuristic modeling to improve farm-level efficiency
- Sensors and data intelligence to provide closed-loop feedback
- Recommendation techniques to provide actionable insights

San Jose, CA Chandrasekar Vuppalapati

Acknowledgments

The author is sincerely thankful to all the government, world trade, banks, food organizations, federal and state government departments, public agencies, private industries, and nongovernmental organizations (NGOs) that are providing services to the underserved and vulnerable communities and salute their efforts and unwavering determination to defeat global poverty and hunger and to ensure food for all. The author is completely indebted to the Department of Pension & Pensioners' Welfare of the Government of India that has helped him and his family survive the family crisis of 1990 when the author, in his teenage, lost his father to a heart attack. Timely pension enabled the author's family to survive through difficulties and ensure sufficient food available to meet day-to-day needs!

The author is deeply indebted to the love, support, and encouragement of his wife Anitha Ilapakurti and the active support of his daughters, Sriya Vuppalapati and Shruti Vuppalapati, who helped draft the first version of the manuscript.

The author is deeply thankful to the support of his elder brother, Rajasekar Vuppalapati, and the joyful support of his younger brother, Jaya Shankar Vuppalapati. Additionally, the author is also very thankful to his elder sisters Padmavati Vuppalapati and Sridevi Vuppalapati.

The author sincerely thanks Santosh Kedari and his team in India for conducting field studies in India for Dairy and Sanjeevani Healthcare analytics products and bringing the market feedback for the betterment of services.

The author is deeply indebted to Sharat Kedari, who helped in setting Cloud and machine learning algorithms and developed economic, climate, and ML models. In addition, the author is thankful for the support, hard work, and dedication of Vanaja Satish who helped develop ML models, climate change models, and pricing models. Finally, the author is sincerely and deeply thankful to Sandhya Vissapragada for her contributions in data science and in creating machine learning models on Microsoft Azure and Google Cloud and developed climate change models, economic models, and deployment models.

Contents

Abbreviations

ABARES	Australian Bureau of Agricultural and Resource Economics and Sciences
AI	Artificial intelligence
AMS	Agricultural Marketing Service
APTMA	All-Pakistan Textile Mills Association
ARIMA	Autoregressive integrated moving average
ASI	Agricultural Stress Index
ASTER	Advanced Spaceborne Thermal Emissions Radiometer
AVHRR	Advanced very-high-resolution radiometer
BBCH	Biologische Bundesanstalt, Bundessortenamt und Chemische Industrie
BEA	US Bureau of Economic Analysis
BLS	Bureau of Labor Statistics
BOP	Balance of payments
CAADP	Comprehensive Africa Agriculture Development Programme
C-ATBD	Climate Algorithm Theoretical Basis Document
CBLD	IMF's Central Bank Legislation Database
CBOT	Chicago Board of Trade
CDF	Cumulative distribution function
CDR	Climate data record
CMIP6	Coupled Model Intercomparison Project Phase 6
CRI	Climate Risk Index
CVT	Computer vision
DMSP	Defense Meteorological Satellite Program
ECMWF	European Centre for Medium-Range Weather Forecasts
ECOSTRESS	ECOsystem Spaceborne Thermal Radiometer Experiment on Space Station
FANTA	Food and Nutrition Technical Assistance
FAO	Food and Agriculture Organization

FCI	Food Corporation of India
FEWS NET	Famine Early Warning Systems Network
FFPI	FAO Food Price Index
FIES	Food Insecurity Experience Scale
FRED	Federal Reserve Economic Data
FSIN	Food Security Information Network
FSNAU	FAO's Food Security and Nutrition Analysis Unit
GAIN	Global Agricultural Information Network
GDP	Gross domestic product
GFCF	Gross fixed capital formation
GHCN	Global Historical Climatology Network
GHCNm	The Global Historical Climatology Network monthly
GIEWS	Global Information and Early Warning System
GNI	Gross national income
GOES	Geostationary Operational Environmental Satellites
GOFI	IGC Grains and Oilseeds Freight Index
GOI	IGC Gains and Oilseeds Index
HER	Electronic Health Records
ICE Futures	Intercontinental Exchange
ICT	Information communication technology
IEA	International Energy Agency
IFAD	International Fund for Agricultural Development
IFPRI	International Food Policy Research Institute
IFS	International Financial Statistics
IGC	International Grains Council
IIP	International Investment Position[1]
ILO	Institute of Labor Organization
ILO	International Labour Organization
IMD	India Meteorological Department
IMF	International Monetary Fund
IPC	Integrated Food Security Phase Classification
IPCC	Intergovernmental Panel on Climate Change
KCBT	Kansas City Board of Trade
KL divergence	Kullback–Leibler divergence
LAC	Latin America and the Caribbean
LACIE	Large Area Crop Inventory Experiment
M.A.S.L	Meters above sea level
MEI	Main economic indicators
METOP	Meteorological operational satellite
MFCC	Mel-frequency cepstral coefficients
MGEX	Minneapolis Grain Exchange
MGP	Minimum guarantee price

[1] The IIP—https://data.imf.org/?sk=7A51304B-6426-40C0-83DD-CA473CA1FD52

ML	Machine learning
MLP	Multilayer perceptron
MODIS	Moderate-resolution imaging spectroradiometer
MSP	Minimum support price
MTBE	methyl tertiary butyl ether
NCEI	National Centers for Environmental Information
NCHS	National Center for Health Statistics
NEBR	National Bureau of Economic Research
NEER	Nominal effective exchange rate
NHANES	National Health and Nutrition Examination Survey
NLP	Natural language processing techniques
NOAA	National Oceanic and Atmospheric Administration
NQF	National Quality Forum
OECD	Organisation for Economic Co-operation and Development
OLS	Ordinary least squares
P.L.480[2] USAID and PL-480, 1961–1969	The administrations of John F. Kennedy and Lyndon B. Johnson marked a revitalization of the US foreign assistance program, signified a growing awareness of the importance of humanitarian aid as a form of diplomacy, and reinforced the belief that American security was linked to the economic progress and stability of other nations
PCA	Principal component analysis
PCE	Personal consumption expenditures
PDF	Probability distribution function
POES	Polar Orbiting Environmental Satellites
PPP	Purchasing power parity
PS&D	Production, supply, and distribution online
Public Debt Act,[3] 1944	The Public Debt Act, 1944, was an act of the Parliament of India which provided a legal framework for the issuance and servicing of government securities in India. The Act oversees government securities and their management by the Reserve Bank of India
RBI	Reserve Bank of India
RMS	Root mean square
S&P 500 Index	Standard & Poor's Financial Index
SCAA	Specialty Coffee Association of America
SDG	Sustainable Development Goals
SDoH	Social determinant of health

[2]USAID & PL-480—https://history.state.gov/milestones/1961-1968/pl-480

[3]Public Debt Act, 1944—https://www.indiacode.nic.in/handle/123456789/2405?view_type=browse&sam_handle=123456789/1362

SDR[4]	Special Drawing Right[5]—the SDR is an international reserve asset, created by the IMF in 1969 to supplement its member countries' official reserves. The value of the SDR is based on a basket of five currencies—the US dollar, the euro, the Chinese renminbi, the Japanese yen, and the British pound sterling
SMR	Steam–methane reforming
SNA	System of National Accounts
SOLAW	The state of the world's land and water resources for food and agriculture: systems at breaking point
SSP	Shared Socioeconomic Pathways
STAR	NOAA Satellite Application Research
SVM	Support vector machines
The HUNT Study	The Trøndelag Health Study (The HUNT Study)
TS	Time series
UN	The United Nations
UNFCCC	United Nations Framework Convention on Climate Change
UNFPA	United Nations Population Fund
UNICEF	United Nations International Children's Emergency Fund
UNSD	United Nations Statistical Division
USDA	US Department of Agriculture
USDA FAS	USDA Foreign Agricultural Service
USDA GAIN	Global Agricultural Information Network
VAR	Vector autoregression
Vicofa	Vietnam Coffee and Cocoa Association
VIIRS	Visible Infrared Imaging Radiometer Suite
WAOB	USDA's World Agricultural Outlook Board
WASDE	World Agricultural Supply and Demand Estimates
WDI	World Development Indicators
WFP	World Food Programme
WGS84	World Geodetic System 1984
WHO	World Health Organization
WPS	World Bank Policy Research Working Paper
WTO	World Trade Organization

[4] SDR Valuation—https://www.imf.org/external/np/fin/data/rms_sdrv.aspx

[5] SDR—https://www.imf.org/en/About/Factsheets/Sheets/2016/08/01/14/51/Special-Drawing-Right-SDR

Part I
Introduction to Artificial Intelligence and Heuristics

Chapter 1
Introduction

This chapter covers:

- What is Artificial Intelligence (AI) and Heuristics?
- Machine Learning (ML)
- Heuristics
- Exploratory Data Analysis (EDA)
- Machine Learning Technique: Association Mining
  - Egypt, Risk Modeling: Wheat Import Origins—Food Security
- Machine Learning Technique: Clustering
  - Egypt, Insecurity in the World's Food Supply: New Import Countries Based on Cluster Technique
- Machine Learning Cluster Model: Wheat Import New Origins
- Machine Learning Insights

This chapter introduces Artificial Intelligence (AI) and covers techniques that constitute AI and heuristics. Next, it introduces techniques to process agricultural and climatology datasets and walks through satellite radiometer data processing. Next, the chapter introduces association mining and clustering techniques. The chapter concludes with importer origin for wheat commodity risk modeling for Egypt.

Classical information computing systems perform actions through software instructions or code. These instructions are either compiled and executed on software stack or embedded as part of the hardware instructions. The system's capacity to learn and behave autonomously is controlled by the developer of the software or

C. Vuppalapati, *Artificial Intelligence and Heuristics for Enhanced Food Security*, International Series in Operations Research & Management Science 331, https://doi.org/10.1007/978-3-031-08743-1_1

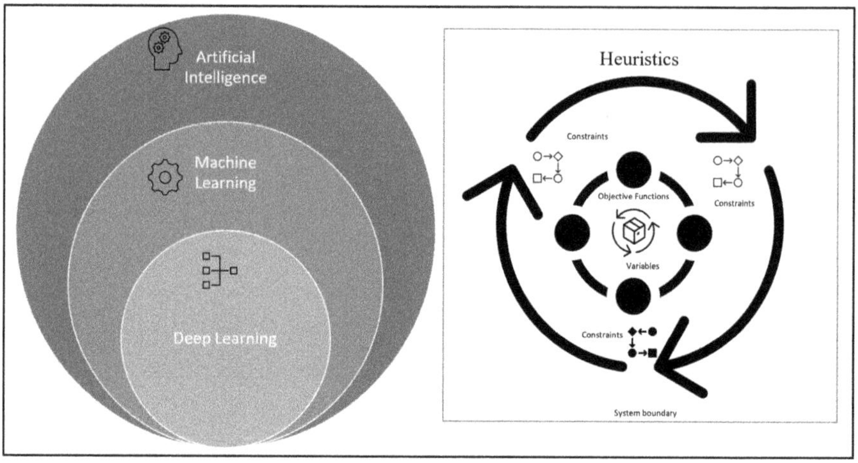

Fig. 1.1 Artificial Intelligence and heuristics

hardware of the classical information systems. These systems' scope is very limited to the specification it is built to. Any deviation from the specification is error, not meeting the published service level agreement (SLA), and, hence, defective. Artificial Intelligence systems, on the other hand, learn from data and act based on the data patterns or data signals. Learning from the data is unlimited so as AI systems capabilities. The self-learning of the AI systems is what makes it a transformational technology of our digital age—transformational as it can be adaptive, autonomous, and augmentative to human needs. In the truest sense, AI systems are designed to help humans and carry out tasks in a very much human cognitive manner, albeit it can function without wear and tear for 24/7 and 365 days. One of the chief reasons why AI applications are getting prominence and industry acceptance is in its software's ability to learn, albeit continuously, from real-world use and experience and its capability to improve its performance. It is no wonder that the applications of AI span from complex high-technology equipment manufacturing to personalized exclusive recommendations.

The Artificial Intelligence (AI) system is comprised of machine learning and deep learning. AI systems are any systems that power through data and enable computers to mimic human behavior. The AI systems morph into the following patterns (please see Fig. 1.1):

- Language understanding and generating systems such as natural language processing (NLP) and natural language generating (NLG) systems
- System that can sense—see/vision—like human,—for example, computer vision (CV) cognitive systems
- Systems that can understand and translate different human languages—cognitive speech to text (speech detection)
- Emotionally understand and derive sentiments

To mimic human behavior, the AI systems need contextualization and transfer of knowledge of humans to perceive and mimic human behavior. It is complex and difficult to understand human behavior holistically. To aid machines to cognitively behave like humans, the AI systems are trained using the following techniques:

Supervised—instructed and trained with annotated behavior data
Unsupervised—learning from the signatures of the data to behave
Reinforcement-based—incentive-driven for actions performed correctly
Mix of supervised and unsupervised—a process called semi-supervised
Network-driven—like human neurons, a process called neural network-based deep learning

AI systems learn from data. So as the data signature changes, the model to retrain to learn new signals could be applied to all the facets of AI.

The AI systems can leverage techniques from statistical methods to improve machines with experience. As described above, supervised, unsupervised, and semi-supervised techniques enable machines to understand and learn signals from the data. During machine learning process, data preparation is generally the most time-consuming process. It takes several iterations to get the base data that could be fed to machines to prepare the models. The process of preparation of data from raw mode to curated form is undertaken by data engineering systems. A commonly used technique such as exploratory data analysis is applied extensively. More will be described in the next section.

Deep learning systems are a subset of machine learning that make computation on multilayered networks that synthesize the way the human brain learns and makes decisions. Deep learning employs network-based learning.

Heuristics (please see Fig. 1.2) are specific to a system that is being modeled; the same goes for the AI systems. The system boundary is a process that is being modeled. Examples include optimization of medical equipment dispensing and just-in-time supply chain analytics, automobile line balancing, agricultural tractor manufacturing and line balance, optimization of agricultural commodity transportation, and food and nutrition diet problems. Heuristics are applied in many industrial cases, and they are, in most cases, an adjunct of machine learning and Artificial Intelligence. Nonetheless, in many cases heuristics are not driven through data but modeled via datasets. The seed or initial model for heuristics is process- or rule-driven. Heuristics are derived from the tacit understanding of a system over a period and modeled with the help of a human agent or a subject matter expertise team. Most of the domain experts generally call heuristics a gut call. Nonetheless, mathematically, heuristics are rule codified systems with a set of constraints defining the system behavior. In heuristics, the definition of system variables is critical. Both constraints and variables are modeled using objective functions with the goal of maximization of the system. The maximization could be improved: just-in-time delivery, manufacturing floor space optimization, optimization of the number of manufacturing agents per workspace, a lower cost of agricultural equipment, less utilization of

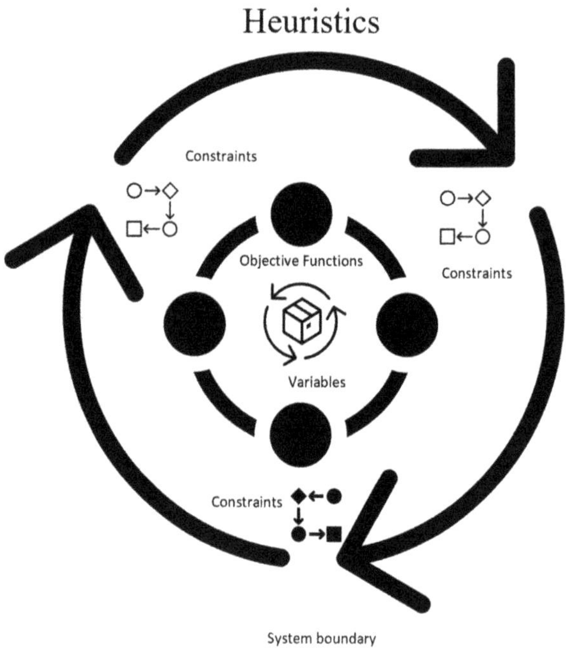

Fig. 1.2 Heuristics

pesticides to improve crop performance, or optimization of routing of commodity delivery from farm to fork, of course through retail value chain.

Data plays an important role in the development of AI systems and heuristics. AI models learn from the data whereas heuristic models are codification of knowledge or a tacit experience of a subject matter expert (SME). Put it succinctly, heuristics models are developed using a set of rules or design processes, constraints, or optimization functions. The heuristic model building involves understanding the variables of the model, understanding constraints, understanding objective functions, and finally solving the heuristics model for optimal solution. Another difference I would like to draw upon was AI models, based on the model type, generate output. The output could be mathematical evaluation of AI model or recommendation for an actionable insight or identification of outlier of a system. Heuristics model does not guarantee output, that is, the model may or may not solve it sometimes to have an optimal solution. Another important distinction is that AI models are deterministic. That is, with similar or the same inputs, the model will always generate the same output, assuming the model equation did not change. Heuristic model is nondeterministic. That is, for the same inputs, the model could result in different solutions. Nevertheless, the models from AI systems could be fed to heuristics and could establish a *learning curve* that AI systems could be enhanced with the new data that heuristics generate (please see Fig. 1.3). In a way, both AI systems and heuristics form a symbiotic process of learning, sharing, and building models.

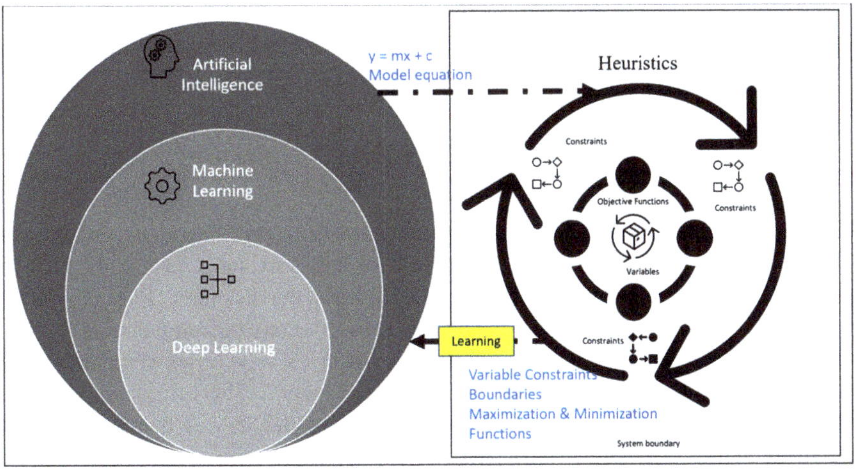

Fig. 1.3 Artificial Intelligence and heuristics feedback loop

Table 1.1 AI and heuristics

	AI models	Heuristics
Models	Data-driven	Rule-driven
Repeatable	Yes	No
Solution	Deterministic	Nondeterministic
Output	Produce results	Produce optimal solution or do not produce solution at all
Model types	Machine learning, deep learning with statistical models	Linear programming, constraint programming, and mixed-integer programming
Feature variables	Exogenous variables, independent variables, and synthetic variable play a role in constructing the model	System variables, constraints, and objective functions define the model
Mathematical formulations	Statistical	Linear and nonlinear algebraic
Model refreshment	Key feature variables or data signature changes	Rules or constraints change

The bidirectional relationship is learning via data, constraints, objective functions, and evaluation of optimal solution. The driver from the AI systems is a new mathematical relationship that describes the system. Here is a summary of AI models and heuristics (please see Table 1.1):

For more details on machine learning techniques, I would highly recommend referring to Machine Learning[1] and Artificial Intelligence for Agricultural Economics [1]. More on heuristics will be covered in the next chapters.

1.1 Exploratory Data Analysis (EDA)

The main purpose of the exploratory data analysis (EDA) is to see signals within the data and understand it. Additionally, findings of the EDA and data signals must be confirmed with subject matter expert (SME) from the business. It is crucial to validate with SME to confirm understanding as well as any hidden business signals. For example, I was handed over the following historical time series of a commodity price series to develop a predictive time series model. The time series was of monthly frequency. When I used the time series model, it was complying about duplicate values of price entry (please see Fig. 1.4):

SME on the data team has confirmed that due to the nature of the business, they had two price issuances for December 2020 and price ceiling (most recent in the month) for the time series model.

2020M11	190.38
2020M12	198.77
2020M12	220.34
2021M01	234.47
2021M02	245.24
2021M03	245.17
2021M04	268.23
2021M05	305.31

The specified time column contains rows with duplicate time stamps. If your data contains multiple time series, review the time series identifier column setting to define the time series identifiers for your data. If the dataset needs to be aggregated, please provide freq and target_aggregation_function parameters and run AutoML again.

Fig. 1.4 Corn prices

[1]Machine Learning and Artificial Intelligence for Agricultural Economics—https://rd.springer.com/book/10.1007/978-3-030-77485-1

The exploratory data analysis (EDA) is knowledge discovery chasm to understand data attributes of the data as to understand business or use case for which the model development is being undertaken. It is crucial to uncover all the understandings. Here are high-level steps [2][2]:

1. Maximize insight into a dataset.
2. Uncover underlying structure.
3. Extract important variables.
4. Detect outliers and anomalies.
5. Test underlying assumptions.
6. Develop parsimonious models.
7. Determine optimal factor settings.

The following contains the decision flow of the EDA (please see Fig. 1.5):

Data exploration and analysis is typically an iterative process, in which the data scientist takes a sample of data and performs the following kinds of task to analyze it and test hypotheses:

- Clean data to handle errors, missing values, and other issues.
- Apply statistical techniques to better understand the data and how the sample might be expected to represent the real-world population of data, allowing random variation.
- Visualize data to determine relationships between variables and, in the case of a machine learning project, identify features that are potentially predictive of the label.
- Derive new features from existing ones that might better encapsulate relationships within the data.
- Revise the hypothesis and repeat the process.

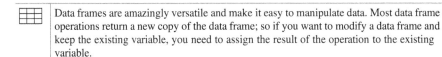

> Data frames are amazingly versatile and make it easy to manipulate data. Most data frame operations return a new copy of the data frame; so if you want to modify a data frame and keep the existing variable, you need to assign the result of the operation to the existing variable.

1.1.1 Vegetation Health Index: Exploratory Data Analysis

The following code explains the entire decision flow. The Vegetation Health Index has a duration of 10 days, a dekad frequency da. Seasonal and weather-related indices are sampled in dekad frequency. For instance, accumulated rainfall during the most recent dekad, which has been aggregated from daily estimates. Every month has three dekads, such that the first two dekads have 10 days (i.e., 1–10,

[2]EDA—https://www.itl.nist.gov/div898/handbook/eda/section1/eda11.htm

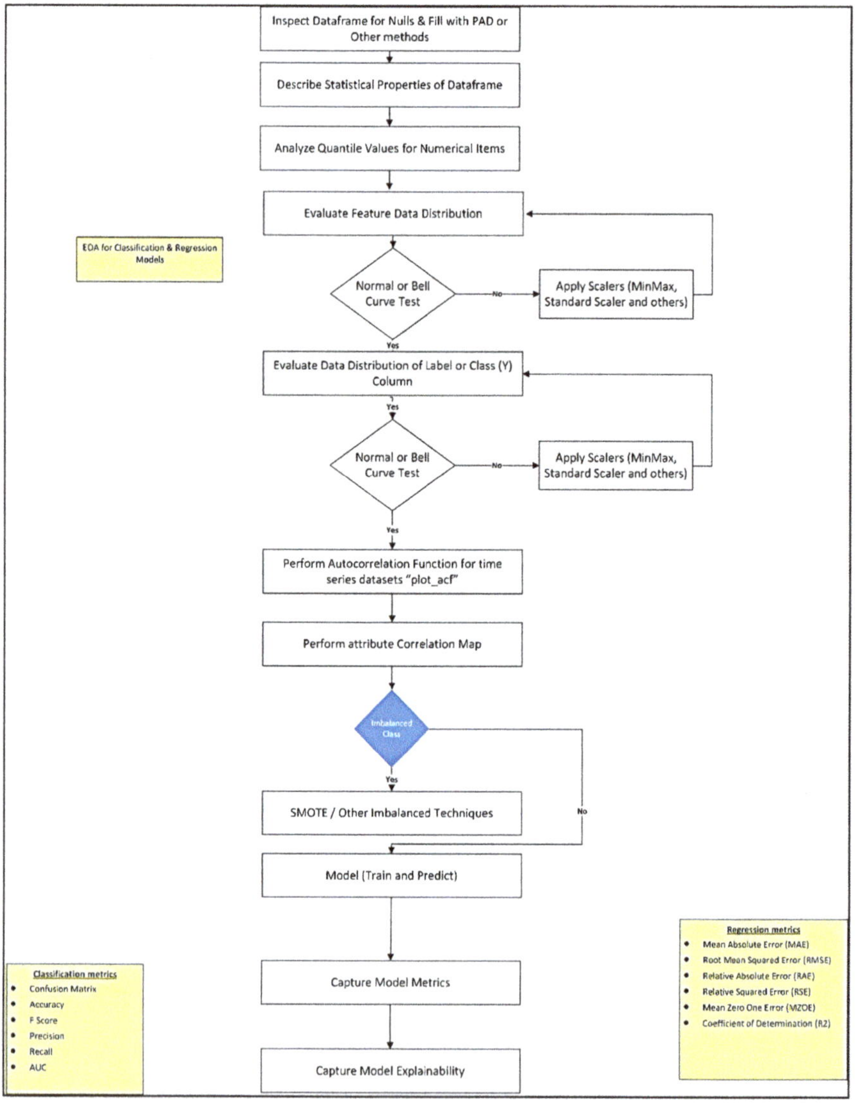

Fig. 1.5 EDA flow chart

11–20), and the third is comprised of the remaining days of the month. Therefore, the length of the third dekad of each month is not consistent and varies from 8 to 11 days, depending on the length of the month.[3] Similarly, earth vegetation and crop health indexes are accumulated in dekad.

[3]Dekadal Rainfall Estimates—http://iridl.ldeo.columbia.edu/maproom/Food_Security/Locusts/Regional/Dekadal_Rainfall/index.html

As can be seen, the northwestern parts of the country have experienced more snow and cloud during dekad 1 January 2022 compared to January 2021 (please see Table 1.2). Similarly, the same assessment can be drawn for dekad 2. Finally, in dekad 3 January 2022, it is hotter in Texas than during the same period January 2021. The sources of the data are the Global Information and Early Warning System (GIEWS), the Meteorological Operational Satellite-Advanced Very-High-Resolution Radiometer (METOP-AVHRR), the World Geodetic System 1984 (WGS84), and the Geographic Lat/Lon. Please see Vegetation Health Index (VHI) legend—values less than 0.15 are critical drought, and values greater than 0.85 are healthy (please see Fig. 1.6).

The World's Food Supply is Made Insecure by Climate Change. Future projections in global yield trends of both maize and wheat[a] indicate a significant decline; these declines can be attributed to the negative impacts of climate change arising from increasing greenhouse gas emissions [3].
The United Nations

[a]The United Nations The World's Food Supply is Made Insecure by Climate Change—https://www.un.org/en/academic-impact/worlds-food-supply-made-insecure-climate-change

The seasonal indicators are designed to allow easy identification of areas of cropped land with a high likelihood of water stress (drought). The indices are based on remote sensing data of vegetation and land surface temperature combined with information on agricultural cropping cycles derived from historical data and a global crop mask.

The raw data[4] is from the Food and Agriculture Organization-Agricultural Stress Index System[5] (FAO-ASIS) and would look as follows:

Indicator	Country	ADM1_ CODE	Province	Date	Data	Year	Month	Dekad	Unit	Source
Vegetation Health Index (VHI)	United States of America	3218	California	1/1/ 1984	0.589	1984	1	1		FAO-ASIS
Vegetation Health Index (VHI)	United States of America	3218	California	1/11/ 1984	0.561	1984	1	2		FAO-ASIS
Vegetation Health Index (VHI)	United States of America	3218	California	1/21/ 1984	0.532	1984	1	3		FAO-ASIS
Vegetation Health Index (VHI)	United States of America	3218	California	2/1/ 1984	0.512	1984	2	1		FAO-ASIS
	United States of America	3218	California	2/11/ 1984	0.519	1984	2	2		FAO-ASIS

(continued)

[4]Spatial Aggregated Data—https://www.fao.org/giews/earthobservation/asis/data/country/USA/MAP_NDVI_ANOMALY/DATA/vhi_adm1_dekad_data.csv

[5]FAO-ASIS—https://www.fao.org/giews/earthobservation/asis/index_1.jsp?lang=en

Indicator	Country	ADM1_ CODE	Province	Date	Data	Year	Month	Dekad	Unit	Source
Vegetation Health Index (VHI)										
Vegetation Health Index (VHI)	United States of America	3218	California	2/21/ 1984	0.538	1984	2	3		FAO-ASIS
Vegetation Health Index (VHI)	United States of America	3218	California	3/1/ 1984	0.566	1984	3	1		FAO-ASIS
Vegetation Health Index (VHI)	United States of America	3218	California	3/11/ 1984	0.594	1984	3	2		FAO-ASIS
Vegetation Health Index (VHI)	United States of America	3218	California	3/21/ 1984	0.64	1984	3	3		FAO-ASIS
Vegetation Health Index (VHI)	United States of America	3218	California	4/1/ 1984	0.657	1984	4	1		FAO-ASIS
Vegetation Health Index (VHI)	United States of America	3218	California	4/11/ 1984	0.648	1984	4	2		FAO-ASIS
Vegetation Health Index (VHI)	United States of America	3218	California	4/21/ 1984	0.642	1984	4	3		FAO-ASIS
Vegetation Health Index (VHI)	United States of America	3218	California	5/1/ 1984	0.629	1984	5	1		FAO-ASIS

1.1.2 US Vegetation Health Index Analysis

The goal of the experiment was to analyze the Vegetation Index of the country from 1982 to 2022. Analyze the Vegetation Health Index by region.

1.1.2.1 Step 1: Load Required Libraries

```
import pandas as pd
import numpy as np
import plotly.graph_objects as go
import seaborn as sns; sns.set()
import matplotlib.pyplot as plt
from scipy import stats
from scipy.stats import pearsonr
```

(continued)

Table 1.2 Spatial aggregated

Dekad[a] 1—January 2022	Dekad[b] 2—January 2022	Dekad[c] 3—January 2022
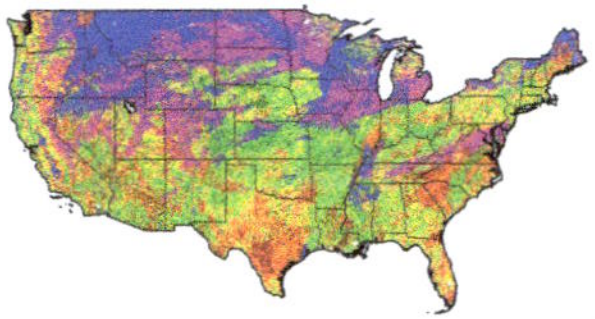	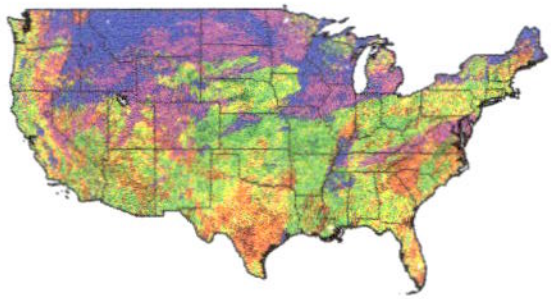	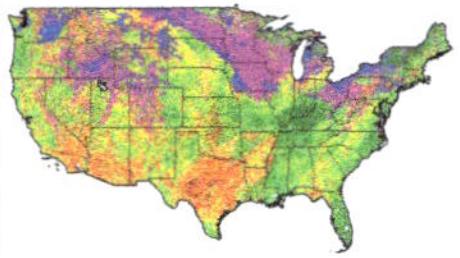
Dekad[d] 1—January 2021	Dekad 2—January 2021	Dekad[e] 3—January 2021
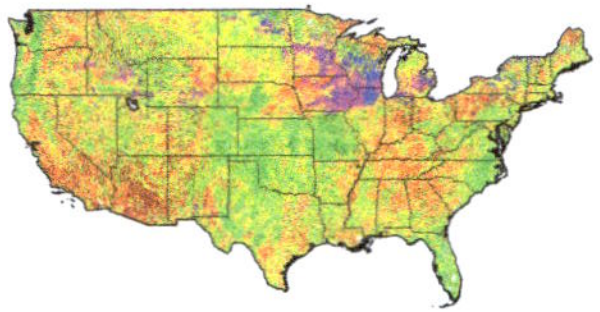	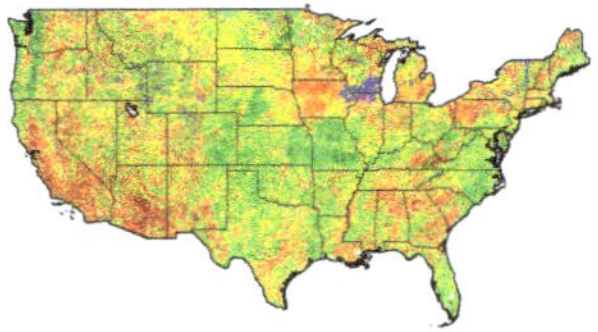	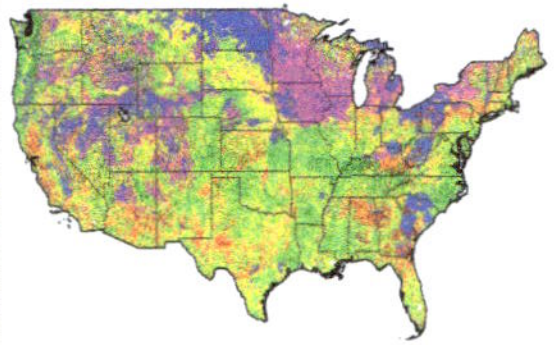

[a]VHI—https://www.fao.org/giews/earthobservation/country/index.jsp?lang=en&code=USA#
[b]Dekad 2 January 2022—https://www.fao.org/giews/earthobservation/asis/data/country/USA/MAP_NDVI_ANOMALY/HR/ot2202h.png
[c]Dekad 3 January 2022—https://www.fao.org/giews/earthobservation/asis/data/country/USA/MAP_NDVI_ANOMALY/HR/ot2203h.png
[d]Dekad 1 January 2021—https://www.fao.org/giews/earthobservation/asis/data/country/USA/MAP_NDVI_ANOMALY/HR/ot2101h.png
[e]Dekad 3 January 2021—https://www.fao.org/giews/earthobservation/asis/data/country/USA/MAP_NDVI_ANOMALY/HR/ot2103h.png
Source: Data source for spatial-aggregated (national and subnational) time series data (Spatial Aggregated Data—https://www.fao.org/giews/earthobservation/asis/data/country/USA/MAP_NDVI_ANOMALY/DATA/vhi_adm1_dekad_data.csv)

Fig. 1.6 VHI legend

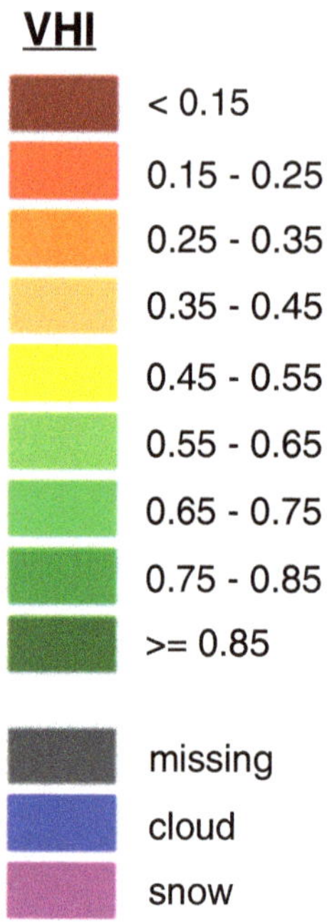


```
import matplotlib.pyplot as plt
from sklearn.preprocessing import MinMaxScaler
from sklearn.model_selection import train_test_split
from sklearn.linear_model import LinearRegression
from sklearn.ensemble import RandomForestClassifier
from sklearn.metrics import r2_score
from sklearn import metrics

import matplotlib.pyplot as plt
import matplotlib as mpl
import numpy as np
import scipy.stats as spstats
import seaborn as sns

%matplotlib inline
mpl.style.reload_library()
mpl.style.use('classic')
mpl.rcParams['figure.facecolor'] = (1, 1, 1, 0)
mpl.rcParams['figure.figsize'] = [6.0, 4.0]
mpl.rcParams['figure.dpi'] = 100
```

1.1.2.2 Step 2: Load Vegetation Health Index Data

Download Vegetation Health Index from FAO Earth Observation-GIEWS.

```
# Importing data

vhiDekadDataforUSADF = pd.read_csv("USA_vhi_adm1_dekad_data.
csv")
vhiDekadDataforUSADF
```

Output:

	Indicator	Country	ADM1_CODE	Province	Date	Data	Year	Month	Dekad	Unit	Source
0	Vegetation Health Index (VHI)	United States of America	3214	Alabama	1/1/1984	0.427	1984	1	1		FAO-ASIS
1	Vegetation Health Index (VHI)	United States of America	3214	Alabama	1/11/1984	0.452	1984	1	2		FAO-ASIS
2	Vegetation Health Index (VHI)	United States of America	3214	Alabama	1/21/1984	0.509	1984	1	3		FAO-ASIS
3	Vegetation Health Index (VHI)	United States of America	3214	Alabama	2/1/1984	0.575	1984	2	1		FAO-ASIS
4	Vegetation Health Index (VHI)	United States of America	3214	Alabama	2/11/1984	0.577	1984	2	2		FAO-ASIS
...	...	...	...	...	...	...	...	...	...	...	...
69209	Vegetation Health Index (VHI)	United States of America	3260	Virginia	2/11/2022	0.260	2022	2	2		FAO-ASIS
69210	Vegetation Health Index (VHI)	United States of America	3261	Washington	2/11/2022	0.386	2022	2	2		FAO-ASIS
69211	Vegetation Health Index (VHI)	United States of America	3262	West Virginia	2/11/2022	0.495	2022	2	2		FAO-ASIS
69212	Vegetation Health Index (VHI)	United States of America	3263	Wisconsin	2/11/2022	0.518	2022	2	2		FAO-ASIS
69213	Vegetation Health Index (VHI)	United States of America	3264	Wyoming	2/11/2022	0.478	2022	2	2		FAO-ASIS

The dataset contains Province, VHI Data, Month, Dekad, and Source (FAO-ASIS). Construct Month Dekad Label so year-over-year changes can be analyzed.

```
vhiDekadDataforUSADF['month_dekad_label'] = 'month_' +
vhiDekadDataforUSADF['Month'].apply(lambda x : str(x)) +
'_dekad_' + vhiDekadDataforUSADF['Dekad'].apply(lambda y : str(y))
```

The above code constructs dekad label.

```
vhiDekadDataforUSADF= vhiDekadDataforUSADF.rename
({'Data':'Vegetation Health Index (VHI)'}, axis=1)
vhiDekadDataforUSADF.set_index("Year", inplace = True)
vhiDekadDataforUSADF
```

Set the data frame, year, as index of the data frame for further data analysis.

Year	Indicator	Country	ADM1_CODE	Province	Date	Vegetation Health Index (VHI)	Month	Dekad	Unit	Source	month_dekad_label	Region
1984	Vegetation Health Index (VHI)	United States of America	3214	Alabama	1/1/1984	0.427	1	1		FAO-ASIS	month_1_dekad_1	South
1984	Vegetation Health Index (VHI)	United States of America	3214	Alabama	1/11/1984	0.452	1	2		FAO-ASIS	month_1_dekad_2	South
1984	Vegetation Health Index (VHI)	United States of America	3214	Alabama	1/21/1984	0.509	1	3		FAO-ASIS	month_1_dekad_3	South
1984	Vegetation Health Index (VHI)	United States of America	3214	Alabama	2/1/1984	0.575	2	1		FAO-ASIS	month_2_dekad_1	South

1.1.2.3 Step 3: Observe Vegetation Index Distribution

Let's analyze the VHI data distribution.

```
import pandas as pd
import matplotlib.pyplot as plt

# This ensures plots are displayed inline in the Jupyter notebook
%matplotlib inline

# Get the label column
label = vhiDekadDataforUSADF['Vegetation Health Index (VHI)']

# Create a figure for 2 subplots (2 rows, 1 column)
fig, ax = plt.subplots(2, 1, figsize = (9,12))

# Plot the histogram
ax[0].hist(label, bins=100)
ax[0].set_ylabel('Frequency')

# Add lines for the mean, median, and mode
ax[0].axvline(label.mean(), color='magenta',
linestyle='dashed', linewidth=2)
ax[0].axvline(label.median(), color='cyan',
linestyle='dashed', linewidth=2)

# Plot the boxplot
ax[1].boxplot(label, vert=False)
ax[1].set_xlabel('Vegetation Health Index (VHI)')

# Add a title to the Figure
fig.suptitle('Vegetation Health Index (VHI)')

# Show the figure
fig.show()
```

Output:

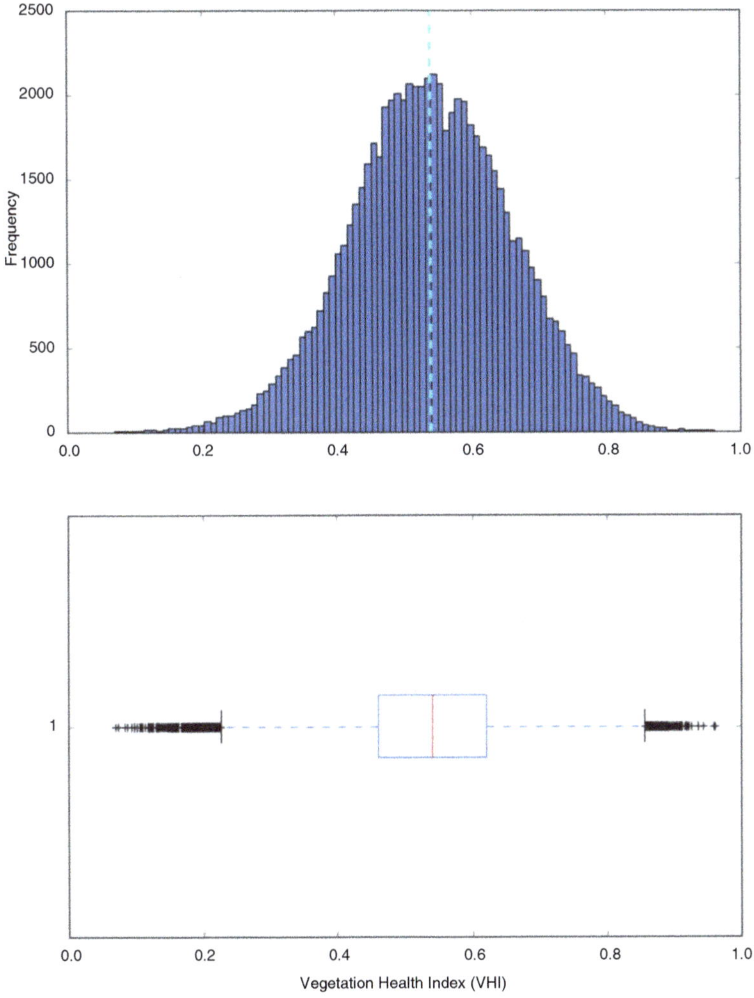

The VHI exhibits normal distribution. The data distribution is as follows: minimum 0.07, mean 0.54, median 0.54, mode 0.51, and maximum 0.96. These values indicate that the region of the country has various levels of Vegetation Health Index.

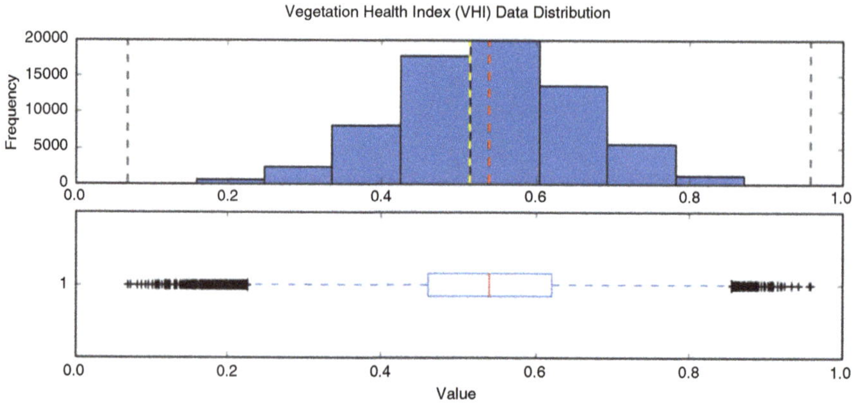

1.1.2.4 Step 4: VHI by Province

To observe VHI by providence, i.e., by states, box plot the values.

```
vhiDekadDataforUSADF.boxplot(column='Vegetation Health Index
(VHI)', by='Province', figsize=(8,5))
```

Output:

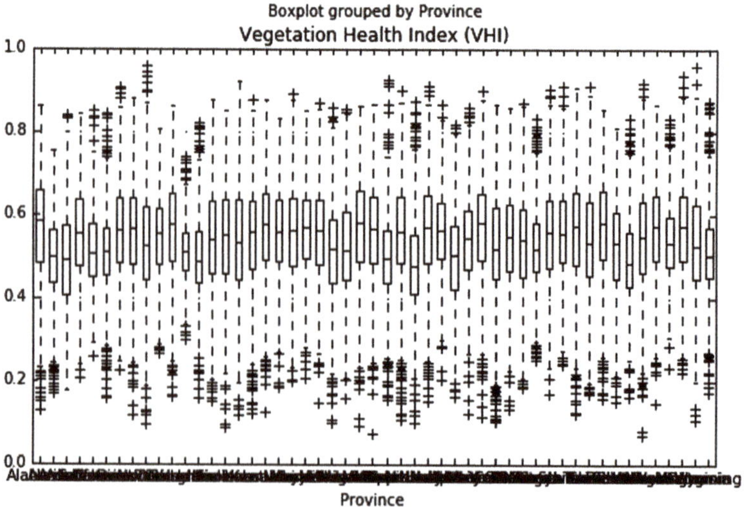

The above box plot is dense. However, for almost all states, the VHI varies between 0.4 and 0.7.

Now, box plot by the region.

```
vhiDekadDataforUSADF.boxplot(column='Vegetation Health Index
(VHI)', by='Region', figsize=(8,5))
```

Output:

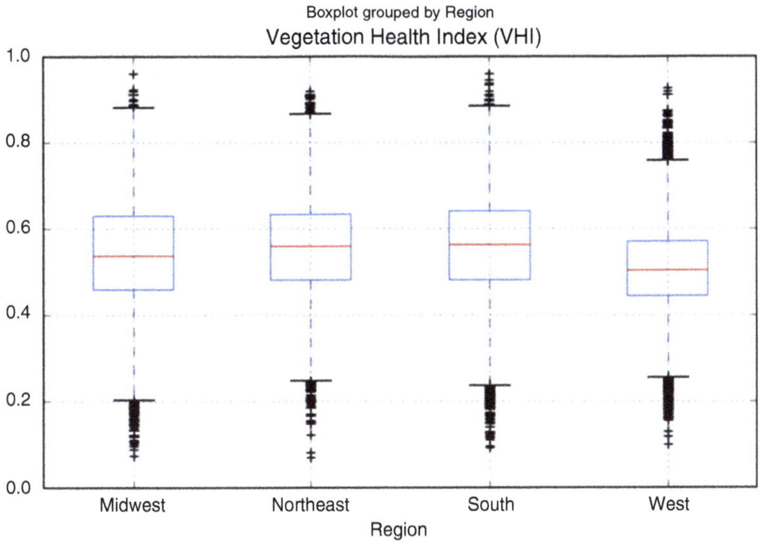

As you can see, the South region exhibits higher values of VHI over the West region.

Bar plot also provides similar results.

```
# Ensure plots are displayed inline in the notebook
%matplotlib inline

from matplotlib import pyplot as plt

# Create a bar plot of name vs grade
plt.bar(x=vhiDekadDataforUSADF['Region'],
height=vhiDekadDataforUSADF['Vegetation Health Index (VHI)'])

# Display the plot
plt.show()
```

Output:

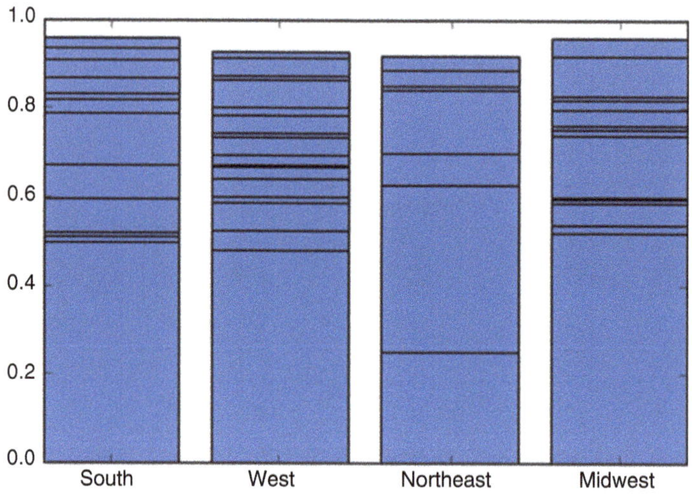

The final maps highlight anomalous vegetation growth and potential drought in crop zones during the growing season. The satellite data used in the calculation of the mean Vegetation Health Index (VHI) and the ASI is the 10-day (dekadal) vegetation data from the METOP-AVHRR (Advanced Very-High-Resolution Radiometer) sensor at 1 km resolution (2007 and after). Data at 1 km resolution for the period 1984–2006 was derived from the NOAA-AVHRR dataset at 16 km resolution. The crop mask is FAO GLC-SHARE.

ML provides pattern analysis and outlier analysis. The pattern analysis would provide lineage of agricultural flow and a holistic view on the food security.

1.2 Machine Learning Technique: Association Mining

Machine learning can learn from data through the application of statistical techniques such as regression, conditional Bayesian models, information gain-based decision trees, ensemble models, and deep learning-based techniques. To learn relationships that are hidden in a set of a transaction database, association mining is applied. Association rules are an important class of regularities in data. Mining of association rules is a fundamental data mining task. It is perhaps the most important model invented and extensively studied by the database and data mining community. Its objective is to find all co-occurrence relationships, called associations, among data items. Since it was first introduced in 1993 by Agrawal et al. [4], it has attracted a great deal of attention. Many efficient algorithms, extensions, and applications have been reported.[6]

Association rule mining, however, does not consider the sequence in which the items are purchase or transactions have occurred. Sequential pattern mining takes care of that. An example of a sequential pattern is "Egypt imports wheat from Ukraine first, then United States of America, then Russia." The items are not purchased at the same time, but one after another. Such patterns are useful in web usage mining for analyzing clickstreams in server logs. They are also useful for finding language or linguistic patterns from natural language texts.

1.2.1 Support

Support for ARM is introduced by Agrawal et al. [4, 5]. It measures the frequency of association, i.e., how many times the specific item has occurred in a dataset. An item with greater support is called frequent or large itemset. In terms of probability theory, we can express support as

$$\text{Support} = \text{P}\ (\text{A} \cap \text{B}) = \frac{\text{Number of Transactions containing both A and B}}{\text{Total Number of Transactions}} \tag{1.1}$$

If support is too low, the rule may occur due to chance. In export and import policies of countries, a low support value can be meaningful. For import origin use case, we can have a minimum support. Here we're looking at available import choices.

1.2.2 Confidence

Confidence measures the strength of the association. It determines how frequently item B occurs in the transaction that contains A [4, 5]. Confidence expresses the conditional probability of an item. The definition of confidence is

[6]Bing Liu, Web Data Mining: Exploring Hyperlinks, Contents, and Usage Data (Data-Centric Systems and Applications), Springer; 2nd ed. 2011 edition (June 26, 2011), ISBN-13 : 978-3642194597.

$$\text{Confidence} = P(A \mid B) = \frac{P(A \cap B)}{P(A)}$$
$$= \frac{\text{Number of Transactions containing both A and B}}{\text{Number of Transactions containing A}} \quad (1.2)$$

Confidence thus determines the predictability of the rule. If confidence of a rule is too low, one cannot reliably infer or predict B from A.

1.2.3 *Frequent Itemset*

In general, a dataset that contains k items can potentially generate up to 2^k -1 frequent itemsets, excluding null set. For importer country origins of Egypt, k is the list of countries that export wheat—more precisely, the list of countries with a higher production and capacity to export. The value of k could be anywhere from 10 to 15. And thus, it could generate many import relationships. For illustration (please see Fig. 1.7), we can start with three countries on the list {United States of America (USA), Canada (C), and France (F)}.

The lattice structure holds relationships among three countries. We have seven possibilities assuming one country can support the import needs of wheat of Egypt. Otherwise, it is distributed across the three countries.

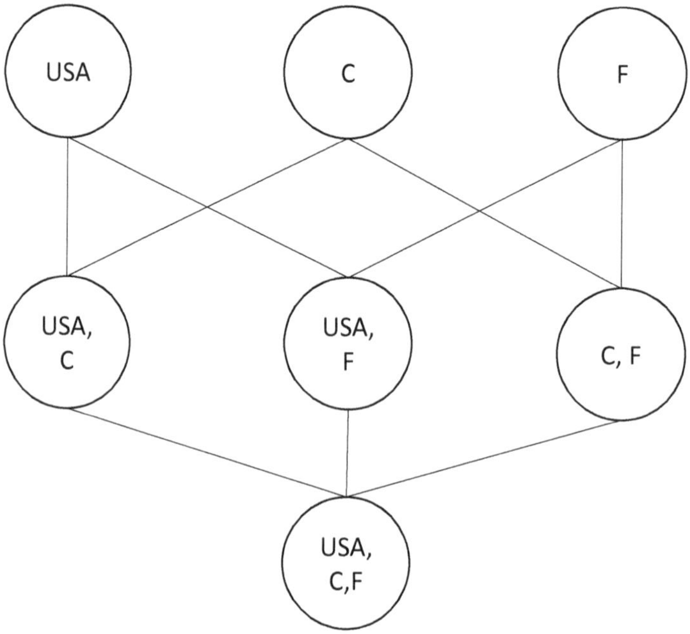

Fig. 1.7 Frequency itemset

A brute force approach for finding frequent itemsets is to determine the support count for every candidate item in the lattice structure. To do this, we need to compare each candidate with every transaction or apply Apriori rule mining.

1.3 Risk Modeling: Wheat Import Origins: Food Security

Artificial Intelligence enhances food security, a measure of availability and access to food. The rational is simple. Artificial Intelligence learns from the data and as the data emerges from multifield sources, the interactive nature of these data signals in terms their influence, causation, severity, length of time, ample/magnitude, spikiness, and composite pattern, the analysis requires sophisticated techniques that Artificial Intelligence offers and detects insights from these signals relatively smoothly. Food security is such a multifaceted-multifield complex issue, and the reasons for lack of food security, or food insecurity, can originate from any of these domains that include economics, geopolitical, agriculture, climate and weather, natural resources, and unexpected shocks.

Egypt is one of the largest importers of wheat to cover domestic needs. Despite having 3835.9688[7] (1000 ha) of agricultural area (main exports are citrus fruits, potatoes, and other types of fruits and vegetables[8]), Egypt imports wheat to support food subsidy programs. Wheat imports from Egypt, the world's largest wheat importer, are projected at 13 million tonnes, like in 2019/2020 despite a good domestic harvest, given the government's recently announced intention to increase reserves in response to threats posed to global supplies by COVID-19.[9] Egypt is one of the largest importers of wheat and continues to operate a bread subsidy system, with recipients getting 150 loaves of traditional baladi bread each month at a subsidized price of Egyptian pound EGP 0.05 per loaf ($0.01). Egypt has a large and growing population, and its production of grains and oilseeds falls well short of its food needs, making large-scale imports necessary.[10] It continues to run a bread subsidy scheme, with loaves provided to much of the population at prices well below the cost of production. At the same time, the Egyptian authorities are working to increase agricultural output, notably by focusing on increasing the uptake of new, higher-yielding varieties while providing information on best practices to the country's farmers.

Consider the wheat import of Egypt in 2020. The overall cereal import requirements in the 2020/2021 marketing year (July/June) are forecasted at about 24.2 million tonnes,[11] close to the previous year's level and 10% higher than the 5-year

[7] Egypt country profile—https://www.fao.org/countryprofiles/index/en/?iso3=EGY

[8] Egypt—https://www.fao.org/giews/countrybrief/country.jsp?lang=en&code=EGY

[9] Wheat Imports—https://www.opportimes.com/the-10-countries-with-the-most-wheat-imports/

[10] Focus on Egypt—https://www.world-grain.com/articles/15959-focus-on-egypt

[11] Egypt—https://www.fao.org/giews/countrybrief/country.jsp?lang=en&code=EGY

Table 1.3 Wheat imports

Partner countries	Wheat import quantity (tonnes), 2020
Australia	203,640
Bulgaria	46,593
Canada	8391
France	593,422
Hungary	19,171
Italy	0
Lithuania	23,518
Romania	294,522
Russian Federation	5,460,508
Ukraine	2,317,534
United Kingdom of Great Britain and Northern Ireland	2
United States of America	75,282

average. Currently, the three largest suppliers are the *Russian Federation*, *Ukraine*, *and Romania*. However, the export tariffs imposed by the *Russian Federation* are likely to result in some changes in import origins.

Please see Table 1.3 for wheat imports from different partner countries.[12]

In terms of percentages, the Russian Federation has supplied 60.39%, Ukraine has shipped 25.63 %, France 6.56 %, Australia 2.52%, and the rest of the world 27.54% (please see Fig. 1.8).

Egypt imports of essential commodity such as wheat are as dynamic as people's needs are. For instance, a decade back, the major import of wheat came from the Russian Federation with 38%, the United States 28%, Australia 10%, and France 10%, and the rest of the world provided approximately 14%.

In 2000, the United States (38.9%), Australia (8.6%), France (2.3%), Argentina (1.6%), and Turkey (1.3%) fulfilled 50% of the imports with the rest of the world filling the others. The import needs of the country change (please see Fig. 1.9). Egyptian imports of common wheat are projected to increase by 24.0%, from 10.3 million metric tonnes in 2009–2011 to 12.8 million metric tonnes in 2021.[13] In 2000, Egypt was the largest importer of common wheat in the North Africa region [6].

As we can see the association of different countries has provided the necessary import needs of Egypt to fulfill food security. One could ask, is there an association of country list that could fulfill import needs to fulfill food security? And the answer rests in the Sankey diagrams below (please see Fig. 1.10):

For instance, between the 2020 and 2011 country list, is there an association pattern (please see Fig. 1.11)?

Similarly, is there a relationship between 2011 and 2000 (please see Fig. 1.12)?

[12] FAO detailed trade matrix—https://www.fao.org/faostat/en/#data/TM

[13] 2012 Outlook of the U.S. and World Wheat Industries—https://ageconsearch.umn.edu/record/133393/files/AAE696.pdf

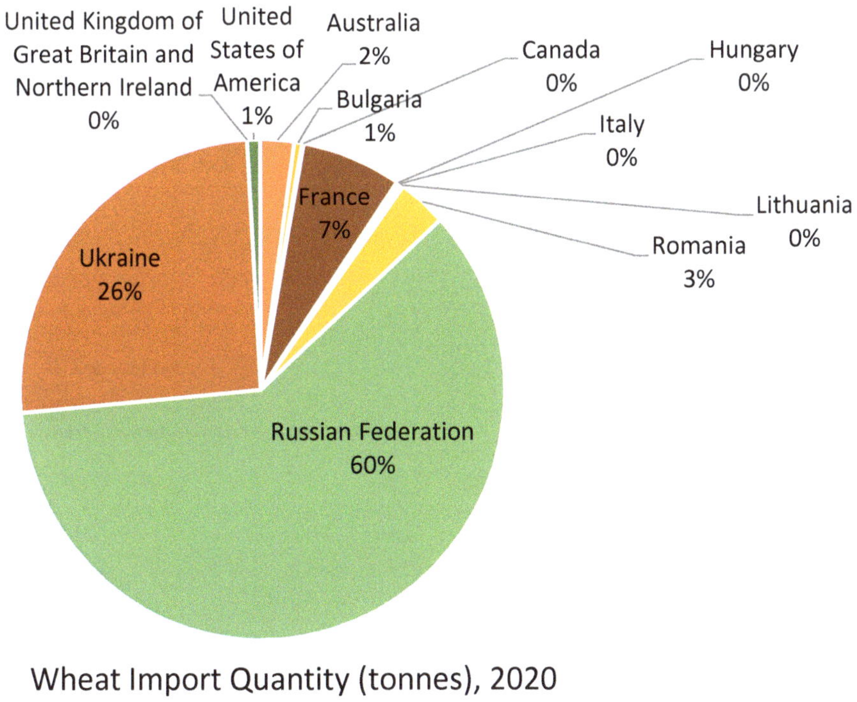

Fig. 1.8 Wheat import quantity: 2020

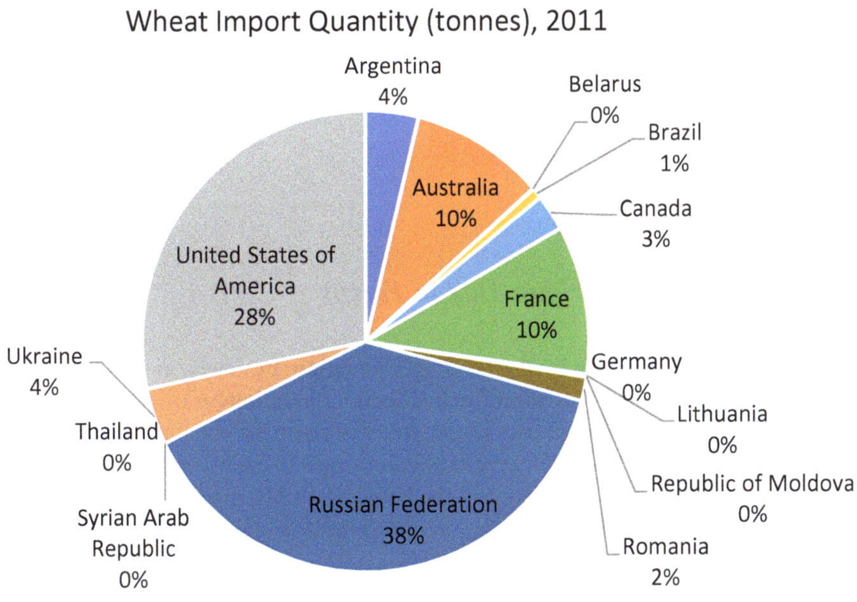

Fig. 1.9 Wheat import quantity: 2011

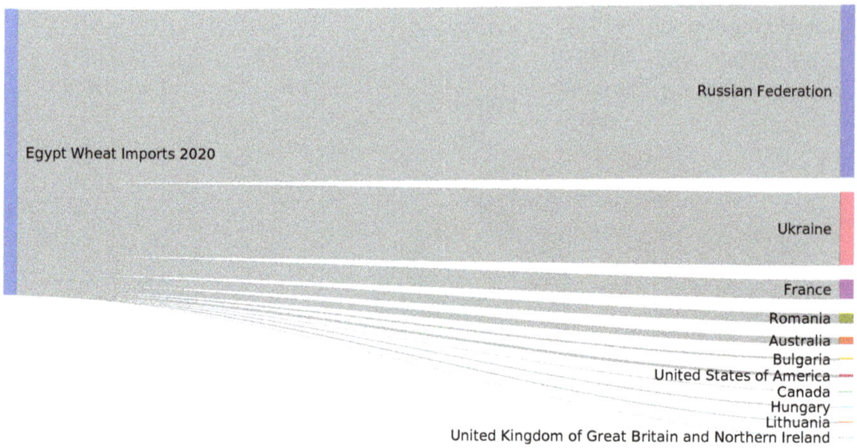

Fig. 1.10 Egypt wheat imports 2020 Sankey

Fig. 1.11 Egypt wheat imports: 2011

1.3.1 New Import Origins Due to Tariffs

The export tariffs imposed by the Russian Federation are likely to result in some changes in import origins (as shown below, before 2022 import origin countries—please see Fig. 1.13). New import origin patterns can be derived with the help of machine learning techniques. Of course, the dynamics of export and import policies of a government is beyond the scope of the book. What ML models could provide is updated import origins that can account for tariffs and provide actionable insights to address the food security issues.

Fig. 1.12 Egypt wheat imports: 2000

Fig. 1.13 Text cloud—import origins: 2020

Association rule manning is applied where two or more than two items have a relationship established as part of a transaction. The core principle of association rule mining is data mining of interactions between two or more items in a dataset. Association rule mining (ARM) is one of the most important and substantial techniques in machine learning [4, 5]. It is particularly important for extracting knowledge from large databases by discovering frequent itemsets and associating item relationships between or among items of a data file. ARM is a powerful exploratory technique with a wide range of applications such as marketing policies, medical diagnosis, financial forecast, credit fraud detection, public administration and actuarial studies, text mining, various kinds of scientific research,

Fig. 1.14 Wheat import origins in 2020 with Russia

telecommunication services, and many other research areas. The emerging importance of ARM encourages machine learning researchers to develop a magnificent number of algorithms such as Apriori, Predictive Apriori, and many others.

New import origin could establish with historical established countries or new ones (refer to Clustering for more details) (please see Fig. 1.14).

Egypt wheat import origins

Software code for this model: Egypt_ImporterOrgin_Wheat_Model.ipynb (Jupyter Notebook Code)

1.3.2 *Step 1: Data Mine Historical Import Transaction Records*

Collect historical import origin transaction records for Egypt. The data is available as part of FAOSTAT trade matrix. The data table would be as follows:

Load libraries:

```
import pandas as pd
import numpy as np
from sklearn.linear_model import LinearRegression
from sklearn.model_selection import StratifiedKFold
from sklearn.metrics import mean_squared_error
from math import sqrt
import random
from sklearn.impute import SimpleImputer
from sklearn.experimental import enable_iterative_imputer
from sklearn.impute import IterativeImputer
import seaborn as sns
import matplotlib.pyplot as plt


dfEgyptWheatImportOrgin = pd.read_csv
('Egypt_Wheat_Import_Orgin_FAOSTAT_data_3-17-2022.csv',
encoding= 'unicode_escape')
dfEgyptWheatImportOrgin.tail()
```

Output: Egypt wheat from partner countries

	Domain Code	Domain	Reporter Country Code (FAO)	Reporter Countries	Partner Country Code (FAO)	Partner Countries	Element Code	Element	Item Code (FAO)	Item	Year Code	Year	Unit	Value	Flag	Flag Description
2463	TM	Detailed trade matrix	59	Egypt	248	Yugoslav SFR	5610	Import Quantity	268	Oil, sunflower	1988	1988	tonnes	12012.0	NaN	Official data
2464	TM	Detailed trade matrix	59	Egypt	231	United States of America	5610	Import Quantity	15	Wheat	1986	1986	tonnes	1293909.0	NaN	Official data
2465	TM	Detailed trade matrix	59	Egypt	251	Zambia	5610	Import Quantity	56	Maize	2006	2006	tonnes	0.0	NaN	Official data
2466	TM	Detailed trade matrix	59	Egypt	181	Zimbabwe	5610	Import Quantity	56	Maize	1992	1992	tonnes	4.0	NaN	Official data
2467	TM	Detailed trade matrix	59	Egypt	181	Zimbabwe	5610	Import Quantity	882	Milk, whole fresh cow	1987	1987	tonnes	6.0	NaN	Official data

Inspect columns of data frame:

```
dfEgyptWheatImportOrigin.columns
```

```
Index(['Domain Code', 'Domain', 'Reporter Country Code (FAO)',
       'Reporter Countries', 'Partner Country Code (FAO)', 'Partner Countries',
       'Element Code', 'Element', 'Item Code (FAO)', 'Item', 'Year Code',
       'Year', 'Unit', 'Value', 'Flag', 'Flag Description'],
      dtype='object')
```

Not all columns are required to perform pattern analysis. Derive necessary columns to perform pattern analysis:

1.3.3 Step 2: Construct Data Frame to Perform Pattern Analysis

```
dfEgyptWheatImportOriginARM = dfEgyptWheatImportOrigin
[['Partner Countries','Year
Code','Year','Unit','Value','Item']]

dfEgyptWheatImportOriginARM.set_index("Year",inplace = True)
dfEgyptWheatImportOriginARM
```

Output:

	Partner Countries	Year Code	Unit	Value	Item
Year					
2002	Afghanistan	2002	tonnes	54.0	Butter, cow milk
2007	Albania	2007	tonnes	110.0	Butter, cow milk
1988	Albania	1988	tonnes	1.0	Milk, whole fresh cow
2020	Algeria	2020	tonnes	0.0	Oil, vegetable origin nes
2020	Australia	2020	tonnes	203640.0	Wheat
...	...	...	...	...	...
1988	Yugoslav SFR	1988	tonnes	12012.0	Oil, sunflower
1986	United States of America	1986	tonnes	1293909.0	Wheat
2006	Zambia	2006	tonnes	0.0	Maize
1992	Zimbabwe	1992	tonnes	4.0	Maize
1987	Zimbabwe	1987	tonnes	6.0	Milk, whole fresh cow

2468 rows × 5 columns

1.3.4 Step 3: Select Wheat Commodity as Part of the Data Frame

```
dfEgyptWheatImportOriginARM = dfEgyptWheatImportOriginARM
[dfEgyptWheatImportOriginARM.Item == 'Wheat']
dfEgyptWheatImportOriginARM
```

Output: Please note that the number of rows has been reduced from 2468 to 416 rows.

	Partner Countries	Year Code	Unit	Value	Item
Year					
2020	Australia	2020	tonnes	203640.0	Wheat
2020	Bulgaria	2020	tonnes	46593.0	Wheat
2020	Canada	2020	tonnes	8391.0	Wheat
2020	France	2020	tonnes	593422.0	Wheat
2020	Hungary	2020	tonnes	19171.0	Wheat
...	...	...	...	...	...
1987	Saudi Arabia	1987	tonnes	34704.0	Wheat
1986	Australia	1986	tonnes	1805812.0	Wheat
1986	Canada	1986	tonnes	231500.0	Wheat
1986	Saudi Arabia	1986	tonnes	74243.0	Wheat
1986	United States of America	1986	tonnes	1293909.0	Wheat

416 rows × 5 columns

Check for nulls.

```
dfEgyptWheatImportOriginARM.isna().sum()
```

If records are null, drop them.

```
Partner Countries    0
Year Code         0
Unit          0
Value         12
Item          0
dtype: int64
```

1.3.5 Step 4: Perform EDA (Exploratory Data Analysis)

This process includes removal of nulls, spaces, and other data issues.

```
dfEgyptWheatImportOriginARM['Partner Countries']
=dfEgyptWheatImportOriginARM['Partner Countries'].apply
(lambda x: x.replace(",",""))

dfEgyptWheatImportOriginARM.head(90)
```

Output:

Year	Partner Countries	Year Code	Unit	Value	Item
2020	Australia	2020	tonnes	203640.0	Wheat
2020	Bulgaria	2020	tonnes	46593.0	Wheat
2020	Canada	2020	tonnes	8391.0	Wheat
2020	France	2020	tonnes	593422.0	Wheat
2020	Hungary	2020	tonnes	19171.0	Wheat
...	...	...	...	...	...
2014	Romania	2014	tonnes	1809852.0	Wheat
2014	Russian Federation	2014	tonnes	4057024.0	Wheat
2014	Ukraine	2014	tonnes	2843061.0	Wheat
2014	United States of America	2014	tonnes	432086.0	Wheat
2013	Argentina	2013	tonnes	12344.0	Wheat

90 rows × 5 columns

1.3.6 Step 5: Pivot the Table to Get Transaction Record for the Association Analysis

```
data_ddfEgyptWheatImportOriginARM =
dfEgyptWheatImportOriginARM.pivot_table( 'Value' , ['Year
Code'], 'Partner Countries')
data_ddfEgyptWheatImportOriginARM
```

Output:

Please note that the countries that have NULLs against a year imply no import of wheat.

Partner Countries	Algeria	Argentina	Australia	Austria	Azerbaijan	Bangladesh	Belarus	Belgium	Belgium-Luxembourg	Brazil	...	Syrian Arab Republic	Thailand	Turkey
Year Code														
1986	NaN	NaN	1805812.0	NaN	NaN	NaN	NaN	NaN	NaN	NaN	...	NaN	NaN	NaN
1987	NaN	NaN	1405498.0	NaN	NaN	NaN	NaN	NaN	NaN	NaN	...	NaN	NaN	NaN
1988	NaN	NaN	1591093.0	NaN	NaN	NaN	NaN	NaN	NaN	NaN	...	NaN	NaN	NaN
1989	NaN	NaN	658692.0	NaN	NaN	4104.0	NaN	NaN	47226.0	NaN	...	NaN	NaN	NaN
1990	NaN	NaN	1805459.0	NaN	NaN	NaN	NaN	NaN	NaN	NaN	...	NaN	NaN	NaN
1991	NaN	NaN	1912096.0	NaN	NaN	NaN	NaN	NaN	NaN	NaN	...	NaN	NaN	12600.0
1992	NaN	NaN	1242054.0	NaN	NaN	NaN	NaN	NaN	NaN	NaN	...	NaN	NaN	NaN
1993	NaN	NaN	755395.0	NaN	NaN	NaN	NaN	NaN	576.0	NaN	...	NaN	NaN	NaN
1994	NaN	NaN	1827624.0	NaN	NaN	NaN	NaN	NaN	NaN	NaN	...	NaN	NaN	NaN
1995	NaN	24809.0	NaN	NaN	NaN	NaN	NaN	NaN	NaN	NaN	...	NaN	NaN	NaN
1996	NaN	NaN	711660.0	NaN	NaN	NaN	NaN	NaN	NaN	NaN	...	NaN	NaN	NaN
1998	NaN	283091.0	625667.0	NaN	NaN	NaN	NaN	NaN	NaN	NaN	...	NaN	NaN	NaN
1999	NaN	25447.0	525043.0	NaN	NaN	NaN	NaN	NaN	NaN	NaN	...	NaN	NaN	172877.0
2000	NaN	147610.0	775559.0	NaN	NaN	NaN	NaN	NaN	NaN	NaN	...	0.0	NaN	123638.0
2001	NaN	4575.0	741843.0	NaN	NaN	NaN	NaN	NaN	NaN	5000.0	...	NaN	NaN	326833.0
2002	935.0	51850.0	930245.0	NaN	NaN	NaN	NaN	NaN	NaN	30807.0	...	105.0	5270.0	NaN
2003	NaN	2000.0	350102.0	NaN	NaN	NaN	NaN	NaN	NaN	3462.0	...	144455.0	NaN	NaN

In the above example, in 1986, Egypt has imported wheat from Australia, Canada, Saudi Arabia, and the United State.

Clean NaNs, to generate record set for association analysis.

```
def row_to_dict(row):
  return row.dropna().to_dict()

data_ddfEgyptWheatImport
OriginARMnew=data_ddfEgyptWheatImportOriginARMnew.apply
(row_to_dict, axis=1)
```

Output:

```
1986  {'Australia': 1805812.0, 'Canada': 231500.0, '...
1987  {'Australia': 1405498.0, 'Canada': 206222.0, '...
1988  {'Australia': 1591093.0, 'Canada': 74940.0, 'F...
1989  {'Australia': 658692.0, 'Bangladesh': 4104.0, ...
1990  {'Australia': 1805459.0, 'Denmark': 43750.0, '...
1991  {'Australia': 1912096.0, 'Canada': 26400.0, 'F...
1992  {'Australia': 1242054.0, 'Canada': 60025.0, 'D...
1993  {'Australia': 755395.0, 'Belgium-Luxembourg': ...
1994  {'Australia': 1827624.0, 'Canada': 39912.0, 'F...
1995  {'Argentina': 24809.0, 'Bulgaria': 5428.0, 'Fr...
1996  {'Australia': 711660.0, 'Canada': 123129.0, 'F...
1998  {'Argentina': 283091.0, 'Australia': 625667.0,...
1999  {'Argentina': 25447.0, 'Australia': 525043.0, ...
```

(continued)

```
2000   {'Argentina': 147610.0, 'Australia': 775559.0,...
2001   {'Argentina': 4575.0, 'Australia': 741843.0, '...
2002   {'Algeria': 935.0, 'Argentina': 51850.0, 'Aust...
2003   {'Argentina': 2000.0, 'Australia': 350102.0, '...
2004   {'Argentina': 123512.0, 'Australia': 1029617.0...
2005   {'Algeria': 10600.0, 'Argentina': 703000.0, 'A...
2006   {'Algeria': 35144.0, 'Argentina': 12450.0, 'Au...
2007   {'Argentina': 73095.0, 'Australia': 288582.0, ...
2009   {'Argentina': 13400.0, 'Australia': 182181.0, ...
2011   {'Argentina': 374933.0, 'Australia': 947185.0,...
2013   {'Argentina': 12344.0, 'Australia': 72961.0, '...
2014   {'Australia': 276184.0, 'Bulgaria': 33000.0, '...
2015   {'Australia': 472182.0, 'Belgium': 0.0, 'Bulga...
2016   {'Argentina': 189000.0, 'Australia': 171548.0,...
2017   {'Argentina': 121989.0, 'Australia': 202026.0,...
2018   {'Australia': 264309.0, 'Belarus': 67604.0, 'B...
2019   {'Australia': 32724.0, 'Belarus': 57750.0, 'Br...
2020   {'Australia': 203640.0, 'Bulgaria': 46593.0, '...
```

Finally, save the record to file to perform association analysis. In the below code, split the record based on the token ":", and split country and export wheat quantity.

```
def retrieveCountryWheatImportDetil(strWheatImportRow):
  strCountrylist = strWheatImportRow #"{'Australia': 1805812.0,
'Canada': 231500.0, 'Saudi Arabia': 74243.0, 'United States of
America': 1293909.0}"
  strCountrylist=strCountrylist.replace("{","")
  strCountrylist=strCountrylist.replace("'","")
  strCountrylist=strCountrylist.replace("}","")
  strCountrylist=strCountrylist.replace("'","")
  countryTokens = strCountrylist.split(",")
  country= []
  yearImport= []
  importWheatTonnes= []
  for countrytoekn in countryTokens:
    countryinportdata=countrytoekn.split(":")
    country.append(countryinportdata[0])
    importWheatTonnes.append(float(countryinportdata[1]))
    yearImport.append(key)

  return country,importWheatTonnes,yearImport

dfCountries = pd.DataFrame(importCountries)

dfWheatQuantity = pd.DataFrame(importCountryQuantities)
```

(continued)

```
dfCountries.to_csv('EgyptWheatImportCountries.csv')
dfWheatQuantity.to_csv('EgyptWheatImportQuantities.csv')


import json
importCountries=[]
importCountryQuantities=[]
importYear=[]
for key, value in data_ddfEgyptWheatImportOriginARMnew.items():

  country=[]
  importWheatTonnes=[]
  wheatimportYear=[]
  countrylist = json.dumps(value)
  country,wheatimport,
wheatimportYear=retrieveCountryWheatImportDetil(countrylist)
  country.append(key)
  wheatimport.append(key)
  importCountries.append(country)
  importCountryQuantities.append(wheatimport)
  importYear.append(wheatimportYear)
```

Call the "retrieveCountryWheatImportDetil" for each record and write it to file. The records would look like:

```
[['"Australia"',
 ' "Canada"',
 ' "Saudi Arabia"',
 ' "United States of America"',
 1986],
 ['"Australia"', ' "Canada"', ' "France"', ' "Saudi Arabia"',
1987],
 ['"Australia"',
 ' "Canada"',
 ' "France"',
 ' "United States of America"',
 1988],
 ['"Australia"',
 ' "Bangladesh"',
 ' "Belgium-Luxembourg"',
 ' "Canada"',
 ' "France"',
 ' "United States of America"',
 1989],
 ['"Australia"',
 ' "Denmark"',
 ' "France"',
```

(continued)

```
 ' "Italy"',
 ' "Sweden"',
 ' "United Kingdom of Great Britain and Northern Ireland"',
 ' "United States of America"',
 ' "Yugoslav SFR"',
 1990],
 ['"Australia"',
 ' "Canada"',
 ' "France"',
 ' "Italy"',
 ' "Libya"',
 ' "Saudi Arabia"',
 ' "Turkey"',
 ' "United States of America"',
 1991],
 ['"Australia"',
 ' "Canada"',
 ' "Denmark"',
 ' "United Kingdom of Great Britain and Northern Ireland"',
 ' "United States of America"',
 1992],
 ['"Australia"', ' "Belgium-Luxembourg"', ' "Denmark"', '
"France"', 1993],
 ['"Australia"',
 ' "Canada"',
 ' "France"',
 ' "Germany"',
 ' "Sweden"',
 ' "United States of America"',
 1994],
 ['"Argentina"',
 ' "Bulgaria"',
 ' "France"',
 ' "United States of America"',
 1995],
```

1.3.7 *Step 6: Save the Import Origin Records*

Finally, save the import origin of countries to feed the association rule mining.

```
dfCountries = pd.DataFrame(importCountries)

dfWheatQuantity = pd.DataFrame(importCountryQuantities)

dfCountries.to_csv('EgyptWheatImportCountries.csv')
dfWheatQuantity.to_csv('EgyptWheatImportQuantities.csv')
```

Year	Import partner country list
2020	{Australia Bulgaria Canada France Hungary Italy Lithuania Romania Russian Federation Ukraine United Kingdom of Great Britain and Northern Ireland United States of America}
2019	{Australia Belarus Brazil Bulgaria Canada France Netherlands Poland Republic of Moldova Romania Russian Federation Ukraine United States of America}
2011	{Argentina Australia Belarus Brazil Canada France Germany Lithuania Republic of Moldova Romania Russian Federation Syrian Arab Republic Thailand Ukraine United States of America}
1986	{Australia Canada Saudi Arabia United States of America}

The number of people living in food insecurity grows every year. And the pandemic has made the problem worse.[a] But is it possible to give everyone access to the food they need?
Before it is too late, the answer rests in the steps that we take as humanity [7].
Author

[a]FOLLOW THE FOOD Why we still haven't solved global food insecurity—https://www.bbc.com/future/bespoke/follow-the-food/the-race-to-improve-food-security/

1.4 Risk Modeling: Association Rules to Overcome Food Insecurity

Software code for this model: Egypt_fooditems_importorigin_apriori.py (Jupyter Notebook Code)

1.4.1 *Step 1: Load Egypt Import Origins 1986–2020*

The following code reads and loads the historical Egypt import origin country data from file.

```
def aprioriFromFile(fname, minSup, minConf):
    C1ItemSet, itemSetList = getFromFile(fname)

    # Final result global frequent itemset
    globalFreqItemSet = dict()
    # Storing global itemset with support count
    globalItemSetWithSup = defaultdict(int)

    L1ItemSet = getAboveMinSup(
        C1ItemSet, itemSetList, minSup, globalItemSetWithSup)
    currentLSet = L1ItemSet
    k = 2

    # Calculating frequent item set
    while(currentLSet):
        # Storing frequent itemset
        globalFreqItemSet[k-1] = currentLSet
        # Self-joining Lk
        candidateSet = getUnion(currentLSet, k)
        # Perform subset testing and remove pruned supersets
        candidateSet = pruning(candidateSet, currentLSet, k-1)
        # Scanning itemSet for counting support
        currentLSet = getAboveMinSup(
            candidateSet, itemSetList, minSup, globalItemSetWithSup)
        k += 1

    rules = associationRule(globalFreqItemSet,
globalItemSetWithSup, minConf)

    rules.sort(key=lambda x: x[2])
```

(continued)

```
    return globalFreqItemSet, rules

if __name__ == "__main__":
    optparser = OptionParser()
    optparser.add_option('-f', '--inputFile',
                dest='inputFile',
                help='CSV filename',
                default=None)
    optparser.add_option('-s', '--minSupport',
                dest='minSup',
                help='Min support (float)',
                default=0.5,
                type='float')
    optparser.add_option('-c', '--minConfidence',
                dest='minConf',
                help='Min confidence (float)',
                default=0.5,
                type='float')

    (options, args) = optparser.parse_args()

    freqItemSet, rules = aprioriFromFile(options.inputFile,
options.minSup, options.minConf)
    print(list(freqItemSet))
    print(len(list(freqItemSet)))

     print('--------------------------------------------------------
----------------------')
    print(list(rules))
    print('---------------------------------------------------------
---------------------')
    for ruleCnt in range(len(rules)):
      print('---------Rule--------------' + str(ruleCnt+1))
      print(rules[ruleCnt])

    print('---------------------------------------------------------
---------------------')
```

1.4.2 Step 2: Perform Association Rules

The purpose of the association rules is to identify pattern of import origin countries and create a frequent itemset.

```
def associationRule(freqItemSet, itemSetWithSup, minConf):
  rules = []
  for k, itemSet in freqItemSet.items():
    for item in itemSet:
      subsets = powerset(item)
      for s in subsets:
        confidence = float(
          itemSetWithSup[item] / itemSetWithSup[frozenset(s)])
        if(confidence > minConf):
          rules.append([set(s), set(item.difference(s)),
confidence])
  return rules
```

1.4.3 Step 3: Perform Above-Minimum Support and Pruning Data

Integral of association rule mining is to identify the minimum support and prune the relationships that are below the minimum support.

```
def getAboveMinSup(itemSet, itemSetList, minSup,
globalItemSetWithSup):
  freqItemSet = set()
  localItemSetWithSup = defaultdict(int)

  for item in itemSet:
    for itemSet in itemSetList:
      if item.issubset(itemSet):
        globalItemSetWithSup[item] += 1
        localItemSetWithSup[item] += 1

  for item, supCount in localItemSetWithSup.items():
    support = float(supCount / len(itemSetList))
    if(support >= minSup):
      freqItemSet.add(item)

  return freqItemSet

def getUnion(itemSet, length):
  return set([i.union(j) for i in itemSet for j in itemSet if len(i.
union(j)) == length])
```

(continued)

```
def pruning(candidateSet, prevFreqSet, length):
  tempCandidateSet = candidateSet.copy()
  for item in candidateSet:
    subsets = combinations(item, length)
    for subset in subsets:
      # if the subset is not in previous K-frequent get, then remove
the set
      if(frozenset(subset) not in prevFreqSet):
        tempCandidateSet.remove(item)
        break
  return tempCandidateSet

def associationRule(freqItemSet, itemSetWithSup, minConf):
  rules = []
  for k, itemSet in freqItemSet.items():
    for item in itemSet:
      subsets = powerset(item)
      for s in subsets:
        confidence = float(
          itemSetWithSup[item] / itemSetWithSup[frozenset(s)])
        if(confidence > minConf):
          rules.append([set(s), set(item.difference(s)),
confidence])
  return rules
```

1.4.4 Step 4: Run the Association Rules

```
python   Egypt_fooditems_importorigin_apriori.py   --inputFile
Egypy_Import_Origins1986-2020.csv --minSupport 0.37 --
minConfidence 0.1
```

Output:

The minimum support and minimum confidence would identify all import origin relationships that could provide wheat imports to Egypt.

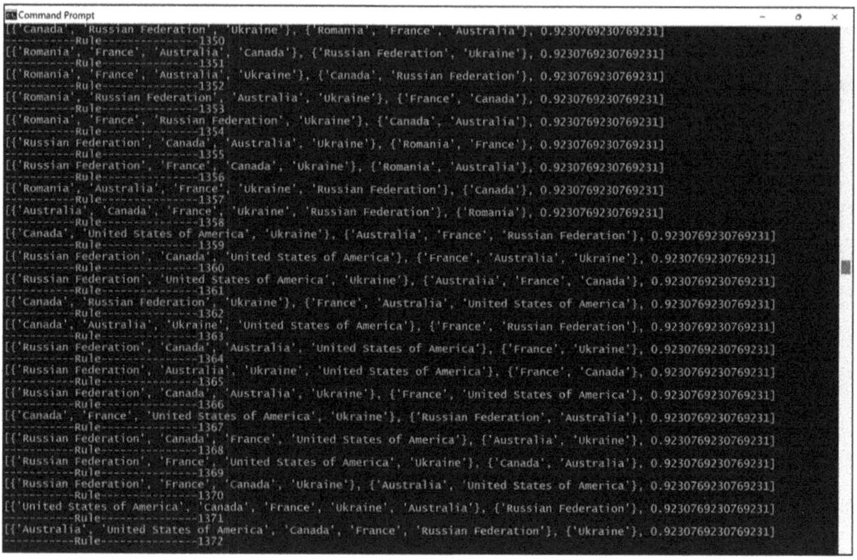

```
Command Prompt
[{'Canada', 'Russian Federation', 'Ukraine'}, {'Romania', 'France', 'Australia'}, 0.923076923076923l]
----------Rule---------------1350
[{'Romania', 'France', 'Australia', 'Canada'}, {'Russian Federation', 'Ukraine'}, 0.923076923076923l]
----------Rule---------------1351
[{'Romania', 'France', 'Australia', 'Ukraine'}, {'Canada', 'Russian Federation'}, 0.923076923076923l]
----------Rule---------------1352
[{'Romania', 'Russian Federation', 'Australia', 'Ukraine'}, {'France', 'Canada'}, 0.923076923076923l]
----------Rule---------------1353
[{'Romania', 'France', 'Russian Federation', 'Ukraine'}, {'Canada', 'Australia'}, 0.923076923076923l]
----------Rule---------------1354
[{'Russian Federation', 'Canada', 'Australia', 'Ukraine'}, {'Romania', 'France'}, 0.923076923076923l]
----------Rule---------------1355
[{'Russian Federation', 'France', 'Canada', 'Ukraine'}, {'Romania', 'Australia'}, 0.923076923076923l]
----------Rule---------------1356
[{'Romania', 'Australia', 'France', 'Ukraine', 'Russian Federation'}, {'Canada'}, 0.923076923076923l]
----------Rule---------------1357
[{'Australia', 'Canada', 'France', 'Ukraine', 'Russian Federation'}, {'Romania'}, 0.923076923076923l]
----------Rule---------------1358
[{'Canada', 'United States of America', 'Ukraine'}, {'Australia', 'France', 'Russian Federation'}, 0.923076923076923l]
----------Rule---------------1359
[{'Russian Federation', 'Canada', 'United States of America'}, {'France', 'Australia', 'Ukraine'}, 0.923076923076923l]
----------Rule---------------1360
[{'Russian Federation', 'United States of America', 'Ukraine'}, {'Australia', 'France', 'Canada'}, 0.923076923076923l]
----------Rule---------------1361
[{'Canada', 'Russian Federation', 'Ukraine'}, {'France', 'Australia', 'United States of America'}, 0.923076923076923l]
----------Rule---------------1362
[{'Canada', 'Australia', 'Ukraine', 'United States of America'}, {'France', 'Russian Federation'}, 0.923076923076923l]
----------Rule---------------1363
[{'Russian Federation', 'Canada', 'Australia', 'United States of America'}, {'France', 'Ukraine'}, 0.923076923076923l]
----------Rule---------------1364
[{'Russian Federation', 'Australia', 'Ukraine', 'United States of America'}, {'France', 'Canada'}, 0.923076923076923l]
----------Rule---------------1365
[{'Russian Federation', 'Canada', 'Australia', 'Ukraine'}, {'France', 'United States of America'}, 0.923076923076923l]
----------Rule---------------1366
[{'Canada', 'France', 'United States of America', 'Ukraine'}, {'Russian Federation', 'Australia'}, 0.923076923076923l]
----------Rule---------------1367
[{'Russian Federation', 'Canada', 'France', 'United States of America'}, {'Australia', 'Ukraine'}, 0.923076923076923l]
----------Rule---------------1368
[{'Russian Federation', 'France', 'United States of America', 'Ukraine'}, {'Canada', 'Australia'}, 0.923076923076923l]
----------Rule---------------1369
[{'Russian Federation', 'France', 'Canada', 'Ukraine'}, {'Australia', 'United States of America'}, 0.923076923076923l]
----------Rule---------------1370
[{'United States of America', 'Canada', 'France', 'Ukraine', 'Australia'}, {'Russian Federation'}, 0.923076923076923l]
----------Rule---------------1371
[{'Australia', 'United States of America', 'Canada', 'France', 'Russian Federation'}, {'Ukraine'}, 0.923076923076923l]
----------Rule---------------1372
```

```
----------Rule---------------4473706
[{'France', 'Slovenia', 'Argentina', 'Australia', 'Syrian Arab
Republic', 'Jordan', 'Romania', 'Russian Federation', 'Canada',
'Czechia', 'Bulgaria', 'Kazakhstan', 'Ukraine'}, {'United States of
America'}, 1.0]
----------Rule-------------4473707
[{'France', 'Slovenia', 'Argentina', 'Australia', 'Kazakhstan',
'Jordan', 'Romania', 'Russian Federation', 'Canada', 'Czechia',
'Bulgaria', 'United States of America', 'Ukraine'}, {'Syrian Arab
Republic'}, 1.0]
----------Rule-------------4473708
[{'France', 'Slovenia', 'Argentina', 'Australia', 'Kazakhstan',
'Syrian Arab Republic', 'Jordan', 'Romania', 'Canada', 'Czechia',
'Bulgaria', 'United States of America', 'Ukraine'}, {'Russian
Federation'}, 1.0]
----------Rule-------------4473709
[{'France', 'Slovenia', 'Argentina', 'Australia', 'Kazakhstan',
'Jordan', 'Romania', 'Russian Federation', 'Canada', 'Syrian Arab
Republic', 'Bulgaria', 'United States of America', 'Ukraine'},
{'Czechia'}, 1.0]
----------Rule-------------4473710
[{'France', 'Slovenia', 'Argentina', 'Kazakhstan', 'Syrian Arab
Republic', 'Jordan', 'Romania', 'Russian Federation', 'Canada',
'Czechia', 'Bulgaria', 'United States of America', 'Ukraine'},
{'Australia'}, 1.0]
----------Rule------------4473711
[{'France', 'Argentina', 'Australia', 'Kazakhstan', 'Syrian Arab
Republic', 'Jordan', 'Romania', 'Russian Federation', 'Canada',
'Czechia', 'Bulgaria', 'United States of America', 'Ukraine'},
{'Slovenia'}, 1.0]
```

```
----------Rule---------------4470714
[{'France', 'Slovenia', 'Australia', 'Jordan', 'Romania', 'Czechia',
'Bulgaria', 'Kazakhstan', 'Ukraine'}, {'Russian Federation',
'Canada', 'Argentina', 'Syrian Arab Republic', 'United States of
America'}, 1.0]
-------------------------------------------------------------
--------------
```

Without { Russia & Ukraine } imports would look like:

```
C:\Hanumayamma\Clustering>python
Egypt_fooditems_importorigin_apriori.py --inputFile
store_data_Copy_withoutRU_csv.csv --minSupport 0.3 --minConfidence
0.1
```

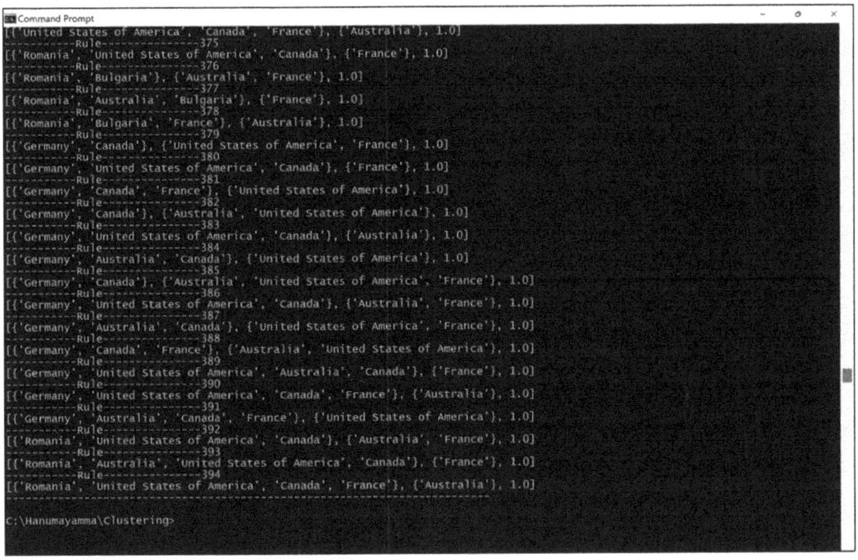


```
----------Rule-------------551750
[{'Jordan', 'Kazakhstan', 'United States of America', 'Slovenia',
'Syrian Arab Republic'}, {'Czechia', 'Canada', 'Bulgaria', 'Romania',
'Australia', 'Argentina', 'France'}, 1.0]
----------Rule-------------554746
[{'Czechia', 'Canada', 'Bulgaria', 'Jordan', 'Kazakhstan', 'United
States of America', 'Australia', 'Argentina', 'France', 'Slovenia',
'Syrian Arab Republic'}, {'Romania'}, 1.0]
----------Rule-------------554747
[{'Czechia', 'Canada', 'Bulgaria', 'Jordan', 'United States of
America', 'Romania', 'Australia', 'Argentina', 'France', 'Slovenia',
'Syrian Arab Republic'}, {'Kazakhstan'}, 1.0]
----------Rule-------------554748
```

```
[{'Czechia', 'Canada', 'Bulgaria', 'Kazakhstan', 'United States of
America', 'Romania', 'Australia', 'Argentina', 'France', 'Slovenia',
'Syrian Arab Republic'}, {'Jordan'}, 1.0]
```

As we can see imports of wheat to Egypt has built-in association rules. For instance, Egyptian government transacts with partner countries on the amount and grade of wheat to be imported.

Association rule assumes that the deficit wheat import could be fulfilled by other countries that have supplied prior to Egypt.

1.5 Machine Learning Technique: Clustering

Clustering is the process of arranging items of similar characteristics. A cluster is a collection of data objects that are like one another within the same cluster and are dissimilar to the objects in other clusters. A cluster of data objects can be treated collectively as one group, and so may be considered as a form of data compression. Although classification is an effective means for distinguishing groups or classes of objects, it requires the often-costly collection and labeling of a large set of training tuples or patterns, which the classifier uses to model each group. The characteristics could be physical or derived through computational, for instance, cluster of agriculture crops of similar phenological life cycles, cluster of agricultural crops with similar soil and water requirements, or cluster of agricultural equipment with similar range of heat/humidity thresholds. Clustering could also be used to mitigate import origins of the countries. Import and export policies of agricultural products are beyond the scope of the book.

Clustering is also called *data segmentation* in some applications because clustering partitions large datasets into groups according to their similarity. Clustering can also be used for outlier detection, where outliers (values that are "far away" from any cluster) may be more interesting than common cases.

As a branch of statistics, cluster analysis has been extensively studied for many years, focusing mainly on distance-based cluster analysis. Cluster analysis tools based on k-means, k-medoids, and several other methods have also been built into many statistical analysis software packages or systems, such as S-Plus, SPSS, and SAS.

In machine learning, clustering is an example of unsupervised learning. Unlike classification, clustering and unsupervised learning do not rely on predefined classes and class-labeled training examples. For this reason, clustering is a form of learning by observation rather than learning by examples. In data mining, efforts have focused on finding methods for efficient and effective cluster analysis in large databases. Active themes of research focus on the scalability of clustering methods, the effectiveness of methods for clustering complex shapes and types of data, high-dimensional clustering techniques, and methods for clustering *mixed numerical and categorical data in large databases*.

In summary [4]:

- Clustering is a technique for *finding similarity groups* in data, called *clusters*.
- Clustering is often called an unsupervised learning task as no class values denoting an a priori grouping of the data instances are given, which is the case in supervised learning.

1.5.1 Types of Data in Cluster

Suppose that a dataset to be clustered contains objects, which may represent persons, houses, documents, countries, agricultural import origins, and so on. The main memory-based clustering algorithms typically operate on either of the following two data structures.

1.5.1.1 Data Matrix (Or Object-by-variable Structure)

This represents objects, such as agricultural commodity import origins, with o variables (also called measurements or attributes), such as country production, country export or import values, and so on. The structure is in the form of a relational table or n-by-p matrix (n objects, p variables) [1, 4, 5]:

$$\begin{bmatrix} x_{11} & \cdots & x_{1f} & \cdots & x_{1p} \\ \cdots & \cdots & \cdots & \cdots & \cdots \\ x_{i1} & \cdots & x_{if} & \cdots & x_{ip} \\ \cdots & \cdots & \cdots & \cdots & \cdots \\ x_{n1} & \cdots & x_{nf} & \cdots & x_{np} \end{bmatrix}$$

1.5.1.2 Dissimilarity Matrix (Or Object-by-object Structure)

This stores a collection of proximities that are available for all pairs of n objects. It is often represented by an n-by-n table [1, 4, 5]:

$$\begin{bmatrix} 0 & & & & \\ d(2,1) & 0 & & & \\ d(3,1) & d(3,2) & 0 & & \\ \vdots & \vdots & \vdots & & \\ d(n,1) & d(n,2) & \cdots & \cdots & 0 \end{bmatrix}$$

where d(i, j) is the measured difference or dissimilarity between objects i and j. In general, d(i, j) is a non-negative number that is close to 0 when objects i and j are highly similar or "near" each other and becomes larger the more they differ.

1.5.2 Distance Calculations

The dissimilarity (or similarity) between the objects described by interval-scaled variables is typically computed based on the distance between each pair of objects. The most popular distance measure is *Euclidean distance*, which is defined as (please see Eqs. 1.3 and 1.4)

$$d(i,j) = \sqrt{(x_{i1} - x_{j1})^2 + x_{i2} - x_{j2})^2 + \ldots + x_{in} - x_{jn})^2} \quad (1.3)$$

Another well-known metric is Manhattan (or city block) distance, defined as

$$d(i,j) = \mid x_{i1} - x_{j1} \mid + \mid x_{i2} - x_{j2} \mid + \ldots + \mid x_{in} - x_{jn} \mid \quad (1.4)$$

Both the Euclidean distance and Manhattan distance satisfy the following mathematic requirements of a distance function:

- $d(i,j) \geq 0$: distance is non-negative number.
- $d(i,i) = 0$: the distance of an object to itself is 0.
- $d(i,j) = d(j,i)$: distance is a symmetric function.
- $d(i,j) \leq (d(i,h) + d(h,j)$: going directly from object I to object j in space is no more than making a detour over any other object h (triangular inequality).

1.5.3 Clustering Approaches

Two of the most common data clustering algorithms are hierarchical clustering and non-hierarchical clustering. In hierarchical clustering, successive clusters are discovered using previously established clusters, while clusters are established all at once using non-hierarchical algorithms (please see Fig. 1.15).

Hierarchical algorithms can be further divided into divisive algorithms and agglomerative algorithms. Divisive clustering begins with all data elements in one cluster, which is later successively divided into smaller ones. On the other hand, agglomerative clustering begins with single-element clusters and works by successively merging them into larger ones [1, 4, 5].

Flowchart of the agglomerative hierarchical clustering algorithm. Agglomerative clustering considers each data point as a cluster in the beginning. Two clusters are then merged in each step until all objects are forced into the same group.

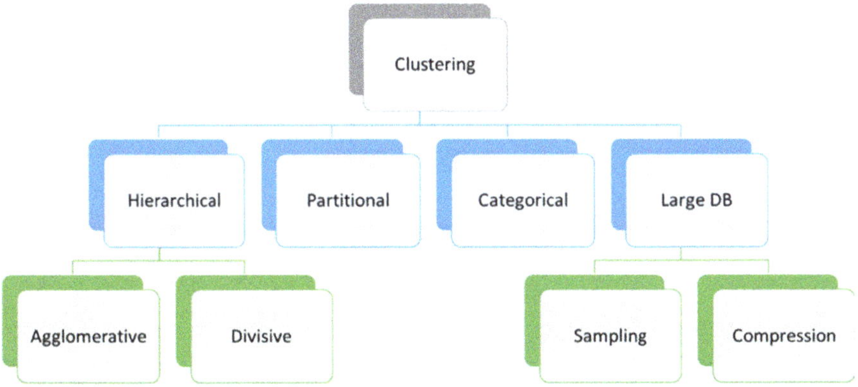

Fig. 1.15 Clustering

Hierarchical clustering is an agglomerative (top-down) clustering method. As its name suggests, the idea of this method is to build a hierarchy of clusters, showing relations between the individual members and merging clusters of data based on similarity. In the first step of clustering, the algorithm will look for the two most similar data points and merge them to create a new "pseudo-data point," which represents the average of the two merged data points. Each iterative step takes the next two closest data points (or pseudo-data points) and merges them. This process is generally continued until there is one large cluster containing all the original data points. Hierarchical clustering results in a "tree," showing the relationship of all the original points [8].

- Hierarchical clustering takes as input a set of points [8].
- It creates a tree in which the points are leaves, and the internal nodes reveal the similarity structure of the points.
- The tree is often called a "dendrogram."
- The method is summarized below:
  - Place all points into their own clusters while there are more than one cluster; do merge the closest pair of clusters.

- The behavior of the algorithm depends on how "closest pair of clusters" is defined.
- Clusters are created in levels creating sets of clusters at each level.
- Agglomerative (please see Fig. 1.16):
  - Initially each item in its own cluster.
  - Iteratively clusters are merged.
  - Bottom up.

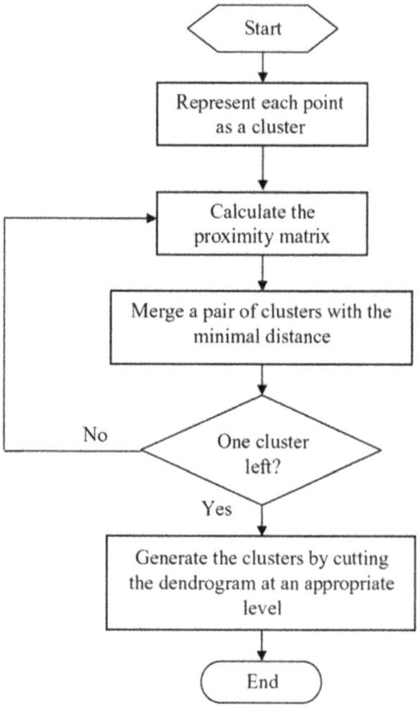

Fig. 1.16 Agglomerative clustering flowchart

- Divisive:
  - Initially all items in one cluster.
  - Large clusters are successively divided.
  - Top down.

- Dendrogram (please see Fig. 1.17): a tree data structure which illustrates hierarchical clustering techniques.
- Each level shows clusters for that level.
  - Leaf—individual clusters
  - Root—one cluster

- A cluster at level i is the union of its children clusters at level i+1.

Hierarchical clusters compute the following distance metrics to evaluate the clusters [1, 4, 5, 8]:

Fig. 1.17 Dendrogram

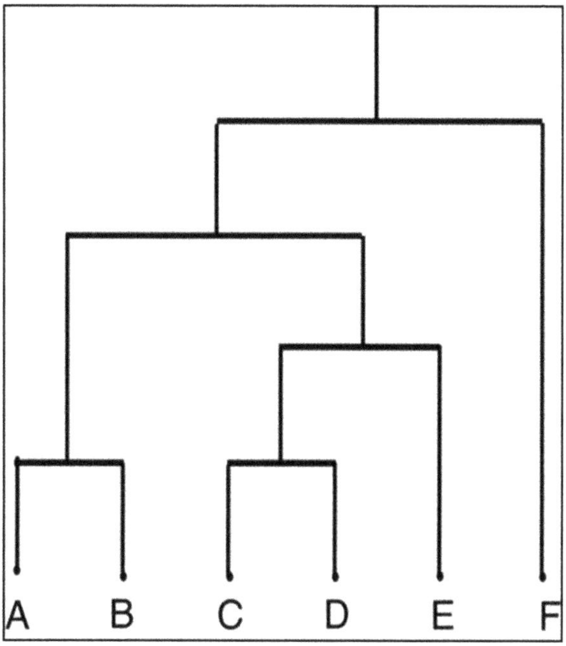

$$\text{Centroid} = C_m = \frac{\sum_{i=1}^{N} (tmi)}{N} \tag{1.5}$$

The centroid of a cluster is equal to the sum of the distance between all points divided by the total number of points.

$$\text{Radius} = R_m = \sqrt{\frac{\sum_{i=1}^{N} (tmi - C)2}{N}} \tag{1.6}$$

The radius of a cluster is the maximum distance between all the points and the centroid. Combine the two clusters whose resulting cluster has the lowest radius. A slight modification is to combine the clusters whose result has the lowest average distance between a point and the centroid. Another modification is to use the sum of the squares of the distances between the points and the centroid. In some algorithms, we shall find these variant definitions of "radius" referred to as "the radius."

Fig. 1.18 Sample cluster

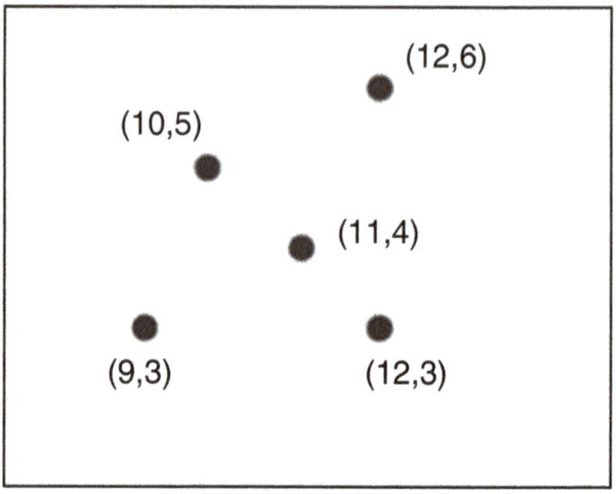

$$\text{Diameter} = D_m = \sqrt{\frac{\sum_{i=1}^{N} \sum_{i=1}^{N} (tmi - tmj)2}{N\,(N-1)}} \tag{1.7}$$

The diameter of a cluster is the maximum distance between any two points of the cluster. Note that the radius and diameter of a cluster are not related directly, as they are in a circle, but there is a tendency for them to be proportional. We may choose to merge those clusters whose resulting cluster has the smallest diameter. Variants of this rule, analogous to the rule for radius, are possible. For instance, to calculate the radius and diameter of the following cluster (please see Fig. 1.18):

Step 1: Calculate the centroid of the cluster.

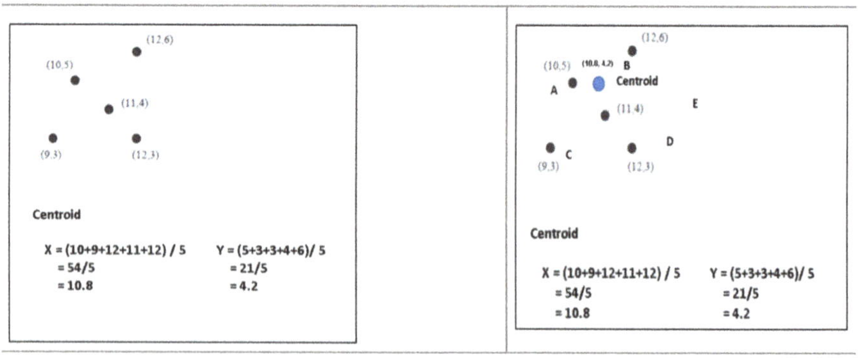

Here is the centroid of the cluster (10.8,4.2).

Step 2: Radius of the cluster. Here is the calculation of distance between the points and the centroid:

Distance between Point A and Centroid	0.8 * SQRT (2) = 1.131371
Distance between Point B and Centroid	2.1633
Distance between Point C and Centroid	2.1633
Distance between Point D and Centroid	1.2 * SQRT (2) = 1.697056
Distance between Point E and Centroid	0.2 * SQRT (2) = 0.282843

There is a tie for the two furthest points from the centroid, (9,3) and (12,6), both at distance $\sqrt{4.68} = 2.16$. Thus, the radius is *2.16* (please see Eqs. 1.5, 1.6, and 1.7). For the diameter, we find the two points in the cluster having the greatest distance. These are again (9,3) and (12,6). Their distance is $\sqrt{18} = 4.24$, so that is the diameter. Notice that the diameter is not exactly twice the radius, although it is close in this case. The reason is that the centroid is not on the line between (9,3) and (12,6).

1.5.3.1 Hierarchical Clustering: Merging Clusters

- *Single link*: The distance between two clusters is the distance between the closest points, also called "neighbor joining." In the figure below, the shortest distance between Cluster A points and Cluster B points is depicted. Similarly, Cluster A and Cluster C points and Cluster B and Cluster C points that are the shortest are linkage mode (please see Fig. 1.19).

- *Average link*: The distance between clusters is the distance between the cluster centroids. In the figure below, the distance between Cluster A and Cluster B

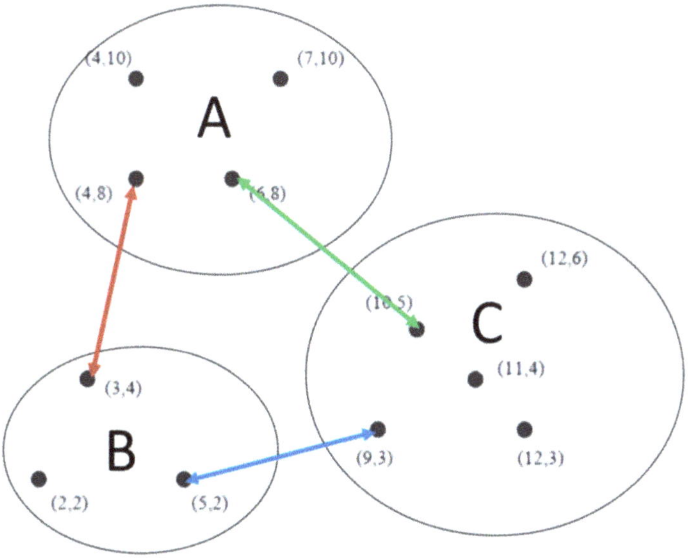

Fig. 1.19 Single link

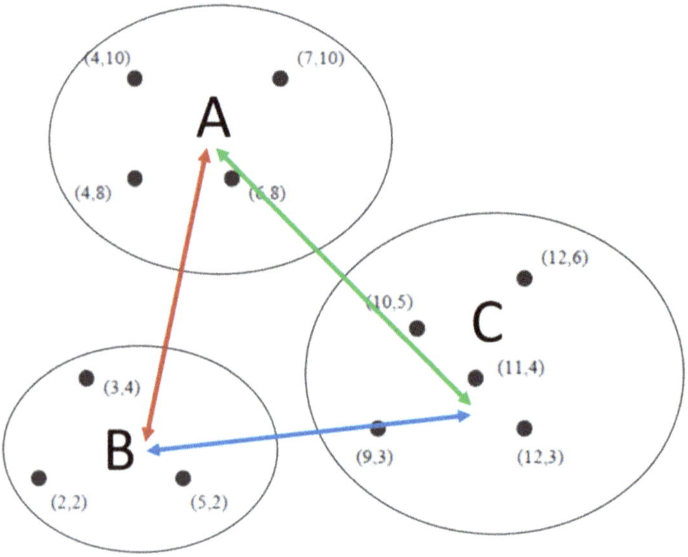

Fig. 1.20 Average link

centroid is depicted. Similarly, Cluster A and Cluster C points and Cluster B and Cluster C centroids are average linkage mode (please see Fig. 1.20).

- F23*Complete link*: The distance between clusters is the distance between the farthest pair of points. In the figure below, the longest distance between Cluster A points and Cluster B points is depicted. Similarly, Cluster A and Cluster C points and Cluster B and Cluster C points that are the longest are linkage mode (please see Fig. 1.21).

1.5.3.2 Non-hierarchical Clustering

As for non-hierarchical clustering, some of the common algorithms are k-means and fuzzy C-means. Other less conventional clustering algorithms include ant-based clustering, which is inspired by collective behavior in decentralized, self-organized systems. In addition to the clustering algorithm, a function or measure for determining the similarity, distance, or relatedness between elements is essential. In conventional clustering, feature-based measures such as cosine similarity, Jaccard distance, or Dice similarity are employed to support the use of hierarchical and non-hierarchical cluster.

The k-means algorithm searches for a predetermined number of clusters within an unlabeled multidimensional dataset. It accomplishes this using a simple conception of what optimal clustering looks like:

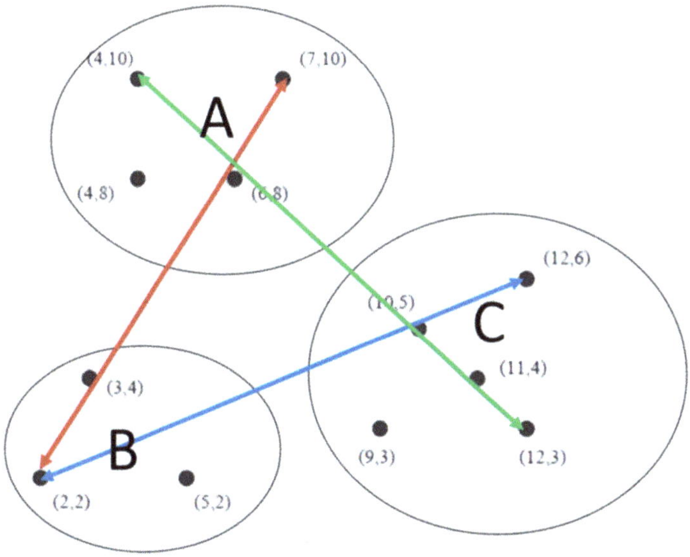

Fig. 1.21 Complete link

- The "cluster center" is the arithmetic means of all the points belonging to the cluster. Each point is closer to its own cluster center than to other cluster centers.

Those two assumptions are the basis of the k-means model. We will soon dive into exactly how the algorithm reaches this solution, but for now let's look at a simple dataset and see the k-means result.

1.6 Insecurity in the World's Food Supply: New Import Countries Based on Cluster Technique

International food markets, year 2022–2023, will probably face shortages due to the conflict. Russia and Ukraine together supply almost one-third of the world's wheat, a quarter of its barley, and nearly three-quarters of its sunflower oil, according to the International Food Policy Research Institute.[14] The latest shock comes on top of two years of disruption related to COVID-19. After panicked shoppers stripped

[14]Ukraine War Stokes Insecurity in the World's Food Supply—https://www.wsj.com/articles/ukraine-war-stokes-insecurity-in-the-worlds-food-supply-11647343035#:~:text=International%20food%20markets%20will%20probably,International%20Food%20Policy%20Research%20Institute

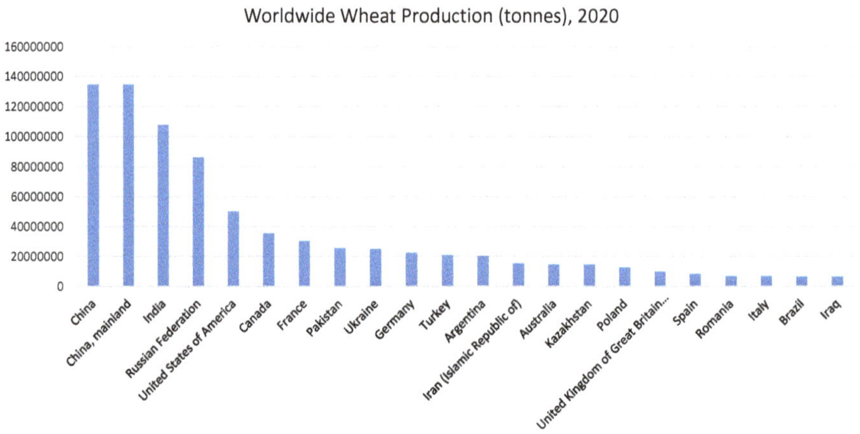

Fig. 1.22 Worldwide wheat production

supermarket shelves in the early days of the pandemic, world food supplies got back to normal surprisingly quickly, but they remained stressed [9].

To establish new import origin countries to overcome deficit by a major supplier issue, we can apply clustering technique to solve it. To prepare clustering data points, firstly, we need to consider the following:

- Current production of wheat
- Exports of wheat
- Percentage of experts of wheat production
- Capacity available to take up new export needs—that is, capacity availability to export (after meeting the internal consumption & stocks needed by local government due to COVID-19).

1.6.1 Step 1: Wheat-Producing Countries

The following figure contains the top wheat-producing countries in the world (FAOSTAT 2020).[15] China (134,254,710 tonnes), India (107,590,000 tonnes), the Russian Federation (85,896,326 tonnes), the United States (49,690,680), and Canada (35,183,000 tonnes) are the top five wheat-producing countries in 2020 (please see Fig. 1.22).

[15]FAOSTAT—https://www.fao.org/faostat/en/#data/QCL

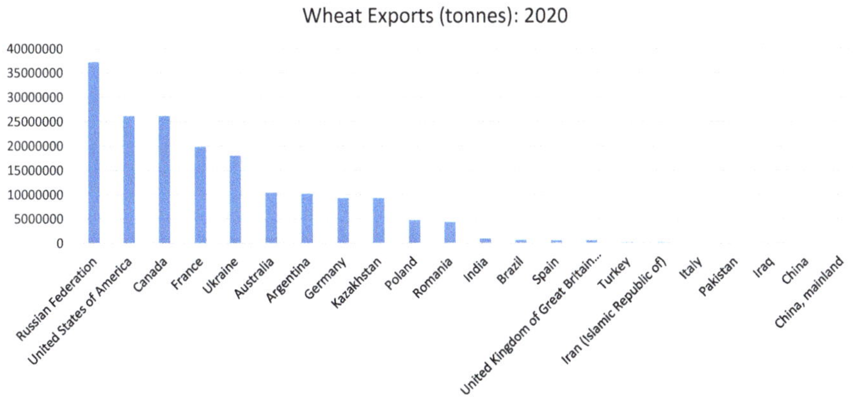

Fig. 1.23 Wheat exports

1.6.2 Step 2: Top Exporting Countries

Get the list of the top wheat-exporting countries in 2020. Of course, there is no rule that the top wheat-producing countries are also the top exporting countries. There could be many government and other policies that limit the export of wheat from a country.

As per FAOSTAT 2020, the following are the top wheat-exporting countries: leading the pack are the Russian Federation (37,267,014 tonnes), the United States (26,131,626), Canada (26,110,509), France (19,792,597), and Ukraine (18,055,673) (please see Fig. 1.23).

1.6.3 Step 3: Available Capacity to Export

If we rearrange production to exports on a Cartesian (X-Y) plane, we can observe countries such as the Russian Federation (37.26%), Canada (26.11%), the United States (26.13%), France (19.79%), and Ukraine (18.05%) utilizing more than 15% of their production to exports (please see Fig. 1.24).

Countries along the bottom right (India with 0.92% and China with 0.00%) are the potential candidates that could step up to meet the deficit in Egypt. Other big grain producers can try to make up some of the Russian and Ukrainian shortfalls. After five consecutive record crops, India's government has plenty of wheat inventories that could be exported. Australia also had a very good harvest this year, but shipping capacity is tight, with slots booked up months in advance.

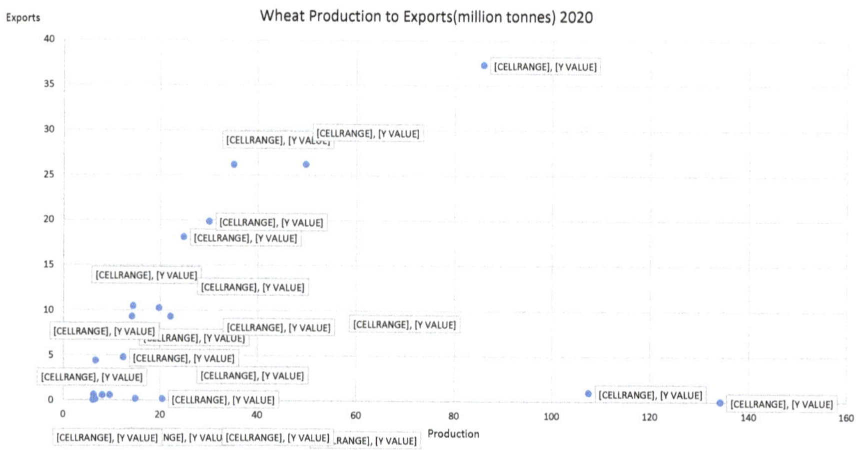

Fig. 1.24 Wheat production vs. export

1.7 Machine Learning Cluster Model: Wheat Import New Origins

Egypt wheat import origins	

	Software code for this model: Egypt_whatimport_countries_agglomerative.py (Jupyter Notebook Code)

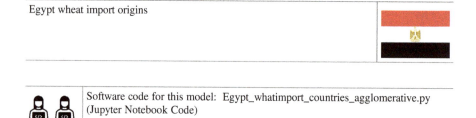

1.7.1 *Step 1: Load Worldwide Wheat Production (Million Tonnes) and Export (Million Tonnes) Data*

As prepared in the above section, load worldwide wheat production and export data.

```
import matplotlib.pyplot as plt
import pandas as pd
#%matplotlib inline
'exec(%matplotlib inline)'
```

(continued)

```
import numpy as np
Egypt_wheatimport_countries_data = pd.read_csv
('Egypt_wheatimport_countries_data.csv')
Egypt_wheatimport_countries_data.shape
print ('\nEgypt_wheatimport_countries_data\n')
print(Egypt_wheatimport_countries_data.head())
```

Output:

```
Command Prompt
C:\Hanumayamma\Clustering>python Egypt_wheatimport_countries_agglomerative.py

Egypt_wheatimport_countries_data

   ImporterID      Country  Production (tonnes)  Production ( million tonnes)  Export (million tonnes)
0           1       Canada             35183000                     35.183000                26.110509
1           3    Australia             14480217                     14.480217                10.400418
2           4       France             30144110                     30.144110                19.792597
3           5   Kazakhstan             14257950                     14.257950                 9.259493
4           6      Romania              6754530                      6.754530                 4.304783
```

1.7.2 Step 2: Construct Dendrogram

The following code constructs a hierarchical dendrogram.

```
data = Egypt_wheatimport_countries_data.iloc[:, 3:5].values
import scipy.cluster.hierarchy as shc

plt.figure(figsize=(10, 7))
plt.title("Egypt_wheatimport_countries Dendograms")
dend = shc.dendrogram(shc.linkage(data, method='average'))
print ('\npring dend \n')
print(dend)
plt.show()
```

Output:

```
{'icoord': [[15.0, 15.0, 25.0, 25.0], [5.0, 5.0, 20.0, 20.0],
[55.0, 55.0, 65.0, 65.0], [45.0, 45.0, 60.0, 60.0], [75.0, 75.0,
85.0, 85.0], [52.5, 52.5, 80.0, 80.0], [35.0, 35.0, 66.25, 66.25],
[95.0, 95.0, 105.0, 105.0], [50.625, 50.625, 100.0, 100.0],
[115.0, 115.0, 125.0, 125.0], [135.0, 135.0, 145.0, 145.0],
[155.0, 155.0, 165.0, 165.0], [140.0, 140.0, 160.0, 160.0],
[120.0, 120.0, 150.0, 150.0], [75.3125, 75.3125, 135.0, 135.0],
```

(continued)

```
[185.0, 185.0, 195.0, 195.0], [175.0, 175.0, 190.0, 190.0],
[105.15625, 105.15625, 182.5, 182.5], [12.5, 12.5, 143.828125,
143.828125]], 'dcoord': [[0.0, 0.004710000000017089,
0.004710000000017089, 0.0], [0.0, 26.67852257589513,
26.67852257589513, 0.004710000000017089], [0.0,
0.4815745912400282 7, 0.4815745912400282 7, 0.0], [0.0,
0.5957667225371448, 0.5957667225371448, 0.48157459124002827],
[0.0, 1.5144950266356094, 1.5144950266356094, 0.0],
[0.5957667225371448, 2.503097841988069, 2.503097841988069,
1.5144950266356094], [0.0, 4.228532784487995,
4.228532784487995, 2.503097841988069], [0.0, 5.236692895069273,
5.236692895069273, 0.0], [4.228532784487995, 7.194643869220741,
7.194643869220741, 5.236692895069273], [0.0,
4.7484945577066835, 4.7484945577066835, 0.0], [0.0,
1.1623736382566507, 1.1623736382566507, 0.0], [0.0,
2.5720753971857033, 2.5720753971857033, 0.0],
[1.1623736382566507, 6.647215427161645, 6.647215427161645,
2.5720753971857033], [4.7484945577066835, 11.61728192201094,
11.61728192201094, 6.647215427161645], [7.194643869220741,
13.17854293346081, 13.17854293346081, 11.61728192201094], [0.0,
8.081239043107438, 8.081239043107438, 0.0], [0.0,
17.52822950792522, 17.52822950792522, 8.081239043107438],
[13.17854293346081, 32.78996152788013, 32.78996152788013,
17.52822950792522], [26.67852257589513, 108.2557902514295 2,
108.2557902514295 2, 32.78996152788013]], 'ivl': ['13', '18',
'19', '4', '9', '12', '17', '10', '11', '8', '14', '15', '16', '1',
'3', '6', '7', '5', '0', '2'], 'leaves': [13, 18, 19, 4, 9, 12, 17,
10, 11, 8, 14, 15, 16, 1, 3, 6, 7, 5, 0, 2], 'color_list': ['g', 'g',
'r', 'r', 'r', 'r', 'r', 'r', 'r', 'r', 'r', 'r', 'r', 'r', 'r', 'r',
'r', 'r', 'b']}
[2 0 2 0 0 4 0 0 0 0 0 0 0 3 0 0 0 0 1 1]
```


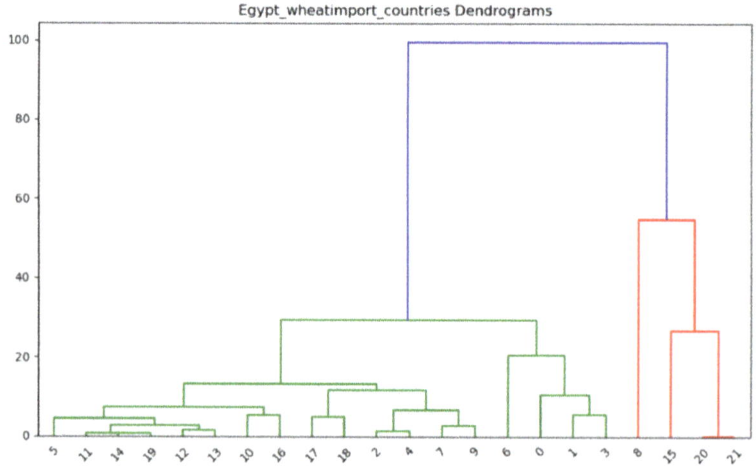

As it can be seen, Egypt wheat import dataset has two clusters constructed—as depicted in the above dendrogram.

1.7.3 Step 3: Construct Hierarchical Cluster

The purpose of the code is to construct the cluster.

```
from sklearn.cluster import AgglomerativeClustering
cluster = AgglomerativeClustering(n_clusters=5,
affinity='manhattan', linkage='average')
print(cluster.fit_predict(data))
plt.figure(figsize=(10, 7))
plt.scatter(data[:,0], data[:,1], c=cluster.labels_,
cmap='rainbow')

plt.title("Export (million tonnes) vs. Production ( million
tonnes) ")
plt.xlabel("Export (million tonnes)")
plt.ylabel("Production ( million tonnes)")

plt.show()
import matplotlib.pyplot as plt
y = Egypt_wheatimport_countries_data['Export (million tonnes)']
z = Egypt_wheatimport_countries_data['Production ( million
tonnes)']
n = Egypt_wheatimport_countries_data['Country']

fig, ax = plt.subplots()
ax.scatter(z, y, c=cluster.labels_, cmap='rainbow')

for i, txt in enumerate(n):
  ax.annotate(txt, (z[i], y[i]))

plt.title("Export (million tonnes) vs. Production ( million
tonnes) ")
plt.xlabel("Export (million tonnes)")
plt.ylabel("Production ( million tonnes)")

plt.show()
```

Output:

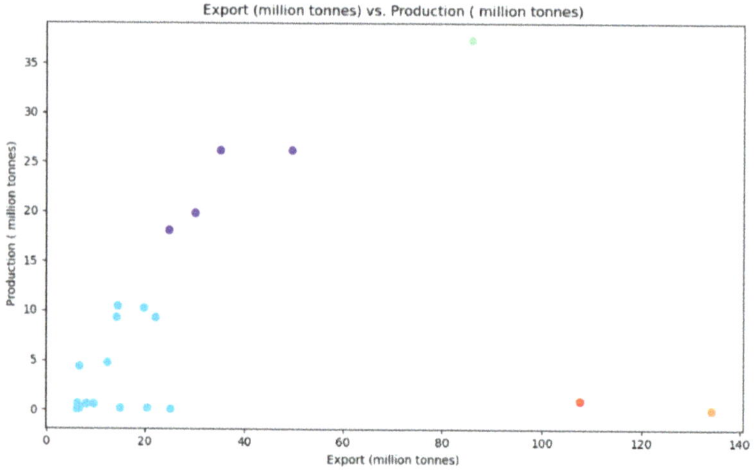

Each color dot represents an importer country. Five clusters are depicted by five colors.

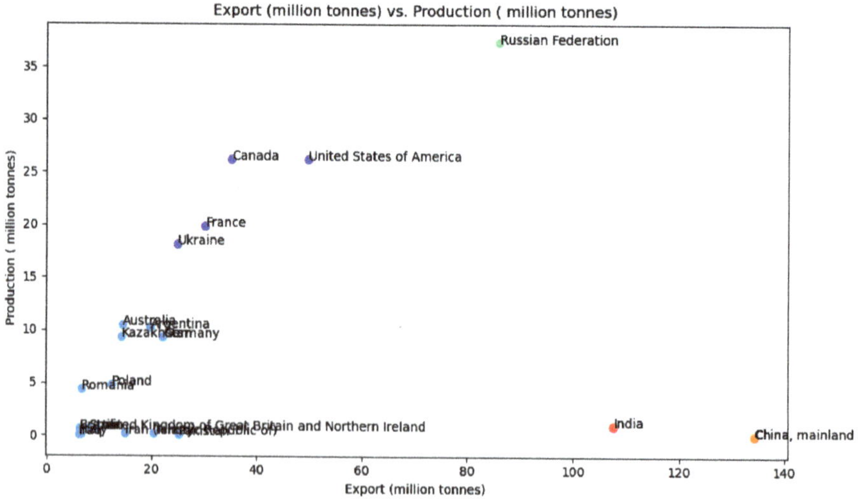

If you run the same cluster code with "average" linkage method, it would keep both China and India in different clusters, signifying Egypt could reach these two countries for filling imports.

1.7.4 Step 4: What-If Modeling

Given tariffs and restrictions on the Russian Federation, what if, removal of the Russian Federation from supply chain and Ukraine from the imports.

The resultant dendrogram and cluster are different, shifting more towards other countries.

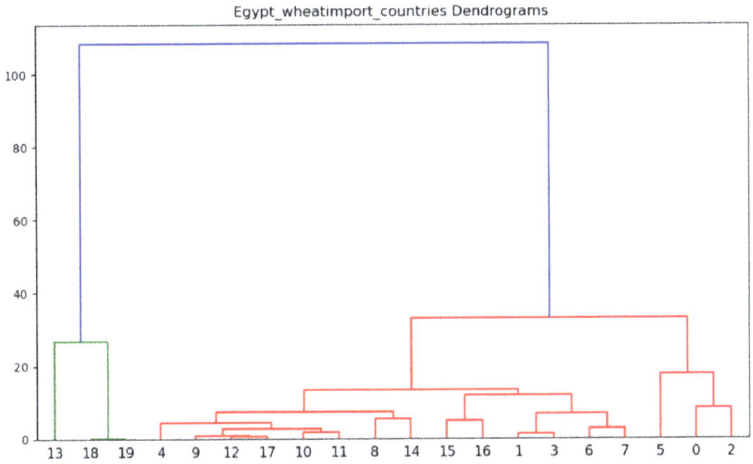

Cluster diagram:

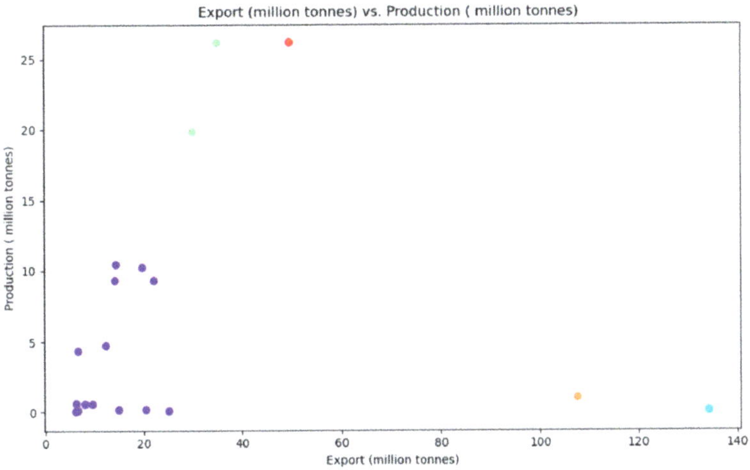

Cluster with labels:

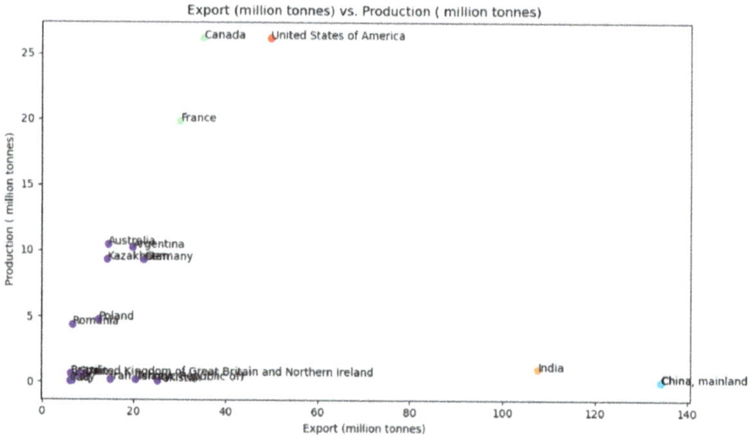

Without the Russian Federation, other major wheat-exporting countries such as the United States have to play a significant role.

No doubt wars and conflicts are the major source of food insecurity as can be seen with Ukraine war. Another major source of food insecurity is climate change. In the next 30 years, food supply and food security will be severely threatened if little or no action is taken to address climate change and the food system's vulnerability to climate change. According to the Intergovernmental Panel on Climate Change (IPCC), the extent of climate change impacts on individual regions will vary over time, and different societal and environmental systems will have varied abilities to mitigate or adapt to change. Negative effects of climate change include the continued rise of global temperatures, changes in precipitation patterns, an increased frequency of droughts and heat waves, sea-level rise, melting of sea ice, and a higher risk of more intense natural disasters.[16] I will cover in subsequent chapters the production vs. food security due to climate change [3].

1.8 Machine Learning Insights

Machine learning typically outputs the following types of analytics:

- Descriptive
- Predictive
- Prescriptive

[16]The World's Food Supply is Made Insecure by Climate Change—https://www.un.org/en/academic-impact/worlds-food-supply-made-insecure-climate-change

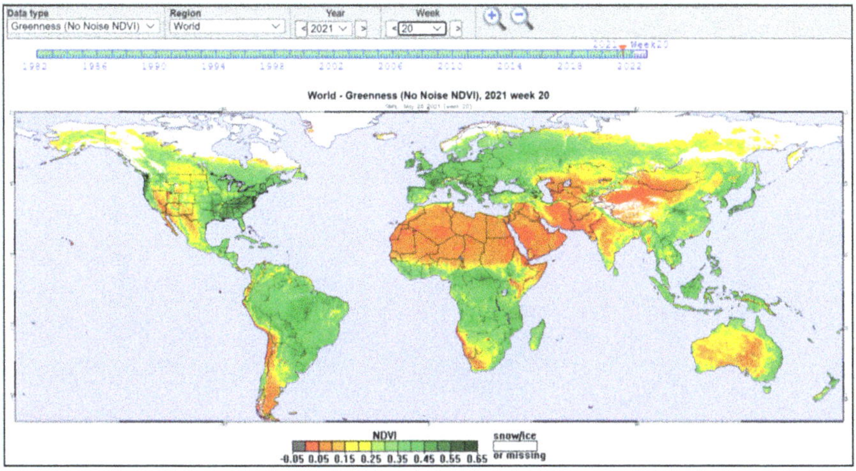

Fig. 1.25 World Vegetation Index,—2021, week 20

1.8.1 Descriptive Analytics: Satellite Vegetation Index

Descriptive analytics helps answer the question "What happened? (Or what is happening?)" in all its form: What were the world agricultural supply and demand[17] last year, and how would these affect food security [10]? What was the Vegetation Index in a particular subnational level last year, and how has it affected[18] food security [11]? Which countries are under severe Vegetation Index—drought and others? What were the countries that are in food insecurity [12]?[19] Descriptive analytics [13] is characterized by traditional business intelligence (BI) and visualizations such as pie charts, bar charts, dendograms, geolocation maps, heatmap, line graphs, tables, or generated narratives. As an example of descriptive analytics (please see Fig. 1.25), consider the following dashboard that displays the World Vegetation Index for week 20 of 2021 (updated April 2021).

No noise (smoothed) Normalized Difference Vegetation Index (SMN) (please see Fig. 1.25): global, 4 km, 7-day composite, validated. The SMN is derived from no noise NDVI, components of which were pre- and post-launch calibrated. SMN can be used to estimate the start and senescence of vegetation, the start of the growing season, and phenological phases.[20]

[17] USDA Data Products—https://www.ers.usda.gov/data-products/

[18] Ukraine war 'catastrophic for global food'—https://www.bbc.com/news/business-60623941

[19] Crisis in Ukraine Drives Food Prices Higher Around World—https://www.voanews.com/a/hold-for-wknd-crisis-in-ukraine-drives-food-prices-higher-around-world/6471261.html

[20] STAR—Global Vegetation Health Products: Browse Archived Images—https://www.star.nesdis.noaa.gov/smcd/emb/vci/VH/vh_browse.php

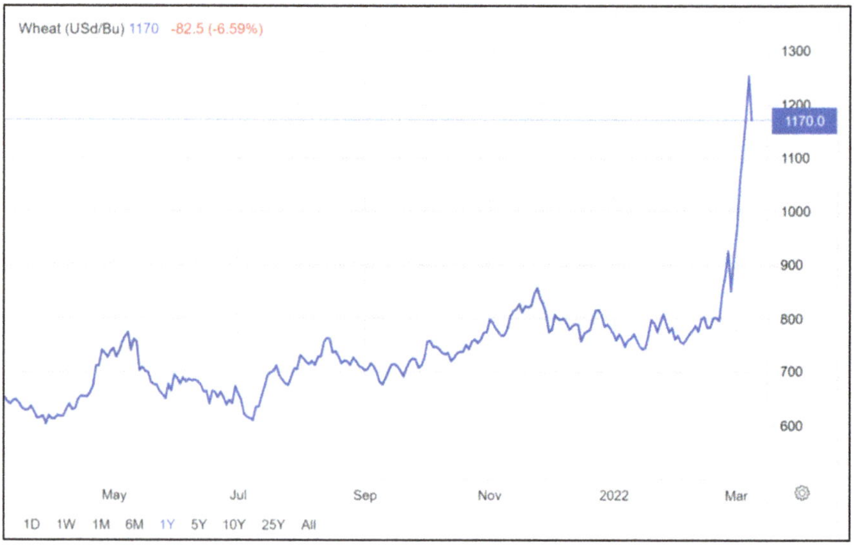

Fig. 1.26 Wheat futures

1.8.2 Predictive Analytics: Food Commodities and Food Insecurity Linkage

Predictive analytics answers, "What will happen?" or "What is likely to happen next?" Predictive analytics is inherently probabilistic (regression & forecasting), and it is the domain of data scientists. Predictive examples of time series forecasting use cases include crop yield prediction, commodity pricing, commodity future movements, export of commodities, and many more. In the following, please find wheat future[21] movement details and price spike due to recent events in the market (please see Fig. 1.26).

Wheat futures are available for trading in the Chicago Board of Trade[22] (CBOT), Euronext,[23] the Kansas City Board of Trade[24] (KCBT), and the Minneapolis Grain Exchange[25] (MGEX). The standard contract unit is 5000 bushels. The United States is the biggest exporter of wheat followed by the European Union, Australia, and Canada. Chicago wheat futures eased from a 14-year high peak of $12.8 per bushel

[21] Wheat Futures—https://tradingeconomics.com/commodity/wheat

[22] CBOT—https://www.cmegroup.com/company/cbot.html

[23] Euronext futures—https://www.euronext.com/en

[24] KC HRW Wheat—https://www.cmegroup.com/markets/agriculture/grains/kc-wheat.html

[25] Minneapolis Grain Exchange—https://www.mgex.com/

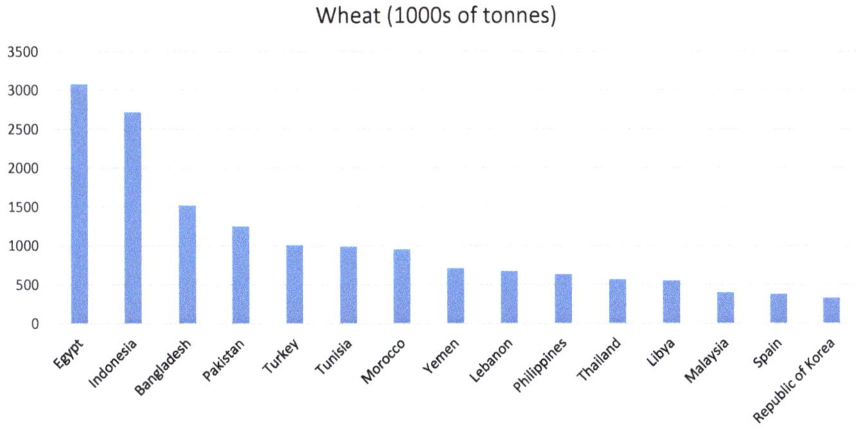

Fig. 1.27 Wheat-importing countries: 2020

to around $11.4, cooling after a run of six daily gains, as investors grappled with supply disruption caused by Russia's invasion of fellow grain exporter Ukraine.[26,27] Amid heavy sanctions and restrictive measures from western economies, exports from the Black Sea have nearly halted. Multinational food companies such as Bunge and ADM have closed facilities in the region, while the world's biggest container ship operator, Maersk/MSC, suspended service to Russian ports. At the same time, the Ukrainian military suspended all commercial operations from Ukrainian ports [10]. Such supply disruptions came on the heels of an already tight market with wheat stocks in major exporting countries at low levels.

Predictive analytics plays an important role in ensuring sustainable food security by mining signals that could have a negative influence on food security, for instance,[28] fears in the Middle East as Ukraine war hits wheat imports[29] (please see Fig. 1.27). The increase in wheat prices will drive food prices higher around the world—the "ripple effect"—as soaring wheat prices hit countries already facing inflation, food insecurity, and conflict [11, 12].

Nearly half of Tunisia's wheat imports come from Ukraine, and the Russian invasion has sent prices to a 14-year high. Even though the Tunisian state controls the price of bread, people fear they will inevitably feel the crunch. "If the price of bread goes up, it'll mean cutbacks elsewhere," said a resident in Tunisia. "We need

[26] Crisis in Ukraine Drives Food Prices Higher Around World—https://www.voanews.com/a/hold-for-wknd-crisis-in-ukraine-drives-food-prices-higher-around-world/6471261.html

[27] Ukraine war 'catastrophic for global food'—https://www.bbc.com/news/business-60623941

[28] Crisis in Ukraine Drives Food Prices Higher Around World—https://www.voanews.com/a/hold-for-wknd-crisis-in-ukraine-drives-food-prices-higher-around-world/6471261.html

[29] 'We need bread': fears in Middle East as Ukraine war hits wheat imports—https://www.theguardian.com/global-development/2022/mar/07/we-need-bread-fears-in-middle-east-as-ukraine-russia-war-hits-wheat-imports

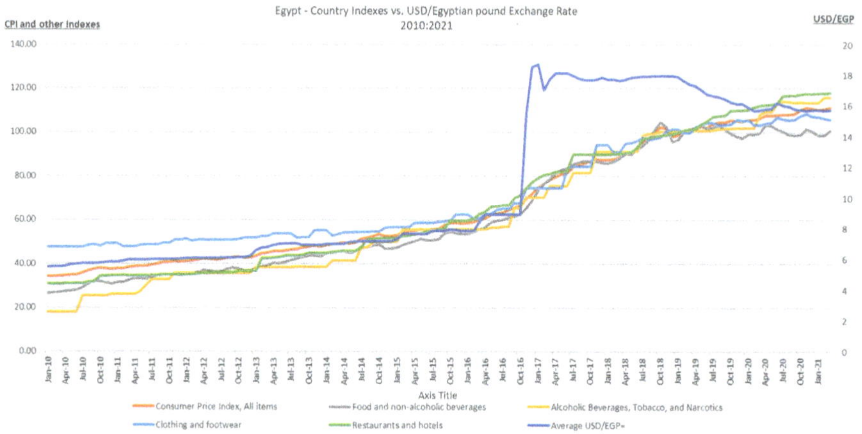

Fig. 1.28 Egypt USD exchange rate

the bread." Tunisia is highly vulnerable to such aftershocks, with a fragile economy battered in recent years by inflation and high unemployment and saddled with large amounts of public debt. But it is far from the only country in the Middle East and North Africa that would face difficulties in the event of prolonged supply-chain disruption and price hikes. The same applies to Yemen, Lebanon, and Egypt.

Please find historical country indexes for Egypt (please see Fig. 1.28), where you can see Consumer Price Index[30]; food and non-alcoholic beverages; alcoholic beverages, tobacco, and narcotics; clothing and footwear; and restaurants and hotels. The price index for the essential items has been trending higher since 2018—a trend precursor of food insecurity. According to the 2019 Global Hunger Index,[31] Egypt suffers from a moderate level of hunger, ranking 61 of 117 countries, compared to 61 of 119 countries in 2018. Food affordability, quality, and safety remain as challenges as Egypt continues to rely on global markets for more than half of its staples. Malnutrition is another growing public health concern, with a 21.4% stunting rate, 16% overweight and/or obesity rate, and 5.5% underweight rate of children under 5 years of age (Fig. 1.29).[32]

A closer look at the above graph (please see Fig. 1.30) reveals the CPI from October 2016, and other indexes have been trending higher. Additionally, as reported by WFP, food insecurity has also grown. The signal for this change was the drop in Egyptian pound (USD/Egyptian pound = [33])as it hits record low in

[30]Country Indexes and Weights—https://data.imf.org/regular.aspx?key=61015892

[31]WFP Countries list—https://www.wfp.org/countries

[32]Egypt—https://www.wfp.org/countries/egypt#:~:text=According%20to%20the%202019%20Global,than%20half%20of%20its%20staples

[33]USD/Egyptian Pound rate—https://fxtop.com/en/historical-exchange-rates.php?A=1&C1=USD&C2=EGP&MA=1&DD1=01&MM1=01&YYYY1=1970&B=1&P=&I=1&DD2=25&MM2=02&YYYY2=2022&btnOK=Go%21

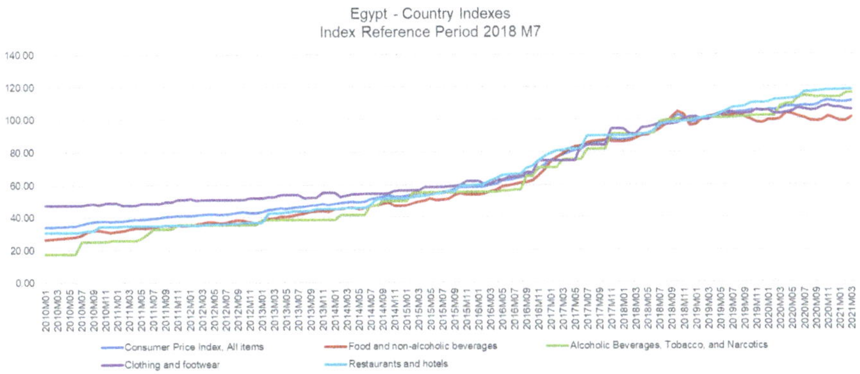

Fig. 1.29 Egypt country indexes

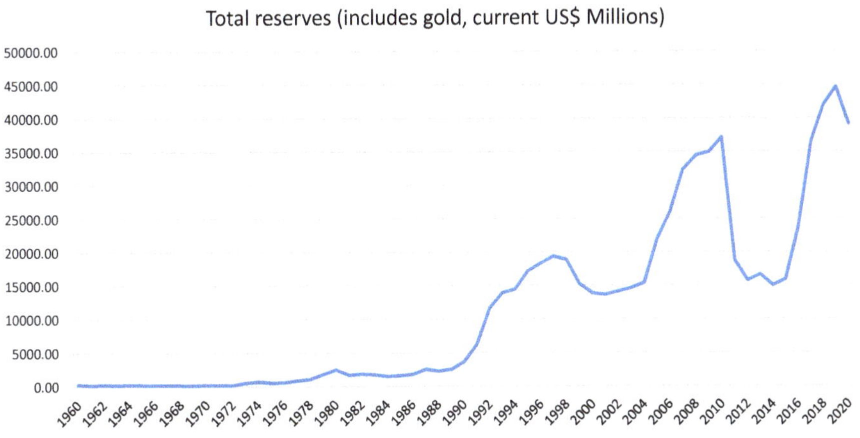

Fig. 1.30 Total reserves (includes gold)

October 2016.[34] Egypt has struggled to earn dollars since a 2011 uprising drove away tourists and foreign investors—the country's main sources of foreign currency. The central bank has been rationing dollars and imposing strict capital controls while maintaining the pound at an artificially strong official rate of 8.8 to the dollar.

Total reserves comprise holdings of monetary gold (please see Fig. 1.8), special drawing rights, reserves of IMF members held by the IMF, and holdings of foreign exchange under the control of monetary authorities. Total reserves play an important role in stabilizing price spikes and ensuring a higher food security, and it plays as a

[34] Egyptian pound hits record low on parallel market amid devaluation talk—https://www.reuters.com/article/egypt-currency/egyptian-pound-hits-record-low-on-parallel-market-amid-devaluation-talk-idUSL1N1CX0TV

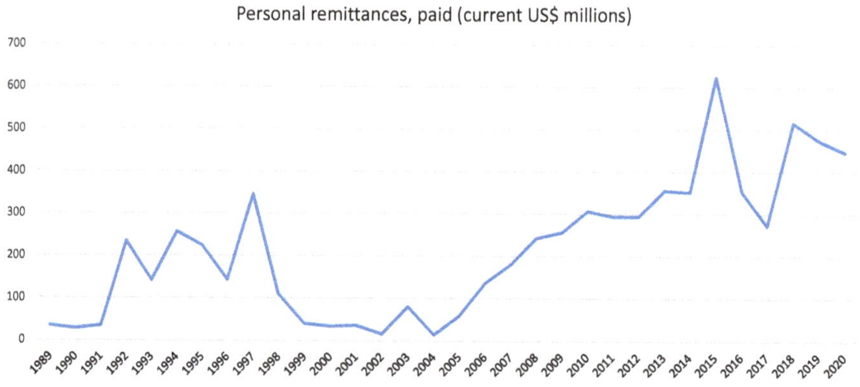

Fig. 1.31 Personal remittances

leading signal to food insecurity prognosis. Egypt,[35] in 2016–2017 (as shown in the below graph), was depleted of foreign reserves!

Another important signal that reflects a drop in purchasing power is personal remittances,[36] paid (current US$) (please see Fig. 1.31). Personal remittances comprise personal transfers and compensation of employees. Personal transfers consist of all current transfers in cash or in kind made or received by resident households to or from nonresident households. Personal transfers thus include all current transfers between resident and nonresident individuals. A drop in personal remittance renders a lower purchasing power, a factor that significantly impacts food security.

> "War leads to greater food insecurity, and food insecurity increases the chance of unrest and violence. So, a conflict in Ukraine leading to hunger and pushing people into food insecurity elsewhere could have [the] potential for unrest and violence in other areas. And really, the world cannot afford another conflict."
> Abeer Etefa
> Cairo, Senior Regional Communications Officer for Middle East, North Africa, Eastern Europe and Central Asia[a]

[a]WFP—https://www.wfp.org/media-contacts

1.8.3 Prescriptive Analytics: Economic Cycles and Gold Price Linkage

Prescriptive analytics takes the inputs from prediction and—combined with rules and constraint-based optimization—enables better decisions about what to do. The

[35]Total Foreign Reserves—FI.RES.TOTL.CD—https://data.worldbank.org/indicator/FI.RES.TOTL.CD

[36]Remittance—https://data.worldbank.org/indicator/BM.TRF.PWKR.CD.DT

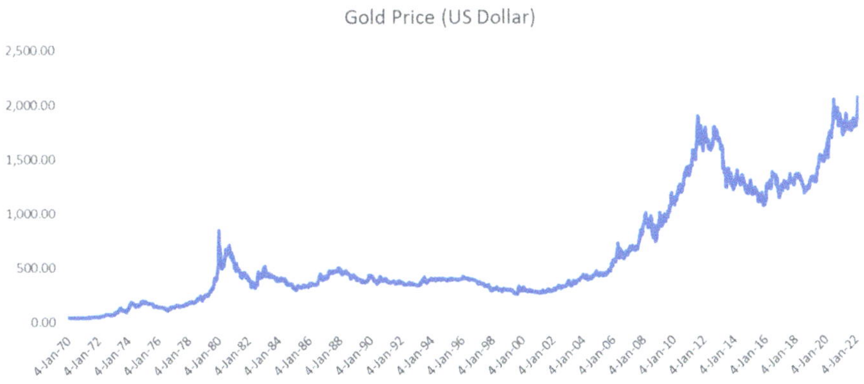

Fig. 1.32 Gold price

decision might be to send an automated task to a human decision-maker along with a set of next action recommendations or to send a precise next action command to another system. Prescriptive analytics is therefore best suited for situations where constraints are precise [14].

Please consider gold commodity pricing. Gold (please see Fig. 1.32) is mostly traded on the OTC London market, the US futures market (COMEX), and the Shanghai Gold Exchange (SGE). The standard futures contract is 100 troy ounces. Gold is an attractive investment during periods of political and economic uncertainty. Half of the gold consumption in the world is in jewelry, 40% in investments, and 10% in industry. The biggest producers of gold are China, Australia, the United States, South Africa, Russia, Peru, and Indonesia. The biggest consumers of gold jewelry are India, China, the United States, Turkey, Saudi Arabia, Russia, and the UAE.[37]

Gold price surged to as high as $2,066 an ounce on Tuesday, March 8, 2022, just 9 dollars shy of its record peak hit in August 2020, as geopolitical and economic uncertainties stemming from the Russia-Ukraine war lifted the demand for the safe-haven metal. In the latest developments, the United States announced a ban on imports of Russian oil, a move that threatened supply chains and heaped further inflationary pressure on economies worldwide. Soaring commodity prices fueled the fears of stagflation for the global economy, bolstering the metal's appeal as an inflation hedge.

Historical analysis of gold provides signals related to market economic events and macroeconomic stressors such as recessions, ongoing debt troubles, monetary easing by the US Federal Reserve, lower interest rates, a weaker US dollar, and onset of the COVID-19 pandemic. As can be seen during economic contraction events (please see Fig. 1.33), gold prices have peaked. For example, during the great

[37] Gold—https://tradingeconomics.com/commodity/gold

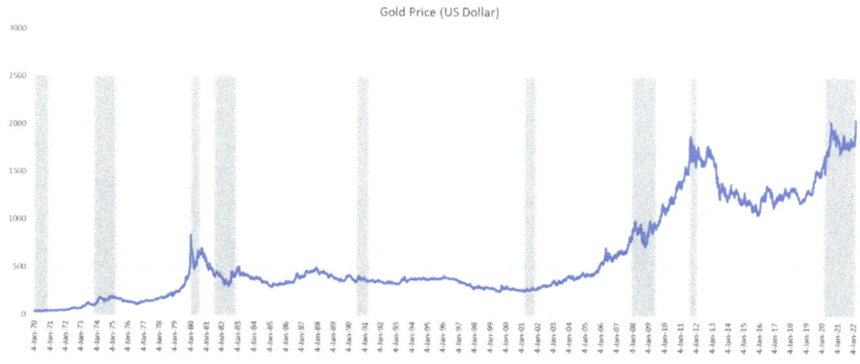

Fig. 1.33 Gold price and recessions

recessions (December 2007 to June 2009),[38] the gold price jumped from $795.6 per troy ounce, December 2007, to $932.5 per troy ounce, June 2009. Of course, gold price is influenced by several other factors, but economic events would have a higher influence as gold is considered one of the safest investments against market events and inflation hedges. Please see Appendix—Economic Frameworks and Macroeconomics section the US Recessions.

Prescriptive analytics enables us to tackle food insecurity issues. Prescriptive analytics enables us to learn from historical food insecurity events and forestall any future occurrence by signal mining or learning. The goal is to serve people by eliminating food insecurity. Consider the following table that has past food insecurity event and the main reasons or signals from the events.

	Main reasons or signals	Event
	Conflict, high food prices, depreciation of local currency and disrupted livelihoods are the major drivers of acute food insecurity. Injection of *Foreign Currency Reserves.*	Yemen[a]: Acute Food Insecurity Situation due to money access. Time period: October to December 2020 and projection for January to June 2021
	Scale-up of livelihood assistance for the winter wheat season, the spring season crops, and vulnerable herding households is essential to prevent further deterioration of household food production capacity in rural areas.	Afghanistan[b]: Acute Food Insecurity Situation due to Crop Failures. Time period: September to October 2021 and projection for November 2021 to March 2022
	Economic reactivation measures: Economic Opportunity creations for localities where a higher proportion of households have depleted their reserves and are employing Crisis or Emergency strategies.	Honduras[c]: Acute Food Insecurity Situation due to economic opportunity loss caused by the COVID-19. Time period: December 2020 to March 2021 and projections for April to June 2021 and July to September 2021

(continued)

[38] The Great Recession—https://www.federalreservehistory.org/essays/great-recession-of-200709

	Main reasons or signals	Event
	Employment: through economic reactivation plan, identify population that got affected—a behavior of the demand for agricultural and non-agricultural employment; temporary and permanent employment, loss of employment and reduction of wages should be monitored. The price of maize, beans and other foods should be monitored in the most affected departments, as they could increase due to storm damage or if the restrictions on mobilization due to the COVID-19 pandemic become strict again.	Guatemala[d]: Acute Food Insecurity Situation due to storm damage of essential commodities , spike in essential food commodities , and COVID-19 induced employment issues. Time period: projection update for November 2020 to March 2021
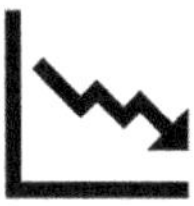	*Economic decline*: Inflation, exchange rate deterioration, reduction in remittances. *Poor harvests*: Poor agricultural harvests due to below-normal rainfall.	Haiti[e]: Acute Food Insecurity Situation due to Economic decline and poor harvest. Time period: August 2020 to February 2021 and projection for March to June 2021
	Recommendations: improvement in economic conditions to improve purchasing power and climate smart technologies for reducing animal diseases	Tajikistan[f]: Acute Food Insecurity Situation due to low purchasing capacity, fewer harvest and low livestock asset holding. Time period: September to December 2012

[a]Yemen: Acute Food Insecurity Situation October–December 2020 and Projection for January–June 2021—https://www.ipcinfo.org/ipc-country-analysis/details-map/en/c/1152947/
[b]Afghanistan: Acute Food Insecurity Situation September–October 2021 and Projection for November 2021–March 2022—https://www.ipcinfo.org/ipc-country-analysis/details-map/en/c/1155210/?iso3=AFG
[c]Honduras: Acute Food Insecurity Situation December 2020–March 2021 and Projections for April–June 2021 and July–September 2021—https://www.ipcinfo.org/ipc-country-analysis/details-map/en/c/1153046/?iso3=HND
[d]Guatemala: Acute Food Insecurity Situation Projection Update for November 2020–March 2021—https://www.ipcinfo.org/ipc-country-analysis/details-map/en/c/1152979/?iso3=GTMHaiti: Acute Food Insecurity Situation August 2020–February 2021 and Projection for March–June 2021—https://www.ipcinfo.org/ipc-country-analysis/details-map/en/c/1152816/?iso3=HTI
[e]Haiti: Acute Food Insecurity Situation August 2020–February 2021 and Projection for March–June 2021—https://www.ipcinfo.org/ipc-country-analysis/details-map/en/c/1152816/?iso3=HTI
[f]Tajikistan: Acute Food Insecurity Situation September–December 2012—https://www.ipcinfo.org/ipc-country-analysis/details-map/en/c/459517/?iso3=TJK

Machine learning plays an important role in understanding signals related to food insecurity events and provides a prescriptive insight.

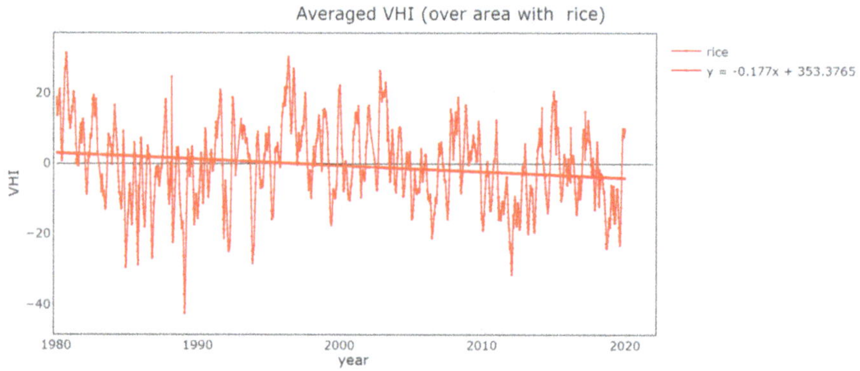

Fig. 1.34 VHI

1.8.4 Prognostics Analytics: Food Security and Vegetation Stress Indicators

Prognostics analytics plays an important role in food security. Prognostic variables such as the Moderate Resolution Imaging Spectroradiometer (MODIS) land surface temperature (LST) product provide temperature data crop land and play an important role in assessing temperature stress. LST and soil moisture are the prognostic variables[39] of the surface energy and water balance model. They are assimilated separately or simultaneously using different assimilation methods. Please find Weekly Averaged Time Series for Province #5: California of United States[40] (please see Fig. 1.34).

Consider another use case—Kansas[41] (please see Fig. 1.35). On average, Kansas is the largest wheat-producing state. Nearly one-fifth of all wheat grown in the United States is grown in Kansas. We are so well known for this major crop that Kansas is identified around the world as the "Wheat State" and "Breadbasket of the World." Somewhere in the world, wheat is being harvested every month of the year. In Kansas, we produce hard red winter and hard white wheat. Hard wheat flour provides a variety of bread products. Durum semolina and flour are used to make pasta.

Wheat is an important crop for Kansas farmers, the state's economy, and our daily diet. The wheat crop supports thousands of producers; their communities; and large industries for supplies, transportation, processing, and marketing. In Kansas, the

[39] Assimilating Satellite Land Surface States Data from Fengyun-4A—https://www.nature.com/articles/s41598-019-55733-3

[40] Averaged VHI—https://www.star.nesdis.noaa.gov/smcd/emb/vci/VH/vh_adminMeanByCrop.php?type=Province_Weekly_MeanPlot

[41] What is growing in Kansas? Our Main Grain Wheat—https://www.kfb.org/WebsitePageFile/file/d35474eb-cac9-4547-b1ed-61b4d89143dd/part%202%20-%20What's%20Growing%20in%20Kansas%20(1).pdf

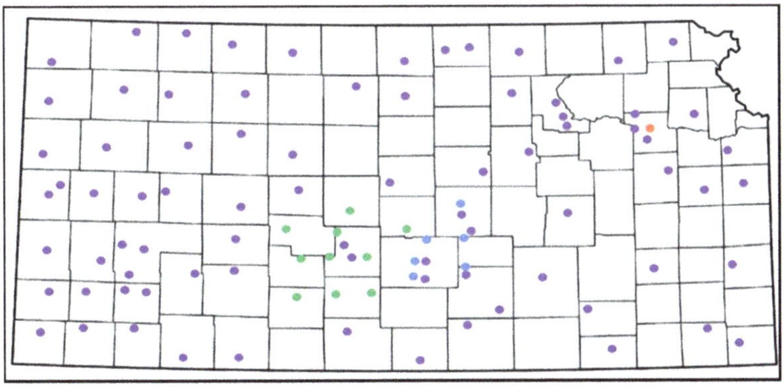

Fig. 1.35 Kansas weather stations

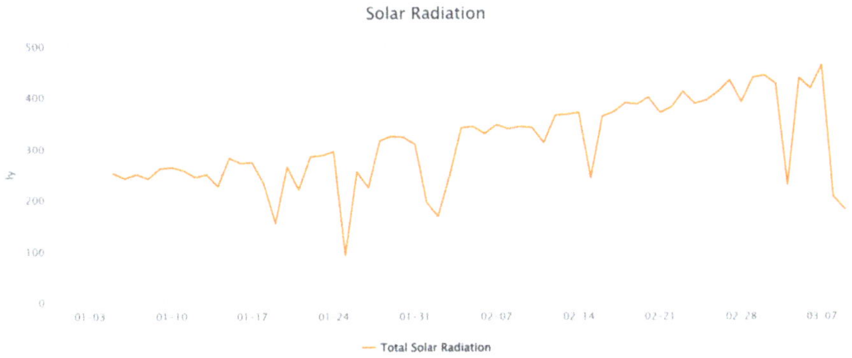

Fig. 1.36 Solar radiation

county that harvests the most wheat is Sumner County, Kansas. The other top ten wheat-producing counties in Kansas are Mitchell, Saline, Dickinson, Rice, McPherson, Reno, Kingman, Harper, and Sedgwick. There are 6 classes and more than 30,000 varieties of wheat. Please see Mesonet data map.[42]

Weather pattern: please look at solar radiation (Fig. 1.36), Garden City, 2022-01-01 to 2022-03-09.[43]

Vegetation Health Index[44] (VHI)—March 2021 METOP—AVHRR (please see Fig. 1.15).[45]

---

[42] Kansas Mesonet, 2022: Kansas Mesonet Historical Data. Accessed 10 March 2022, http://mesonet.k-state.edu/weather/historical

[43] Solar Radiation—http://mesonet.k-state.edu/weather/historical/#!

[44] Earth Observation—https://www.fao.org/giews/earthobservation/country/index.jsp?lang=en&code=USA#

[45] Vegetation Health Index—(VHI)—March 2021 METOP—AVHRR—https://www.fao.org/giews/earthobservation/asis/data/country/USA/MAP_NDVI_ANOMALY/HR/om2103h.png

The solar radiation has a huge impact on the wheat yield.[46] The effects of temperature and solar radiation have a strong influence on Spring wheat. This model uses relatively few, conservative relationships to define leaf-area development as a function of temperature, biomass accumulation as a function of intercepted radiation, and seed growth as calculated from a linear increase in harvest index with time. Temperature especially influenced the duration of ontogenetic events, with cool temperature being clearly advantageous for increasing the environmental yield potential of wheat. For unstressed wheat crops, the simple, mechanistic model accounted for much of the variability in grain yield among seasons. Additionally, climate change and dryland agriculture acreage are projected to expand due to climate changes. The Vegetation Health Index (VHI) illustrates the severity of drought based on vegetation health and the influence of temperature on plant conditions. The VHI is a composite index and the elementary indicator used to compute the ASI. It combines both the Vegetation Condition Index (VCI) and the Temperature Condition Index (TCI). The TCI is calculated using a similar equation to the VCI but relates the current temperature to the long-term maximum and minimum, as it is assumed that higher temperatures tend to cause a deterioration in vegetation conditions. A decrease in the VHI would, for example, indicate relatively poor vegetation conditions and warmer temperatures, signifying stressed vegetation conditions, and over a longer period would be indicative of drought. The VHI images are computed for the two main seasons and in three modalities: dekadal, monthly, and annual. The Vegetation Health stress Index is a prognostic variable that provides health of the cropland.[47] Examining typical dryland yield potentials and yield improvement measures is crucial for developing future dryland crop production systems. Weather data and the VHI are critical for agriculture (please see Fig. 1.37).[48]

After reading this chapter, you should be able to answer Artificial Intelligence, machine learning, and heuristics. You should be able to answer machine learning techniques including association rule mining (ARM) and clustering techniques. You should be able to apply AI techniques to solve importer risk origin for wheat commodities and, finally, to extend statistical techniques to address risk models for importer origin using ARM and cluster techniques.

---

[46] A model of the temperature and solar-radiation effects on spring wheat growth and yield—https://www.sciencedirect.com/science/article/pii/0378429091900735

[47] Evaluating sources of an apparent cold bias in MODIS land surface temperatures in the St. Elias Mountains, Yukon, Canada—https://tc.copernicus.org/preprints/tc-2021-211/

[48] Dryland maize yield potentials and constraints: A case study in western Kansas—https://onlinelibrary.wiley.com/doi/pdfdirect/10.1002/fes3.328

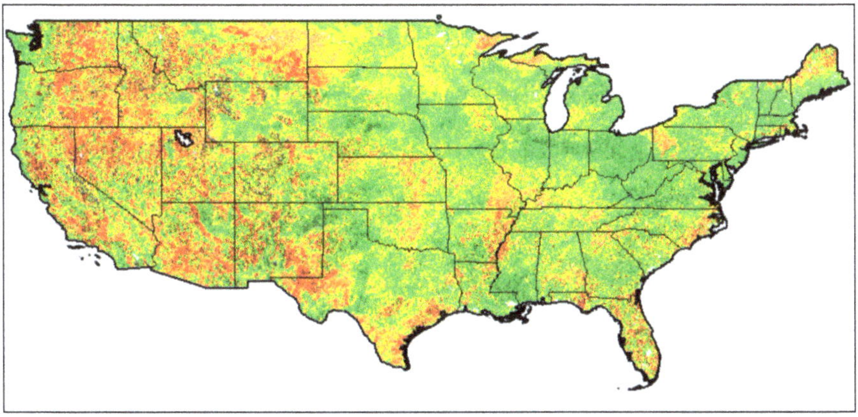

Fig. 1.37 US VHI. Disclaimers: The boundaries and names shown and the designations used on the maps do not imply the expression of any opinion whatsoever on the part of book or FAO concerning the legal status of any country, territory, city or area or of its authorities, or concerning the delimitation of its frontiers and boundaries

References

1. Chandrasekar Vuppalapati, Machine Learning and Artificial Intelligence for Agricultural Economics: Prognostic Data Analytics to Serve Small Scale Farmers Worldwide, Springer; 1st ed. 2021 edition (October 5, 2021), ISBN-13:978-3030774844
2. NIST/SEMATECH e-Handbook of Statistical Methods, http://www.itl.nist.gov/div898/handbook/, March 20, 2022.
3. Tilman D, Balzer C, Hill J, Befort BL (2011). Global food demand and the sustainable intensification of agriculture. PNAS, 108(50) pp. 20260-64.
4. Bing Liu, Web Data Mining: Exploring Hyperlinks, Contents, and Usage Data (Data-Centric Systems and Applications), Springer; 2nd ed. 2011 edition (June 26, 2011), ISBN-13:978-3642194597
5. Jiawei Han ,Micheline Kamber and Jian Pei, Data Mining: Concepts and Techniques, Publisher: Morgan Kaufmann; 3 edition (June 15, 2011), ISBN-10: 9780123814791
6. Richard D. Taylor and Won W. Koo, 2012 Outlook of the U.S. and World Wheat Industries, 2012-2021, June 2012, https://ageconsearch.umn.edu/record/133393/files/AAE696.pdf, Access Date: March 10, 2022
7. William Park, FOLLOW THE FOOD - Why we still haven't solved global food insecurity, https://www.bbc.com/future/bespoke/follow-the-food/the-race-to-improve-food-security/, Access Date: March 10, 2022
8. Don Wunsch and Rui Xu, Clustering, Wiley-IEEE Press (October 24, 2008), ISBN-13: 978-0470276808
9. Carol Ryan, Ukraine War Stokes Insecurity in the World's Food Supply, March 15, 2022 7:17 am ET, https://www.wsj.com/articles/ukraine-war-stokes-insecurity-in-the-worlds-food-supply-11647343035#:~:text=International%20food%20markets%20will%20probably,International%20Food%20Policy%20Research%20Institute., Access Date: March 19, 2022
10. Emma Simpson, Ukraine war 'catastrophic for global food', March 7 2022, https://www.bbc.com/news/business-60623941, Access Date: March 15, 2022

11. Rob Garver, Crisis in Ukraine Drives Food Prices Higher Around World, March 06, 2022 3:26 PM, https://www.voanews.com/a/hold-for-wknd-crisis-in-ukraine-drives-food-prices-higher-around-world/6471261.html, Access Date: March 15, 2022
12. Simon Speakman Cordall in Tunis and Lizzy Davies, 'We need bread': fears in Middle East as Ukraine war hits wheat imports, Mon 7 Mar 2022 08.38 EST, Access Date: March 15, 2022
13. Dan Vesset, "Descriptive analytics 101: What happened?", Publication Date: May 10, 2018, https://www.ibm.com/blogs/business-analytics/descriptive-analytics-101-what-happened/ , Access Date: April 30, 2019
14. Dan Vesset, "Prescriptive analytics 101: What should be done about it?", Publication Date: May 15, 2018, https://www.ibm.com/blogs/business-analytics/prescriptive-analytics-done/ , Access Date: Sep 18, 2019

Chapter 2
Heuristics

This chapter covers:

- Heuristics
- Mathematical Solvers
- Lear Programming or Linear Optimization
- Mixed-Integer Programming or Optimization
- Constraint Optimization or Constraint Programming
  - Knapsack
  - Risk Reduced Knapsack
- Mathematical Optimization Solver Use Cases
  - Manufacturing: General Assembly Line Balancing
  - Healthcare and Supply Chain : Surgical Suture Management
- Model and Simulation Process
- Choosing Solver Technology

This chapter introduces heuristics and mathematical optimization. It introduces the framework to formulate an optimization process. Next, it introduces linear programming, mixed-integer programming, and constraint programming techniques. To run the optimization code, the chapter introduces several types of solvers and compares the solvers. The chapter salutes and sincerely thanks George Dantzig for creating trillions of dollars of value and saving countless years of life across the globe through the power of optimization. This chapter showcases two industrial use cases: manufacturing general assembly line balancing and surgical suture management. Finally, the chapter concludes with model and simulation process and choosing solver technology.

C. Vuppalapati, *Artificial Intelligence and Heuristics for Enhanced Food Security*, International Series in Operations Research & Management Science 331,
https://doi.org/10.1007/978-3-031-08743-1_2

I would like to sincerely thank the contribution of George Dantzig who introduced the world to the power of optimization, creating trillions of dollars of value and saving countless years of life across the globe.[1]

From the industry use case point of view, I have seen heuristics applied to many engineering and science products/projects, including construction, transportation, manufacturing, healthcare, agriculture, machinery movement, and others. Regular data science and heuristics project phases are highly similar, except that there are differences in project scoping and model design phases. During the project initiation phase of Knowledge Discovery in Databases (KDD), developing an application understanding is critical to progress the subsequent phases of KDD. In my experience, the more time that data engineering, data science, and product/project teams spend during this phase, the clearer the picture and the higher the possible customer satisfaction when the model gets productized. The other phases of KDD, creating target dataset, data cleaning and preprocessing, and data mining task, are very similar. I would say understanding data and data signatures is pretty much the same for regular data science and heuristics. Finally, during the tail end of KDD, i.e., choosing data mining algorithms, validating and presenting insights are noticeably different between data science and heuristics projects. In data science projects, the choice of Artificial Intelligence algorithms—supervised, unsupervised, or deep learning—is based on the signature of the data and potential outcomes that are chartered during the project scoping phase. For heuristics projects, the solver design and translation of solver inputs such as decision variables, constraints, and objective functions into mathematical formulation and into feasible or infeasible solutions are very specific to the solver. The learning from the data and the codifying heuristics into mathematical formation are both data verified ultimately, a convergence of data-driven approach, if you will.

Project scoping with cost optimization, Resource efficiencies, and profit maximization goal	Define Data Needs, Build Data Collection, Perform Quality assessment	Define Constraints and Define Objective Functions	Translate Constraints and Objective functions into Solver mathematical formulation	Build and solve the model

Project scoping with cost optimization, resource efficiencies, and profit maximization goal	Define data needs, build data collection, perform quality assessment	Define constraints and define objective functions	Translate constraints and objective functions into solver mathematical formulation	Build and solve the model

[1] George Bernard Dantzig: The Pioneer of Linear Optimization—https://mbrjournal.com/2021/01/26/george-bernard-dantzig-the-pioneer-of-linear-optimization/

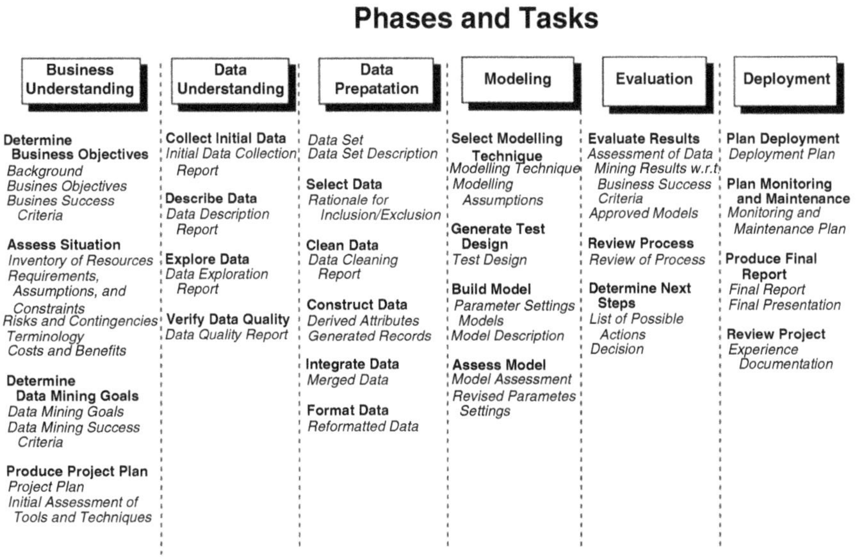
Phases and Tasks
Business Understanding
Data Understanding
Data Prepatation
Modeling
Evaluation
Deployment
Determine Business Objectives
Background
Busines Objectives
Busines Success Criteria
Assess Situation
Inventory of Resources
Requirements, Assumptions, and Constraints
Risks and Contingencies
Terminology
Costs and Benefits
Determine Data Mining Goals
Data Mining Goals
Data Mining Success Criteria
Produce Project Plan
Project Plan
Initial Assessment of Tools and Techniques
Collect Initial Data
Initial Data Collection Report
Describe Data
Data Description Report
Explore Data
Data Exploration Report
Verify Data Quality
Data Quality Report
Data Set
Data Set Description
Select Data
Rationale for Inclusion/Exclusion
Clean Data
Data Cleaning Report
Construct Data
Derived Attributes
Generated Records
Integrate Data
Merged Data
Format Data
Reformatted Data
Select Modelling Technique
Modelling Technique
Modelling Assumptions
Generate Test Design
Test Design
Build Model
Parameter Settings
Models
Model Description
Assess Model
Model Assessment
Revised Parametes Settings
Evaluate Results
Assessment of Data Mining Results w.r.t. Business Success Criteria
Approved Models
Review Process
Review of Process
Determine Next Steps
List of Possible Actions
Decision
Plan Deployment
Deployment Plan
Plan Monitoring and Maintenance
Monitoring and Maintenance Plan
Produce Final Report
Final Report
Final Presentation
Review Project
Experience Documentation

Fig. 2.1 CRISP-DM phases [1]

Artificial Intelligence (AI) and heuristics both deal with the data: in AI the models are built using data, and in the latter ones, the models are constructed from understanding optimizations that the human user embodies through the process of learning and transferring the process in mathematical formulations. Nevertheless, data and its validation are as important in heuristics models as in machine learning (ML) and AI models.

In data science during the initial phases (please see Fig. 2.1) of cross-industry standard process for data mining (CRISP-DM[2]) models, it is very vital to understand the business or use case expectations from the practitioner, stakeholders, sponsors, and/or end users. As a data science team, we generally approach with the model centric view [1]. That is, we focus on our specialty in data science models and try to fit prospective use cases into one of the data science models. I have seen numerous AI/ML projects that did not see the light of the day despite the huge amount of research and development and project dollars invested due to this limited view. The successful AI/ML programs emphasize business understandings and data understandings as critical and important as the design of the AI/ML algorithms.

Moreover, it is imperative that we take a stance and guide the stakeholders and project users if they are unsure about what kind of models are right for the opportunity at hand or problem they sought to solve. For example, I was pulled into a customer discussion during the project charter phase for an electric automobile company in California. The customer manufacturing team was under the impression

[2]CRISP-DM Help Overview—https://www.ibm.com/docs/en/spss-modeler/SaaS?topic=dm-crisp-help-overview

that the general assembly (GA) line balancing optimization, the main line that has workstations and operators to assemble electric automobiles, was an AI project. My sales account team was convinced and pulled data scientists to model AI. The experience of modeling GA line balancing optimization using AI was fitting square objects into circular holes. After several discussions, both the account team and customers are not advancing any phases of CRISP-DM. I could explain to customers that the use case they sought to solve is the classical mathematical optimization process that aims to model their current manufacturing process. In other words, we are modeling heuristics. Fast forward, one and a half years later, GA line balancing heuristic model is in production and used by the manufacturing team.

The Assembly Line Balancing Problem (ALB)

The assembly line balancing problem (ALB) is a production planning problem concerned with allocating tasks to the stations on the assembly line, first proposed and formulated as a mathematical programming problem in 1955[3]. A solution to the ALB is a set of decisions that determine which tasks are assigned to each station [2].

The first usage of the line production system for assembly was realized in 1901 by the Olds Motor Vehicle Company, and the concept was patented as an "assembly line" by the company owner Ransom Olds[4]. The Olds assembly line did not use a conveyor, however, as the vehicles were simply rolled on wheels from one workstation to the next. In 1913, Henry Ford's Model T assembly line first integrated conveyance with the assembly line concept, an innovation which achieved vast industrial success as well as historical acclaim [3].

In the diagram below (please see Fig. 2.2), the illustration of a GA line with operators, assembly processes, and tasks are depicted. Task[5] is the smallest possible indivisible operations work performed during assembly [5]. Consider the process of embodiment of several tasks performed while assembling an automobile or agriculture tractor.[6] For example, the assembly of wheels of agriculture tractor process (P_{wheel}) has following tasks: i) move to the left front end of the agriculture tractor ($T_{\text{move front}}$), ii) attach tractor wheels (T_{attach}), iii) fasten the bolts (T_{bolts}), and iv) close the wheel covers (T_{covers}). Operator has to account for front and rear wheel

[3]Salveson, M. E. (1955). The assembly line balancing problem. Journal of Industrial Engineering, 6, 18-25.

[4]Domm, Robert. Michigan Yesterday and Today, Publisher : Voyageur Press; First edition (October 1, 2009), ISBN-13 : 978-0760333853.

[5]A Study on General Assembly Line Balancing Methods and Techniques—https://tigerprints.clemson.edu/cgi/viewcontent.cgi?referer=&httpsredir=1&article=2550&context=all_dissertations

[6]Agricultural Tractors—https://www.postrock.k-state.edu/crops/tractorsafety/tractor-safety-book-module-4.pdf

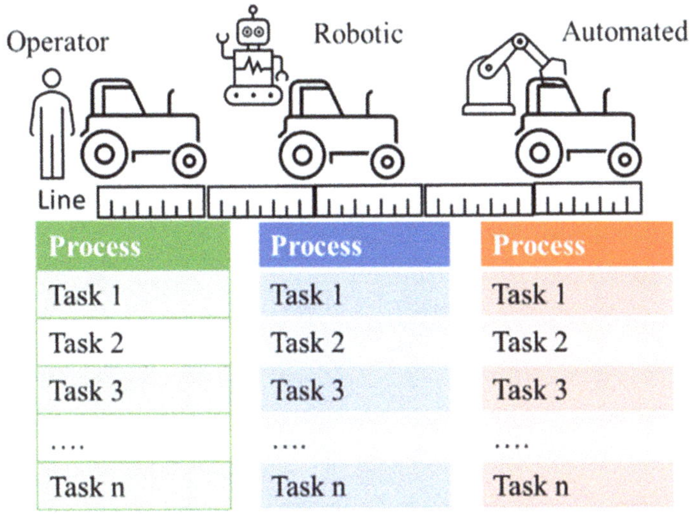

Fig. 2.2 GA line—process and task [4]

size differences when assembling the wheels. These processes must be repeated to all sections of the tractor: front right, rear left, and rear right. The tasks have time to complete, and summation of all individual task ($\sum all\ tasks$) time represents the amount of time to complete a process (please see Table 2.1). For some manual processes, there is break or rest time. And tasks also have order and relationships. For instance, precedence relationships between two individual tasks are used to codify these constraints, with the task that must come first labeled the predecessor and the latter task called the successor. Some of the tasks are manual, and some of them are automatic and robotic. For example, moving the wheels to the workstation is a manual process. Automatic tasks include testing communication antennas, fitting farm equipment, and so on. Process also has other parameters such as the zone on the manufacturing line to be performed and ergonomics (e.g., height of the workstation or underground task) [4]. I will discuss modeling in the subsequent sections.

This process was codified in the object model to facilitate solvers to find the solution.[7] More technical details will be covered in the subsequent sections of the chapter. The manufacturing and GA line team released several greenfield and brownfield factories models developed using heuristics [4].

Similar modeling steps can be applied to other GA line optimizations. For instance, consider the agriculture equipment advanced farm tractor GA line. Farm tractors[8] haven't changed much since internal combustion engines started replacing

[7] N. Bjørner, M. Levatich, N. P. Lopes, A. Rybalchenko, C. Vuppalapati. Supercharging Plant Configurations Using Z3. In Proc. of the 18th International Conference on Integration of Constraint Programming, Artificial Intelligence, and Operations Research (CPAIOR), July 2021.

[8] Assembling Tractors and Equipment—https://www.assemblymag.com/articles/89827-assembling-tractors-and-equipment

Table 2.1 Processes and tasks

Process	Tasks	Time (seconds)	Order
$P_{wheel\ attach}$	$T_{move\ front\ left}$	1	1
	T_{attach}	1	2
	T_{bolts}	1	3
	T_{covers}	2	4
	$T_{move\ rear\ left}$	1	5
	T_{attach}	1	6
	T_{bolts}	1	7
	T_{covers}	2	8
Process	Tasks	Time (seconds)	Order
$P_{wheel\ attach}$	$T_{move\ front\ right}$	1	9
	T_{attach}	1	10
	T_{bolts}	1	11
	T_{covers}	2	12
	$T_{move\ rear\ right}$	1	13
	T_{attach}	1	14
	T_{bolts}	1	15
	T_{covers}	2	16

steam power 100 years ago. But today's machines are faster, more powerful, lighter, cleaner, and smarter than ever. In fact, some large tractors boast more lines of software code than the early space shuttles [6].

Processes and tasks for tractor equipment give more emphasis on operational effectiveness of the tractor under harsh environments (please see Fig. 2.3). Tractor components are heavier, thicker, and stiffer than automotive parts, and they must endure higher loads and harsher work environments. Hence, GA line workstations are equipped with more power equipment for generating higher torque needs. That is, more demand for high-torque tooling that is flexible enough to be used on several applications in one station. Although similar design processes of GA line optimization could be applied, nevertheless, variation of certain tasks needed to address equipment-specific engineering characteristics.

2.1 Heuristics

A heuristic is a method drawn from experience, common sense, or an educated guess that aims at providing or contributing to providing a practical solution to a problem that is usually very difficult to solve (NP-Hard), and consequently an optimal or good, feasible solution is too complicated to obtain[9] [7].

[9]Practical Artificial Intelligence: Machine Learning, Bots, and Agent Solutions Using C# by Arnaldo Pérez Castaño Apress © 2018

Fig. 2.3 Tractor assembly [6]

Heuristic methods can be used to speed up the process of finding a good, feasible solution by providing us with a shortcut. This speed-up process is usually carried out via search algorithms where we traverse a tree representing the space of possible solutions. The application of certain problem-specific heuristics can significantly reduce tree search.

More than half a century ago, the research community began analyzing formally the quality of the solutions generated by heuristics.[10] The heuristics with guaranteed performance bounds eventually became known as approximation algorithms [8]. The idea behind approximation algorithms was to develop procedures to generate provable near-optimal solutions to optimization problems that could not be solved efficiently by the computational techniques available at that time. With the advent of the theory of NP-completeness in the early 1970s, approximation algorithms became more prominent as the need to generate near-optimal solutions for NP-hard optimization problems became the most important avenue for dealing with computational intractability. In general, NP-completeness is a class of computational problems[11] for which no efficient solution algorithm has been found. Many significant computer-science problems belong to this class, e.g., the traveling salesman problem, satisfiability problems, and graph-covering problems [9].

NP (nondeterministic polynomial) is a set of decision problems that can be solved by a **n**ondeterministic Turing machine in **p**olynomial time. P is a subset of NP (any problem that can be solved by a deterministic machine in polynomial time can also be solved by a nondeterministic machine in polynomial time). Informally, NP is a set of decision problems that can be solved by a polynomial time via a "lucky

[10] Handbook of Approximation Algorithms and Metaheuristics: Contemporary and Emerging Applications, Volume 2, Second Edition.

[11] NP Complete—https://www.britannica.com/science/NP-complete-problem

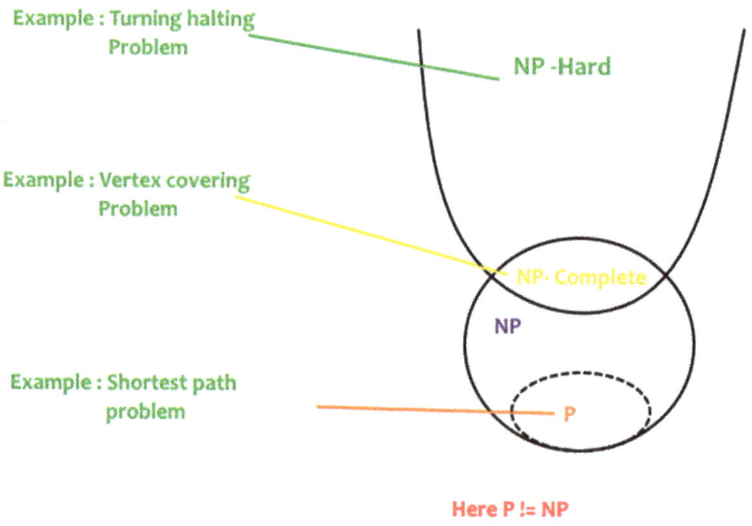

Fig. 2.4 NP

algorithm," a magical algorithm that always makes a right guess among the given set of choices (please see Fig. 2.4).

As it was established in the 1970s, for some problems one could generate near-optimal solutions quickly, while for other problems it was established that generating provably good suboptimal solutions was as difficult as generating optimal ones. Other approaches based on probabilistic analysis and randomized algorithms became popular in the 1980s. The introduction of new techniques to solve linear programming problems started a new wave for developing approximation algorithms that matured and saw tremendous growth in the 1990s. To deal with the inapproximable problems, in a practical sense, there were a few techniques introduced in the 1980s and 1990s. These methodologies have been referred to as metaheuristics and may be viewed as problem-independent methodologies that can be applied to sets of problems. Metaheuristics techniques include simulated annealing (SA), ant colony optimization (ACO), evolutionary computation (EC), Tabu search (TS), memetic algorithms (MAs), and so on. In a simple sense, a metaheuristic[12] is a combination of heuristic methods that aims to effectively promote the exploration of the search space [10]. Thus, it is possible to overcome the local search trap in a complex space. It can also be understood as a general heuristic method that is developed to guide a specific heuristic (a larger strategy for developing smaller or heuristic methods). A metaheuristic is inspired by several topics that can be highlighted, such as analogies with physical, chemical, biological, semantic, and social phenomena.

---

[12] Advanced Computational Methods for Agriculture Machinery Movement Optimization with Applications in Sugarcane Production—https://www.mdpi.com/2077-0472/10/10/434#cite

George Bernard Dantzig: The Pioneer of Linear Optimization[13]

"Linear programming was developed by George B. Dantzig in 1947 as a technique for planning the

diversified activities of the U.S. Air Force."[14,15] Robert Dorfman, Paul A. Samuelson and Robert M. Solow (1958; 3). Dorfman, Samuelson and Solow go on to note that Dantzig's fundamental paper was circulated privately for several years and finally published as Dantzig (1951). A complete listing of Dantzig's early contributions to developing the theory of linear programming can be found in Dantzig (1963; 597-597) [11–13].

George Dantzig introduced the world to the power of optimization, creating trillions of dollars of value and saving countless years of life across the globe. In this laudation, John Birge describes the fascinating life and incredible accomplishments of a scholar whose footprints led the way to almost everything the global economy produces [11].

Linear programs and Dantzig's many other contributions to optimization have driven enormous increases in productivity throughout the global economy. Linear programming has become a vital tool in advancing artificial intelligence and machine learning, and it is used in electrical stimulation therapy, chemotherapy plans, drug discovery, radiation therapy designs, and finding optimal diets. Linear programming and its various extensions continue to play an influential role in the economy and in all our lives [14].

Air Force Project SCOOP (Scientific Computation of Optimal Programs)[16]

Project SCOOP (Scientific Computation of Optimal Programs) officially started during this period [14]. In August 1948 Marshall Wood and I briefed the Air Staff on the use of electronic computers in military planning. We stated, I quote [14]:

1. "The primary objective of Project SCOOP is the development of an advanced design for an Integrated and comprehensive system for the planning and control of all Air Force activities."
2. "The recent development of high-speed digital electronic computers presages an extensive application of mathematics to large-scale management

(continued)

[13] George Bernard Dantzig: The Pioneer of Linear Optimization—https://mbrjournal.com/2021/01/26/george-bernard-dantzig-the-pioneer-of-linear-optimization/

[14] APPLIED ECONOMICS—Chapter 10: Linear Programming—https://www.economics.ubc.ca/files/2014/02/pdf_course_erwin-diewert-ECON594Ch10.pdf

[15] George Dantzig, 90; Created Linear Programming—https://www.latimes.com/archives/la-xpm-2005-may-22-me-dantzig22-story.html

[16] Impact of Linear Programming on Computer Development—https://apps.dtic.mil/dtic/tr/fulltext/u2/a157659.pdf

of problems of the quantitative type. Project SCOOP Is designed to prepare the Air Force to take maximum advantage of these developments."

Stigler's diet for 1939 data cost $39.93 per year (daily cost of $0.1093) and included varying amounts of wheat flour, evaporated milk, cabbage, spinach, and dried navy beans (Stigler 1945). Stigler's 1939 diet problem was the first "large-scale" problem that was solved using the simplex method (Dantzig 1963, 1990)[17]

In this book, we will limit heuristics to solving linear programming. Combinatorial optimization is the area of mathematical optimization that seeks to find the best solution to a problem out of a very large set of possible solutions.[18] Some of the examples include vehicle routing, farm equipment optimal use, fertilizer mix, manufacturing general line optimization, and supply chain. The possible number of solutions is numerous, too many for a computer to search them all.

Mathematical optimization[19] underpins many applications in science and engineering, as it provides a set of formal tools to compute the "best" action, design, control, or model from a set of possibilities [15]. In data science and machine learning, mathematical optimization is the engine of model fitting. The key elements include unconstrained, constrained, convex optimization, and optimization for model fitting. The chapter also provides a model to formulate and solve optimization problems early and often using standard modeling languages and solvers.

2.1.1 Model Formulation

Formulation consists[20] of: [15]

1. Model parameters: the parameters are fixed (e.g., constants or fixed) or varied (e.g., calculated at runtime).

[17] Author(s): George J. Stigler, The Cost of Subsistence, Source: Journal of Farm Economics, Vol. 27, No. 2 (May, 1945), pp. 303-314, Published by: Oxford University Press on behalf of the Agricultural & Applied Economics Association, Stable URL: http://www.jstor.org/stable/1231810 ; https://math.berkeley.edu/~mgu/MA170F2015/Diet.pdf. Accessed: 30/09/2013 18:56.

[18] About OR Tools—https://developers.google.com/optimization/introduction/overview

[19] C. Vuppalapati, A. Ilapakurti, S. Kedari, R. Vuppalapati, J. Vuppalapati and S. Kedari, "The Role of Combinatorial Mathematical Optimization and Heuristics to improve Small Farmers to Veterinarian access and to create a Sustainable Food Future for the World," 2020 Fourth World Conference on Smart Trends in Systems, Security and Sustainability (WorldS4), 2020, pp. 214-221, doi: https://doi.org/10.1109/WorldS450073.2020.9210339.

[20] Introduction to Optimization Models—https://www.usna.edu/Users/math/uhan/sa305/lessons/02.pdf

2. Model transition or decision variables: mathematical formulation evaluation will result in a decision path (UML refers it as outcome of Diamond), and the variables are centric to model formulation.
3. Aim or objective function: this is central to mathematical formulation with evaluation of decision variables—either to be maximized or minimized. In the below code, the purpose of objective function is to minimize travel time and distance.
4. Constraints: these are rules of use case, phenomena, or problem to be solved by the mathematical formulation. Constraints are bound to variables or general.

To solve a mathematical model, the solution must satisfy all the values of model decision variables and must validate solving all the model-defined constraints. The value of the model is the value of the objective function. The optimized solution shall solve the value that is as good as the value of all other feasible solutions. To narrow down and identify optimal solutions, following techniques are used:

- *Linear optimization (or linear programming)*[21]: linear optimization (or linear programming) is the name given to computing the best solution to a problem modeled as a set of linear relationships. These problems arise in many scientific and engineering disciplines. (The word "programming" is a bit of a misnomer, like how "computer" once meant "a person who computes." Here, "programming" refers to the arrangement of a plan rather than programming in a computer language.)
- *Mixed-integer programming (MIPs)*[22]: linear optimization problems that require some of the variables to be integers are called mixed-integer programs (MIPs). These variables can arise in a couple of ways—integer variables and Boolean variables.
- *Constraint programming (CP) or constraint satisfaction problem (CSP)*[23]: constraint optimization, or constraint programming (CP), is the name given to identifying feasible solutions out of a very large set of candidates, where the problem can be modeled in terms of arbitrary constraints. CP problems arise in many scientific and engineering disciplines. (The word "programming" is a bit of a misnomer, like how "computer" once meant "a person who computes." Here, "programming" refers to the arrangement of a plan, rather than programming in a computer language.) A constraint programming solver that uses SAT (satisfiability) methods.
- *Other optimizations*: assignment, routing, bin packing, and network flows use LP and CSP. In this book, I would consider them as an extension of basic optimization.

[21] Linear Optimization—https://developers.google.com/optimization/lp?hl=en
[22] Integer Optimization—https://developers.google.com/optimization/mip?hl=en
[23] Constraint Optimization—https://developers.google.com/optimization/cp?hl=en

The theory of linear programming provides a good introduction to the study of constrained maximization (and minimization) problems where some or all the constraints are in the form of inequalities rather than equalities. Many models in economics and agriculture can be expressed as inequality constrained optimization problems. A linear program is a special case of this general class of problems where both the objective function and the constraint functions are linear in the decision variables.

Linear programming problems are important for several reasons [12]:

- Many general constrained optimization problems can be approximated by a linear program.
- The mathematical prerequisites for studying linear programming are minimal; only knowledge of matrix algebra is required.
- Linear programming theory provides a good introduction to the theory of duality in nonlinear programming.

Our approach to heuristics modeling is very much influenced by concepts and methodologies honed and developed in software engineering and most specifically formal methods communities. Thus, a starting point is to describe using logical notation a set of domains, functions over the domains, and constraints over the signature. This is standard for any domain constraint engineering problem. Here are important high-level steps to solve a LP problem:

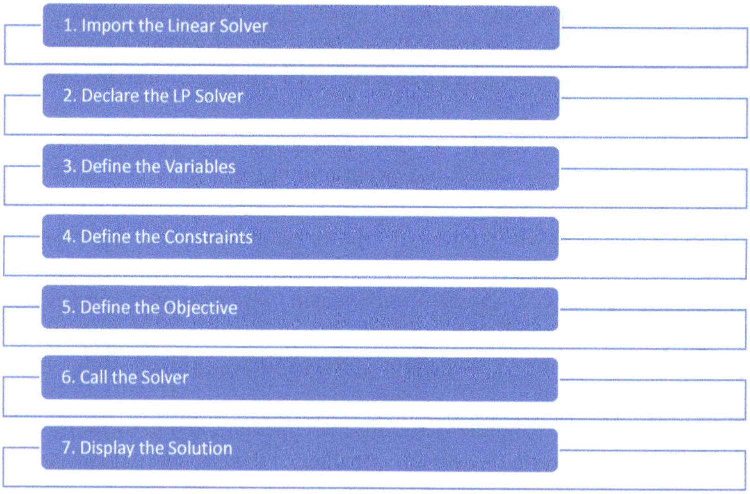

2.1.2 Solvers

To solve a mathematical model, the solution must satisfy all the values of model decision variables and must validate solving all the model-defined constraints. The

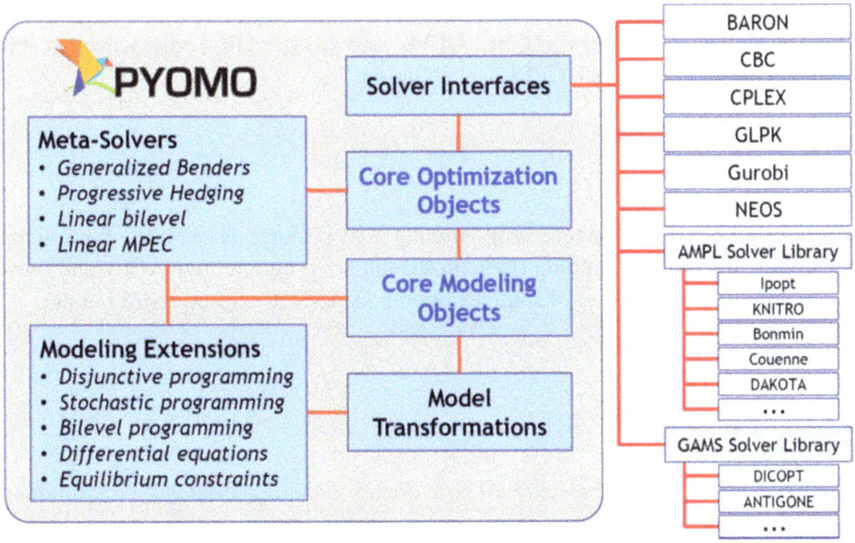

Fig. 2.5 Solvers [15]

value of the model is the value of the objective function. The optimized solution shall solve the value that is as good as the value of all other feasible solutions (please see Fig. 2.5).

2.1.2.1 Pyomo

Pyomo[24] is an open-source software package that supports a diverse set of mathematical optimization capabilities (LP, QP, NP, MILP) for formulating, solving, and analyzing optimization models. It is a Python-based package. At its core, Pyomo supports modeling of structured optimization applications. Pyomo is used to define and create general symbolic specific problem instances utilizing commercial (CPLEX, GUROBI) and open-source solvers (GLPK, CBC, and others).

Pyomo supports a wide range of problem types, including: [15]

- Linear programming (LP)
- Quadratic programming (QP)
- Nonlinear programming (NP)
- Mixed-integer linear programming (MILP)
- And others[25,26,27]

[24] Pyomo—http://www.pyomo.org/about

[25] About Pyomo—http://www.pyomo.org/about

[26] Pyomo Documentation—http://www.pyomo.org/documentation

[27] ND-Pyomo-Cookbook—https://github.com/jckantor/ND-Pyomo-Cookbook

To run and solve optimization, Pyomo provides interfaces to several solvers. In this article, we will emphasize CBC, GLPK, and Google OR-Tools. Other notable commercial solvers include CPLEX and Gurobi.

2.1.2.2 GLPK Solver

The GLPK12[28] (GNU Linear Programming Kit) package is intended for solving large-scale linear programming (LP), mixed-integer programming (MIP), and other related problems. It is a set of routines written in ANSI C and organized in the form of a callable library.

2.1.2.3 CBC Solver

The CBC[29] (Coin-or branch and cut) is an open-source mixed-integer programming solver written in C++. It can be used as a callable library or using a stand-alone executable. It can be called through AMPL (natively), GAMS.

2.1.2.4 Z3 Solver

Z3[30] is a theorem prover from Microsoft Research. Z3 integrates a custom optimized consequence finding module built tightly with its conflict-driven clause learning, CDCL, engine.

2.1.2.5 Google Operation Research (OR) Tools

The OR-Tools[31] is an open-source software for combinatorial optimization, which seeks to find the best solution to a problem out of a very large set of possible solutions. Here are some examples of problems that OR-Tools solve:

- Vehicle routing
- Scheduling
- Bin packaging

The OR-Tools includes solvers for:

---

[28] GLPK—https://www.gnu.org/software/glpk/
[29] The CBC Solver—https://projects.coin-or.org/Cbc
[30] Z3 Solver—https://github.com/Z3Prover/z3
[31] Google OR Tools—https://developers.google.com/optimization/introduction/overview

- Constraint programming
- Linear and mixed-integer programming
- Vehicle routing
- Graph algorithm

The primary OR-Tools linear optimization solver is GLOP, Google's linear programming system. It's fast, memory efficient, and numerically stable. The next section shows how to use GLOP to solve a simple linear problem in all the supported languages.[32]

2.1.3 Decision Variables or Variables

The decision variables are the variables that will decide the solver output. They represent a feasible or infeasible solution. To solve any problem, we first need to identify the decision variables. In the figure x and y are decision variables.[33]

```
# Import the linear solver wrapper
from ortools.linear_solver import pywraplp

# Declare the LP solver

solver = pywraplp.Solver.CreateSolver('GLOP')

# Create the variables
x = solver.NumVar(0, solver.infinity(), 'x')
y = solver.NumVar(0, solver.infinity(), 'y')

print('Number of variables =', solver.NumVariables())
```

2.1.4 Constraints

The constraints are the restrictions or limitations on the decision variables. They usually limit the value of the decision variables. In the above example, the limit on the availability of resources x and y is my constraints.

[32] GLOP—https://developers.google.com/optimization/lp/glop

[33] Linear Programming—https://developers.google.com/optimization/lp/lp_example

```
# Constraint 0: x + 2y <= 14.
solver.Add(x + 2 * y <= 14.0)

# Constraint 1: 3x - y >= 0.
solver.Add(3 * x - y >= 0.0)

# Constraint 2: x - y <= 2.
solver.Add(x - y <= 2.0)

print('Number of constraints =', solver.NumConstraints())
```

2.1.5 *Objective Function*

It is defined as the objective of making decisions. In the above example, the company wishes to increase the total profit represented by Z. So, profit is my objective function.

The following code defines the objective function, 3x + 4y, and specifies that this is a maximization problem.

```
# Objective function: 3x + 4y.
solver.Maximize(3 * x + 4 * y)
```

2.2 Use Case: Surgical Sutures

Supply chain teams are the major benefactors of heuristics modeling, in my experience. Mathematics modeling that developed as part of heuristics is not confined to one domain. Supply chain teams applied heuristic modeling to solve optimization of routing, just-in-time (JIT) replenishment of essential and critical material to reduce waste, and truck routing to improve customer experience. I have implanted a comprehensive system to store, replenish, and dispense the sutures in a hospital to reduce wastage of surgical suture. For different types of surgical sutures [16], please refer to the below figure[34] (please see Fig. 2.6). The system is comprised of kiosk, suture hub, and smart shelf.

[34] Jonathan Kantor, Atlas of Suturing Techniques: Approaches to Surgical Wound, Laceration, and Cosmetic Repair, Publisher : McGraw-Hill Education / Medical; 1st edition (January 29, 2016), ISBN-13 : 978-0071836579.

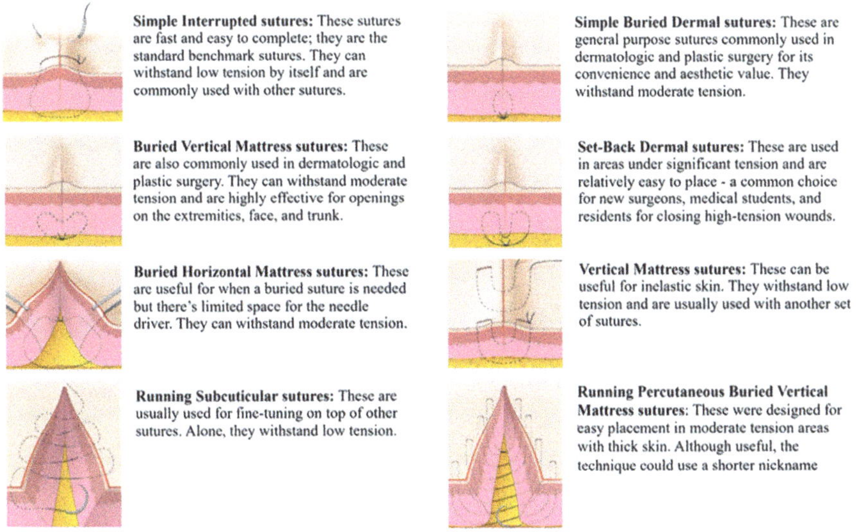
Simple Interrupted sutures: These sutures are fast and easy to complete; they are the standard benchmark sutures. They can withstand low tension by itself and are commonly used with other sutures.
Simple Buried Dermal sutures: These are general purpose sutures commonly used in dermatologic and plastic surgery for its convenience and aesthetic value. They withstand moderate tension.
Buried Vertical Mattress sutures: These are also commonly used in dermatologic and plastic surgery. They can withstand moderate tension and are highly effective for openings on the extremities, face, and trunk.
Set-Back Dermal sutures: These are used in areas under significant tension and are relatively easy to place - a common choice for new surgeons, medical students, and residents for closing high-tension wounds.
Buried Horizontal Mattress sutures: These are useful for when a buried suture is needed but there's limited space for the needle driver. They can withstand moderate tension.
Vertical Mattress sutures: These can be useful for inelastic skin. They withstand low tension and are usually used with another set of sutures.
Running Subcuticular sutures: These are usually used for fine-tuning on top of other sutures. Alone, they withstand low tension.
Running Percutaneous Buried Vertical Mattress sutures: These were designed for easy placement in moderate tension areas with thick skin. Although useful, the technique could use a shorter nickname

Fig. 2.6 Suture types [16]

The kiosk consists of functionalities that enable the hospital operating staff, head nurse, and operating surgeon to retrieve sutures from the suture hub. The suture hub is a shelf that consists of various suture types and sizes stacked together. These shelf boxes are operated electronically and open and dispense appropriate suture based on the user request. Finally, the smart shelf, as its name implies, understands what suture it holds and its count. As and when sutures are dispensed, the smart shelf understands the type of dispensed suture and its running count. In essence, the shelf has its own understanding and smartness. This was possible by the integration of machine learning and sensor technologies that connect the smart shelf with the central server.

The heart of heuristics modeling is implemented using linear programming model engine, a configurator tool, which is used to identify hospital needs based on the usage of suture for various surgical processes and methods and issues of replenishment. The model engine utilizes historical and scheduled case data extracted from a customer's surgical database to identify configuration and replenishment. The algorithm inputs include:

- Historical procedure data including suture SKU consumption, date, and location (e.g., OR)
- Scheduled procedure data, doctor suture preference choices
- Dimensions
- Priority

The output included sutures needed for the scheduled procedures and the number of stacks needed for each suture. The goal was to reduce overall wastage of sutures.

The modeling (please see Fig. 2.7):

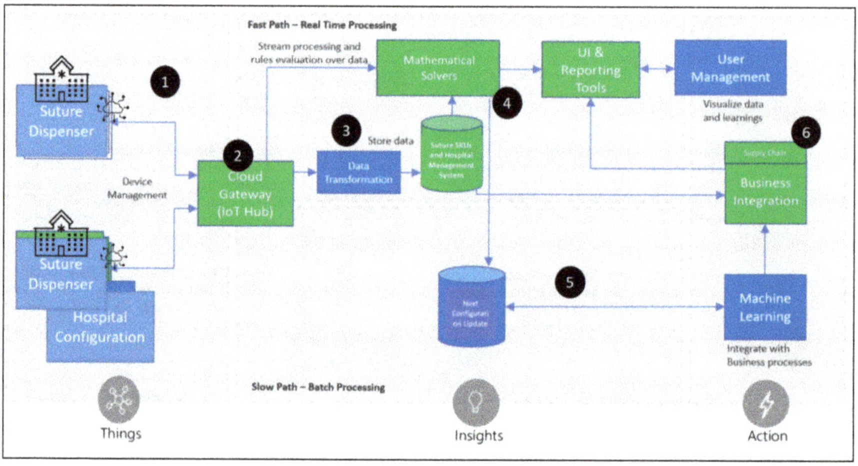

Fig. 2.7 Suture dispenser

The solver models that codify heuristics are deployed in the above architecture.

1	The input to the solver comes from dispensers that are deployed in hospital operating rooms (OR).
2	Through the Internet of Things (IoT) connectivity, the usage-related patterns such as the number of suture boxes that are used by a surgeon, a medical procedure, and hospital information are sent to the central server via IoT payload.
3	The data from suture dispenser is transformed and combined with hospital management system and fed to mathematical solver.
4	Mathematical solver runs constraint modeling and computes the utilization of sutures and calculates the replenishment counts for sutures. It processes the upcoming hospital surgery list, surgeon's raster, and procedure to generate the required list of sutures. The output of solver is persisted into hospital order management systems.
5	The solver computed replenishment list is used for business integration patterns such as the best time to overall reduce the cost. Additionally, the data is used to report the run rate of sutures vs. surgeon utilization.
6	Just-in-time (JIT) orders and supply chain optimization are performed as part of the business operations.

Heuristics modeling: solver needs variable, constraints, and optimization functions to generate suture supply chain order list.

Variables:

I: Surgery procedure index
J: Suture type or product ID—simple interrupted sutures, buried vertical mattress suture, running subcuticular suture, and others
K: Index of doctors—a database that is contextualized for a given hospital
P: Box dimension—some of them are small or large[a]
T: Replenishment cycle—a time series data variable. On cold start, for a new hospital, this value is initially endogenic. Later, as the system collects more data, the variable becomes more adaptive
N_{kit}: Number of methods I by surgeon k scheduled in a replenishment cycle
A_{kij}: The required amount of suture j in procedure I by surgeon K
W_{ki}: The weight of each surgeon procedure in the time period
H_{jp}: Capacity of suture box j in size/dimension of p
M_p: The number of boxes or stack dimension p in the automated dispenser

[a]Suture Storage Rack Dispenser Rack Dispenser Rack 25-3/4 X 5-3/4 X 27 Inch, 20 Modules, 10 Back Plates, 20 Module Clips, 30 Divider Panels, Wall Mount—https://mms.mckesson.com/product/138140/J-J-Healthcare-Systems-MR20

Optimization functions:

The goal is to reduce wastage and improve utilization of suture by the surgeons without wasting. As you may know, once a suture box is opened, it is disposed of irrespective of the usage of all items or not. This is where most of the issue is with respect to the wastage.

$$\text{Max} \sum_{kit} [W]_{ki} \, [Z]_{kit} \, [N]_{kit} \tag{2.1}$$

Where Z_{kit} is binary variable set tp 1 for at replenishment cycle t, procedure i performed by surgeon k and 0 for otherwise.

Constraints:

- Indicate that required suture SKUs are sourced in the dispenser and to determine the number of stacks.
- Suture boxes are less than the maximum number of suture boxes.

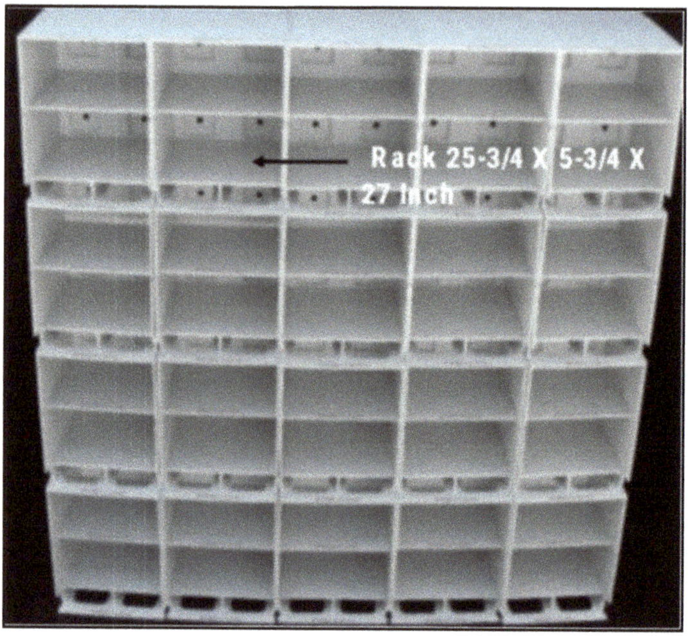

The COVID-19 pandemic, importantly, has changed the phase of usage of heuristics. For example, many of my wafer manufacturing customers are using heuristics in risk modeling, traditionally an area that finance teams apply. Specifically, supplier risk modeling from geographical, health preparedness, and primary vs. secondary supplier types.

Severe Drought Adds to Afghanistan's Woes, Endangering Millions as Economy Collapses

Farmers who survived two decades of war abandon land as water sources dry up [17].

One of the major contributors of food insecurity is conflicts. Conflicts with natural calamities such as drought are the receipt for sustained food insecurity. A clear case can be witnessed in what is happening in Afghanistan.

This year (2021)'s deep drought is compounding the economic crisis that deepened when the Taliban overthrew the previous Afghan government[35], prompting the United States and others to freeze some $9 billion in Afghan central-bank assets and spurring many of the country's professionals to leave.

(continued)

[35] Severe Drought Adds to Afghanistan's Woes, Endangering Millions as Economy Collapses. By Sune Engel Rasmussen—https://www.wsj.com/articles/severe-drought-adds-to-afghanistans-woes-endangering-millions-as-economy-collapses-11633872935?page=1

Now the scarcity of water is slashing farmers' incomes and driving up food bills for people in cities. The United Nations estimates that the drought is threatening the livelihood of up to 9 million Afghans and affecting 25 of the country's 34 provinces.

Tough conditions have already put 14 million people – more than one-third of the Afghan population – in a food-security crisis, according to the UN Food and Agriculture Organization. Afghanistan's current harvest is expected to be 15% below average due to the drought, the FAO said.

Afghanistan is heavily reliant on livelihoods sensitive to fluctuations in weather, such as rain-fed agriculture and cattle farming. Impoverished villagers often lack the capital and technology to switch to more modern and resilient farming methods. Climate change stands to increase the burden.

Around 12% of Afghanistan's land is suitable for agriculture, but around 80% of the population relies on farming for survival, said Samim Hoshmand, a former top climate negotiator under Afghanistan's National Environment Protection Agency [17].

General Assembly (GA) Line [4] Optimization[36]

We describe our experiences using Z3 for synthesizing and optimizing next-generation plant configurations for a car manufacturing company. Our approach leverages unique capabilities of Z3: a combination of specialized solvers for finite domain bit-vectors and uninterpreted functions and a programmable extension that we call constraints as code. To optimize plant configurations using Z3, we identify useful formalisms from Satisfiability Modulo Theories solvers and integrate solving capabilities for the resulting nontrivial optimization problems. [4]

The digital transformation is widely recognized as an ongoing seismic shift in today's industries. For automobile industry, the digital transformations are powering driving experiences. With new models and factories being churned out at a brisk pace, there is an urgent need for automating and optimizing production plants to increase the pace of production while reducing costs and resource requirements. The organization of production assembly lines involves a combination of hundreds of assembly stations and thousands of operators completing tens of thousands of tasks with tens of thousands of different tools available. Some tasks must be completed in sequential order, some stations may not be able to service tasks with conflicting requirements,

(continued)

[36]N. Bjørner, M. Levatich, N. P. Lopes, A. Rybalchenko, C. Vuppalapati. Supercharging Plant Configurations Using Z3. In Proc. of the 18th International Conference on Integration of Constraint Programming, Artificial Intelligence, and Operations Research (CPAIOR), July 2021.

only a subset of available operators may be able to work on a given task, and all tasks are packed into stringent timing bounds on each station.

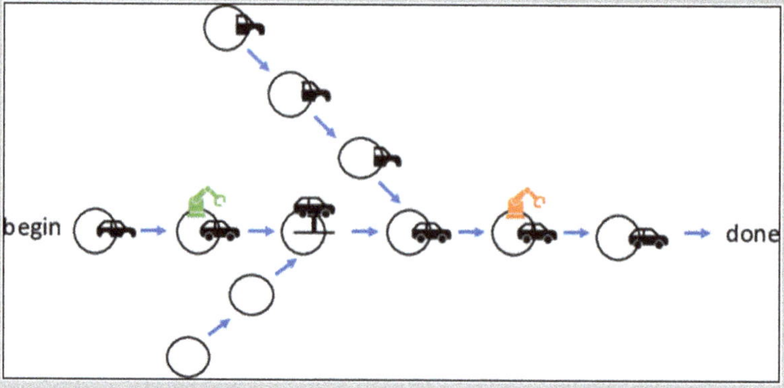

2.3 Solvers and Comparison

While CLP, GLPK, and CBC are drop-in replaceable one of the other,[37] CP-SAT has quite different APIs; therefore, the syntax among the two tests unavoidably changed a bit. Most of Google solver examples could be found on Google solver site.[38]

2.3.1 Variable Creation

In CBC/CLP/GLPK, the following approach is followed:

```
var = self.solver.BoolVar(name=name)
self._or_variables['reconstruction_cost'][reco_id] = var
self._objective.SetCoefficient(var, coeff)
```

CP-SAT counterpart is a little bit different (note the integer rounding).

[37] Solvers Comparison—https://github.com/google/or-tools/issues/1522

[38] Google Solver Examples—https://github.com/google/or-tools/tree/stable/examples/python

```
var = self.model.NewBoolVar(name=name)
self._or_variables['reconstruction_cost'][reco_id] = var
self._objective_vars.append(var)
self._objective_coeff.append(int(coeff * 1e4))
self.model.Minimize(cp_model.LinearExpr.ScalProd(self.
_objective_vars, self._objective_coeff))
```

2.3.2 *Constraint Creation*

In CBC/CLP/GLPK, constraints are created as follows:

```
constraint_out = self.solver.Constraint(0, 0)
for many times:
  var = get_variable(...)
  constraint_out.SetCoefficient(var, 1)
```

CP-SAT has the following syntax:

```
self.model.Add(R_node - exit_edge - sum(out_link_edges) == 0)
... # or sometimes
self.model.AddLinearConstraint(sum(related_nodes), 0, 1)
```

2.3.3 *Solve*

In CBC/CLP/GLPK:

```
self.solver.Solve()
```

CP-SAT:

```
self.solver.parameters.num_search_workers = 8
status = self.solver.Solve(self.model)
```

Linear programming[a] has become a vital tool in advancing Artificial Intelligence and machine learning, and it is used in electrical stimulation therapy, chemotherapy plans, drug discovery, radiation therapy designs, and finding optimal diets.

[a]George Bernard Dantzig: The Pioneer of Linear Optimization—https://mbrjournal.com/2021/01/26/george-bernard-dantzig-the-pioneer-of-linear-optimization/

2.4 Linear Optimization or Linear Programming

Linear optimization (or linear programming) is the name given to computing the best solution to a problem modeled as a set of linear relationships. These problems arise in many scientific and engineering disciplines. This is one of the essential techniques that could reduce food loss and improve food production, both of which lead to a higher food security. Linear programming is used to optimize resource usage—an optimal balance between farm inputs[39] (water, fertilizers) and output (production of commodities such as rice, wheat, corn, and cotton) [18]. Most of the resources, restrictions, aims, and sensitivities of these kinds of matter that can be compiled with developing models based on linear programming are considered and determined as an optimal cropping pattern [18].

In its simplest form, linear programming is a method of determining a profit-maximizing combination of farm enterprises that is feasible with respect to a set fixed farm constraints.[40] Early applications of linear programming in farm planning assumed a profit maximization behavior, a single-period planning horizon (no growth), and a certain environment (no uncertainty about prices, yields, and so forth) [19].

Assumptions of Linear Programming [19]

Several assumptions about the nature of the production process, the resources, and activities are implicit in the linear programming model:

1. Optimization: it is assumed that an appropriate objective function is either maximized or minimized.
2. Fixedness: at least one constraint has a nonzero right-hand side coefficient
3. Finiteness: it is assumed that there are only a finite number of activities and constraints to be considered so that a solution may be sought.

[39]Mohammad Heydari, Faridah Othman, Meysam Salarijazi,Iman Ahmadianfar, Mohammad Sadegh Sadeghian, Predicting the Amount of Fertilizers using Linear Programming for Agricultural Products from Optimum Cropping Pattern, December 2018Journal of Geographical Studies 2(1): 22-29, DOI: https://doi.org/10.21523/gcj5.18020103

[40]Optimization Model for Agricultural Machinery Selection in Elsuki Agricultural Scheme Using linear Programming, http://repository.sustech.edu/bitstream/handle/123456789/13031/Optimization%20Model%20for%20...%20.pdf?sequence=1

4. Determinism: all coefficients in the model are assumed to be known constantly.
5. Continuity: it is assumed that resources can be used and activities produced in quantities that are frictional units.
6. Homogeneity: it is assumed that all units of the same resource or activity are identical.
7. Additively: the activities are assumed to be additive in the sense that when two or more are used, their total product is the sum of their individual products. That is no interaction effects between activities and permitted.
8. Proportionality: the gross margin resource requirements per unit of activities are assumed to be constant regardless of the level of the activities used. A constant gross margin per unit of activity assumes a perfectly elastic demand curve for the product and perfectly elastic supplies of any variable inputs that may be used (Edwin, 1998)

All solvers have good functionality. But it is imperative to apply the right programming for the problem at hand:

Linear programming is applied when[41]:

1. A linear objective function.
2. Linear constraints that can be equalities or inequalities.
3. Bounds on variables that can be positive, negative, finite, or infinite. Specifically, they are all binary.

2.4.1 Crop Land: Profit Maximization and Optimization Using Linear Programming

A farmer has recently acquired a 110 hectares piece of land.[42] He has decided to grow wheat and barley on that land. Due to the quality of the sun and the region's excellent climate, the entire production of wheat and barley can be sold [20]. He wants to know how to plant each variety in the 110 hectares, given the costs, net profits, and labor requirements according to the data shown below:

Variety	Cost (price/hec)	Net profit (price/hec)	Man-days/hec
Wheat	100	50	10
Barley	200	120	30

The farmer has a budget of US$10,000 and availability of 1200 man-days during the planning horizon. Find the optimal solution and the optimal value.

Step 1: Identify the decision variables.

[41] Solver comparison—https://github.com/google/or-tools/issues/1522

[42] Introductory guide on Linear Programming for (aspiring) data scientists—https://www.analyticsvidhya.com/blog/2017/02/lintroductory-guide-on-linear-programming-explained-in-simple-english/

The total area for growing wheat = X (in hectares)
The total area for growing barley = Y (in hectares)
X and Y are my decision variables.

Step 2: Write the objective function.

Since production from the entire land can be sold in the market, the farmer would want to maximize the profit for his total produce. We are given the net profit for both wheat and barley. The farmer earns a net profit of US$50 for each hectare of wheat and US$120 for barley.

Our objective function (given by Z) is Max $Z = 50X + 120Y$.

2.4.2 *Solver Code: Farm Profit Maximization*

Software code for this model can be found on GitHub link: farmProfitMaximizationLinearProgramming.py (Python Code)

In the below code, we have initiated linear solver GLOP.[43,44]—Next define constraints and objective functions.

```
from ortools.linear_solver import pywraplp

def farmProfitMaximizationLinearProgramming():
   """Linear programming sample."""
   # Instantiate a Glop solver, naming it LinearExample.
   solver = pywraplp.Solver.CreateSolver('GLOP')

   # Create the two variables and let them take on any non-negative
value.
   x = solver.NumVar(0, solver.infinity(), 'x')
   y = solver.NumVar(0, solver.infinity(), 'y')

   print('Number of variables =', solver.NumVariables())

   # Constraint 0: X + 2Y ≤ 100.cd
   solver.Add(x + 2 * y <= 100.0)

   # Constraint 1: X + 3Y ≤ 120
   solver.Add(x + 3 * y <= 120.0)
```

(continued)

[43]OR Tools—https://github.com/google/or-tools

[44]Operations Research and Algorithms at Google—http://aixia2015.unife.it/wp-content/uploads/slides/invited/Perron-OR_at_Google.pdf

```
  # Constraint 2: X + Y ≤ 110.
  solver.Add(x + y <= 110.0)

  print('Number of constraints =', solver.NumConstraints())
```

Finally, solve for profit maximization.

```
# Objective function: 50X + 120Y
  solver.Maximize(50 * x + 120 * y)

  # Solve the system.
  status = solver.Solve()

  if status == pywraplp.Solver.OPTIMAL:
    print('Solution:')
    print('Objective value =', solver.Objective().Value())
    print('x =', x.solution_value())
    print('y =', y.solution_value())
  else:
    print('The problem does not have an optimal solution.')

  print('\nAdvanced usage:')
  print('Problem solved in %f milliseconds' % solver.wall_time())
  print('Problem solved in %d iterations' % solver.iterations())

farmProfitMaximizationLinearProgramming()
```

Output:
Run the solver.

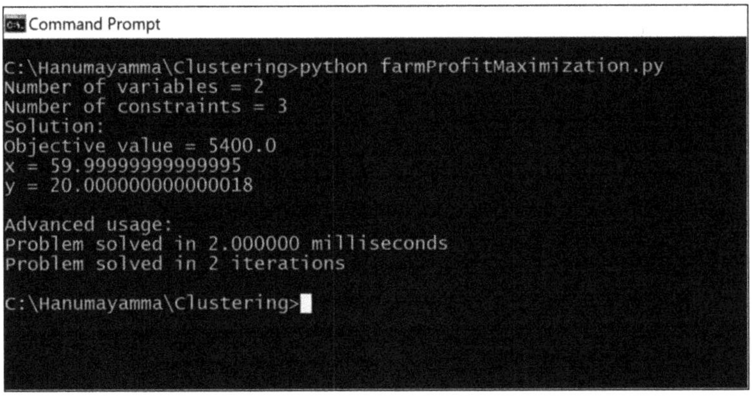
Command Prompt
C:\Hanumayamma\Clustering>python farmProfitMaximization.py
Number of variables = 2
Number of constraints = 3
Solution:
Objective value = 5400.0
x = 59.99999999999995
y = 20.000000000000018
Advanced usage:
Problem solved in 2.000000 milliseconds
Problem solved in 2 iterations
C:\Hanumayamma\Clustering>

2.4.3 Solving Graphically

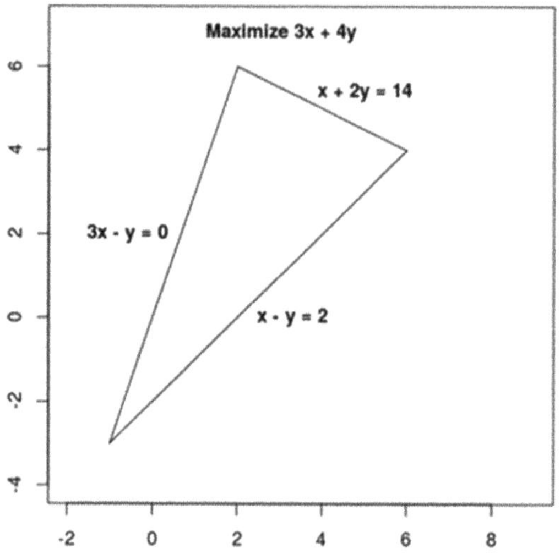

Since we know that $X, Y \geq 0$, we will consider only the first quadrant.[45]

To plot for the graph for the above equations, first I will simplify all the equations [20].

$100X + 200Y \leq 10{,}000$ can be simplified to $X + 2Y \leq 100$ by dividing by 100.

$10X + 30Y \leq 1200$ can be simplified to $X + 3Y \leq 120$ by dividing by 10.

The third equation is in its simplified form, $X + Y \leq 110$ (please see Fig. 2.8).

Plot the first two lines on a graph in the first quadrant (like shown below).

The optimal feasible solution is achieved at the point of intersection where the budget and man-days constraints are active. This means the point at which the equations $X + 2Y \leq 100$ and $X + 3Y \leq 120$ intersect gives us the optimal solution.

The values of X and Y which give the optimal solution are at (60,20).

To maximize profit the farmer should produce wheat and barley in 60 hectares and 20 hectares of land, respectively.

The maximum profit the company will gain is as follows:

$$\text{Max } Z = 50^{*}\ (60) + 120^{*}\ (20) = \text{US\$5400}$$

[45] Introductory guide on Linear Programming for (aspiring) data scientists—https://www.analyticsvidhya.com/blog/2017/02/lintroductory-guide-on-linear-programming-explained-in-simple-english/

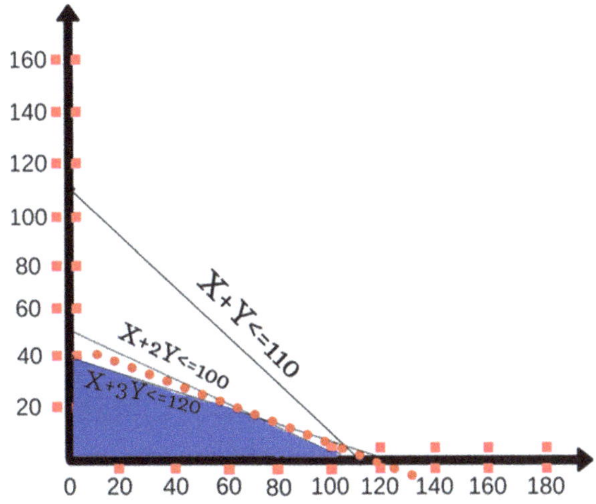

Fig. 2.8 Line equation ($X + Y <= 110$)

Linear programming and its various extensions continue to play an influential role in the economy and in all our lives.[a]

[a]George Bernard Dantzig: The Pioneer of Linear Optimization—https://mbrjournal.com/2021/01/26/george-bernard-dantzig-the-pioneer-of-linear-optimization/

2.4.4 Predicting the Number and Quantity of Fertilizers Using Linear Programming

Fertilizer costs are one of the expensive inputs to agriculture. By optimizing the decision to buy fertilizers, I mean timing-wise, farmers would reduce the cost of agriculture produce and increase the overall affordability of food. This would increase small farm sustainability and enhance food security. In the following example, I would like to apply linear programming to optimize the type of crop with respect to fertilizer usage.

Software code for this model can be found on GitHub Link: farmFertilizerCostOptimizationLinearProgramming.py (Python Code)

Let's say farming is planned on a total cultivation land of 2100 hectares with the following crop details (please see below Table 2.2).

The required data for modeling were prepared in the form of constants, the upper and lower limits values, and computational values in the pre-modeling phase.

Table 2.2 Crop yields [18]

Crops	Phosphate fertilizer (kg/ha)	Nitrate fertilizer (kg/ha)	Potash fertilizer (kg/ha)	Other (kg/ha)	Total (kg/ha)
Wheat	57,072	87,381	6886	3043	154,381
Barley	28,107	33,213	1705	1603	64,629
Husks	9948	15,906	898	458	27,210
Corn	2882	6628	383	139	10,032
Pea	2727	2851	323	22	5924
Lentil	5240	5768	331	1	11,024
Cotton	1614	2059	120	21	3813
Sugar beet	7374	8250	1361	697	17,682
Watermelon	7471	7769	636	1833	17,708
Cucumber	10,170	16,443	2234	2796	31,642
Potato	10,778	14,455	2808	478	28,518
Onions	9325	13,311	1132	1142	24,910
Tomatoes	9531	15,177	636	1339	27,492
Canola	13,239	16,248	1374	369	31,230
Beans	8879	9676	588	311	19,453
Soya bean	6401	9390	2676	171	18,639
Rice	17,766	24,142	2994	369	45,272
Sum	208,522	288,667	27,088	14,791	539,560

$$\text{Minimum land required for production}_{i=(}\text{Minimum tonnage}_i)/ (\text{Average production per hectare}_i) \tag{2.2}$$

$$\text{Minimum water required to provide the desired capacity}_i = (\text{Min land required}_i)^* (\text{Minimum required water}_i) \tag{2.3}$$

$$\text{Value per hectare}_i = (\text{The product value per ton})_i^* (\text{Average production per hectare})_i \tag{2.4}$$

For each crop, availability of optimum land to cultivate to increase profit maximization (please see Table 2.2). Finally, our objective was to increase profit and reduce overall fertilizer costs and water needs—minimum water required. Please refer to the table for the amount of consumed fertilizer for optimized cropping pattern[46] [18]:

[46]Mohammad Heydari, Faridah Othman, Meysam Salarijazi,Iman Ahmadianfar, Mohammad Sadegh Sadeghian, Predicting the Amount of Fertilizers using Linear Programming for Agricultural Products from Optimum Cropping Pattern, December 2018Journal of Geographical Studies 2(1): 22-29, DOI: https://doi.org/10.21523/gcj5.18020103

Linear programming (LP) can be used to solve questions on matching diets to nutritional and other additional constraints with a minimum number of changes. Linear programming is a mathematical technique that allows the generation of optimal solutions that satisfy several constraints at once.[a]

[a]van Dooren C. A Review of the Use of Linear Programming to Optimize Diets, Nutritiously, Economically and Environmentally. Front Nutr. 2018;5:48. Published 2018 Jun 21. doi: https://doi.org/10.3389/fnut.2018.00048

2.4.5 Step 1: Create Variables

Create variables for each crop (please see Table 2.3).

```
solver = pywraplp.Solver.CreateSolver('GLOP')

  # Create the two variables and let them take on any non-negative
value.
  Wheat = solver.NumVar(0, solver.infinity(), 'Wheat')
  Barley = solver.NumVar(0, solver.infinity(), 'Barley')
  Husks = solver.NumVar(0, solver.infinity(), 'Husks')
  Corn = solver.NumVar(0, solver.infinity(), 'Corn')
  Pea = solver.NumVar(0, solver.infinity(), 'Pea')
  Lentil = solver.NumVar(0, solver.infinity(), 'Lentil')
  Cotton = solver.NumVar(0, solver.infinity(), 'Cotton')
  Sugarbeet = solver.NumVar(0, solver.infinity(), 'Sugarbeet')
  Watermelon = solver.NumVar(0, solver.infinity(), 'Watermelon')
  Cucumber = solver.NumVar(0, solver.infinity(), 'Cucumber')
  Potato = solver.NumVar(0, solver.infinity(), 'Potato')
  Onions = solver.NumVar(0, solver.infinity(), 'Onions')
  Tomatoes = solver.NumVar(0, solver.infinity(), 'Tomatoes')
  Canola = solver.NumVar(0, solver.infinity(), 'Canola')
  Beans = solver.NumVar(0, solver.infinity(), 'Beans')
  Soyspring = solver.NumVar(0, solver.infinity(), 'Soyspring')
  Rice = solver.NumVar(0, solver.infinity(), 'Rice')

  print('Number of variables =', solver.NumVariables())
```

2.4.6 Step 2: Constraints

Total land required is the land available to cultivate:

Table 2.3 Agricultural products—min, max, and average

Agricultural products	Units	Wheat	Barley	Husks	Corn	Pea	Lentil	Cotton	Sugar beet	Watermelon	Cucumber	Potato	Onions	Tomatoes	Canola	Beans	Soyspring	Rice
Minimum required water	m3/ ha	4340	3730	4180	5060	3940	4630	9160	4710	11,850	3800	2970	4530	5364	6590	4930	3220	8890
Average production per hectare	ton/ ha	2.68	2.71	4.25	6.39	1.05	1.2	2.37	42.02	27.69	19.48	29.03	37.18	37.69	2.08	1.67	2.34	4.23
Product value per tone	1000 $	1050	780	850	870	1900	2000	2200	210	374	300	300	200	200	1900	1800	1700	2700
Maximum available agricultural land	ha	400	300	40	20	200	200	200	30	40	40	40	40	40	140	200	60	110
Minimum land required for production	ha	374	185	24	8	57	59	8	2	3	2	3	3	5	72	60	56	95
Optimum land	ha	377	185	40	20	57.1	58.6	8.4	30	40	40	40	40	40	72.2	60	60	110

Source: Mohammad Heydari, Faridah Othman, Meysam Salarijazi,Iman Ahmadianfar, Mohammad Sadegh Sadeghian, Predicting the Amount of Fertilizers using Linear Programming for Agricultural Products from Optimum Cropping Pattern, December 2018 Journal of Geographical Studies 2(1):22-29, DOI: https://doi.org/10.21523/gcj5.18020103 [18]

```
solver.Add( Wheat + Barley+ Husks + Corn + Pea + Lentil + Cotton +
Sugarbeet + Watermelon + Cucumber + Potato + Onions + Tomatoes +
Canola + Beans + Soyspring + Rice <= 2100.0)
```

The minimum and maximum land available for each crop: sample details as presented in the table

```
solver.Add(Wheat <= 400)
  solver.Add(Wheat >= 374)
  solver.Add(Barley <= 300)
  solver.Add(Barley >= 185)
  solver.Add(Husks <= 40)
  solver.Add(Husks >= 24)
  solver.Add(Corn <= 20)
  solver.Add(Corn >= 8)
  solver.Add(Pea <= 200)
  solver.Add(Pea >= 57)
  solver.Add(Lentil <= 200)
  solver.Add(Lentil >= 59)
  solver.Add(Cotton <= 200)
  solver.Add(Cotton >= 8)
  solver.Add(Sugarbeet <= 30)
  solver.Add(Sugarbeet >= 2)
  solver.Add(Watermelon <= 40)
  solver.Add(Watermelon >= 3)
  solver.Add(Cucumber <= 40)
  solver.Add(Cucumber >= 2)
  solver.Add(Onions <= 40)
  solver.Add(Onions >= 3)
  solver.Add(Tomatoes <= 40)
  solver.Add(Tomatoes >= 5)
  solver.Add(Canola <= 140)
  solver.Add(Canola >= 72)
  solver.Add(Beans <= 200)
  solver.Add(Beans >= 60)
  solver.Add(Soyspring <= 60)
  solver.Add(Soyspring >= 56)
  solver.Add( Potato <= 40)
  solver.Add(Potato >= 3)
  solver.Add( Rice <= 110)
  solver.Add(Rice >= 95)
```

The optimum land for each crop is required to be specified for the model to yield best optimization:

```
solver.Add( 3.77 * Wheat+ 1.85 * Barley+ 0.4 * Husks +   0.2 * Corn +
0.571 * Pea + 0.586 * Lentil + 0.084 * Cotton + 0.3 * Sugarbeet + 0.4 *
Watermelon + 0.4 * Cucumber + 0.4 * Potato + 0.4 * Onions + 0.4 *
Tomatoes + 0.722 * Canola + 0.6 * Beans + 0.6 * Soyspring + 1.1 * Rice
<= 2100.0)

print ('Number of constraints =', solver.NumConstraints())
```

Now, at this point solver is ready with variables and constraints specified.

2.4.7 Step 3: Objective Function

The goal of optimization is to find the minimum water usage for the crops. To yield the results, run the minimum water required as an objective function: please note that we have divided "minimum water required per hectare" with 1000 to get cubic water required: m^3/ha.

Maximum Z =

$$\sum (\text{Optimal area of agricultural land for production} * \text{Value per hectare})_i \quad \forall i = 1, 2, 3, 4, \quad \ldots \tag{2.5}$$

```
solver.Minimize(4.34 * Wheat+ 3.73 * Barley+ 4.18 * Husks +   05.06
* Corn + 3.94 * Pea + 4.63 * Lentil + 9.16 * Cotton + 4.71 * Sugarbeet +
11.85 * Watermelon + 3.8 * Cucumber + 2.97 * Potato + 4.53 * Onions +
5.36 * Tomatoes + 6.59 * Canola + 4.93 * Beans + 3.22 * Soyspring +
8.89 * Rice)
```

2.4.8 Step 4: Solve

Solve the model formulation:

```
# Solve the system.
  status = solver.Solve()
  if status == pywraplp.Solver.OPTIMAL:
```

(continued)

```
    print('Solution:')
    print('Objective value =', solver.Objective().Value())
    print('Wheat =', Wheat.solution_value())
    print('Barley =', Barley.solution_value())
    print('Husks =', Husks.solution_value())
    print('Corn =', Corn.solution_value())
    print('Pea =', Pea.solution_value())
    print('Barley =', Barley.solution_value())
    print('Lentil =', Lentil.solution_value())
    print('Cotton =', Cotton.solution_value())
    print('Sugarbeet =', Sugarbeet.solution_value())
    print('Watermelon =', Watermelon.solution_value())
    print('Cucumber =', Cucumber.solution_value())
    print('Potato =', Potato.solution_value())
    print('Onions =', Onions.solution_value())
    print('Tomatoes =', Tomatoes.solution_value())
    print('Canola =', Canola.solution_value())
    print('Beans =', Beans.solution_value())
    print('Soyspring =', Soyspring.solution_value())
    print('Rice =', Rice.solution_value())

  else:
    print('The problem does not have an optimal solution.')

  print('\nAdvanced usage:')
  print('Problem solved in %f milliseconds' % solver.wall_time())
  print('Problem solved in %d iterations' % solver.iterations())
```

Output:

```
Command Prompt

C:\Hanumayamma\Clustering>python farmFertilizerCostOptimizationLinearProgramming.py
Number of variables = 17
Number of constraints = 36
Solution:
Objective value = 4922.0599999999995
Wheat = 374.0
Barley = 185.0
Husks = 24.0
Corn = 8.0
Pea = 57.0
Barley = 185.0
Lentil = 59.0
Cotton = 8.0
Sugarbeet = 2.0
Watermelon = 3.0
Cucumber = 2.0
Potato = 3.0
Onions = 3.0
Tomatoes = 5.0
Canola = 72.0
Beans = 60.0
Soyspring = 56.0
Rice = 95.0

Advanced usage:
Problem solved in 4.000000 milliseconds
Problem solved in 0 iterations
```

Table 2.4 Experiment runs

(Maximize Profit [i])	(Minimum required water[±])	(Maximize Optimum Land[Δ])
Objective value = 1900825.62	**Objective value = 7217.08**	**Objective value = 2100.0**
Wheat = 374.0	Wheat = 374.0	Wheat = 374.0
Barley = 185.0	Barley = 185.0	Barley = 185.0
Husks = 24.0	Husks = 24.0	Husks = 24.0
Corn = 20.0	Corn = 20.0	Corn = 8.0
Pea = 57.0	Pea = 57.0	Pea = 57.0
Barley = 185.0	Barley = 185.0	Barley = 185.0
Lentil = 88.29	Lentil = 59.0	Lentil = 59.0
Cotton = 200.0	Cotton = 200.0	Cotton = 8.0
Sugar beet = 2.0	Sugar beet = 9.89	Sugarbeet = 2.0
Watermelon = 3.0	Watermelon = 40.0	Watermelon = 3.0
Cucumber = 2.0	Cucumber = 2.0	Cucumber = 2.0
Potato = 3.0	Potato = 3.0	Potato = 3.0
Onions = 3.0	Onions = 3.0	Onions = 3.0
Tomatoes = 5.0	Tomatoes = 5.0	Tomatoes = 5.0
Canola = 72.0	Canola = 72.0	Canola = 72.0
Beans = 60.0	Beans = 60.0	Beans = 87.99
Soyspring = 56.0	Soyspring = 56.0	Soyspring = 60.0
Rice = 95.0	Rice = 95.0	Rice = 110.0

[i] solver.Maximize(1050 * Wheat+ 780 * Barley+ 850 * Husks + 870 * Corn + 1900 * Pea + 2000 * Lentil + 2200 * Cotton + 210 * Sugarbeet + 374 * Watermelon + 300 * Cucumber + 300 * Potato + 200 * Onions + 200 * Tomatoes + 1900 * Canola + 1800 * Beans + 1700 * Soyspring + 2700 * Rice)

[±] solver.Maximize(4.34 * Wheat+ 3.73 * Barley+ 4.18 * Husks + 05.06 * Corn + 3.94 * Pea + 4.63 * Lentil + 9.16 * Cotton + 4.71 * Sugarbeet + 11.85 * Watermelon + 3.8 * Cucumber + 2.97 * Potato + 4.53 * Onions + 5.36 * Tomatoes + 6.59 * Canola + 4.93 * Beans + 3.22 * Soyspring + 8.89 * Rice)

[Δ] solver.Maximize(3.77 * Wheat+ 1.85 * Barley+ 0.4 * Husks + 0.2 * Corn + 0.571 * Pea + 0.586 * Lentil + 0.084 * Cotton + 0.3 * Sugarbeet + 0.4 * Watermelon + 0.4 * Cucumber + 0.4 * Potato + 0.4 * Onions + 0.4 * Tomatoes + 0.722 * Canola + 0.6 * Beans + 0.6 * Soyspring + 1.1 * Rice)

In the above output, the solver finds the minimum water usage—4922.0 cubic meters—for optimizing water. If you run the same formulation, with the criterion of increasing profit maximization, the solver outputs 7217 cubic meters of water (please see Table 2.4 for all experiment runs).

2.5 Mixed-Integer Programming (MIPs)

Linear optimization problems that require some of the variables to be integers are called mixed-integer programs (MIPs).[47] These variables can arise in a couple of ways:

[47] Google MIPs—https://developers.google.com/optimization/mip

- Integer variables: represent numbers of items, such as workstations, sutures, cars, or television sets, and the problem is to decide how many of each item to manufacture to maximize profit. Typically, such problems can be set up as standard linear optimization problems, with the added requirement that the variables must be integers.
- Boolean variables: represent decisions with 0–1 values. As an example, consider the use case of sutures supply chain optimization. Here the usage of suture variable was Boolean with 0 representing the suture not used and 1 being the suture used.

The basic steps to solve MIP optimization problems are very similar to that of linear programming. Here are the steps:

- Import the linear solver wrapper,
- Declare the MIP solver.
- Define the variables.
- Define the constraints.
- Define the objective.
- Call the MIP solver.
- Display the solution.

2.5.1 Simple MIP Algebraic Line Problem

The following code enables to create mixed-integer program for the following algebraic equation:

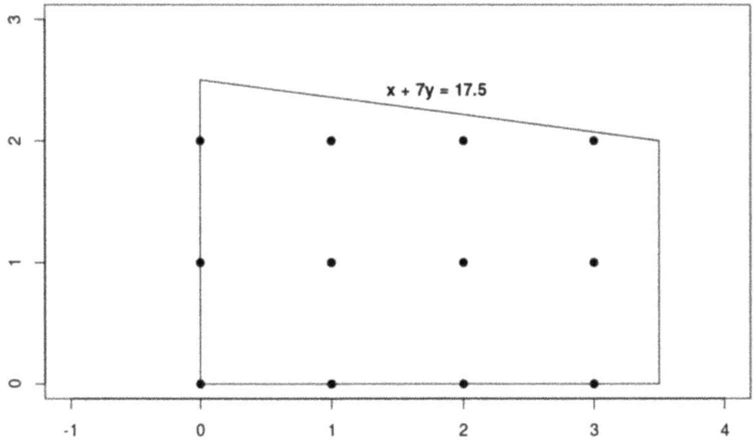

2.5.1.1 Declare the MIP solver

Step 1: declare the MIP solver.

```
# Create the mip solver with the SCIP backend.
solver = pywraplp.Solver.CreateSolver('SCIP')
```

2.5.1.2 Define the Variables

Define model variables. In the following code, we are explicitly creating integer variables.

```
infinity = solver.infinity()
# x and y are integer non-negative variables.
x = solver.IntVar(0.0, infinity, 'x')
y = solver.IntVar(0.0, infinity, 'y')

print('Number of variables =', solver.NumVariables())
```

2.5.1.3 Define the Constraints

The following code defines the constraints for the problem:

```
# x + 7 * y <= 17.5.
solver.Add(x + 7 * y <= 17.5)

# x <= 3.5.
solver.Add(x <= 3.5)

print('Number of constraints =', solver.NumConstraints())
```

2.5.1.4 Define the Objective

Create objective functions.

```
# Maximize x + 10 * y.
solver.Maximize(x + 10 * y)
```

2.5.1.5 Call the Solver

Call the solver to solve the problem.

```
status = solver.Solve()

if status == pywraplp.Solver.OPTIMAL:
  print('Solution:')
  print('Objective value =', solver.Objective().Value())
  print('x =', x.solution_value())
  print('y =', y.solution_value())
else:
  print('The problem does not have an optimal solution.')
```

Output:
The run will yield the following output:

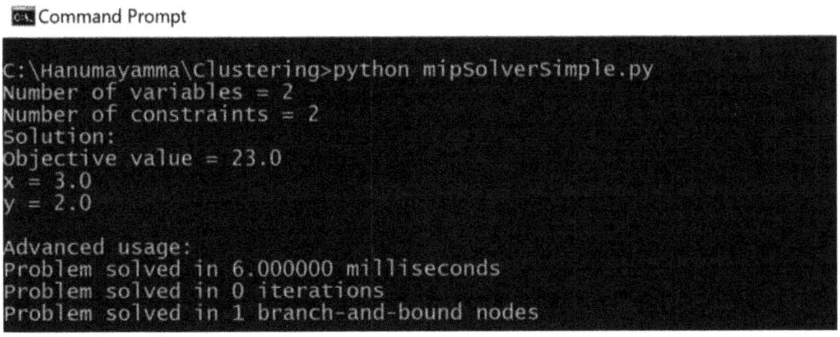
Command Prompt
C:\Hanumayamma\Clustering>python mipSolverSimple.py
Number of variables = 2
Number of constraints = 2
Solution:
Objective value = 23.0
x = 3.0
y = 2.0
Advanced usage:
Problem solved in 6.000000 milliseconds
Problem solved in 0 iterations
Problem solved in 1 branch-and-bound nodes

2.5.2 *Difference Between MIP and Linear Solving*

As you can see above, LP and MIP output is different due to the difference of variable types and solver function applied.

The following code enables us to create linear programming for the following algebraic equation: please note areas of variable, constraints, and objective function definitions.

```
from ortools.linear_solver import pywraplp

def main():
  # Create the linear solver with the GLOP backend.
  solver = pywraplp.Solver.CreateSolver('GLOP')

  infinity = solver.infinity()
  # Create the variables x and y.
  x = solver.NumVar(0.0, infinity, 'x')
  y = solver.NumVar(0.0, infinity, 'y')

  print('Number of variables =', solver.NumVariables())

  # x + 7 * y <= 17.5.
  solver.Add(x + 7 * y <= 17.5)

  # x <= 3.5.
  solver.Add(x <= 3.5)

  print('Number of constraints =', solver.NumConstraints())

  # Maximize x + 10 * y.
  solver.Maximize(x + 10 * y)

  status = solver.Solve()

  if status == pywraplp.Solver.OPTIMAL:
    print('Solution:')
    print('Objective value =', solver.Objective().Value())
    print('x =', x.solution_value())
    print('y =', y.solution_value())
  else:
    print('The problem does not have an optimal solution.')

  print('\nAdvanced usage:')
  print('Problem solved in %f milliseconds' % solver.wall_time())
  print('Problem solved in %d iterations' % solver.iterations())
  print('Problem solved in %d branch-and-bound nodes' % solver.
nodes())

if __name__ == '__main__':
  main()
```

Instead of being integer variables, X and Y are continuous variables. Solver constraints and solver are applied based on the linear programming.

Output:

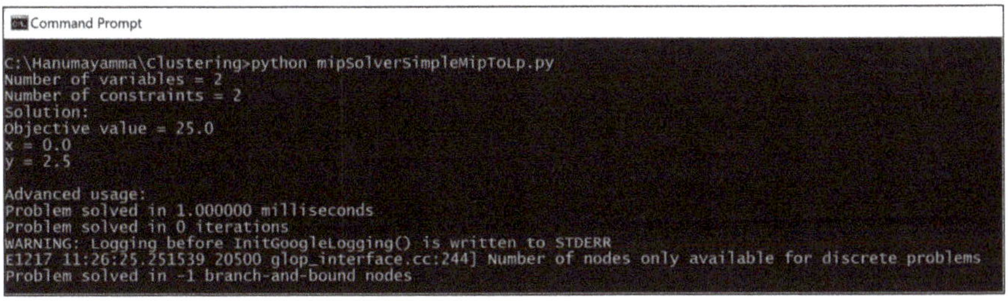
Command Prompt
C:\Hanumayamma\Clustering>python mipSolverSimpleMipToLp.py
Number of variables = 2
Number of constraints = 2
Solution:
Objective value = 25.0
x = 0.0
y = 2.5
Advanced usage:
Problem solved in 1.000000 milliseconds
Problem solved in 0 iterations
WARNING: Logging before InitGoogleLogging() is written to STDERR
E1217 11:26:25.251539 20500 glop_interface.cc:244] Number of nodes only available for discrete problems
Problem solved in -1 branch-and-bound nodes

The solution to the linear problem occurs at the point $x = 0$, $y = 2.5$, where the objective function equals 25. Here's a graph showing the solutions to both the linear and integer problems.

Notice that the integer solution is not close to the linear solution, compared with most other integer points in the feasible region. In general, the solutions to a linear optimization problem and the corresponding integer optimization problems can be far apart. Because of this, the two types of problems require different methods for their solution.[48]

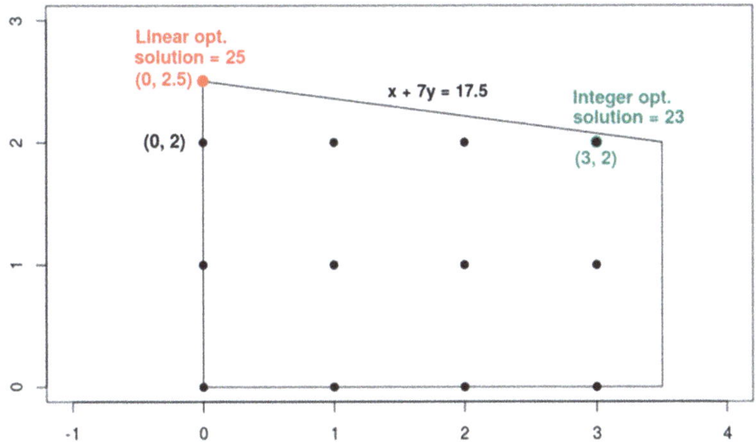

2.5.3 *Predicting the Number and Quantity of Fertilizers Using MIP Programming*

The purpose of the code was what I have described in the above section. The only additional item is when we run the fertilizer optimization in mixed-integer programming, the results are like that of LP but different.

[48] LP and MIP—https://developers.google.com/optimization/mip/mip_example

Software code for this model can be found on GitHub Link:
farmFertilizerCostOptimizationMIP.py (Python Code)

2.5.4 Code

I have highlighted the code difference between LP and MIP:

```
# -*- coding: utf-8 -*-
"""
Created on Tue Nov 16 16:48:13 2021

@author: CHVUPPAL
"""

from ortools.linear_solver import pywraplp

def LinearProgrammingExample():
  """Linear programming sample."""
  # Instantiate a Glop solver, naming it LinearExample.
  solver = pywraplp.Solver.CreateSolver('SCIP')

  # Create the two variables and let them take on any non-negative
value.
  Wheat = solver.IntVar(0, solver.infinity(), 'Wheat')
  Barley = solver.IntVar(0, solver.infinity(), 'Barley')
  Husks = solver.IntVar(0, solver.infinity(), 'Husks')
  Corn = solver.IntVar(0, solver.infinity(), 'Corn')
  Pea = solver.IntVar(0, solver.infinity(), 'Pea')
  Lentil = solver.IntVar(0, solver.infinity(), 'Lentil')
  Cotton = solver.IntVar(0, solver.infinity(), 'Cotton')
  Sugarbeet = solver.IntVar(0, solver.infinity(), 'Sugarbeet')
  Watermelon = solver.IntVar(0, solver.infinity(), 'Watermelon')
  Cucumber = solver.IntVar(0, solver.infinity(), 'Cucumber')
  Potato = solver.IntVar(0, solver.infinity(), 'Potato')
  Onions = solver.IntVar(0, solver.infinity(), 'Onions')
  Tomatoes = solver.IntVar(0, solver.infinity(), 'Tomatoes')
  Canola = solver.IntVar(0, solver.infinity(), 'Canola')
  Beans = solver.IntVar(0, solver.infinity(), 'Beans')
  Soyspring = solver.IntVar(0, solver.infinity(), 'Soyspring')
  Rice = solver.IntVar(0, solver.infinity(), 'Rice')

  print('Number of variables =', solver.NumVariables())

  # Constraint 0:Max Land
```

(continued)

```
  solver.Add( Wheat + Barley+ Husks + Corn + Pea + Lentil + Cotton +
Sugarbeet + Watermelon + Cucumber + Potato + Onions + Tomatoes +
Canola + Beans + Soyspring + Rice <= 2100.0)

   solver.Add(Wheat <= 400)
   solver.Add(Wheat >= 374)
   solver.Add(Barley <= 300)
   solver.Add(Barley >= 185)
   solver.Add(Husks <= 40)
   solver.Add(Husks >= 24)
   solver.Add(Corn <= 20)
   solver.Add(Corn >= 8)
   solver.Add(Pea <= 200)
   solver.Add(Pea >= 57)
   solver.Add(Lentil <= 200)
   solver.Add(Lentil >= 59)
   solver.Add(Cotton <= 200)
   solver.Add(Cotton >= 8)
   solver.Add(Sugarbeet <= 30)
   solver.Add(Sugarbeet >= 2)
   solver.Add(Watermelon <= 40)
   solver.Add(Watermelon >= 3)
   solver.Add(Cucumber <= 40)
   solver.Add(Cucumber >= 2)
   solver.Add(Onions <= 40)
   solver.Add(Onions >= 3)
   solver.Add(Tomatoes <= 40)
   solver.Add(Tomatoes >= 5)
   solver.Add(Canola <= 140)
   solver.Add(Canola >= 72)
   solver.Add(Beans <= 200)
   solver.Add(Beans >= 60)
   solver.Add(Soyspring <= 60)
   solver.Add(Soyspring >= 56)
   solver.Add( Potato <= 40)
   solver.Add(Potato >= 3)
   solver.Add( Rice <= 110)
   solver.Add(Rice >= 95)

   # Optimal Land
  solver.Add( 3.77 * Wheat+ 1.85 * Barley+ 0.4 * Husks +   0.2 * Corn
+ 0.571 * Pea + 0.586 * Lentil + 0.084 * Cotton + 0.3 * Sugarbeet +
0.4 * Watermelon + 0.4 * Cucumber + 0.4 * Potato + 0.4 * Onions + 0.4 *
Tomatoes + 0.722 * Canola + 0.6 * Beans + 0.6 * Soyspring + 1.1 * Rice
<= 2100.0)
  # solver.Add(4.34 * Wheat+ 3.73 * Barley+ 4.18 * Husks +   05.06 *
Corn + 3.94 * Pea + 4.63 * Lentil + 9.16 * Cotton + 4.71 * Sugarbeet +
11.85 * Watermelon + 3.8 * Cucumber + 2.97 * Potato + 4.53 * Onions +
5.36 * Tomatoes + 6.59 * Canola + 4.93 * Beans + 3.22 * Soyspring +
8.89 * Rice <= 4000)
```

(continued)

```
# Number of constraints
print('Number of constraints =', solver.NumConstraints())

# Objective function: The product value per tone

solver.Maximize(1050 * Wheat+ 780  * Barley+ 850 * Husks +  870
* Corn + 1900 * Pea + 2000 * Lentil + 2200 * Cotton + 210 * Sugarbeet +
374 * Watermelon + 300 * Cucumber + 300 * Potato + 200 * Onions + 200
* Tomatoes + 1900 * Canola + 1800 * Beans + 1700  * Soyspring + 2700 *
Rice)

# Solve the system.
status = solver.Solve()

if status == pywraplp.Solver.OPTIMAL:
  print('Solution:')
  print('Objective value =', solver.Objective().Value())
  print('Wheat =', Wheat.solution_value())
  print('Barley =', Barley.solution_value())
  print('Husks =', Husks.solution_value())
  print('Corn =', Corn.solution_value())
  print('Pea =', Pea.solution_value())
  print('Barley =', Barley.solution_value())
  print('Lentil =', Lentil.solution_value())
  print('Cotton =', Cotton.solution_value())
  print('Sugarbeet =', Sugarbeet.solution_value())
  print('Watermelon =', Watermelon.solution_value())
  print('Cucumber =', Cucumber.solution_value())
  print('Potato =', Potato.solution_value())
  print('Onions =', Onions.solution_value())
  print('Tomatoes =', Tomatoes.solution_value())
  print('Canola =', Canola.solution_value())
  print('Beans =', Beans.solution_value())
  print('Soyspring =', Soyspring.solution_value())
  print('Rice =', Rice.solution_value())

else:
  print('The problem does not have an optimal solution.')

print('\nAdvanced usage:')
print('Problem solved in %f milliseconds' % solver.wall_time())
print('Problem solved in %d iterations' % solver.iterations())
```

As you can see, the only difference is import library and variable definitions.

Command Prompt

```
C:\Hanumayamma\Clustering>python farmFertilizerCostOptimizationMIP.py
Number of variables = 17
Number of constraints = 36
Solution:
Objective value = 1900242.0000000002
Wheat = 374.0
Barley = 185.0
Husks = 24.0
Corn = 20.0
Pea = 57.0
Barley = 185.0
Lentil = 88.0
Cotton = 200.0
Sugarbeet = 2.0
Watermelon = 3.0
Cucumber = 2.0
Potato = 3.0
Onions = 3.0
Tomatoes = 5.0
Canola = 72.0
Beans = 60.0
Soyspring = 56.0
Rice = 95.0

Advanced usage:
Problem solved in 100.000000 milliseconds
```

2.6 Constraint Optimization or Constraint Programming (CP)

Constraint optimization, or constraint programming (CP),[49] is the name given to identifying feasible solutions out of a very large set of candidates, where the problem can be modeled in terms of arbitrary constraints. CP problems arise in many scientific and engineering disciplines. (The word "programming" is a bit of a misnomer, like how "computer" once meant "a person who computes." Here, "programming" refers to the arrangement of a plan rather than programming in a computer language.)

CP is based on feasibility[50] (finding a feasible solution) rather than optimization (finding an optimal solution) and focuses on the constraints and variables rather than the objective function. In fact, a CP problem may not even have an objective function—the goal may simply be to narrow down a very large set of possible solutions to a more manageable subset by adding constraints to the problem.

Employee scheduling, bus driver scheduling,[51] agriculture seed selection,[52] and others are good examples of constraint programming [21].

---

[49] Constraint Programming—https://developers.google.com/optimization/cp

[50] Optimization 101 with OR-Tools—https://www.kaggle.com/nicapotato/optimisation-101-with-or-tools

[51] Bus driver scheduling—https://github.com/google/or-tools/blob/stable/examples/python/bus_driver_scheduling_sat.py

[52] Chance-constrained optimization models for agricultural seed development and selection—https://keep.lib.asu.edu/items/157571/view

2.6.1 *Maximize Line Equation:* x + 2y + 2z

Let's consider the following equation: x + 2y + 2z. Our goal is to find optimal solution.

$$x + \frac{7}{2}y + \frac{3}{2}z \leq 40 \tag{2.6}$$

$$3x - 5y + 7z \leq 90 \tag{2.7}$$

$$5x + 2y - 6z \leq 70 \tag{2.8}$$

$$X, y, z >= 0 \tag{2.9}$$

$$X, Y, Z \text{ Integer} \tag{2.10}$$

In order to increase computational speed,[53] the CP-SAT solver works over the integers. Modify Eq. (2.5) to integer coefficients.

$$2x + 7y + 3z \leq 80 \tag{2.11}$$

2.6.1.1 Import the Libraries

Import CP SAT library.

```
from ortools.sat.python import cp_model
```

2.6.1.2 Declare the Model

Declare the model.

```
model = cp_model.CpModel()
```

[53] CP-SAT—https://developers.google.com/optimization/cp/cp_example

2.6.1.3 Create the Variables

Declare the model.

```
var_upper_bound = max(50, 45, 37)
x = model.NewIntVar(0, var_upper_bound, 'x')
y = model.NewIntVar(0, var_upper_bound, 'y')
z = model.NewIntVar(0, var_upper_bound, 'z')
```

2.6.1.4 Define the Constraints

Translate to constraints Eqs. (2.5, 2.6, 2.7, 2.8, and 2.9).

```
model.Add(2 * x + 7 * y + 3 * z <= 80)
model.Add(3 * x - 5 * y + 7 * z <= 90)
model.Add(5 * x + 2 * y - 6 * z <= 70)
```

2.6.1.5 Define the Objective Function

Define the objective function.

```
model.Maximize( x + 2 * y + 3 * z)
```

2.6.1.6 Call the Solver

```
solver = cp_model.CpSolver()
status = solver.Solve(model)
```

Total code:

```
# -*- coding: utf-8 -*-
"""
Created on Sat Dec 18 11:07:01 2021

@author: CHVUPPAL
"""

from ortools.sat.python import cp_model
model = cp_model.CpModel()

# Create the variables
var_upper_bound = max(150, 145, 70)
x = model.NewIntVar(0, var_upper_bound, 'x')
y = model.NewIntVar(0, var_upper_bound, 'y')
z = model.NewIntVar(0, var_upper_bound, 'z')

# Define the constraints
model.Add(2 * x + 7 * y + 3 * z <= 80)
model.Add(3 * x - 5 * y + 7 * z <= 90)
model.Add(5 * x + 2 * y - 6 * z <= 70)

# Define the objective function
model.Maximize(x + 2 * y + 2 * z)

# Call the solver

solver = cp_model.CpSolver()
# Enumerate all solutions.
solver.parameters.enumerate_all_solutions = True

status = solver.Solve(model)

# Display the solution
print('Status = %s' % solver.StatusName(status))

if status == cp_model.OPTIMAL or status == cp_model.FEASIBLE:
  print(f'Maximum of objective function: {solver.ObjectiveValue
()}\n')
  print(f'x = {solver.Value(x)}')
  print(f'y = {solver.Value(y)}')
  print(f'z = {solver.Value(z)}')
else:
  print('No solution found.')
```

Output:

```
Command Prompt

C:\Hanumayamma\Clustering>python CPLineMaximization.py
Status = OPTIMAL
Maximum of objective function: 40.0

x = 0
y = 5
z = 15

C:\Hanumayamma\Clustering>
```

2.6.2 Risk Reduction Knapsack

In the risk reduction knapsack, we consider the potential agricultural yield risk that could be caused by uncertainty and variation. We would again like to select C varieties from a given set, *I* of alternative seeds, each with an estimated yield randomly distributed (according to a known distribution, which in this case is assumed to be normal) with means μ_i and variances σ_i^2. We would like to maximize the total average yield that the selected varieties offer (i.e., total average knapsack yield), subject to a risk constraint. We set a minimum weight level, called "required minimum weight" and denoted by R. We would like to ensure that the probability that the total knapsack yield is greater than or equal to R.

The model can be stated as

$$\max \sum_{i \in I} uixi \tag{2.12}$$

$$s.t. \sum_{i \in I} xi \leq C \tag{2.13}$$

$$P(W \leq R) \leq 1 - \alpha \tag{2.14}$$

$$x_i \in \{0, 1\} \forall_{\mathrm{I}} \in \mathrm{I} \tag{2.15}$$

where W is the total knapsack yield which is defined as the sum of the yields of all selected varieties and x_i indicates if a variety is selected for the knapsack. The variable x_i is 1 if the variety is selected, 0 otherwise. *W* is a random variable given by the sum of the yields of the varieties in the chosen knapsack. When the variety yields are independent and identically distributed, the distribution of the total knapsack yield (which is a random variable that is the sum of the random variety yields of the varieties in the knapsack) is also normally distributed with mean μ and variance σ^2 equal to

$$\mu = \sum_{i \in I} uixi \tag{2.16}$$

$$\sigma^2 = \sum_{i \in I} \sigma 2\ xi \tag{2.17}$$

Hence, when the variety yields are normally distributed and independent, it is possible to rewrite the above model as

$$\max \sum_{i \in I} uixi \tag{2.18}$$

$$s.t. \sum_{i \in I} xi \leq C \tag{2.19}$$

$$P(W \leq R) \leq 1 - \alpha \tag{2.20}$$

$$P(W \leq R) = P\left(Z \leq \frac{R - \mu}{\sqrt{\sigma 2}}\right) = \phi\left(\frac{R - \mu}{\sqrt{\sigma 2}}\right) \leq 1 - \alpha \rightarrow \frac{R - \mu}{\sqrt{\sigma 2}} \leq \phi^{-1}(1 - \alpha) \tag{2.21}$$

This knapsack should give a different result when compared to the naïve knapsack, since the addition of a constraint reduces the feasible region of this problem, and the solution to this problem should be less than that of the naïve knapsack. This also makes intuitive sense, because varieties with high variances but otherwise high means are penalized because of their larger variances and therefore not selected.

Extending the above program:

```
# PREPROCESSING
n_seeds = len(set([entry['Variety'] for entry in seeds]))

# MODEL
model = cp_model.CpModel()
x_select = [model.NewBoolVar('') for i in range(n_seeds)]
denom = model.NewIntVar(1, 1000, '')
division = model.NewIntVar(0, 10, '')
sumX = model.NewIntVar(0, 10000, '')
sumY = model.NewIntVar(1, 10000, '')
scaled_requireYield = model.NewIntVar(0, 1 , '')

# select exactly "capacity"
model.Add(sum(x_select) == capacity)
## model.Add(sum([x_select[i] * int(round(seeds[i]['Mean']*10))
for i in range(n_seeds)]) >= 10*requireYield)
```

(continued)

```
model.Add(sumX == sum([x_select[i] * int(round(seeds[i]['Mean']
*10)) for i in range(n_seeds)]))
model.Add(denom == scaled_requireYield * requireYield)
model.AddDivisionEquality(division, sumX, denom)

model.Add(sumX >= requireYield)

# maximize sum of scores selected -> # ASSUMPTION: * 10 makes all the
values integral
model.Maximize(sum([x_select[i] * int(round(seeds[i]['Mean']
*10)) for i in range(n_seeds)]))

# SOLVE
solver = cp_model.CpSolver()
solver.parameters.log_search_progress = True
model.Proto().objective.scaling_factor = -1./10    # inverse
scaling for solver logging output
status = solver.Solve(model)
```

Required yield	Varieties in the selected knapsack	Average knapsack yield	Knapsack standard deviation
	{ X7, X10, X12, X15, X20, X24, X28, X29, X31, X39}	734.9	53.23
662	{ X7, X12, X15, X20, X24, X28, X29, X31, X39, X43}	728.1	34.21
650	{ X6, X7, X10, X14, X15, X24, X28, X29, X31, X36}	714.9	

The second entry is the results of adding the constraint that the knapsack yield had to be greater than or equal to a value of 662 at least 95% of the time. This resulted in a knapsack yield of 728.1 and a standard deviation of 34.21. Compared to the naïve knapsack, the knapsack exchanges the X10 variety for the X43 variety. This results in a knapsack yield loss of 6.8 bushels; however, it improves the standard deviation by 19.02. The difference in the selected set of varieties can be attributed to the addition of the risk constraint. This risk constraint forces the model to select less risky varieties, and as a result, it attempts to lower the standard deviation of the knapsack. Since the naïve knapsack gives the highest possible knapsack value, it is reasonable to expect to lose some knapsack value in exchange for choosing a less risky knapsack.

2.6.3 Complete Code

```
# -*- coding: utf-8 -*-
"""
Created on Sat Dec 18 14:53:44 2021

@author: CHVUPPAL
"""
import math
from ortools.sat.python import cp_model

# DATA
capacity = 10
requireYield=650
risk=20
seeds = [
 {"Variety":"X1", "Mean":59.95, "Variance":120.52},
  {"Variety":"X2", "Mean":41.81, "Variance":158.45},
  {"Variety":"X3", "Mean":55.1, "Variance":73.44},
  {"Variety":"X4", "Mean":57.36, "Variance":275.64},
  {"Variety":"X5", "Mean":51.03, "Variance":245.45},
  {"Variety":"X6", "Mean":68.84, "Variance":30.79},
  {"Variety":"X7", "Mean":71.39, "Variance":68.64},
  {"Variety":"X8", "Mean":60.06, "Variance":37.17},
  {"Variety":"X9", "Mean":48.12, "Variance":729.44},
  {"Variety":"X10", "Mean":68.93, "Variance":115.99},
  {"Variety":"X11", "Mean":40.48, "Variance":745.22},
  {"Variety":"X12", "Mean":71.21, "Variance":568.47},
  {"Variety":"X13", "Mean":64.61, "Variance":526.54},
  {"Variety":"X14", "Mean":62.76, "Variance":110.07},
  {"Variety":"X15", "Mean":71.76, "Variance":25.68},
  {"Variety":"X16", "Mean":47.72, "Variance":214.78},
  {"Variety":"X17", "Mean":62.59, "Variance":130.38},
  {"Variety":"X18", "Mean":53, "Variance":63.56},
  {"Variety":"X19", "Mean":65.75, "Variance":275.79},
  {"Variety":"X20", "Mean":73.18, "Variance":31.27},
  {"Variety":"X21", "Mean":55.49, "Variance":657.44},
  {"Variety":"X22", "Mean":61.61, "Variance":613.14},
  {"Variety":"X23", "Mean":66.38, "Variance":514.91},
  {"Variety":"X24", "Mean":77.5, "Variance":786.65},
  {"Variety":"X25", "Mean":66.59, "Variance":83.68},
  {"Variety":"X26", "Mean":57.34, "Variance":28.79},
  {"Variety":"X27", "Mean":58.5, "Variance":670.47},
  {"Variety":"X28", "Mean":71.42, "Variance":461.13},
  {"Variety":"X29", "Mean":79.87, "Variance":146.08},
  {"Variety":"X30", "Mean":43.17, "Variance":480.85},
  {"Variety":"X31", "Mean":78, "Variance":321.96},
  {"Variety":"X32", "Mean":54.49, "Variance":42.52},
  {"Variety":"X33", "Mean":41.16, "Variance":656.93},
```

(continued)

```
 {"Variety":"X34", "Mean":42.74,  "Variance":847.96},
 {"Variety":"X35", "Mean":59.07,  "Variance":540.1},
 {"Variety":"X36", "Mean":64.37,  "Variance":312.94},
 {"Variety":"X37", "Mean":43.42,  "Variance":96.6},
 {"Variety":"X38", "Mean":56.95,  "Variance":743.69},
 {"Variety":"X39", "Mean":71.58,  "Variance":307.59},
 {"Variety":"X40", "Mean":62.97,  "Variance":549.18},
 {"Variety":"X41", "Mean":63.32,  "Variance":59.68},
 {"Variety":"X42", "Mean":49.97,  "Variance":37.08},
 {"Variety":"X43", "Mean":62.1,  "Variance":26.82},
 {"Variety":"X44", "Mean":40.33,  "Variance":127.47},
 {"Variety":"X45", "Mean":54.24,  "Variance":187.15},
 {"Variety":"X46", "Mean":60.27,  "Variance":816.93},
 {"Variety":"X47", "Mean":48.44,  "Variance":282.85},
 {"Variety":"X48", "Mean":68.29,  "Variance":38.62},
 {"Variety":"X49", "Mean":40.82,  "Variance":670.32},
 {"Variety":"X50", "Mean":41.17,  "Variance":41.41},

 ]

# PREPROCESSING
n_seeds = len(set([entry['Variety'] for entry in seeds]))

# MODEL
model = cp_model.CpModel()
x_select = [model.NewBoolVar('') for i in range(n_seeds)]
denom = model.NewIntVar(1, 1000, '')
division = model.NewIntVar(0, 10, '')
sumX = model.NewIntVar(0, 10000, '')
sumY = model.NewIntVar(1, 10000, '')
scaled_requireYield = model.NewIntVar(0, 1 , '')

# select exactly "capacity"
model.Add(sum(x_select) == capacity)
## model.Add(sum([x_select[i] * int(round(seeds[i]['Mean']*10))
for i in range(n_seeds)]) >= 10*requireYield)

model.Add(sumX == sum([x_select[i] * int(round(seeds[i]['Mean']
*10)) for i in range(n_seeds)]))
model.Add(denom == scaled_requireYield * requireYield)

model.Add(denom <= sumX)

model.AddDivisionEquality(division,sumX,denom)

model.Add(division >= 0)

model.Add(sumX >= requireYield)
```

(continued)

```
# maximize sum of scores selected -> # ASSUMPTION: * 10 makes all the
values integral
model.Maximize(sum([x_select[i] * int(round(seeds[i]['Mean']
*10)) for i in range(n_seeds)]))

# SOLVE
solver = cp_model.CpSolver()
solver.parameters.log_search_progress = True
model.Proto().objective.scaling_factor = -1./10     # inverse
scaling for solver logging output
status = solver.Solve(model)

if status == cp_model.OPTIMAL:
 print('denom =', solver.Value(denom))
 print('sumX =', solver.Value(sumX))
 print('division =', solver.Value(division))

 selected = [i for i in range(n_seeds) if solver.Value(x_select
[i]) == 1]
 print("\n".join([str(seeds[i]) for i in selected]))
```

In summary the purpose of this work is to establish a tool that breeders can use to make better decisions when it comes to seed progression. This code is a chance-constrained knapsack optimization model, which selects the N varieties which maximize the average yield of the knapsack. This model is constrained by yield requirements and a minimum chance of success. The mean and the variance of the yields of the seed varieties are assumed to be normal and are used to create the chance constraint. However, this method assumes that the breeder has very good knowledge of the variety's distribution, mean, and variance. Since this isn't always the case, the breeders can use sampling to replace the need of calculating a probability distribution.

The code shows the risks of selecting knapsacks based only on the mean yields of the varieties. This method can result in reduced yields due to factors such as weather conditions, gene variability, and soil type. By creating a risk reduction knapsack, we can select varieties that ensure a higher mean yield when uncertain events do occur. Our goal is to ensure a higher yield and zero hunger through improved productivity.

Sustainable Development Goal 2: Zero Hunger
End hunger, achieve food security and improved nutrition and promote sustainable agriculture.[a]

[a]SDG2—https://www.ifc.org/wps/wcm/connect/topics_ext_content/ifc_external_corporate_site/development+impact/sdgs/measure-by-goal/sdg-2

2.7 Model and Simulation (M&S) Process and Industrialization Heuristics

Model and simulation[54] is a proven methodology to industrialize heuristics models. Establish purpose and scope of the project charter, formulate conceptual model, analyze data, and develop simulation model are the steps in developing the model (please see Fig. 2.9).

Validate and verify model, execute model, analyze data, configure-control, and publish model is the industrialization of the model. Analyzing simulate model is performed with the customer to confirm the workings of the model. Other steps are more internal to the heuristics team. In what I have seen with many customers, verify and validate model and simulation are an iterative process. Customers change the data configuration and would like to validate. For instance, in the GA line optimization model, the model was verified with the data from different models and different variants (please see Fig. 2.10). The variants include change in the seating,

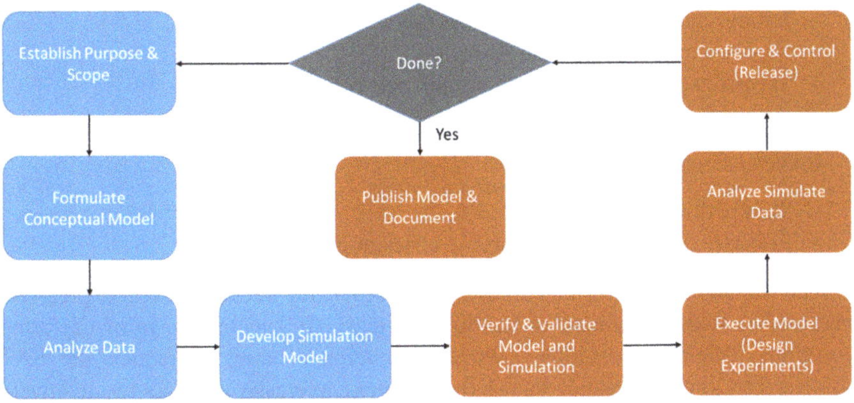

Fig. 2.9 Model simulation process [22]

Fig. 2.10 Automobile variants

[54]Margaret L. Loper, Modeling and Simulation in the Systems Engineering Life Cycle: Core Concepts and Accompanying Lectures, Publisher : Springer; 2015th edition (May 19, 2015), ISBN-13 : 978-1447156338.

change in the steering (left & right), types of tires for different weather conditions, and so on.

2.8 Choosing Solver Technology

Generally, solver technology choice is dependent on the model capabilities[55]: as models get complex,[56] choosing the right model will yield the maximum optimization (please see Fig. 2.11).

- Mixed-integer programming
  - Strong optimization, lower bounding (linear only)
  - Able to handle huge problems 1,000s of variants and constraints

- Finite domain propagation (FD)
  - Strong satisfaction, poor optimization
  - Highly expressive constraints
  - Specialized algorithms for important sub-constraints

- Boolean satisfaction (SAT)
  - Satisfaction principally
  - Effective conflict learning, highly efficient propagation

The way to understand the expressiveness/efficiency trade-off is that expressiveness comes with the benefit of handling increasingly succinct ways of capturing constraints, their propositional encoded counterparts being impractical for SAT solvers. The lower efficiency means that the more expressive solvers use relatively more overhead per succinct constraint than a SAT solver per clause. Each class of domain is labeled by distinguishing features of their mainstream state-of-the-art solvers. Modern SAT solvers are mainly based on a CDCL architecture that alternates a search for a solution to propositional variables with resolution inferences when a search dead-ends. These inferences rely on a limited set of premises. Garbage collection of unused derived clauses is a central ingredient to make CDCL scalable. MIP solvers use interior point methods and primal and dual simplex algorithms. Simplex pivoting performs a Gauss-Jordan elimination, which amounts to globally solving for a selected variable with respect to all constraints. CP solvers have

[55]G12: From Solver Independent Solutions to Efficient Solutions—https://people.eng.unimelb.edu.au/pstuckey/G12-CP-ICLP.pdf

[56]Bus Driver Svheduling—https://github.com/google/or-tools/blob/stable/examples/python/bus_driver_scheduling_sat.py

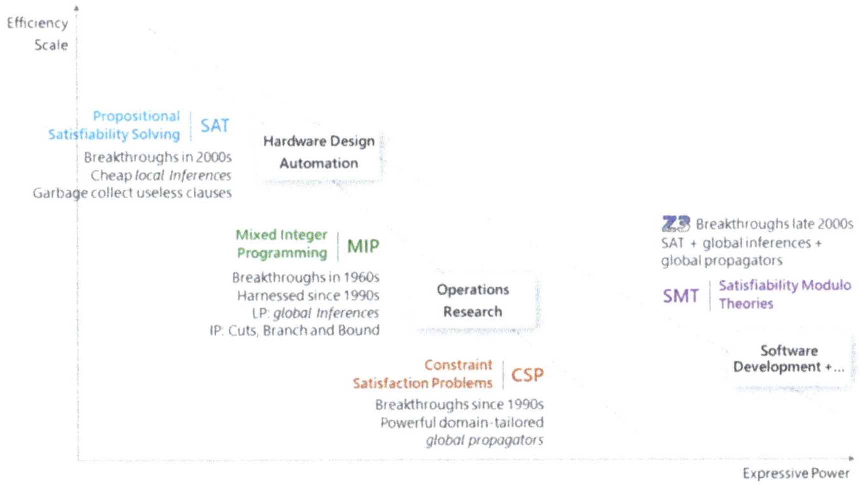

Fig. 2.11 Solver technology [4]

perfected the art of efficient propagators for global constraints. Effective propagators narrow the solution space maximally with minimal overhead.

After reading this chapter, you should be able to answer heuristics and mathematical optimization. You should be able to develop optimization models, feasible solutions using linear programming, constraint programming, and mixed-integer programming. You should be able to answer solver technology and develop heuristics data science models using the model and simulation process.

References

1. Pete Chapman, The CRISP-DM User Guide, https://s2.smu.edu/~mhd/8331f03/crisp.pdf , Data of Access: December 15, 2021.
2. Salveson, M. E. (1955). The assembly line balancing problem. Journal of Industrial Engineering, 6, 18-25.
3. Domm, Robert. Michigan Yesterday and Today, Publisher : Voyageur Press; First edition (October 1, 2009), ISBN-13: 978-0760333853
4. N. Bjørner, M. Levatich, N. P. Lopes, A. Rybalchenko, C. Vuppalapati. Supercharging Plant Configurations Using Z3. In Proc. of the 18th International Conference on Integration of Constraint Programming, Artificial Intelligence, and Operations Research (CPAIOR), July 2021.
5. Bryan Pearcel, A STUDY ON GENERAL ASSEMBLY LINE BALANCING MODELING METHODS AND TECHNIQUES, Publication Date: 12-2015, https://tigerprints.clemson.edu/

cgi/viewcontent.cgi?referer=&httpsredir=1&article=2550&context=all_dissertations, Access Date: December 18 2021
6. Austin Weber, Assembling Tractors and Equipment, Publication Date: February 09 2012, https://www.assemblymag.com/articles/89827-assembling-tractors-and-equipment, Access Date: December 18 2021
7. Arnaldo Pérez Castaño, Practical Artificial Intelligence: Machine Learning, Bots, and Agent Solutions Using C#, Publisher : Apress; 1st ed. edition (May 24, 2018), ISBN-13: 978-1484233566
8. Teofilo F. Gonzalez (Editor), Handbook of Approximation Algorithms and Metaheuristics: Contemporary and Emerging Applications, Volume 2, Publisher: Routledge; 2nd edition (June 30, 2020), ISBN-13: 978-0367571597
9. Richard Karp, Stephen Arthur Cook, Erik Gregersen, NP-complete problem mathematics, https://www.britannica.com/science/NP-complete-problem , Access Date: December 19, 2021
10. Filip M, Zoubek T, Bumbalek R, Cerny P, Batista CE, Olsan P, Bartos P, Kriz P, Xiao M, Dolan A, Findura P. Advanced Computational Methods for Agriculture Machinery Movement Optimization with Applications in Sugarcane Production. Agriculture. 2020; 10(10):434. https://doi.org/10.3390/agriculture10100434
11. John R. Birge, George Bernard Dantzig: The Pioneer of Linear Optimization, Winter 2021, https://mbrjournal.com/2021/01/26/george-bernard-dantzig-the-pioneer-of-linear-optimization/ , Access Date: December 19, 2021
12. W.E. Diewert, APPLIED ECONOMICS - Chapter 10: Linear Programming, Publication Date: July 2013, https://www.economics.ubc.ca/files/2014/02/pdf_course_erwin-diewert-ECON594 Ch10.pdf , Access Date: December 14, 2021
13. George Dantzig, 90; Created Linear Programming L.A. TIMES ARCHIVES, Publication Date: MAY 22, 2005, 12 AM PT, URL: https://www.latimes.com/archives/la-xpm-2005-may-22-me-dantzig22-story.html , Access Date: December 19, 2021
14. G B Dantzig, Impact of Linear Programming on Computer Development, Stanford University CA Systems Optimization Lab, June 1985, https://apps.dtic.mil/dtic/tr/fulltext/u2/a157659.pdf , Access Date: December 19, 2021
15. C. Vuppalapati, A. Ilapakurti, S. Kedari, R. Vuppalapati, J. Vuppalapati and S. Kedari, "The Role of Combinatorial Mathematical Optimization and Heuristics to improve Small Farmers to Veterinarian access and to create a Sustainable Food Future for the World," 2020 Fourth World Conference on Smart Trends in Systems, Security and Sustainability (WorldS4), 2020, pp. 214-221, doi: https://doi.org/10.1109/WorldS450073.2020.9210339.
16. Jonathan Kantor, Atlas of Suturing Techniques: Approaches to Surgical Wound, Laceration, and Cosmetic Repair, Publisher : McGraw-Hill Education / Medical; 1st edition (January 29, 2016), ISBN-13: 978-0071836579
17. Sune Engel Rasmussen, Severe Drought Adds to Afghanistan's Woes, Endangering Millions as Economy Collapses, Publication: Oct. 10, 2021 9:35 am ET, https://www.wsj.com/articles/severe-drought-adds-to-afghanistans-woes-endangering-millions-as-economy-collapses-11633872935?page=1 , Access Date: December 18, 2021
18. Mohammad Heydari, Faridah Othman, Meysam Salarijazi,Iman Ahmadianfar, Mohammad Sadegh Sadeghian, Predicting the Amount of Fertilizers using Linear Programming for Agricultural Products from Optimum Cropping Pattern, December 2018Journal of Geographical Studies 2(1):22-29, DOI: https://doi.org/10.21523/gcj5.18020103

19. Alameen Alwathiq Alameen Mohamed, Optimization Model for Agricultural Machinery Selection in Elsuki Agricultural Scheme Using linear Programming, July 2011, http://repository.sustech.edu/bitstream/handle/123456789/13031/Optimization%20Model%20for%20...%20.pdf?sequence=1, December 12,2021
20. Karthe, Introductory guide on Linear Programming for (aspiring) data scientists, Publication Date: February 28, 2017, https://www.analyticsvidhya.com/blog/2017/02/lintroductory-guide-on-linear-programming-explained-in-simple-english/ , Access Date: 12/12/2021
21. Ozkan Meric Ozcan, Chance-constrained optimization models for agricultural seed development and selection, August 2019, https://keep.lib.asu.edu/items/157571/view , Access Date: December 17, 2021
22. Margaret L. Loper, Modeling and Simulation in the Systems Engineering Life Cycle: Core Concepts and Accompanying Lectures, Publisher : Springer; 2015th edition (May 19, 2015), ISBN-13: 978-1447156338

Chapter 3
Data Engineering Techniques for Machine Learning and Heuristics

This chapter covers:

- Food Security Data
- Heuristics Data and Frequencies
- Data and Food Security Long Tail
- Data Enrichment Strategies
- Food Security Supplier Risk Modeling—Food-Fuel Conundrum
- Saudi Arabia Rice Imports and Food Security
  - Machine Learning Model: Saudi Arabia Rice Import Origin Cluster
  - Machine Learning Cluster Model—Food-Fuel Conundrum

This chapter introduces food security data sources and data engineering attributes for handling agricultural datasets. The chapter starts with heuristics data and frequencies and introduces food security long tail. Next, the chapter introduces data enrichment techniques to fix any data-related issues. Finally, the chapter concludes with food security risk model that was due to potential food-fuel policy change.

The importance of data and data collection and engineering techniques could be summarized with a single statement!

	By the authority vested in me as President by the Constitution and the laws of the United States of America, it is hereby ordered—Executive Order on Ensuring a Data-Driven Response to COVID-19 and Future High-Consequence Public Health Threats [1]. JOSEPH R. BIDEN JR.

(continued)

C. Vuppalapati, *Artificial Intelligence and Heuristics for Enhanced Food Security*, International Series in Operations Research & Management Science 331,
https://doi.org/10.1007/978-3-031-08743-1_3

	THE WHITE HOUSE,[1] January 21, 2021.

Effective data collection and data engineering techniques have advanced us to address the COVID-19 pandemic in an effective manner. The executive order emphasizes five aspects of the data:

- Effective collection and dissemination of data
- Enhancing data collection and collaboration capabilities
- Effectiveness, interoperability, and connectivity of data systems
- Advancing innovation in public health data and analytics
- Data privacy and effective use

Though the executive order is focused on health and public systems to take current and future threats, the principles of the data are still applicable across all data domains: agriculture, food security, cyber security, and public health systems.

Effective data is essential to develop machine learning and heuristic models, and the demand for good-quality statistical data is necessary now than ever before. The argument is simple. Data plays an important role in ensuring food security. Timely and reliable statistics are key inputs to the broad development strategy not only for nations of the world but also very critical for commercial enterprises. At the same token, good-quality statistical data is important for small farmers and agricultural businesses. Improvements in the quality and quantity of data on all aspects of development are essential if we are to achieve the goal of a world without poverty[2] and reach sustainability objectives. Good data are needed to set baselines, identify effective public and private actions, set goals and targets, monitor progress, and evaluate impacts. They are also an essential tool of good government, providing means for people to assess what governments do and helping them to participate directly in the development process. For example, during the Haiti Acute Food Insecurity National Situation in June–July 2013 and Southeastern Department in September 2013, it was recommended that data collection at agro-ecological zones is essential to prevent future food insecurity. The recommendation by the Integrated Food Security Phase Classification (IPC) is to establish a system for the regular collection of disaggregated information (rainfall, market prices, health and nutrition, water and sanitation, agricultural production, food preservation and processing) for future analysis purposes[3] and, similarly, to collect data related to prices in local

[1]Executive Order on Ensuring a Data-Driven Response to COVID-19 and Future High-Consequence Public Health Threats—https://www.whitehouse.gov/briefing-room/presidential-actions/2021/01/21/executive-order-ensuring-a-data-driven-response-to-covid-19-and-future-high-consequence-public-health-threats/

[2]The World Bank—https://data.worldbank.org/about

[3]Haiti: Acute Food Insecurity National Situation in June–July 2013 and Projection for Southeastern Department in September 2013—https://www.ipcinfo.org/ipc-country-analysis/details-map/en/c/459529/?iso3=HTI

markets, weekly wages per household, daily price, type, and frequency of survival strategies employed by households to enhance food security for Honduras.[4]

3.1 Food Security Data

Globally, climate change[5] is expected to threaten food production and certain aspects of food quality, as well as food prices and distribution systems. Many crop yields are predicted to decline because of the combined effects of changes in rainfall, severe weather events, and increasing competition from weeds and pests on crop plants. Livestock and fish production are also projected to decline. Prices are expected to rise in response to declining food production and associated trends such as increasingly expensive petroleum (used for agricultural inputs such as pesticides and fertilizers).

Like many important datasets, food security (please see Fig. 3.1) data is a time series data. The parameters that capture food security are bucketed into four groups: affordability, availability, quality and safety, and natural resources and resilience.[6]

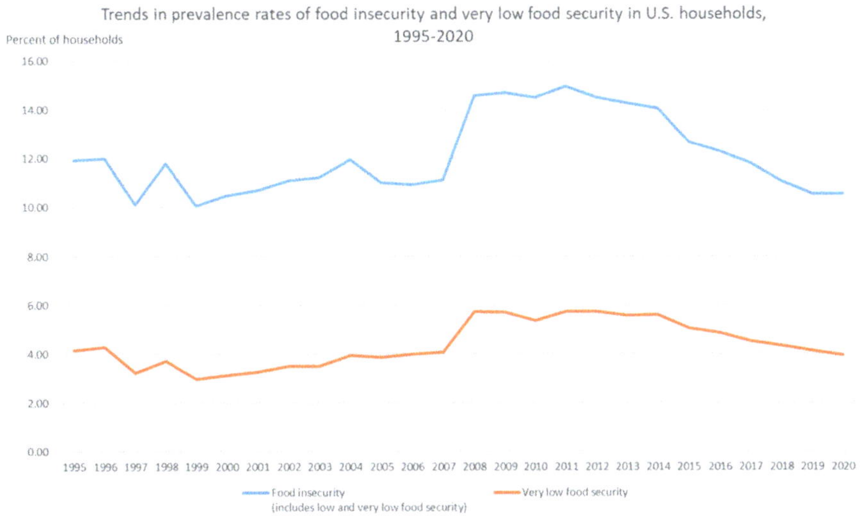

Fig. 3.1 Food insecurity

[4]Honduras: Acute Food Insecurity Situation in Southern Honduras December 2012–January 2013—https://www.ipcinfo.org/ipc-country-analysis/details-map/en/c/459515/?iso3=HND

[5]Food Security—https://www.cdc.gov/climateandhealth/effects/food_security.htm

[6]Agriculture and Food Security—https://www.usaid.gov/what-we-do/agriculture-and-food-security

The dataset for food security is captured as part of FAOSTAT[7] and as part of the survey data of the National Health and Nutrition Examination Survey[8] (NHANES) (though not a direct agriculture data, it can function as a proxy for food security). To review food security data 2017–March 2020 Pre-Pandemic Sample Data Release, please check the National Center for Health Statistics[9] National Health and Nutrition Examination Survey Food Security[10] (FSQ_I). Please find the below graph on trends in prevalence rates of food insecurity and very low food security in US households, 1995–2020 (Source: USDA, Economic Research Service, using data from the Current Population Survey Food Security Supplement[11]). Food security questionnaire provides several categories that describe household food security category (FSDHH): household full food security,[12] no affirmative response in any of these items; household marginal food security; household low food security; and household very low food security. More on food security will be covered in the next chapter. Please refer to Appendix A for Food Security Data Sources.

Agriculture—where the fight against hunger and climate change come together
Agriculture and food systems are partly responsible for climate change, but they are also part of the solution. Appropriate actions in agriculture, forestry and fisheries can mitigate greenhouse gas emission and promote climate adaptation—with efforts to reduce emissions and adjust practices to the new reality often enhancing and supporting one another. For millions of people, especially rural family farmers in developing countries, our actions can make the difference between poverty and prosperity, between hunger and food security and nutrition. Agriculture—where the fight against hunger and climate change come together—can unlock solutions[13] [2].

3.2 Heuristics Data and Frequencies

Heuristics modeling requires data, and it consists of at least two sets of data. One type includes time series data, and the other set includes rules or process optimization input data. In general, it is composite data comprising measured data with the processes that govern it. For example, in line balancing use case that was described in Chap. 2, data requirement to solve line balancing (please see Fig. 3.2 and

[7] FAOSTAT Food Security—https://www.fao.org/faostat/en/#data/FS

[8] NHANES—https://www.cdc.gov/nchs/nhanes/index.htm

[9] National Center for Health Statistics—What's new—https://www.cdc.gov/nchs/nhanes/new_nhanes.htm

[10] Food Security (FSQ_I)—https://wwwn.cdc.gov/Nchs/Nhanes/2015-2016/FSQ_I.htm

[11] Food Insecurity prevalence in US—https://www.ers.usda.gov/media/xtddtqat/trends.xlsx

[12] 2015–2016 Data Documentation, Codebook, and Frequencies Food Security (FSQ_I)—https://wwwn.cdc.gov/Nchs/Nhanes/2015-2016/FSQ_I.htm

[13] FAO'S WORK ON CLIMATE CHANGE—https://www.fao.org/3/i8037e/i8037e.pdf

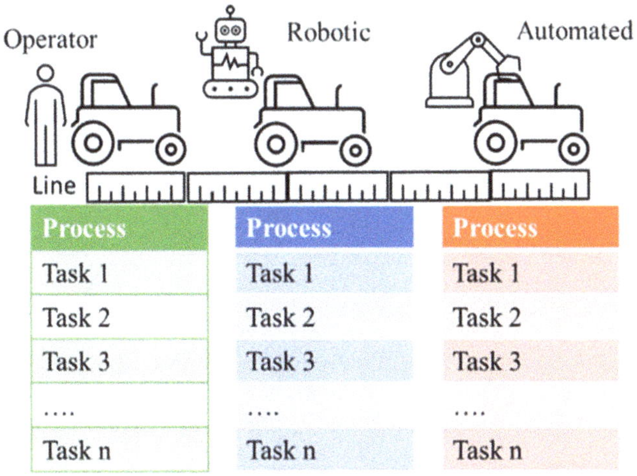

Fig. 3.2 GA line

Table 3.1) for manufacturing requires understanding a number of process steps that needed to be performed at each workstation on the manufacturing general assembly line.

- Number of process steps to perform at each workstation
- Number of tasks related to each process

Second process steps: the objective is to maximize station utilization and minimize the overall number of operators on the manufacturing line. The manufacturing line can be comprised of Sequential stations or a combination of sequential and parallel stations. Station availability for operator assignment is defined in the workstation design table. Workstations are premium and limited resources. The objective is to maximize utilization with the objective to produce more at the least cost.

Some of the rules:

- Workstations can have one or more operators assigned to them. Each workstation has height and zone constraints.
- Multiple zones can be worked on in the same station if there are no overlaps between operators and any automatic process.
- Tasks with process work height range of ±500 mm can be assigned to a single operator. For all operators in a station, the heights need to be in the same range. This constraint is not applicable to robotic process.
- Operators can be assigned only in stations that are unlocked.

Process rules are input data to optimization process. Derivation of process variables and constraints also the data to the heuristics model.

Table 3.1 Processes and tasks

Process	Tasks	Time (s)	Order
$P_{wheel\ attach}$	$T_{move\ front\ left}$	1	1
	T_{attach}	1	2
	T_{bolts}	1	3
	T_{covers}	2	4
	$T_{move\ rear\ left}$	1	5
	T_{attach}	1	6
	T_{bolts}	1	7
	T_{covers}	2	8
Process	Tasks	Time (s)	Order
$P_{wheel\ attach}$	$T_{move\ front\ right}$	1	9
	T_{attach}	1	10
	T_{bolts}	1	11
	T_{covers}	2	12
	$T_{move\ rear\ right}$	1	13
	T_{attach}	1	14
	T_{bolts}	1	15
	T_{covers}	2	16

- Number of processes (P) = 1000
- Number of stations (ST) = 800
- Number of tasks (T) = 8000

$$\text{TaskTimeInSec} \leq \sum_{\text{Task}=1}^{\text{Task All}} \text{Processes} + \sum \text{Any auxiliary task (Utilization of sets and others)}.$$

Consider another use case of heuristics model—the Stigler Diet[14] problem—that addresses nutritional needs given a set of foods. The data inputs are tabular data of nutrient and daily recommended intake (please see Table 3.2), plus the commodity list of food items with pricing details.

The set of foods Stigler evaluated reflected the time (1944). The nutritional data below is per dollar, not per unit, so the objective is to determine how many dollars to spend on each foodstuff. For the complete list, please check Stigler Diet (please see Table 3.3).[15]

[14] Stigler Diet—https://developers.google.com/optimization/lp/stigler_diet

[15] Stigler Diet—https://developers.google.com/optimization/lp/stigler_diet

Table 3.2 Nutrient table

Nutrient	Daily recommended intake
Calories	3000 calories
Protein	70 g
Calcium	.8 g
Iron	12 mg
Vitamin A	5000 IU
Thiamine (vitamin B1)	1.8 mg
Riboflavin (vitamin B2)	2.7 mg
Niacin	18 mg
Ascorbic acid (vitamin C)	75 mg

3.3 Data and Food Security Long Tail

The long tail[16,17] is a business strategy that allows companies to realize significant profits by selling low volumes of hard-to-find items to many customers, instead of only selling large volumes of a reduced number of popular items [3]. The term was first coined in 2004 by Chris Anderson, who argued that products in low demand or with low sales volume can collectively make up market share that rival or exceed the relatively few current bestsellers and blockbusters but only if the store or distribution channel is large enough. In statistics, a long tail of some distributions of numbers is the portion of the distribution having many occurrences far from the "head" or central part of the distribution. In terms of data services available to farmers vs. food security, we could see long tail. Though the data needed is acute, many developing countries' farmers would have no available data sources or only available for commercial farmers. And hence the usage of long-tail vs. food security is justified (please see Fig. 3.3).

For centuries, data has been driving agriculture. Agricultural practitioners are data scientists who have been using naturally available data[18] from the nature to plan the agricultural crops and production [4]. They have been closely studying data from weather, seasons, soil content, availability of natural resources, availability of time in day or shifting the time (through daylight savings) to yield maximum field available time, ambient temperature, crop growing pattern, and phenological details of wide varieties of crops as well as the local conditions, trade, commerce, imports/exports, policies to maximize the yield, replenish cycles of soil, moon phases, labor availability, animal husbandry, and the bio natural fertilizers to plan their crops and successfully harvest. In essence, the food security is assured by agricultural

[16] Fuzzy Logic Infused Intelligent Scent Dispenser For Creating Memorable Customer Experience of Long-Tail Connected Venues—https://ieeexplore.ieee.org/document/8527046

[17] Long Tail—https://www.investopedia.com/terms/l/long-tail.asp

[18] Chandrasekar Vuppalapati, Machine Learning and Artificial Intelligence for Agricultural Economics: Prognostic Data Analytics to Serve Small Scale Farmers Worldwide, Publisher: Springer; 1st ed. 2021 edition (October 5, 2021), ISBN-13: 978-3030774844.

Table 3.3 Commodity—calorie

Commodity	Unit	1939 price (cents)	Calories (kcal)	Protein (g)	Calcium (g)	Iron (mg)	Vitamin A (KIU)	Thiamine (mg)	Riboflavin (mg)	Niacin (mg)	Ascorbic acid (mg)
Wheat flour (enriched)	10 lb.	36	44.7	1411	2	365	0	55.4	33.3	441	0
Macaroni	1 lb.	14.1	11.6	418	0.7	54	0	3.2	1.9	68	0
Wheat cereal (enriched)	28 oz.	24.2	11.8	377	14.4	175	0	14.4	8.8	114	0
Corn flakes	8 oz.	7.1	11.4	252	0.1	56	0	13.5	2.3	68	0
Corn meal	1 lb.	4.6	36.0	897	1.7	99	30.9	17.4	7.9	106	0

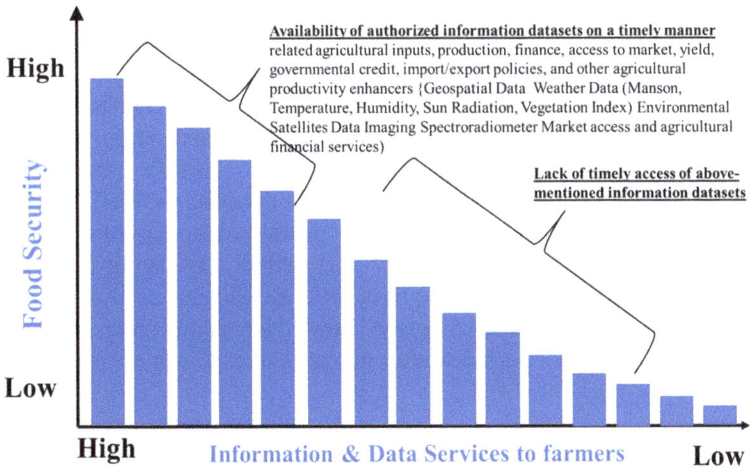

Fig. 3.3 Food security vs. information and data services to farmers

practitioners through the study of data for centuries. And we have around 800 million of such agricultural data practitioners/scientists around the world[19] [5].

Farmers the world over, in dealing with costs, returns, and risks, are calculating economic agents. Within their small individual, allocative domain they are fine-tuning entrepreneurs, tuning so subtly that many experts fail to see how efficient they are[20] [6].

Having data from diverse sources provides the development of climate smart agricultural algorithms that enable repeatable usage patterns for the current and future agricultural needs. Additionally, it enables geo diversity and extend learning-sharing cycle worldwide. Of course, agriculture is heavily localized with datasets that span and overlap with country specific macro environment and needed to account for such in algorithms.

The insights value chain is multiplicative, i.e., you are only as good as the weakest link in the chain. A lack of data source results in no value for the use case and, hence, low level of food security in the figure.

Here are steps:

- Connect multi-scale, multi-domain, or multi-format agricultural data bridge real-time distributed and parallel data systems.
- Create new methodologies and frameworks for tracking and processing data.
- Identify new approaches to data archiving and sharing that support findable, accessible, interoperable, and reusable (FAIR) standards.

[19] Data Science for Food and Agricultural Systems (DSFAS)—https://nifa.usda.gov/program/dsfas

[20] Relationship between milk production and price variations in the EC—http://aei.pitt.edu/36837/1/A67.pdf

The data availability to farmers in terms of information and data services has a direct impact on food security—availability of authorized information datasets on a timely manner related to agricultural inputs, production, finance, access to market, yield, governmental credit, import/export policies, and other agricultural productivity enhancers (geospatial data, weather data (monsoon, temperature, humidity, sun radiation), Vegetation Index, environmental satellite data imaging spectroradiometer, market access, and agricultural financial services). The rich and available authorized data have a higher agricultural productivity that has a direct impact on the four pillars of food security—accessibility, availability, affordability, and future sustainability. Fusing data from different domains catalyzes[21] activities that harness big data for synthesizing new knowledge, making predictive decisions, and fostering data-supported innovation in agriculture, to achieve sustained food security.

Recovery from the Great Depression: The Farm Channel in Spring 1933[22] [7]
"The depression in the manufacturing industry of the country is due chiefly to the fact that agricultural products generally have been selling below the cost of production, and thereby destroyed the purchasing power in the domestic market of nearly half of all our people. We are going to restore the purchasing power of the farmer."
Franklin D. Roosevelt,
Campaign speech[23] in Atlanta, GA, October 24, 1932 [8]

3.4 Data Enrichment

Datasets, in general due to perturbation of inputs or due to non-calibration equipment or due to human errors, come from various lines of sources, businesses, industries, academia, or field operations. Historically, the data sources are from transactional and from the reporting information systems. With the advent of the Internet of Things (IoT) and smart and intelligent agriculture, the data sources, now, are more near real time or real time, sub-milliseconds. As the sources of agriculture have expanded, so as the data issues. One could expect, therefore, that the real-world agricultural datasets contain missing values, due to sensor calibration issues of IoT devices or data transformation issues when injecting into corporate or governmental data lakes, often encoded as NaN, or they can be blank or empty. Such datasets

[21] Data Science for Food and Agricultural Systems (DSFAS)—https://nifa.usda.gov/program/dsfas

[22] Recovery from the Great Depression—https://pubs.aeaweb.org/doi/pdfplus/10.1257/aer.20170237#page=14

[23] Campaign Speech Franklin D. Roosevelt | October 24, 1932—https://teachingamericanhistory.org/library/document/campaign-speech/

however are incompatible with machine learning and artificial intelligent algorithms, for instance, scikit-learn, which assume that all values in an array are numerical and that all have and hold meaning.[24]

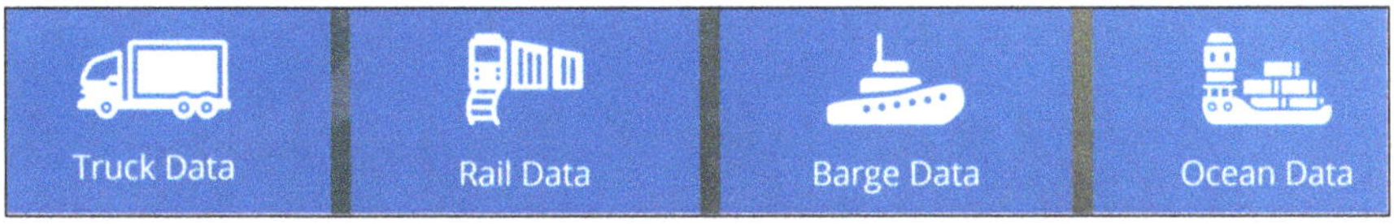

A basic strategy to use incomplete datasets employed by many data scientists is to discard the entire rows and/or columns containing missing values. Despite losing valuable information, despite forgoing underling data signal or a foreseeable prospective trend, this is a general operating practice followed by small and large data science teams. Lacking resources or even a theoretical framework, researchers, methodologists, and software developers resort to editing the data to lend an appearance of completeness. Unfortunately, ad hoc edits may do more harm than good, producing answers that are biased, inefficient (lacking in power), and unreliable[25].

To maximize return on assets of information acquisition systems and to better serve customers, a novel strategy is to impute the missing values, i.e., to infer them from the known part of the data. The strategy for imputation could entail the following:

- Replacing missing value with a placeholder value, mean, or other value[26]
  - Imputation using (mean/median) values
  - Imputation using (most frequent) or (zero/constant) values
  - Imputation using k-NN
  - Imputation using multivariate imputation by chained equation (MICE)
- Completely removing the rows and columns that have missing value
- Inferring value based on statistical methods[27] (probabilistic principal component analysis (PPCA), multivariate imputation by chained equations (MICE)) [9]

[24]Imputation of missing values—https://scikit-learn.org/stable/modules/impute.html

[25]Missing Data: Our View of the State of the Art—https://pubmed.ncbi.nlm.nih.gov/12090408/

[26]Imputation of missing values—https://scikit-learn.org/stable/modules/impute.html

[27]MICE vs PPCA: Missing data imputation in healthcare—https://www.sciencedirect.com/science/article/pii/S2352914819302783

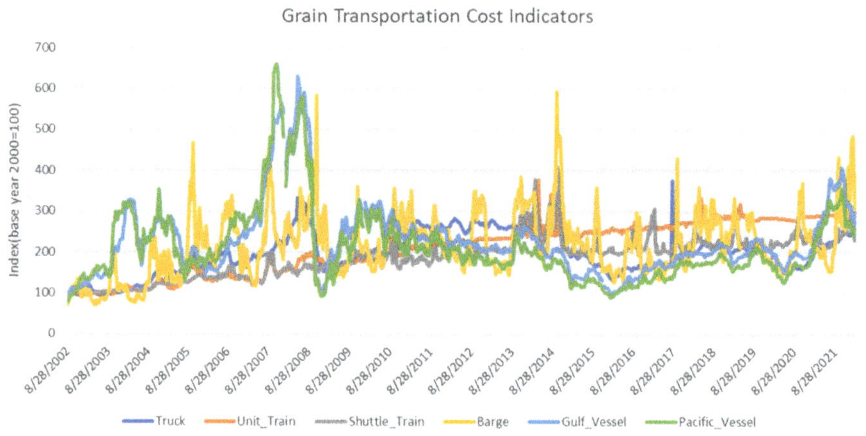

Fig. 3.4 Grain transportation cost indicators

3.4.1 Mean/Median Strategy

This strategy works for column with missing values and is highly effective for small numeric datasets. However, it does not factor in correlation between features and could lead to lower accuracies.

Please consider two important datasets (wheat prices and agriculture grain transportation). Grain Transportation Cost Indicators dataset[28] shows the weekly cost indices of transporting grain by each mode: truck, rail, barge, and ocean-going vessels. The base of each index (set to 100) is the average of monthly costs in 2000. Grain transportation cost indicators (please see Fig. 3.4) are a proxy of agricultural demand and supply health.[29] While transporting grain to the market may be the last input cost in the production of grain, it is a critical decision a producer must make, especially when margins are thin. Determining which market to sell your grain (if you have options) can be a complex decision. Many factors will impact the transportation cost of grain and determine the most profitable option. Those factors include grain price, distance, fuel price, wait time, quality discounts, labor, and truck capacity. It is common for most producers to make their market decision based on only one of these factors[30] [10].

Most producers in Western Kentucky have multiple potential markets to deliver their grain. This leads to the question of, "Should I sell my grain to the closest elevator, or should I transport it a further distance to an elevator offering a higher

[28] Grain Transportation Cost Indicators—https://agtransport.usda.gov/Transportation-Costs/Grain-Transportation-Cost-Indicators/8uye-ieij

[29] Grain Transportation Report Datasets—https://www.ams.usda.gov/services/transportation-analysis/gtr-datasets

[30] Post-Harvest Management Cooperative Extension Service the Economics of Grain Transportation—http://www2.ca.uky.edu/agcomm/pubs/AEC/AEC100/AEC100.pdf

Fig. 3.5 Wheat futures

price?" What market you choose not only will determine the price you receive but will also determine the cost associated with transportation. The market that provides the highest price is not always the most profitable price. The trade-off between maximizing price per bushel received from the buyer and minimizing transportation costs could be the difference between making a profit that year or being in the red. Grain transportation cost along with demand aspects of the transportation routes and vehicles could provide insight into economic signal. Overlaying the essential staple commodity price, for instance wheat, would provide demand and price curve between transportation cost and wheat demand.

"Freight costs are related to fuel prices (marine gas oil and fuel oil) and fuel prices are obviously linked to oil prices. This all means higher prices for food".
Trinidad & Tobago's former Minister[31] of Energy [11]

3.4.1.1 Wheat Futures

Wheat futures[32] are available for trading in the Chicago Board of Trade (CBOT), Euronext, Kansas City Board of Trade (KCBT), and the Minneapolis Grain Exchange (MGEX) (please see Fig. 3.5). The standard contract unit is 5000 bushels. The United States is the biggest exporter of wheat followed by the European Union, Australia, and Canada.

[31] Caribbean Food Security Likely To Be Impacted By Russia-Ukraine Conflict—https://www.forbes.com/sites/daphneewingchow/2022/02/27/caribbean-food-security-likely-to-be-impacted-by-russia-ukraine-conflict/?sh=1ece953739d4

[32] Wheat Futures—https://www.investing.com/commodities/us-wheat-historical-data

Conflicts, Low Stock Levels, and Food Security

Chicago wheat futures surged to $11 per bushel in the first week of March 2022, the highest since 2008 as the Russian invasion of Ukraine brought supply disruptions from two of the world's largest producers. Amid heavy sanctions and restrictive measures from western economies, exports from the Black Sea have nearly halted. Multinational food companies such as Bunge and ADM have closed facilities in the region, while the world's biggest container ship operator Maersk/MSC suspended service to Russian ports.

At the same time, the Ukrainian military suspended all commercial operations from Ukrainian ports. With Russia and Ukraine accounting for roughly 30% of the world's wheat exports, conflict in the region affects crucial supply from an already tight market. Meanwhile, wheat stocks in major exporting countries are already at low levels, and ongoing droughts are reducing maize availability in South America, US, and Canada.[33]

By comparing the grain transportation costs with wheat futures data, we can deduce the movements of prices of both as producers would like to maximize the profit. Profit maximization occurs with the biggest gap between the total revenue and the total costs.

$$\text{Profit} = \text{Total Revenue (TR)Total Costs (TC)} \tag{3.1}$$

However, there are nulls in the dataset that could potentially change the signal. In the following figure, please find the null values of Gulf and Pacific Vessel cost indicators. The same goes with Barge. Some of the nulls are due to public holidays, and some of them are on business days.

To overcome signal mangling issue due to missing data, it could be legitimate that on public holidays the transportation cost for vessels may not be available; we can apply the following imputation strategies to develop missing data filling strategies (please Fig. 3.6).

3.4.2 *Simple Imputation Strategy*

Software code for this model: WheatPricetoGrainTransportationCostIndicators.ipynb (Jupyter Notebook Code)

[33] Wheat Futures—https://tradingeconomics.com/commodity/wheat

	Date	Week	Month	Year	Truck	Unit_Train	Shuttle_Train	Barge	Gulf_Vessel	Pacific_Vessel	Price	Open	High
8	12/29/2021	52	12	2021	243	299.03	303.79	397.22	NaN	NaN	758.5	774	782.25
12	12/1/2021	48	11	2021	250	297.12	250.57	266.67	NaN	NaN	782	799.75	812
60	12/30/2020	52	12	2020	177	286.87	244.62	232.22	NaN	NaN	638.5	642.12	664.38
64	12/2/2020	48	11	2020	168	287.13	222.81	258.33	NaN	NaN	616.8	576.75	619.12
112	1/1/2020	52	12	2019	206	283.22	210.03	175.00	NaN	NaN	64.25	552.25	568.12
116	12/4/2019	48	12	2019	206	283.22	209.34	195.00	NaN	NaN	533	524.62	536.38
140	6/19/2019	24	6	2019	206	NaN	230.20	293.33	193.43	168.44	526	530.25	557.12
141	6/12/2019	23	6	2019	208	287.27	235.07	NaN	195.66	170.21	526	537.75	549.12
142	6/5/2019	22	6	2019	210	289.69	227.22	NaN	196.78	171.99	539.5	505	543.75
143	5/29/2019	21	5	2019	211	NaN	219.12	NaN	194.54	170.21	503.25	507	528.25
145	5/15/2019	19	5	2019	212	286.45	222.56	NaN	190.07	163.12	490.38	463.5	492.62
146	5/8/2019	18	5	2019	213	295.39	216.24	NaN	190.07	163.12	463.12	423.62	472.88
164	1/2/2019	52	12	2018	205	284.06	212.63	215.56	NaN	NaN	19.88	516.38	524.5
169	11/28/2018	47	11	2018	219	283.34	214.21	179.44	NaN	NaN	31.75	527.75	532
216	1/3/2018	53	12	2017	199	272.12	219.01	187.78	NaN	NaN	20.75	430.25	436
221	11/29/2017	48	11	2017	196	272.12	208.25	166.67	NaN	NaN	419	439.5	442.75
321	12/30/2015	52	12	2015	150	253.03	200.64	157.22	NaN	NaN	78.38	472.5	479.63
325	12/2/2015	48	11	2015	162	248.64	196.96	155.56	NaN	NaN	90.38	489.38	497.88
347	7/1/2015	26	6	2015	191	249.67	02.36	NaN	143.11	127.66	575.13	585.5	599.63
348	6/24/2015	25	6	2015	192	248.99	01.64	NaN	145.35	131.21	588.25	565.5	617.13
420	2/5/2014	5	2	2014	265	283.08	62.89	NaN	252.68	198.58	616.88	623.25	628.13
460	5/1/2013	17	4	2013	258	234.09	06.58	NaN	207.96	173.76	806.63	730.5	834.38

Fig. 3.6 Data NaNs

Step 1: Load Wheat Futures and Grain Transportation Cost Data

```
from sklearn.linear_model import LinearRegression
from sklearn.model_selection import StratifiedKFold
from sklearn.datasets import fetch_california_housing
from sklearn.linear_model import LinearRegression
from sklearn.model_selection import StratifiedKFold
from sklearn.metrics import mean_squared_error
from math import sqrt
import random
import numpy as np
random.seed(0)

#Fetching the dataset
import pandas as pd
grainTransportationCostIndicatorsUSWheatDF = pd.read_csv
('US_WHEAT_Grain_Transportation_Cost_IndicatorsRelationship.
csv')

grainTransportationCostIndicatorsUSWheatDF.tail(10)
```

Output:

	Date	Week	Month	Year	Truck	Unit_Train	Shuttle_Train	Barge	Gulf_Vessel	Pacific_Vessel	Price	Open	High
895	12/29/2004	52	12	2004	133	129.77	124.30	166.67	NaN	NaN	382.5	412.25	416
896	12/22/2004	51	12	2004	133	129.89	124.45	192.11	271.61	280.55	413.5	409.5	419
897	12/15/2004	50	12	2004	134	132.67	124.45	167.22	282.79	308.87	410.5	413.5	416.75
898	12/8/2004	49	12	2004	139	134.21	124.45	136.11	310.54	345.23	413	367	414
899	12/1/2004	48	11	2004	142	133.22	124.45	133.33	314.71	353.65	365.5	378	385.5
900	11/24/2004	47	11	2004	142	127.14	121.33	131.11	287.89	319.77	376.25	405	413
901	11/17/2004	46	11	2004	143	127.06	121.28	166.11	273.91	290.48	402.5	401	408
902	11/10/2004	45	11	2004	145	125.77	121.28	206.11	271.95	277.41	400	391	419
903	11/3/2004	44	10	2004	148	123.10	121.28	205.00	270.40	281.45	395	409	434
904	10/27/2004	43	10	2004	148	116.61	118.17	235.00	268.56	283.83	403.5	368	403.5

To find the total number of nulls in the data frame:

```
grainTransportationCostIndicatorsUSWheatDF.isna().sum()
```

Output:

```
Date              0
Week              0
Month             0
Year              0
Truck             0
Unit_Train        2
Shuttle_Train     0
Barge            10
Gulf_Vessel      18
Pacific_Vessel   18
Price             0
Open              0
High              0
dtype: int64
```

Please note that Barge, Gulf_Vessel, and Pacific Vessel cost indicators have null values.

Step 2: List Null Values Within the Cost and Wheat Futures Dataset
To find time occurrences of nulls, please apply the following data frame technique:

```
grainTransportationCostIndicatorsWheatNARows =
grainTransportationCostIndicatorsUSWheatDF
[grainTransportationCostIndicatorsUSWheatDF.isna().any
(axis=1)]
grainTransportationCostIndicatorsWheatNARows
```

Please note that you will find one or more null entries per row below:

	Date	Week	Month	Year	Truck	Unit_Train	Shuttle_Train	Barge	Gulf_Vessel	Pacific_Vessel	Price	Open	High
8	12/29/2021	52	12	2021	243	299.03	303.79	397.22	NaN	NaN	758.5	774	782.25
12	12/1/2021	48	11	2021	250	297.12	250.57	266.67	NaN	NaN	782	799.75	812
60	12/30/2020	52	12	2020	177	286.87	244.62	232.22	NaN	NaN	638.5	642.12	664.38
64	12/2/2020	48	11	2020	168	287.13	222.81	258.33	NaN	NaN	616.8	576.75	619.12
112	1/1/2020	52	12	2019	206	283.22	210.03	175.00	NaN	NaN	564.25	552.25	568.12
116	12/4/2019	48	12	2019	206	283.22	209.34	195.00	NaN	NaN	533	524.62	536.38
140	6/19/2019	24	6	2019	206	NaN	230.20	293.33	193.43	168.44	526	530.25	557.12
141	6/12/2019	23	6	2019	208	287.27	235.07	NaN	195.66	170.21	526	537.75	549.12
142	6/5/2019	22	6	2019	210	289.69	227.22	NaN	196.78	171.99	539.5	505	543.75
143	5/29/2019	21	5	2019	211	NaN	219.12	NaN	194.54	170.21	503.25	507	528.25
145	5/15/2019	19	5	2019	212	286.45	222.56	NaN	190.07	163.12	490.38	463.5	492.62
146	5/8/2019	18	5	2019	213	295.39	216.24	NaN	190.07	163.12	463.12	423.62	472.88
164	1/2/2019	52	12	2018	205	284.06	212.63	215.56	NaN	NaN	519.88	516.38	524.5
169	11/28/2018	47	11	2018	219	283.34	214.21	179.44	NaN	NaN	531.75	527.75	532
216	1/3/2018	53	12	2017	199	272.12	219.01	187.78	NaN	NaN	420.75	430.25	436
221	11/29/2017	48	11	2017	196	272.12	208.25	166.67	NaN	NaN	419	439.5	442.75
321	12/30/2015	52	12	2015	150	253.03	200.64	157.22	NaN	NaN	478.38	472.5	479.63
325	12/2/2015	48	11	2015	162	248.64	196.96	155.56	NaN	NaN	490.38	489.38	497.88
347	7/1/2015	26	6	2015	191	249.67	202.36	NaN	143.11	127.66	575.13	585.5	599.63
348	6/24/2015	25	6	2015	192	248.99	201.64	NaN	145.35	131.21	588.25	565.5	617.13
420	2/5/2014	5	2	2014	265	283.08	262.89	NaN	252.68	198.58	616.88	623.25	628.13
460	5/1/2013	17	4	2013	258	234.09	206.58	NaN	207.96	173.76	806.63	730.5	834.38

Step 3: Create Train and Target Placeholders

```
train = grainTransportationCostIndicatorsUSWheatDF[['Truck',
'Unit_Train', 'Shuttle_Train',
   'Barge', 'Gulf_Vessel', 'Pacific_Vessel']]
target = grainTransportationCostIndicatorsUSWheatDF
[['WheatPrice']]
```

Output:

	Truck	Unit_Train	Shuttle_Train	Barge	Gulf_Vessel	Pacific_Vessel
895	133	129.77	124.30	166.67	NaN	NaN
896	133	129.89	124.45	192.11	271.61	280.55
897	134	132.67	124.45	167.22	282.79	308.87
898	139	134.21	124.45	136.11	310.54	345.23
899	142	133.22	124.45	133.33	314.71	353.65
900	142	127.14	121.33	131.11	287.89	319.77
901	143	127.06	121.28	166.11	273.91	290.48
902	145	125.77	121.28	206.11	271.95	277.41
903	148	123.10	121.28	205.00	270.40	281.45
904	148	116.61	118.17	235.00	268.56	283.83

Step 4: Simple Imputer

In the following code, apply SCIKIT Simple Imputer to mean strategy to replace NaN with median values.

```
#Impute the values using scikit-learn SimpleImpute Class
from sklearn.impute import SimpleImputer
imp_mean = SimpleImputer( strategy='mean') #for median imputation
replace 'mean' with 'median'
imp_mean.fit(train)

imputed_train_array = imp_mean.transform(train)
print(imputed_train_array)
column = imputed_train_array
print(column.size)
```

Output: total column size 6335

```
[[ 272.   297.82 229.33 ... 295.17 257.09 1059.1 ]
 [ 270.   297.82 229.62 ... 290.7  255.32 843.  ]
 [ 265.   297.82 252.85 ... 272.81 241.13 797.  ]
 ...
 [ 145.   125.77 121.28 ... 271.95 277.41 400.  ]
```

(continued)

```
 [148.  123.1 121.28 ... 270.4  281.45 395. ]
 [148.  116.61 118.17 ... 268.56 283.83 403.5]]
6335
```

Convert array into data frame.[34]

```
imputed_traindf = pd.DataFrame(imputed_train_array)
```

Output:

	Truck	Unit_Train	Shuttle_Train	Barge	Gulf_Vessel	Pacific_Vessel	WheatPrice
895	133.0	129.77	124.30	166.67	232.815479	213.859245	382.50
896	133.0	129.89	124.45	192.11	271.610000	280.550000	413.50
897	134.0	132.67	124.45	167.22	282.790000	308.870000	410.50
898	139.0	134.21	124.45	136.11	310.540000	345.230000	413.00
899	142.0	133.22	124.45	133.33	314.710000	353.650000	365.50
900	142.0	127.14	121.33	131.11	287.890000	319.770000	376.25
901	143.0	127.06	121.28	166.11	273.910000	290.480000	402.50
902	145.0	125.77	121.28	206.11	271.950000	277.410000	400.00
903	148.0	123.10	121.28	205.00	270.400000	281.450000	395.00

Please note Gulf and Pacific Vessel values are imputed with mean strategy.

3.4.3 *Most Frequent or (Zero/Constant) Strategy*

The most frequent imputation strategy[35] fits with numerical/categorical (mostly recommended) values. That is, missing categorical column values can be filled with this strategy.[36]

[34] SimpleImputer—https://scikit-learn.org/stable/modules/generated/sklearn.impute.SimpleImputer.html

[35] sklearn SimpleImputer too slow for categorical data represented as string values—https://datascience.stackexchange.com/questions/66034/sklearn-simpleimputer-too-slow-for-categorical-data-represented-as-string-values

[36] 6 Different Ways to Compensate for Missing Values In a Dataset (Data Imputation with examples)—https://towardsdatascience.com/6-different-ways-to-compensate-for-missing-values-data-imputation-with-examples-6022d9ca0779

Software code for this model can be found on GitHub Link: WheatPricetoGrainTransportationCostIndicators.ipynb (Jupyter Notebook Code)

Step 1: Create the Most Frequent Imputer

```
#Impute the values using scikit-learn SimpleImpute Class
from sklearn.impute import SimpleImputer
imp_mean = SimpleImputer(missing_values=np.nan,
strategy='most_frequent')
imp_mean.fit(train)

imputed_train_mf_array = imp_mean.transform(train)
print(imputed_train_mf_array)
column = imputed_train_mf_array
print(column.size)
```

Step 2: Imputed Most Frequent Data Frame

```
imputed_trainMFdf = pd.DataFrame(imputed_train_mf_array,
        columns=['Truck', 'Unit_Train', 'Shuttle_Train',
    'Barge', 'Gulf_Vessel', 'Pacific_Vessel','WheatPrice'])

imputed_trainMFdf
```

	Truck	Unit_Train	Shuttle_Train	Barge	Gulf_Vessel	Pacific_Vessel	WheatPrice
895	133.0	129.77	124.30	166.67	205.72	170.21	382.50
896	133.0	129.89	124.45	192.11	271.61	280.55	413.50
897	134.0	132.67	124.45	167.22	282.79	308.87	410.50
898	139.0	134.21	124.45	136.11	310.54	345.23	413.00
899	142.0	133.22	124.45	133.33	314.71	353.65	365.50
900	142.0	127.14	121.33	131.11	287.89	319.77	376.25
901	143.0	127.06	121.28	166.11	273.91	290.48	402.50
902	145.0	125.77	121.28	206.11	271.95	277.41	400.00
903	148.0	123.10	121.28	205.00	270.40	281.45	395.00
904	148.0	116.61	118.17	235.00	268.56	283.83	403.50

The imputed strategy, bounding rectangle, replaces NAN values with converted values. Unlike simple impute strategy that replaced Gulf Vessel and Pacific Vessel

Grain Cost indicator with 232.82 and 213.85, the most frequent strategy has replaced the cost indicators by using the most frequent values from within the dataset (205.72 & 170.21). A cost differential can yield a higher food security[37] or a lower food insecurity [11].

Categorical values take long time to fit.

3.4.4 *k-NN Strategy*

The k nearest neighbors is an algorithm that is used for simple classification purposes. The algorithm uses "feature similarity" to predict the values of any new data points.

The idea is to impute array with a passed in initial impute function (mean impute) and then use the resulting complete array to construct a K-D tree.[38] Use this K-D tree to compute the nearest neighbors. After finding "k" nearest neighbors, take the weighted average of them.

K-D tree is a partitioning data structure that arranges points in k-dimensional space. K-D tree performs analysis in memory.

Let's see some example code using impute library which provides a simple and easy way to use KNN for imputation:

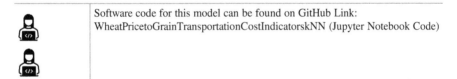

Software code for this model can be found on GitHub Link: WheatPricetoGrainTransportationCostIndicatorskNN (Jupyter Notebook Code)

Steps 1 and 2 are the same: load the Bahamas Dataset and creation of WheatPricetoGrainTransportationCostIndicatorskNNDF

Step 3:
Apply kNN = 1. That is, the average (by distance) of the nearest neighbor 1 fills all null cells with the nearest cell value.

[37] Caribbean Food Security Likely To Be Impacted By Russia-Ukraine Conflict—https://www.forbes.com/sites/daphneewingchow/2022/02/27/caribbean-food-security-likely-to-be-impacted-by-russia-ukraine-conflict/?sh=1ece953739d4

[38] Source code for impyute.imputation.cs.fast_knn—https://impyute.readthedocs.io/en/master/_modules/impyute/imputation/cs/fast_knn.html

```
import sys
from impyute.imputation.cs import fast_knn
from math import sqrt
import random
import numpy as np

sys.setrecursionlimit(100000) #Increase the recursion limit of
the OS

data = grainTransportationCostIndicatorsUSWheatkNNDF.to_numpy
()
# Weighted average (by distance) of nearest 1 neighbour
print('\naverage (by distance) of nearest 1 neighbour')
imputed_TransportationCostIndicatorsUSWheat_knn1_array
=fast_knn(data, k=1)
print(imputed_TransportationCostIndicatorsUSWheat_knn1_array)
```

Output:

```
average (by distance) of nearest 1 neighbour
[[ 297.82 229.33 308.89 295.17 257.09 1059.1 ]
 [ 297.82 229.62 385.  290.7  255.32 843. ]
 [ 297.82 252.85 400.56 272.81 241.13 797. ]
 ...
 [ 125.77 121.28 206.11 271.95 277.41 400. ]
 [ 123.1  121.28 205.  270.4  281.45 395. ]
 [ 116.61 118.17 235.  268.56 283.83 403.5 ]]
```

Step 4: Convert kNN Array Back into Data Frame

```
imputed_kNNtraindf = pd.DataFrame
(imputed_TransportationCostIndicatorsUSWheat_knn1_array, columns =
['Unit_Train', 'Shuttle_Train',
   'Barge', 'Gulf_Vessel', 'Pacific_Vessel', 'WheatPrice'])
```

Output:

	Unit_Train	Shuttle_Train	Barge	Gulf_Vessel	Pacific_Vessel	WheatPrice
895	129.77	124.30	166.67	263.05	242.40	382.50
896	129.89	124.45	192.11	271.61	280.55	413.50
897	132.67	124.45	167.22	282.79	308.87	410.50
898	134.21	124.45	136.11	310.54	345.23	413.00
899	133.22	124.45	133.33	314.71	353.65	365.50
900	127.14	121.33	131.11	287.89	319.77	376.25
901	127.06	121.28	166.11	273.91	290.48	402.50
902	125.77	121.28	206.11	271.95	277.41	400.00
903	123.10	121.28	205.00	270.40	281.45	395.00
904	116.61	118.17	235.00	268.56	283.83	403.50

For K = 2 and higher, the weighted average values of the data are as below. As you can see, the weighted average values are computed for K is greater than 1.

```
# Weighted average of nearest 2 neighbour
print('\nWeighted average of nearest 2 neighbour')
imputed_TransportationCostIndicatorsUSWheat_knn2_array
=fast_knn(data, k=2)
print(imputed_TransportationCostIndicatorsUSWheat_knn2_array)

# Weighted average of nearest 3 neighbour
print('\nWeighted average of nearest 3 neighbour')
imputed_TransportationCostIndicatorsUSWheat_knn3_array
=fast_knn(data, k=3)
print(imputed_TransportationCostIndicatorsUSWheat_knn3_array)
```

Output:
K = 2

	Unit_Train	Shuttle_Train	Barge	Gulf_Vessel	Pacific_Vessel	WheatPrice
895	129.77	124.30	166.67	261.792759	250.735044	382.50
896	129.89	124.45	192.11	271.610000	280.550000	413.50
897	132.67	124.45	167.22	282.790000	308.870000	410.50
898	134.21	124.45	136.11	310.540000	345.230000	413.00
899	133.22	124.45	133.33	314.710000	353.650000	365.50
900	127.14	121.33	131.11	287.890000	319.770000	376.25
901	127.06	121.28	166.11	273.910000	290.480000	402.50
902	125.77	121.28	206.11	271.950000	277.410000	400.00
903	123.10	121.28	205.00	270.400000	281.450000	395.00
904	116.61	118.17	235.00	268.560000	283.830000	403.50

$K = 3$

	Unit_Train	Shuttle_Train	Barge	Gulf_Vessel	Pacific_Vessel	WheatPrice
895	129.77	124.30	166.67	261.792759	250.735044	382.50
896	129.89	124.45	192.11	271.610000	280.550000	413.50
897	132.67	124.45	167.22	282.790000	308.870000	410.50
898	134.21	124.45	136.11	310.540000	345.230000	413.00
899	133.22	124.45	133.33	314.710000	353.650000	365.50
900	127.14	121.33	131.11	287.890000	319.770000	376.25
901	127.06	121.28	166.11	273.910000	290.480000	402.50
902	125.77	121.28	206.11	271.950000	277.410000	400.00
903	123.10	121.28	205.00	270.400000	281.450000	395.00
904	116.61	118.17	235.00	268.560000	283.830000	403.50

k-NN imputation is much more accurate than simple imputer with mean/median or frequent imputation strategies. Nonetheless, computation is as expensive as k-NN does in memory data wrangling and training. If dataset has quite higher outliers, the performance and accuracy of k-NN are affected immensely. Expense would be much higher as compared to datasets with least outliers.

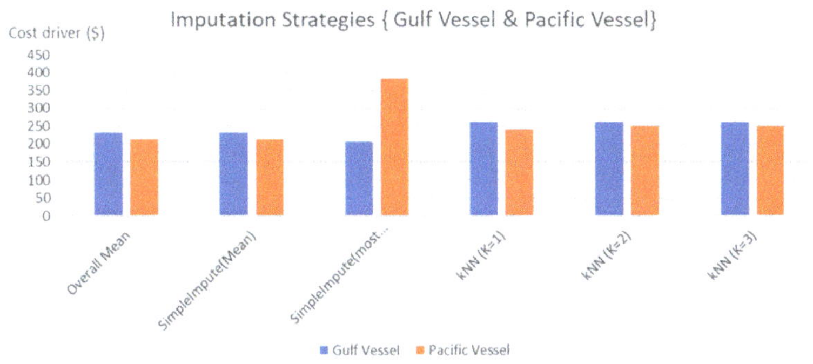

3.4.5 *Multivariate Imputation by Chained Equation (MICE) Strategy*

Multivariate imputation using chained equations or "multiple imputation by chained equations" (MICE) accounts for the process that created missing data; preserves the relations in the data, a stark contrast to what constant or mean/median simple imputer; and preserves the uncertainty about relations. Please find the main steps used in MICE:[39] MICE performs various regressions to predict the missing values from the dataset [12]. The type of regression is based on the type of missing column. With a multiple imputation method, each variable with missing data is modeled conditionally using the other variables in the data before filling in the missing values. MICE,[40] sometimes called "fully conditional specification" or "sequential regression multiple imputation," has emerged in the statistical literature as one principled method of addressing missing data [13]. Creating multiple imputations, as opposed to single imputations, accounts for the statistical uncertainty in the imputations. In addition, the chained equations approach is very flexible and can handle variables of varying types (e.g., continuous, or binary) as well as complexities such as bounds or survey skip patterns.

Software code for this model can be found on GitHub Link: WheatPricetoGrainTransportationCostIndicatorsMICE (Jupyter Notebook Code)

[39] MICE—https://www.jstatsoft.org/article/view/v045i03/v45i03.pdf

[40] Multiple imputation by chained equations: what is it and how does it work?—https://www.ncbi.nlm.nih.gov/pmc/articles/PMC3074241/

Step 1: Wheat Futures and Grain Transportation Cost Indicators

```
from sklearn.datasets import fetch_california_housing
from sklearn.linear_model import LinearRegression
from sklearn.model_selection import StratifiedKFold
from sklearn.metrics import mean_squared_error
from math import sqrt
import random
import numpy as np
random.seed(0)

#Fetching the dataset
import pandas as pd
grainTransportationCostIndicatorsUSWheatDF = pd.read_csv
('US_WHEAT_Grain_Transportation_Cost_IndicatorsRelationship.
csv')

grainTransportationCostIndicatorsUSWheatDF.tail(10)
```

Output:

	Date	Week	Month	Year	Truck	Unit_Train	Shuttle_Train	Barge	Gulf_Vessel	Pacific_Vessel	Price	Open	High
895	12/29/2004	52	12	2004	133	129.77	124.30	166.67	NaN	NaN	382.5	412.25	416
896	12/22/2004	51	12	2004	133	129.89	124.45	192.11	271.61	280.55	413.5	409.5	419
897	12/15/2004	50	12	2004	134	132.67	124.45	167.22	282.79	308.87	410.5	413.5	416.75
898	12/8/2004	49	12	2004	139	134.21	124.45	136.11	310.54	345.23	413	367	414
899	12/1/2004	48	11	2004	142	133.22	124.45	133.33	314.71	353.65	365.5	378	385.5
900	11/24/2004	47	11	2004	142	127.14	121.33	131.11	287.89	319.77	376.25	405	413
901	11/17/2004	46	11	2004	143	127.06	121.28	166.11	273.91	290.48	402.5	401	408
902	11/10/2004	45	11	2004	145	125.77	121.28	206.11	271.95	277.41	400	391	419
903	11/3/2004	44	10	2004	148	123.10	121.28	205.00	270.40	281.45	395	409	434
904	10/27/2004	43	10	2004	148	116.61	118.17	235.00	268.56	283.83	403.5	368	403.5

Step 2: Load the Entire Dataset

```
grainTransportationCostIndicatorsUSWheatMICENDF                    =
grainTransportationCostIndicatorsUSWheatDF
[['Date','Truck','Unit_Train', 'Shuttle_Train',
   'Barge', 'Gulf_Vessel', 'Pacific_Vessel', 'Price']]
```

Output:

	Truck	Unit_Train	Shuttle_Train	Barge	Gulf_Vessel	Pacific_Vessel
895	133	129.77	124.30	166.67	NaN	NaN
896	133	129.89	124.45	192.11	271.61	280.55
897	134	132.67	124.45	167.22	282.79	308.87
898	139	134.21	124.45	136.11	310.54	345.23
899	142	133.22	124.45	133.33	314.71	353.65
900	142	127.14	121.33	131.11	287.89	319.77
901	143	127.06	121.28	166.11	273.91	290.48
902	145	125.77	121.28	206.11	271.95	277.41
903	148	123.10	121.28	205.00	270.40	281.45
904	148	116.61	118.17	235.00	268.56	283.83

Step 3: Apply the Multivariate Imputation by Chained Equation (MICE)

```
import pandas as pd
import numpy as np
# importing the MICE from fancyimpute library
from fancyimpute import IterativeImputer

df = grainTransportationCostIndicatorsUSWheatMICENDF

# printing the dataframe
print(df)

# calling the MICE class
mice_imputer = IterativeImputer()
# imputing the missing value with mice imputer
imputed_train_array = mice_imputer.fit_transform(df)

# printing dataframe
print(imputed_train_array)
```

Output:

```
     Truck  Unit_Train  Shuttle_Train   Barge  Gulf_Vessel  Pacific_Vessel  \
0      272      297.82         229.33  308.89       295.17          257.09
1      270      297.82         229.62  385.00       290.70          255.32
2      265      297.82         252.85  400.56       272.81          241.13
3      258      299.14         271.61  483.89       279.52          244.68
4      254      299.14         273.80  477.78       288.46          248.23
..     ...         ...            ...     ...          ...             ...
900    142      127.14         121.33  131.11       287.89          319.77
901    143      127.06         121.28  166.11       273.91          290.48
902    145      125.77         121.28  206.11       271.95          277.41
903    148      123.10         121.28  205.00       270.40          281.45
904    148      116.61         118.17  235.00       268.56          283.83

     WheatPrice
0       1059.10
1        843.00
2        797.00
3        797.75
4        763.25
..          ...
900      376.25
901      402.50
902      400.00
903      395.00
904      403.50

[905 rows x 7 columns]
[[ 272.    297.82  229.33 ...  295.17  257.09 1059.1 ]
 [ 270.    297.82  229.62 ...  290.7   255.32  843.  ]
 [ 265.    297.82  252.85 ...  272.81  241.13  797.  ]
 ...
 [ 145.    125.77  121.28 ...  271.95  277.41  400.  ]
 [ 148.    123.1   121.28 ...  270.4   281.45  395.  ]
```

Step 4: Convert Imputed Array to Data Frame

```
imputed_traindf = pd.DataFrame(imputed_train_array,
        columns=['Truck', 'Unit_Train', 'Shuttle_Train',
   'Barge', 'Gulf_Vessel', 'Pacific_Vessel','WheatPrice'])
imputed_traindf.tail(10)
```

Output:

	Truck	Unit_Train	Shuttle_Train	Barge	Gulf_Vessel	Pacific_Vessel	WheatPrice
895	133.0	129.77	124.30	166.67	209.161179	216.732253	382.50
896	133.0	129.89	124.45	192.11	271.610000	280.550000	413.50
897	134.0	132.67	124.45	167.22	282.790000	308.870000	410.50
898	139.0	134.21	124.45	136.11	310.540000	345.230000	413.00
899	142.0	133.22	124.45	133.33	314.710000	353.650000	365.50
900	142.0	127.14	121.33	131.11	287.890000	319.770000	376.25
901	143.0	127.06	121.28	166.11	273.910000	290.480000	402.50
902	145.0	125.77	121.28	206.11	271.950000	277.410000	400.00
903	148.0	123.10	121.28	205.00	270.400000	281.450000	395.00
904	148.0	116.61	118.17	235.00	268.560000	283.830000	403.50

Here is the summary of all imputation techniques: MICE fairs very close to the dataset mean and considers all the attributes of the data frame.

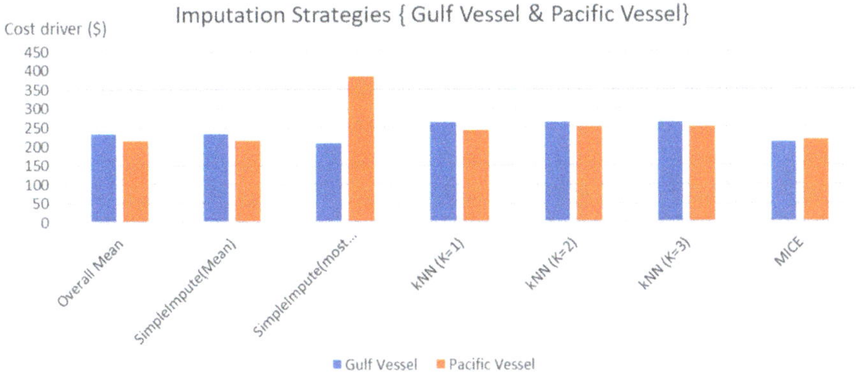

3.5 Food Security Supplier Risk Modeling: Food-Fuel Conundrum

Food security is tightly coupled with many factors such as monsoons, economic factors, oil prices, inflation, local government policies, and food tastes, which are either historical or current or could emerge from the future. Historical factors such as droughts, floods, macroeconomic economic events, and crop failure that have occurred continue to affect food security. Historical datasets provide data signatures that could be mined to provide predictive and prescriptive recommendations.

Current events, for example, lower rains, conflicts, wars, economic depreciation of currencies, and economic tariffs/trade wars, tend to increase or worsen food insecurity. Trade wars are huge threats to food security.[41] International trade has proven to be a critical mechanism for growth and development. It helped build stronger value chains, mitigate conflict, and provide access to higher quality and quantities of goods and services. International trade has also provided consumers with access to a more diversified and nutritious food basket [14]. However, for trade to improve food security to the greatest number of people across the globe, greater international cooperation is necessary. Trade wars and conflicts decrease cooperation (a factor we can revisit as part of the Coupled Model Intercomparison Project 6 (CMPI6) Shared Socioeconomic Pathways (SSP3)).

Future food insecurity events can be forestalled and completely eliminated, provided necessary actions are pursued. This is the great advantage of machine learning and Artificial Intelligence models. New emerging policies and governmental decree could sometimes lead into food security issues. The food security equation, under such circumstances, has two parts. Government regulations or policies tend to increase threat to food importers. What-if risk modeling is a way to mitigate the risk.

3.5.1 Saudi Arabia Rice Imports and Food Security

Saudi Arabia has the largest economy in the Arab world with a GDP and per capita income of $785 billion and $22,953, respectively, in 2019. Saudi's population is currently 34.2 million and is expected to exceed 40 million by 2030. According to the UN trade data, Saudi imported more than $18 billion worth of food and agricultural products in 2018, and the US market share was more than 8%. The country has a growing population, a strong food service sector, a rapidly maturing food retail sector, and a new and developing traditional tourism industry and hosts millions of religious pilgrims each year. Saudi also imports most of its food and has become more open to business and tourism over the past several years. Post anticipates these factors will continue to support robust demand[42] for imported food despite volatile oil prices and the effects of COVID-19 [15].

[41] Trade Wars are huge threat to food security—https://unctad.org/news/trade-wars-are-huge-threats-food-security

[42] Top exporters to Saudi—https://agriexchange.apeda.gov.in/MarketReport/Reports/Exporter_Guide_Riyadh_Saudi_Arabia_12-31-2020.pdf

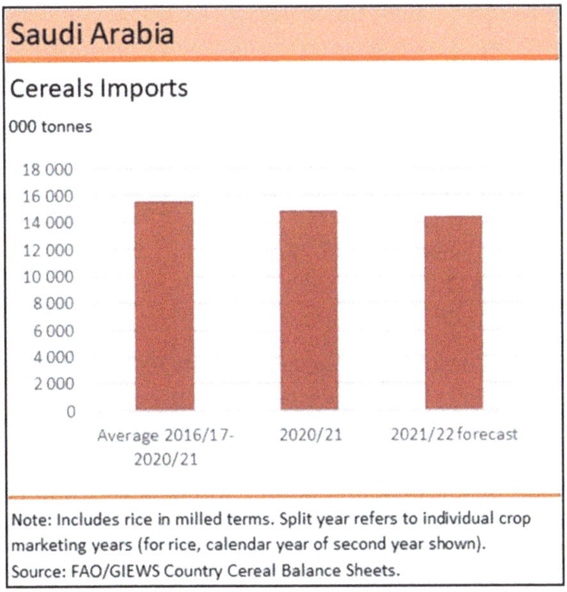

Fig. 3.7 Saudi Arabia

In the Middle East, especially Saudi Arabia, food security will be a matter of concern for policy makers (please see Fig. 3.7). In desert countries the food commodities are mostly imported from other countries as their local production is not enough to meet the domestic needs, and Saudi Arabia[43] is no exception [16]. Most of the cereals and red meat are imported. Saudi Arabia, a "fossil fuel superpower," is 1 of the top 20 economies in the world, a country with a business-friendly and trade-oriented environment. In 2019 its imports grew by 6.5% from the previous year and reached US$144.3 billion. Food imports also continued to rise especially for cereals, rice, meat, and dairy products.[44] With only 1.5% of its overall land area classified as arable, Saudi Arabia is unable to domestically produce sufficient output of agri-food products to meet local market demand, let alone a growing appetite for a wide range of foods and gourmet products. The major limiting factors in agriculture production are land and water, such that by 2050, Saudi Arabia is expected to import all of its domestic needs. Owing to natural and geographic conditions, crop production is limited to crops irrigated using underground water reserves. Although some areas receive rainfall, the amounts are not sufficient to grow any rainfed crops.[45]

Saudi Arabia is a major importer of cereals, with a value of around US$$4.3 billion. Among the most imported cereals, rice places first, followed by barley, corn,

[43] Achieving food security in the Kingdom of Saudi Arabia through innovation: Potential role of agricultural extension—https://www.sciencedirect.com/science/article/pii/S1658077X16300996

[44] Food Importers and Food Import Trends in Saudi Arabia 2020—https://bestfoodimporters.com/food-importers-and-food-import-trends-in-saudi-arabia-2020/

[45] Saudi Arabia Country Profile—https://www.fao.org/giews/countrybrief/country.jsp?code=SAU&lang=ar

and wheat. Saudi Arabia's rice per capita consumption is estimated at 35 kg/year. *The major supplier of rice until 2017 was India, but due to the shortage of production, it couldn't supply all the kingdom's needs.* Nowadays, Pakistan and Thailand supply rice, along with India.

Top five suppliers of high-value food[46] products to Saudi [15]:

1. UAE (13.2%)
2. Brazil (11.1%)
3. The United States (5.9%)
4. Egypt (5.5%)
5. India (5%)

The United States annually exports more than $1.5 billion (USD) of agricultural and related products directly and indirectly to the Kingdom of Saudi Arabia. Saudi Arabia is reliant on imports to meet up to 75% of its food consumption needs, and US retail food exports to Saudi Arabia have grown rapidly over the past several years. In 2021, retail food exports are projected to reach more than $570 million, which is nearly 47% of the total US agricultural and related product[47] exports to Saudi Arabia [17].

Saudi Arabia has taken important steps towards improving agricultural development for enhancing food security amid climate changes, heatwaves, and droughts. Together, it has contributed in improving indicators of food access, patterns of healthy consumption, reduction of waste, and realization of high rates of self-sufficiency.[48] As it is clear, Saudi Arabia is charting a path towards food security, and usage of technologies such as smart climate agriculture sensors and analytics would play an important role[49] [18, 19].

[46] Top exporters to Saudi—https://agriexchange.apeda.gov.in/MarketReport/Reports/Exporter_Guide_Riyadh_Saudi_Arabia_12-31-2020.pdf

[47] Food and Agricultural Import Regulations and Standards Country Report—https://apps.fas.usda.gov/newgainapi/api/Report/DownloadReportByFileName?fileName=Food%20and%20Agricultural%20Import%20Regulations%20and%20Standards%20Country%20Report_Riyadh_Saudi%20Arabia_12-31-2021.pdf

[48] Saudi Arabia Succeeded in Improving Food Security, Reducing Waste and Achieving Self-sufficiency, Reports Deputy Minister of Environment—https://www.spa.gov.sa/viewstory.php?lang=en&newsid=2292762

[49] How Saudi Arabia is charting a path toward food security—https://www.arabnews.com/node/1870111/saudi-arabia

3.5.2 National Policy on Biofuels and Food-Fuel Conundrum

India imports almost 80% of its oil requirements and has seen prices spiralling upwards in recent weeks. To reduce imports, the Indian government has looked towards renewable energy, electric vehicles, and hybrid or blended fuel. The government has notified the consideration of National Policy on Biofuels[50]-2018 (NPB-2018) to boost[51] the use of mixed gasoline in June 2018 [20]. The impact of biofuels on food security to be studied as ethanol production using food crops is a double-edged sword,[52] especially if the consideration of ethanol production permits the use of sugarcane, molasses, sugar, maize, damaged food grains, and surplus rice lying with the Food Corporation of India (FCI). This initiative will have a considerable impact on the food security across the world as India is one of the major exporters of rice.[53]

India is one of the largest exporters of rice across the world (please see Fig. 3.8). Additionally, rice is being imported by African countries that are experiencing huge food insecurity.

The National Policy on Biofuels includes the following potential domestic raw materials for the production of biofuels or for ethanol production: B molasses; sugarcane juice; biomass in the form of grasses; agriculture residues (rice straw, cotton stalk, corn cobs, sawdust, bagasse, etc.); sugar containing materials like sugar beet, sweet sorghum, etc.; starch containing materials such as corn, cassava, rotten potatoes, etc.; damaged food grains like wheat, broken rice, etc. which are unfit for human consumption; and food grains during the surplus phase. Algal feedstock and cultivation of seaweeds can also be a potential feedstock for ethanol production.[54] The risk of food security does exist; given climate change and uncertainty of monsoons seen recently, any low rain monsoon in the seeding and harvest season would result in a high risk for farmers and cause heavy food insecurity. High-import-dependent countries can feel the pinch of food insecurity [20].

[50] National Policy on Biofuels—https://s3-us-west-2.amazonaws.com/visionresources/infographics/National-Policy-on-Biofuels-2018.pdf

[51] National Policy on Biofuels 2018—https://mopng.gov.in/files/uploads/NATIONAL_POLICY_ON_BIOFUELS-2018.pdf

[52] The food-fuel conundrum—https://www.thehindubusinessline.com/opinion/the-food-fuel-conundrum/article35348867.ece

[53] National Policy on Biofuels 2018: Here are key things you should know-
https://economictimes.indiatimes.com//small-biz/productline/power-generation/national-policy-on-biofuels-2018-here-are-key-things-you-should-know/articleshow/71922729.cms?utm_source=contentofinterest&utm_medium=text&utm_campaign=cppst

[54] National Policy on Biofuels 2018—https://mopng.gov.in/files/uploads/NATIONAL_POLICY_ON_BIOFUELS-2018.pdf

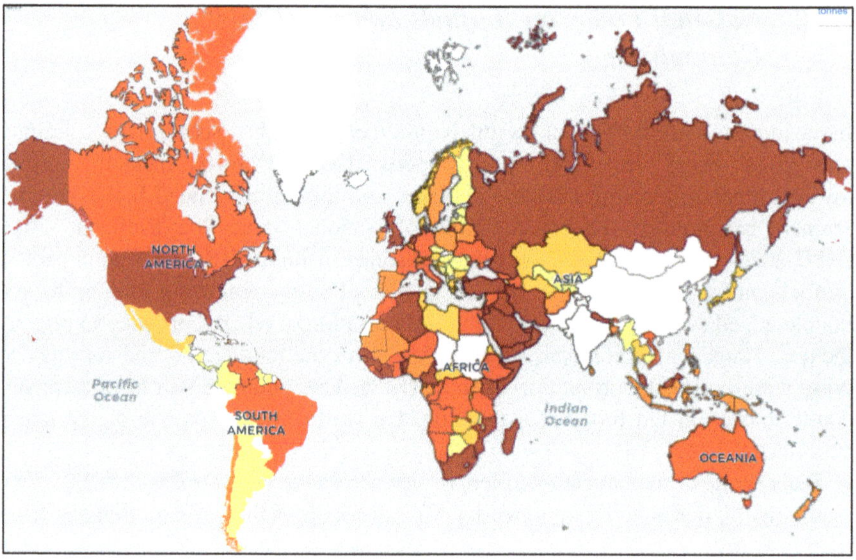

Fig. 3.8 Rice exports

3.5.2.1 Milled Rice

In the 2020/2021 crop year, China produced over 148 million metric tons of milled rice, a higher volume than any other country.[55] India came in second place with 122 million metric tons of milled rice in that crop year. Milled rice is a labor-intensive process. After the rice is grown and harvested from rice paddies, it undergoes a process to prepare it for human consumption, known as milling. Milling removes the rice husk and bran layers, and the product is the rice that can be found in your local grocery store or supermarket. The total volume of milled rice produced worldwide reached 497.7 million metric tons in the 2019/2020 crop year.

3.5.2.2 Rice Exports in 2020

India has produced 118.93 million tonnes and has exported 12.99 million tonnes (10.92%) of milled rice in 2020. Saudi Arabia (1.28 million tonnes) is the major importer of milled rice (please see Fig. 3.9), followed by Iran (Islamic Republic of) with 1.10 million tonnes and Benin with 962,555 tonnes (FAOSTAT 2020).

[55] Rice Exports—https://www.statista.com/statistics/255945/top-countries-of-destination-for-us-rice-exports-2011/

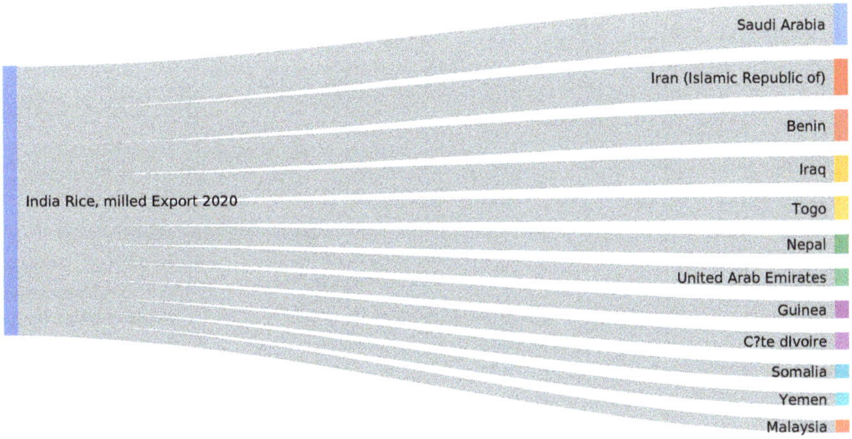

Fig. 3.9 Milled rice exports, 2020, India

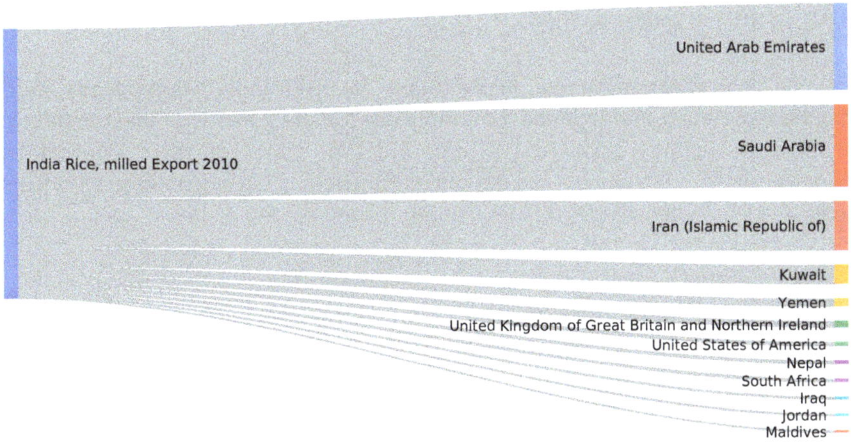

Fig. 3.10 Milled rice exports, 2010, India

3.5.2.3 Rice Exports in 2010

India has produced 96,023,326 tonnes of milled rice and has exported 2,204,535 tonnes (2.29%) in 2010 with the United Arab Emirates importing 661,481 tonnes (please see Fig. 3.10), followed by Saudi Arabia (621,763 tonnes) and Iran (Islamic Republic of) in the third position with 374,474 tonnes (FAOSTAT 2020).

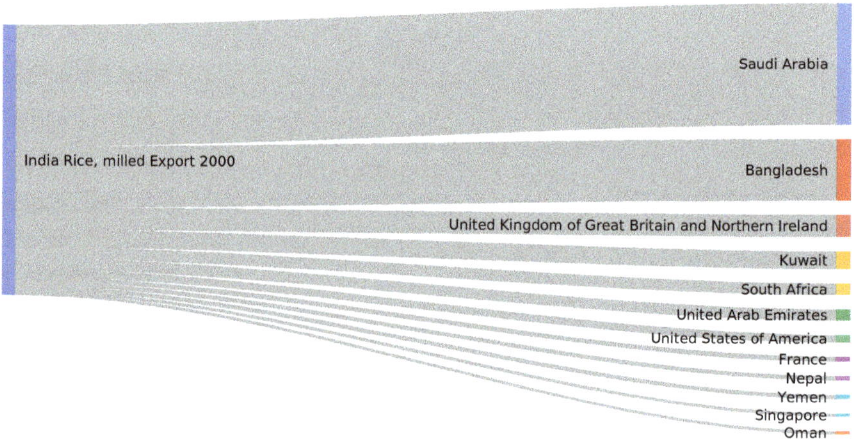

Fig. 3.11 Milled rice exports, 2000, India

3.5.2.4 Rice Exports in 2000

We can observe similar trends in 2000 also. In 2000, India has produced 85.01 million tonnes and has exported 1.52 million tonnes (1.79%) with major exported countries including Saudi Arabia with 631,406 tonnes of rice, followed by Bangladesh (318,416 tonnes) and the United Kingdom of Great Britain and Northern Ireland (113,879 tonnes) (FAOSTAT 2020) (please see Fig. 3.11).

3.5.2.5 Rice Exports in 1990

In 1990 India has produced 74.38 million tonnes of milled rice and has exported 502,323 tonnes (0.67%) of milled rice. A similar trend can be observed in 1990 with the USSR as a major importer of rice (266,389 tonnes), followed by Saudi Arabia (125,549 tonnes) and the United Kingdom of Great Britain and Northern Ireland (28,426 tonnes) (FAOSTAT 2020) (please see Fig. 3.12).

India Agriculture: Trade and Trade Policy[56]
Since 1970, India's trade in cereals has shown a trend from net imports to net exports of both wheat and rice—a trend that reflects shifts in trade policy, as well as longer term changes in supply and demand. Through the 1980s and early 1990s, Indian agriculture had export restrictions and overvalued

(continued)

[56] USDA—Trade—https://www.ers.usda.gov/webdocs/publications/45802/11577_err41d_1_.pdf?v=0

exchange rates that resulted in net taxation of the farm sector. Exports of agricultural goods, including wheat and rice, were restricted through various regulations to bolster India's domestic food security. For wheat and rice, quantitative controls on imports and exports were administered through the Food Corporation of India (FCI).

In the mid-1990s, trade policies were changed when quantitative restrictions on imports were lifted and replaced by tariffs. The wheat tariff was initially set at zero, but was raised to 50% in 1999 to curb imports into southern India at a time when surpluses were growing in the north. The rice tariff has remained at 70%, a level that prohibits trade from occurring.

In 2000, India began to provide budgetary subsidies to support exports of surplus wheat and rice when the combination of declining world prices and higher domestic prices made Indian wheat and rice uncompetitive in world markets. In 2005, the Government halted export subsidies because of tightening domestic supplies and reduced Indian competitiveness in international markets, although private traders remain free to export wheat and rice.

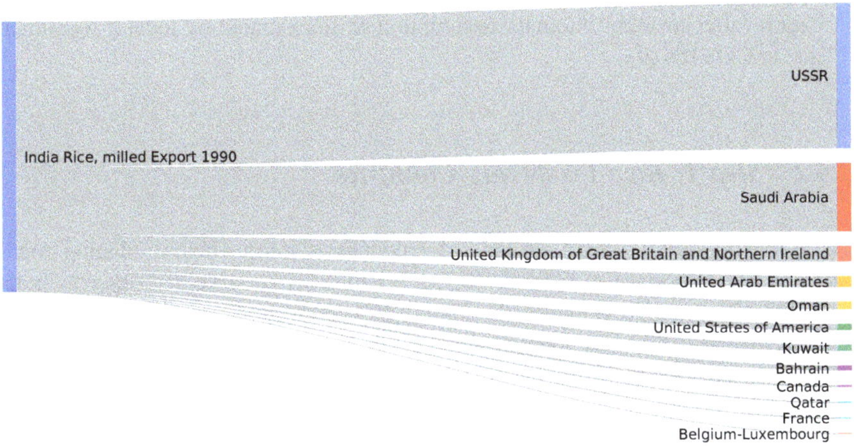

Fig. 3.12 Milled rice exports, 1990, India

3.6 Machine Learning Model: Saudi Arabia Rice Import Origin Cluster

Saudi Arabia rice import origins

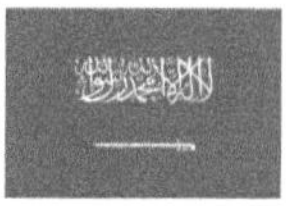

Software code for this model: SaudiArbia_ImporterOrgin_Rice_countries_agglomerative.py

To establish new import origin countries to overcome deficit by a major supplier issue, we can apply clustering technique to solve it. To prepare clustering data points, firstly, we need to consider the following:

- Current production of rice
- Exports of rice
- Percentage of experts of rice production
- Capacity available to take up new export needs—that is, capacity availability to export (after meeting internal consumption & stocks need by local government due to COVID-19).

3.6.1 Step 1: Rice-Producing Countries

The following figure (please see Fig. 3.13) contains the top wheat-producing countries in the world (FAOSTAT 2020).[57] China (141,310,620 tonnes), India (118,929,435 tonnes), Bangladesh (36,622,229 tonnes), Indonesia (36,451,018), and Vietnam (28,520,184 tonnes) are the top five wheat-producing countries in 2020.

3.6.2 Step 2: Top Exporting Countries

Get the list of the top rice-exporting countries in 2020.[58] Of course, there is no rule that the top rice-producing countries are also the top exporting countries. There

[57] FAOSTAT—https://www.fao.org/faostat/en/#data/QCL

[58] Rice export countries—https://www.fao.org/faostat/en/#data/TCL

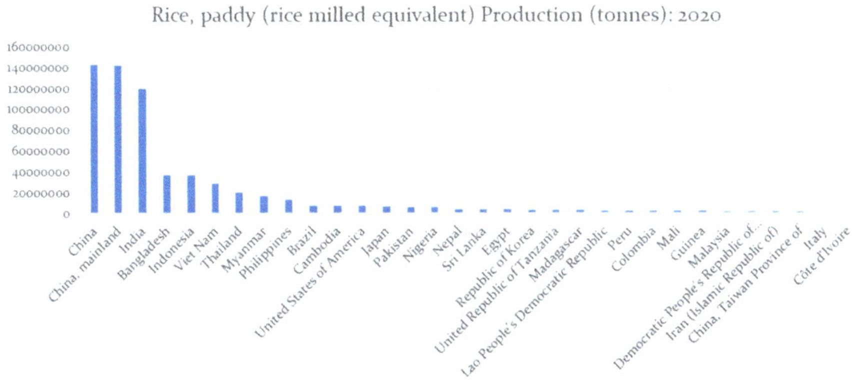

Fig. 3.13 Rice, paddy production, 2020

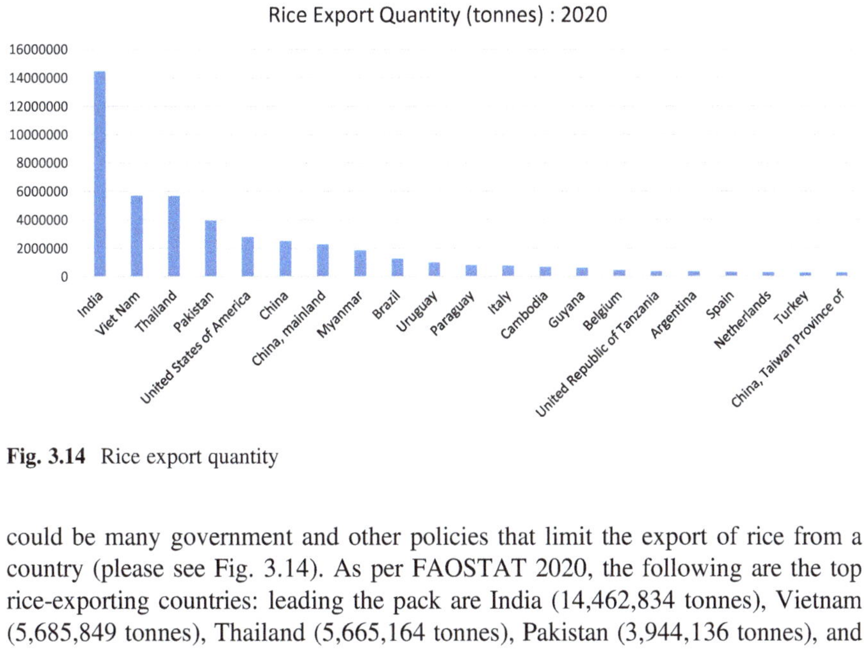

Fig. 3.14 Rice export quantity

could be many government and other policies that limit the export of rice from a country (please see Fig. 3.14). As per FAOSTAT 2020, the following are the top rice-exporting countries: leading the pack are India (14,462,834 tonnes), Vietnam (5,685,849 tonnes), Thailand (5,665,164 tonnes), Pakistan (3,944,136 tonnes), and the United States (2,791,901 tonnes).

3.6.3 Step 3: Available Capacity to Export

If we rearrange the production to exports on a Cartesian (X–Y) plane, we can observe that countries such Pakistan (70.23%), the United States (40.54%),

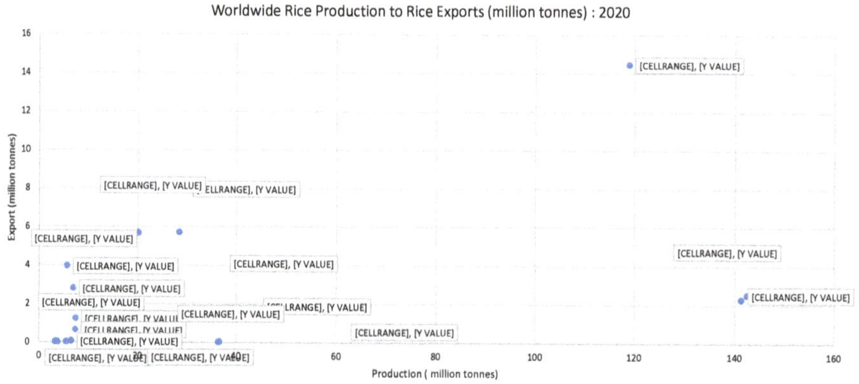

Fig. 3.15 Worldwide rice production

Table 3.4 Export percentage

Country	Production (million tonnes)	Export (million tonnes)	Export percentage (%)
China	142.4784	2.506497	1.759212466
China, mainland	141.3106	2.270515	1.60675468
India	118.9294	14.46283	12.1608532
Bangladesh	36.62223	0.012998	0.035492105
Indonesia	36.45102	0.000352	0.000965679
Vietnam	28.52018	5.685849	19.93622832
Thailand	20.16409	5.665164	28.09530644
Myanmar	16.7417	1.845199	11.02157487
Philippines	12.86967	0.000287	0.00223005
Brazil	7.397704	1.217888	16.46305394
Cambodia	7.31032	0.630941	8.63082601
United States	6.885434	2.791901	40.5479306
Japan	6.474069	0.040394	0.623935272
Pakistan	5.615657	3.944136	70.2346315
Nigeria	5.450724	0.000054	0.000990694
Nepal	3.702436	0.000159	0.00429447
Sri Lanka	3.415656	0.009214	0.269757844
Egypt	3.263969	0.000006	0.000183825

Thailand (28.09%), Vietnam (19.93%), and Brazil (16.46%) utilize more than 15% of their production to exports (please see Fig. 3.15).

Countries along the bottom right (China with 1.75%) are the potential candidates that could step up to meet the deficit in Saudi Arabia. Other big grain producers can try to make up some of India's potential shortfalls (please see Table 3.4).

$$\text{Exported percentage} = \frac{\text{Export rice (tonnes)}}{\text{Production in tonnes}}$$

3.6.4 *Step 4: Export Capacity After Meeting Local Government Mandates and Food Export Polices*

Now the available capacity to export is 100%—exported percentage. However, it could not be practical for any country to export all of its production. Stocks and local consumption levels are mandated by local governments (due to COVID-19, the numbers are increased. In general, the WTO needs to be notified in advance[59]). By considering *mandated percentage of 97% of production to be local consumption*, the resultant percentage can be exported (please see Table 3.5).

Table 3.5 Export capacity available

Country	Production (million tonnes)	Export (million tonnes)	Export percentage (%)	Available capacity	Export capacity after local mandates (%)
China	142.4784	2.506497	1.759212466	98.24078753	28.24078753
China, mainland	141.3106	2.270515	1.60675468	98.39324532	28.39324532
India	118.9294	14.46283	12.1608532	87.8391468	17.8391468
Bangladesh	36.62223	0.012998	0.035492105	99.96450789	29.96450789
Indonesia	36.45102	0.000352	0.000965679	99.99903432	29.99903432
Viet Nam	28.52018	5.685849	19.93622832	80.06377168	10.06377168
Thailand	20.16409	5.665164	28.09530644	71.90469356	1.904693561
Myanmar	16.7417	1.845199	11.02157487	88.97842513	18.97842513
Philippines	12.86967	0.000287	0.00223005	99.99776995	29.99776995
Brazil	7.397704	1.217888	16.46305394	83.53694606	13.53694606
Cambodia	7.31032	0.630941	8.63082601	91.36917399	21.36917399
United States of America	6.885434	2.791901	40.5479306	59.4520694	**−10.5479306**
Japan	6.474069	0.040394	0.623935272	99.37606473	29.37606473
Pakistan	5.615657	3.944136	70.2346315	29.7653685	**−40.2346315**
Nigeria	5.450724	0.000054	0.000990694	99.99900931	29.99900931
Nepal	3.702436	0.000159	0.00429447	99.99570553	29.99570553
Sri Lanka	3.415656	0.009214	0.269757844	99.73024216	29.73024216
Egypt	3.263969	0.000006	0.000183825	99.99981617	29.99981617

[59] How to address agricultural export restrictions?—https://www.wto.org/english/forums_e/debates_e/debate33_e.htm

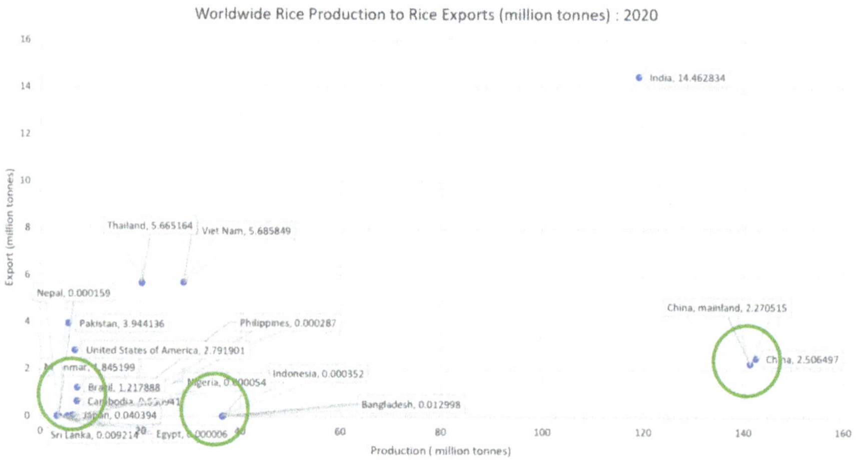

Fig. 3.16 Worldwide rice production, 2020

WORLD TRADE ORGANIZATION	*How to address agricultural export restrictions?* Under WTO rules, countries can restrict exports of agricultural products but only temporarily and they have to comply with GATT Article XI (ie, 11), in this case paragraph 2(a), and with Article 12 of the Agriculture Agreement. These require the restricting country to take into account the impact on importing countries' food security, to notify the WTO as soon as possible, and as far in advance as possible, to be prepared to discuss the restriction with importing countries and to supply them with detailed information when asked for it.

Countries with export capacity after local governmental tariffs and mandates are the candidates that Saudi Arabia could approach to fulfill import needs, if there is rice shortage due to food-fuel issues (please see Fig. 3.16). Only Bangladesh, Indonesia, the Philippines, Japan, Nigeria, Nepal, Sri Lanka, and Egypt could be considered provided political, social, and economic policies are met and bilateral cooperation conditions are fulfilled. Please note that Saudi Arabia could meet only 25% of its rice imports!

3.7 Machine Learning Cluster Model: Food-Fuel Conundrum

Saudi Arabia rice import origins	

Software code for this model: SaudiArbia_RiceImport_Cluster_Agglomerative.py (Jupyter Notebook Code)

3.7.1 Step 1: Load Worldwide Rice Production (Million Tonnes) and Export (Million Tonnes) Data

As prepared in the above section, load worldwide rice production and export data.

```
"""
Created on Sun Mar 20 20:29:51 2022

@author: CHVUPPAL
"""

import matplotlib.pyplot as plt
import pandas as pd
#%matplotlib inline
'exec(%matplotlib inline)'

import numpy as np
SaudiArbia_RiceImport_countries_data = pd.read_csv
('SaudiArbia_RiceImport_countries_data.csv')
SaudiArbia_RiceImport_countries_data.shape
print ('\nSaudiArbia_RiceImport_countries_data\n')
print(SaudiArbia_RiceImport_countries_data.head())

data = SaudiArbia_RiceImport_countries_data.iloc[:, 3:5].values
#SaudiArbia_RiceImport_countries_data['Current Export
Percentage (%)', 'AvailableCapacitytoExport(%)']
#SaudiArbia_RiceImport_countries_data.iloc[:, 7:8].values
print(data)
```

Output:

```
Command Prompt - python SaudiArbia_RiceImport_Cluster_Agglomerative.py

C:\Hanumayamma\Clustering>python SaudiArbia_RiceImport_Cluster_Agglomerative.py

SaudiArbia_RiceImport_countries_data

   ImporterID            Country  Production ( million tonnes)  Current Export Percentage (%)  AvailableCapacitytoExport(%)
0           1              China                    142.478356                       1.759212                     28.240788
1           2    China, mainland                    141.310620                       1.606755                     28.393245
2           4         Bangladesh                     36.622229                       0.035492                     29.964508
3           5          Indonesia                     36.451018                       0.000966                     29.999034
4           6           Viet Nam                     28.520184                      19.936228                     10.063772
[[1.75921247e+00 2.82407875e+01]
 [1.60675468e+00 2.83932453e+01]
 [3.54921050e-02 2.99645079e+01]
 [9.65679000e-04 2.99990343e+01]
 [1.99362283e+01 1.00637717e+01]
 [2.80953064e+01 1.90469356e+00]
 [1.10215749e+01 1.89784251e+01]
 [2.23005000e-03 2.99977699e+01]
 [1.64630539e+01 1.35369461e+01]
 [8.63082601e+00 2.13691740e+01]
 [6.23935272e-01 2.93760647e+01]
 [9.90694000e-04 2.99990093e+01]
 [4.29447000e-03 2.99957055e+01]
 [2.69757844e-01 2.97302422e+01]
 [1.83825000e-04 2.99998162e+01]]
```

3.7.2 Step 2: Construct Dendrogram

The following code constructs a hierarchical dendrogram.

```
import scipy.cluster.hierarchy as shc

plt.figure(figsize=(10, 7))
plt.title("SaudiArbia_RiceImport_countries Dendograms")
dend = shc.dendrogram(shc.linkage(data, method='average'))
print ('\npring dend \n')
print(dend)

plt.show()
```

Output:

```
{'icoord': [[15.0, 15.0, 25.0, 25.0], [5.0, 5.0, 20.0, 20.0],
[55.0, 55.0, 65.0, 65.0], [45.0, 45.0, 60.0, 60.0], [75.0, 75.0,
85.0, 85.0], [52.5, 52.5, 80.0, 80.0], [35.0, 35.0, 66.25, 66.25],
[95.0, 95.0, 105.0, 105.0], [50.625, 50.625, 100.0, 100.0],
[115.0, 115.0, 125.0, 125.0], [135.0, 135.0, 145.0, 145.0],
[155.0, 155.0, 165.0, 165.0], [140.0, 140.0, 160.0, 160.0],
[120.0, 120.0, 150.0, 150.0], [75.3125, 75.3125, 135.0, 135.0],
[185.0, 185.0, 195.0, 195.0], [175.0, 175.0, 190.0, 190.0],
[105.15625, 105.15625, 182.5, 182.5], [12.5, 12.5, 143.828125,
143.828125]], 'dcoord': [[0.0, 0.004710000000017089,
0.004710000000017089, 0.0], [0.0, 26.67852257589513,
26.67852257589513, 0.004710000000017089], [0.0,
0.48157459124002827, 0.48157459124002827, 0.0], [0.0,
```

(continued)

```
0.5957667225371448, 0.5957667225371448, 0.48157459124002827],
[0.0, 1.5144950266356094, 1.5144950266356094, 0.0],
[0.5957667225371448, 2.503097841988069, 2.503097841988069,
1.5144950266356094], [0.0, 4.228532784487995,
4.228532784487995, 2.503097841988069], [0.0, 5.236692895069273,
5.236692895069273, 0.0], [4.228532784487995, 7.194643869220741,
7.194643869220741, 5.236692895069273], [0.0,
4.7484945577066835, 4.7484945577066835, 0.0], [0.0,
1.1623736382566507, 1.1623736382566507, 0.0], [0.0,
2.5720753971857033, 2.5720753971857033, 0.0],
[1.1623736382566507, 6.647215427161645, 6.647215427161645,
2.5720753971857033], [4.7484945577066835, 11.61728192201094,
11.61728192201094, 6.647215427161645], [7.194643869220741,
13.17854293346081, 13.17854293346081, 11.61728192201094], [0.0,
8.081239043107438, 8.081239043107438, 0.0], [0.0,
17.52822950792522, 17.52822950792522, 8.081239043107438],
[13.17854293346081, 32.78996152788013, 32.78996152788013,
17.52822950792522], [26.67852257589513, 108.25579025142952,
108.25579025142952, 32.78996152788013]], 'ivl': ['13', '18',
'19', '4', '9', '12', '17', '10', '11', '8', '14', '15', '16', '1',
'3', '6', '7', '5', '0', '2'], 'leaves': [13, 18, 19, 4, 9, 12, 17,
10, 11, 8, 14, 15, 16, 1, 3, 6, 7, 5, 0, 2], 'color_list': ['g', 'g',
'r', 'r', 'r', 'r', 'r', 'r', 'r', 'r', 'r', 'r', 'r', 'r', 'r', 'r',
'r', 'r', 'b']}
[2 0 2 0 0 4 0 0 0 0 0 0 0 3 0 0 0 0 1 1]
```

As it can be seen, Saudi Arabia rice import dataset has two clusters constructed—as depicted in the above dendrogram.

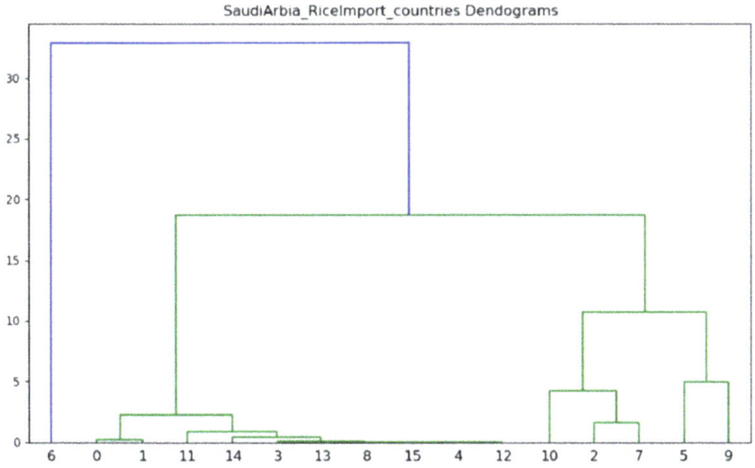

3.7.3 Step 3: Construct Hierarchical Cluster

The purpose of the code is to construct the cluster.

```
from sklearn.cluster import AgglomerativeClustering
x = SaudiArbia_RiceImport_countries_data['Current Export
Percentage (%)'] #'y - x
y = SaudiArbia_RiceImport_countries_data
['AvailableCapacitytoExport(%)'] #' z - y
cluster = AgglomerativeClustering(n_clusters=5,
affinity='manhattan', linkage='average')
print(cluster.fit_predict(data))
plt.figure(figsize=(10, 7))
plt.scatter(x, y, c=cluster.labels_, cmap='rainbow')
plt.title("Current Export Percentage (%) vs.Available Capacity to
Export(%) ")
plt.xlabel("Current Export Percentage (%)")
plt.ylabel("AvailableCapacitytoExport(%)")
plt.show()
```

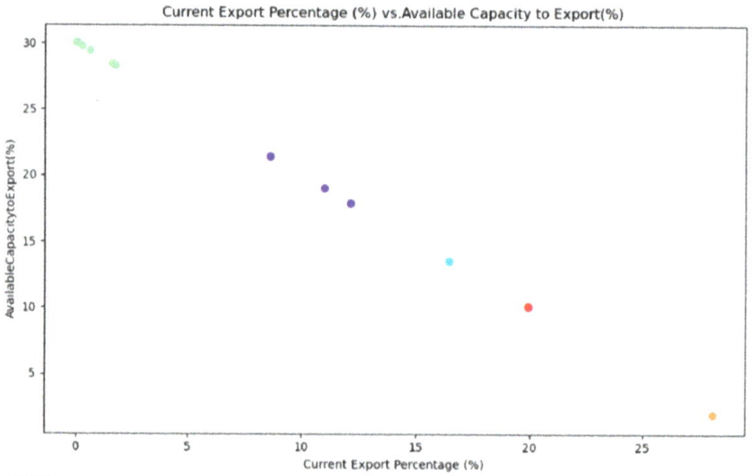

Output:

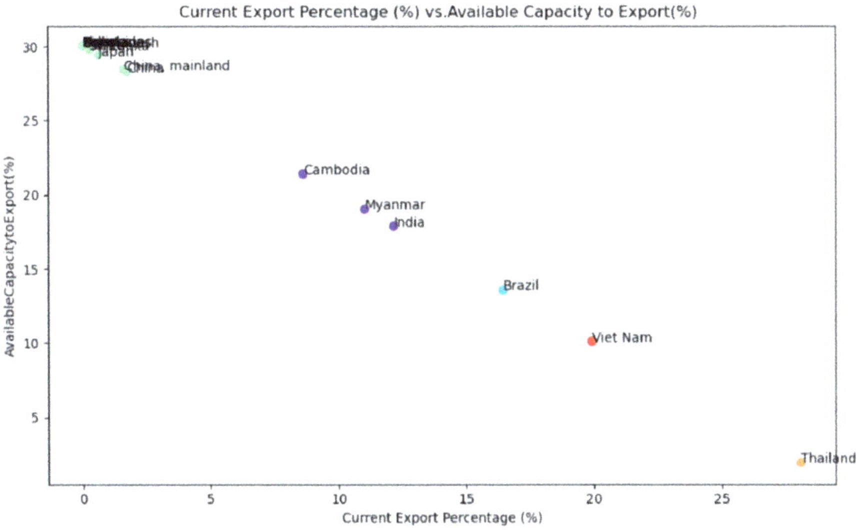

Each color dot represents an importer country. The five clusters are depicted by five colors.

3.7.4 Step 4: What-If Modeling

Given fuel vs. food, what if India stops the exports of rice. The following is the resultant cluster shift right. The resultant dendrogram and cluster are different, shifted more towards other countries.

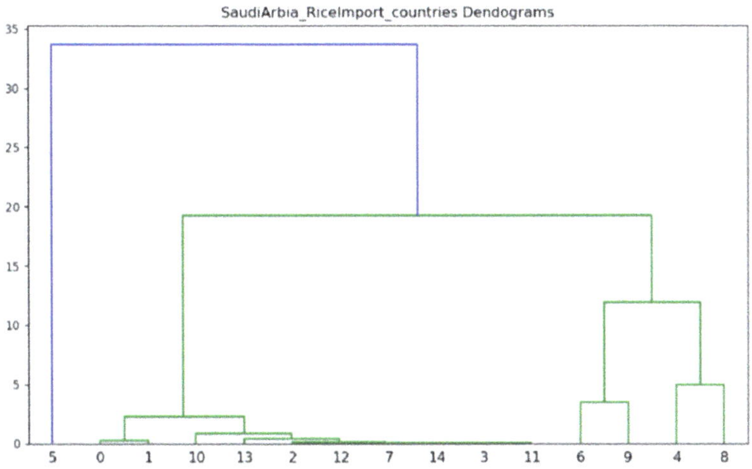

Cluster diagram:

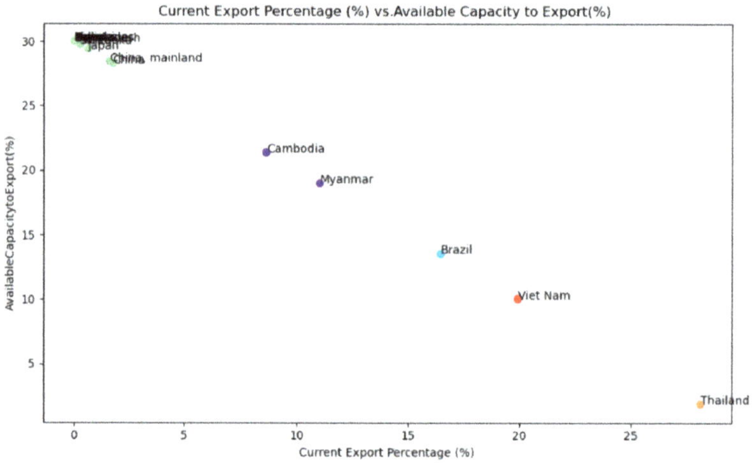

Without India, other major rice-exporting countries such as Brazil, China, Thailand, Myanmar, Cambodia, and Vietnam must contribute more.

No doubt Fuel-Food policy shifts more towards Fuel are the major source of food insecurity as can be seen with India moving to crop production to oil. Another major source of food insecurity is due to climate change. In the next 30 years, food supply and food security will be severely threatened if little or no action is taken to address climate change and the food system's vulnerability to climate change. According to the Intergovernmental Panel on Climate Change (IPCC), the extent of climate change impacts on individual regions will vary over time, and different societal and environmental systems will have varied abilities to mitigate or adapt to change. Negative effects of climate change include the continued rise of global temperatures, changes in precipitation patterns, an increased frequency of droughts and heat waves, sea-level rise, melting of sea ice, and a higher risk of more intense natural disasters.[60] I will cover in the subsequent chapters the production vs. food security due to climate change.

After reading this chapter, you should be able to answer food security data sources and data engineering attributes for handling agricultural datasets. You should be able to answer heuristics data and frequencies, data and food security long tail, and data enrichment strategies. Finally, you should be able to answer food security risk model that was due to the potential food-fuel policy change.

[60] The World's Food Supply is Made Insecure by Climate Change—https://www.un.org/en/academic-impact/worlds-food-supply-made-insecure-climate-change

References

1. THE WHITE HOUSE, Executive Order on Ensuring a Data-Driven Response to COVID-19 and Future High-Consequence Public Health Threats, January 21, 2021, https://www.whitehouse.gov/briefing-room/presidential-actions/2021/01/21/executive-order-ensuring-a-data-driven-response-to-covid-19-and-future-high-consequence-public-health-threats/, Access Date: March 18, 2022
2. FAO, FAO's Work on Climate Change – Conference 2017, 2017, https://www.fao.org/3/i8037e/i8037e.pdf, Access Date: September 18, 2020
3. C. Vuppalapati, R. Vuppalapati, S. Kedari, A. Ilapakurti, J. S. Vuppalapati and S. Kedari, "Fuzzy Logic Infused Intelligent Scent Dispenser For Creating Memorable Customer Experienceof Long-Tail Connected Venues," *2018 International Conference on Machine Learning and Cybernetics (ICMLC)*, 2018, pp. 149-154, https://doi.org/10.1109/ICMLC.2018.8527046
4. Chandrasekar Vuppalapati, Machine Learning and Artificial Intelligence for Agricultural Economics: Prognostic Data Analytics to Serve Small Scale Farmers Worldwide, Publisher: Springer; 1st ed. 2021 edition (October 5, 2021), ISBN-13: 978-3030774844
5. Ann E. Stapleton, Data Science for Food and Agricultural Systems (DSFAS), https://nifa.usda.gov/grants/programs/data-science-food-agricultural-systems-dsfas, Access Date: February 27, 2022
6. T.W. Schultz Distortions of Agricultural Incentives (1978), Relationship between milk production and price variations in the EC, July 1981, http://aei.pitt.edu/36837/1/A67.pdf, Access Date: November 22, 2019
7. Joshua K. Hausman, Paul W. Rhode, and Johannes F. Wieland, Recovery from the Great Depression: The Farm Channel in Spring 1933, 2019, https://pubs.aeaweb.org/doi/pdfplus/10.1257/aer.20170237#page=14, Access Date: November 22, 2020
8. Franklin D. Roosevelt, Campaign Speech, October 24, 1932, https://teachingamericanhistory.org/document/campaign-speech/, Access Date: September 18, 2019
9. Harshad Hegde, Neel Shimpi, Aloksagar Panny, Ingrid Glurich, Pamela Christie, Amit Acharya, MICE vs PPCA: Missing data imputation in healthcare, Informatics in Medicine Unlocked, Volume 17, 2019, 100275, ISSN 2352-9148, https://doi.org/10.1016/j.imu.2019.100275
10. Jordan Shockley, Post-Harvest Management Cooperative Extension Service the Economics of Grain Transportation, 2016, http://www2.ca.uky.edu/agcomm/pubs/AEC/AEC100/AEC100.pdf, Access Date: September 18, 2019
11. Daphne Ewing-Chow, Caribbean Food Security Likely To Be Impacted By Russia-Ukraine Conflict, Feb 27, 2022,03:29 am EST, https://www.forbes.com/sites/daphneewingchow/2022/02/27/caribbean-food-security-likely-to-be-impacted-by-russia-ukraine-conflict/?sh=1ece953739d4, Access Date: 10 March 2022
12. Stef van Buuren and Karin Groothuis-Oudshoorn, mice: Multivariate Imputation by Chained Equations in R, December 2011, https://www.jstatsoft.org/article/view/v045i03/v45i03.pdf, Access Date: January 08, 2022
13. Melissa J. Azur, Elizabeth A. Stuart, Constantine Frangakis, and Philip J. Leaf, Multiple imputation by chained equations: what is it and how does it work? 2011 Mar, https://www.ncbi.nlm.nih.gov/pmc/articles/PMC3074241/, Access Date: May 2020
14. Dr. Mukhisa Kituyi, Secretary-General of UNCTAD, Trade wars are huge threats to food security, 22 January 2020, https://unctad.org/news/trade-wars-are-huge-threats-food-security, Access Date: March 01, 2022
15. Mark Ford and Hussein Mousa, Report Name: Exporter Guide, November 19, 2020, https://agriexchange.apeda.gov.in/MarketReport/Reports/Exporter_Guide_Riyadh_Saudi_Arabia_12-31-2020.pdf, Access Date: January 08, 2021

16. Sajid Fiaz, Mehmood Ali Noor, Fahad Owis Aldosri, Achieving food security in the Kingdom of Saudi Arabia through innovation: Potential role of agricultural extension, Journal of the Saudi Society of Agricultural Sciences, Volume 17, Issue 4, 2018, Pages 365-375, ISSN 1658-077X, https://doi.org/10.1016/j.jssas.2016.09.001
17. Hussein Mousa and Mark Ford, Food and Agricultural Import Regulations and Standards Country Report, January 03, 2022, https://apps.fas.usda.gov/newgainapi/api/Report/DownloadReportByFileName?fileName=Food%20and%20Agricultural%20Import%20Regulations%20and%20Standards%20Country%20Report_Riyadh_Saudi%20Arabia_12-31-2021.pdf, Access Date: March 10, 2022
18. Saudi Press Agency, Saudi Arabia Succeeded in Improving Food Security, Reducing Waste and Achieving Self-sufficiency, Reports Deputy Minister of Environment, Tuesday 1443/2/28 – 2021/10/05, https://www.spa.gov.sa/viewstory.php?lang=en&newsid=2292762, Access Date: March 10, 2022
19. GEORGE CHARLES DARLEY, How Saudi Arabia is charting a path toward food security, Updated 19 October 2021, https://www.arabnews.com/node/1870111/saudi-arabia, Access Date: March 10, 2022
20. Indian Government, National Policy in Biofuels – 2018, 2018, https://mopng.gov.in/files/uploads/NATIONAL_POLICY_ON_BIOFUELS-2018.pdf, Access Date: September 18, 2021

Part II
Food Security Machine Learning and Heuristics Models

Chapter 4
Food Security

This Chapter Covers:

- Food Security
- Linkages of Agriculture (A), Food (F), and Nutrition (N)
- Key Drivers of Food Security
- Food Security Indicators and Drivers
- The Global Food Security Index (GFSI)
- Machine Learning Models
  - Economic Access and Food Security Predictive Model
  - Commodity Price Prediction and Linear Programming
  - Food Affordability Predictive Model
- Food Security and Technological Innovation
- Small Farm Sustainability
- Data-Driven Food Security Models

The chapter introduces food security, key drivers of food insecurity, and food security indicators and drivers. Additionally, it introduces the Global Food Security Index (GFSI) framework and the UN Suite of Food Security Indicators and also machine learning and linear programming to develop commodity price prediction, economic access and food security, and food affordability models. Finally, the chapter concludes with the role of technological innovations, data-driven models, and small farm sustainability to enhance food security.

C. Vuppalapati, *Artificial Intelligence and Heuristics for Enhanced Food Security*,
International Series in Operations Research & Management Science 331,
https://doi.org/10.1007/978-3-031-08743-1_4

A household is not "food secure" unless it "feels" food secure[1]. Food security exists when all people, always, have physical and economic access to sufficient, safe, and nutritious food that meets their dietary needs and food preferences for an active and healthy life[2]. Food security is a global need, and every country in the world is actively working to ensure enhanced food security [1]. For instance, the US Department of Agriculture (USDA) Economic Research Service (ERS) monitors the food security of US households through an annual, nationally representative survey. While most US households are food secure, a minority of US households experience food insecurity at times during the year, meaning that their access to adequate food for active, healthy living is limited by lack of money and other resources[3]. Some experience very low food security, a more severe range of food insecurity where food intake of one or more members is reduced and normal eating patterns are disrupted [2]. In 2020, 89.5 percent of US households were food secure throughout the year. The remaining 10.5 percent of households were food insecure at least some time during the year, including 3.9 percent (5.1 million households) that had very low food security[4]. Prevalence[5] of food insecurity (please see Figs. 4.1 and 4.2) is not uniform across the country due to both the characteristics of populations and state-level policies and economic conditions [2–4].

Level of food security in the European Union (EU) is not uniform[6]. The 2021 edition of the Global Food Security Index (GFSI) [7] report has reported the EU as the biggest concentration of food security leaders in the world. The list of the top 20 food secure nations includes 13 EU countries, with Ireland, Austria, the United Kingdom, Finland, Switzerland, and the Netherlands securing the top 6 ranks with even France and the United States in the 9th position. Developed by the Economist Intelligence Unit (EIU), the Global Food Security Index (GFSI) highlights drivers and factors that affect food security in more than a hundred countries across the world[8]. One reason the level of food security in the EU is not uniform is that the performances of

[1]Issues and Challenges of Inclusive Development: Essays in Honor of Prof. R. Radhakrishna by R. Maria Saleth, S. Galab and E. Revathi, Publisher: Springer; 1st ed. 2020 edition (June 19, 2020), ISBN-13: 978-9811522284

[2]Food Security—https://www.fao.org/fileadmin/templates/faoitaly/documents/pdf/pdf_Food_Security_Cocept_Note.pdf

[3]Food Security and Nutrition Assistance—https://www.ers.usda.gov/data-products/ag-and-food-statistics-charting-the-essentials/food-security-and-nutrition-assistance/

[4]The prevalence of food insecurity in 2020 is unchanged from 2019—https://www.ers.usda.gov/data-products/chart-gallery/gallery/chart-detail/?chartId=58378

[5]Prevalence of food insecurity is not uniform across the country—https://www.ers.usda.gov/data-products/chart-gallery/gallery/chart-detail/?chartId=58392

[6]Level of food security in the EU is not uniform, report shows—https://www.euractiv.com/section/agriculture-food/news/level-of-food-security-in-the-eu-is-not-uniform-report-shows/

[7]The Global Food Security Index—https://impact.economist.com/sustainability/project/food-security-index/Home/Methodology

[8]Country Rankings 2021—https://impact.economist.com/sustainability/project/food-security-index/Index

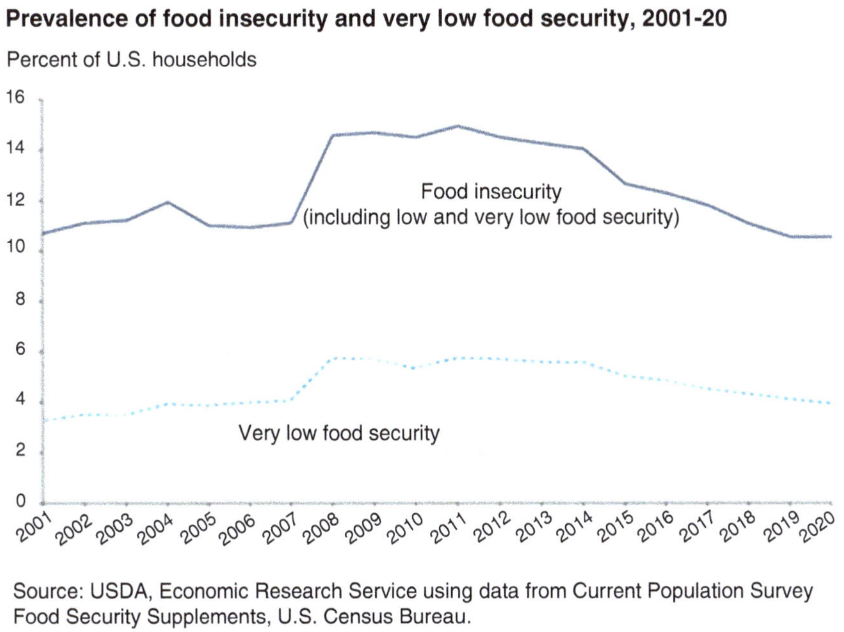

Fig. 4.1 Prevalence of food security and very low food security, 2001–2020

both Mediterranean countries and Central and Eastern Europe are significantly lower than in Western and Northern Europe. Slovakia, Ukraine, and Serbia are the lowest-ranked countries in the EU, getting respectively the 42nd, 58th, and 60th place in the food security standings worldwide. Food security could be measured using one of the important indicators: the real household gross disposable income per capita. The following provides countries within the European Union (source: OECD Household Dashboard 2007:2021[9]). In Greece, the real gross disposable income of households per capita (index = 2007) reached a record high in Q2 of 2007 and a record low in 2013[10,11] as can be seen in the above figure. In the Czech Republic, the real gross disposable income of households per capita (index = 2007) reached a record high in 2020-Q4 and a record low in 2007-Q4[12].

---

[9] OECD: Household Dashboard—https://stats.oecd.org/Index.aspx?DataSetCode=HH_DASH%20#

[10] Greece—The real gross disposable income of households per capita—https://tradingeconomics.com/greece/the-real-gross-disposable-income-of-households-per-capita-idx-2008-eurostat-data.html#:~:text=(index%20%3D%202008)-,Greece%20%2D%20The%20real%20gross%20disposable%20income%20of%20households%20per%20capita,a%20record%20high%20of%20EUR100.

[11] Greece—https://www.oecdbetterlifeindex.org/countries/greece/

[12] First quarter of 2019 Household real income per capita up in both euro area and EU28—https://ec.europa.eu/eurostat/documents/2995521/10012317/2-26072019-AP-EN.pdf/fb30e9c2-23a1-4293-89c5-5f42e98aa46f

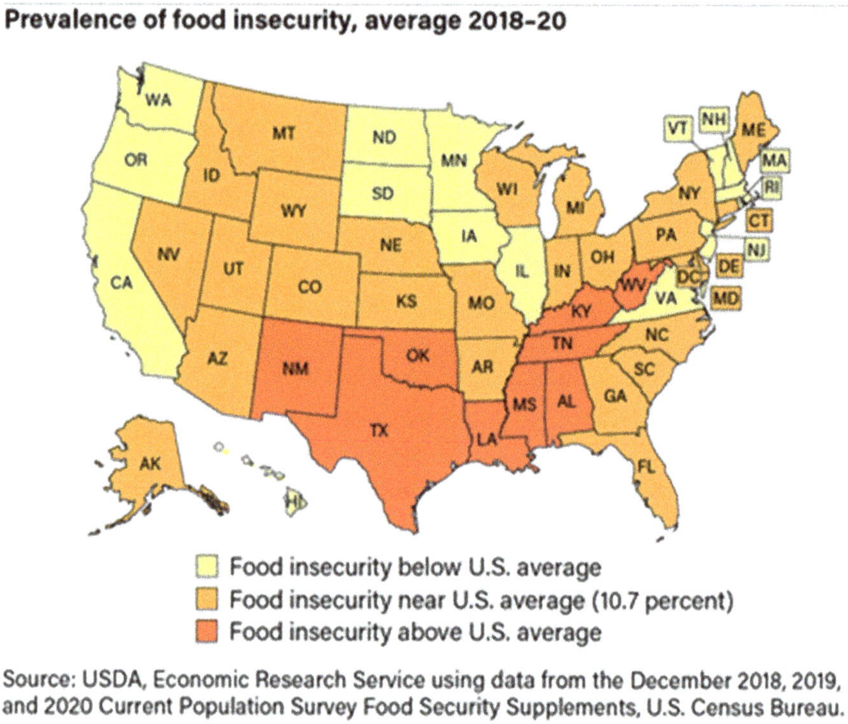

Fig. 4.2 Prevalence of food security, average 2018–2020

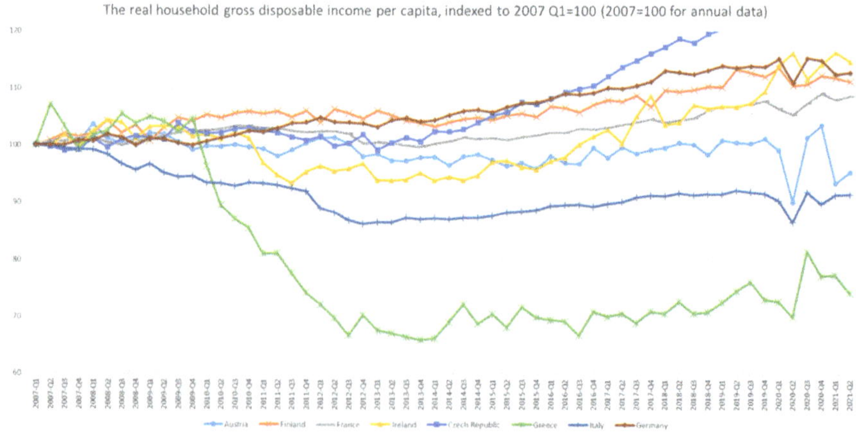

Real Household Gross Disposable Income Per Capita

This indicator shows the real household gross disposable income per capita, indexed to 2007 Q1=100 (2007=100 for annual data). As such, it shows how much

(continued)

	households' income has grown or shrunk after adjusting for how much purchasing power the money has from the beginning of 2007. For example, if money income increases more than consumer prices, real income increases. If money income increases less than consumer prices, real income declines. Household disposable income equals the total income received, after deduction of taxes on income and wealth and social contributions and includes monetary social benefits (such as unemployment benefits). It does not include in-kind transfers, such as those related to health and education provided free or at economically insignificant prices by government and non-profit institutions serving households (NPISHs).

Food security in Asia and the Pacific region has "two faces."[13] With Japan and Singapore making the top 20 in the Global Food Security Index, Tajikistan, Bangladesh, and Laos are in the 83rd, 84th, and 91st positions. While Asia's economic growth and ongoing structural transformation deepen the complexity in managing the limited natural resources required for food security, many pockets of Asia continue to struggle with high levels of poverty and poor nutrition [5]. The COVID-19 pandemic has increased food security risks in Asia and the Pacific as strict quarantine measures and export bans on basic food items have affected all stages of food supply chains[14]. Household food consumption and nutrition have been significantly affected by loss of jobs and income and limited access to food. Informal sector workers—70% of total employment in the region—in particular are at a higher risk. Some scenarios present figures of up to 130 million additional people at risk of becoming acutely food insecure, with up to 24 million in the Asia and Pacific region.

Latin America and the Caribbean managed to reduce the number of undernourished by 20 million compared to the year 2000. However, 2018 marks the fourth consecutive year in which hunger shows a continuous increase[15]. Moderate or severe food insecurity in Latin America increased considerably. This increase caused more than 32 million people to join the almost 155 million who lived in food insecurity in the region in 2014–2016. The region has shown significant progress in reducing child malnutrition, and it is significantly distant and below the global prevalence of malnutrition in girls and boys. However, malnutrition due to excessive weight in the region is one of the highest in the world, and it continues to increase. With Costa Rica 24th and Chile 28th in leading GFSI, Latin America has lower food security index scores as listed with Venezuela 102nd and Haiti 106th [6]. Chile has a drastic

[13] Food Security in Asia and the Pacific, https://www.adb.org/publications/food-security-asia-and-pacific

[14] Food Security in Asia and the Pacific amid the COVID-19 Pandemic—https://www.adb.org/sites/default/files/publication/611671/adb-brief-139-food-security-asia-pacific-covid-19.pdf

[15] Regional Overview of Food Security in Latin America and the Caribbean—https://www.fao.org/documents/card/en/c/ca6979en/

shift in real household income in 2020 and 2021 due to the government's pandemic-era cash handouts; suddenly, Chileans are flushed with cash[16] (please see below figure—Source OECD[17] Chile 2007:Q1–2021:Q4).

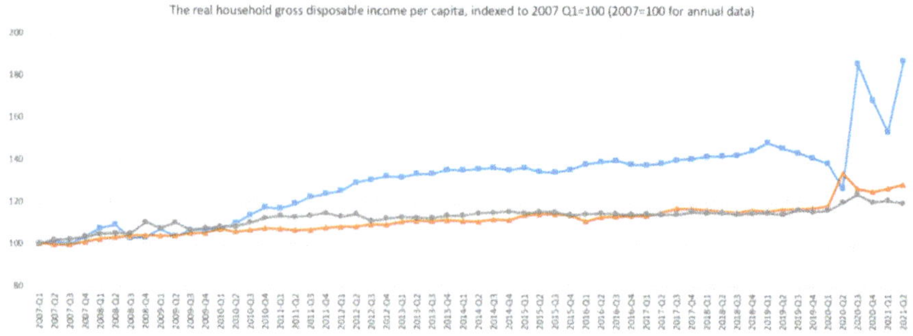

Food insecurity in the Middle East and North Africa (MENA) is a growing challenge. Even before COVID-19, UN agencies estimated that over 55 million of its population of 456.7 million was undernourished. The pandemic, protracted conflict, and other factors make hunger more common. In 2020, MENA's share of the world's acutely food insecure people was 20%, disproportionately high compared to its 6% share of the population[18]. The number one driver of hunger on the planet is man-made conflict[19], and as reported by the World Bank, the food security issue has worsened with conflict, such as in Yemen and Syria. The UN estimates the number of Yemenis afflicted by food insecurity reached 24 million—~83% of the population—in 2021, with 16.2 million needing emergency food. The war in Syria has had devastating consequences: over 12 million Syrians are food insecure, an increase of 4.5 million in 2020 alone [7].

[16]The World's Hottest Economy—https://www.bloomberg.com/news/newsletters/2021-09-21/what-s-happening-in-the-world-economy-chile-s-economy-is-red-hot

[17]Household data—https://stats.oecd.org/Index.aspx?DataSetCode=HH_DASH%20#

[18]MENA Has a Food Security Problem, But There Are Ways to Address It—https://www.worldbank.org/en/news/opinion/2021/09/24/mena-has-a-food-security-problem-but-there-are-ways-to-address-it

[19]Nearly 60% of the World's Hungriest People Live in Just Ten Countries. Why?—https://www.wfpusa.org/articles/60-percent-of-the-worlds-hungry-live-in-just-8-countries-why/

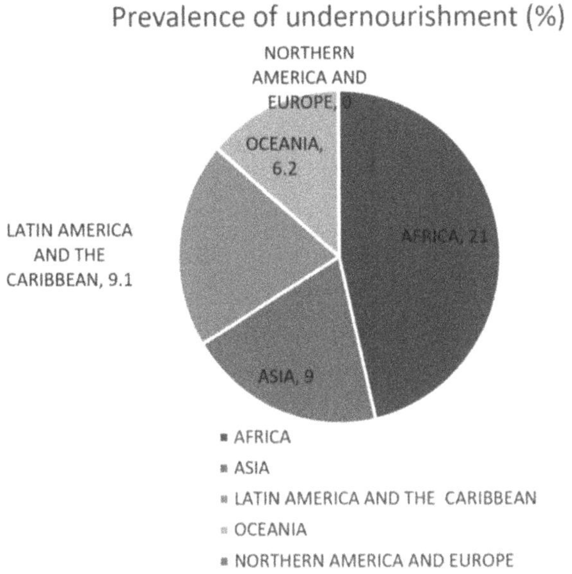

As of 2019, 234 million sub-Saharan Africans were chronically undernourished, more than in any other region. In the whole of Africa[20], 250 million people were experiencing hunger, which is nearly 20% of the population. In Africa, hunger is increasing at an alarming rate. The COVID-19 pandemic, conflict, drought, economic woes, and extreme weather are reversing years of progress [8]. It is evident that food security is important to the world, and food insecurity is one of the biggest issues that every continent of the world is grappling with. The numbers show enduring and troubling regional inequalities. About one in five people (21 percent of the population) was facing hunger in Africa in 2020—more than double the proportion of any other region. This represents an increase of 3 percentage points in 1 year. This is followed by Latin America and the Caribbean (9.1 percent) and Asia (9.0 percent), with increases of 2.0 and 1.1 percentage points, respectively, between 2019 and 2020[21] (please see Table—Source FAO).

	2005	2010	2015	2016	2017	2018	2019	2020*
World	12.4	9.2	8.3	8.3	8.1	8.3	8.4	9.9
Africa	21.3	18	16.9	17.5	17.1	17.8	18	21
Northern Africa	8.5	7.3	6.1	6.2	6.5	6.4	6.4	7.1

(continued)

[20] Africa hunger, famine: Facts, FAQs, and how to help—https://www.worldvision.org/hunger-news-stories/africa-hunger-famine-facts

[21] FAO, IFAD, UNICEF, WFP and WHO. 2021. The State of Food Security and Nutrition in the World 2021. Transforming food systems for food security, improved nutrition and affordable healthy diets for all. Rome, FAO. https://doi.org/10.4060/cb4474en

	2005	2010	2015	2016	2017	2018	2019	2020*
Sub-Saharan Africa	24.6	20.6	19.4	20.1	19.5	20.4	20.6	24.1
Eastern Africa	33	28.4	24.8	25.6	24.9	25.9	25.6	28.1
Middle Africa	36.8	28.9	28.7	29.6	28.4	29.4	30.3	31.8
Southern Africa	5	6.2	7.5	7.9	7.3	7.6	7.6	10.1
Western Africa	14.2	11.3	11.5	11.9	11.8	12.5	12.9	18.7
Asia	13.9	9.5	8.3	8	7.8	7.8	7.9	9
Central Asia	10.6	4.4	2.9	3.2	3.2	3.1	3	3.4
Eastern Asia	6.8	<2.5	<2.5	<2.5	<2.5	<2.5	<2.5	<2.5
South-eastern Asia	17.3	11.6	8.3	7.8	7.4	6.9	7	7.3
Southern Asia	20.5	15.6	14.1	13.2	13	13.1	13.3	15.8
Western Asia and	9	9.1	14.3	15	14.5	14.4	14.4	15.1
Western Asia and Northern Africa	8.8	8.2	10.5	10.9	10.7	10.6	10.7	11.3
Latin America and the Caribbean	9.3	6.9	5.8	6.8	6.6	6.8	7.1	9.1
Caribbean	19.2	15.9	15.2	15.4	15.3	16.1	15.8	16.1
Latin America	8.6	6.2	5.1	6.2	6	6.1	6.5	8.6
Central America	8	7.4	7.5	8.1	7.9	8	8.1	10.6
South America	8.8	5.7	4.2	5.4	5.2	5.4	5.8	7.8
Oceania	6.9	5.3	6.1	6.2	6.3	6.2	6.2	6.2
Northern America and Europe	<2.5	<2.5	<2.5	<2.5	<2.5	<2.5	<2.5	<2.5

The most frequently cited official definition, however, of food security comes from 1996 UN World Food Summit[22]: "Food security[23] means having, always, both physical and economic access to sufficient food to meet dietary needs for a productive and healthy life." Prior to the 1996 UN Summit, academicians and policy makers referred to food hunger as lack of food security. The absence of operational definition has complicated policy effectiveness as for any government to undertake investments that citizens see as effective. In most Asian countries, the operational definition of food security has taken the form of domestic price stability relative to world prices. This divergence between domestic prices and world prices, at least on a day-to-day basis, then requires state control over trade flows in commodities, especially stable commodities such as rice and wheat. That is, the state must intervene in commodity marketing. To avoid food insecurity, keep finance flowing[24]. [9]

In line with the official UN definition of food security, the Global Food Security Index (GFSI) bases the drivers of food security on affordability, availability, quality and safety, and natural resources and resilience. The GFSI is a dynamic quantitative and

[22] World Food Summit, Rome, 1996—https://www.fao.org/3/w3548e/w3548e00.htm

[23] Agriculture and Food Security—https://www.usaid.gov/what-we-do/agriculture-and-food-security

[24] To avoid food insecurity, keep finance flowing—https://blogs.worldbank.org/psd/avoid-food-insecurity-keep-finance-flowing

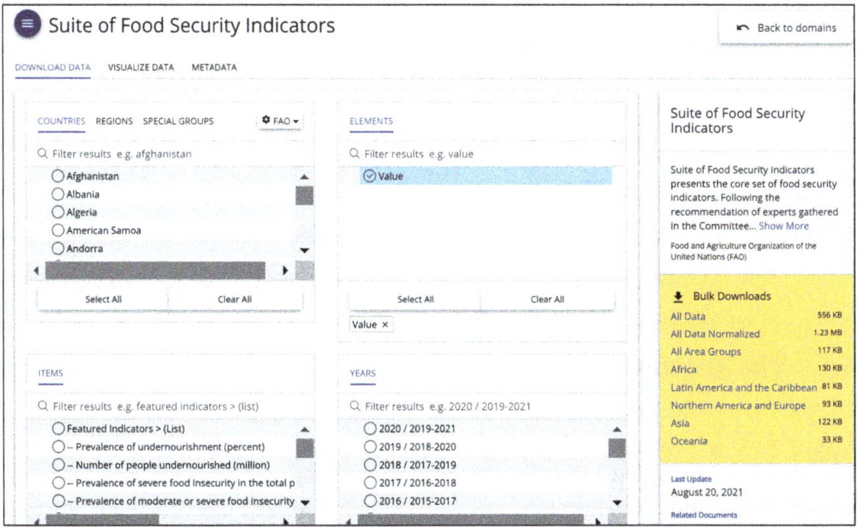

Fig. 4.3 Suite of food security indicators. Food Security Indicators—https://www.fao.org/faostat/en/#data/FS

qualitative benchmarking model constructed from 58 unique indicators that measure the drivers of food security across both developing and developed countries [25].

Similarly, the UN Food and Agricultural Organization (FAO) core food security indicators (please see Fig. 4.3), aiming to capture various aspects of food insecurity, are identified following the recommendation of experts gathered in the Committee on World Food Security (CFS) Round Table on hunger measurement, hosted at FAO headquarters in September 2011. The food security indicators are categorized in four major categories: availability, access, stability, and utilization.

Tracking progress towards global food security is critical for designing and evaluating policies and programs. Nonetheless, finding appropriate indicators is challenging. It has been discussed widely that the concept of food security is

[25] The GFSI Methodology—https://impact.economist.com/sustainability/project/food-security-index/Home/Methodology

multidimensional, dynamic, and even context specific. The complexity of the concept, compounded by the challenge of collecting data, led to a veritable proliferation of indicators in the last two decades[26] [10]. Please refer to Appendix A for Food Security Data Sources.

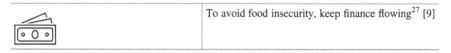

To avoid food insecurity, keep finance flowing[27] [9]

4.1 The Human Problem of Hunger

Ensuring food security is the government's responsibility. For instance, the US Federal Government spending on USDA's food and nutrition assistance programs totaled $122.1 billion in fiscal year (FY) 2020 or 32 percent more than the previous fiscal year. Food and nutrition assistance accounts for almost 20% of the EU's total humanitarian budget. EU humanitarian food assistance in 2020 is around €500 million[28].

Governments across the world, however, are at a critical juncture. From war zones to oil-rich Gulf states, hunger is a threat[29] (please see Fig. 4.4—Hunger Analytics Dashboard). There are 957 million people across 93 countries who do not have enough to eat[30] and 239 million people in need of life-saving humanitarian action and protection [11]. Compared with 2019[31], 46 million more people in Africa, almost 57 million more in Asia, and about 14 million more in Latin America and the Caribbean were affected by hunger.

Global extreme poverty, as per the World Bank[32], is expected to rise in 2020 for the first time in over 20 years as the disruption of the COVID-19 pandemic compounds the forces of conflict and climate change, which were already slowing poverty reduction progress [12]. Much of this poverty, hunger, and malnutrition is

[26] The use of the Global Food Security Index to inform the situation in food insecure countries—https://publications.jrc.ec.europa.eu/repository/bitstream/JRC108638/jrc108638-revised_version.pdf

[27] To avoid food insecurity, keep finance flowing—https://blogs.worldbank.org/psd/avoid-food-insecurity-keep-finance-flowing

[28] EU Food Assistance—https://ec.europa.eu/echo/what/humanitarian-aid/food-assistance_en

[29] Arab governments are worried about food security—https://www.economist.com/middle-east-and-africa/2021/04/15/arab-governments-are-worried-about-food-security

[30] 2021 is going to be a bad year for world hunger—https://www.un.org/en/food-systems-summit/news/2021-going-be-bad-year-world-hunger

[31] Food Security—https://www.fao.org/state-of-food-security-nutrition

[32] COVID-19 to Add as Many as 150 Million Extreme Poor by 2021—https://www.worldbank.org/en/news/press-release/2020/10/07/covid-19-to-add-as-many-as-150-million-extreme-poor-by-2021

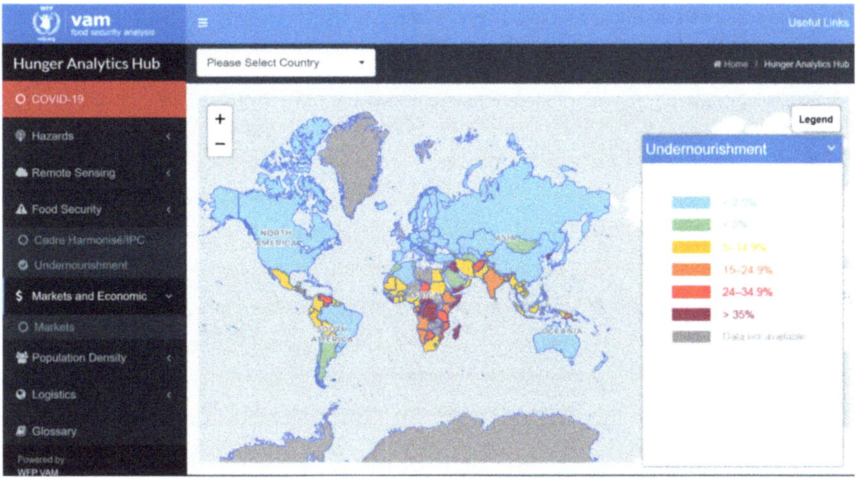

Fig. 4.4 Hunger analytics. Source: https://dataviz.vam.wfp.org/Hunger-Analytics-Hub

already concentrated in rural areas in developing countries which will experience severe economic and opportunity contractions, where most people rely on agriculture for their livelihoods. These challenges are likely to worsen in the years to come: the global population is expected to swell from 7.3 to 8.5 billion by 2030, and again to 9.7 billion by 2050, placing unprecedented pressure on food systems. Rising incomes will further increase the demand for food—particularly foods such as meat—that requires more resources to produce. With the increased demand for greener fuel and biodiesel, food security will remain an important economic development issue over the next several decades. As food-versus-fuel tension becomes more intense[33] [13], the day will come when more agricultural products will be used for energy than food. Adding to the conundrum, the COVID-19 pandemic has changed the face of the earth in terms of supply chain, resource availability, and human labor and has exposed our vulnerabilities in food security to an even greater extent (please see Fig. 4.5). These changes, together with the COVID-19 pandemic and widespread environmental shifts and variability, will exert increasing pressure on the natural resources on which food production relies.

[33] Cargill CEO Says Global Food Prices to Stay High on Labor Crunch—https://finance.yahoo.com/news/cargill-ceo-says-global-food-125939772.html

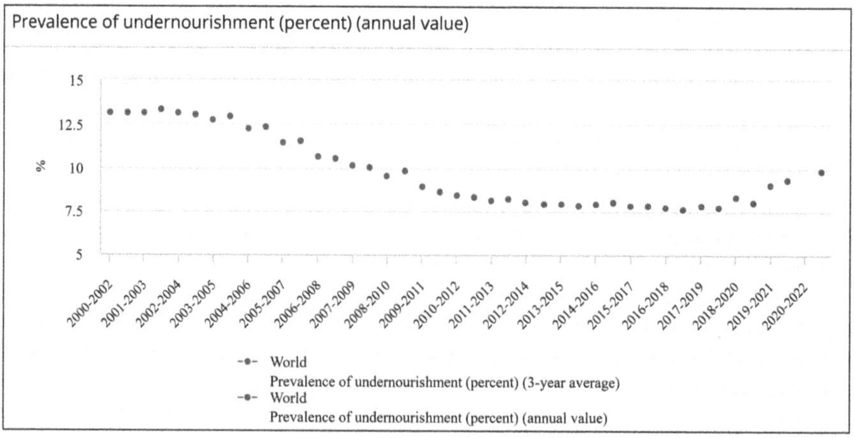

Fig. 4.5 Prevalence of severe food insecurity in the total population (percent) (annual value)

4.1.1 Linkages Among Agriculture (A), Food (F), and Nutrition (N)[34]

The world produces sufficient food[35, 36]. There is enough food for everyone[37]. According to the Food and Agriculture Organization of the United Nations (2009a, 2009b), the world produces more than 1 1/2 times enough food to feed everyone on the planet (please Fig. 4.6 for food linkage). That's already enough to feed ten billion people, the world's 2050 projected population peak. Yet millions (worldwide more than 10 percent of people are hungry) of people sleep hungry, and the main reason is due to three factors: conflicts, food loss, and economic access [14–16].

[34] C. Peter Timmer, Walter P. Falcon, and Scott R. Pearson, FOOD POLICY ANALYSIS, Published for The World Bank—The Johns Hopkins University Press, 1983, https://documents1.worldbank.org/curated/en/308741468762347702/pdf/multi0page.pdf

[35] We Already Grow Enough Food for 10 Billion People… and Still Can't End Hunger—https://foodfirst.org/publication/we-already-grow-enough-food-for-10-billion-people-and-still-cant-end-hunger/

[36] We produce enough food to feed 10 billion people. So why does hunger still exist?—https://medium.com/@jeremyerdman/we-produce-enough-food-to-feed-10-billion-people-so-why-does-hunger-still-exist-8086d2657539

[37] How to feed 10 billion people—https://www.unep.org/news-and-stories/story/how-feed-10-billion-people

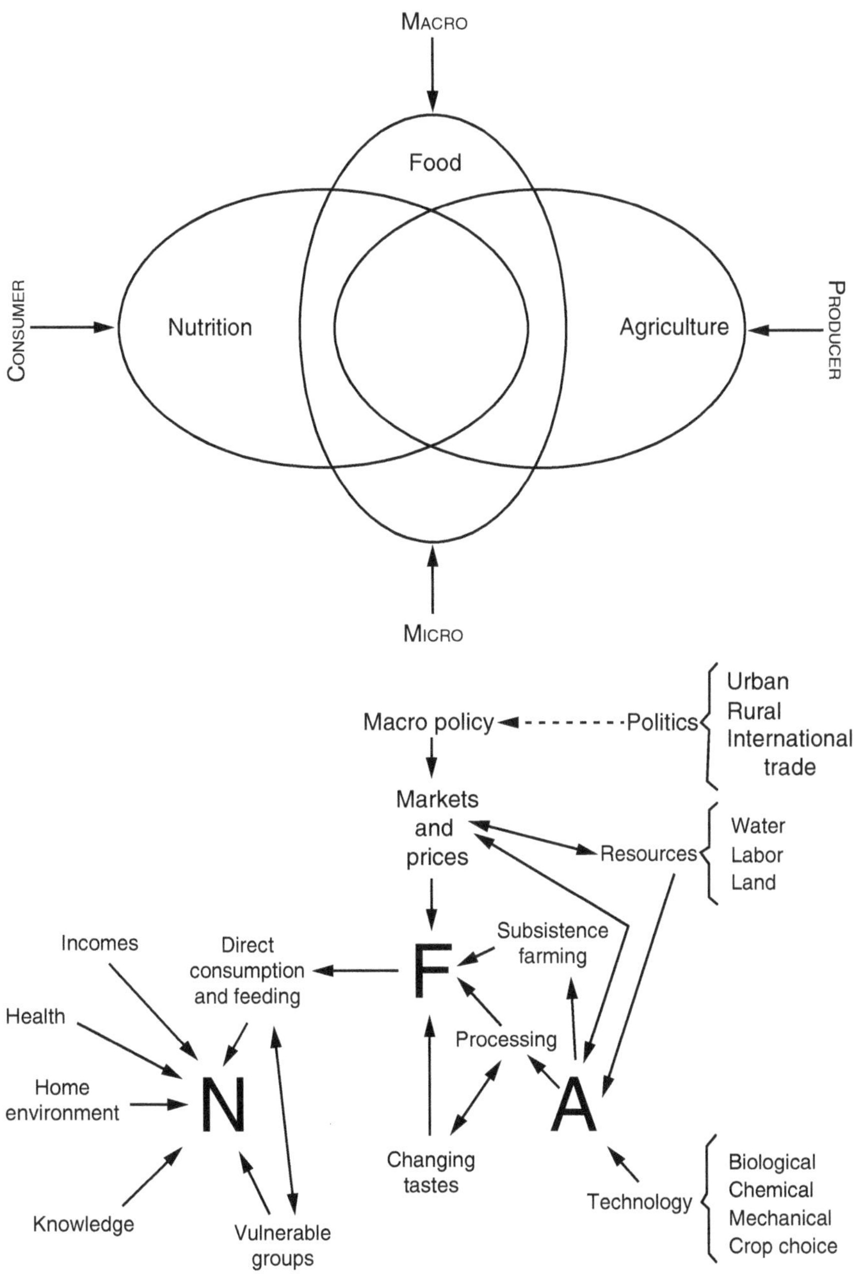

Fig. 4.6 Food linkage

4.1.2 Man-Made Conflicts

The number one driver of hunger on the planet is man-made conflict[38]. Conflict tears families apart, forces entire communities from their homes, destroys infrastructure, and disrupts food production. It's a vicious force and one that has pushed 80 million innocent civilians to the most extreme levels of hunger imaginable [17]. As UN World Food Program Executive Director David Beasley says, "Wars and conflicts are driving hunger in a way we've never seen before." Conflict analysis is beyond the scope of the book.

4.1.3 Food Loss

Our inability to feed the entirety of the world's population is mostly due to food waste. Millions of Americans struggle to put food on the table, while millions of pounds of excess food are wasted every year[39] [18]. Globally, 30–40% of all food is wasted. In less developed countries, this loss is due to the lack of infrastructure and knowledge to keep food fresh. For example, India loses 30–40% of its produce because retail and wholesalers lack cold storage [16]. Reducing food waste can help address food insecurity.

The terms food loss and food waste are often used interchangeably and grounded in legal jurisdictions, but they are quite different in terms of origin and scope. Food "waste" is most readily defined at the retail and consumer stages[40], where outputs of the agricultural system are self-evidently "food" for human consumption [19].

However, food "loss" refers to the decrease in food quantity or quality, which makes it unfit for human consumption, occurring throughout the supply chain, from harvest through to processing and distribution. This wasted food, which is still potentially fit for human consumption, could potentially feed those in need and thereby contribute to enhancing food security. By reducing food losses and waste, more food could be made available for consumption without the need for more farm output[41] [20]. If only one-fourth of the food wasted could be saved, it would be

[38] Nearly 60% of the World's Hungriest People Live in Just Ten Countries. Why?—https://www.wfpusa.org/articles/60-percent-of-the-worlds-hungry-live-in-just-8-countries-why/

[39] Reducing Food Waste Can Help Address Food Insecurity—https://www.usnews.com/news/healthiest-communities/articles/2019-10-22/commentary-the-link-between-food-waste-and-food-insecurity

[40] J. Parfitt, M. Barthel, S. Macnaughton Food waste within food supply chains: quantification and potential for change to 2050 Philosoph. Trans. R. Soc. B: Biol. Sci., 365 (1554) (2010), pp. 3065–3081

[41] Z. Babar, S. Mirgani S. Food Security in the Middle East Oxford University Press, Oxford: UK (2014) ISBN-13: 978-1849043021

sufficient to feed all currently undernourished people[42,43,44] [21–24]. Generally, as can be seen, the major food loss occurs during the farm, storage, and transport value chain stages[45] (please see Fig. 4.7).

Sustainable Development Goal 2: Zero Hunger
End hunger, achieve food security and improved nutrition and promote sustainable agriculture[46].

4.1.4 Economic Access

Food insecurity is defined as the disruption of food intake or eating patterns because of a lack of money and other resources. The number one driver of hunger on the planet is man-made conflict[47]. Conflict tears families apart and forces entire communities from their homes [17]. The great majority of the world's hungry people are the very poor, the landless and nearly landless, the vulnerable groups of young children, pregnant and lactating women, and the elderly. Food insecurity is a key issue in the economic stability domain[48]. Economic access drives food security or a lack thereof, and the following labels refer to the degree of food security (these labels are outcome of changes by the USDA in response to recommendations of an expert panel convened at USDA's request by the Committee on National Statistics (CNSTAT) of the National Academies[49]):

[42] Basher, Syed Abul and Raboy, David and Kaitibie, Simeon and Hossain, Ishrat, Understanding Challenges to Food Security in Dry Arab Micro-States: Evidence from Qatari Micro-Data (December 22, 2013). Journal of Agricultural and Food Industrial Organization 2013 (11) 1, Available at SSRN: https://ssrn.com/abstract=2499890

[43] Giallombardo, G.; Mirabelli, G.; Solina, V. An Integrated Model for the Harvest, Storage, and Distribution of Perishable Crops. Appl. Sci. 2021, 11, 6855. https://doi.org/10.3390/app11156855

[44] Smart Delivery Systems: Solving Complex Vehicle Routing Problems (Intelligent Data-Centric Systems: Sensor Collected Intelligence) 1st Edition by Jakub Nalepa (Editor) ISBN-13: 978-0128157152

[45] Food Loss and Waste Database—https://www.fao.org/platform-food-loss-waste/flw-data/en/

[46] SDG2—https://www.ifc.org/wps/wcm/connect/topics_ext_content/ifc_external_corporate_site/development+impact/sdgs/measure-by-goal/sdg-2

[47] Nearly 60% of the World's Hungriest People Live in Just Ten Countries. Why?—https://www.wfpusa.org/articles/60-percent-of-the-worlds-hungry-live-in-just-8-countries-why/

[48] Food Insecurity—https://www.healthypeople.gov/2020/topics-objectives/topic/social-determinants-health/interventions-resources/food-insecurity

[49] Definitions of Food Security—https://www.ers.usda.gov/topics/food-nutrition-assistance/food-security-in-the-us/definitions-of-food-security.aspx

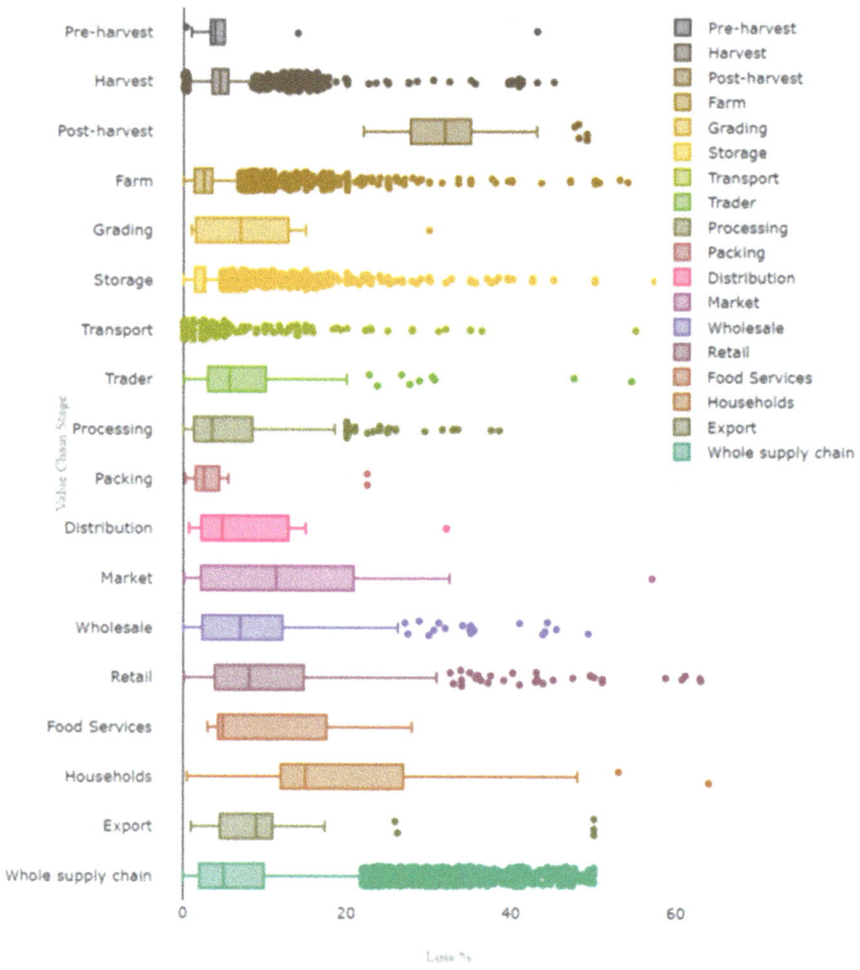

Fig. 4.7 Food loss

Food security	• High food security (old label = food security): no reported indications of food-access problems or limitations • Marginal food security (old label = food security): one or two reported indications—typically of anxiety over food sufficiency or shortage of food in the house. Little or no indication of changes in diets or food intake
Food insecurity	• Low food security (old label = food insecurity without hunger): reports of reduced quality, variety, or desirability of diet. Little or no indication of reduced food intake • Very low food security (old label = food insecurity with hunger): reports of multiple indications of disrupted eating patterns and reduced food intake

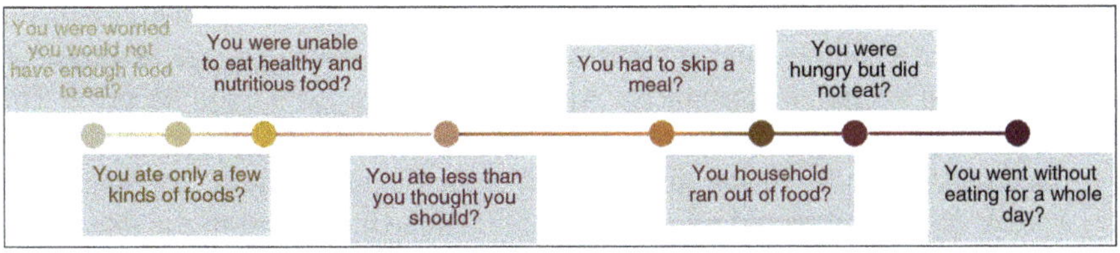

Fig. 4.8 FIES global reference scale

4.1.5 Food Insecurity Experience Scale (FIES)

An experience-based food security scale is used to produce a measure of access to food at different levels of severity that can be compared across contexts. It relies on data obtained by asking people, directly in surveys, about the occurrence of conditions and behaviors that are known to reflect constrained access to food[50] (Fig. 4.8).

The Food Insecurity Experience Scale (FIES[51]) is developed based on responses that, during the last 12 months, there was a time when, because of lack of money or other resources:

1. You were worried you would not have enough food to eat?
2. You were unable to eat healthy and nutritious food?
3. You ate only a few kinds of food?
4. You had to skip a meal?
5. You ate less than you thought you should?
6. Your household ran out of food?
7. You were hungry but did not eat?
8. You went without eating for a whole day?

4.2 Key Drivers of Food Insecurity

The key drivers of global food insecurity are food prices, food price instability, oil prices, grain stock levels, exchange rate volatility, rising demand for food due to population growth, speculation and commodity futures, climate change, political instability/conflicts[52], and changing consumption patterns and pressure on food production rate from climate change, natural resource availability as affected by land degradation and water scarcity, biofuel production, and a lack of public and

[50] The State of Food Security and Nutrition in the World 2021—https://www.fao.org/3/cb4474en/cb4474en.pdf

[51] The Food Insecurity Experience Scale—https://www.fao.org/in-action/voices-of-the-hungry/fies/en/

[52] Macron tells French farmers: Ukraine war will weigh on you, and it will last—https://www.yahoo.com/news/macron-tells-french-farmers-ukraine-075753530.html

private investment in infrastructure. The degree of importance of each key driver varies between countries and regions according to their unique set of physical, economic, and social circumstances. Agricultural prices drive food security. The agricultural price increase[53] could be due to (a) crop failure, (b) tight global supply, (c) trade restrictions by major suppliers, (d) panic buying by large importers, and (e) a weak dollar, and record oil prices were the immediate cause of the rise in rice prices. Please find food insecurities:

- In Kenya[54], about 2.4 million people are estimated to be severely food insecure between November 2021 and January 2022, reflecting consecutive poor rainy seasons since late 2020 that affected crop and livestock production, mainly in northern and eastern pastoral, agropastoral, and marginal agricultural areas.
- In Mozambique[55], food security conditions are expected to worsen in 2018 in southern and some central provinces due to the unfavorable weather conditions that are anticipated to cause a reduction in the 2018 cereal harvest.
- In Chad[56], about 990 000 are projected to be food insecure from June 2013 to August due to the serious deterioration of pastoral conditions in the Sahel.
- In Djibouti[57], due to the impact of consecutive unfavorable rainy seasons on pastoral livelihoods, about 197 000 people are severely food insecure, mainly concentrated in pastoral areas.
- In Nigeria[58], in December 2016, due to economic downturn, steep depreciation of the local currency, population displacements, and severe insecurity in northern areas, more than eight million people are estimated to be food insecure.
- In Myanmar[59], income losses due to the impact of the COVID-19 pandemic have affected the food security situation of vulnerable households.
- In Pakistan[60], 2016, population displacement and localized cereal production shortfalls had contributed to food insecurity.

[53] Factors Behind the Rise in Global Rice Prices in 2008— https://www.ers.usda.gov/publications/pub-details/?pubid=38490

[54] CROP PROSPECTS and Quarterly Global Report FOOD SITUATION—https://www.fao.org/3/cb7877en/cb7877en.pdf

[55] Crop Prospects and Food Situation, June 2018—https://www.fao.org/3/I9666EN/i9666en.pdf

[56] Crop Prospects and Food Situation, June 2018—https://www.fao.org/3/I9666EN/i9666en.pdf

[57] Crop Prospects and Food Situation, June 2018—https://www.fao.org/3/I9666EN/i9666en.pdf

[58] Crops and Food, December 2016,—https://www.fao.org/3/i6558e/i6558e.pdf

[59] Food Security and COVID-19—https://www.worldbank.org/en/topic/agriculture/brief/food-security-and-covid-19

[60] Crops and Food—https://www.fao.org/3/i6558e/i6558e.pdf

4.2.1 Food Price Instability

Food price instability[61] has very serious consequences in developing countries for it impacts on [25]:

- *Food security*, with some poor households being obliged to reduce their consumption when prices rise.
- *Agricultural modernization*, as producers do not invest if prices are too unstable; this brings green revolutions to a standstill and subsequently obstructs economic development.
- *Political stability* if price rises spark urban riots.
- *Macroeconomic stability*, as food price instability may in certain cases affect the state's budget, the trade balance, exchange rates, or even growth and inflation rates.

Price instability means that production remains highly sensitive to climatic hazards and is little responsive to price incentives (and this in turn maintains price instability). Price instability generates a risk for farmers, which leads them to invest very little (both because they are risk-averse, an example we can see from Minimum Support Price of Indian Farmers vs. Government[62] [26], and because banks are reluctant to lend while the price risk is high). These low levels of agricultural investment mean that production remains highly sensitive to climatic hazards (with very little use of irrigation or drought-resistant varieties) and responds only sluggishly to price incentives, with producers finding it difficult to boost production when prices rise. But production sensitivity to climatic hazards and its poor responsiveness to price rises further accentuate price instability.

> Food price instability is a self-sustaining phenomenon as it prompts behaviors that tend to maintain high levels of instability (vicious circles).
> Price instability and low agricultural investment therefore form a vicious circle[63] [25]

It should also be noted that regardless of which instruments are used, all policies [14] designed (please see Fig. 4.9) to provide protection against imported instability are liable to be rendered ineffective by spillover effects. This occurs if the borders with neighboring countries are porous. If country A manages to protect itself from soaring international prices, then the price on its domestic market will be far lower than that in neighboring countries. This is likely to induce (legal or illegal) exports to these countries, thus raising the price in country A and compromising the policy implemented by this country to mitigate imported instability. A symmetrical

[61] Managing food price instability in developing countries: A critical analysis of strategies and instruments (2013)—https://europa.eu/capacity4dev/file/14518/download?token=gpMqQy9Y

[62] India's farmers faced down a popular prime minister and won. What will they do now?—https://www.npr.org/2021/11/26/1059200463/india-farmer-protests-modi-farm-laws

[63] Managing food price instability in developing countries: A critical analysis of strategies and instruments (2013)—https://europa.eu/capacity4dev/file/14518/download?token=gpMqQy9Y

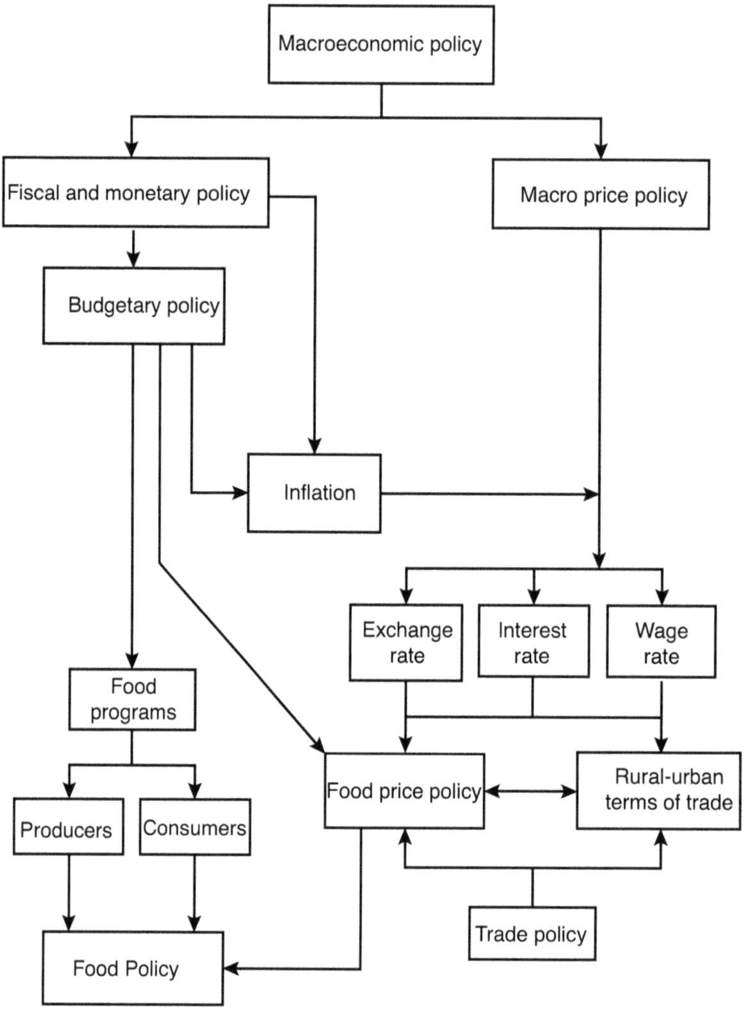

Fig. 4.9 Food policy

situation may arise if the country manages to maintain a relatively high producer price despite falling international prices. Here, the product is liable to flood into the country from neighbors (capitalizing on the high price), driving prices down. The answer here lies in better border control (when possible) or in choosing to tackle imported instability by implementing measures on a regional scale.

Despite all these difficulties, it is possible for a country that deploys adequate resources to control imported instability. Shining examples of this are China and India during the 2008 price crisis.

4.2.2 High Correlation Between Agricultural Commodities and Oil Prices

Demand and supply, energy policy, and weather have shown linkage between agricultural commodities and oil. In 2005, the US energy policy changed to increase the use of biofuels in gasoline production, increasing demand for ethanol[64] [27]. Major agricultural commodity prices generally trend together due to common price determinants[65]. Food commodity prices are also linked with those for crude oil. However, more conventional supply (e.g., weather events) and demand (e.g., the value of the US dollar) factors are important determinants of the recent price movements for corn, soybeans, and wheat. Favorable weather conditions resulted in strong corn and soybean supply in 2017, driving prices downwards. The positive weather conditions persisted for soybeans in 2018, further decreasing prices for soybeans despite crude oil price advances. These price data suggest that supply and demand factors dominate when food commodity prices and crude oil prices diverge.

Agricultural commodity prices are becoming increasingly correlated with oil prices[66]. Oil prices affect agricultural input prices directly and indirectly (e.g., through the price of fuel and fertilizer). In addition, depending on the relative prices of agricultural crops and oil, biofuel production may become profitable (without government support) in some OECD countries. Financial investment in commodities may also have contributed to an increasing correlation between oil and non-oil commodity prices because of the significant share of such investment that tracks indexes containing a basket of different commodities. High and volatile oil prices (if that is what is expected) could therefore contribute to higher and more volatile agricultural prices, through higher input costs, higher demand for the commodities used in the production of biofuels (sugar, maize, vegetable oils), through competition for land with commodities that are not used directly to produce fuel, and possibly through financial investment in commodity baskets [28].

4.2.3 Food Security and Grain Stock Levels

Low stocks relative to use and uncertainty about stock levels in some parts of the world contributed to the 2007/2008 price spike. Stocks can be drawn down in response to a supply or demand shock, but once they have been depleted, supply

[64] Do Rises in Oil Prices Mean Rises in Food Prices?, https://www.stlouisfed.org/on-the-economy/2016/november/rises-oil-prices-mean-rises-food-prices

[65] The relationship between crude oil prices and export prices of major agricultural commodities—https://www.bls.gov/opub/btn/volume-8/the-relationship-between-crude-oil-and-export-prices-of-major-agricultural-commodities.htm

[66] Price Volatility in Food and Agricultural Markets: Policy Responses—https://www.fao.org/fileadmin/templates/est/Volatility/Interagency_Report_to_the_G20_on_Food_Price_Volatility.pdf

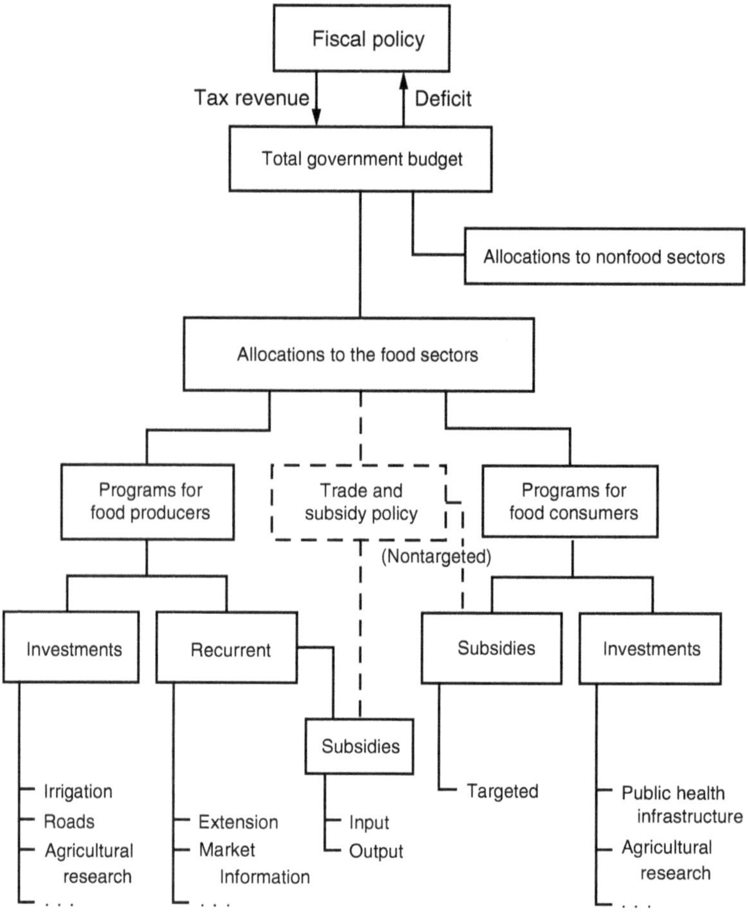

Fig. 4.10 Fiscal policy

can no longer be increased until new production comes on board. Even expectations of depleted stocks may lead prices to rise sharply. The low stock levels observed in recent years have been attributed to the partial dismantling of price support and intervention purchase schemes in some OECD countries, as well as to the correction of the quality of information on private- and government-held stocks in important producing and consuming countries. Stocks were rebuilt in 2009 and the first part of 2010, but currently stocks are again being depleted. If stock levels remain low in major markets, and projections based on existing knowledge of market conditions and policy settings suggest that they may, the risk of volatility in prices will remain high [14] (please see Fig. 4.10).

4.2.4 *Stronger Demand and Slow Growth in Agricultural Productivity*

Stronger demand for food crops and animal products in conjunction with slow growth in agricultural productivity and low stocks results in upward pressure on prices. Recent years have also seen some shift in production patterns, particularly of food and feed grains, and world markets are more dependent on supplies from the Black Sea region and other, newer, agricultural production regions than in the past. Yields in these regions are less stable and supply more variable than in some other parts of the world where natural conditions are better and where application of the most up-to-date technologies and management practices have increased and stabilized yields. As the geographical distribution of production changes, supply may therefore become more variable, in turn leading to increased price volatility.

The same underlying factors that are leading to increased demand for food—growth in population, affluence leading to increased demand for animal protein, urbanization, and biofuels—are also increasing pressure on finite resources such as land and water[67]. While such resource constraints are, thus far, more local than global in nature, a growing concern is evident, and the associated uncertainty may imply upward pressure on prices and continuing or increased volatility.

4.2.5 *Exchange Rate Volatility*

Trade in many agricultural commodities is denominated in USD. A depreciating USD causes dollar-denominated international commodity prices to rise, although not to the full extent of the depreciation. The opposite occurs when the dollar appreciates. For example, consider below the Sugar Futures (US Sugar #11 Futures[68]) graph. The Sugar No. 11 contract is the world benchmark contract for raw sugar trading and is available on the Intercontinental Exchange[69] (ICE). The biggest producer and exporter of sugar in the world is Brazil (21% of total production and 45% of total exports). Now, let's analyze signal between USD/Brazilian real[70] exchange rate and ICE 11 futures. As indicated in the graph, higher USD/BRL values (i.e., appreciate in USD) coincided with a drop in ICE 11 future sugar price. Vice versa is also true. That is, drop in USD/BRL (depreciate in USD) shows an increase in sugar futures. These currency movements added to the amplitude of the price changes observed. (They also help to explain why demand remained strong in

[67] AQUASTAT—https://www.fao.org/aquastat/statistics/query/index.html?lang=en

[68] US Sugar #11 Futures—https://www.investing.com/commodities/us-sugar-no11-historical-data

[69] The ICE Sugar Contract 11—https://tradingeconomics.com/commodity/sugar

[70] USD/BRL—https://fxtop.com/en/historical-exchange-rates.php?A=1&C1=USD&C2=BRL&MA=1&DD1=01&MM1=01&YYYY1=1970&B=1&P=&I=1&DD2=25&MM2=02&YYYY2=2022&btnOK=Go%21

countries where the currency was appreciating against the dollar and why falling prices were not fully felt in the same countries once the dollar began to appreciate again.) Exchange rate volatility per se is beyond the scope of this book, but if the future is marked by increased exchange rate volatility, this will also have repercussions for the volatility of international prices of commodities.

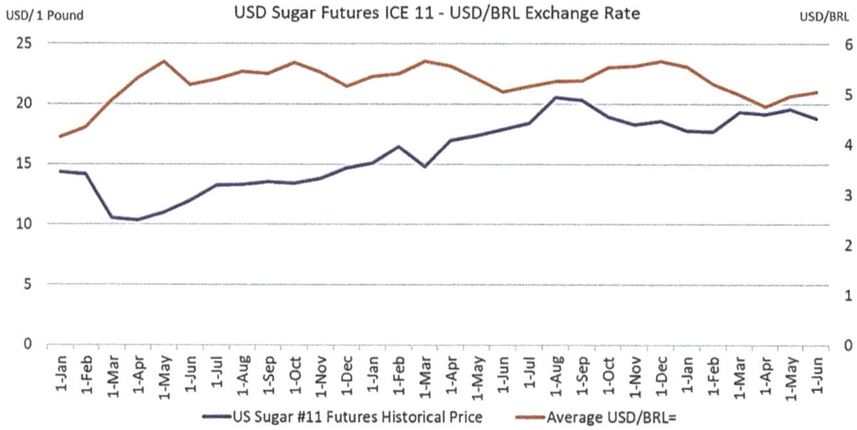

4.2.6 *Speculations and Commodity Futures*

There is no doubt that investment in financial derivatives markets for agricultural commodities increased strongly in the mid-2000s, but there is disagreement about the role of financial speculation as a driver of agricultural commodity price increases and volatility. While analysts argue about whether financial speculation has been a major factor, most agree that increased participation by noncommercial actors such as index funds, swap dealers, and money managers in financial markets probably acted to amplify short-term price swings and could have contributed to the formation of price bubbles in some situations. Against this background the extent to which financial speculation might be a determinant of agricultural price volatility in the future is also subject to disagreement. It is clear, however, that well-functioning derivative markets for agricultural commodities could play a significant role in reducing or smoothing price fluctuations—indeed, this is one of the primary functions of commodity futures markets.

4.3 Food Security Indicators and Drivers

The United States leads efforts to improve global food security[71], providing about half of global food aid [29]. Global food security has improved over the past 20 years, but both challenges and opportunities remain. The USDA ERS researchers analyze the roles of trade, agricultural productivity, safety nets, incomes, prices, and better data and measurement in achieving these gains. Please refer to Appendix A for Food Security Data Sources.

The COVID-19 pandemic has unprecedented impact on the food security[72]. Revisions to food security have new revelations [30]:

- The number of food-insecure people in 2020 is estimated at 921 million, an increase of 160 million (21 percent) from pre-pandemic estimates. The increase in the number of food-insecure people in 2020 due to COVID-19 is almost double the original estimate in the 2020 International Food Security Assessment (IFSA) report.
- The sharp increase in food insecurity due to COVID-19 reflects a sharper GDP decline in 2020 than originally projected and a slower recovery in 2021 across the 76 countries covered in the assessment. Asia and Latin America and the Caribbean (LAC) are the two regions with the largest revisions to 2020 GDP, with their GDP growth projections revised downwards by about 6 percentage points, on average, from the projection in the 2020 IFSA report.

Feed the Future[73]

The U.S. Government launched the Feed the Future initiative in the wake of the 2007/2008 global food price spikes to reduce global hunger, undernutrition, and extreme poverty. Feed the Future's results and critical contributions to the U.S.' economy, security and leadership have garnered broad bipartisan support, culminating in the enactment of the Global Food Security Act (GFSA) of 2016.

The GFSA called for a new whole-of-government global food security strategy that the 11 Feed the Future partner agencies and departments worked

(continued)

[71] Assessing Food Security—https://www.ers.usda.gov/topics/international-markets-us-trade/global-food-security/readings/

[72] COVID-19 Working Paper: International Food Security Assessment, 2020–2030: COVID-19 Update and Impacts on Food Insecurity—https://www.ers.usda.gov/webdocs/publications/100276/ap-087.pdf?v=6037.9

[73] THE U.S. GOVERNMENT'S GLOBAL FOOD SECURITY RESEARCH STRATEGY—https://www.usaid.gov/sites/default/files/documents/1867/GFS_2017_Research_Strategy_508C.pdf

together to create, along with department and agency specific implementation plans.

The resulting 2017–2021 Global Food Security Strategy (GFSS) describes in detail how the U.S. intends to direct Feed the Future resources and programming to advance three strategic objectives: promoting inclusive, sustainable agriculture-led economic growth; building resilience among vulnerable populations and households; and improving nutritional outcomes, especially among women and children.

To achieve these objectives, the GFSS highlighted that Feed the Future research investments should "ensure a pipeline of innovations, tools and approaches designed to improve agriculture, food security, resilience and nutrition priorities in the face of complex, dynamic challenges."

4.4 The Global Food Security Index (GFSI) Framework

The GFSI is based on contributing factors rather than outcomes of food security. It tends to measure the conditions for food security or an enabling environment for food security instead of actual food security level. The GFSI does not capture the entire spectrum of food security. It reflects specific aspects chosen by the team of experts that designed the index. The index focuses on the GDP as well as poverty and on the agricultural production side. The GFSI, like any other composite indicator, does not allow to draw any causal inference between the dimensions of the indicators (affordability, availability, quality, and safety), or the individual indicators included, and food security.

Here are the major indicators[74]: 

1)	Affordability[75]
1.1)	Change in average food costs

(continued)

[74]Global Food Security Index 2021: The 10-year anniversary October 12, 2021—https://impact.economist.com/sustainability/project/food-security-index/Resources

[75]The GFSI Methodology—https://my.corteva.com/GFSI?file=ei21_gfsir

1)	Affordability
1.2)	Proportion of population under global poverty line
1.3)	Inequality-adjusted income index
1.4)	Agricultural import tariffs
1.5)	Food safety-net programs
1.5.1)	Presence of food safety-net programs
1.5.2)	Funding for food safety-net programs
1.5.3)	Coverage of food safety-net programs
1.5.4)	Operation of food safety-net programs
1.6)	Market access and agricultural financial services
1.6.1)	Access to finance and financial products for farmers
1.6.2)	Access to diversified financial products
1.6.3)	Access to market data and mobile banking

The United Nations FAO looks availability from the point of daily intake: average dietary energy supply adequacy; average value of food production; share of dietary energy supply derived from cereals, roots, and tubers; average protein supply; and average supply of protein of animal origin consist in this section.

4.4.1 Change Is Average Food Costs

Food prices represent an important component of cost of living that affects households' ability to afford food. Higher food prices lead to greater food insecurity[76]. The USDA estimates that 40 million people, including more than 12 million children, in the United States are food insecure as of 2017. For a family struggling to afford housing, utilities, transportation, and other necessities, the additional burden of high food prices can have a significant impact on their household budget. Seven counties fall into the top 10% for both food insecurity and meal cost (please see Fig. 4.11).

COVID-19 has exacerbated food prices. With the onset of the pandemic in March 2020, many consumers experienced, for the first time in their lifetimes, empty grocery store shelves. More than a year into the pandemic, they are facing another unfamiliar trend when it comes to accessing food: notably higher prices. Increases in wages in the food sector, rising agricultural commodity prices, transportation bottlenecks, and strong consumer demand have led to the highest annual grocery price increases in a decade and the highest annual restaurant price increases since the early 1980s[77] [31]. As depicted in the above figure, from May 2019 to May 2020, consumer prices for meats, poultry, fish, and eggs rose 10.0 percent, the largest

[76] Food Price Variation—https://www.feedingamerica.org/sites/default/files/2019-05/2017-map-the-meal-gap-food-price-variation_0.pdf

[77] What Is Driving the Increase in Food Prices?—https://econofact.org/what-is-driving-the-increase-in-food-prices

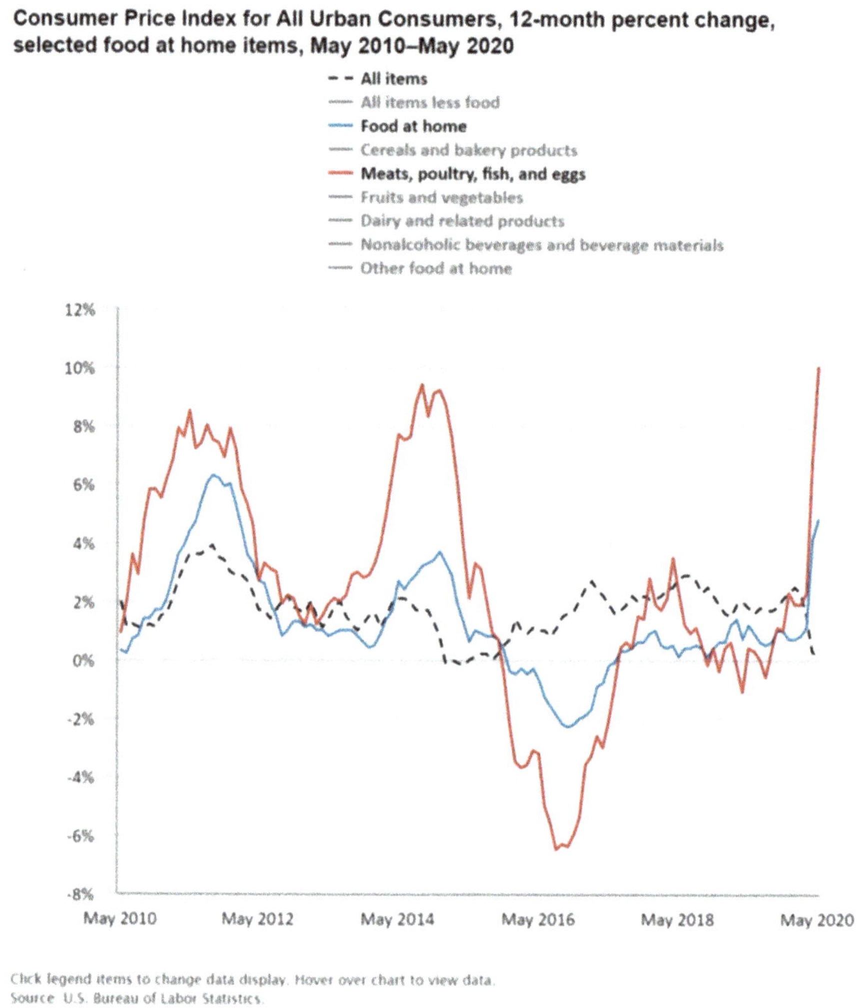

Fig. 4.11 CPI [31, 32]

12-month percentage increase since the year ended May 2004. Prices for dairy and related products increased by 5.7 percent, and prices for non-alcoholic beverages rose 4.1 percent[78] [32].

Heuristics play an important role in optimizing food basket. In OECD countries micronutrient inadequacy can coexist with excess calorie intake. Vulnerable groups, especially pregnant women within low socioeconomic groups and their families, are

[78]Consumer prices for food at home increased 4.8 percent for year ended May 2020—https://www.bls.gov/opub/ted/2020/consumer-prices-for-food-at-home-increased-4-point-8-percent-for-year-ended-may-2020.htm

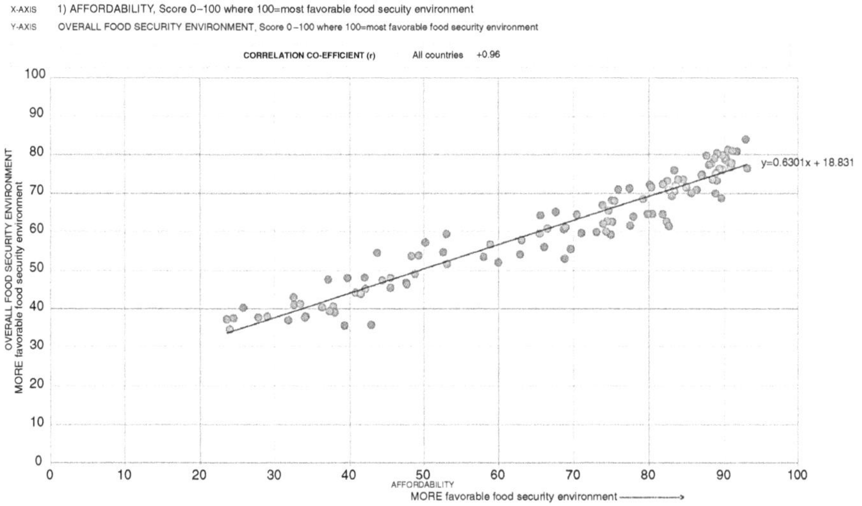

Fig. 4.12 Affordability and overall food security [GSFI model]

at a high risk. Evidence suggests that increased intake of micronutrient-dense foods with low energy density can help prevent nutrition-related noncommunicable diseases (NCD) along with micronutrient inadequacies, and corresponding national food-based dietary guidelines (FBDGs) have been developed in many countries[79]. However, micronutrient-dense foods are relatively expensive, so people especially those on low incomes buy less, and this increases the risk of micronutrient inadequacies. Even in high-income countries, economic constraints and actual lifestyles lead people to consume diets with a low micronutrient-energy ratio. Both micronutrient inadequacy and excess weight gain are expected to increase along with inequalities during economic crises even in high-income countries [33].

There is a need for mathematical modeling to help calculate which foods can supply the optimum nutrient recommendations for low cost, especially for income-strapped households and authorities, e.g., catering services within the public sector. The method of LP has been used to optimize the average daily nutrient intake of children and adults since the 1950s [33].

The correlation coefficient of relationship between affordability and overall food security is +0.96. That is, affordability drives 96% of food security (please see Fig. 4.12).

In terms of relationship, affordability and food safety are governed by:

$$\text{Over all Food Security (Y)} = 0.6301 \times \text{Affordability} + 18.831 \qquad (1)$$

[79]Parlesak A, Tetens I, Dejgård Jensen J, et al. Use of Linear Programming to Develop Cost-Minimized Nutritionally Adequate Health Promoting Food Baskets. PLoS One. 2016;11(10): e0163411. Published 2016 Oct 19. doi:10.1371/journal.pone.0163411

4.5 Economic Access: Food Security Predictive Model

The goal of the ML model is to predict poverty in the United States. The data contains socioeconomic, demographic, and geographic data in counties in the United States, enabling the development of ML model to predict the poverty rate of counties[80].

Software code for this model can be found on GitHub Link: foodSecurityModel.ipynb (Jupyter Notebook Python Code)

Dataset: the dataset for the ML model is downloaded from Microsoft Data Science Capstone[81]. The original dataset is compiled from a wide range of sources and made publicly available by the US Department of Agriculture, Economic Research Service (USDA ERS).

4.5.1 *Step 1: Load Dataset*

```
# Importing data
foodSecurityData = pd.read_csv("PovertyDataPreparation.csv")
```

	row_id	area__rucc	area__urban_influence	econ__economic_typology	econ__pct_civilian_labor	econ__pct_unemployment	econ__pc
0	0	Nonmetro - Completely rural or less than 2,500...	Noncore adjacent to a large metro area	Federal/State government-dependent	0.358	0.089	
1	2	Nonmetro - Urban population of 2,500 to 19,999...	Micropolitan adjacent to a large metro area	Manufacturing-dependent	0.503	0.057	
2	5	Nonmetro - Urban population of 2,500 to 19,999...	Noncore adjacent to micro area and contains a ...	Federal/State government-dependent	0.578	0.049	
		Metro - Counties in					

[80]NJ Food Bank Poverty Rate Prediction—https://gallery.azure.ai/Experiment/NJ-Food-Bank-Poverty-Rate-Prediction

[81]Capstone Project—https://www.datasciencecapstone.org/

4.5.2 Step 2: Inspect Data from a Statistical Point of View

	row_id	econ__pct_civilian_labor	econ__pct_unemployment	econ__pct_uninsured_adults	econ__pct_uninsured_children	demo__pct_female	demo__pct_below_18_years_of_age	demo_
count	2836.000000	2836.000000	2836.000000	2836.000000	2836.000000	2836.000000	2836.000000	
mean	3158.406206	0.464087	0.060207	0.214484	0.081478	0.500608	0.228295	
std	1814.463440	0.066527	0.021677	0.065821	0.036371	0.020984	0.032882	
min	0.000000	0.258000	0.011000	0.046000	0.009000	0.338000	0.127000	
25%	1578.500000	0.419750	0.045000	0.164000	0.055000	0.495000	0.208000	
50%	3198.500000	0.465000	0.058000	0.213000	0.074000	0.504000	0.227000	
75%	4724.250000	0.511000	0.072000	0.260000	0.100000	0.512000	0.245000	
max	6276.000000	0.936000	0.240000	0.495000	0.249000	0.576000	0.417000	

4.5.3 Step 3: Inspect Food Security Data Types

```
foodSecurityData.dtypes
```

Data frame type call lists all the data frame column names and data types. As can be seen from the output, we have integer, float, and object types.

```
row_id                                        int64
area__rucc                                   object
area__urban_influence                        object
econ__economic_typology                      object
econ__pct_civilian_labor                    float64
econ__pct_unemployment                      float64
econ__pct_uninsured_adults                  float64
econ__pct_uninsured_children                float64
demo__pct_female                            float64
demo__pct_below_18_years_of_age             float64
demo__pct_aged_65_years_and_older           float64
demo__pct_hispanic                          float64
demo__pct_non_hispanic_african_american     float64
demo__pct_non_hispanic_white                float64
demo__pct_american_indian_or_alaskan_native float64
demo__pct_asian                             float64
demo__pct_adults_less_than_a_high_school_diploma float64
demo__pct_adults_with_high_school_diploma   float64
demo__pct_adults_with_some_college          float64
demo__pct_adults_bachelors_or_higher        float64
demo__birth_rate_per_1k                       int64
demo__death_rate_per_1k                       int64
health__pct_adult_obesity                   float64
health__pct_adult_smoking                   float64
health__pct_diabetes                        float64
health__pct_low_birthweight                 float64
```

4.5.4 Step 4

Since the data has both descriptive and numerical data, group all numerical data attributes into numerical features list.

In the code below, we are grouping both float and integer column types into numerical features list.

```
#A list that record the name of numerical features
numerical_features=foodSecurityData.dtypes[foodSecurityData.
dtypes=='float64'].index.tolist() + foodSecurityData.dtypes
[foodSecurityData.dtypes=='int64'].index.tolist()
numerical_features
```

To see all column statistical properties, describe the data frame, and transpose it so that we can list data based on statistical column based:

```
foodSecurityData.describe().transpose()
```

We have a total of 2836 rows with varying mean and standard deviations.

To see graphical density view:

```
def show_density(var_data,colname):
  from matplotlib import pyplot as plt

  fig = plt.figure(figsize=(10,4))

  # Plot density
  var_data.plot.density()

  # Add titles and labels
  plt.title(colname + ' Density')

  # Show the mean, median, and mode
  plt.axvline(x=var_data.mean(), color = 'cyan',
linestyle='dashed', linewidth = 2)
  plt.axvline(x=var_data.median(), color = 'red',
linestyle='dashed', linewidth = 2)
  plt.axvline(x=var_data.mode()[0], color = 'yellow',
linestyle='dashed', linewidth = 2)

  # Show the figure
  plt.show()

for col in numerical_features:
  show_density(foodSecurityData[col],col)
```

	count	mean	std	min	25%	50%	75%
row_id	2836.0	3158.406206	1814.463440	0.000000	1578.500000	3198.500000	4724.250000
econ__pct_civilian_labor	2836.0	0.464087	0.066527	0.258000	0.419750	0.465000	0.511000
econ__pct_unemployment	2836.0	0.060207	0.021677	0.011000	0.045000	0.058000	0.072000
econ__pct_uninsured_adults	2836.0	0.214484	0.065821	0.046000	0.164000	0.213000	0.260000
econ__pct_uninsured_children	2836.0	0.081478	0.036371	0.009000	0.055000	0.074000	0.100000
demo__pct_female	2836.0	0.500608	0.020984	0.338000	0.495000	0.504000	0.512000
o__pct_below_18_years_of_age	2836.0	0.228295	0.032882	0.127000	0.208000	0.227000	0.245000
_pct_aged_65_years_and_older	2836.0	0.167891	0.041406	0.069000	0.141000	0.165000	0.190000
demo__pct_hispanic	2836.0	0.087996	0.138234	0.000000	0.019000	0.035000	0.088000
non_hispanic_african_american	2836.0	0.093995	0.146095	0.000000	0.008000	0.026000	0.103000
lemo__pct_non_hispanic_white	2836.0	0.772697	0.202078	0.060000	0.656000	0.853000	0.934000
rican_indian_or_alaskan_native	2836.0	0.021295	0.071813	0.000000	0.002000	0.007000	0.013000
demo__pct_asian	2836.0	0.013668	0.024117	0.000000	0.003000	0.008000	0.014000
s_than_a_high_school_diploma	2836.0	0.148526	0.066529	0.016129	0.098616	0.133902	0.193661
ults_with_high_school_diploma	2836.0	0.349109	0.071190	0.072821	0.304554	0.354157	0.399604
pct_adults_with_some_college	2836.0	0.300006	0.050237	0.112821	0.265084	0.299700	0.334004
ct_adults_bachelors_or_higher	2836.0	0.202359	0.091434	0.049116	0.139365	0.177958	0.237525

Density view: percentage of civilian labor, unemployment, uninsured adults, poverty rates, and uninsured children exhibit normal distribution. Female density graph (demo_pct_female) has shifted right—this is due to a greater number of female populations in the dataset.

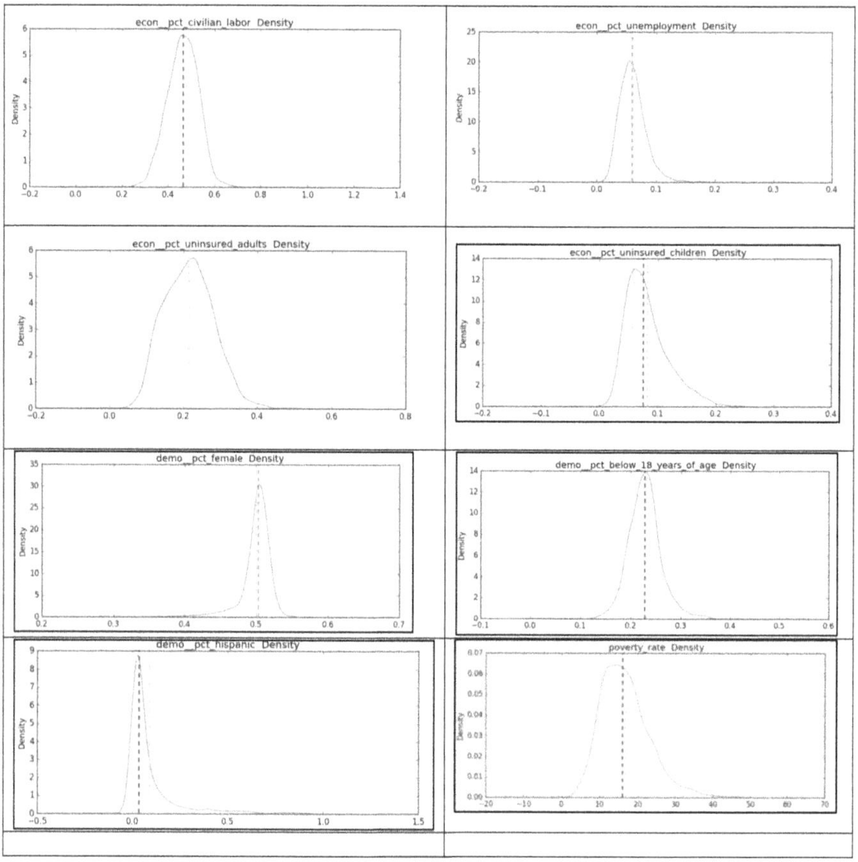

4.5.5 Step 5: Economy Typology

Economic topology affected the poverty rate. The following is a table of median poverty rate given the economic topology:

```
foodSecurityData['econ__economic_typology'].value_counts()
```

```
Nonspecialized                        1207
Manufacturing-dependent                469
Federal/State government-dependent     349
Farm-dependent                         318
Recreation                             284
Mining-dependent                       209
Name: econ__economic_typology, dtype: int64
```

Economic typology[82]: the 2015 County Typology Codes classify all US counties according to six mutually exclusive categories of economic dependence and six overlapping categories of policy-relevant themes. The economic dependencies include farming, mining, manufacturing, Federal/State government, recreation, and nonspecialized counties. The policy-relevant types include low education, low employment, persistent poverty, persistent child poverty, population loss, and retirement destination.

The government-dependent economic topology has the highest median poverty rate. This is consistent with the finding that percentage civilian labor is negatively correlated with poverty rate according to the linear regression model. This suggests that the poverty rate is smaller for counties when more people work in the civilian sector as opposed to the public sector.

Urban Influence Category[83]

The 2013 Urban Influence Codes form a classification scheme that distinguishes metropolitan counties by population size of their metro area and nonmetropolitan counties by the size of the largest city or town and proximity to metro and micropolitan areas. The standard Office of Management and Budget (OMB) metro and nonmetro categories have been subdivided into two metro and ten nonmetro categories, resulting in a 12-part county classification. The following table lists all major influencers.

```
foodSecurityData['area__urban_influence'].value_counts()
```

Small-in a metro area with fewer than one million residents	650
Large-in a metro area with at least one million residents or more	425
Noncore adjacent to a small metro with town of at least 2500 residents	331
Micropolitan adjacent to a small metro area	258
Micropolitan not adjacent to a metro area	225
Noncore adjacent to micro area and contains a town of 2500–19,999 residents	194
Noncore adjacent to a large metro area	150

(continued)

[82]County Typology Codes—https://www.ers.usda.gov/data-products/county-typology-codes/

[83]Urban Influence Codes—https://www.ers.usda.gov/data-products/urban-influence-codes/

Noncore adjacent to micro area and does not contain a town of at least 2500 residents	131
Noncore adjacent to a small metro and does not contain a town of at least 2500 residents	127
Micropolitan adjacent to a large metro area	122
Noncore not adjacent to a metro/micro area and does not contain a town of at least 2500 residents	113
Noncore not adjacent to a metro/micro area and contains a town of 2500 or more residents	110

4.5.6 Step 6: Poverty Rate—Class Variable

The Census Bureau uses a set of money income thresholds that vary by family size and composition to determine who is in poverty. If a family's total income is less than the family's threshold, then that family and every individual in it are considered in poverty. The official poverty thresholds do not vary geographically, but they are updated for inflation using the Consumer Price Index (CPI-U)[84]. Here are the rates[85]:

- $13,465 for an individual under 65 years of age
- $12,413 for those 65 years or older
- $20,591 for a family of three (two adults and one child)
- $31,417 for a family of five (two adults and three children)

```
sns.distplot(foodSecurityData['poverty_rate'],
     label="Skewness : %.2f"%(foodSecurityData['poverty_rate']\
                         .skew())).legend(loc="best")
```

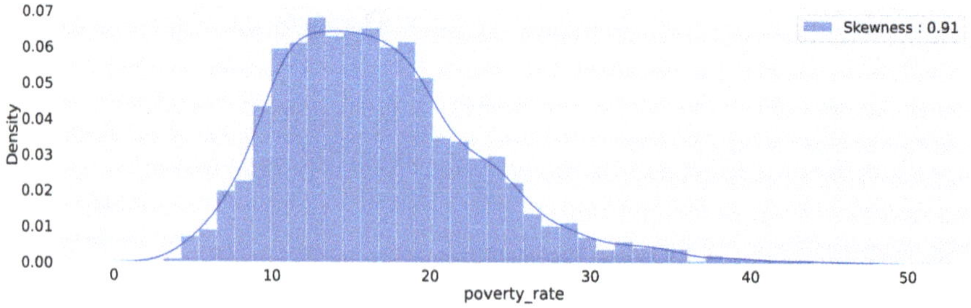

84 How the Census Bureau Measures Poverty—https://www.census.gov/topics/income-poverty/poverty/guidance/poverty-measures.html

85 Poverty Rates—https://www.census.gov/data/tables/time-series/demo/income-poverty/historical-poverty-thresholds.html

4.5.7 Step 7: Correlation Graphs

```
f, axes = plt.subplots(10, 3,figsize=(6*2,20*2))
plt.subplots_adjust( wspace=0.35, hspace=0.35)

for i,ax in zip(numerical_features,axes.flat):

    ax.scatter(x = foodSecurityData[i], y = foodSecurityData
['poverty_rate'],s=1.5,color=['red'])
    ax.set_xlabel(i,size=5.5*2)
    ax.set_ylabel('poverty_rate',size=5.5*2)
    ax.tick_params(labelsize=8)
```

As can be seen below the graph:

- Poverty rates are positively correlated (68.94%) with adults with less than a high school diploma. In fact, adults with less than a high school diploma are the major contributor of poverty.

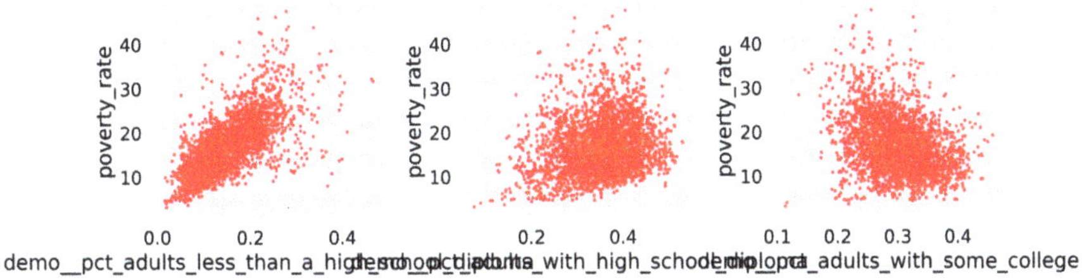

- Poverty rates are positively correlated (61.25%) with unemployment rate. That is, the more the unemployment rates, the higher the poverty rates.
- Poverty rates are positively correlated (56.45%) with uninsured adults.

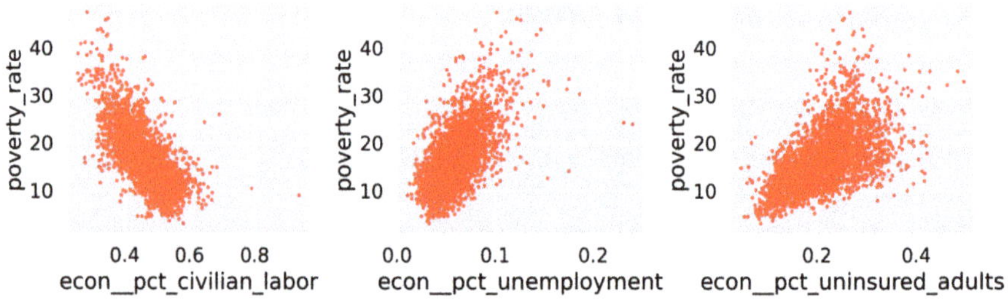

- Poverty rates are positively correlated (13.21%) with uninsured children.
- Poverty rates are negatively correlated (−3.6%) with percentage of female.

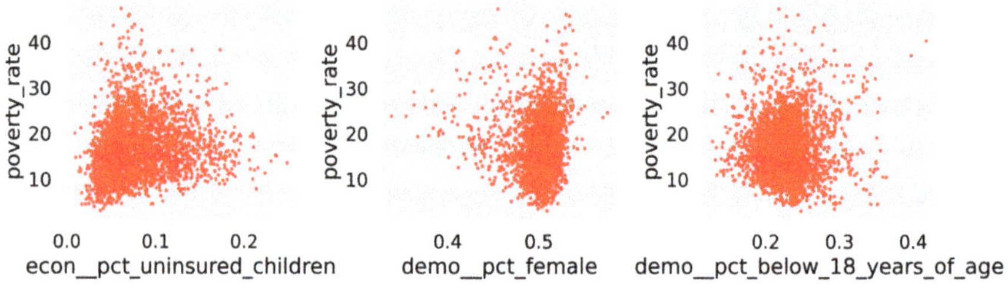

4.5.8 Step 8: Correlation Graphs (Numerically) Computed

```
temp=foodSecurityData.corr().loc[['poverty_rate']]
temp.transpose().sort_values(by='poverty_rate',
ascending=False)
```

	poverty_rate
poverty_rate	1.000000
demo__pct_adults_less_than_a_high_school_diploma	0.689451
econ__pct_unemployment	0.612574
econ__pct_uninsured_adults	0.564513
health__pct_diabetes	0.550976
health__pct_low_birthweight	0.546240
demo__pct_non_hispanic_african_american	0.502056
health__pct_physical_inacticity	0.464225
health__pct_adult_obesity	0.455410
health__homicides_per_100k	0.405997
health__motor_vehicle_crash_deaths_per_100k	0.398524
demo__death_rate_per_1k	0.282489
health__pop_per_dentist	0.264164
health__pct_adult_smoking	0.243161
demo__pct_american_indian_or_alaskan_native	0.227126
demo__pct_adults_with_high_school_diploma	0.196028
health__pop_per_primary_care_physician	0.157423
econ__pct_uninsured_children	0.132193
demo__pct_hispanic	0.098439
health__air_pollution_particulate_matter	0.040138
demo__pct_below_18_years_of_age	0.024054
row_id	-0.004722
demo__pct_female	-0.036073
demo__pct_aged_65_years_and_older	-0.062110
demo__pct_asian	-0.170318
demo__pct_adults_with_some_college	-0.328941
health__pct_excessive_drinking	-0.347744
demo__pct_adults_bachelors_or_higher	-0.473552
demo__pct_non_hispanic_white	-0.483955
econ__pct_civilian_labor	-0.697299

4.5.9 Step 9: Categorical Variables and Plots

For categorical values, check data features.

```
# For categorical features, we would plot violinplot.
sns.set(font_scale = 1.6)
g=sns.violinplot(x='econ__economic_typology',
y='poverty_rate',hue='yr',data=foodSecurityData)
g1=g.set_xticklabels(g.get_xticklabels(),rotation=90)
```

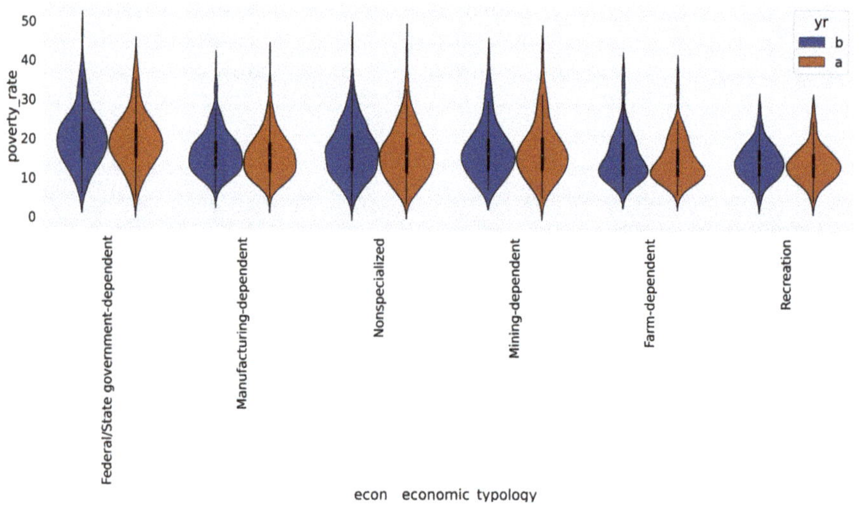

4.5.10 Step 10: Split Data

Split the data into training and testing data.

```
x=data.drop('poverty_rate',axis=1)
y=data[['poverty_rate']]
from sklearn.model_selection import train_test_split
x_train,x_test,y_train,y_test=train_test_split(x,y,
test_size=0.3)
```

4.5.11 Step 11: Machine Learning Regressive Model

```
from sklearn.pipeline import make_pipeline
from sklearn.feature_selection import SelectKBest
from sklearn.model_selection import StratifiedKFold
from sklearn.model_selection import GridSearchCV
from sklearn.model_selection import cross_val_score
from sklearn.feature_selection import SelectFromModel
from sklearn.ensemble import RandomForestRegressor,
AdaBoostRegressor, ExtraTreesRegressor,
GradientBoostingRegressor
```

(continued)

```
from sklearn.linear_model import LinearRegression,LassoCV,
Ridge, LassoLarsCV,ElasticNetCV
import xgboost as xgb
import lightgbm as lgb
from sklearn.metrics import mean_squared_error

def compute_score(model,x,y,
scoring='neg_mean_squared_error'):
  y=y.values.ravel()
  xval=cross_val_score(model,x,y,cv=5,scoring=scoring)
  return xval
```

Create random forest regressor:

```
#turn run_gs to True if you want to run the gridsearch again.
run_gs= False

if run_gs:
  parameter_grid={
        'max_depth' : [40],
        'n_estimators': [200,250],
        'max_features': ['sqrt', 'auto', 'log2'],
        'min_samples_split': [2],
        'min_samples_leaf': [1],
        'bootstrap': [True, False],}

  forest=RandomForestRegressor()
  cross_validation=StratifiedKFold(n_splits=5)

  grid_search=GridSearchCV(forest,
            scoring='neg_mean_squared_error',
            param_grid=parameter_grid,
            cv=cross_validation,
            verbose=1)
  grid_search.fit(x_train,y_train.values.ravel())

  model= grid_search
  parameters=grid_search.best_params_

  print('Best score: {}'.format(grid_search.best_score_))
  print('Best parameters: {}'.format(grid_search.best_params_))

else:
  parameters = {'bootstrap': False, 'max_depth':
40, 'max_features': 'sqrt', 'min_samples_leaf':
1, 'min_samples_split': 2, 'n_estimators': 200}
```

(continued)

```
RFG_model=RandomForestRegressor(**parameters)
RFG_model.fit(x_train,y_train)

print(compute_score(RFG_model,x_train,y_train))
```

Output:
[-0.00275396 -0.00342545 -0.00290793 -0.00304581 -0.00340464]
R^2 score:

```
r2_score(y_test,y_pred)
```

Output: 0.8390685259704225
Mean squared error:

```
mean_squared_error(y_test,RFG_model.predict(x_test))
```

Output: 0.0009572003811444977

4.5.12 Step 12: Feature Importance

```
features = pd.DataFrame()
features['feature'] = x_train.columns
features['importance'] = RFG_model.feature_importances_
features.sort_values(by=['importance'], ascending=True,
inplace=True)
features=features.iloc[-10:,:]
f, ax = plt.subplots(1, 1,figsize=(5,5))
features.set_index('feature', inplace=True)
features.plot(kind='barh', ax=ax,fontsize=5)
ax.legend(prop={'size': 6})
RFG_top10=features.index.tolist()
```

The feature importance, as predicted by the random forest regressor, lists civilian labor as the major influencer. Next, adults with less than a high school diploma and unemployment are listed as the major causes of poverty.

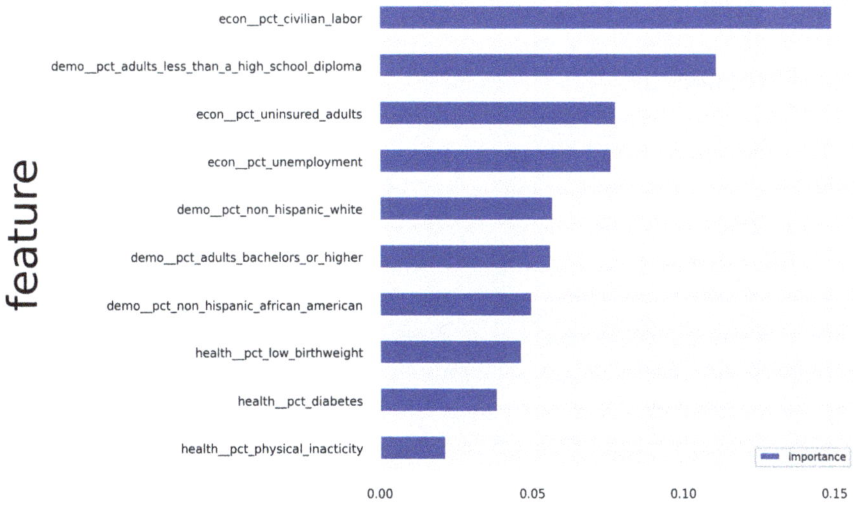

There are many features that correlate with the poverty rate in a US county, but the ones that standout are the following:

- Percent civilian labor—civilian labor force, annual average, as percent of population. Poverty rate is negatively correlated with civilian labor percentage.
- Percent unemployment—unemployment, annual average, as percent of population. Poverty rate is positively correlated with unemployment.
- Percent uninsured adults—percent of adults without health insurance. Poverty rate is positively correlated with percentage of uninsured adults.
- Percent uninsured children—percent of children without health insurance. Poverty rate is negatively correlated with percentage of uninsured children.
- Percent below 18 years of age—percent of population that is below 18 years of age. Poverty rate is negatively correlated with percentage of population below 18 years of age.
- Percent non-Hispanic African American—percent of population that identifies as African American. Poverty rate is positively correlated with percentage of non-Hispanic African Americans.
- Percent American Indian or Alaskan Native—percent of population that identifies as Native American. Poverty rate is positively correlated with percentage of American Indian or Alaskan Native.
- Percent adults with less than a high school diploma—percent of adult population that does not have a high school diploma. Poverty rate is positively correlated with percentage of adults with less than a high school diploma.
- Percent diabetes—percent of population with diabetes. Poverty rate is negatively correlated with percentage of population with diabetes.

4.5.13 Step 13: XGB Regressor

Create XGB regressor.

```
run_gs= False

if run_gs:
  parameter_grid={

        'max_depth' : [20],
        'min_child_weight': [4,6,8],
        'gamma': [0.3,0.5,0.7 ],
                  }

  forest=xgb.XGBRegressor()
  cross_validation=StratifiedKFold(n_splits=5)

  grid_search=GridSearchCV(forest,
             scoring='neg_mean_squared_error',
             param_grid=parameter_grid,
             cv=cross_validation,
             verbose=1)

  grid_search.fit(x_train,y_train.values.ravel())
  model= grid_search

  parameters=grid_search.best_params_

  print('Best score: {}'.format(grid_search.best_score_))
 print('Best parameters: {}'.format(grid_search.best_params_))

else:
  parameters = {'gamma': 0.5, 'max_depth':
20, 'min_child_weight': 6}
  xgb_model=xgb.XGBRegressor()
  xgb_model.fit(x_train,y_train)
```

R^2 score:

```
r2_score(y_test,y_pred)
```

Output: 0.7936 or 79.36%
Mean squared error:

```
mean_squared_error(y_test,xgb_model.predict(x_test))
```

Output: 0.003200374170111798

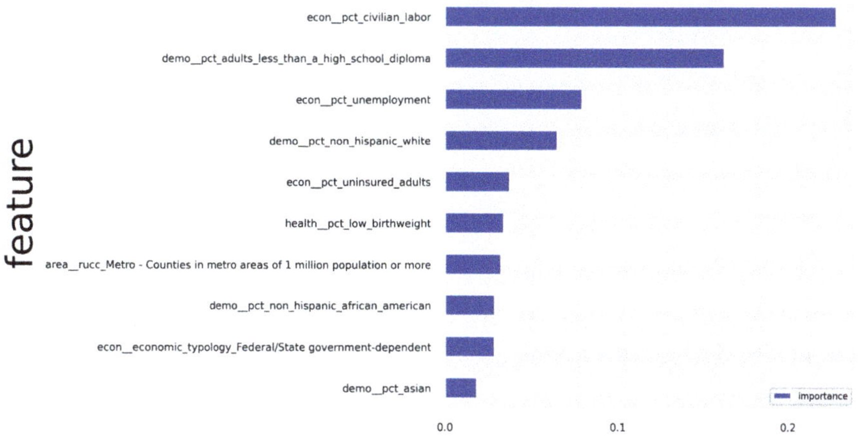

4.5.14 *Agricultural Import Tariff*

Agricultural import tariffs: Tariffs are the largest source of global economic costs generated from agricultural policy distortions. Countries apply tariffs to protect domestic industries against price competition from imports. Tariffs are higher on agricultural products than they are on non-agricultural goods in more than 90 percent of countries. That is because in some regions, agricultural products are often deemed "sensitive," meaning they are important for national security reasons (please see Fig. 4.13).

Agricultural tariff has a negative correlation with overall food security. That is, increase in agricultural tariffs reduces the food security by 29.9%.

$$\text{Over all Food Security (Y)} = -0.2991 \times \text{Agricultural Tarif} + 65.444 \quad (2)$$

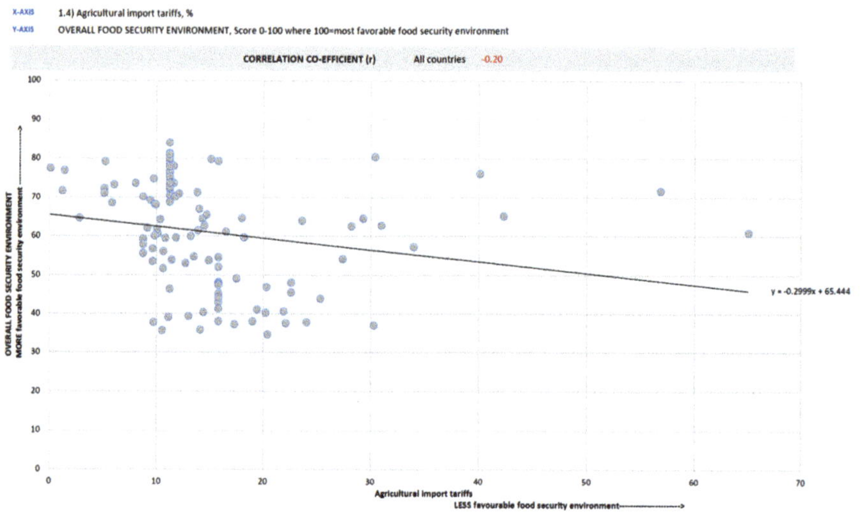

Fig. 4.13 GFSI—agricultural tariff

4.5.15 Market Access and Agricultural Financial Services

Inadequate food and nutrition affect human well-being, particularly for many poor subpopulations living in rural areas. Market access and agricultural financial services improve food access. Market access influences dietary diversity and food security for smallholder households in many ways. In Ethiopia, most smallholders are subsistence farmers who have poor access to markets. This study used primary data from a household survey to examine the relationship between market access and the dietary diversity and food security for 324 smallholder households in the Yayu area of southwestern Ethiopia in early 2018. Multivariate regression analysis showed that households located far from market centers consumed not only less diverse foods but also spend less on food consumption than households located close to market centers[86] [34].

A study conducted in Ecuador with a size of 383 surveys[87] was determined by a stratified random sampling method with proportional affixation. Dietary diversity was measured through the Household Dietary Diversity Score (HDDS), with 12 food groups (cereals; roots and tubers; fruits; sugar/honey; meat and eggs; legumes or grains; vegetables; oils/fats; milk and dairy products; meats; miscellaneous; fish and shellfish) over a recall period of 7 days. A Poisson regression model was used to determine the relationship between the HDDS and sociodemographic variables. The

[86]Does market access improve dietary diversity and food security—https://agrifoodecon.springeropen.com/articles/10.1186/s40100-021-00190-8#Sec14

[87]Factors That Determine the Dietary Diversity Score in Rural Households: The Case of the Paute River Basin of Azuay Province—https://www.ncbi.nlm.nih.gov/pmc/articles/PMC7923421/

results show that the average HDDS of food consumption is 10.89 foods. Of the analyzed food groups, the most consumed are cereals; roots and tubers; fruits; and sugar/honey. In addition, the determinants that best explain the HDDS in the predictive model were housing size, household size, per capita food expenditure, area of cultivated land, level of education, and marital status of the head of household [35].

Dietary diversity could be increased by participating to food markets. Research has examined relationships between food market participation and household dietary diversity in populations of rural Malawi facing hunger and poor nutrition. Using Poisson regression and survey data from 400 households in 2 districts of rural Malawi in post-harvest and lean seasons of 2017/18, it was clearly established associations between food purchase diversity and household dietary diversity that households engaging more with food markets are more likely to have diversified diets and better nutrition[88].

Consider the Household Dietary Diversity Score (HDDS)[89] which contains data crowdsourced daily from Venezuelans using the Premise data mobile application. The data collected allows the fast measurement of Household Dietary Diversity Score (HDDS), with the goal of providing context around the food security in vulnerable communities [36].

4.6 Machine Learning Model and Linear Programming: Fertilizer Price Prediction Using Commodity (Rice, Sorghum, Maize, and Wheat) and Oil Prices

The ML model to predict fertilizer cost is based on the below demand and supply components[90], and linear programming incorporates objective functions based on the ML model [37].

- Demand components
  - The demand for corn (as demand goes high, so is the need to produce more and hence to apply more fertilizers to get optimal yield)
  - Demand for wheat
  - Demand for rice
  - Demand for soybeans
- Supply: crude (on supply side)

[88] Does household participation in food markets increase dietary diversity? Evidence from rural Malawi—https://www.sciencedirect.com/science/article/pii/S2211912420301395

[89] HDD—https://data.humdata.org/dataset/open_daily_food

[90] Predicting Fertilizer Prices—https://www.agmanager.info/sites/default/files/pdf/FertilizerProjection_2021-2022.pdf

- On the supply side, consider the cost of crude.
- Natural gas is complementing crude—the same behavior could be expected from demand swings of natural gas

4.6.1 Data Sources

Data retrieved from the World Bank Commodity Market's Pink Sheet data that gets published every month[91].

	Software code for this model can be found on GitHub Link: UREA_EE_BULK_Fertilizer_PricePredict.Ipynb (Jupyter Notebook)

4.6.2 Step 1: Load Required Libraries

The first step in predicting urea fertilizer cost is to load all libraries and Pink Sheet data.

```
# Importing data

CMOHistoricalDataDF = pd.read_csv("CMO-Historical-Data-
MonthlyRegressionFinal.csv")

CMOHistoricalDataDF
```

[91] Pink Sheet Data—https://www.worldbank.org/en/research/commodity-markets

Output:

	Year	NGAS_US US ($/mmbtu)	SOYBEANS ($/mt)	CRUDE_PETRO ($/bbl)	MAIZE ($/mt)	SORGHUM ($/mt)	RICE_05 ($/mt)	WHEAT_US_SRW ($/mt)	WHEAT_US_HRW ($/mt)	UREA_EE_BULK ($/mt)
0	197901	1.02	284.00	17.45	105.41	97.06	278.52	140.36	137.79	143.00
1	197902	1.05	298.00	20.75	107.01	97.83	279.50	144.77	140.36	143.00
2	197903	1.10	310.00	22.02	109.52	99.21	293.00	144.04	139.99	131.88
3	197904	1.11	300.00	22.43	111.49	98.27	295.46	142.20	139.99	135.00
4	197905	1.15	300.00	33.50	112.57	100.49	296.78	141.43	143.30	141.75
...	...	...	...	...	...	...	...	...	...	...
501	202010	2.25	454.25	39.90	186.75	NaN	471.00	245.20	NaN	245.00
502	202011	2.59	499.98	42.30	190.38	NaN	489.00	247.95	NaN	245.00
503	202012	2.54	510.94	48.73	198.77	NaN	520.00	251.15	NaN	245.00
504	202101	2.67	576.30	53.60	234.47	NaN	545.00	276.45	NaN	265.00
505	202102	5.07	574.80	60.46	245.24	NaN	557.00	276.63	NaN	335.00

The Pink Sheet data is loaded from January 1979 to December 2020.

4.6.3 *Step 2: Statistical Properties of Data*

```
CMOHistoricalDataDF.describe()
```

Output:

	Year	NGAS_US US ($/mmbtu)	SOYBEANS ($/mt)	CRUDE_PETRO ($/bbl)	MAIZE ($/mt)	SORGHUM ($/mt)	RICE_05 ($/mt)	WHEAT_US_SRW ($/mt)	WHEAT_US_HRW ($/mt)	UREA_EE_BULK ($/mt)
count	506.000000	506.000000	506.000000	506.000000	506.000000	500.000000	506.000000	506.000000	498.000000	506.000000
mean	199964.978261	3.233636	322.219960	42.101403	139.801344	134.967520	337.408834	168.935119	180.688574	185.365059
std	1218.367423	2.014272	109.708207	28.549572	54.484496	51.105978	123.794435	57.446569	62.307556	109.216730
min	197901.000000	1.020000	183.000000	9.620000	65.310000	62.530000	163.750000	85.300000	101.780000	62.500000
25%	198907.250000	1.900000	241.000000	18.907500	103.667500	98.935000	251.265000	130.090000	140.005000	103.500000
50%	200001.500000	2.580000	283.500000	30.890000	118.835000	114.470000	303.750000	153.165000	164.240000	155.200000
75%	201007.750000	3.895000	383.560000	58.147500	163.300000	161.930000	418.940000	197.472500	199.865000	242.972500
max	202102.000000	13.520000	684.020000	132.830000	333.050000	302.530000	907.000000	419.610000	439.720000	785.000000

Count of records 506.

We might get a clearer idea of the distribution of production (millions of tonnes) values by visualizing the data. The common plot types for visualizing numeric data distributions are histograms and box plots.

```
import pandas as pd
import matplotlib.pyplot as plt

# This ensures plots are displayed inline in the Jupyter notebook
%matplotlib inline

# Get the label column
label = CMOHistoricalDataDF['UREA_EE_BULK ($/mt)']

# Create a figure for 2 subplots (2 rows, 1 column)
fig, ax = plt.subplots(2, 1, figsize = (9,12))

# Plot the histogram
ax[0].hist(label, bins=100)
ax[0].set_ylabel('Frequency')

# Add lines for the mean, median, and mode
ax[0].axvline(label.mean(), color='magenta',
linestyle='dashed', linewidth=2)
ax[0].axvline(label.median(), color='cyan',
linestyle='dashed', linewidth=2)

# Plot the boxplot
ax[1].boxplot(label, vert=False)
ax[1].set_xlabel('UREA_EE_BULK ($/mt)')

# Add a title to the Figure
fig.suptitle('UREA_EE_BULK ($/mt)')

# Show the figure
fig.show()
```

Output:

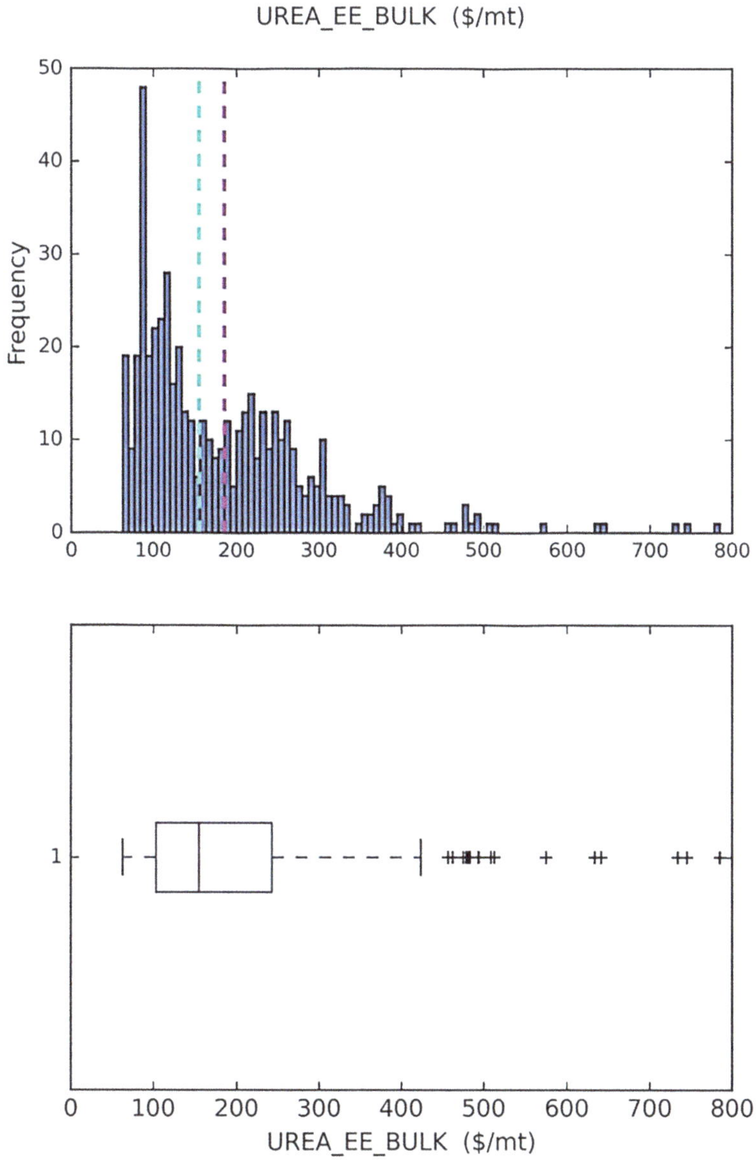

As you can see, the usage of urea is right skewed—evidently, peaks in the production and price of urea are seen in the 2000s and 2010s.

Below are the price movements (please see Fig. 4.14) of urea[92]:

[92] UREA Price Movements—https://www.netcials.com/commodity-monthly-price-history/UREA-EE-BULK/

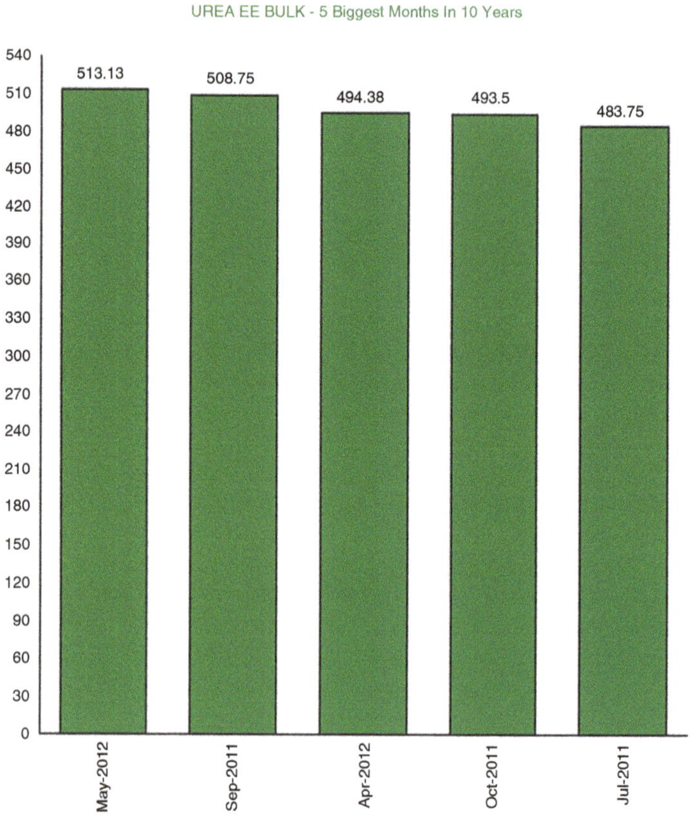

Fig. 4.14 Urea prices

And the following are the best prices for farmers, that is, cost of buying urea is lowest[93]:

Month-year	Price ($)
Jun-2016	142.63
May-2017	178.75
Jul-2016	181
Jul-2017	181
Aug-2016	186.25

Correlation analysis of production to all other features provides the following details:

[93] Lower Prices—https://www.netcials.com/commodity-monthly-price-history/UREA-EE-BULK/

```
plt.figure(figsize=(12, 9))
sns.heatmap(CMOHistoricalDataDF.corr())
```

Output:

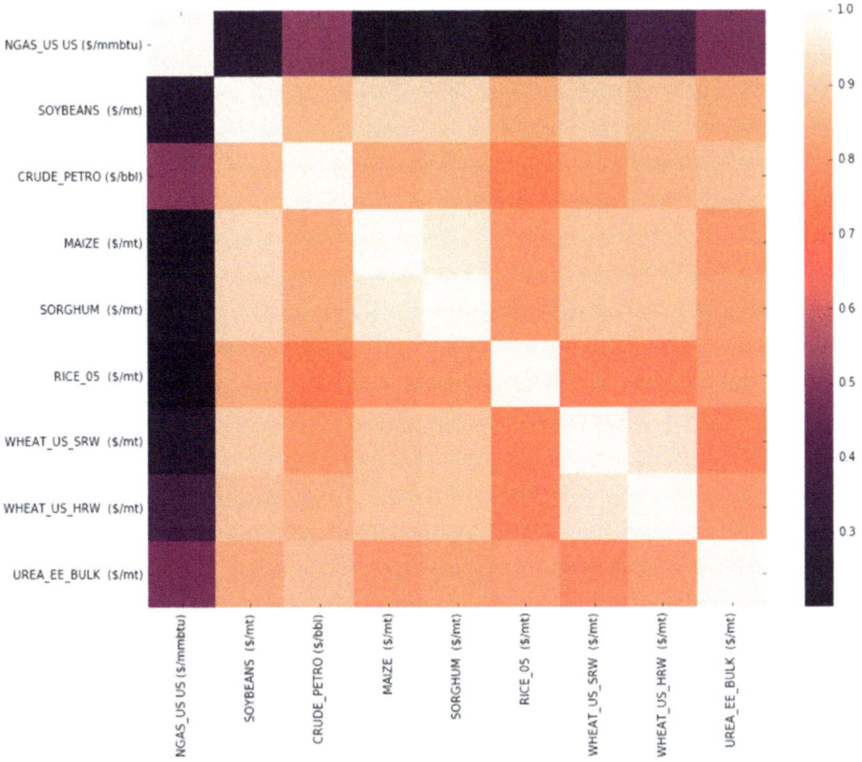

Highly correlated map between urea and commodity prices.
Scatter plot provides the distribution of feature and urea usage:

```
fig = go.Figure()

# Add traces
fig.add_trace(go.Scatter(x=CMOHistoricalDataDF.index,
y=CMOHistoricalDataDF['UREA_EE_BULK ($/mt)'],
          mode='lines',
          name='UREA_EE_BULK ($/mt) '))
fig.update_layout(
  title="UREA_EE_BULK ($/mt)")

fig.show()
```

Output:

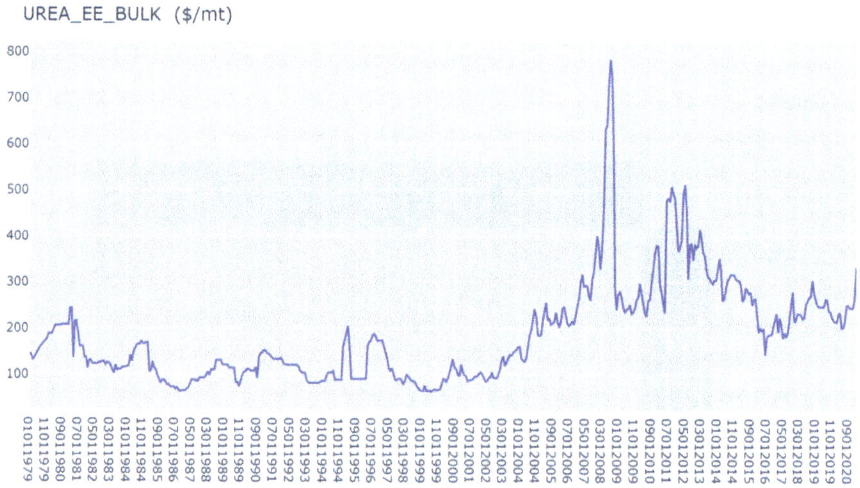

Urea peaked in 2008 at $780 per metric tonnes.

```
fig = go.Figure()

# Add traces
fig.add_trace(go.Scatter(x=CMOHistoricalDataDF.index,
y=CMOHistoricalDataDF['CRUDE_PETRO ($/bbl)'],
         mode='lines',
         name='CRUDE_PETRO ($/bbl)'))
fig.update_layout(
  title='CRUDE_PETRO ($/bbl)')

fig.show()
```

Crude petrol had more fluctuations:

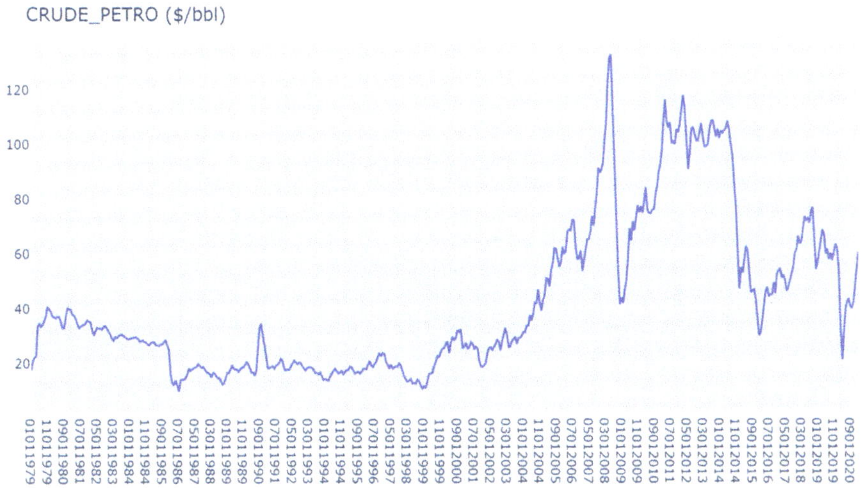


```
fig = go.Figure()

# Add traces
fig.add_trace(go.Scatter(x=CMOHistoricalDataDF.index,
y=CMOHistoricalDataDF['MAIZE ($/mt)'],
          mode='lines',
          name='MAIZE ($/mt)'))
fig.update_layout(
  title="MAIZE ($/mt)")

fig.show()
```

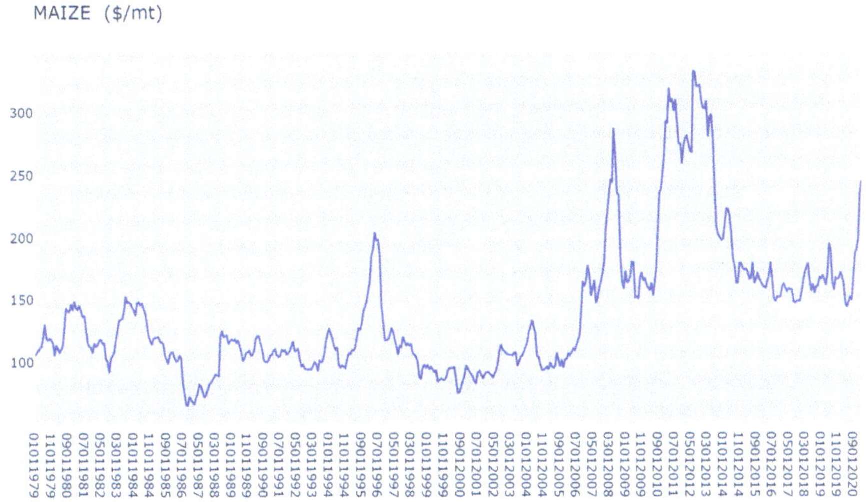

4.6.4 Perform Linear Regression

Load feature and label data frames for performing regression task:

```
X, y = CMOHistoricalDataDF[['SOYBEANS ($/mt)', 'CRUDE_PETRO
($/bbl)',
   'MAIZE ($/mt)', 'SORGHUM ($/mt)', 'RICE_05 ($/mt)',
   'WHEAT_US_SRW ($/mt)', 'WHEAT_US_HRW ($/mt)']].values,
CMOHistoricalDataDF['UREA_EE_BULK ($/mt)'].values
print('Features:',X[:10], '\nLabels:', y[:10], sep='\n')
```

Output:

```
Features:
[[284.   17.45 105.41  97.06 278.52 140.36 137.79]
 [298.   20.75 107.01  97.83 279.5  144.77 140.36]
 [310.   22.02 109.52  99.21 293.   144.04 139.99]
 [300.   22.43 111.49  98.27 295.46 142.2  139.99]
 [300.   33.5  112.57 100.49 296.78 141.43 143.3 ]
 [322.   34.67 120.35 109.46 303.31 163.88 167.92]
 [322.   33.5  130.51 118.9  304.05 168.29 174.53]
 [302.   34.   118.85 111.38 328.43 162.77 169.02]
 [292.   35.5  117.96 112.6  338.17 164.98 174.16]
 [283.   37.25 119.1  116.2  340.38 163.1  178.21]]

Labels:
[143.  143.  131.88 135.  141.75 148.5  155.4  160.  163.75 167.5 ]
```

Perform data distribution analysis.

```
# Create a function that we can re-use
def show_distribution(var_data,colname):
  from matplotlib import pyplot as plt

  # Get statistics
  min_val = var_data.min()
  max_val = var_data.max()
  mean_val = var_data.mean()
  med_val = var_data.median()
  mod_val = var_data.mode()[0]

  print('Minimum:{:.2f}\nMean:{:.2f}\nMedian:{:.2f}\nMode:
{:.2f}\nMaximum:{:.2f}\n'.format(min_val,
```

(continued)

```
                                                            mean_val,
                                                            med_val,
                                                            mod_val,
                                                            max_val))

  # Create a figure for 2 subplots (2 rows, 1 column)
  fig, ax = plt.subplots(2, 1, figsize = (10,4))
  # Plot the histogram
  ax[0].hist(var_data)
  ax[0].set_ylabel('Frequency')
  # Add lines for the mean, median, and mode
  ax[0].axvline(x=min_val, color = 'gray', linestyle='dashed',
linewidth = 2)
  ax[0].axvline(x=mean_val, color = 'cyan', linestyle='dashed',
linewidth = 2)
  ax[0].axvline(x=med_val, color = 'red', linestyle='dashed',
linewidth = 2)
  ax[0].axvline(x=mod_val, color = 'yellow', linestyle='dashed',
linewidth = 2)
  ax[0].axvline(x=max_val, color = 'gray', linestyle='dashed',
linewidth = 2)

  # Plot the boxplot
  ax[1].boxplot(var_data, vert=False)
  ax[1].set_xlabel('Value')

  # Add a title to the Figure
  fig.suptitle(colname + ' Data Distribution')

  # Show the figure
  fig.show()
# Get the variable to examine

for col in ['SOYBEANS ($/mt)', 'CRUDE_PETRO ($/bbl)',
    'MAIZE ($/mt)', 'SORGHUM ($/mt)', 'RICE_05 ($/mt)',
    'WHEAT_US_SRW ($/mt)', 'WHEAT_US_HRW ($/mt)', 'UREA_EE_BULK
($/mt)']:
  show_distribution(CMOHistoricalDataDF[col],col)

# Call the function
```

Output:

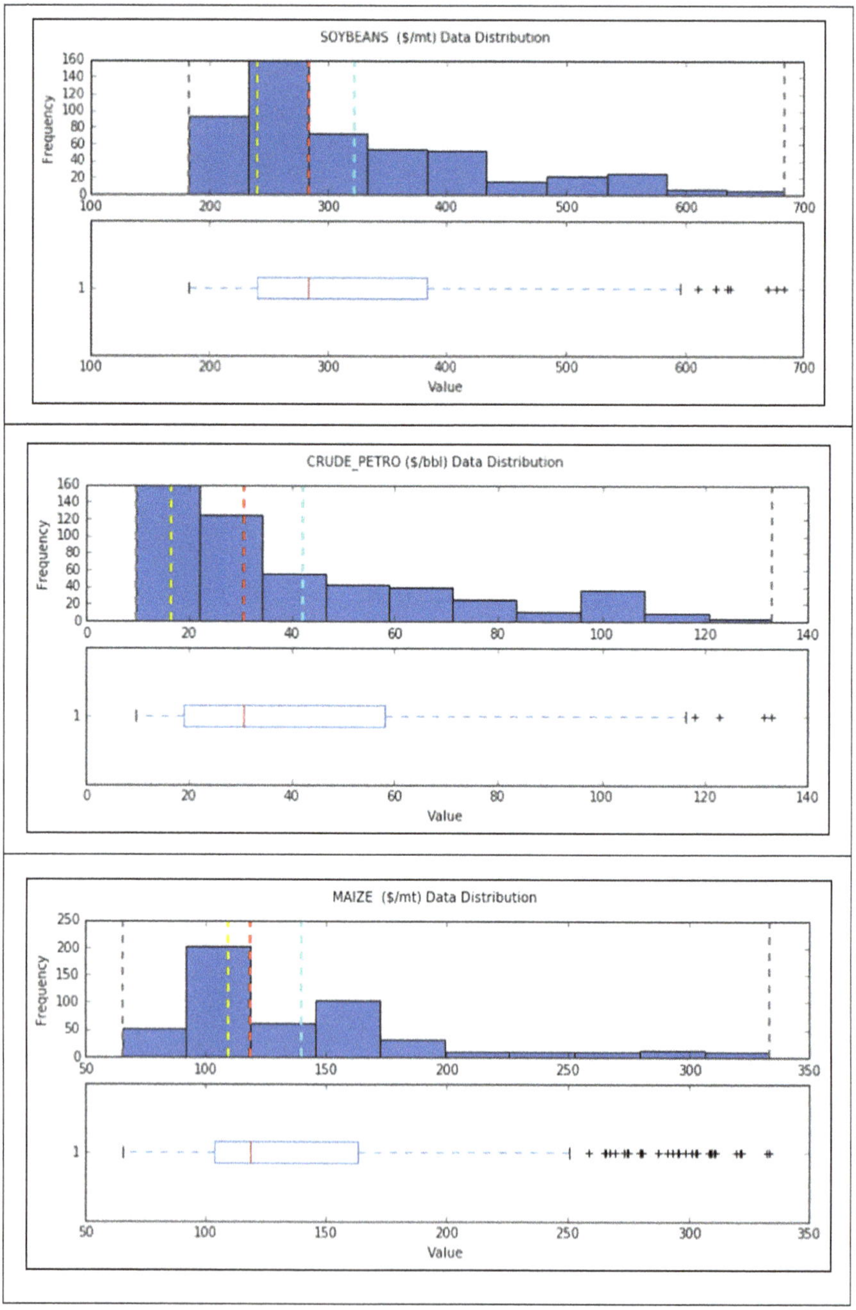
SOYBEANS ($/mt) Data Distribution
Frequency
Value
CRUDE_PETRO ($/bbl) Data Distribution
Frequency
Value
MAIZE ($/mt) Data Distribution
Frequency
Value

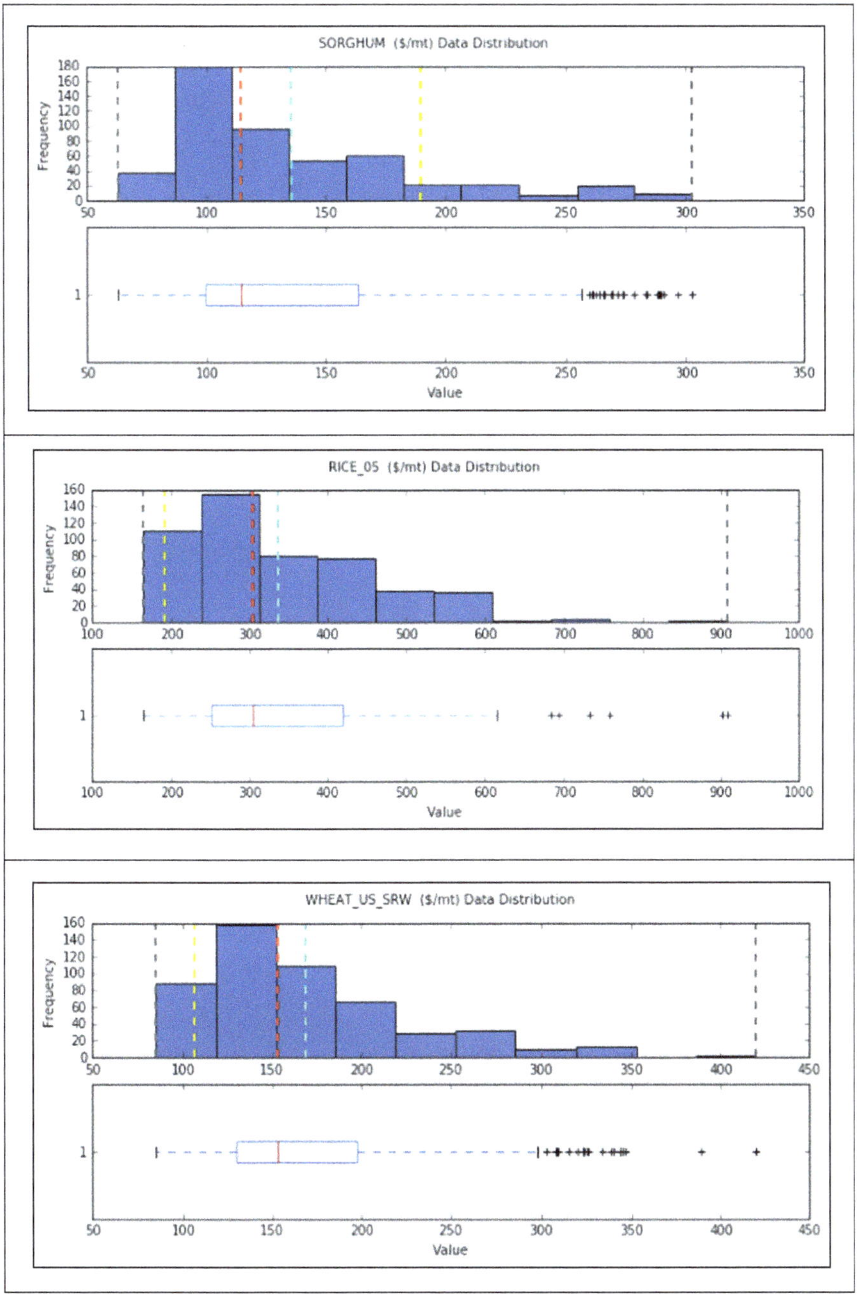
SORGHUM ($/mt) Data Distribution
Frequency
Value
RICE_05 ($/mt) Data Distribution
Frequency
Value
WHEAT_US_SRW ($/mt) Data Distribution
Frequency
Value

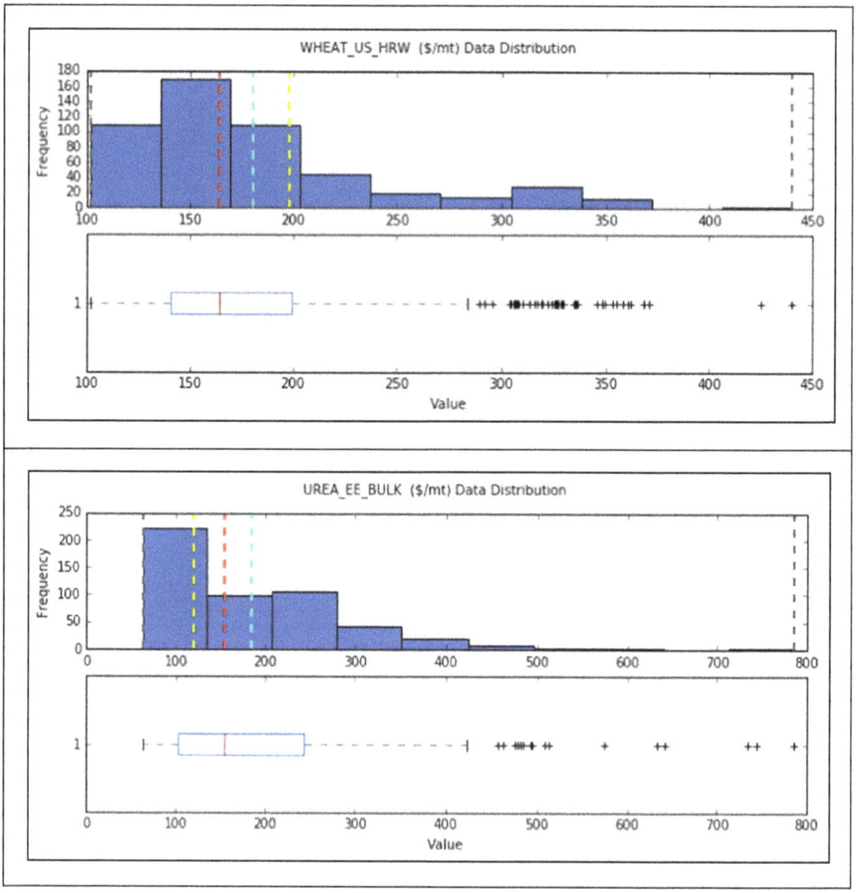

All the feature variables are skewed right of the graph.
Perform distribution analysis:

```
def show_density(var_data,colname):
  from matplotlib import pyplot as plt

  fig = plt.figure(figsize=(10,4))

  # Plot density
  var_data.plot.density()

  # Add titles and labels
  plt.title(colname + ' Density')
```

(continued)

```
  # Show the mean, median, and mode
  plt.axvline(x=var_data.mean(), color = 'cyan',
linestyle='dashed', linewidth = 2)
  plt.axvline(x=var_data.median(), color = 'red',
linestyle='dashed', linewidth = 2)
  plt.axvline(x=var_data.mode()[0], color = 'yellow',
linestyle='dashed', linewidth = 2)

  # Show the figure
  plt.show()

# Get the density of Grade

for col in ['NGAS_US US ($/mmbtu)', 'SOYBEANS ($/mt)',
'CRUDE_PETRO ($/bbl)',
    'MAIZE ($/mt)', 'SORGHUM ($/mt)', 'RICE_05 ($/mt)',
    'WHEAT_US_SRW ($/mt)', 'WHEAT_US_HRW ($/mt)', 'UREA_EE_BULK
($/mt)']:
  show_density(CMOHistoricalDataDF[col],col)
```

Output:

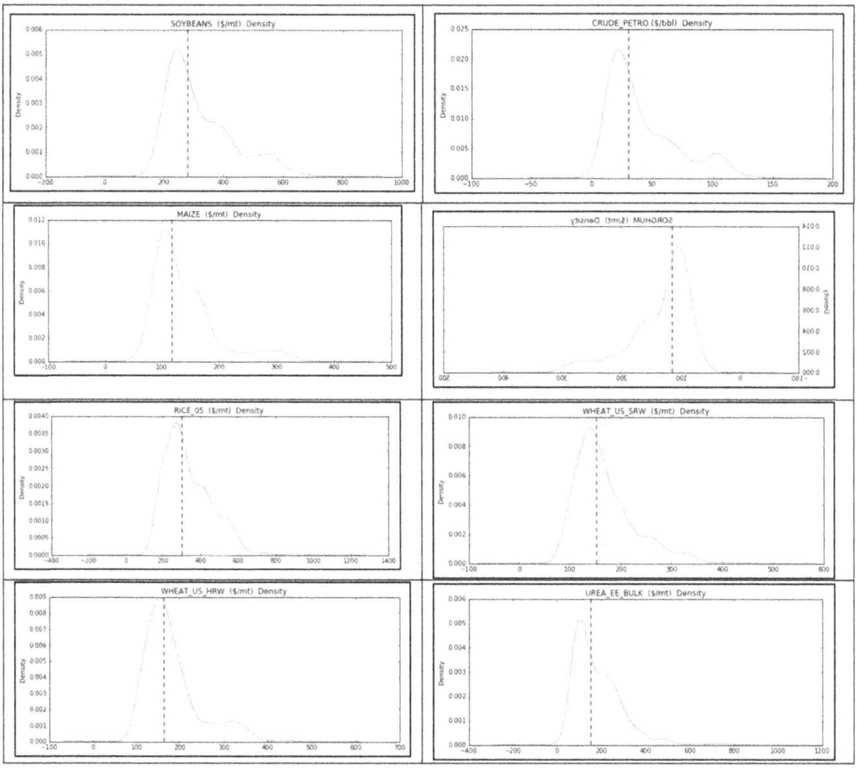

Prepare regression data by train and split:

```
X_train, X_test, y_train, y_test = train_test_split(X, y,
test_size = 0.2,random_state = 0)

regressor = LinearRegression()
regressor.fit(X_train,y_train)

# predicting the test
y_pred = regressor.predict(X_test)
y_pred
```

Plot the prediction:

```
import matplotlib.pyplot as plt

%matplotlib inline

plt.scatter(y_test, y_pred)
plt.xlabel('Actual Labels')
plt.ylabel('Predicted Labels')
plt.title('Milk Production (in Million Tonnes)')
# overlay the regression line
z = np.polyfit(y_test, y_pred, 1)
p = np.poly1d(z)
plt.plot(y_test,p(y_test), color='magenta')
plt.show()
```

Output:

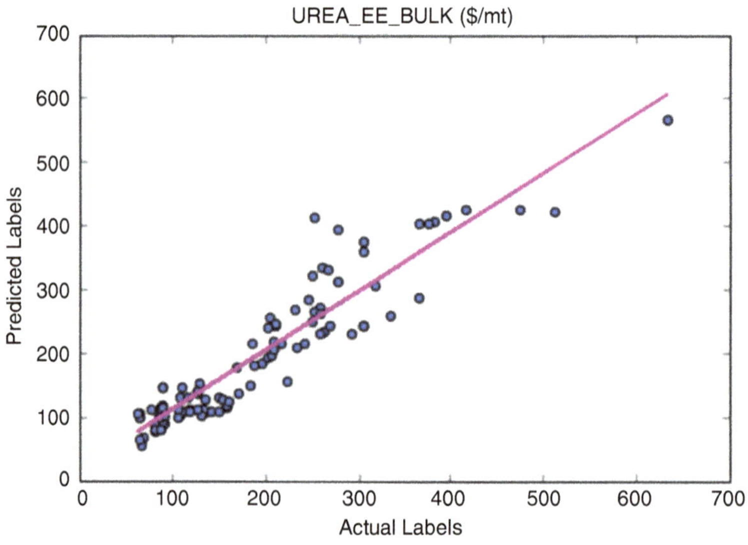

As you can see, the actual and predicted values follow the linear prediction curve. Perform prediction to actual summary:

```
pred_y_df = pd.DataFrame({'Actual Value':y_test, 'Predicted
value':y_pred,'Difference': y_test-y_pred})
pred_y_df
```

Output:

	Actual Value	Predicted value	Difference
0	210.00	245.005245	-35.005245
1	262.00	333.660915	-71.660915
2	135.00	126.877903	8.122097
3	382.50	408.368254	-25.868254
4	109.50	102.039228	7.460772
...	...	...	...
97	110.50	147.619785	-37.119785
98	217.75	216.860138	0.889862
99	116.50	130.170142	-13.670142
100	172.25	136.048094	36.201906
101	92.50	104.190909	-11.690909

102 rows × 3 columns

Evaluate intercept values:

```
'''
'SOYBEANS ($/mt)', 'CRUDE_PETRO ($/bbl)',
   'MAIZE ($/mt)', 'SORGHUM ($/mt)', 'RICE_05 ($/mt)',
   'WHEAT_US_SRW ($/mt)', 'WHEAT_US_HRW ($/mt)'
   '''
regressor.coef_
```

Output:

```
array([0.0280269, 2.09727059, -0.11114963, 0.37699977,
0.29655391, -0.46832559, 0.34333731])
```

Contribution?	Feature
+82.887	RICE_05 ($/mt)
+48.191	WHEAT_US_HRW ($/mt)
+43.518	CRUDE_PETRO ($/bbl)
+36.882	SORGHUM ($/mt)
+8.352	SOYBEANS ($/mt)
-11.894	MAIZE ($/mt)
-29.767	<BIAS>
-67.799	WHEAT_US_SRW ($/mt)

Fig. 4.15 Model explainability

Please note: Except for MAIZE and WHEAT_US_SRW, other coefficients are positive, implies any change in the positive demands will have increase in price for urea.

Check model score:

```
from sklearn.metrics import mean_squared_error, r2_score

mse = mean_squared_error(y_test, y_pred)
print("MSE:", mse)

rmse = np.sqrt(mse)
print("RMSE:", rmse)

r2 = r2_score(y_test, y_pred)
print("R2:", r2)
```

Output:

```
MSE: 1564.520028729579
   RMSE: 39.554014065952636
   R2: 0.8669524827821189
```

Perform explainability of the model (please see Figs. 4.15 and 4.16):

Following are the major global and local influencers when explainability is performed: CRUDE PETRO, RICE_05, WHEAT_US_SRW, and WHEAT_US_HRW. Rice fields have a huge influence on urea usage[94] [37].

[94]Urea fertilization and the N-cycle of rice-fields in the Camargue (S. France)—https://link.springer.com/article/10.1023/A:1003247327586#citeas

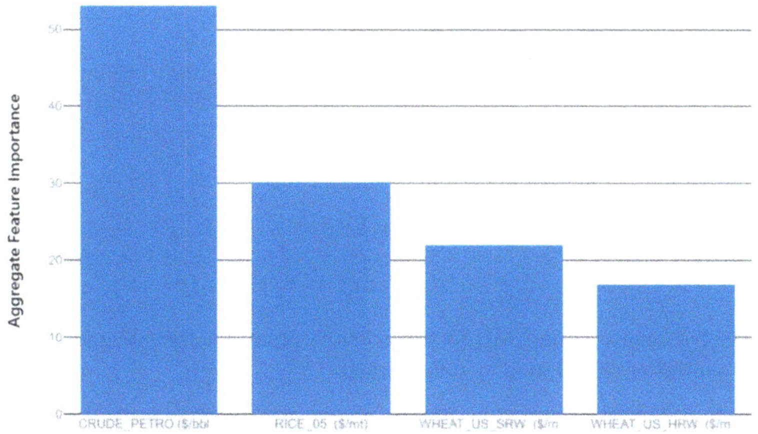

Fig. 4.16 Model explainability

4.6.5 *What-If Analysis*

Let's look how the following what-if analysis would help small farmers:

4.6.5.1 Demand Spike for Crude Oil

What if the demand for crude oil goes up to $150.29 $/bbl in the next 3 months? Let us input increase in demand for major commodities and predict the price of urea:

```
WIF_X_test = [[267.  , 150.29, 109.4, 111.88, 512.5, 140.14,
195.16],
    [449.35, 74.88, 164.56, 166.35, 591.  , 206.5 , 206.25],
    [338.  , 17.88, 124.4 , 118.  , 304.5 , 158.62, 183.49],
    [684.02, 105.27, 331.99, 273.34, 567.75, 333.73, 349.4 ],
    [223.  , 26.38, 114.91,  99.01, 191.12, 117.77, 128.6 ],
    [252.  , 36.58, 108.2 , 114.5 , 395.97, 155.8 , 156.53],
    [360.93, 57.67, 163.59, 147.93, 430.  , 197.52, 181.15],
    [334.  , 65.16, 160.23, 150.01, 317.6 , 180.78, 195.72]]
```

Output: urea cost goes up to $476.29.

```
array([476.29221529, 333.66091462, 126.87790326, 408.36825409,
    102.03922783, 183.35776574, 236.09484681, 231.71579693])
```

The cost would go up to $476.

4.6.5.2 Demand Goes Down

What if the demand for crude oil goes down in next 3 months? The urea price goes up to $ 224.61.

Output:

```
array([224.61974434, 333.66091462, 126.87790326, 408.36825409,
    102.03922783, 183.35776574, 236.09484681, 231.71579693])
```

The advantage of using ML to predict urea price is easy integration with the market and faster closed loop to small farmers.

> Able to predict fertilizer price using machine learning and Artificial Intelligence will save the lives of small farmers in the developing countries. Having the ability to analyze the historical trends in price data of oil, staple crops, and major fertilizers would provide ML model all the avenue of predicting the price of fertilizer in a lead of one or two agricultural calendars. This is the easy part! Disseminating these insights on a timely basis and enabling small farmers to get these details and, importantly, having these insights embedded into policy makers' manual of governance are the most difficult part, the ones that require help from human counterparts.

2)	Availability[95]
2.1)	Sufficiency of supply
2.1.1)	Food supply adequacy
2.1.2)	Dependency on chronic food aid
2.2)	Agricultural research and development
2.2.1)	Public expenditure on agricultural research and development
2.1.2)	Access to agricultural technology, education, and resources
2.3)	Agricultural infrastructure
2.3.1)	Crop storage facilities
2.3.2)	Road infrastructure
2.3.3)	Air, port, and rail infrastructure
2.3.4)	Irrigation infrastructure
2.4)	Volatility of agricultural production
2.5)	Political and social barriers to access
2.5.1)	Armed conflict
2.5.2)	Political stability risk
2.5.3)	Corruption
2.5.4)	Gender inequality
2.6)	Food loss
2.7)	Food security and access policy commitments
2.7.1)	Food security strategy

[95] The GFSI Methodology—https://my.corteva.com/GFSI?file=ei21_gfsir

Agricultural research and development: Research and innovation is a major driver for increasing the availability of food[96]. Its role in ensuring food security, particularly concerning genetic improvement and biotechnologies, is evaluated. Food security can be improved thanks to research and innovation, for example, for strengthening the adaptation to climate change and improving the tolerances of crops and livestock to stresses, drought, and floods. Innovation needs to be accessible to farmers and adapted to their needs.

Agricultural infrastructure: In low-income economies and less industrialized countries, food security relies on small-scale farming systems in a dual structure. A description of the agricultural potential of key areas such as the Black Sea area or sub-Saharan Africa is carried out, as well as specific farm level modeling. At farm level, developing countries are characterized by specific features such a small scale and a household subsistence character of producers[97]: therefore, specific models need to be developed to capture in low-income economies and at farm level, and from there at regional or national level, the impacts of different policy options—penetration of ICT.

There is a need for mathematical modeling to help calculate which foods can supply the optimum nutrient recommendations for low cost, especially for income-strapped households and authorities, e.g., catering services within the public sector. The method of LP has been used to optimize the average daily nutrient intake of children and adults since the 1950s.

4.7 California Health and Human Services (CHHS) Food Affordability Predictive Model

An adequate, nutritious diet is a necessity at all stages of life. Inadequate diets can impair intellectual performance and have been linked to more frequent school absence and poorer educational achievement in children. Nutrition also plays a significant role in causing or preventing several illnesses, such as cardiovascular disease, some cancers, obesity, type 2 diabetes, and anemia.

At least two factors influence the affordability of food and the dietary choices of families—*the cost of food and family income.* The inability to afford food is a major factor in food insecurity (please see Fig. 4.17), which has a spectrum of effects including anxiety over food sufficiency or food shortages, reduced quality or desirability of diet, disrupted eating patterns, and reduced food intake. This dataset

[96] Global food security—https://ec.europa.eu/jrc/en/research-topic/global-food-security

[97] Food security and agriculture in low income and transition economies—https://ec.europa.eu/jrc/en/research-topic/global-food-security

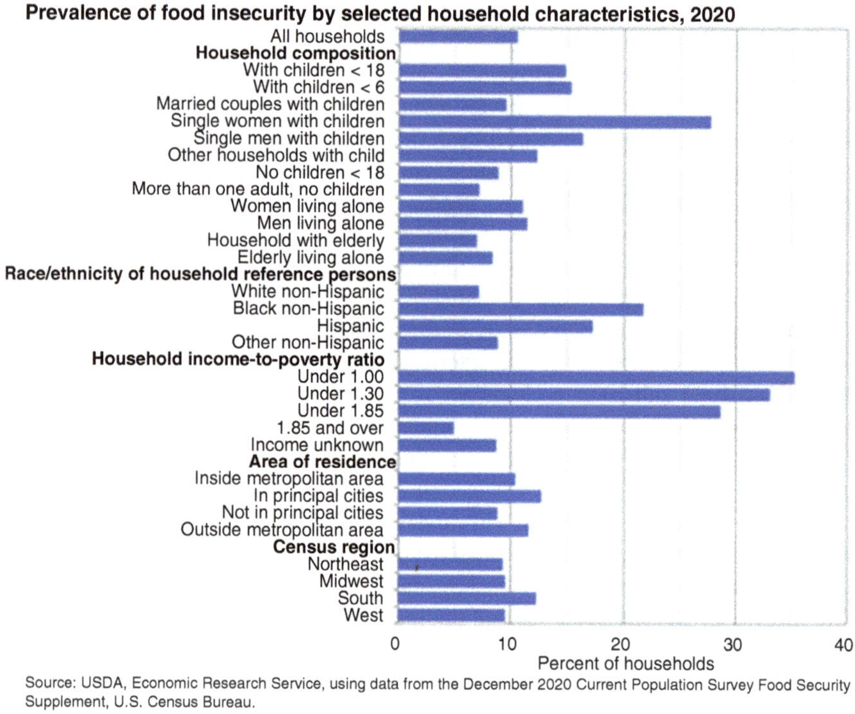

Fig. 4.17 Prevalence of food insecurity

covers the data from 2006 to 2010; "Indicator ID" that has been used is "757" (food affordability for female-headed household with children under 18 years)[98] [38].

This table contains data on the average cost of a market basket of nutritious food items relative to income for female-headed households with children, for California, its regions, counties, and cities/towns. The ratio uses data from the US Department of Agriculture and the US Census Bureau. The table is part of a series of indicators in the Healthy Communities Data and Indicators Project of the Office of Health Equity. An adequate, nutritious diet is a necessity at all stages of life. At least two factors influence the affordability of food and the dietary choices of families—the cost of food and family income. The inability to afford food is a major factor in food insecurity, which has a spectrum of effects including anxiety over food sufficiency or food shortages; reduced quality or desirability of diet; and disrupted eating patterns and reduced food intake. More information about the data table and a data dictionary can be found in the Attachments.

[98] Food Insecurity by Household Characteristics—https://www.ers.usda.gov/topics/food-nutrition-assistance/food-security-in-the-us/key-statistics-graphics.aspx

4.7.1 Dataset: California Health and Human Services Food Affordability[99]

Company: Department of Health Care Services[100] (DHCS) and California Department of Public Health (CDPH)[101]

Software code for this model can be found on GitHub Link: CHHS_Food_Affordability_PredictiveModel.ipynb (Jupyter Notebook Code)

4.7.2 Food Insecurity by Household Characteristics [38]

The prevalence of food insecurity varied considerably among household types. Rates of food insecurity were higher than the national average (10.5 percent) for the following groups (please see Table 4.1):

- All households with children (14.8 percent)
- Households with children under age 6 (15.3 percent)
- Households with children headed by a single woman (27.7 percent) or a single man (16.3 percent)
- Households with Black, non-Hispanic (21.7 percent), and Hispanic reference persons (17.2 percent; a household reference person is an adult household member in whose name the housing unit is owned or rented)
- Households with incomes below 185 percent of the poverty threshold (28.6 percent; the Federal poverty line was $26,246 for a family of four in 2020)

[99] CHHS Model—https://data.chhs.ca.gov/dataset/food-affordability-2006-2010
[100] DEPARTMENT OF HEALTH CARE SERVICES—https://www.dhcs.ca.gov/
[101] California Department of Public Health (CDPH)—https://www.cdph.ca.gov/

Table 4.1 food_afford_cdp_co_region_ca4-14-13-ada.csv. Data Dictionary—https://data.chhs.ca.gov/dataset/food-affordability-2006-2010/resource/da6d5b83-bbde-4192-92cb-e575fc6417ca

Name	Description	Type	Constraint
Race_Ethnicity_Code	Code for a race/ethnicity group: 1 = American Indian or Alaska Native (AIAN), 2 = Asian, 3 = Black or African American (AfricanAm), 4 = Latino, 5 = Native Hawaiian or Other Pacific Islander (NHOPI), 6 = White, 7 = Multiple, 8 = Other, 9 = Total	Integer	Level: nominal Required: 1
Race_Ethnicity_Name	Name of race/ethnic group: AIAN (American Indian or Alaska Native), Asian, African American, Latino, NHOPI (Native Hawaiian or Other Pacific Islander), White, Multiple, Other, Total	String	Required: 1
Geographic_Unit	Type of geographic unit: PL = Place (includes cities, towns, and census-designated places (CDP). It does not include unincorporated communities), CO = County, RE = region, CA = State	String	Required:1
Geographic_Unit_Value	Value of geographic unit: 5-digit FIPS place code, 5-digit FIPS county code, 2-digit region id, 2-digit FIPS state code	Integer	Level: nominal Required: 1
Geographic_Unit_Name	Name of geographic unit: place name, county name, region name, or state name	String	Required: 1
County	Name of county that geotype is in. Not available for geotypes RE and CA	String	
FIPS_Code	FIPS (Federal Information Processing Standards) code of county that geotype is in: 2-digit census state code (06) plus 3-digit census county code	Integer	Level: nominal
Region_Name	Metropolitan Planning Organization (MPO) regions as reported in the 2010 California Regional progress report	String	Required: 1
Region_Code	Metropolitan Planning Organization (MPO)-based region code: 01 = Bay Area, 02 = Butte, 03 = Central/Southeast Sierra, 04 = Monterey Bay, 05 = North Coast, 06 = Northeast Sierra, 07 = Northern Sacramento Valley,	Integer	Level: nominal

(continued)

Table 4.1 (continued)

Name	Description	Type	Constraint
	08 = Sacramento Area, 09 = San Diego, 10 = San Joaquin Valley, 11 = San Luis Obispo, 12 = Santa Barbara, 13 = Shasta, 14 = Southern California		
Annual_Food_Cost	The annual cost of food is based on the USDA's low-cost food plan, which includes a market basket of items that families would have to purchase to provide a nutritious diet for each family member	Number	
Median_Income	Median income of female-headed family with children <18 years	Number	Level: ratio
Affordability_Ratio	Ratio of food cost to income for female-headed family with children <18 years	Number	Level: ratio
LL95_Affordability_Ratio	Lower limit of 95% confidence interval	Number	Level: ratio
UL95_Affordability_Ratio	Upper limit of 95% confidence interval	Number	Level: ratio
Percent_Standard_Error	Standard error of percentage	Number	Level: ratio
Percent_Relative_Standard_Error	Relative standard error (se/percent * 100) expressed as a percent	Number	Level: ratio
California_Decile_Group	California places and/or census tracts into 10 groups (or deciles) according to the distribution of values of the affordability ratio. Equal values or "ties" are assigned the mean decile rank	Integer	Level: nominal
California_Rate_Ratio	This indicates how many times the local percentage is higher or lower than the state percentage. Values higher than 1 indicate that local rates are higher than state rates	Number	Level: ratio
Average_Family_Size	Average family size for a female-headed family with children <18 years, specific to a geography, all races combined	Integer	Level: ratio
Version_Datetime	Date and time stamp of version of data	Datetime	Required: 1

4.7.3 Step 1: Load Required Libraries

The first step in predicting food affordability is to load the required libraries:

```
import pandas as pd
import numpy as np
import plotly.graph_objects as go
import seaborn as sns; sns.set()
import matplotlib.pyplot as plt
from scipy import stats
from scipy.stats import pearsonr
import matplotlib.pyplot as plt
from sklearn.preprocessing import MinMaxScaler
from sklearn.model_selection import train_test_split
from sklearn.linear_model import LinearRegression
from sklearn.ensemble import RandomForestClassifier
from sklearn.metrics import r2_score
from sklearn import metrics

import eli5
import matplotlib.pyplot as plt
import matplotlib as mpl
import numpy as np
import scipy.stats as spstats
import seaborn as sns

%matplotlib inline
import pandas as pd
pd.options.display.max_columns=100

from matplotlib import pyplot as plt
import numpy as np
import seaborn as sns
import pylab as plot
```

Load scalers—min max and other scalers.

4.7.4 Step 2: Load Food Affordability Data

The first step in predicting food affordability is to load the required libraries:

```
# Importing data
foodAffordabilityData = pd.read_csv("Food_Affordability__2006-
2010_v2.csv")
```

code	race_eth_name	geotype	geotypevalue	geoname	county_name	county_fips	region_name	region_code	cost_yr	median_income	affordability_ratio
1	AIAN	CA	6	California	NaN	NaN	California	NaN	7508.0	23777.0	0.315779
2	Asian	CA	6	California	NaN	NaN	California	NaN	7508.0	38508.0	0.194980

4.7.5 Step 3: Inspect Data Frame for Column and Types

Inspect data frame for column types[102]:

```
foodAffordabilityData.dtypes
```

Output:

```
ind_id                           object
ind_definition                   object
reportyear                       object
race_eth_code                   float64
race_eth_name                    object
geotype                          object
geotypevalue                    float64
geoname                          object
county_name                      object
county_fips                     float64
region_name                      object
region_code                     float64
cost_yr                         float64
median_income                   float64
affordability_ratio             float64
LL95_affordability_ratio        float64
UL95_affordability_ratio        float64
se_food_afford                  float64
rse_food_afford                 float64
food_afford_decile              float64
CA_RR_Affordability             float64
ave_fam_size                    float64
version                          object
dtype: object
```

[102] ICA 10 & 12—https://gallery.azure.ai/Experiment/ICA-2

4.7.6 Step 4: Clean Columns (Drop Columns)

Drop columns that have numeric representation of categorical columns: ind_definition, reace_eth_name, LL95_affordability_ratio, UL95_affordability_ratio, and version.

```
foodAffordabilityData.drop('ind_id',axis=1,inplace=True)
foodAffordabilityData.drop('ind_definition',axis=1,
inplace=True)
foodAffordabilityData.drop('race_eth_name',axis=1,
inplace=True)
foodAffordabilityData.drop('LL95_affordability_ratio',axis=1,
inplace=True)
foodAffordabilityData.drop('UL95_affordability_ratio',axis=1,
inplace=True)
foodAffordabilityData.drop('version',axis=1,inplace=True)
foodAffordabilityData.drop('se_food_afford',axis=1,
inplace=True)
foodAffordabilityData.drop('rse_food_afford',axis=1,
inplace=True)
```

Describe the affordability data frame:

```
foodAffordabilityData.describe()
```

Output:

	race_eth_code	geotypevalue	county_fips	region_code	cost_yr	median_income	affordability_ratio	food_afford_decile	CA_RR_Affordability	ave_fam_size
count	14364.000000	14364.000000	14229.000000	14355.000000	11043.000000	3473.000000	3473.000000	960.000000	3473.000000	12096.000000
mean	5.000000	40680.393484	6057.977862	7.930408	7269.147371	35985.685081	0.357114	5.500000	1.340507	3.175714
std	2.582079	25834.492705	31.048709	4.564384	1596.524986	27436.558125	0.451169	2.873778	1.693561	0.762813
min	1.000000	1.000000	6001.000000	1.000000	3095.425200	2500.000000	0.021258	1.000000	0.079797	1.360000
25%	3.000000	17480.500000	6035.000000	4.000000	6253.057530	20219.000000	0.158028	3.000000	0.593193	2.660000
50%	5.000000	40382.000000	6059.000000	8.000000	7148.640000	30371.000000	0.245429	5.500000	0.921273	3.130000
75%	7.000000	60609.500000	6083.000000	12.000000	8083.378272	44083.000000	0.381940	8.000000	1.433696	3.550000
max	9.000000	87090.000000	6115.000000	14.000000	16872.049540	250000.000000	4.852371	10.000000	18.214432	7.200000

List numeric feature columns:

```
#A list that record the name of numerical features
numerical_features=foodAffordabilityData.dtypes
[foodAffordabilityData.dtypes=='float64'].index.tolist() +
foodAffordabilityData.dtypes[foodAffordabilityData.
dtypes=='int64'].index.tolist()
numerical_features
```

```
['county_fips',
 'region_code',
 'cost_yr',
 'median_income',
 'affordability_ratio',
 'CA_decile',
 'CA_RR',
 'ave_fam_size',
 'race_eth_code',
 'geotypevalue']
```

```
foodAffordabilityData.describe().transpose()
```

	count	mean	std	min	25%	50%	75%	max
race_eth_code	14364.0	5.000000	2.582079	1.000000	3.000000	5.000000	7.000000	9.000000
geotypevalue	14364.0	40680.393484	25834.492705	1.000000	17480.500000	40382.000000	60609.500000	87090.000000
county_fips	14229.0	6057.977862	31.048709	6001.000000	6035.000000	6059.000000	6083.000000	6115.000000
region_code	14355.0	7.930408	4.564384	1.000000	4.000000	8.000000	12.000000	14.000000
cost_yr	11043.0	7269.147371	1596.524986	3095.425200	6253.057530	7148.640000	8083.378272	16872.049540
median_income	3473.0	35985.685081	27436.558125	2500.000000	20219.000000	30371.000000	44083.000000	250000.000000
affordability_ratio	3473.0	0.357114	0.451169	0.021258	0.158028	0.245429	0.381940	4.852371
food_afford_decile	960.0	5.500000	2.873778	1.000000	3.000000	5.500000	8.000000	10.000000
CA_RR_Affordability	3473.0	1.340507	1.693561	0.079797	0.593193	0.921273	1.433696	18.214432
ave_fam_size	12096.0	3.175714	0.762813	1.360000	2.660000	3.130000	3.550000	7.200000

4.7.7 Step 5: Density Function for Numerical Columns

To visualize data centrality or bell curve nature distribution and skewness (tail), let's display column data on the density function.

```
def show_density(var_data,colname):
    from matplotlib import pyplot as plt

    fig = plt.figure(figsize=(10,4))

    # Plot density
    var_data.plot.density()

    # Add titles and labels
    plt.title(colname + ' Density')
```

(continued)

```
  # Show the mean, median, and mode
  plt.axvline(x=var_data.mean(), color = 'cyan',
linestyle='dashed', linewidth = 2)
  plt.axvline(x=var_data.median(), color = 'red',
linestyle='dashed', linewidth = 2)
  plt.axvline(x=var_data.mode()[0], color = 'yellow',
linestyle='dashed', linewidth = 2)

  # Show the figure
  plt.show()

for col in numerical_features:
  show_density(foodAffordabilityData[col],col)
```

We are utilizing data frame plot density.

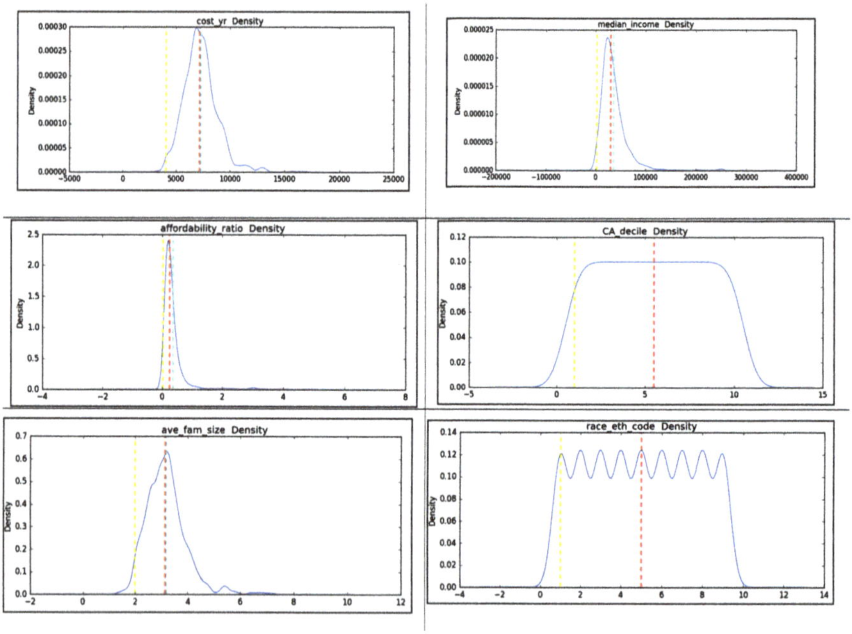

Except CA decile density and race ETH code, other feature engineer parameters exhibit normal data distribution.

4.7.8 Step 6: Check for Null Rows

Check null rows and drop. Later, I will introduce application of data filling:

```
foodAffordabilityData.isnull().sum()
```

Output: medium income, affordability ratio, food affordability ratio, food afford decile, and CA RR affordability columns have null values.

```
reportyear                  1
race_eth_code               1
geotype                     1
geotypevalue                1
geoname                     1
county_name               136
county_fips               136
region_name                 1
region_code                10
cost_yr                  3322
median_income           10892
affordability_ratio     10892
food_afford_decile      13405
CA_RR_Affordability     10892
ave_fam_size             2269
dtype: int64
```

Categorical value count provides column tuple values.

```
foodAffordabilityData['region_name'].value_counts()
```

```
Southern California              316
Bay Area                         165
San Joaquin Valley               165
Sacramento Area                   68
North Coast                       45
Monterey Bay                      44
San Diego                         40
Northeast Sierra                  26
Central/Southeast Sierra          25
Santa Barbara                     19
San Luis Obispo                   15
Butte                             13
Northern Sacramento Valley        11
Shasta                             8
Name: region_name, dtype: int64
```

Southern California, the Bay Area, San Joaquin Valley, and Sacramento Area have the highest number of data rows.

Distribution of median income and affordability ratio provide data distribution details.

```
sns.distplot(foodAffordabilityData['median_income'],
      label="Skewness : %.2f"%(foodAffordabilityData
['median_income']\
                    .skew())).legend(loc="best")
```

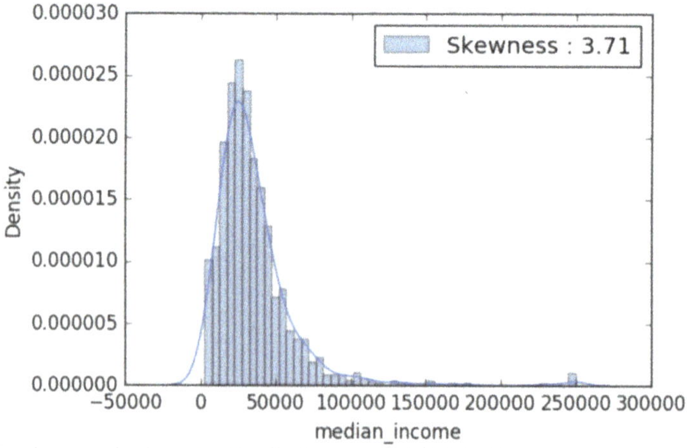

Median income is densely distributed around $50,000.

4.7.9 Step 7: Correlation of Affordability Ratio with Feature Parameters

Let's check the spread of affordability with key attributes.

```
f, axes = plt.subplots(4, 3,figsize=(6*2,7*2))
plt.subplots_adjust( wspace=0.35, hspace=0.35)
for i,ax in zip(numerical_features,axes.flat):
    ax.scatter(x = foodAffordabilityData[i], y =
foodAffordabilityData['affordability_ratio'],s=1.5,color=
['red'])
    ax.set_xlabel(i,size=5.5*2)
    ax.set_ylabel('affordability_ratio',size=5.5*2)
    ax.tick_params(labelsize=8)
```

Affordability ratio was distributed evenly with geotype and county FIPS.

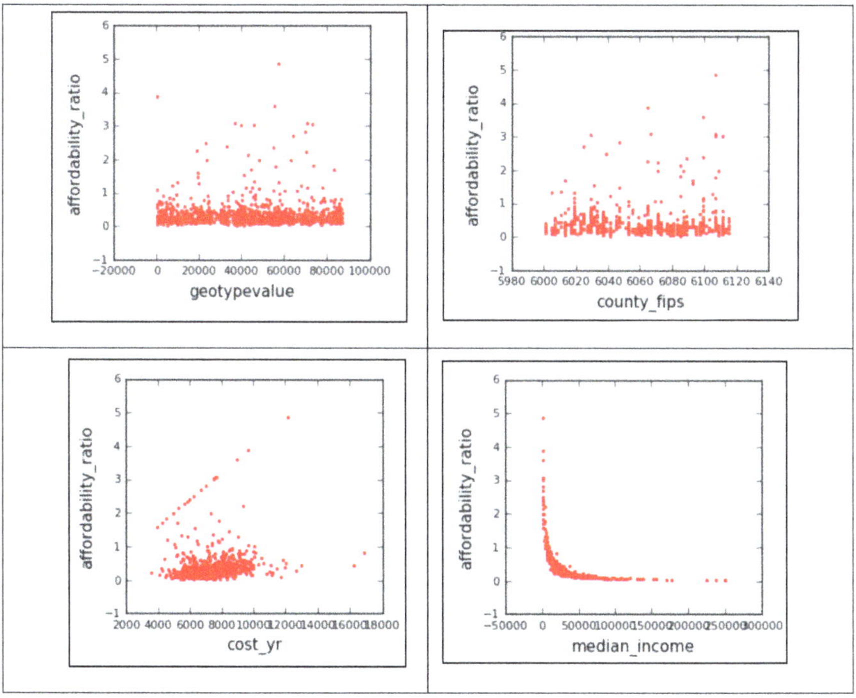

With low yearly costs, the affordability ratios are higher. As cost increased, the affordability ratio has declined. Median income is densely located between no income and $50,000. Affordability ratio and median income were negatively correlated with −43.95%.

For average farm size, the affordability ratio is lower for family sizes of more than two. As family size increased, the affordability ratio was at a minimum.

```
temp=foodAffordabilityData.corr().loc
[['affordability_ratio']]
temp.transpose().sort_values(by='affordability_ratio',
ascending=False)
```

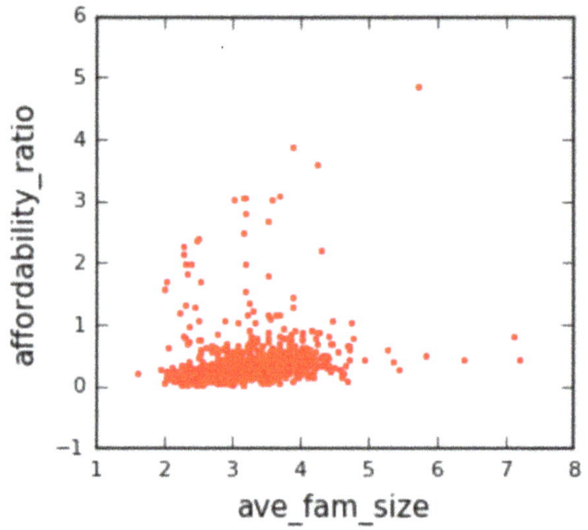

Output:

	affordability_ratio
affordability_ratio	1.000000
CA_RR_Affordability	1.000000
ave_fam_size	0.202765
cost_yr	0.142685
county_fips	0.079695
region_code	0.075913
geotypevalue	0.019441
median_income	-0.439577
food_afford_decile	-0.619112
race_eth_code	NaN

4.7.10 Step 8: Scalers

Prepare scalers list to transform data.

```
from sklearn.preprocessing import MinMaxScaler
from sklearn.preprocessing import MaxAbsScaler
from sklearn.preprocessing import StandardScaler
from sklearn.preprocessing import RobustScaler
from sklearn.preprocessing import Normalizer

scalers=[]
X={}
temp=numerical_features.copy()

scalers.append(MinMaxScaler())
scalers.append(MaxAbsScaler())
scalers.append(StandardScaler())
scalers.append(RobustScaler())
scalers.append(Normalizer())
for scaler in scalers:
  tdata=foodAffordabilityData.copy()
  tdata[temp]=scaler.fit_transform(tdata[temp])
  X[str(scaler)]=tdata
del temp
```

4.7.11 Step 8: Split Data

Split the data.

```
x=data.drop('affordability_ratio',axis=1)
y=data[['affordability_ratio']]
from sklearn.model_selection import train_test_split
x_train,x_test,y_train,y_test=train_test_split(x,y,
test_size=0.3)
```

4.7.12 Step 9: Random Forest Regressor Model

Random forest regressor

```
#turn run_gs to True if you want to run the gridsearch again.
run_gs= False

if run_gs:
  parameter_grid={
         'max_depth' : [40],
         'n_estimators': [200,250],
         'max_features': ['sqrt', 'auto', 'log2'],
         'min_samples_split': [2 ],
         'min_samples_leaf': [1],
         'bootstrap': [True, False],}

  forest=RandomForestRegressor()
  cross_validation=StratifiedKFold(n_splits=5)

  grid_search=GridSearchCV(forest,
             scoring='neg_mean_squared_error',
             param_grid=parameter_grid,
             cv=cross_validation,
             verbose=1)
  grid_search.fit(x_train,y_train.values.ravel())

  model= grid_search
  parameters=grid_search.best_params_

  print('Best score: {}'.format(grid_search.best_score_))
  print('Best parameters: {}'.format(grid_search.best_params_))

else:
  parameters = {'bootstrap': False, 'max_depth':
40, 'max_features': 'sqrt', 'min_samples_leaf':
1, 'min_samples_split': 2, 'n_estimators': 200}
  RFG_model=RandomForestRegressor(**parameters)
  RFG_model.fit(x_train,y_train)

  print(compute_score(RFG_model,x_train,y_train))
```

Output: [-8.56862185e-05 -2.08871089e-04 -5.47815387e-05 -1.18127668e-04 -9.07219038e-04]

R^2 value

```
r2_score(y_test,y_pred)
```

0.9853825976195444
Mean squared error

```
mean_squared_error(y_test,RFG_model.predict(x_test))
```

Output: 0.00010735815265748431
Very low error

4.7.13 Step 10: Feature Importance

```
features = pd.DataFrame()
features['feature'] = x_train.columns
features['importance'] = RFG_model.feature_importances_
features.sort_values(by=['importance'], ascending=True,
inplace=True)
features=features.iloc[-10:,:]
f, ax = plt.subplots(1, 1,figsize=(5,5))
features.set_index('feature', inplace=True)
features.plot(kind='barh', ax=ax,fontsize=5)
ax.legend(prop={'size': 6})
RFG_top10=features.index.tolist()
```

Features of the highest importance:

- CA_RR: a higher local percentage compared to the state percentage has influence on affordability ratio.
- Median income: median income of female-headed family with children <18 years has the second most explainability with affordability ratio.
- CA decile.
- Cost year: the annual cost of food is based on the USDA's low-cost food plan, which includes a market basket of items that families would have to purchase to provide a nutritious diet for each family member and significantly influences affordability ratio.
- Average farm size: the average family size for a female-headed family with children <18 years, specific to a geography, all races combined influences affordability ratio.

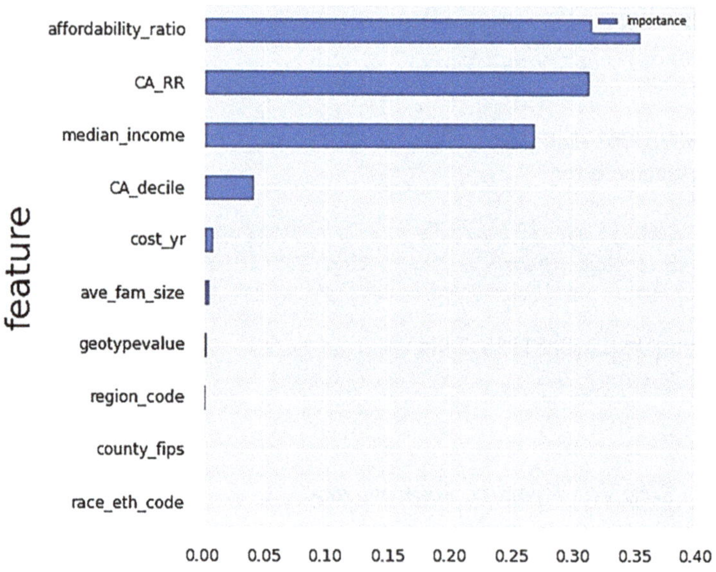

4.7.14 Step 11: Other Regressor Performance Table

Regressor name	R^2	Mean squared error	Feature graph
XGBRegressor	95.13%	0.000507124 2929758312	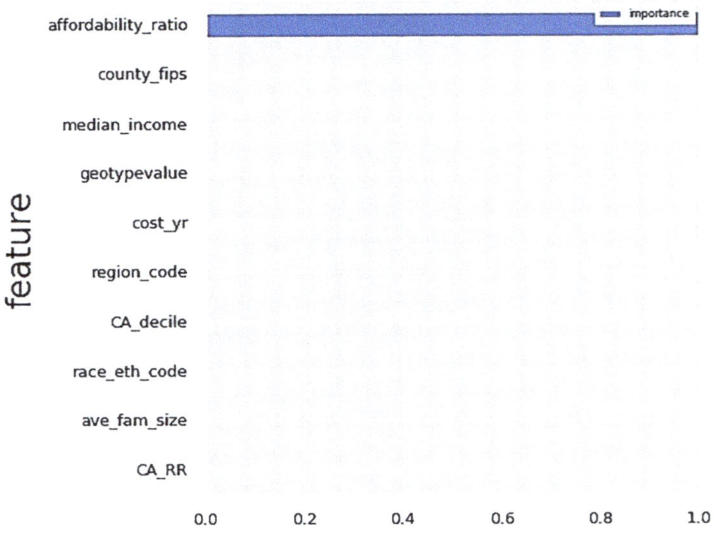

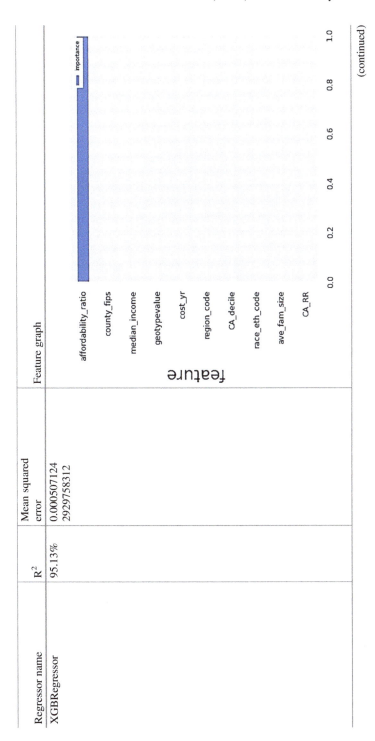

(continued)

(continued)

Regressor name	R^2	Mean squared error	Feature graph
GradientBoostingRegressor	95.44%	0.0004755553823553814	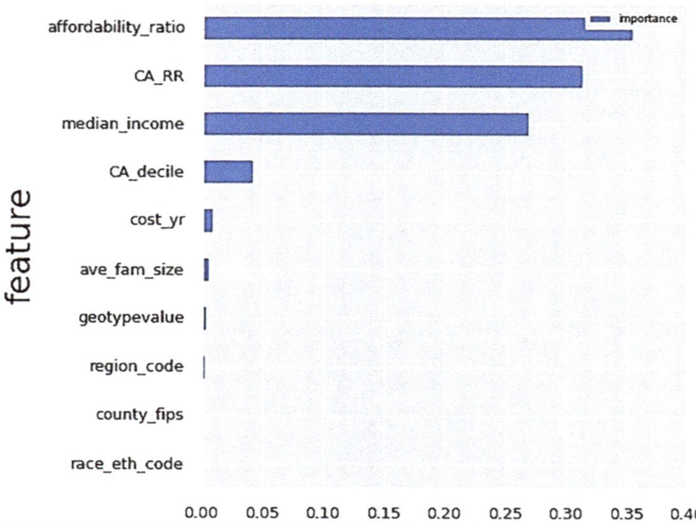

LGBMRegressor	77.44%	0.0023481012114087846	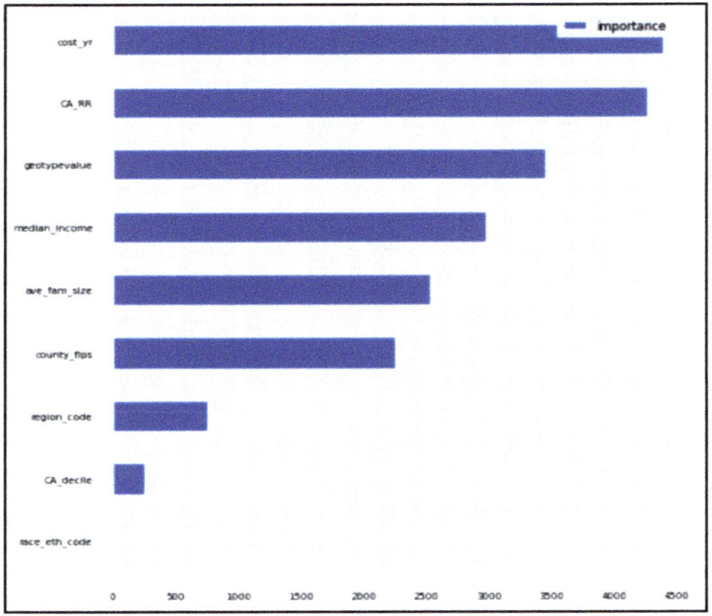

4.8 Food Security and Technological Innovations

Technological innovations that enhance food production are the key underpinning and the best antidote to reduce food insecurity. Given that food insecurity is more prevalent in rural parts of the countries, the creation and improvement of economic opportunities at rural levels will alleviate food insecurity to most of the world's population. The technological strategies should align with the following major initiatives that specifically improve food production at rural levels:

1. Technologies and practices that advance the productivity frontier to drive income growth, improve diets, and promote natural resource conservation.
2. Technologies and practices that reduce, manage, and mitigate risk to support resilient, prosperous, and well-nourished individuals, households, and communities.
3. Improved knowledge of how to achieve human outcomes: train farmers to make the most of new technology; invest in skills and training of women; and generate evidence on how to improve economic opportunity, nutrition, and resilience sustainably and equitably.
4. Data-driven insights and approaches: in addition to using the agricultural survey data, data sources that collect real-time electronic sensor data of critical agriculture and dairy enhance better decision-making on farms. These advances need to be applied more directly on farm providing better data and analysis on agricultural systems, helping farmers to choose the most efficient practices and inputs and respond to changes in weather and markets with greater effectiveness[103] [39].
5. Supply chain efficiencies in delivering farm to fork; create more efficient food chains.

> "The pandemic and global recession may cause over 1.4% of the world's population to fall into extreme poverty," said World Bank Group President David Malpass. "In order to reverse this serious setback to development progress and poverty reduction, countries will need to prepare for a different economy post-COVID, by allowing capital, labor, skills, and innovation to move into new businesses and sectors. World Bank Group support—across IBRD, IDA, IFC and MIGA—will help developing countries resume growth and respond to the health, social, and economic impacts of COVID-19 as they work toward a sustainable and inclusive recovery."

[103] 10 ideas to boost global food security—https://www.fwi.co.uk/news/environment/10-ideas-boost-global-food-security

4.9 Small Farm Sustainability

The small farmers make a big contribution to agriculture and dairy production in the world. There are more than 608 million small farms around the world, occupying between 70 and 80 percent of the world's farmland and producing around 80 percent of the world's food in value terms, according to detailed new research by the Food and Agriculture Organization of the United Nations (FAO). With 65 percent of poor working adults making a living, small farm development is one of the most powerful tools to end extreme poverty, boost shared prosperity, and feed a projected 9.7 billion people by 2050.

Despite their importance, ironically, small farms are disappearing at an alarming rate. For instance, small farms represented 46 percent of production in the United States in 1991. But by 2015, that share had fallen to under 25 percent, according to the US Department of Agriculture. In the European Union, three million farms (around 20%) have disappeared during the last 8 years, mainly small farms. The same trend can be witnessed across other continents. The chief reason for small farm disappearance is the lack of economic sustainability, and causes of it include (a) the lack of sufficient capital and availability of low-cost credit from the government and/or nationalized banks; (b) unstable farm incomes due to poor price realization, a high transaction cost, and poor bargaining power due to small-marketed surplus; (c) lack of off-farm employment opportunities at rural level; and (d) exposure to highly volatile speculative stock markets, abrupt crude oil price changes, steep cross-border trade tensions, and unprecedented international tariffs and trade wars. As a result, many small farmers have committed self-destruction and suicides due to skyrocketing debt and burden of high-interest loans.

Part of the reason smaller family farms can't produce enough to feed more than their own families or communities, and themselves may struggle with food insecurity[104], is that they lack technological solutions to optimize and infuse agricultural innovations, monitor soil conditions, improve livestock management, and streamline operations [40]. The problems that require solutions are frequently connected to the outcomes of generations of unsustainable farming practices. These life-taking issues could be avoided if small-scale farmers are equipped with the same macroeconomic information and analytical capabilities [40] as currently available to mid- and large-scale corporate farms, and we can eradicate marginalization and save our small farmers.

[104]How AI-Driven Technology Is Increasing Food Security, And Improving The Lives Of Farmers Worldwide—https://www.forbes.com/sites/anniebrown/2021/08/12/how-ai-driven-technology-is-increasing-food-security-and-improving-the-lives-of-farmers-worldwide/?sh=2baf41963d4f

4.10 Data

In today's world of plenty, one in nine people is hungry and unable to lead a healthy, active, and productive life[105]. Millions rely on humanitarian assistance just so that they can meet their daily food requirements. Conflict, climate shocks, and economic volatility—hunger's main drivers—not only erode people's livelihoods but also hinder the future development of countries.

Data (please see Fig. 4.18) is a lifeline to overcome and eradicate food insecurity. Targeted action to eradicate hunger, food insecurity, and malnutrition is only possible if actors understand why people are deprived[106]. Such understanding requires the availability of reliable data, statistics, and information, adequate capacity to analyze the available information, and good communication skills to inform decision-makers.

Data drives innovations and optimizes processes, the core pillars of achieving food security. A cited[107] challenge in achieving development goals aimed at poverty and hunger reduction is the lack of reliable on-the-ground data. The ground data could manifest in terms of output of data mining from satellite image data, economic survey data, farm feed real-time data, and/or policy initiative data from the government [41–43].

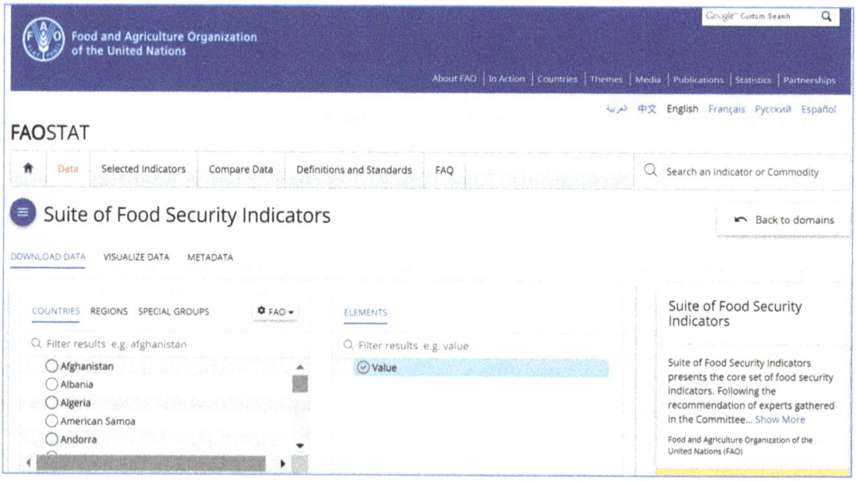

Fig. 4.18 FAOSTAT (https://www.fao.org/faostat/en/#data/FS)

[105] Food Security Information Network (FSIN)—https://www.fsinplatform.org/about-us

[106] Information Systems for Food Security and Nutrition—https://www.fao.org/3/au836e/au836e.pdf

[107] Data Science For Food Security—https://woodsinstitute.stanford.edu/system/files/publications/FSE%20Research%20Brief%20Final.pdf

Limited or insufficient data makes it difficult to establish baseline conditions and to assess the effectiveness of various aid programs. In the past, researchers and policy makers had to rely on ground surveys, which are expensive, time-consuming, and rarely conducted. This has led to large data gaps in mapping sustainable development goal progress, such as in agricultural and poverty statistics. There are early warning systems[108] available to alert any food insecurity. Please refer to Appendix A for Food Security Data Sources.

In the next chapter, I will cover the other two important categories of GFSI: Quality and Safety and Natural Resources and Resilience. Additionally, I would like to introduce machine learning and mathematical optimization models that enhance food security.

After reading this chapter, you should be able to answer the key drivers of food insecurity and food security indicators and drivers. Additionally, you should be able to apply the Global Food Security Index (GFSI) framework to develop machine learning and mathematical optimization models. You should be able to mathematically develop food affordability vs. median income and yearly food cost regressive model. You should be able to answer the role of technological innovations, data-driven models, and small farm sustainability to enhance food security.

References

1. Issues and Challenges of Inclusive Development: Essays in Honor of Prof. R. Radhakrishna by R. Maria Saleth, S. Galab and E. Revathi, Publisher: Springer; 1st ed. 2020 edition (June 19, 2020), ISBN-13: 978-9811522284
2. Anikka Martin, Food Security and Nutrition Assistance, Last updated: Monday, November 08, 2021, https://www.ers.usda.gov/data-products/ag-and-food-statistics-charting-the-essentials/food-security-and-nutrition-assistance/, Access Date: December 12, 2021
3. Alisha Coleman-Jensen, Matthew P. Rabbitt, Laura Hales, and Christian A. Gregory, The prevalence of food insecurity in 2020 is unchanged from 2019, https://www.ers.usda.gov/data-products/chart-gallery/gallery/chart-detail/?chartId=58378, Access Date: December 20, 2021
4. lisha Coleman-Jensen, Matthew P. Rabbitt, Laura Hales, and Christian A. Gregory, Prevalence of food insecurity is not uniform across the country, Last updated: Monday, November 08, 2021, https://www.ers.usda.gov/data-products/chart-gallery/gallery/chart-detail/?chartId=58392, Access Date: December 21, 2021
5. Asian Development Bank. 2013. Food Security in Asia and the Pacific. © Asian Development Bank. http://hdl.handle.net/11540/1435. License: CC BY 3.0 IGO.
6. FAO, RAHO, WFP, UNICEF, Regional Overview of Food Security in Latin America and the Caribbean, https://www.fao.org/documents/card/en/c/ca6979en/, ISBN: 978-92-5-132446-2, Year of publication: 2020

[108] Tool Early Warning Hub—https://www.foodsecurityportal.org/tools/early-warning-hub

7. Ferid Belhaj & Ayat Soliman, MENA Has a Food Security Problem, But There Are Ways to Address It, Publication Date: SEPTEMBER 25, 2021, https://www.worldbank.org/en/news/opinion/2021/09/24/mena-has-a-food-security-problem-but-there-are-ways-to-address-it, Access Date: December 20, 2021
8. Kathryn Reid, Africa hunger, famine: Facts, FAQs, and how to help, Publication Date: May 21, 2021, https://www.worldvision.org/hunger-news-stories/africa-hunger-famine-facts, Access Date: December 20, 2021
9. PANOS VARANGISJUAN BUCHENAUDIEGO ARIASTOSHIAKI ONO, To avoid food insecurity, keep finance flowing, Publication Date: MAY 27, 2020, https://blogs.worldbank.org/psd/avoid-food-insecurity-keep-finance-flowing, Access Date: December 21, 2021
10. Thomas, A., D`hombres, B., Casubolo, C., Saisana, M. and Kayitakire, F., The use of the Global Food Security Index to inform the situation in food insecure countries, EUR 28885 EN, Publications Office of the European Union, Luxembourg, 2017, ISBN 978-92-79-76681-7, doi:10.2760/83356, JRC108638
11. GERNOT LAGANDA, PROC, CHIEF OF CLIMATE AND DISASTER RISK REDUCTION PROGRAMMES, UN WORLD FOOD PROGRAMME, 2021 is going to be a bad year for world hunger, 2021, https://www.un.org/en/food-systems-summit/news/2021-going-be-bad-year-world-hunger, December 20, 2021
12. Elizabeth Howton, Mark Felsenthal, David W. Young, COVID-19 to Add as Many as 150 Million Extreme Poor by 2021, PRESS RELEASE NO: 2021/024/DEC-GPV, https://www.worldbank.org/en/news/press-release/2020/10/07/covid-19-to-add-as-many-as-150-million-extreme-poor-by-2021, Access Date: December 20, 2021
13. Alfred Cang, Cargill CEO Says Global Food Prices to Stay High on Labor Crunch, November 17, 2021, 4:59 AM PST, https://www.bloomberg.com/news/articles/2021-11-17/cargill-ceo-says-global-food-prices-to-stay-high-on-labor-crunch, Access Date: December 10, 2021
14. Peter Timmer, Walter P. Falcon, and Scott R. Pearson, FOOD POLICY ANALYSIS, Published for The World Bank - The Johns Hopkins University Press, 1983, https://documents1.worldbank.org/curated/en/308741468762347702/pdf/multi0page.pdf
15. Eric Holt-Giménez, Annie Shattuck, Miguel Altieri, Hans Herren & Steve Gliessman (2012) We Already Grow Enough Food for 10 Billion People . . . and Still Can't End Hunger, Journal of Sustainable Agriculture, 36:6, 595–598, https://doi.org/10.1080/10440046.2012.695331
16. Jeremy Erdman, We produce enough food to feed 10 billion people. So why does hunger still exist?, Publication Date: Feb 1, 2018, https://medium.com/@jeremyerdman/we-produce-enough-food-to-feed-10-billion-people-so-why-does-hunger-still-exist-8086d2657539, Access Date: December 10, 2021
17. World Food Program USA, Nearly 60% of the World's Hungriest People Live in Just Ten Countries. Why?, November 8, 2021, https://www.wfpusa.org/articles/60-percent-of-the-worlds-hungry-live-in-just-8-countries-why/, Access Date: December 10, 2021
18. Janice Phillips, Reducing Food Waste Can Help Address Food Insecurity, Publication Date: Oct. 22, 2019, at 6:00 a.m, URL: https://www.usnews.com/news/healthiest-communities/articles/2019-10-22/commentary-the-link-between-food-waste-and-food-insecurity, Access Date: December 10, 2021
19. J. Parfitt, M. Barthel, S. Macnaughton Food waste within food supply chains: quantification and potential for change to 2050 Philosoph. Trans. R. Soc. B: Biol. Sci., 365 (1554) (2010), pp. 3065–3081
20. Z. Babar, S. Mirgani S. Food Security in the Middle East Oxford University Press, Oxford: UK (2014) ISBN-13: 978-1849043021
21. Zahir Irani, Amir M. Sharif, Habin Lee, Emel Aktas, Zeynep Topaloğlu, Tamara van't Wout, Samsul Huda, Managing food security through food waste and loss: Small data to big data, Computers & Operations Research, Volume 98, 2018, Pages 367–383, ISSN 0305-0548, https://doi.org/10.1016/j.cor.2017.10.007. (https://www.sciencedirect.com/science/article/pii/S030505481730271X)

22. Dana Marsetiya Utama, Shanty Kusuma Dewi, Abdul Wahid & Imam Santoso (2020) The vehicle routing problem for perishable goods: A systematic review, Cogent Engineering, 7:1, 1816148, https://doi.org/10.1080/23311916.2020.1816148
23. Giallombardo, G.; Mirabelli, G.; Solina, V. An Integrated Model for the Harvest, Storage, and Distribution of Perishable Crops. Appl. Sci. 2021, 11, 6855. https://doi.org/10.3390/app11156855
24. Smart Delivery Systems: Solving Complex Vehicle Routing Problems (Intelligent Data-Centric Systems: Sensor Collected Intelligence) 1st Edition by Jakub Nalepa (Editor) ISBN-13: 978-0128157152
25. Franck Galtier Cirad with the collaboration of Bruno Vindel, Agence Française de Développement., Managing food price instability in developing countries: A critical analysis of strategies and instruments (2013)" - https://europa.eu/capacity4dev/file/14518/download?token=gpMqQy9Y
26. LAUREN FRAYER, India's farmers faced down a popular prime minister and won. What will they do now?, November 26, 20216:22 PM ET, https://www.npr.org/2021/11/26/1059200463/india-farmer-protests-modi-farm-laws, Access Date: December 14, 2021
27. Michael T Owyang, Hannah Shell, Do Rises in Oil Prices Mean Rises in Food Prices?, November 24, 2016, https://www.stlouisfed.org/on-the-economy/2016/november/rises-oil-prices-mean-rises-food-prices, Access Date: December 01, 2021
28. FAO, IFAD, IMF, OECD, UNCTAD, WFP, the World Bank, the WTO, IFPRI and the UN HLTF, Price Volatility in Food and Agricultural Markets: Policy Responses (2 June 2011): https://www.oecd.org/tad/agricultural-trade/48152638.pdf [accessed 5 May 2016]
29. Yacob Abrehe Zereyesus, Assessing Food Security, Last updated: Wednesday, December 01, 2021, https://www.ers.usda.gov/topics/international-markets-us-trade/global-food-security/readings/, Access Date: December 20, 2021
30. Felix Baquedano, Yacob Abrehe Zereyesus, Cheryl Christensen, and Constanza Valdes, COVID-19 Working Paper: International Food Security Assessment, 2020–2030: COVID-19 Update and Impacts on Food Insecurity https://www.ers.usda.gov/webdocs/publications/100276/ap-087.pdf?v=6037.9, Access Date: December 01, 2021
31. Jayson L. Lusk, What Is Driving the Increase in Food Prices?, Publication Date: November 12, 2021, https://econofact.org/what-is-driving-the-increase-in-food-prices, Access Date: December 21, 2021
32. Bureau of Labor Statistics, U.S. Department of Labor, The Economics Daily, Consumer prices for food at home increased 4.8 percent for year ended May 2020 at https://www.bls.gov/opub/ted/2020/consumer-prices-for-food-at-home-increased-4-point-8-percent-for-year-ended-may-2020.htm (visited December 24, 2021)
33. Parlesak A, Tetens I, Dejgård Jensen J, et al. Use of Linear Programming to Develop Cost-Minimized Nutritionally Adequate Health Promoting Food Baskets. PLoS One. 2016;11(10): e0163411. Published 2016 Oct 19. https://doi.org/10.1371/journal.pone.0163411
34. Usman, M.A., Callo-Concha, D. Does market access improve dietary diversity and food security? Evidence from Southwestern Ethiopian smallholder coffee producers. Agric Econ 9, 18 (2021). https://doi.org/10.1186/s40100-021-00190-8
35. Cordero-Ahiman OV, Vanegas JL, Franco-Crespo C, Beltrán-Romero P, Quinde-Lituma ME. Factors That Determine the Dietary Diversity Score in Rural Households: The Case of the Paute River Basin of Azuay Province, Ecuador. Int J Environ Res Public Health. 2021;18(4):2059. Published 2021 Feb 20. https://doi.org/10.3390/ijerph18042059
36. Mirriam Matita, Ephraim W. Chirwa, Deborah Johnston, Jacob Mazalale, Richard Smith, Helen Walls, Does household participation in food markets increase dietary diversity? Evidence from rural Malawi, Global Food Security, Volume 28, 2021, 100486, ISSN 2211 9124, https://doi.org/10.1016/j.gfs.2020.100486. (https://www.sciencedirect.com/science/article/pii/S2211912420301395)

37. Golterman, H.L., Bruijn, P., Schouffoer, J.G.M. et al. Urea fertilization and the N-cycle of rice-fields in the Camargue (S. France). Hydrobiologia 384, 7–20 (1998). https://doi.org/10.1023/A:1003247327586
38. Alisha Coleman-Jensen, Matthew P. Rabbitt, Laura Hales, and Christian A. Gregory, Wednesday, September 08, 2021, Food Insecurity by Household Characteristics, https://www.ers.usda.gov/topics/food-nutrition-assistance/food-security-in-the-us/key-statistics-graphics.aspx, Access Date: December 23, 2021
39. William Frazer, 10 ideas to boost global food security, May 2015, https://www.fwi.co.uk/news/environment/10-ideas-boost-global-food-security, Access Date: December 01, 2021
40. Annie Brown, How AI-Driven Technology Is Increasing Food Security, And Improving The Lives Of Farmers Worldwide, Aug 12, 2021, 04:18pm EDT, https://www.forbes.com/sites/anniebrown/2021/08/12/how-ai-driven-technology-is-increasing-food-security-and-improving-the-lives-of-farmers-worldwide/?sh=2baf41963d4f, Access Date: September 18, 2021
41. CENTER ON FOOD SECURITY AND THE ENVIRONMENT, Data Science For Food Security, Summer 2017, https://woodsinstitute.stanford.edu/system/files/publications/FSE%20Research%20Brief%20Final.pdf, Access Date: September 18, 2021
42. Emerging Technologies for Promoting Food Security: Overcoming the World Food Crisis, Publisher: Woodhead Publishing; 1st edition (December 18, 2015), ISBN-13: 978-1782423355
43. Jiawei Han, Micheline Kamber and Jian Pei, Data Mining: Concepts and Techniques, Publisher: Morgan Kaufmann; 3 edition (June 15, 2011), ISBN-10: 9780123814791

Chapter 5
Food Security: Quality and Safety Drivers

This Chapter Covers:

- Food Security
- Quality and Safety Metric
- Dietary Diversity
  - Machine Learning Model: Household Dietary Diversity Score (HDDS)
- US National Surveys of Dietary Intake and Nutritional Status
  - National Health and Nutrition Examination Survey (NHANES)
- NHANES Food Security
  - Machine Learning Model Development: Body Measure associations between body weight and the health and nutritional status (BPXSY2 vs. BPXSY1)
  - NHANES 2019-2020 Questionnaire Instruments FSQ—Prevalence of Food Security
- Micronutrients Availability
- Data Driven Food Security Models

The chapter introduces food security quality and safety drivers. It introduces the components of dietary diversity and key indicators of dietary diversity and provides in-depth analysis on Household Dietary Diversity Score (HDDS) and Food Consumption Score (FCS). Next, the chapter introduces the food security and nutrition data framework. As part of the framework, it analyzes macroeconomic drivers and its impact on nutrition. Next, the chapter introduces National Surveys of Dietary Intake and Nutritional Status and develops machine learning regression models

C. Vuppalapati, *Artificial Intelligence and Heuristics for Enhanced Food Security*, International Series in Operations Research & Management Science 331,
https://doi.org/10.1007/978-3-031-08743-1_5

using HDDS on Venezuela data to predict who is most at risk of future hunger. Finally, the chapter concludes with machine learning models that leverage the National Health and Nutrition Examination Survey (NHANES) data to learn national health profile and food security among surveyed populations.

The Global Food Security Index (GFSI) is an annual assessment measuring food security through affordability, availability, and quality and safety and natural resources and resilience metrics. The drivers of food affordability include food cost or food inflation, population income (proportion of population under global poverty line), agricultural import tariffs, food safety-net programs, and market access & agricultural financial services. The drivers of availability relay on sufficiency of supply of food, agricultural research and development, agricultural infrastructure, volatility of agricultural production, food loss, and food security policy commitment. Both affordability and availability are factors in economic and agricultural production. Quality and safety, on the other hand, look at nutritional standards and dietary guidelines. This chapter focuses on this important pillar of food security.

Food security is one of the prerequisites for a healthy individual, a well-functioning family, and high-achieving societies. Lack of food security is a humanitarian, social, political, and public health issue. Higher food prices could compound unrest[1] [1]. Rising food prices have resulted in both food insecurity and improvisation[2] [2]. Lack of food security could cause social unrest, regime changes, civil war, and economic collapse. A succession of alarming events—including excessive food price volatility, financial crises, and climate change and related weather shocks threatening food production—has elevated food security to a top priority for governments and the global development community[3] [3]. Food producers have struggled with shortages, bottlenecks, transportation, weather, and labor woes, all of which have caused food prices to rise[4] [4], a fact that we can see from the below FAO Food Price Index (FFPI) (please see Fig. 5.1).

The FAO Food Price Index (FFPI) is an important indicator of global food commodity price movements. It is a measure of the monthly change in international prices of a basket of food commodities. It consists of the average of five commodities (meat, dairy, cereals, oil, and sugar) group price indices weighted by the average export shares of each of the groups over 2014–2016. The same five commodities are the major dietary indicators of key food security indicators that will be covered in subsequent sections.

---

[1] Middle East, North Africa Risk Unrest on High Food Prices—https://finance.yahoo.com/news/middle-east-north-africa-risk-050000342.html

[2] Rising food prices have resulted in both food insecurity and improvisation—https://www.npr.org/2021/11/09/1054032209/rising-food-prices-have-resulted-in-both-food-insecurity-and-improvisation

[3] Measuring the Food Access Dimension of Food Security: A Critical Review and Mapping of Indicators—https://journals.sagepub.com/doi/pdf/10.1177/0379572115587274

[4] Here's why your food prices keep going up—https://www.washingtonpost.com/business/2021/09/15/food-inflation-faq/

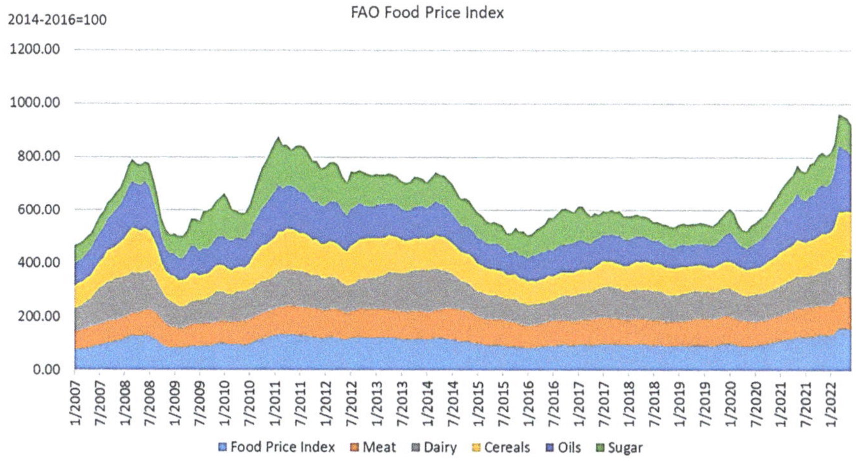

Fig. 5.1 FAO food price index

Global food prices rose "sharply" during 2021[5] [5]. Since the end of 2019, the United Nations' gauge of food prices has risen (a fact that can be witnessed in nominal and real terms[6]) by about a third, with the causes of the surge—bad weather, higher shipping costs, worker shortages, the COVID-19, an energy crunch (increase in biofuels), and rising fertilizer costs—meaning high prices could persist for years to come. Countries in the Middle East and North Africa such as Sudan, Yemen, Lebanon, Tunisia, and Egypt are the most exposed given their imports of wheat and sugar, according to Bloomberg Economics.

Five food group prices have surged resulting in food price index surge (2008 and 2011 spikes[7] [6]; please see Fig. 5.2). This would have an impact on the accessibility and availability of food and directly impact overall food security of the world. For 2021, the FAO Cereal Price Index averaged 131.2 points, up 28.0 points (27.2%) from 2020, and the highest annual average registered since 2012. Demand for corn in China, an ongoing drought in Brazil, and increased global use of vegetable oils, sugar, and cereals has caused prices to surge rapidly around the globe[8] [7]. For 2021, the FAO Vegetable Oil Price Index averaged 164.8 points, up as much as 65.4 points (or 65.8%) from 2020, and marking an all-time annual high (please see Fig. 5.3). The demand was due to weakening palm and sunflower oil prices[9], while soy and

[5]Global food prices rose 'sharply' during 2021—https://news.un.org/en/story/2022/01/1109212

[6]World Food Situation—https://www.fao.org/worldfoodsituation/foodpricesindex/en/

[7]Chapter: Volatile Volatility: Conceptual and Measurement Issues Related to Price Trends and Volatility—Book: Food Price Volatility and Its Implications for Food Security and Policy, ISBN-13: 978-3319281995

[8]Global food prices surge to their highest level in a decade—https://www.cnn.com/2021/06/04/business/inflation-food-prices/index.html

[9]India's sunflower oil imports could jump to record as prices dip below soyoil—https://finance.yahoo.com/news/indias-sunflower-oil-imports-could-092259915.html

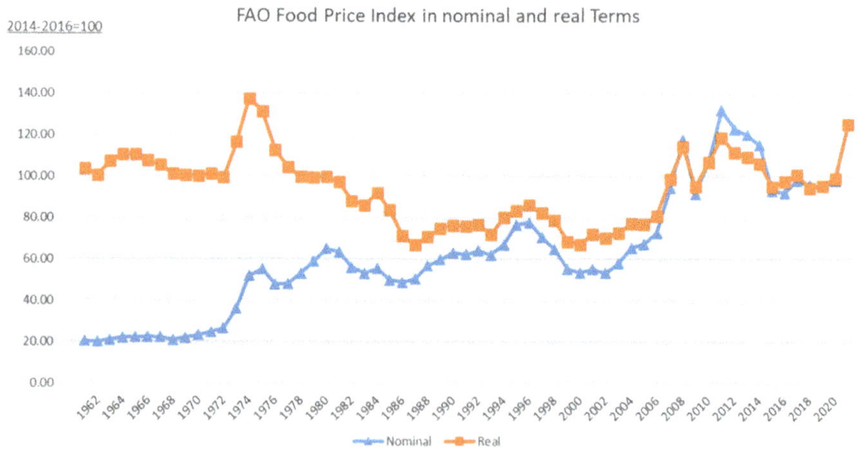

Fig. 5.2 FAO Food Price Index in nominal and real terms

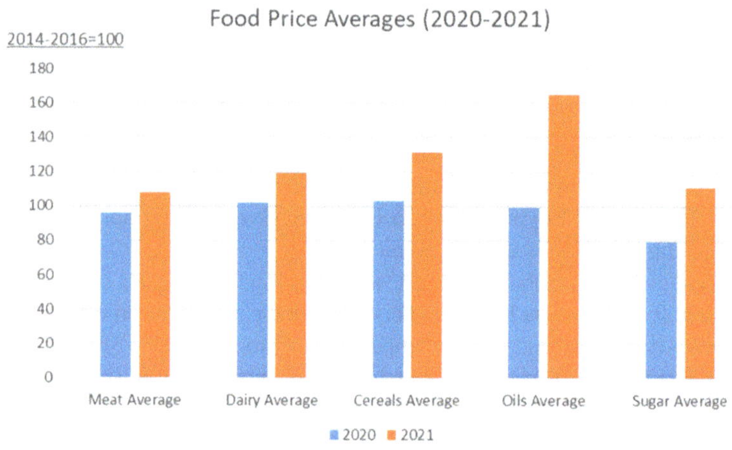

Fig. 5.3 Food price averages

rapeseed oil values remained virtually unchanged month on month [9]. Softening demand due to COVID-19 cases also affected the prices. In 2021, the FAO Dairy Price Index averaged 119.0 points, up 17.2 points (16.9%) from 2020, reflecting sustained import demand throughout the year, especially from Asia (One of the bigger buyers is China, and they use a lot of wheat as a milk replacer for piglets[10]. [10]), and tight exportable supplies from the leading producing regions. In 2021, the FAO Meat Price Index averaged 107.6 points, up 12.1 points (12.7%) from 2020. Across the different categories, ovine meat registered the sharpest increase in prices,

[10]Dairy farmers could see $20 milk in 2022—https://www.yahoo.com/news/dairy-farmers-could-see-20-120321307.html

followed by bovine and poultry meats, while pig meat prices fell marginally. For year 2021, the FAO Sugar Price Index averaged 110.7 points, up 31.2 points (or 37.5%) from 2020, and the highest since 2016[11]. Throughout the year concerns over the reduced output in Brazil amid stronger global demand for sugar underpinned the increase in prices. The weakening of the Brazilian Real against the US dollar and lower ethanol prices also contributed to lowering of world sugar prices in December. The world's record food bill is hitting poorer countries hardest[12] and raising the concerns of global food security [11]. Higher usage of biodiesels and moderate global price increases for meat and dairy products also contributed to the steep rise in global food prices[13] [7].

> "COVID-19 is potentially catastrophic for millions who are already hanging by a thread. It is a hammer blow for millions more who can only eat if they earn a wage. Lockdowns and global economic recession have already decimated their nest eggs. It only takes one more shock – like COVID-19 – to push them over the edge. We must collectively act now to mitigate the impact of this global catastrophe" [8].
> Arif Husain, WFP's Chief Economist

5.1 Food Security: Quality and Safety Metric

Food insecurity is a multi-million-dollar problem in the United States of America, impacting nearly 14.3 million individuals at just over $1800 per patient who experiences food insecurity[14] [12]. And the biggest trouble is that there's no clear path to resolution (both the data and food policy level). With millions of people across the country suffering from food insecurity, 11 million of whom the US Department of Agriculture (USDA) says are children, it's hard not to see the impacts the issue can have on public health. Heads of households are often put under the stress of determining where their next meals will come from, rationing food between family members, or going without food so their children or other dependents can eat.

COVID-19 has exacerbated food insecurity. Before the pandemic, the number of families (13.7 million households or 10.5% of all US households[15]; please see pie

---

[11] World Food Situation—https://www.fao.org/worldfoodsituation/foodpricesindex/en/

[12] World's record food bill is hitting poorer countries hardest—https://www.business-standard.com/article/international/world-s-record-food-bill-is-hitting-poorer-countries-hardest-121111200065_1.html

[13] Global food prices surge to their highest level in a decade—https://www.cnn.com/2021/06/04/business/inflation-food-prices/index.html

[14] Using Quality Measures in Food Security, SDOH Programming—https://patientengagementhit.com/news/using-quality-measures-in-food-security-sdoh-programming

[15] Food Security in the U.S.—Key Statistics & Graphics, https://www.ers.usda.gov/topics/food-nutrition-assistance/food-security-in-the-us/key-statistics-graphics.aspx

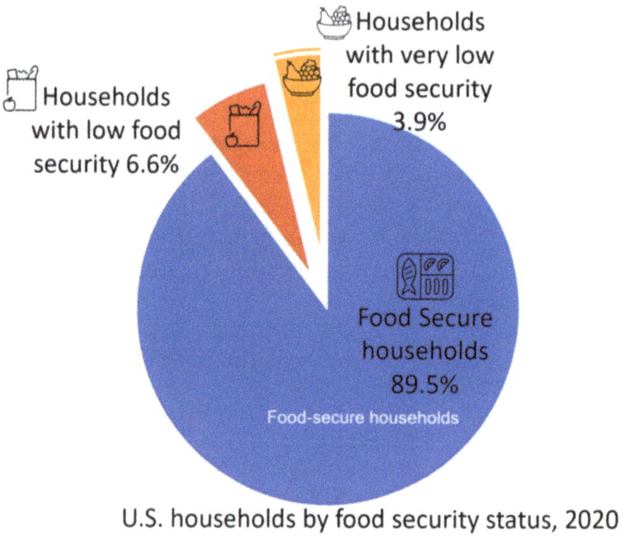

Fig. 5.4 Households by food security status (2020)

chart in Fig. 5.4) experiencing food insecurity [13]—defined as a lack of consistent access to enough food for an active, healthy life—had been steadily falling[16] (food insecurity (includes low and very low food security) in 2015 was 12.66%, and it has reduced to 10.54% in 2020; please see the below figure). But now, as economic instability and a health crisis take over, new estimates point to some of the worst rates of food insecurity (35 million Americans[17] [14]) in the United States of America in years. According to one estimate by researchers at Northwestern University[18] [15], food insecurity more than doubled because of the economic crisis brought on by the outbreak, hitting as many as 23% of households earlier this year (please see Fig. 5.5). Globally, as per the United Nations World Food Programme[19] [12], the global pandemic has the chance to double the number of people experiencing acute food insecurity, from 135 million in 2019 to 265 million in 2020[20] [16]. Food security is only one of the top social determinants of health.

---

[16] Trends in Prevalence rates—https://www.ers.usda.gov/media/xtddtqat/trends.xlsx

[17] Food Insecurity In The U.S. By The Numbers—https://www.npr.org/2020/09/27/912486921/food-insecurity-in-the-u-s-by-the-numbers

[18] How Much Has Food Insecurity Risen? Evidence from the Census Household Pulse Survey—https://www.ipr.northwestern.edu/documents/reports/ipr-rapid-research-reports-pulse-hh-data-10-june-2020.pdf

[19] COVID-19 will double number of people facing food crises unless swift action is taken—https://www.wfp.org/news/covid-19-will-double-number-people-facing-food-crises-unless-swift-action-taken

[20] How local organizations are addressing food insecurity in greater Sacramento area—https://www.abc10.com/article/news/local/sacramento/food-insecurity-greater-sacramento-area/103-dcb949b5-8f7f-46d8-85d9-16fc5817a206

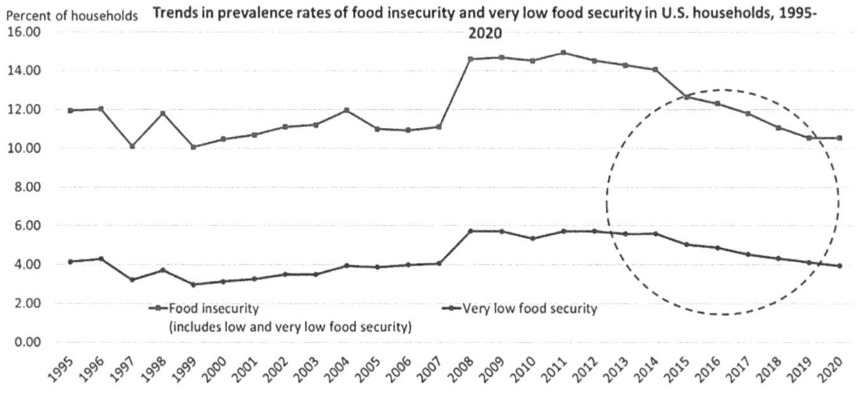

Fig. 5.5 Food security vs. poverty rates

Food security is a complex and tightly intervened framework that holistically touches many verticals of the society. The below figure illustrates the key dimensions[21] (availability, access, utilization, and stability) [3], levels (from global to individual), and components (quantity, quality, safety, cultural acceptability, and preferences) of food security as usually delineated. Availability and access can be measured at all levels from the global to the individual, whereas utilization refers to the ability of individuals to absorb and effectively use the nutrients ingested for normal body functions. Availability and access include several components: quantity (i.e., enough food and energy), quality (i.e., foods that provide all essential nutrients), safety (i.e., food that is free of contaminants and does not pose health risks), and cultural acceptability and preferences (i.e., foods that people like and that fit into traditional or preferred diets). Stability is a cross-cutting dimension that refers to food being available and accessible and utilization being always adequate, so that people do not have to worry about the risk of being food insecure during certain seasons or due to external events.

When assessing food security, it is important to differentiate the level at which data are collected and the level at which food security statements are made. The data collected across the food security ecosystem span across the national surveys, population behavior, preference, macroeconomic and microeconomic indicators, healthcare qualitative metrics, and important drivers that influence food security. The data with advanced analytics provide prognostics as well as guidance for governmental agencies to counter food security issues (please see Fig. 5.6).

Quality and safety metric has five main indicators including diet diversification, nutritional standards (with three sub-indicators including national dietary guidelines, national nutrition plan or strategy, and nutrition monitoring and surveillance), micronutrient availability (with three sub-indicators including dietary availability

[21] Measuring the Food Access Dimension of Food Security: A Critical Review and Mapping of Indicators—https://journals.sagepub.com/doi/pdf/10.1177/0379572115587274

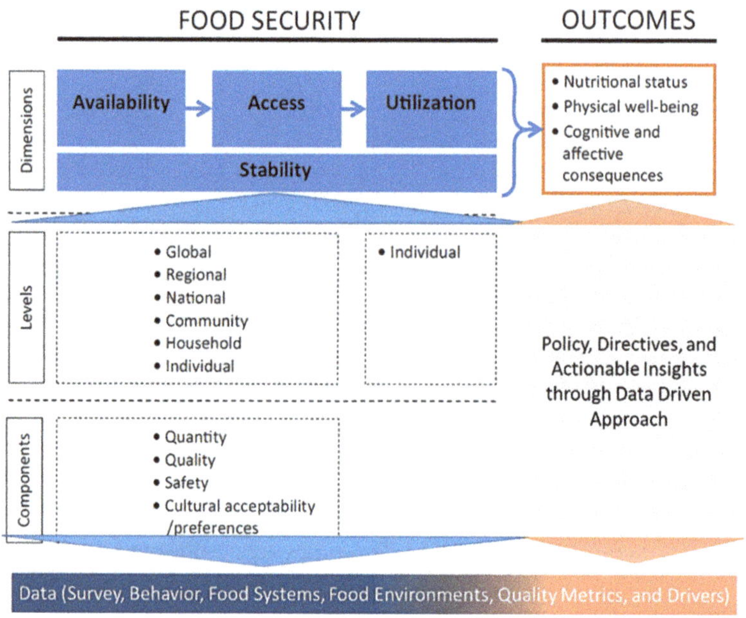

Fig. 5.6 Food security and outcomes

of vitamin A, dietary availability of animal iron, and dietary availability of vegetal iron), protein quality, and food safety (with three sub-indicators including agency to ensure the safety and health of food, percentage of population with access to potable water, and presence of formal grocery sector). The following table summarizes quality and safety parameters:

3)	Quality and safety[22]
3.1)	Dietary diversity
3.2)	Nutritional standards
3.2.1)	National dietary guidelines
3.2.2)	National nutrition plan or strategy
3.2.3)	Nutrition labeling
3.2.4)	Nutrition monitoring and surveillance
3.3)	Micronutrient availability
3.3.1)	Dietary availability of vitamin A
3.3.2)	Dietary availability of iron
3.3.3)	Dietary availability of zinc
3.4)	Protein quality
3.5)	Food safety
3.5.1)	Food safety mechanisms

(continued)

[22] The GFSI Methodology - https://my.corteva.com/GFSI?file=ei21_gfsir

3)	Quality and safety
3.5.2)	Access to drinking water
3.5.3)	Ability to store food safely

Natural resources and resilience: It has seven main indicators including exposure, water, land, oceans, sensitivity, adaptive capacity, and demographic stresses. I will be covering them in the next chapter.

In 2018, the overall food quality and safety scores have declined in almost all countries compared with scores in 2017, a fact that is reflected in the decline of average (mean and median) values (please see Fig. 5.7).

In 2021, the minimum values have declined, indicating the overall thresholds of quality and safety have reduced. This would have a direct impact on the quality of life across the globe. The prognosis for 2022, given COVID-19 spread, very similarity to that of the 2021 (please see Table 5.1).

The GFSI should be used as scientific evidence to convince policy makers, especially in developing countries, to invest in food safety systems. This is also a guide to inform donors of the most needed developing countries for their investments. The international donor community should target more investments to promote food safety at domestic levels in developing countries instead of focusing

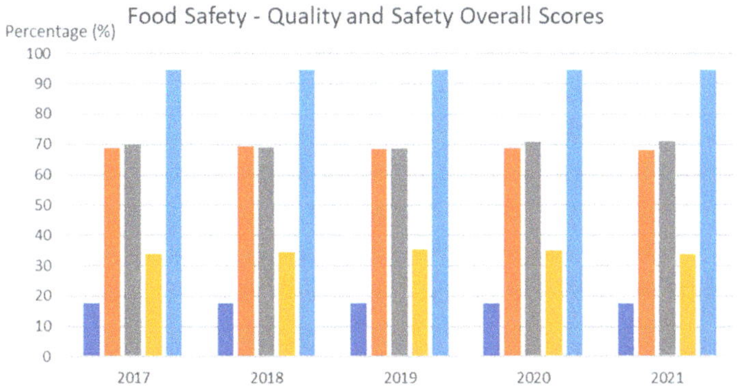

Fig. 5.7 Food safety quality scores (2017–2021)

Table 5.1 Quality and safety

Quality and safety					
	Weight	Average (mean)	Average (median)	Minimum	Maximum
2017	17.6	68.6	69.9	33.7	94.4
2018	17.6	69.3	69.1	34.5	94.4
2019	17.6	68.5	68.8	35.3	94.5
2020	17.6	68.6	70.7	34.9	94.5
2021	17.6	68	71.2	33.8	94.5

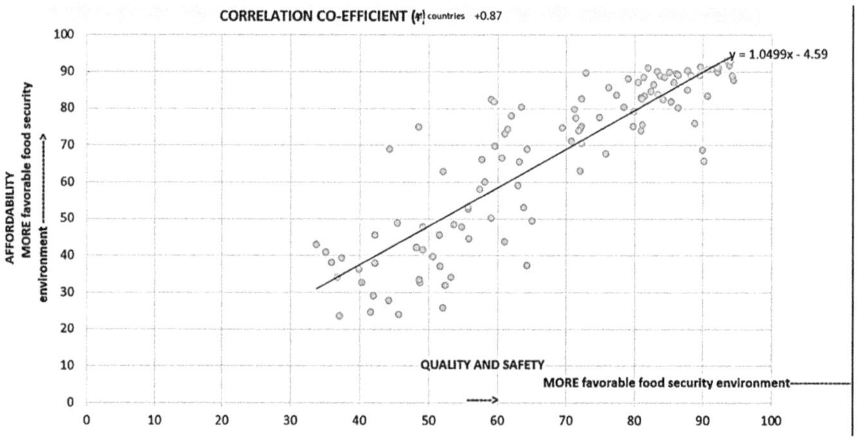

Fig. 5.8 GFSI

on exports (international markets). It is time for a "globalized food safety system" and "one food safety system for all."

The overall correlation between quality and safety driver to food safety is +0.85 (85%). That is, the improvement of quality and safety will have a very positive influence on overall world (please see Fig. 5.8).

$$\text{Quality and safety (Y)} = 0.85 \times \text{food safety} + 18.831 \qquad (5.1)$$

5.1.1 Dietary Diversity

Dietary diversity[23] [17] is defined as the number of different foods or food groups consumed over a given reference period[24] [18]. Different foods and food groups are good sources for various macro- and micronutrients, so a diverse diet best ensures nutrient adequacy[25]. The principle of dietary diversity is embedded in evidence-

[23] Operationalizing Dietary Diversity: A Review of Measurement Issues and Research Priorities—https://academic.oup.com/jn/article/133/11/3911S/4818042

[24] Dietary Diversity as a Food Security Indicator—https://ageconsearch.umn.edu/record/16474/

[25] Minimum Dietary Diversity for Women (MDD-W)—https://www.fantaproject.org/monitoring-and-evaluation/minimum-dietary-diversity-women-indicator-mddw

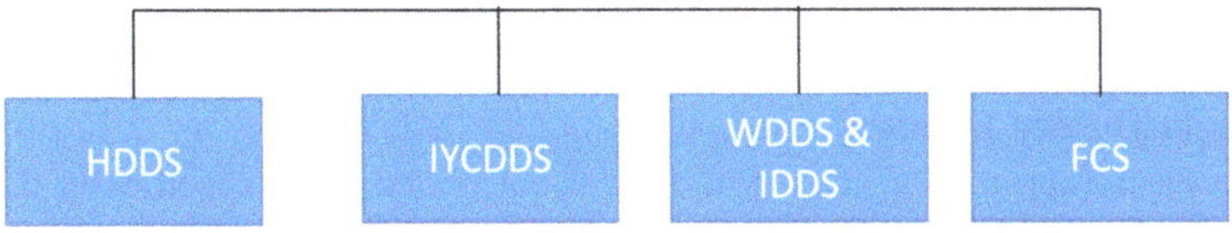

Fig. 5.9 Dietary diversity indicators

based healthy diet patterns, such as the Mediterranean diet and the "DASH" (Dietary Approaches to Stop Hypertension) diet, and is affirmed in all national food-based dietary guidelines. The World Health Organization (WHO) notes that a healthy diet contains fruits, vegetables, legumes, nuts, and whole grains[26]. Dietary diversity could be improved by household participation in food markets, proximity to food markets, and improvement in diverse food cultivation[27] [19].

Lack of dietary diversity is a particularly severe problem among poor populations from the developing world because their diets are predominantly based on starchy staples and often include little or no animal products and few fresh fruits and vegetables[28] [17]. These plant-based diets tend to be low in several micronutrients, and the micronutrients they contain are often in a form that is not easily absorbed. Although other aspects of dietary quality such as high intakes of fat, salt, and refined sugar have not typically been a concern for developing countries, recent shifts in global dietary and activity patterns resulting from increases in income and urbanization are making these problems increasingly relevant for countries in transition as well. Despite the well-recognized importance of dietary diversity, there is still a lack of consensus about what dietary diversity represents. There is also a lack of uniformity in methods to measure dietary diversity and in approaches to develop and validate indicators. In this chapter, the following are the summary of dietary diversity indicators[29] which are studied (please see Fig. 5.9) [3]:

- Household Dietary Diversity Score (HDDS)
- Infant and Young Child Dietary Diversity Score (IYCDDS)
- Women's and Individual Dietary Diversity Score (WDDS and IDDS)
- Food Consumption Score (FCS)

[26] Healthy diet—https://www.who.int/en/news-room/fact-sheets/detail/healthy-diet

[27] Does household participation in food markets increase dietary diversity? Evidence from rural Malawi, Global Food Security—https://www.sciencedirect.com/science/article/pii/S2211912420301395

[28] Operationalizing Dietary Diversity: A Review of Measurement Issues and Research Priorities, The Journal of Nutrition—https://academic.oup.com/jn/article/133/11/3911S/4818042

[29] Measuring the Food Access Dimension of Food Security: A Critical Review and Mapping of Indicators—https://journals.sagepub.com/doi/pdf/10.1177/0379572115587274

5.1.1.1 Household Dietary Diversity Score (HDDS)

Household Dietary Diversity Score was initially developed as an indicator of the food access component of household food security and more specifically the quantity and quality of food access at the household level. Food access was defined as the ability to acquire sufficient quantity and quality of food to meet all household members' nutritional requirements for a productive life[30] [20]:

- FANTA[31]: 12 food groups. 2 food groups capture consumption of staple foods (cereals; roots and tubers); 8 food groups capture consumption of micronutrient-rich foods (vegetables; fruits; meat; eggs; fish; legumes, nuts and seeds; dairy); 3 food groups capture consumption of energy-rich foods (oils and fats; sweets; spices, condiments, and beverages) [21]
- FAO: uses a list of 16 food groups, which are then aggregated into the 12 food groups of the FANTA indicator (cereals; white roots and tubers; vitamin A-rich vegetables and tubers; dark-green leafy vegetables; other vegetables; vitamin A-rich fruits; other fruits; organ meat; flesh meat; eggs; fish and seafood; legumes, nuts, and seeds; milk and milk products; oils and fats; sweets; spices, condiments, and beverages)

  Recall period: 24 h
  Calculation: Simple count of the number of food groups consumed[32] [22]:

$$\text{HDDS} = \sum_{i=1}^{15} \text{Food Groups} \tag{5.2}$$

  Rage: 0–12

5.1.1.2 Infant and Young Child Dietary Diversity Score (IYCDDS)

Indicator[33] was designed to assess dietary diversity in complementary foods for children 6–23 months (as a measure of micronutrient density of complementary foods). It is one of the eight WHO-recommended indicators to measure infant and

[30] Guidelines for Measuring Household and Individual Dietary Diversity—https://www.fao.org/3/i1983e/i1983e.pdf

[31] Household Dietary Diversity Score (HDDS) for Measurement of Household Food Access: Indicator Guide—https://www.fantaproject.org/sites/default/files/resources/HDDS_v2_Sep06_0.pdf

[32] Effect of Remittances on Food Security in Venezuelan Households—https://publications.iadb.org/publications/english/document/Effect-of-Remittances-on-Food-Security-in-Venezuelan-Households.pdf

[33] Indicators for assessing infant and young child feeding practices: definitions and measurement methods—https://www.who.int/publications/i/item/9789240018389

young child feeding practices. It measures the quality of food access at the individual level [23].

Item list: food groups (grains, roots, and tubers; legumes and nuts; dairy products; flesh foods; eggs; vitamin A-rich fruits and vegetables; other fruits and vegetables)

Recall period: 24 h

Cutoff: WHO guidelines on indicators for assessing infant and young child feeding practices use four food groups to define minimum dietary diversity, based on findings from multicounty study and extensive stakeholder consultation.

5.1.1.3 Women's and Individual Dietary Diversity Score (WDDS and IDDS)

This indicator was designed[34] to assess an individual's access to a variety of foods, a key dimension of dietary quality (meant to reflect micronutrient adequacy of the diet) [24]. It was originally developed for use in women of reproductive age (WDDS) to reflect the mean probability of micronutrient adequacy; now it is also used for individuals $>$ 2 years (IDDS). It is a measure of the quality of food access at the individual level.

Item list: 16 food groups (cereals; vitamin A-rich vegetables and tubers; white roots and tubers; dark-green leafy vegetables; other vegetables; vitamin A-rich fruits; other fruits; organ meat; flesh meat; eggs; fish; legumes, nuts, and seeds; dairy; oils and fats; sweets; and condiments), which are then aggregated into 9 food groups (starchy staples; dark-green leafy vegetables; other vitamin A-rich fruits and vegetables; other fruits and vegetables; organ meat; meat and fish; eggs; legumes, nuts, and seeds; and milk and milk products)

The disaggregated version is used to look at specific foods or nutrients of interest (iron, vitamin A, animal-source foods, etc.)

Recall period: 24 h

5.1.1.4 Food Consumption Score (FCS)

Food Consumption Score is a composite score based on household dietary diversity, frequency of household food group consumption, and relative nutritional importance of different food groups[35]. It is thus meant to reflect the quality and quantity of food access at the household level. FCS item list includes country-specific foods grouped into eight standard food groups each with food group-specific weight (w).

---

[34] Guidelines for Measuring Household and Individual Dietary Diversity—https://www.fao.org/publications/card/en/c/5aacbe39-068f-513b-b17d-1d92959654ea/

[35] Zero Hunger—https://sustainabledevelopment-rwanda.github.io/2-1-1/

Item	Weight (w)
Staples	$W = 2.0$
Pulses	$W = 3.0$
Vegetables	$W = 1.0$
Fruit	$W = 1.0$
Meat and fish	$W = 4.0$
Milk	$W = 4.0$
Sugar	$W = 0.5$
Oil	$W = 0.5$

FCS calculation:

$$\begin{aligned}\text{FCS} = {} & w_{\text{staple}}{}^{*} f_{\text{staple}} + w_{\text{pulse}}{}^{*} f_{\text{pulse}} + w_{\text{vegetables}}{}^{*} f_{\text{vegetables}} + w_{\text{fruit}}{}^{*} f_{\text{fruit}} \\ & + w_{\text{meat and fist}}{}^{*} f_{\text{meat and fish}} + w_{\text{milk}}{}^{*} f_{\text{milk}} + w_{\text{sugar}}{}^{*} f_{\text{sugar}} + w_{\text{oil}}{}^{*} f_{\text{oil}} \end{aligned} \tag{5.3}$$

where w_{i} is the weight of each food group and fi is the frequency of consumption (number of days, out of 7). Range value for FCS is 0–112.

Typical cutoff values are:

FCS	Profile
0–21	Poor food consumption
21.5–35	Borderline food consumption
>35	Acceptable food consumption

Cutoffs are higher in locations where oil and sugar are eaten daily (0–28; 28.5–42; > 42).

The food items listed are an example from the Regional Bureau for Southern Africa (ODJ) region.

	Food items (examples)	Food groups (definitive)	Weight (definitive)
1	Maize, maize porridge, rice, sorghum, millet pasta, bread, and other cereals	Main staples	2
	Cassava, potatoes and sweet potatoes, other tubers, and plantains	Main staples	2
2	Beans, peas, groundnuts, and cashew nuts	Pulses	3
3	Vegetables, leaves	Vegetables	1
4	Fruits	Fruit	1
5	Beef, goat, poultry, pork, eggs, and fish	Meat and fish	4
6	Milk yogurt and other diary	Milk	4
7	Sugar and sugar products and honey	Sugar	0.5
8	Oils, fats, and butter	Oil	0.5
9	Spices, tea, coffee, salt, fish power, and small amounts of milk for tea	Condiments	0

Dietary diversity is a major requirement for all age groups, especially for children and pregnant women who get all the essential nutrients, and it can thus be used as one of the core indicators when assessing feeding practices and nutrition of children. Lack of dietary diversity is a major cause of undernutrition, and it poses a serious health challenge across the world, especially in developing countries[36] [25]. Therefore, adequate information on the association between dietary diversity and undernutrition to identify potential strategies for the prevention of undernutrition is critical. Lack of dietary diversity has negative implications for the public health[37] [26].

Dairy products play a significant role in ensuring food security. As clearly indicated through the assigned weights, FCS, improving and accessing dairy products and market is one of the best ways to overcome food insecurity.

Monitoring livestock's health, including vital signs, daily activity levels, and food intake, ensures their health is one of the fastest-growing aspects of AI and machine learning in agriculture. Cow necklace[38] is a Class 10: Wearable veterinary sensor for use in capturing a cow's vital signs, providing data to the farmer to monitor the cow's milk productivity, and improving its overall health.

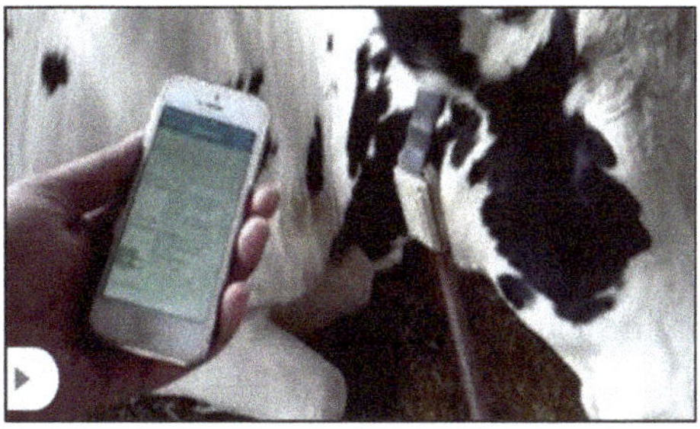

Economic recessions negatively affect dietary diversity. For instance, the 2008 recession was associated with diverse impacts on diets. Calorie intake decreased in high-income countries but increased in middle-income countries[39] [27]. Fruit and vegetable consumption are reduced, especially for more disadvantaged individuals, which may negatively affect health. When the economy goes down, so does the

[36] The influence of dietary diversity on the nutritional status of children between 6 and 23 months of age in Tanzania—https://bmcpediatr.biomedcentral.com/articles/10.1186/s12887-019-1897-5

[37] Dietary diversity and associated factors among adolescents in eastern Uganda: a cross-sectional study—https://bmcpublichealth.biomedcentral.com/articles/10.1186/s12889-020-08669-7

[38] Class 10—Hanumayamma Innovations and Technologies, Inc. - Trade Mark—https://www.trademark247.com/cow%2Bnecklace-87655622-1.html

[39] Impacts of the 2008 Great Recession on dietary intake: a systematic review and meta-analysis—https://www.ncbi.nlm.nih.gov/pmc/articles/PMC8084260/

quality of our diets[40] [28]. Dietary quality plummeted along with the economy. According to the study examining dietary trends during the Great Recession, adults overall ate more refined grains and solid fats, and children increased their intake of added sugar during the recession. The impacts of the downturn were especially pronounced in food-insecure households[41], where individuals significantly reduced their intake of protein and dark-green vegetables while increasing total sugars [29]. During the economic cycles, the farm production diversity is shunned due to economics. This translates to lower household dietary diversity as there is a stronger linkage between farm production diversity and household dietary diversity[42] [30]. Understanding Food Insecurity During the Great Recession[43] [31] is important learning from preventing food insecurity during economic contractions and a helpful policy tool for public health and governmental policies. Finally, favorable international banking policies that would favor maximization of currency funds availability for families in need would improve food security and should be taken as a comprehensive economic and banking policy of the country that is going through economic crisis[44] [22].

Dietary trends during the Great Recession
"The Great Recession had a negative impact on dietary behaviors in both adults and children. Economic downturn impacts household income, employment status and subsequent household food security levels."
Jacqueline Vernarelli, PhD, director of research education and associate professor of public health at Sacred Heart University

To view the overall influence of other drivers on food security, please see Fig. 5.10. It presents a food systems diagram to illustrate how the drivers behind food security and nutrition trends specifically create multiple impacts throughout the food systems (food systems, including food environments), leading to impacts on the four dimensions of food security (availability, access, utilization, and stability), as well as the two additional dimensions of agency and sustainability. These drivers have impacts on attributes of diets (quantity, quality, diversity, safety, and adequacy) and nutrition and health outcomes (nutrition and health). The major drivers are

[40] When the economy goes down, so does the quality of our diets—https://nutrition.org/when-the-economy-goes-down-so-does-the-quality-of-our-diets/

[41] Nutritional Impact of Economic Downturn: How the Great Recession Shaped Dietary Behaviors in US Adults—https://academic.oup.com/cdn/article/5/Supplement_2/191/6293416

[42] Agricultural transformation and food and nutrition security in Ghana: Does farm production diversity (still) matter for household dietary diversity?—https://www.sciencedirect.com/science/article/pii/S0306919217308758

[43] Understanding Food Insecurity During the Great Recession—https://web.stanford.edu/group/recessiontrends-dev/cgi-bin/web/resources/research-project/understanding-food-insecurity-during-great-recession

[44] Effect of Remittances on Food Security in Venezuelan Households—https://publications.iadb.org/publications/english/document/Effect-of-Remittances-on-Food-Security-in-Venezuelan-Households.pdf

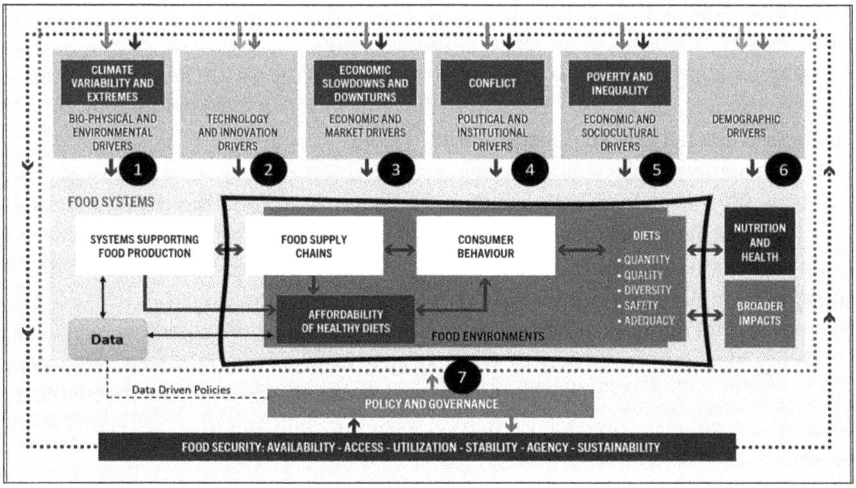

Fig. 5.10 Food security influencer. Source: Adapted from HLPE (2020). Food security and nutrition: building a global narrative towards 2030. A report by the High-Level Panel of Experts on Food Security and Nutrition of the Committee on World Food Security, Rome [32]. The State of Food Security and Nutrition in the World—https://www.fao.org/3/cb4474en/cb4474en.pdf

indicated in the dark boxes that are behind the recent rise in hunger and slowdown in progress in reducing all forms of malnutrition:

1. Climate variability and extremes
2. Technology and innovation drivers[a]
3. Economic slowdown and downturns
4. Conflict
5. Poverty and inequality
6. Demographic drivers[a]
7. Policy and governance

[a]Positive drivers

In Fig. 5.10, conflict (political and institutional drivers), climate variability and extremes (biophysical and environmental drivers), economic slowdowns and downturns (economic and market drivers), and poverty and inequality (economic and sociocultural drivers) are external drivers that act upon food systems (curved box). These drivers tend to create multiple, compounding impacts on food systems that negatively affect food security and nutrition. Because the drivers coexist and interact, this complexity must be fully understood and addressed when designing program and policy responses. The unaffordability of healthy diets is regarded here as an internal driver resulting from the effect of other drivers or factors that directly affects the cost of nutritious foods throughout the food system.

Efficient food production systems and food supply chains contribute to higher food security. Higher farm production diversity significantly contributes to dietary diversity in some situations, but not in all. Improving small farmers' access to markets seems to be a more effective strategy to improve nutrition than promoting production diversity on subsistence farms[45] [33].

Data component plays an important role between food systems and food environments [32]. It collects and utilizes information from food drivers (one through seven) and facilitates optimization by feeding systems supporting food prediction. It feeds back food environments data (supply chain, consumer behavior, and affordability of healthy foods) to improve overall food system effectiveness to drive higher food security. Finally, it enables analytics and heuristics to provide a forecasting capability that improves overall food system efficiencies plus feeds into food policies and governance to have effective policies to overcome food security issues. Data-driven policies, derived from food systems, help improve governance and policy.

Affordability of a diet is determined by the cost of food relative to people's income. Therefore, the dietary diversity is an economic issue. Forecasting economic cycles, recessionary trends, and agricultural commodity price movements could provide a lead-in to create food policies that could overall reduce the impact on marginalized and economic disadvantage communities. This would not only improve the food security to overall population level but also save public health-related expenses due to lack of food security and food diversity-related issues. Implementing effective policies to mitigate adverse nutritional changes and encourage positive changes during the COVID-19 pandemic and other major economic shocks should be prioritized.

In addition to economic prognosis, dietary diversity could be increased by participating to food markets. Food and Nutrition Database for Dietary Studies (FNDDS)[46] is a database that provides the nutrient values for foods and beverages reported in What We Eat in America, the dietary intake component of the National Health and Nutrition Examination Survey (NHANES). Because FNDDS is used to generate the nutrient intake data files for What We Eat in America, NHANES[47] (National Health and Nutrition Examination Survey), it is not required to estimate nutrient intakes from the survey. FNDDS is made available for researchers to review the nutrient profiles for specific foods and beverages as well as their associated portions and recipes. Such detailed information makes it possible for researchers to conduct enhanced analysis of dietary intakes.

---

[45] Production diversity and dietary diversity in smallholder farm households—https://www.pnas.org/content/112/34/10657

[46] FDDS—https://www.ars.usda.gov/northeast-area/beltsville-md-bhnrc/beltsville-human-nutrition-research-center/food-surveys-research-group/docs/fndds/

[47] NHANES—https://www.cdc.gov/nchs/nhanes/index.htm

Difference between dietary diversity and diet quality[48]:
Dietary diversity is one dimension of diet quality. Diverse diets can still lack macronutrient balance and moderation, which are other dimensions of diet quality. Diets lack balance when they are too high or too low in fat, protein, or carbohydrate. Diets lack moderation when they include excessive consumption of energy (calories), salt, or free sugars. Food group diversity does not ensure balance or moderation. Food group diversity also does not in itself ensure that the carbohydrates, proteins, and fats consumed are of high quality. Dietary diversity is, however, associated with better micronutrient density (micronutrients per 100 calories) and micronutrient adequacy of diets.

5.1.2 Rates of Food Insecurity vs. the Poverty Rate

Access to adequate food is critical for health and well-being, and lack of food may have lasting consequences for health and development, especially for children. Since 2000, rates of poverty and food insecurity in the United States of America have been rising, and both spiked dramatically in 2008 with the onset of the Great Recession. As it is evident in the below figure, food insecurity was 11.11% in 2007 and in 2008 rose to 14.59%. In terms of the number of poor families[49], 7,623 families in 2007 were having food insecurity, and that number grew to 8,792[50] [13] in 2009[51]. In 2009, 23.2%[52] of children lived in food-insecure households, up from 16.9% in 1999, while the fraction of children living in households with very low food security has almost doubled, reaching 1.3%[53]. The Great Recession has not only increased poverty and hunger in the United States, but it has also changed the relationship between the two (please see Fig. 5.11). Traditionally, rates of poverty and food insecurity have closely tracked each other, with poverty rates remaining slightly higher. However, during the Great Recession, rates of food insecurity grew beyond the poverty rate[54] [31].

[48] Difference between dietary diversity and diet quality—https://www.fantaproject.org/monitoring-and-evaluation/minimum-dietary-diversity-women-indicator-mddw

[49] Current Population Survey—https://www2.census.gov/programs-surveys/cps/techdocs/cpsmar21.pdf

[50] Food Security in the U.S. Key Statistics & Graphics, Last updated: Wednesday—https://www.ers.usda.gov/topics/food-nutrition-assistance/food-security-in-the-us/key-statistics-graphics.aspx

[51] Trends in Prevalence rates—https://www.ers.usda.gov/media/xtddtqat/trends.xlsx

[52] U.S. Poverty Rate By State In 2021—https://www.forbes.com/sites/andrewdepietro/2021/11/04/us-poverty-rate-by-state-in-2021/?sh=4db0523b1b38

[53] Table 13. Number of Families Below the Poverty Level and Poverty Rate: 1959 to 2020—https://www2.census.gov/programs-surveys/cps/tables/time-series/historical-poverty-people/hstpov13.xlsx

[54] Understanding Food Insecurity During the Great Recession—https://web.stanford.edu/group/recessiontrends-dev/cgi-bin/web/resources/research-project/understanding-food-insecurity-during-great-recession

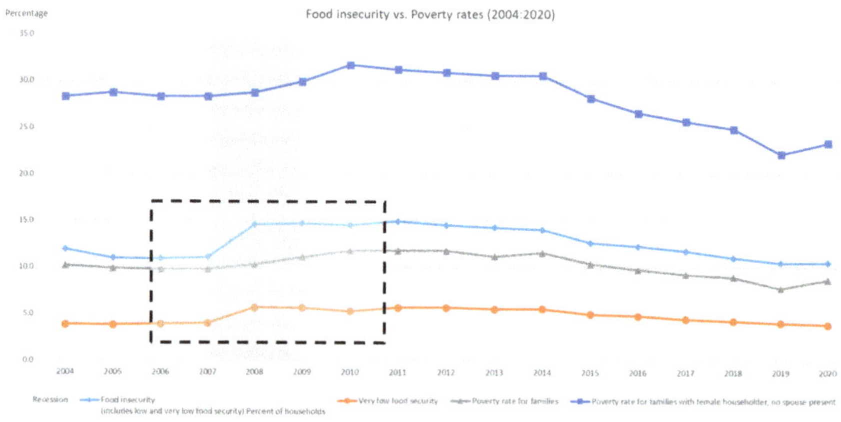

Fig. 5.11 Food security vs. poverty rates

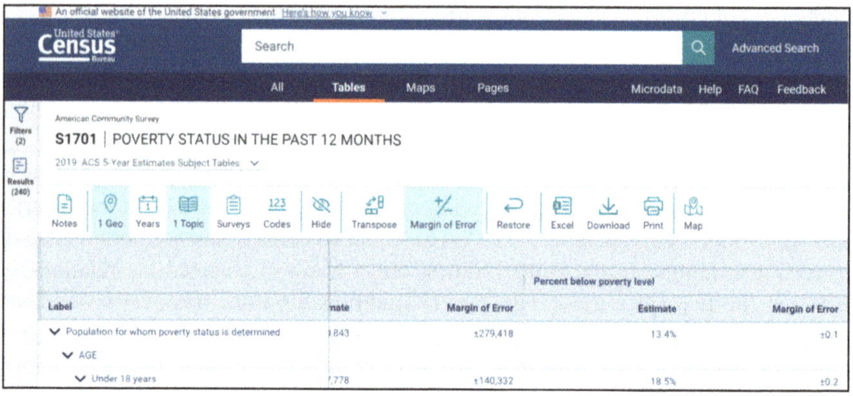

Fig. 5.12 Census UI. Source: https://data.census.gov/cedsci/table?q=&t=Poverty&g=0100000US&tid=ACSST5Y2019.S1701

According to the latest data from the US Census Bureau—namely, the 2019 American Community Survey, 5-Year Estimates—the US poverty rate nationally is 13.4%[55]. This means that 13.4% of the national population lives below the poverty line. This is equal to more than approximately 42.5 million Americans living below the poverty line. While there's still room for improvement, the poverty rate in America has gotten better over the last 5 years. In 2014, the share of the US population living below the poverty line was 15.6%, equivalent to more than 47.7 million Americans. Fortunately, both on the national level and on the state level (for most states), poverty rates have declined from 2014 to 2019 (please see Fig. 5.12).

[55] Poverty Rates in U.S.—https://data.census.gov/cedsci/table?q=&t=Poverty&g=0100000US&tid=ACSST5Y2019.S1701

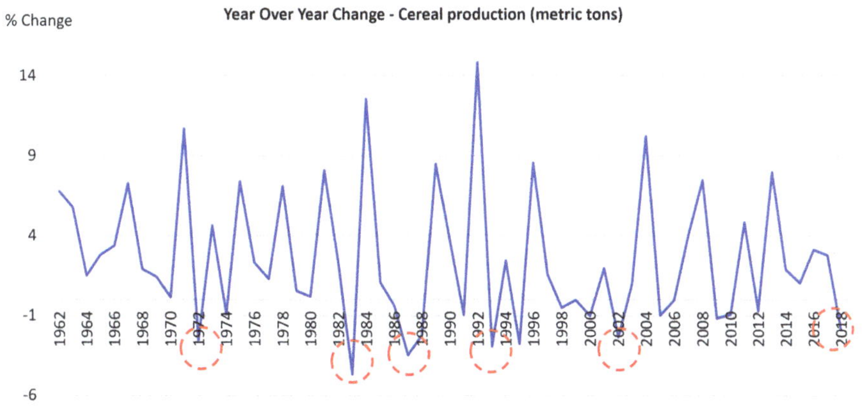

Fig. 5.13 Cereal production. Source data: The World Bank and FAO—Cereal Production [metric tons (World Cereal Production (Metric Tons)—https://data.worldbank.org/indicator/AG.PRD.CREL.MT)]

5.2 Signal Mining: Cereal Productions vs. Food Prices vs. Food Insecurity [34, 35][56,57]

The 2021 Global Report on Food Crises (GRFC 2021) highlights the remarkably high severity and the number of people in crisis or worse (IPC/CH Phase 3 or above) or equivalent in 55 countries/territories, driven by persistent conflict, pre-existing and COVID-19-related economic shocks, and weather extremes. As indicative of the severity of the crisis, the number identified in the 2021 is the highest in the last 5 years[58] [36]. As per chief economist at the United Nations' Food and Agriculture Organization , a global food crisis could be approaching[59] but we're not there [37].

As the world economy, we have seen significant food production drop (please see Fig. 5.13) in 1972–1973[60] [34] (−28 million tons), 1983 (−69 million), 1987 (−56 million tons), 1988 (−34.6 million tons), 1993 (−57 million tons), 2010 (−57.2 million tons), 2012 (−18.2 million tons), and 2018 (−57.2 million tons). Current food crisis is due to mix of drought, conflict, and the COVID-19 pandemic. By

[56] Issues and Challenges of Inclusive Development: Essays in Honor of Prof. R. Radhakrishna, ISBN-13: 978-9811522284

[57] The 1972-73 Food Price Spiral—https://www.brookings.edu/wp-content/uploads/1973/06/1973b_bpea_schnittker.pdf

[58] Global Report on Food Crises—2021—https://www.wfp.org/publications/global-report-food-crises-2021

[59] Amid drought, conflict and rocketing prices, a global food crisis could be approaching, top expert warns—https://www.washingtonpost.com/world/2021/12/15/global-food-crisis-pandemic/

[60] Issues and Challenges of Inclusive Development: Essays in Honor of Prof. R. Radhakrishna, ISBN-13: 978-9811522284

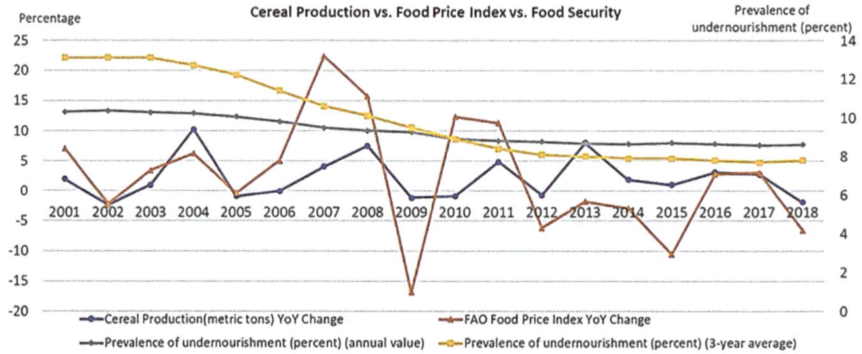

Fig. 5.14 Cereal production vs. food price index vs. food insecurity

assuring the next season so that farmers can do their planting so that the next harvest is assured is the way to fix the issue, per UN chief economist.

If we take signal analysis of three important indicators—for instance, cereal productions, FAO Food Price Index, and food security data, we can see a clear signal between each parameter and among all.

During the years when cereal production dropped, the food price index or food prices shown increase:

- For years 2004–2005, a drop of 21.9 million tons of cereal production exhibited 5.03% increase in Food Price Index. Consecutive lower cereal production of 687 K metric tons for the year (2006–2007) has increased prices by 22%. Both very low food security and food insecurity numbers were increased by 1.55% and 1.75%, respectively (please see Fig. 5.14). As noted in food security in the United States of America[61] [13], the prevalence of very low food security increased from 3.3% in 2001 to 3.9% in 2004 and remained essentially unchanged through 2007 (4.1%).
- For years 2008–2009, a drop of 29 million tons of cereal production exhibited Food Price Index increase by 12.33%. Very low food security was increased by 0.68% and food insecurity numbers decreased by 0.17%.

The converse is also true:

- For years 2003–2004, a 10% increase of cereal production has resulted in a drop of 0.35% price change from the previous year. The same trend can be observed in food insecurity (−7.9%) and very low food security (−1.77%) percentages. In the previous decade, food insecurity increased from 10.7% in 2001 to 11.9% in 2004, declined to about 11% in 2005–2007, then increased significantly in 2008

[61] Food Security & Trends in Prevalence Rates—https://www.ers.usda.gov/media/xtddtqat/trends.xlsx

(to 14.6%), and remained essentially unchanged (i.e., the difference was not statistically significant) at that level in 2009 and 2010.

- The same trend can be observed for years 2014–2015. Year-to-year declines in food insecurity from 2014 to 2015 and 2016 through 2018 were also statistically significant[62].

Although food insecurity depends on several factors, the agricultural productions and commodity prices have a huge influence—"Hungry Season[63]" [13] or "hungry trio factors."

Data sources:

Source	Link	
Food Security—Trends in Prevalence Rates	https://www.ers.usda.gov/media/xtddtqat/trends.xlsx	USDA
FAO Food Price Index	https://www.fao.org/fileadmin/templates/worldfood/Reports_and_docs/Food_price_indices_data_jan62.xls	Food and Agriculture Organization of the United Nations
FAO Food Price Index—Nominal and Real Terms	https://www.fao.org/fileadmin/templates/worldfood/Reports_and_docs/food_price_index_nominal_real_jan652.xls	
Cereal Production	https://data.worldbank.org/indicator/AG.PRD.CREL.MT	THE WORLD BANK

5.3 National Surveys of Dietary Intake and Nutritional Status

National surveys include surveys conducted by the US Department of Agriculture (USDA) and surveys conducted by the US Department of Health and Human Services (DHHS). The Food Supply: Historical Data and the Nationwide Food Consumption Surveys (NFCS) are conducted by the USDA. The Total Diet Study and National Health and Nutrition Examination Survey (NHANES) are undertakings of the US Department of Health and Human Services[64]. We will use the NHANES data to analyze the health trends [38].

Nutrition monitoring in the United States of America is conducted as part of the National Nutrition Monitoring System (NNMS) and Coordinated State Surveillance

[62] Food Insecurity—https://www.ers.usda.gov/topics/food-nutrition-assistance/food-security-in-the-us/key-statistics-graphics.aspx

[63] Issues and Challenges of Inclusive Development: Essays in Honor of Prof. R. Radhakrishna, ISBN-13: 978-9811522284

[64] Dietary Intake and Nutritional Status: Trends and Assessment—https://www.ncbi.nlm.nih.gov/books/NBK218765/

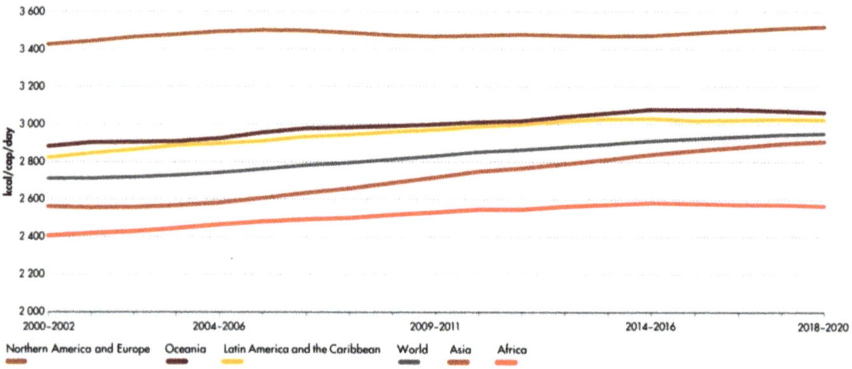

Fig. 5.15 Food supply

System (CSSS). The Centers for Disease Control and Prevention (CDC) contributes to nutrition monitoring through CSSS, in which the nutritional status of the high-risk pediatric population and pregnant women is monitored based on information obtained from service delivery programs operated by selected state and metropolitan health jurisdictions.

Nutritional status has been defined as an individual's health condition as it is influenced by the intake and utilization of nutrients[65]. In theory, optimal nutritional status should be attained by consuming sufficient, but not excessive, sources of energy, essential nutrients, and other food components (such as dietary fiber) not containing toxins or contaminants [39].

Food supply[66]

The world average dietary energy supply (DES), measured as calories per capita per day, has been increasing steadily to 2,950 kcal per person per day over the period from 2018 to 2020, up 9% compared with 2000 to 2002 (please see Fig. 5.15). It is the highest in Northern America and Europe at about 3,520 kcal per person per day; the gap with Oceania and Latin America and the Caribbean, slightly above 3,000 kcal per person per day, is substantial. The fastest increase took place in Asia where DES went up 14% over the last two decades, while the lowest among all regions, Africa, has also witnessed a steady increase in DES followed by a slight decline in recent years, probably due to the drought in 2016 and political conflicts in some countries that adversely affected agricultural production[67].

[65] Diet and Health: Implications for Reducing Chronic Disease Risk—https://www.ncbi.nlm.nih.gov/books/NBK218765/

[66] Food Supply—https://www.fao.org/3/cb4477en/online/cb4477en.html#chapter-3_2

[67] Drought in the Horn of Africa [online]. Rome. [cited September 2021]. http://www.fao.org/emergencies/crisis/drought-hoa/intro/en

5.3.1 *Composition of Foods Raw, Processed, Prepared USDA National Nutrient Database for Standard Reference, Release 27*

The USDA National Nutrient Database for Standard Reference[68] (SR) is the major source of food composition data in the United States of America. It provides the foundation for most food composition databases in the public and private sectors. As information is updated, new versions of the database are released. This version, Release 27 (SR27), contains data on 8,618 food items and up to 150 food components. It replaces SR26 issued in August 2013. The database consists of several sets of data: food descriptions, nutrients, weights and measures, footnotes, and sources of data.

5.3.2 *Data Dictionary*

Data dictionary defines four principal and six support file (table) formats and their parameters for the relational database containing all food, nutrient, and related data and an abbreviated flat file with all the food items, but fewer nutrients, and not all the other related information[69]. Nutrition data is ASCII file—Abbreviated file (please see Table 5.2).

Software code for this model can be found on GitHub Link: CompositionofFoodsRawProcessedPrepared_USDA_NationalNutrientDatabase.ipynb (Jupyter Notebook Code)

Load database: ABBREV.csv (download – sr27abxl.zip)[70]

---

[68] Composition of Foods Raw, Processed, Prepared USDA National Nutrient Database for Standard Reference, Release 27—https://data.nal.usda.gov/dataset/composition-foods-raw-processed-prepared-usda-national-nutrient-database-standard-referenc-4

[69] Data Dictionary—https://data.nal.usda.gov/dataset/composition-foods-raw-processed-prepared-usda-national-nutrient-database-standard-1#{}

[70] Database Zip—https://data.nal.usda.gov/system/files/sr27abxl.zip

Table 5.2 SR27 data dictionary

Field name	Description	Data type
NDB_No	Five-digit Nutrient Data Bank number that uniquely identifies a food item. If this field is defined as numeric, the leading zero will be lost	Character
Shrt_Desc	60-character abbreviated description of food item. Generated from the 200-character description using abbreviations in Appendix A. If short description is longer than 60 characters, additional abbreviations are made	Character
Water	Water (g/100 g)	Decimal
Energ_Kcal	Food energy (kcal/100 g)	Numeric
Protein	Protein (g/100 g)	Decimal
Lipid_Tot	Total lipid (fat)(g/100 g)	Decimal
Ash	Ash (g/100 g)	Decimal
Carbohydrt	Carbohydrate by difference (g/100 g)	Decimal
Fiber_TD	Total dietary fiber (g/100 g)	Decimal
Sugar_Tot	Total sugars (g/100 g)	Decimal
Calcium	Calcium (mg/100 g)	Numeric
Iron	Iron (mg/100 g)	Decimal
Magnesium	Magnesium (mg/100 g)	Numeric
Phosphorus	Phosphorus (mg/100 g)	Numeric
Potassium	Potassium (mg/100 g)	Numeric
Sodium	Sodium (mg/100 g)	Numeric
Zinc	Zinc (mg/100 g)	Decimal
Copper	Copper (mg/100 g)	Decimal
Manganese	Manganese (mg/100 g)	Decimal
Selenium	Selenium (Âµg/100 g)	Decimal
Vit_C	Vitamin C (mg/100 g)	Decimal
Thiamin	Thiamin (mg/100 g)	Decimal
Riboflavin	Riboflavin (mg/100 g)	Decimal
Niacin	Niacin (mg/100 g)	Decimal
Panto_acid	Pantothenic acid (mg/100 g)	Decimal
Vit_B6	Vitamin B6 (mg/100 g)	Decimal
Folate_Tot	Folate total (Âµg/100 g)	Numeric
Folic_acid	Folic acid (Âµg/100 g)	Numeric
Food_Folate	Food folate (Âµg/100 g)	Numeric
Folate_DFE	Folate (Âµg dietary folate equivalents/100 g)	Numeric
Choline_Tot	Choline total (mg/100 g)	Numeric
Vit_B12	Vitamin B12 (Âµg/100 g)	Decimal
Vit_A_IU	Vitamin A (IU/100 g)	Numeric
Vit_A_RAE	Vitamin A (Âµg retinol activity equivalents/100 g)	Numeric
Retinol	Retinol (Âµg/100 g)	Numeric
Alpha_Carot	Alpha-carotene (Âµg/100 g)	Numeric
Beta_Carot	Beta-carotene (Âµg/100 g)	Numeric
Beta_Crypt	Beta-cryptoxanthin (Âµg/100 g)	Numeric
Lycopene	Lycopene (Âµg/100 g)	Numeric

(continued)

Table 5.2 (continued)

Field name	Description	Data type
Lut+Zea	Lutein+zeazanthin (Âµg/100 g)	Numeric
Vit_E	Vitamin E (alpha-tocopherol) (mg/100 g)	Decimal
Vit_D_mcg	Vitamin D (Âµg/100 g)	Decimal
Vit_D_IU	Vitamin D (IU/100 g)	Numeric
Vit_K	Vitamin K (phylloquinone) (Âµg/100 g)	Decimal
FA_Sat	Saturated fatty acid (g/100 g)	Decimal
FA_Mono	Monounsaturated fatty acids (g/100 g)	Decimal
FA_Poly	Polyunsaturated fatty acids (g/100 g)	Decimal
Cholestrl	Cholesterol (mg/100 g)	Decimal
GmWt_1	First household weight for this item from the weight file	Decimal
GmWt_Desc1	Description of household weight number 1	Character
GmWt_2	Second household weight for this item from the weight file	Decimal
GmWt_Desc2	Description of household weight number 2	Character
Refuse_Pct	Percent refuse	Numeric

5.3.2.1 Step 1: Load Data

Load the USDA National Nutrient Database for Standard Reference file. Please select CP1252 encoding to overcome load character errors[71].

```
nRowsRead = 1000 # specify 'None' if want to read whole file
# test.csv may have more rows in reality, but we are only loading/
previewing the first 1000 rows
dfUSDANND = pd.read_csv('ABBREV.csv', delimiter=',', nrows =
nRowsRead, encoding='cp1252')
dfUSDANND.dataframeName = 'ABBREV.csv'
nRow, nCol = dfUSDANND.shape
print(f'There are {nRow} rows and {nCol} columns')
dfUSDANND.head(50)
```

[71] Read CSV Errors—https://stackoverflow.com/questions/42339876/error-unicodedecodeerror-utf-8-codec-cant-decode-byte-0xff-in-position-0-in

Output:

NDB_No	Shrt_Desc	Water_(g)	Energ_Kcal	Protein_(g)	Lipid_Tot_(g)	Ash_(g)	Carbohydrt_(g)	Fiber_TD_(g)	Sugar_Tot_(g)	...	Vit_K_(µg)
1001	BUTTER,WITH SALT	15.87	717	0.85	81.11	2.11	0.06	0.0	0.06	...	7.0
1002	BUTTER,WHIPPED,WITH SALT	15.87	717	0.85	81.11	2.11	0.06	0.0	0.06	...	7.0
1003	BUTTER OIL,ANHYDROUS	0.24	876	0.28	99.48	0.00	0.00	0.0	0.00	...	8.6
1004	CHEESE,BLUE	42.41	353	21.40	28.74	5.11	2.34	0.0	0.50	...	2.4
1005	CHEESE,BRICK	41.11	371	23.24	29.68	3.18	2.79	0.0	0.51	...	2.5
1006	CHEESE,BRIE	48.42	334	20.75	27.68	2.70	0.45	0.0	0.45	...	2.3
1007	CHEESE,CAMEMBERT	51.80	300	19.80	24.26	3.68	0.46	0.0	0.46	...	2.0
1008	CHEESE,CARAWAY	39.28	376	25.18	29.20	3.28	3.06	0.0	NaN	...	NaN
1009	CHEESE,CHEDDAR	37.10	406	24.04	33.82	3.71	1.33	0.0	0.28	...	2.9

5.3.2.2 Step 2: Statistical Properties of Different Food Items

To find statistical values of each food item, that is, mean value of protein; fiber; calcium; vitamins C, A, and D; and others:

```
dfUSDANND.describe()
```

	NDB_No	Water_(g)	Energ_Kcal	Protein_(g)	Lipid_Tot_(g)	Ash_(g)	Carbohydrt_(g)	Fiber_TD_(g)	Sugar_Tot_(g)	Calcium_(mg)
count	1000.000000	1000.000000	1000.000000	1000.000000	1000.000000	1000.00000	1000.000000	958.000000	810.000000	991.000000
mean	3351.181000	53.404470	283.503000	10.032420	21.723860	1.93224	12.926770	1.639457	7.620346	166.944501
std	1474.112347	32.325082	257.923358	11.528427	30.944452	5.11056	19.287237	6.131702	13.263861	320.019059
min	1001.000000	0.000000	0.000000	0.000000	0.000000	0.00000	0.000000	0.000000	0.000000	0.000000
25%	2019.750000	17.070000	78.000000	0.827500	2.117500	0.44500	0.060000	0.000000	0.000000	10.000000
50%	3805.500000	65.115000	189.500000	3.880000	7.355000	0.94000	6.180000	0.000000	2.290000	24.000000
75%	4655.250000	81.255000	376.000000	18.740000	26.560000	1.90250	14.210000	0.400000	8.947500	133.000000
max	5236.000000	94.780000	902.000000	84.630000	100.000000	99.80000	86.680000	53.200000	74.460000	2240.000000

8 rows × 50 columns

5.3.2.3 Step 3: Nutrient Analysis: Food Items with Energy K Calories Above the Mean and Protein (g) with Max Values

To perform food nutrient and energy analysis, query the data frame with subset data collection: for instance, to retrieve all food items that have maximum energy K calories:

```
dfUSDANND[dfUSDANND['Energ_Kcal'] == dfUSDANND['Energ_Kcal'].
max()]
```

As you can see fish oil, beef, lard, and mutton tallow have the maximum K caloric food.

Output:

	NDB_No	Shrt_Desc	Water_(g)	Energ_Kcal	Protein_(g)	Lipid_Tot_(g)	Ash_(g)	Carbohydrt_(g)	Fiber_TD_(g)	Sugar_Tot_(g)
610	4001	FAT,BEEF TALLOW	0.0	902	0.0	100.0	0.0	0.0	0.0	0.0
611	4002	LARD	0.0	902	0.0	100.0	0.0	0.0	0.0	0.0
664	4520	FAT,MUTTON TALLOW	0.0	902	0.0	100.0	0.0	0.0	0.0	0.0
701	4589	FISH OIL,COD LIVER	0.0	902	0.0	100.0	0.0	0.0	0.0	NaN
702	4590	FISH OIL,HERRING	0.0	902	0.0	100.0	0.0	0.0	0.0	NaN
703	4591	FISH OIL,MENHADEN	0.0	902	0.0	100.0	0.0	0.0	0.0	NaN
704	4592	FISH OIL,MENHADEN,FULLY HYDR	0.0	902	0.0	100.0	0.0	0.0	0.0	NaN
705	4593	FISH OIL,SALMON	0.0	902	0.0	100.0	0.0	0.0	0.0	NaN
706	4594	FISH OIL,SARDINE	0.0	902	0.0	100.0	0.0	0.0	0.0	NaN

The fish oil, cod liver is extracted from the livers of Atlantic cod and then filtered for purity. The leading producers of cod liver oil supplements are northern countries like Norway and Iceland, but you can find it in most supermarkets and health food stores. People take cod liver oil capsules to help with arthritic joint pain and cardiovascular disease prevention[72].

Biodiesel: Mutton tallow, along with other animal fats, can be converted to biodiesel to power vehicles. The process of producing biodiesel from tallow is very similar to that of making it from plants. Tallow-based biodiesel burns more efficiently and more cleanly than plant biodiesel but crystallizes at higher temperatures, making it unsuitable for use in very cold climates[73].

To retrieve high cholesterol food item, get food items that have max cholesterol:

```
dfUSDANND[dfUSDANND['Cholestrl_(mg)'] == dfUSDANND['Cholestrl_
(mg)'].max()]
```

Output: Egg, yolk, dried has the highest cholesterol value of 2307 grams.

	NDB_No	Shrt_Desc	Water_(g)	Energ_Kcal	Protein_(g)	Lipid_Tot_(g)	Ash_(g)	Carbohydrt_(g)	Fiber_TD_(g)	Sugar_Tot_(g)	...
125	1137	EGG,YOLK,DRIED	2.73	679	33.73	58.13	3.3	2.1	0.0	0.07	...

1 rows × 53 columns

To get food items that have cholesterol greater than mean value (60.96 mg):

[72] Fish Oil, Cod Liver - https://www.webmd.com/diet/cod-liver-oil-health-benefits#1

[73] Tallow - https://sciencing.com/differences-between-glycolic-acid-glycerin-8062972.html

```
dfUSDANND[dfUSDANND['Cholestrl_(mg)'] >= (dfUSDANND
['Cholestrl_(mg)'].mean())]
```

	NDB_No	Shrt_Desc	Water_(g)	Energ_Kcal	Protein_(g)	Lipid_Tot_(g)	Ash_(g)	Carbohydrt_(g)	Fiber_TD_(g)	Sugar_Tot_(g)
0	1001	BUTTER,WITH SALT	15.87	717	0.85	81.11	2.11	0.06	0.0	0.06
1	1002	BUTTER,WHIPPED,WITH SALT	15.87	717	0.85	81.11	2.11	0.06	0.0	0.06
2	1003	BUTTER OIL,ANHYDROUS	0.24	876	0.28	99.48	0.00	0.00	0.0	0.00
3	1004	CHEESE,BLUE	42.41	353	21.40	28.74	5.11	2.34	0.0	0.50
4	1005	CHEESE,BRICK	41.11	371	23.24	29.68	3.18	2.79	0.0	0.51
...	...	...	...	...	...	...	...	...	...	...
993	5215	TURKEY,BACK,FROM WHL BIRD,NON-ENHANCED,MEAT ON...	76.01	113	21.28	2.50	1.03	0.15	0.0	0.10
994	5216	TURKEY,BACK,FROM WHL BIRD,NON-ENHANCED,MEAT ON...	65.23	173	27.71	6.04	1.12	0.00	0.0	0.00
996	5220	TURKEY,BREAST,FROM WHL BIRD,NON-ENHANCED,MEAT ...	67.88	147	30.13	2.08	1.22	0.00	0.0	0.00
998	5228	TURKEY,WING,FROM WHL BIRD,NON-ENHANCED,MEAT ON...	67.88	147	30.13	2.08	1.22	0.00	0.0	0.00

5.3.2.4 Step 4: Correlation Data

To retrieve correlation data:

```
plotCorrelationMatrix(dfUSDANND, 9)
```

Output:

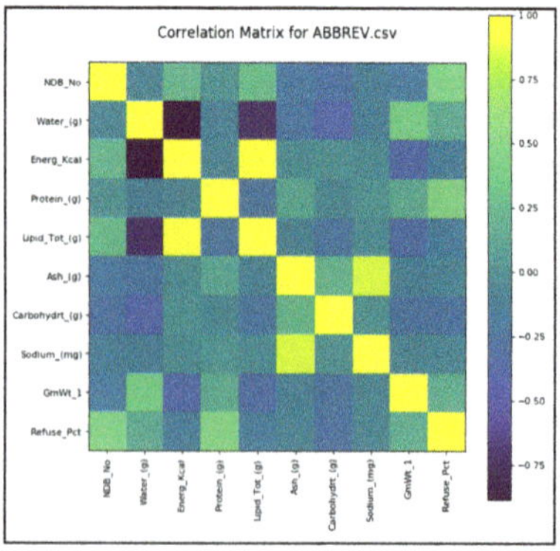

5.3.3 Exploratory Data Analysis: USDA National Nutrient Data

FoodData Central is an integrated data system that provides expanded nutrient[74] profile data and links to related agricultural and experimental research. The data can be downloaded from FoodData Central Data[75]. FoodData Central takes the analysis, compilation, and presentation of nutrient that can be used by, and has benefits for, a variety of users, including researchers, policy makers, academicians and educators, nutrition and health professionals, product developers, and others. It includes five distinct data types that provide information on food and nutrient profiles: Foundation Foods, Food and Nutrient Database for Dietary Studies 2017–2018 (FNDDS 2017–2018), National Nutrient Database for Standard Reference Legacy Release (SR Legacy), USDA Global Branded Food Products Database (Branded Foods), and Experimental Foods. Each of these data types has a unique purpose and unique attributes. FoodData Central links these distinct data types in one location, thus strengthening the ability of researchers, policy makers, and others to address vital issues related to food, nutrition, and diet-health interactions. It provides a broad snapshot in time of the nutrients and other components found in a wide variety of foods and food products[76].

5.3.3.1 Step 1: Load Libraries

To begin this exploratory analysis, first import libraries and define functions for plotting the data using matplotlib. Depending on the data, not all plots will be made:

```
from mpl_toolkits.mplot3d import Axes3D
from sklearn.preprocessing import StandardScaler
import matplotlib.pyplot as plt # plotting
import numpy as np # linear algebra
import os # accessing directory structure
import pandas as pd # data processing, CSV file I/O (e.g. pd.
read_csv)
```

[74] FoodData Central—https://fdc.nal.usda.gov/

[75] Data Download—https://fdc.nal.usda.gov/download-datasets.html

[76] USDA—https://fdc.nal.usda.gov/

Step 2: Helper Functions to Plot per Column Distribution

Plot column-level distribution:

```
# Distribution graphs (histogram/bar graph) of column data
def plotPerColumnDistribution(df, nGraphShown, nGraphPerRow):
  nunique = df.nunique()
  df = df[[col for col in df if nunique[col] > 1 and nunique[col] <
50]] # For displaying purposes, pick columns that have between 1 and
50 unique values
  nRow, nCol = df.shape
  columnNames = list(df)
  nGraphRow = (nCol + nGraphPerRow - 1) / nGraphPerRow
  plt.figure(num = None, figsize = (6 * nGraphPerRow, 8 * nGraphRow),
dpi = 80, facecolor = 'w', edgecolor = 'k')
  for i in range(min(nCol, nGraphShown)):
    plt.subplot(nGraphRow, nGraphPerRow, i + 1)
    columnDf = df.iloc[:, i]
    if (not np.issubdtype(type(columnDf.iloc[0]), np.number)):
      valueCounts = columnDf.value_counts()
      valueCounts.plot.bar()
    else:
      columnDf.hist()
    plt.ylabel('counts')
    plt.xticks(rotation = 90)
    plt.title(f'{columnNames[i]} (column {i})')
  plt.tight_layout(pad = 1.0, w_pad = 1.0, h_pad = 1.0)
  plt.show()
```

Plot correlation matrix:

```
# Correlation matrix
def plotCorrelationMatrix(df, graphWidth):
  filename = df.dataframeName
  df = df.dropna('columns') # drop columns with NaN
  df = df[[col for col in df if df[col].nunique() > 1]] # keep columns
where there are more than 1 unique values
  if df.shape[1] < 2:
    print(f'No correlation plots shown: The number of non-NaN or
constant columns ({df.shape[1]}) is less than 2')
    return
  corr = df.corr()
  plt.figure(num=None, figsize=(graphWidth, graphWidth), dpi=80,
facecolor='w', edgecolor='k')
  corrMat = plt.matshow(corr, fignum = 1)
  plt.xticks(range(len(corr.columns)), corr.columns,
rotation=90)
```

(continued)

```
plt.yticks(range(len(corr.columns)), corr.columns)
plt.gca().xaxis.tick_bottom()
plt.colorbar(corrMat)
plt.title(f'Correlation Matrix for {filename}', fontsize=15)
plt.show()
```

Plot scatter matrix:

```
# Scatter and density plots
def plotScatterMatrix(df, plotSize, textSize):
  df = df.select_dtypes(include = [np.number]) # keep only
numerical columns
  # Remove rows and columns that would lead to df being singular
  df = df.dropna('columns')
  df = df[[col for col in df if df[col].nunique() > 1]] # keep columns
where there are more than 1 unique values
  columnNames = list(df)
  if len(columnNames) > 10: # reduce the number of columns for matrix
inversion of kernel density plots
    columnNames = columnNames[:10]
  df = df[columnNames]
  ax = pd.plotting.scatter_matrix(df, alpha=0.75, figsize=
[plotSize, plotSize], diagonal='kde')
  corrs = df.corr().values
  for i, j in zip(*plt.np.triu_indices_from(ax, k = 1)):
    ax[i, j].annotate('Corr. coef = %.3f' % corrs[i, j], (0.8, 0.2),
xycoords='axes fraction', ha='center', va='center',
size=textSize)
  plt.suptitle('Scatter and Density Plot')
  plt.show()
```

5.3.3.2 Step 3: Load USDA Nutrition Data

Load USDA Nutrition Data[77]. USDA Dataset can be found at USDA site[78]:

[77] USDA National Nutrient—https://www.kaggle.com/kerneler/starter-usda-national-nutrient-e9255702-6/data

[78] USDA Foundation Data—https://fdc.nal.usda.gov/fdc-datasets/FoodData_Central_foundation_food_csv_2021-10-28.zip

```
nRowsRead = 1000 # specify 'None' if want to read whole file
nRowsRead = 1000 # specify 'None' if want to read whole file
# test.csv may have more rows in reality, but we are only loading/
previewing the first 1000 rows
dfNutrientTest = pd.read_csv('Nutrient_test.csv',
delimiter=',', nrows = nRowsRead)
dfNutrientTest.dataframeName = 'Nutrient_test.csv'
nRow, nCol = dfNutrientTest.shape
print(f'There are {nRow} rows and {nCol} columns')
```

Output: There are 1000 rows and 41 columns:

```
dfNutrientTest.head(5)
```

	ID	FoodGroup	Descrip	Energy_kcal	Protein_g	Fat_g	Carb_g	Sugar_g	Fiber_g	VitA_mcg	...	Folate_USRDA	Niacin_USRDA	Riboflavin_USI
0	23116	Beef Products	Beef, chuck, under blade steak, boneless, sepa...	275.0	28.23	18.00	0.00	0.00	0.0	8.0	...	0.0175	0.235750	0.17
1	10047	Pork Products	Pork, fresh, loin, center rib (roasts), bone-i...	248.0	26.99	14.68	0.00	0.00	0.0	5.0	...	0.0000	0.593125	0.22
2	15270	Finfish and Shellfish Products	Crustaceans, shrimp, untreated, raw	85.0	20.10	0.51	0.00	0.00	0.0	0.0	...	0.0000	0.000000	0.00
3	1259	Dairy and Egg Products	Cheese spread, American or Cheddar cheese base...	176.0	13.41	8.88	10.71	7.06	0.0	185.0	...	0.0000	0.009562	0.34
4	19100	Sweets	Candies, fudge, chocolate, prepared-from-recipe	411.0	2.39	10.41	76.44	73.12	1.7	44.0	...	0.0100	0.011000	0.06

Print the first five header rows:

5.3.3.3 Step 4: Plot per Column Distribution

```
plotPerColumnDistribution(dfNutrientTest, 10, 5)
```

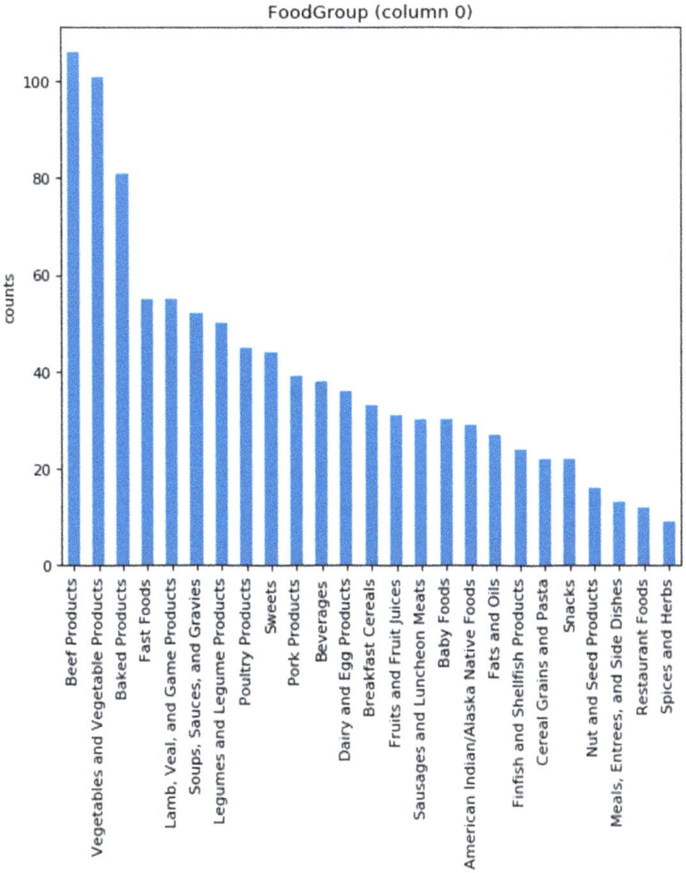

5.3.3.4 Step 5: Plot Correlation Matrix

```
plotCorrelationMatrix(dfNutrientTrain, 9)
```

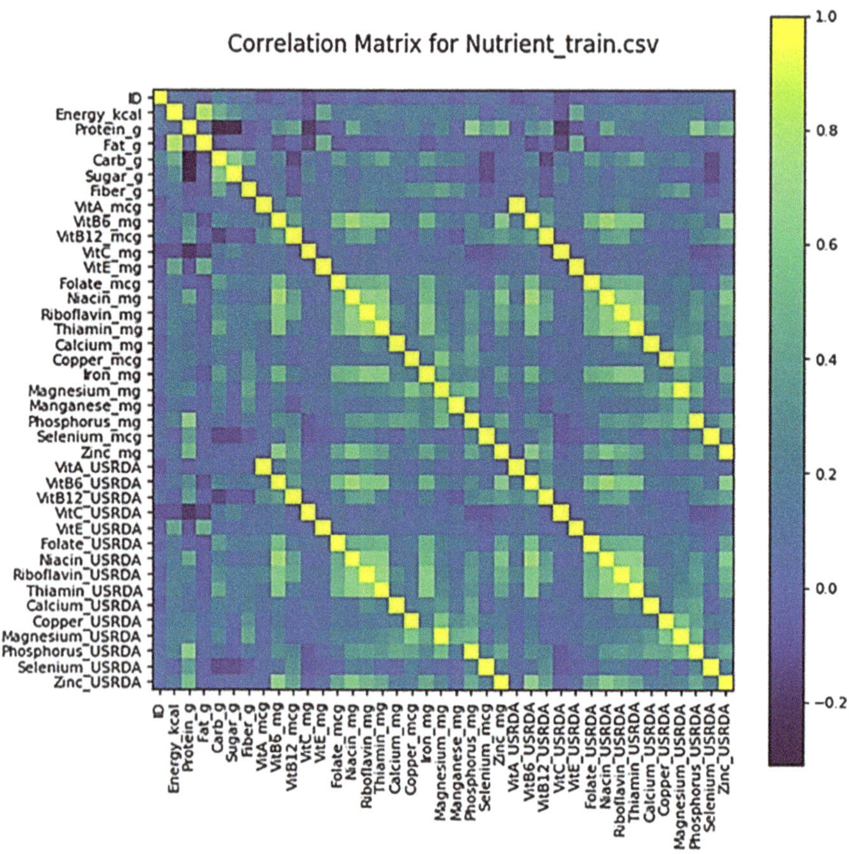

5.4 Household Dietary Diversity Score (HDDS): Venezuela

The data collected in Venezuela, as part of the JuntosEsMejor Challenge funded by the US Agency for International Development (USAID) and Inter-American Development Bank (IDB). Venezuela seen unprecedented displacement of people in the western hemisphere for various reasons, and the purpose of the dataset is to access food security[79].

[79]Better Together Challenge—Innovating with Venezuela for a brighter future - https://juntosesmejorve.org/en/

5.4.1 Dataset

Premise[80] organization belonging to the Food Security Humanitarian Cluster in Venezuela uses data to influence decision-making within the cluster. Since the beginning of the Venezuelan crisis back in 2014, this organization has been tracking and documenting the failing delivery of basic public services and the impact of the crisis on the food security of the affected population in Zulia state [40]. The organization used to collect data once a year (one point in time) with enumerators and with a limited regional coverage (only in Zulia and Maracaibo). They have been using the same dataset for the past 10 months because the conditions in Venezuela have worsened and collecting on-the-ground data is challenging[81] for a variety of reasons, including the onset of COVID-19 [41].

The data collected in the humanitarian information management space designed two surveys to launch on the Premise app and crowdsource data to have timely information about dietary daily consumption and needs as well as weekly coping mechanisms.

5.4.2 Survey #1: Daily Food Consumption

The key objective of this survey is to collect data that helps organizations in the food security space identify where and who needs food assistance. To do this, the survey is composed of a set of questions that derive from the Household Dietary Diversity Score.

The Household Dietary Diversity Score (HDDS) is a population-level indicator of household food access. The questions represent a rapid assessment tool and a simple count of food groups that a household has consumed over the preceding 24 h. The data collected can also be analyzed to provide information on specific food groups of interest. The HDDS is meant to reflect the economic ability of a household to access

[80] Solving The Need For Up-To-Date Food Consumption Data In Venezuela - https://www.premise.com/solving-the-need-for-up-to-date-food-consumption-data-in-venezuela/

[81] Known unknowns: The challenge of collecting COVID-19 data in Venezuela - https://www.thenewhumanitarian.org/news-feature/2020/06/04/coronavirus-data-Venezuela

a variety of foods in the past 24 h. The survey collected specifically in the regions of Caracas, Miranda, Tachira, and Zulia a sample size of 95 submissions per region and made sure that at least 50% of the respondents are women.

5.4.3 Survey #2: Weekly Food Consumption

The key objective of this survey is to collect data that helps organizations in the food security space identify where and who needs food assistance. To do this, the survey is composed of a set of questions that derive in two key scores: the Food Consumption Score (FCS) and the Coping Strategy Index (CSI). At least 50% of the respondents are women.

The Food Consumption Score (FCS) is used by the World Food Programme (WFP) to measure the frequency of consumption of different food groups by a household over the previous 7 days. There are standard weights for each food group (more nutritionally dense foods are given higher weights), and overall scores can range from 0 to 112.

The Coping Strategy Index (CSI) is another WFP indicator that helps assess overall food security by measuring the extent to which households may use harmful coping strategies when there isn't enough food or money to buy food.

5.4.4 Machine Learning Model: Household Dietary Diversity Score (HDDS)—Venezuela

Software code for this model can be found on GitHub Link:
Final_OpenDailyFood.ipynb (Jupyter Notebook Code)

HDDS contains data crowdsourced[82] daily from Venezuelans (please see Fig. 5.16 for the Internet search engine-extracted map) using the Premise Data mobile application. The data collected allows the fast measurement of Household Dietary Diversity Score (HDDS), with the goal of providing context around the food security in vulnerable communities [34].

[82] HDD—https://data.humdata.org/dataset/open_daily_food

Fig. 5.16 Venezuelan map

5.4.4.1 Step 1: Load Data

```
import pandas as pd

# load the training dataset
opendailyfooddfwithlabelsandnumericals = pd.read_csv
("open_daily_food.csv")
opendailyfooddfwithlabelsandnumericals.head()
```

Output:

	submission_id	submission_date	gender	age	geography	financial_situation	education	employment_status	submission_state	cereal_24hrs	...
0	5635627144708096	2021-09-10	Female	26 to 35 years old	Rural	Prefer not to answer	Prefer not to answer	None of the above	Zulia	1	...
1	5627486738841600	2021-09-07	Female	26 to 35 years old	Rural	Prefer not to answer	Prefer not to answer	None of the above	Zulia	1	...
2	5142630263685120	2020-07-21	Male	Not Available	Rural	I can afford food, but nothing else	College or university	Student and work part-time	Zulia	1	...
3	6094210198667264	2020-07-19	Male	Not Available	Rural	I can afford food, but nothing else	College or university	Student and work part-time	Zulia	1	...
4	6272566634479616	2020-07-25	Male	Not Available	Rural	I can afford food, but nothing else	College or university	Student and work part-time	Zulia	1	...

5.4.4.2 Step 2: Calculate FCS Score

As per Eq. (5.2), calculate the FCS score for the dataset. Apply weights and compute the FCS score:

```
opendailyfooddfwithlabelsandnumericals['wt_score'] = 2*
opendailyfooddfwithlabelsandnumericals.cereal_24hrs +
2*opendailyfooddfwithlabelsandnumericals.plantains_24hrs  +
1*opendailyfooddfwithlabelsandnumericals.vegs_24hrs  +
1*opendailyfooddfwithlabelsandnumericals.fruits_24hrs  +
4*opendailyfooddfwithlabelsandnumericals.meat_24hrs  +
4*opendailyfooddfwithlabelsandnumericals.eggs_24hrs  +
4*opendailyfooddfwithlabelsandnumericals.fish_24hrs  +
3*opendailyfooddfwithlabelsandnumericals.grains_24hrs  +
4*opendailyfooddfwithlabelsandnumericals.dairy_24hrs
+  0.5* opendailyfooddfwithlabelsandnumericals.oils_fats_24hrs
+  0.5 * opendailyfooddfwithlabelsandnumericals.sugars_24hrs

opendailyfooddfwithlabelsandnumericals
[['wt_score','fcs','cereal_24hrs' ,  'plantains_24hrs' ,
'vegs_24hrs' ,   'fruits_24hrs' ,  'meat_24hrs'   ,'eggs_24hrs' ,
'fish_24hrs' ,  'grains_24hrs' ,  'dairy_24hrs'   ,
'oils_fats_24hrs'   , 'sugars_24hrs']]
```

Output:

	wt_score	fcs	cereal_24hrs	plantains_24hrs	vegs_24hrs	fruits_24hrs	meat_24hrs	eggs_24hrs	fish_24hrs	grains_24hrs	dairy_24hrs	oils_fats_24
0	21	16.0	1	1	1	1	1	0	1	1	1	
1	25	16.0	1	1	1	1	1	1	1	1	1	
2	18	13.0	1	1	1	1	1	1	0	0	1	
3	16	13.0	1	0	1	1	1	0	1	0	1	
4	17	12.0	1	1	1	0	1	0	1	0	1	
...	...	...	...	...	...	...	...	...	...	...	...	
199508	14	12.5	1	1	1	1	0	1	0	0	1	
199509	16	15.0	1	1	1	0	0	1	0	1	1	
199510	10	10.5	1	0	1	0	0	0	0	1	1	
199511	18	13.0	1	1	1	1	1	1	0	0	1	
199512	7	8.0	1	0	1	0	1	0	0	0	0	

199513 rows × 13 columns

As can be seen, wt_score and fcs are populated with the computed values.

5.4.4.3 Step 3: FCS Score

Calculate FCS score based on the HDDS and FCS[83] [42]:

[83]Measuring Household Food Security Index for High Hill Tribal Community of Nagaland—https://pdfs.semanticscholar.org/2c3c/a9e28afc515f710693aba714681e43d32fd4.pdf

```
def calcFSI(hdds, fcs):
  if hdds <=5 and fcs <=7:
    ret = 'low food security'
  elif hdds >7 and fcs >10 :
    ret = 'full food security'
  elif (hdds>5 and hdds<=7) and (fcs >7 and fcs <=10) :
    ret = 'marginal food security'
  else:
    ret = 'not applicable'
  return ret
opendailyfooddfwithlabelsandnumericals['fsi'] =
opendailyfooddfwithlabelsandnumericals.apply(lambda x: calcFSI
(x['hdds'], x['fcs']), axis=1)
```

Output:

us	submission_state	cereal_24hrs	...	fish_24hrs	grains_24hrs	dairy_24hrs	oils_fats_24hrs	sugars_24hrs	head_household	hdds	fcs	wt_score	fsi
ve	Zulia	1	...	1	1	1	1	1	Yes	10	16.0	21	full food security
ve	Zulia	1	...	1	1	1	1	1	Yes	11	16.0	25	full food security
rk ne	Zulia	1	...	0	0	1	1	1	No	9	13.0	18	full food security
rk ne	Zulia	1	...	1	0	1	1	1	No	8	13.0	16	full food security
rk ne	Zulia	1	...	1	0	1	1	1	No	8	12.0	17	full food security

5.4.4.4 Step 4: Visualize Food Security Score Distribution Histogram

Prepare the histogram of the FSI scores. As it can be seen, the graph exhibits normal distribution:

```
import pandas as pd
import matplotlib.pyplot as plt
# This ensures plots are
displayed inline in the
Jupyter notebook
%matplotlib inline
# Get the label column
label = opendailyfooddfwithlabel
sandnumericals['wt_score']
# Create a figure for 2 subplots
(2 rows, 1 column)
fig, ax = plt.subplots(2, 1,
figsize = (9,12))
# Plot the histogram
ax[0].hist(label, bins=100)
ax[0].set_ylabel('Frequency')
# Add lines for the mean, median, and
mode
ax[0].axvline(label.mean(),
color='magenta',
linestyle='dashed', linewidth=2)
ax[0].axvline(label.median(),
color='cyan', linestyle='dashed',
linewidth=2)
# Plot the boxplot
ax[1].boxplot(label, vert=False)
ax[1].set_xlabel('wt_score')
# Add a title to the Figure
fig.suptitle('wt_score Distribu-
tion')
# Show the figure
fig.show()
```

Output:

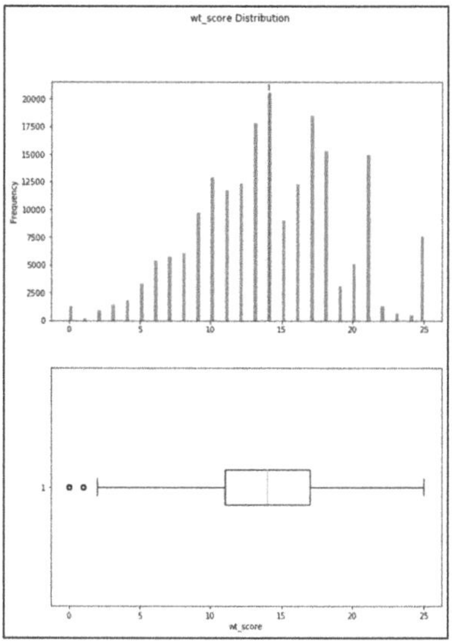

5.4.4.5 Step 5: Value Counts of Categorical Features

Plot the categorical features of Food Score HDDS data. The features include gender, age, geography, financial situation, education, and employment status:

- As it can be seen, the crowdsource data has the highest age counts for over a 45-year-old group, followed by 26–35 years old and 36–45 years old. Below 18 represents the smallest numbers.
- The highest survey responses came from city center or metropolitan area followed by suburban/peri-urban and rural.

```
import numpy as np
# plot a bar plot for each
categorical feature count
categorical_features =
['gender', 'age',
'geography','financial_situation',
'education',
'employment_status','fsi']
for col in categorical_features:
  counts = opendailyfooddfwithlabel
sandnumericals[col].value_counts
().sort_index()
  fig = plt.figure(figsize=(9, 6))
  ax = fig.gca()
  counts.plot.bar(ax = ax,
color='steelblue')
  ax.set_title(col + ' counts')
  ax.set_xlabel(col)
  ax.set_ylabel("Frequency")
plt.show()
```

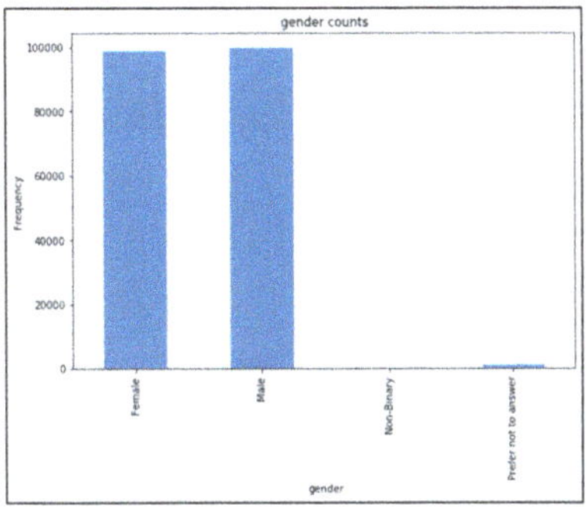

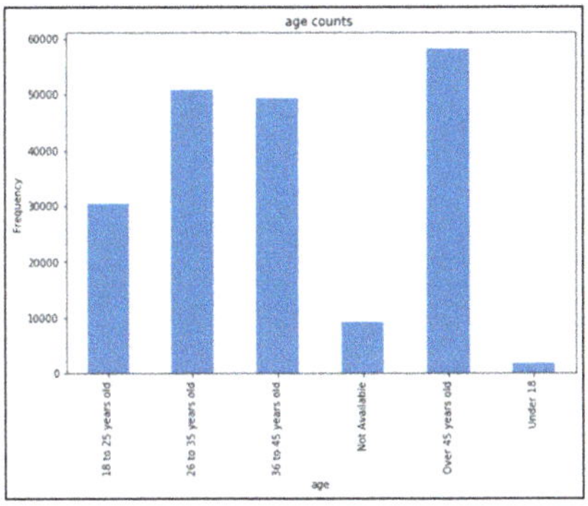

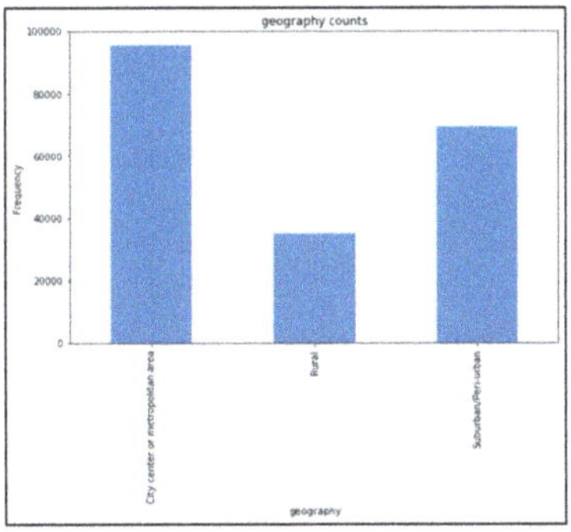

- Financial situation of survey respondents shows significant distribution with the highest number of survey respondents' category including first "can afford food but nothing else," next "afford food and regular expenses," and third "cannot afford enough food for the family."

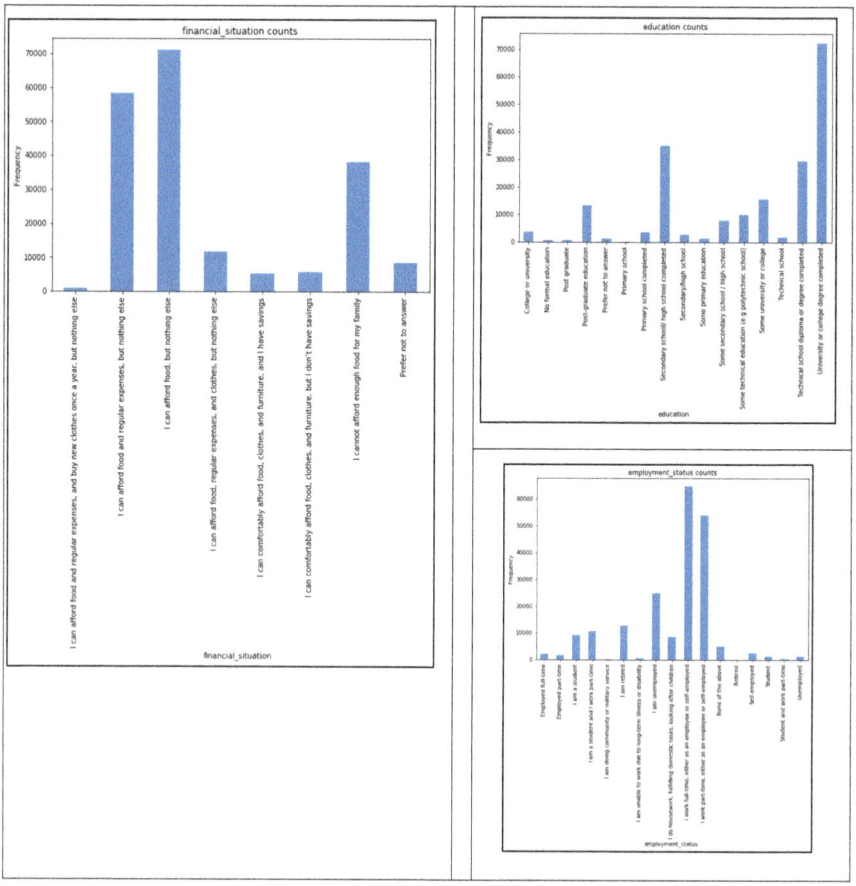

- Education-wise the highest number of survey respondents is first "university or college degree completed," next "secondary or high school completed," followed by "technical school diploma or degree completed," and finally "postgraduate" education. "No formal school," "technical school," and "some primary education" remain a very small portion of survey respondents.
- Majority of survey respondents are employed "full time" or "self-employed," followed by "part-time" employment. Unemployed, retired, student, and self-employed categories represent a small share of survey respondents.

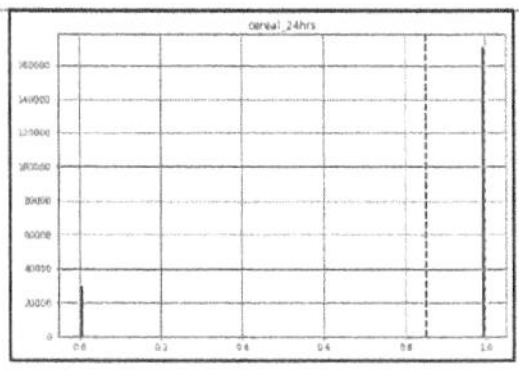

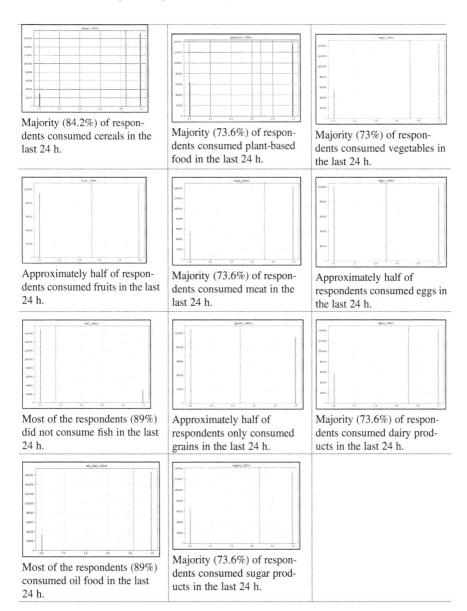

Majority (84.2%) of respondents consumed cereals in the last 24 h.

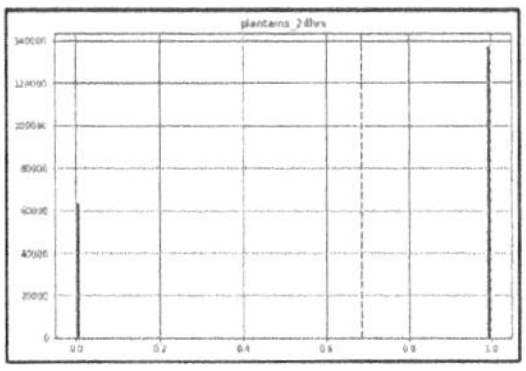

Majority (73.6%) of respondents consumed plant-based food in the last 24 h.

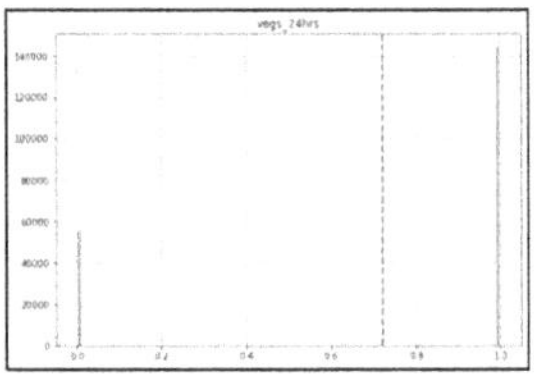

Majority (73%) of respondents consumed vegetables in the last 24 h.

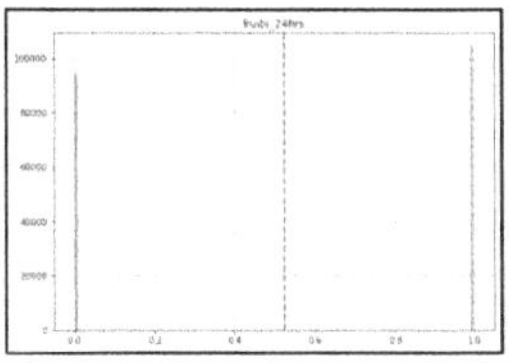

Approximately half of respondents consumed fruits in the last 24 h.

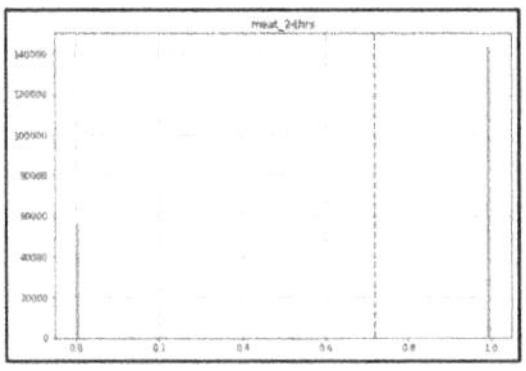

Majority (73.6%) of respondents consumed meat in the last 24 h.

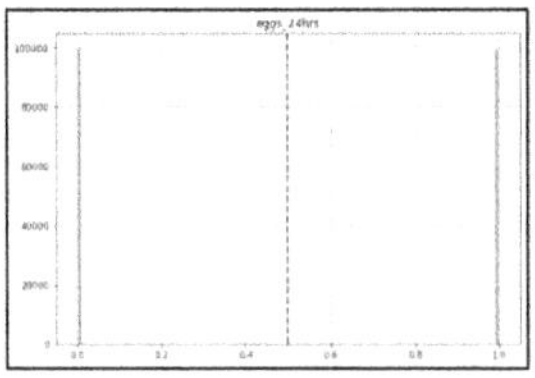

Approximately half of respondents consumed eggs in the last 24 h.

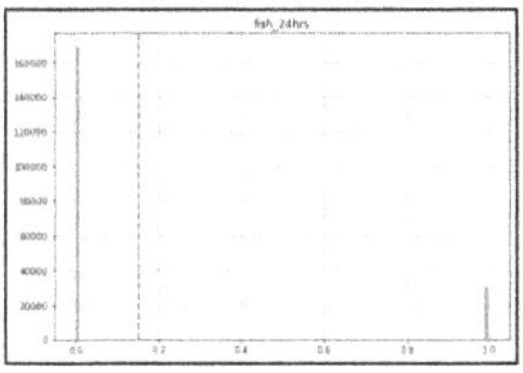

Most of the respondents (89%) did not consume fish in the last 24 h.

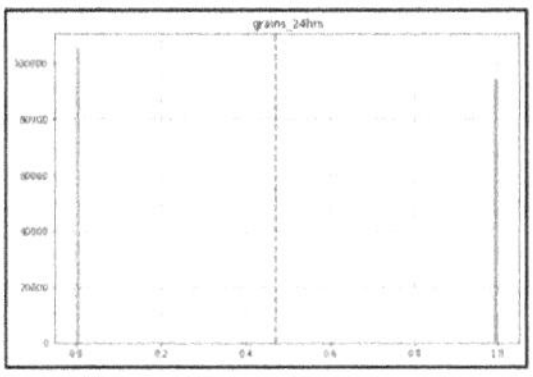

Approximately half of respondents only consumed grains in the last 24 h.

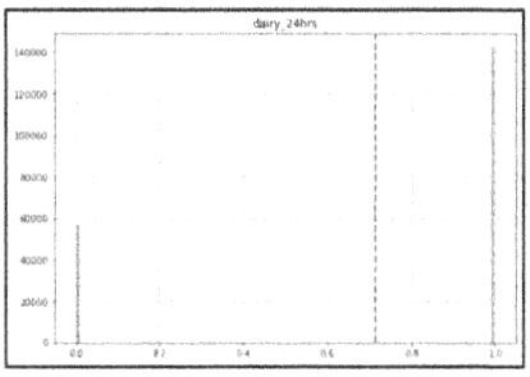

Majority (73.6%) of respondents consumed dairy products in the last 24 h.

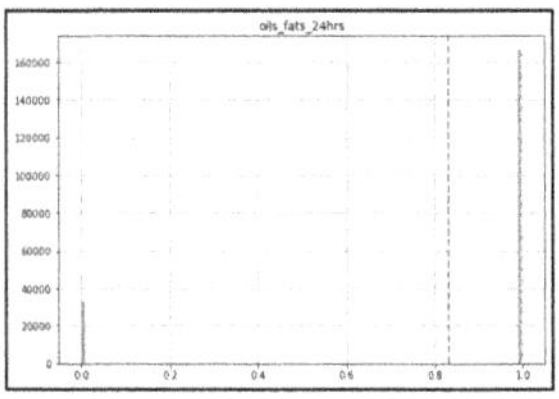

Most of the respondents (89%) consumed oil food in the last 24 h.

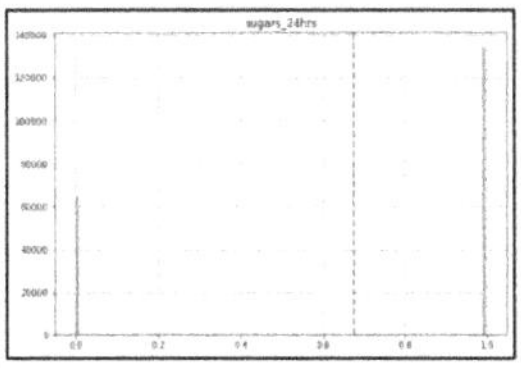

Majority (73.6%) of respondents consumed sugar products in the last 24 h.

5.4.4.6 Step 6: Correlation

The correlation between HDDS and FCS shows a considerable influence:

```
pd.set_option("display.max_rows", None, "display.max_columns",
None)
corr = opendailyfoodwithallnumericalDF.corr() # df is the pandas
dataframe
c1 = corr.abs().unstack()
c1.sort_values(ascending = False)
```

Output:

dairy_24hrs	fsicode	0.621652
fsicode	dairy_24hrs	0.621652
hdds	fruits_24hrs	0.573350
fruits_24hrs	hdds	0.573350
dairy_24hrs	wt_score	0.571910
wt_score	dairy_24hrs	0.571910
plantains_24hrs	hdds	0.534908
hdds	plantains_24hrs	0.534908
vegs_24hrs	hdds	0.531808

What the correlation values tell is that dairy, cereals, and fruits play an important role in food security.

5.4.4.7 Step 7: Prepare Data Frame for Decision Tree

```
opendailyfoodDT=opendailyfooddfwithlabelsandnumericals
[['cereal_24hrs', 'plantains_24hrs', 'vegs_24hrs',
'fruits_24hrs',
    'meat_24hrs', 'eggs_24hrs', 'fish_24hrs', 'grains_24hrs',
'dairy_24hrs',
    'oils_fats_24hrs', 'sugars_24hrs', 'agecode',
'geographycode',
    'educationcode', 'financial_situationcode',
'employment_statuscode',
    'head_householdcode','fsi']]

X = opendailyfoodDT.iloc[:, :-1].values
y = opendailyfoodDT.iloc[:, -1].values

from sklearn.model_selection import train_test_split
X_train, X_test, y_train, y_test = train_test_split(X, y,
test_size = 0.25, random_state = 0)
```

Output: X:

```
array([[1., 1., 1., ..., 1., 1., 1.],
    [1., 1., 1., ..., 1., 1., 1.],
    [1., 1., 1., ..., 2., 2., 1.],
    ...,
    [1., 0., 1., ..., 9., 1., 2.],
    [1., 1., 1., ..., 9., 1., 1.],
    [1., 0., 1., ..., 9., 1., 2.]])
```

Y:

```
array(['full food security', 'full food security', 'full food
security',
   ..., 'not applicable', 'full food security', 'not applicable'],
   dtype=object)
```

5.4.4.8 Step 8: Construct Decision Tree

Prepare the decision tree with a maximum size of 4:

```
from sklearn.tree import DecisionTreeClassifier

from sklearn.tree import plot_tree
import matplotlib.pyplot as plt

fn=['cereal_24hrs', 'plantains_24hrs', 'vegs_24hrs',
    'fruits_24hrs', 'meat_24hrs', 'eggs_24hrs', 'fish_24hrs',
    'grains_24hrs', 'dairy_24hrs', 'oils_fats_24hrs',
'sugars_24hrs',
    'agecode',
    'geographycode', 'educationcode', 'financial_situationcode',
    'employment_statuscode', 'head_householdcode']

target=class_labels

clf = DecisionTreeClassifier(max_depth=4) #max_depth is maximum
number of levels in the tree
clf.fit(X_train, y_train)
```

(continued)

```
plt.figure(figsize=(30,40))
a = plot_tree(clf,
        feature_names= fn,
        class_names=target,
        filled=True,
        rounded=True,
        fontsize=12)
```

Output:
Please find decision rules:

```
|--- fruits_24hrs <= 0.50
|   |--- dairy_24hrs <= 0.50
|   |   |--- vegs_24hrs <= 0.50
|   |   |   |--- grains_24hrs <= 0.50
|   |   |   |   |--- class: low food security
|   |   |   |--- grains_24hrs > 0.50
|   |   |   |   |--- class: not applicable
|   |   |--- vegs_24hrs > 0.50
|   |   |   |--- grains_24hrs <= 0.50
|   |   |   |   |--- class: not applicable
|   |   |   |--- grains_24hrs > 0.50
|   |   |   |   |--- class: not applicable
|   |--- dairy_24hrs > 0.50
|   |   |--- plantains_24hrs <= 0.50
|   |   |   |--- meat_24hrs <= 0.50
|   |   |   |   |--- class: not applicable
|   |   |   |--- meat_24hrs > 0.50
|   |   |   |   |--- class: not applicable
|   |   |--- plantains_24hrs > 0.50
|   |   |   |--- eggs_24hrs <= 0.50
|   |   |   |   |--- class: not applicable
|   |   |   |--- eggs_24hrs > 0.50
|   |   |   |   |--- class: full food security
|--- fruits_24hrs > 0.50
|   |--- plantains_24hrs <= 0.50
|   |   |--- dairy_24hrs <= 0.50
|   |   |   |--- grains_24hrs <= 0.50
|   |   |   |   |--- class: low food security
|   |   |   |--- grains_24hrs > 0.50
|   |   |   |   |--- class: not applicable
|   |   |--- dairy_24hrs > 0.50
|   |   |   |--- vegs_24hrs <= 0.50
|   |   |   |   |--- class: not applicable
|   |   |   |--- vegs_24hrs > 0.50
|   |   |   |   |--- class: not applicable
|   |--- plantains_24hrs > 0.50
```

(continued)

```
| | |--- dairy_24hrs <= 0.50
| | | |--- grains_24hrs <= 0.50
| | | | |--- class: marginal food security
| | | |--- grains_24hrs > 0.50
| | | | |--- class: full food security
| | |--- dairy_24hrs > 0.50
| | | |--- sugars_24hrs <= 0.50
| | | | |--- class: not applicable
| | | |--- sugars_24hrs > 0.50
| | | | |--- class: full food security
```

5.4.4.9 Step 9: Construct OLS Regression Model

The purpose of the regression model is to evaluate and construct the influence of feature variables on food security (additionally, to evaluate coefficients of variables):

```
#Using OLS method

from statsmodels.formula.api import ols

formula='fsicode ~ cereal_24hrs + plantains_24hrs + vegs_24hrs +
fruits_24hrs + meat_24hrs + eggs_24hrs + fish_24hrs + grains_24hrs +
dairy_24hrs + oils_fats_24hrs + sugars_24hrs + geographycode +
financial_situationcode + employment_statuscode'
model=ols(formula,dataols).fit()
model.summary()
```

Output:

$$
\begin{aligned}
\text{FSI} = 4.277 &- 0.254128184^{*}\ \text{cereal\_24hrs} - 0.20105063^{*}\ \text{plantains\_24hrs} \\
&- 0.33186527^{*}\ \text{vegs\_24hrs} - 0.297613375^{*}\ \text{fruits\_24hrs} \\
&- 0.395512963^{*}\ \text{meat\_24hrs} - 0.340520904^{*}\ \text{eggs\_24hrs} \\
&- 0.155376389^{*}\ \text{fish\_24hrs} - 0.466889269^{*}\ \text{grains\_24hrs} \\
&- 0.894577464^{*}\ \text{dairy\_24hrs} - 0.203706591^{*}\ \text{oils\_fats\_24hrs} \\
&- 0.180889263^{*}\ \text{sugars\_24hrs} - 0.00834227^{*}\ \text{geographycode} \\
&- 0.001507924^{*}\ \text{financial\_situationcode} \\
&+ 0.001308516^{*}\ \text{employment\_statuscode}
\end{aligned} \tag{5.4}
$$

OLS Regression Results

Dep. Variable:	fsicode	**R-squared:**	0.730
Model:	OLS	**Adj. R-squared:**	0.730
Method:	Least Squares	**F-statistic:**	3.862e+04
Date:	Sun, 16 Jan 2022	**Prob (F-statistic):**	0.00
Time:	07:42:08	**Log-Likelihood:**	−1.4005e+05
No. Observations:	199496	**AIC:**	2.801e+05
Df Residuals:	199481	**BIC:**	2.803e+05
Df Model:	14		
Covariance Type:	nonrobust		

	coef	**std err**	**t**	**P>|t|**	**[0.025**	**0.975]**
Intercept	4.2771	0.006	703.417	0.000	4.265	4.289
cereal_24hrs	−0.2541	0.003	−78.722	0.000	−0.260	−0.248
plantains_24hrs	−0.2011	0.003	−77.724	0.000	−0.206	−0.196
vegs_24hrs	−0.3319	0.003	−123.269	0.000	−0.337	−0.327
fruits_24hrs	−0.2976	0.002	−123.605	0.000	−0.302	−0.293
meat_24hrs	−0.3955	0.003	−150.716	0.000	−0.401	−0.390
eggs_24hrs	−0.3405	0.002	−152.886	0.000	−0.345	−0.336
fish_24hrs	−0.1554	0.003	−49.186	0.000	−0.162	−0.149
grains_24hrs	−0.4669	0.002	−205.557	0.000	−0.471	−0.462
dairy_24hrs	−0.8946	0.003	−337.283	0.000	−0.900	−0.889
oils_fats_24hrs	−0.2037	0.003	−63.991	0.000	−0.210	−0.197
sugars_24hrs	−0.1809	0.003	−70.297	0.000	−0.186	−0.176
geographycode	−0.0083	0.002	−5.546	0.000	−0.011	−0.005
financial_situationcode	−0.0015	0.001	−1.958	0.050	−0.003	1.59e−06
employment_statuscode	0.0013	0.000	3.366	0.001	0.001	0.002

Omnibus:	138.963	**Durbin-Watson:**	1.802
Prob(Omnibus):	0.000	**Jarque-Bera (JB):**	130.299
Skew:	0.038	**Prob(JB):**	5.08e−29
Kurtosis:	2.901	**Cond. No.**	51.0

As it can be seen, all the feature variables influence food security. Additional details on residuals can provide more details[84].

[84] Regression Plots—https://www.statsmodels.org/stable/examples/notebooks/generated/regression_plots.html

5.4.4.10 Step 10: Excel Model

The same model developed in Excel provides very similar results:

SUMMARY OUTPUT

Regression Statistics	
Multiple R	0.854686745
R Square	0.730489432
Adjusted R	0.730470518
Standard E	0.488262608
Observatio	199496

ANOVA

	df	SS	MS	F	ignificance F
Regression	14	128898.1277	9207.009121	38619.94404	0
Residual	199481	47556.3451	0.238400374		
Total	199495	176454.4728			

	Coefficients	Standard Error	t Stat	P-value	Lower 95%	Upper 95%	Lower 95.0%	Upper 95.0%
Intercept	4.277051233	0.006080392	703.4169766	0	4.265134	4.288969	4.265133811	4.288968656
cereal_24h	-0.254128184	0.003228154	-78.72245287	0	-0.26046	-0.2478	-0.260455288	-0.247801081
plantains_	-0.20105063	0.00258671	-77.72447055	0	-0.20612	-0.19598	-0.206120519	-0.195980742
vegs_24hr	-0.33186527	0.002692207	-123.2688727	0	-0.33714	-0.32659	-0.33714193	-0.32658861
fruits_24h	-0.297613375	0.002407769	-123.6054514	0	-0.30233	-0.29289	-0.302332544	-0.292894206
meat_24h	-0.395512963	0.002624234	-150.7156075	0	-0.40066	-0.39037	-0.400656397	-0.390369528
eggs_24hr	-0.340520904	0.00222729	-152.8857504	0	-0.34489	-0.33616	-0.344886338	-0.336155469
fish_24hrs	-0.155376389	0.003158983	-49.18556704	0	-0.16157	-0.14918	-0.161567921	-0.149184858
grains_24h	-0.466889269	0.002271334	-205.5572607	0	-0.47134	-0.46244	-0.471341029	-0.462437508
dairy_24hr	-0.894577464	0.002652309	-337.2825044	0	-0.89978	-0.88938	-0.899775926	-0.889379002
oils_fats_2	-0.203706591	0.003183364	-63.9909916	0	-0.20995	-0.19747	-0.209945907	-0.197467276
sugars_24	-0.180889263	0.002573198	-70.29745039	0	-0.18593	-0.17585	-0.185932669	-0.175845857
geography	-0.00834227	0.001504117	-5.546290202	2.92163E-08	-0.01129	-0.00539	-0.011290303	-0.005394237
financial_s	-0.001507924	0.000770172	-1.957906044	0.05024243	-0.00302	1.59E-06	-0.003017441	1.59413E-06
employme	0.001308516	0.000388764	3.365838003	0.000763259	0.000547	0.00207	0.000546548	0.002070484

FSI= 4.277 - 0.254128184* cereal_24hrs - 0.20105063 * plantains_24hrs - 0.33186527 * vegs_24hrs - 0.297613375 * fruits_24hrs - 0.395512963 * meat_24hrs - 0.340520904 * eggs_24hrs - 0.155376389 * fish_24hrs - 0.466889269 * grains_24hrs - 0.894577464 * dairy_24hrs - 0.203706591 * oils_fats_24hrs - 0.180889263 * sugars_24hrs - 0.00834227 * geographycode - 0.001507924 * financial_situationcode + 0.001308516 * employment_statuscode

5.4.4.11 Step 11: Logistic Regression

Construct logic regression to understand and be able to predict who can hunger or experience food insecurity:

```
# Train the model
from sklearn.linear_model import LogisticRegression

# Set regularization rate
reg = 0.01

# train a logistic regression model on the training set
model = LogisticRegression(C=1/reg, solver="liblinear").fit
(X_train, y_train)
print (model)
```

Output:

```
LogisticRegression(C=100.0, solver='liblinear')
```

5.4.4.12 Step 12: Predict

```
predictions = model.predict(X_test)
print('Predicted labels: ', predictions)
print('Actual labels:    ' ,y_test)
```

Output:

```
Predicted labels:  ['full food security' 'not applicable' 'full
food security' ...
 'not applicable' 'not applicable' 'not applicable']
Actual labels:    ['full food security' 'not applicable' 'full food
security' ...
 'not applicable' 'not applicable' 'not applicable']
```

5.4.4.13 Step 13: Evaluate the Model

```
from sklearn.metrics import accuracy_score

print('Accuracy: ', accuracy_score(y_test, predictions))
```

Output:
Accuracy: 0.9206600633596663

5.4.4.14 Step 14: Classification Report

```
from sklearn. metrics import classification_report

print(classification_report(y_test, predictions))
```

Output:

```
                          precision    recall  f1-score   support

    full food security       1.00      1.00      1.00     19949
     low food security       0.88      0.77      0.82      5480
marginal food security       0.71      0.64      0.68      3391
        not applicable       0.89      0.93      0.91     21054

              accuracy                           0.92     49874
             macro avg       0.87      0.84      0.85     49874
          weighted avg       0.92      0.92      0.92     49874
```

5.4.4.15 Step 15: Develop Confusion Matrix

```
from sklearn.metrics import confusion_matrix

# Print the confusion matrix
mcm = confusion_matrix(y_test, predictions)
print(mcm)
```

Output:

```
[[19949     0     0     0]
 [    0  4221    28  1231]
 [    0     0  2184  1207]
 [   47   584   860 19563]]


import numpy as np
import matplotlib.pyplot as plt
%matplotlib inline
plt.imshow(mcm, interpolation="nearest", cmap=plt.cm.Blues)
plt.colorbar()
tick_marks = np.arange(len(fsi_classes))
plt.xticks(tick_marks, fsi_classes, rotation=45)
plt.yticks(tick_marks, fsi_classes)
plt.xlabel("Actual fsi")
plt.ylabel("Predicted fsi")
```

Output:

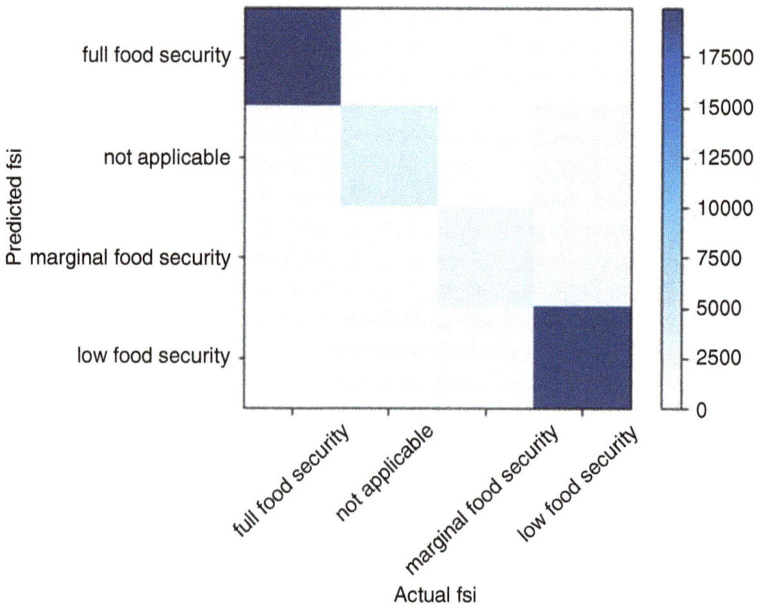

5.4.4.16 Step 16: Construct Ensemble model

```
X, y = opendailyfooddfwithlabelsandnumericals[['cereal_24hrs',
'plantains_24hrs', 'vegs_24hrs','fruits_24hrs', 'meat_24hrs',
'eggs_24hrs', 'fish_24hrs','grains_24hrs', 'dairy_24hrs',
'oils_fats_24hrs', 'sugars_24hrs','age','geography',
'education', 'financial_situation','employment_status',
'head_household']].values,
opendailyfooddfwithlabelsandnumericals['fsicode'].values
print('Features:',X[:10], '\nLabels:', y[:10], sep='\n')
```

Output:

```
Features:
[[1 1 1 1 1 0 1 1 1 1 1 '26 to 35 years old' 'Rural'
 'Prefer not to answer' 'Prefer not to answer' 'None of the above'
'Yes']
 [1 1 1 1 1 1 1 1 1 1 1 '26 to 35 years old' 'Rural'
```

(continued)

```
 'Prefer not to answer' 'Prefer not to answer' 'None of the above'
'Yes']
 [1 1 1 1 1 1 0 0 1 1 1 'Not Available' 'Rural' 'College or university'
 'I can afford food, but nothing else' 'Student and work part-time'
'No']
 [1 0 1 1 1 0 1 0 1 1 1 'Not Available' 'Rural' 'College or university'
 'I can afford food, but nothing else' 'Student and work part-time'
'No']
 [1 1 1 0 1 0 1 0 1 1 1 'Not Available' 'Rural' 'College or university'
 'I can afford food, but nothing else' 'Student and work part-time'
'No']
 [1 0 1 1 1 1 0 0 1 1 1 'Not Available' 'Rural' 'College or university'
 'I can afford food, but nothing else' 'Student and work part-time'
'No']
 [1 1 1 1 0 1 1 1 1 1 0 'Not Available' 'Rural' 'College or university'
 'I can afford food, but nothing else' 'Student and work part-time'
'No']
 [1 1 1 1 0 0 1 1 1 1 1 'Not Available' 'Rural' 'College or university'
 'I can afford food, but nothing else' 'Student and work part-time'
'No']
 [1 1 1 0 0 0 1 1 1 1 1 'Not Available' 'Rural' 'College or university'
 'I can afford food, but nothing else' 'Student and work part-time'
'No']
 [1 1 1 0 1 1 0 0 1 1 1 'Not Available' 'Rural' 'College or university'
 'I can afford food, but nothing else' 'Student and work part-time'
'No']]

Labels:
[1 1 1 1 1 1 1 1 1 1]


from sklearn.model_selection import train_test_split

# Split data 70%-30% into training set and test set
X_train, X_test, y_train, y_test = train_test_split(X, y,
test_size=0.30, random_state=0)

print ('Training Set: %d, rows\nTest Set: %d rows' % (X_train.size,
X_test.size))
```

Output:
Training set: 2373999 rows
Test set: 1017433 rows

5.4.4.17 Step 17: Train the Model

```
# Train the model
from sklearn.compose import ColumnTransformer
from sklearn.pipeline import Pipeline
from sklearn.impute import SimpleImputer
from sklearn.preprocessing import StandardScaler, OneHotEncoder
from sklearn.linear_model import LinearRegression
import numpy as np

# Define preprocessing for numeric columns (scale them)
# numeric_features = [6,7,8,9]
numeric_features_columns=[0,1,2,3,4,5,6,7,8,9,10]
numeric_transformer = Pipeline(steps=[
   ('scaler', StandardScaler())])

# Define preprocessing for categorical features (encode them)
categorical_features = [11,12,13,14,15,16]
categorical_transformer = Pipeline(steps=[
   ('onehot', OneHotEncoder(handle_unknown='ignore'))])

# Combine preprocessing steps
preprocessor = ColumnTransformer(
   transformers=[
      ('num', numeric_transformer, numeric_features_columns),
      ('cat', categorical_transformer, categorical_features)])

# Create preprocessing and training pipeline
pipeline = Pipeline(steps=[('preprocessor', preprocessor),
                ('regressor', LinearRegression(normalize=False))])

# fit the pipeline to train a linear regression model on the training
set
model = pipeline.fit(X_train, (y_train))
print (model)
```

In the above, the code constructs categorical and numerical pipelines and applies numerical transformation and label one-hot encoding. Finally, prepare the model.

Output:

```
Pipeline(steps=[('preprocessor',
        ColumnTransformer(transformers=[('num',
                        Pipeline(steps=[('scaler',
                              StandardScaler())]),
                        [0, 1, 2, 3, 4, 5, 6, 7, 8, 9,
                        10]),
```

(continued)

```
                    ('cat',
                    Pipeline(steps=[('onehot',
                              OneHotEncoder
(handle_unknown='ignore'))]),
                    [11, 12, 13, 14, 15, 16])])),
        ('regressor', LinearRegression())])
```

5.4.4.18 Step 18: Train and Predict

Construct mean square error, root mean square error, and R^2 values:

```
import seaborn as sns
import matplotlib.pyplot as plt

import plotly.graph_objects as go
import seaborn as sns; sns.set()
import matplotlib.pyplot as plt
from scipy import stats
from scipy.stats import pearsonr
import matplotlib.pyplot as plt
from sklearn.preprocessing import MinMaxScaler
from sklearn.model_selection import train_test_split
from sklearn.linear_model import LinearRegression
from sklearn.ensemble import RandomForestClassifier
from sklearn.metrics import r2_score
from sklearn import metrics
from sklearn.metrics import mean_squared_error, r2_score
predictions = model.predict(X_test)
np.set_printoptions(suppress=True)
print('Predicted labels: ', np.round(predictions)[:10])
print('Actual labels: ' ,y_test[:10])

mse = mean_squared_error(y_test, predictions)
print("MSE:", mse)
rmse = np.sqrt(mse)
print("RMSE:", rmse)
r2 = r2_score(y_test, predictions)
print("R2:", r2)

plt.scatter(y_test, predictions)
plt.xlabel('Actual Labels')
plt.ylabel('Predicted Labels')
plt.title('FSI Predictions - Preprocessed')
z = np.polyfit(y_test, predictions, 1)
```

(continued)

```
p = np.poly1d(z)
plt.plot(y_test,p(y_test), color='magenta')
plt.show()
```

Output:

```
Predicted labels: [1. 2. 1. 3. 2. 1. 2. 1. 2. 3.]
Actual labels: [1 2 1 2 3 1 2 1 2 3]
MSE: 0.14869348225756376
RMSE: 0.3856079385302691
R2: 0.833583164291577
```

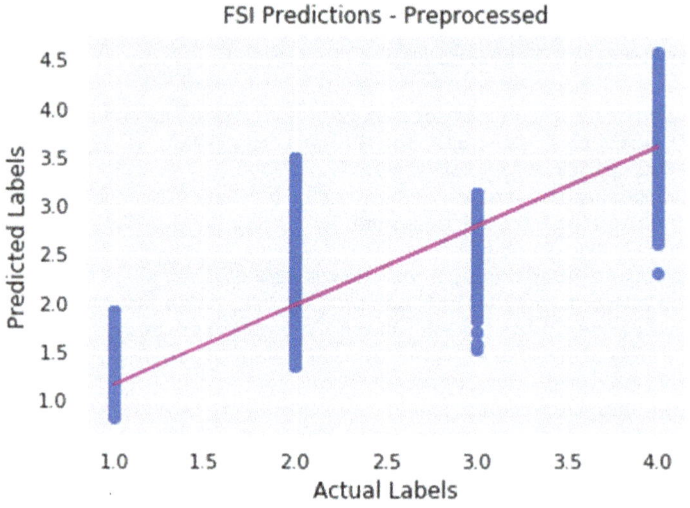

5.4.4.19 Step 19: Construct Gradient Boost Regressor and Predict the Model

Using ensemble model libraries, construct Gradient Boost Regressor.

```
from sklearn.pipeline import Pipeline
from sklearn.ensemble import GradientBoostingRegressor
from sklearn import preprocessing
pipeline = Pipeline(steps=[('preprocessor', preprocessor),
            ('regressor', GradientBoostingRegressor())])
```

(continued)

```
# train a logistic regression model on the training set
model = pipeline.fit(X_train, (y_train))
print (model)
```

Output:

```
Pipeline(steps=[('preprocessor',
        ColumnTransformer(transformers=[('num',
                        Pipeline(steps=[('scaler',
                                StandardScaler())]),
                        [0, 1, 2, 3, 4, 5, 6, 7, 8, 9,
                        10]),
                        ('cat',
                        Pipeline(steps=[('onehot',
                                OneHotEncoder
(handle_unknown='ignore'))]),
                        [11, 12, 13, 14, 15, 16])])),
        ('regressor', GradientBoostingRegressor())])


predictions = model.predict(X_test)

mse = mean_squared_error(y_test, predictions)
print("MSE:", mse)
rmse = np.sqrt(mse)
print("RMSE:", rmse)
r2 = r2_score(y_test, predictions)
print("R2:", r2)

plt.scatter(y_test, predictions)
plt.xlabel('Actual Labels')
plt.ylabel('Predicted Labels')
plt.title('FSI Predictions - Ensemble')
z = np.polyfit(y_test, predictions, 1)
p = np.poly1d(z)
plt.plot(y_test,p(y_test), color='magenta')
plt.show()
```

Output:

```
MSE: 0.14869348225756376
RMSE: 0.3856079385302691
R2: 0.833583164291577
```

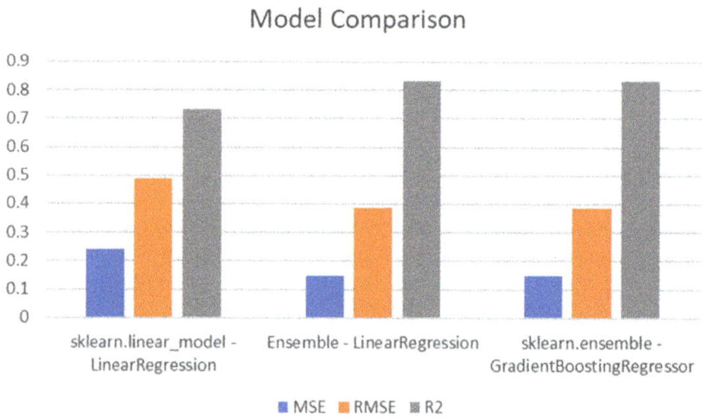

Fig. 5.17 Model performance

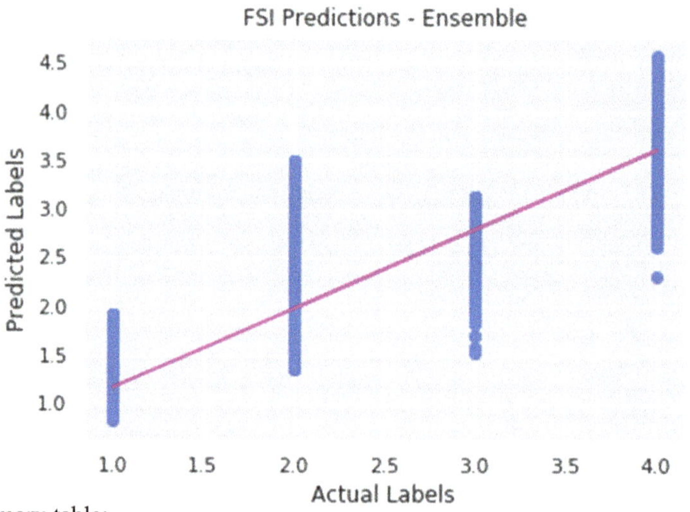

Summary table:

Model	Type	MSE	RMSE	R^2
LinearRegression	sklearn.linear_model	0.238602	0.4884	0.7333
LinearRegression	Ensemble	0.1487	0.3856	0.8335
GradientBoostingRegressor	sklearn.ensemble	0.14967	0.38688	0.8327

In conclusion (please see Fig. 5.17 for model performance), household food security is an important measure of well-being. Although it may not encapsulate all dimensions of poverty, the inability of households to obtain access to enough food for an active, healthy life is surely an important component of their poverty. Accordingly, devising an appropriate measure of food security outcomes is useful to identify the food insecure, assess the severity of their food shortfall, characterize the nature of their insecurity (e.g., seasonal vs. chronic), predict who is most at risk of future hunger, monitor changes in circumstances, and assess the impact of interventions[85]. However, obtaining detailed data on food security status—such as 24-h recall data on caloric intakes—can be time-consuming and expensive and requires a high level of technical skill both in data collection and analysis. In the above analysis, crowdsource data was used to identify dietary diversity in Venezuela and developed a predictive model for who can go potential hunger (equation 5.3).

Production diversity and dietary diversity in smallholder farm households[86] [3]:
Given that hunger and malnutrition are still widespread problems in many developing countries, the question of how to make agriculture and food systems more nutrition-sensitive is of high relevance for research and policy. Many of the undernourished people in Africa and Asia are small-scale subsistence farmers. Diversifying production on these farms is often perceived as a promising strategy to improve dietary quality and diversity. This hypothesis is tested with data from smallholder farm households in Indonesia, Kenya, Ethiopia, and Malawi:

- Higher farm production diversity significantly contributes to dietary diversity in some situations, but not in all.
- Improving small farmers' access to markets seems to be a more effective strategy to improve nutrition than promoting production diversity on subsistence farms.
- Undernutrition and micronutrient malnutrition remain problems of significant magnitude in large parts of the developing world. Improved nutrition requires not only better access to food for poor population segments but also higher dietary quality and diversity. Because many of the poor and undernourished people are smallholder farmers, diversifying production on these smallholder farms is widely perceived as a useful approach to improve dietary diversity.

Market participation and market access are driven through higher productivity. Technology plans an important role.

5.5 National Quality Forum: Food Insecurity Measures

Working through the Measure Incubator®, NQF is facilitating multistakeholder feedback on food insecurity quality measures[87]. Food insecurity is a social determinant of health (SDoH) that affects one in nine American households. And the same is equally applicable to the other parts of the world. The social determinants of health[88] (SDoH) are the conditions in the environments where people are born, live, learn,

[85] Dietary Diversity as a Food Security Indicator—https://ageconsearch.umn.edu/record/16474/

[86] Measuring the Food Access Dimension of Food Security: A Critical Review and Mapping of Indicators—https://journals.sagepub.com/doi/pdf/10.1177/0379572115587274

[87] NQF—https://www.qualityforum.org/Food_Insecurity_Measures.aspx

[88] Social Determinants of Health—https://health.gov/healthypeople/objectives-and-data/social-determinants-health#:~:text=Social%20determinants%20of%20health%20(SDOH,of%2Dlife%20outcomes%20and%20risks.

Fig. 5.18 SDoH

work, play, worship, and age that affect a wide range of health, functioning, and quality-of-life outcomes and risks. SDoH can be grouped into five categories: economic stability, educational access and quality, healthcare access and quality, neighborhood and build environment, and social and community context. Please see image for SDOH[89] (please see Fig. 5.18).

Food insecurity and other SDoH have a significant impact on disparities in health, healthcare, and health outcomes. Food insecurity contributes to diabetes, stroke, cancer, hypertension, asthma, chronic kidney disease, and other chronic conditions. Addressing and accounting for SDoH is critical for improving healthcare quality and achieving health equity. Yet, healthcare quality improvement efforts have largely excluded food insecurity and SDoH. Establishing performance measures that address food insecurity within the clinical setting is critical to improving health and health outcomes for individuals with food insecurity. Please refer to Appendix A for Food Security Data Sources.

[89] SDOH—Healthy People 2030, U.S. Department of Health and Human Services, Office of Disease Prevention and Health Promotion. Retrieved [January 16, 2022], from https://health.gov/healthypeople/objectives-and-data/social-determinants-health

Table 5.3 Food security measures

Measure title and description	Type	Source
1. Screening for Food Insecurity: The percentage of patients that have been screened for food insecurity	Process	Electronic health records (EHR)
2. Appropriate Clinical Action After Screening: Percentage of patients that screened positive for food insecurity using the US Household Food Security Module: Six-Item Short Form of the Food Security Module, US Adult Food Security Survey Module (US AFSSM), US Household Food Security Module (US HFSSM), or Hunger Vital Signs (HVS) screening tool that were assessed for food insecurity severity and appropriate clinical action taken	Process	Electronic health records (EHR)
3. A Change in Severity of Food Insecurity: Percentage of patients with a decrease in severity of food insecurity after appropriate clinical action	Outcome	EHR

5.5.1 Social Determinant of Health (SDoH) and Population Criteria

Food insecurity within the clinical setting is critical to improving health and health outcomes for individuals with food insecurity (please see Table 5.3). The patient population criteria involve identifying patient groups based on the following proposed measures[90].

5.5.2 Food Security Measure Stratification

The stratification of food security measures includes the following:

- Household size and composition
- Race-ethnicity
- Gender
- Age
- Education completed
- Employment status
- Marital status
- Income level
- Payer source
- Sexual orientation
- Disabilities
- Transportation status
- Immigration/nativity status (US born vs. non-US born)

[90] Food Security Measures—https://www.qualityforum.org/WorkArea/linkit.aspx?LinkIdentifier=id&ItemID=91827

5.6 National Health and Nutrition Examination Survey (NHANES)

The National Health and Nutrition Examination Survey[91] (NHANES) is a program of studies designed to assess the health and nutritional status of adults and children in the United States of America. The survey is unique in that it combines interviews and physical examinations. NHANES is a major program of the National Center for Health Statistics (NCHS). NCHS is part of the Centers for Disease Control and Prevention (CDC) and has the responsibility for producing vital and health statistics for the nation. NHANES is a comprehensive survey, including physical examinations, laboratory analyses, questionnaires, and demographic information, and representativeness of the population including minority and underrepresented groups are ensured with weighting and sampling methods[92].

The NHANES program began in the early 1960s and has been conducted as a series of surveys focusing on different population groups or health topics. In 1999, the survey became a continuous program that has a changing focus on a variety of health and nutrition measurements to meet emerging needs. The survey examines a nationally representative sample of about 5,000 persons each year. These persons are in counties across the country, 15 of which are visited each year.

The NHANES interview includes demographic, socioeconomic, dietary, and health-related questions. The examination component consists of medical, dental, and physiological measurements, as well as laboratory tests administered by highly trained medical personnel.

Findings from this survey will be used to determine the prevalence of major diseases and risk factors for diseases. Information will be used to assess nutritional status and its association with health promotion and disease prevention. NHANES findings are also the basis for national standards for such measurements as height, weight, and blood pressure. Data from this survey will be used in epidemiological studies and health sciences research, which help develop sound public health policy, direct and design health programs and services, and expand the health knowledge for the nation.

Nutritional examination data functions as a barometer in identifying the prevalence of critical diseases through the survey mechanism, for example, Prevalence of Depression Among Adults Aged 20 and Over: United States[93]. The data from the National Health and Nutrition Examination Survey provide important insights [43]:

[91] About NHANES—https://www.cdc.gov/nchs/nhanes/about_nhanes.htm

[92] BMIN503 Final Project Methods—https://rstudio-pubs-static.s3.amazonaws.com/557639_762de2760a0d46859d04f1b2821bbc63.html#linear_regression

[93] Prevalence of Depression Among Adults Aged 20 and Over: United States of America, 2013–2016—https://www.cdc.gov/nchs/products/databriefs/db303.htm#depression_prevalence_20_and_over

- During 2013–2016, 8.1% of American adults aged 20 and over had depression in a given 2-week period.
- Women (10.4%) were almost twice as likely as were men (5.5%) to have had depression.
- Depression was lower among non-Hispanic Asian adults, compared with Hispanic, non-Hispanic black, or non-Hispanic white adults.
- The prevalence of depression decreased as family income levels increased.

Results of NHANES benefit people in the United States of America in important ways. Facts about the distribution of health problems and risk factors in the population give researchers important clues to the causes of disease. Information collected from the current survey is compared with information collected in previous surveys. This allows health planners to detect the extent various health problems, and risk factors have changed in the US population over time. By identifying the healthcare needs of the population, government agencies and private sector organizations can establish policies and plan research, education, and health promotion programs that help improve present health status and will prevent future health problems. NHANES interviews provide food exposure of the US population, for instance, blood total mercury levels and depression in the United States of America [94] [44].

5.6.1 *Family Questionnaire*

Household- and family-level information is collected as part of family questionnaire. NHANES 2019–2020 Questionnaire Instruments[95] include Consumer Behavior, Demographic Background, Family Questionnaire Hand Cards, Food Security, Housing Characteristics, Income, Salt Sample Collection, and Smoking.

5.6.2 *NHANES Food Security*

NHANES has assessed household food security with the US Food Security Survey Module (US FSSM) since 1999[96]. Similar modules have been used in various surveys, including the Census Bureau's Current Population Survey (CPS), and a growing number of state, local, and regional studies, such as the California Health Interview Survey (CHIS). USDA's food security statistics are based on a national

---

[94] Blood total mercury levels and depression in the United States of America: National Health and Nutrition Examination Survey 2005-2008, June 2012—https://idea.library.drexel.edu/islandora/object/idea%3A5374/datastream/OBJ/view

[95] NHANES 2019-2020 Questionnaire Instruments—https://wwwn.cdc.gov/nchs/nhanes/continuousnhanes/questionnaires.aspx?Cycle=2019-2020

[96] FSQ: https://wwwn.cdc.gov/Nchs/Nhanes/2017-2018/FSQ_J.htm

food security survey conducted as an annual supplement to the monthly Current Population Survey (CPS). The CPS is a nationally representative survey conducted by the Bureau of the Census for the Bureau of Labor Statistics. The CPS provides data for the nation's monthly unemployment statistics, annual income, and poverty statistics. In December of each year, after completing the labor force interview, about 40,000 households respond to the food security questions—and to questions about food spending and about the use of federal and community food assistance programs. The households interviewed in the CPS are selected to be representative of all civilian households at the state and national levels[97] [13].

The NHANES household interview included two questionnaires: a sample participant questionnaire to collect information regarding each individual survey participant's personal health status and a family questionnaire to collect information at the family level, such as total income for the family, food availability in the family, etc.

5.7 NHANES 2017–2018 Questionnaire Data Food Security Data

Software code for this model can be found on GitHub Link: NHANES_FSQ_Data.py (Python Code)

5.7.1 Dataset[98]: FSQ_J.XPT[99]

Code:

```
"""
Created on Sun Jan 2 18:25:00 2022

@author: CHVUPPAL
"""
```

(continued)

[97] Food Security in the U.S.—Measurements—https://www.ers.usda.gov/topics/food-nutrition-assistance/food-security-in-the-us/measurement.aspx

[98] NHANES 2017–2018 Questionnaire Data Food Security—https://wwwn.cdc.gov/Nchs/Nhanes/Search/DataPage.aspx?Component=Questionnaire&Cycle=2017-2018

[99] FSQ_J.XPT—https://wwwn.cdc.gov/Nchs/Nhanes/2017-2018/FSQ_J.XPT

```
# https://wwwn.cdc.gov/nchs/nhanes/tutorials/samplecode.aspx
# https://wwwn.cdc.gov/nchs/data/tutorials/
DB303_Fig1_Stata_LOG.log

import pandas as pd
from functools import reduce
import matplotlib.pyplot as plt

## ** Download Demographic (DEMO) Data and Keep Variables Of
Interest **
# Load DEMO_I.XPT
dfFSQ_J = pd.read_sas('FSQ_J.XPT')
#dfDEMO_I.set_index('SEQN') # Respondent sequence number
print('--------------------------------------')
print('dfFSQ_J - keep seqn riagendr ridageyr sdmvstra sdmvpsu
wtmec2yr')
print('--------------------------------------')
print(dfFSQ_J)

print(list(dfFSQ_J.columns))
```

Output:

```
C:\Hanumayamma\Clustering>python NHANES_FSQ_Data.py
--------------------------------------
dfDEMO_J - keep seqn riagendr ridageyr sdmvstra sdmvpsu wtmec2yr
--------------------------------------
          SEQN  SDDSRVYR  RIDSTATR  RIAGENDR  RIDAGEYR  ...  SDMVPSU  SDMVSTRA  INDHHIN2  INDFMIN2  INDFMPIR
0      93703.0      10.0       2.0       2.0       2.0  ...      2.0     145.0      15.0      15.0      5.00
1      93704.0      10.0       2.0       1.0       2.0  ...      1.0     143.0      15.0      15.0      5.00
2      93705.0      10.0       2.0       2.0      66.0  ...      2.0     145.0       3.0       3.0      0.82
3      93706.0      10.0       2.0       1.0      18.0  ...      2.0     134.0       NaN       NaN       NaN
4      93707.0      10.0       2.0       1.0      13.0  ...      1.0     138.0      10.0      10.0      1.88
...        ...       ...       ...       ...       ...  ...      ...       ...       ...       ...       ...
9249  102952.0      10.0       2.0       2.0      70.0  ...      2.0     138.0       4.0       4.0      0.95
9250  102953.0      10.0       2.0       1.0      42.0  ...      2.0     137.0      12.0      12.0       NaN
9251  102954.0      10.0       2.0       2.0      41.0  ...      1.0     144.0      10.0      10.0      1.18
9252  102955.0      10.0       2.0       2.0      14.0  ...      1.0     136.0       9.0       9.0      2.24
9253  102956.0      10.0       2.0       1.0      38.0  ...      1.0     142.0       7.0       7.0      1.56

[9254 rows x 46 columns]
['SEQN', 'SDDSRVYR', 'RIDSTATR', 'RIAGENDR', 'RIDAGEYR', 'RIDAGEMN', 'RIDRETH1', 'RIDRETH3', 'RIDEXMON', 'RIDEXAG
M', 'DMQMILIZ', 'DMQADFC', 'DMDBORN4', 'DMDCITZN', 'DMDYRSUS', 'DMDEDUC3', 'DMDEDUC2', 'DMDMARTL', 'RIDEXPRG', 'S
IALANG', 'SIAPROXY', 'SIAINTRP', 'FIALANG', 'FIAPROXY', 'FIAINTRP', 'MIALANG', 'MIAPROXY', 'MIAINTRP', 'AIALANGA'
, 'DMDHHSIZ', 'DMDFMSIZ', 'DMDHHSZA', 'DMDHHSZB', 'DMDHHSZE', 'DMDHRGND', 'DMDHRAGZ', 'DMDHREDZ', 'DMDHRMAZ', 'DM
DHSEDZ', 'WTINT2YR', 'WTMEC2YR', 'SDMVPSU', 'SDMVSTRA', 'INDHHIN2', 'INDFMIN2', 'INDFMPIR']
```

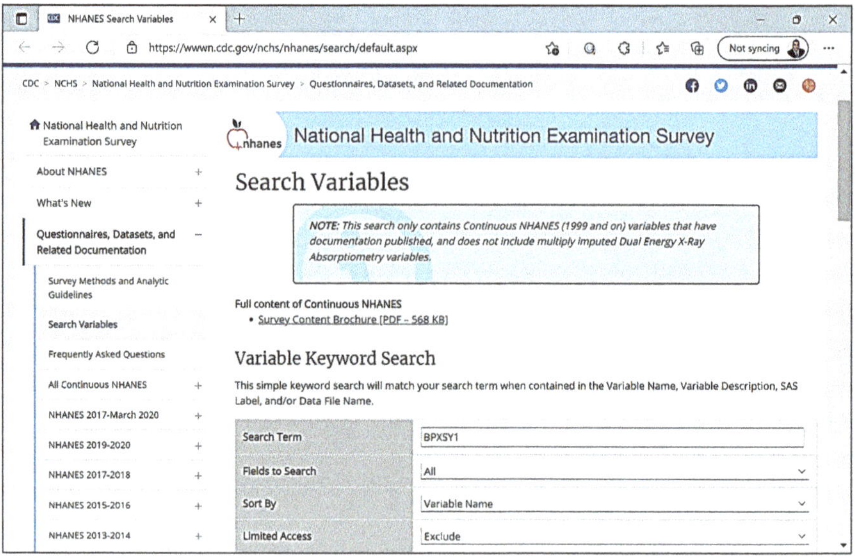

Fig. 5.19 Search

5.7.2 *Search NHANES Dataset*

The NHANES provides huge and valuable dataset. The easiest way to search for a feature variable (please see Fig. 5.19) is to use a search tool[100]:

5.8 Use Case: National Center for Health Statistics (NCHS) Data Brief No. 303—Prevalence of Depression Among Adults Aged 20 and Over: United States, 2013–2016

Major depression is a common and treatable mental disorder characterized by changes in mood and cognitive and physical symptoms over a 2-week period. It is associated with high societal costs and greater functional impairment than many other chronic diseases, including diabetes and arthritis. Depression rates differ by age, sex, income, and health behaviors [43].

[100] Search Variables—https://wwwn.cdc.gov/nchs/nhanes/search/default.aspx

Software code for this model can be found on GitHub Link:
NHANES_Percentageofpersonsaged20andoverwithdepression.py
(Python Code)
NHANES_NHANES_Percentageofpersonsaged20andoverwithde
pressionModel.ipynb
(Jupyter Notebook Code)

5.8.1 Step 1: Demographics Dataset—Percentage of Persons Aged 20 and over with Depression, by Age and Sex: United States, 2013–2016

Load the following datasets as per Brody DJ, Pratt LA, and Hughes J. Prevalence of depression among adults aged 20 and over: United States, 2013–2016. NCHS Data Brief, No. 303. Hyattsville, MD: National Center for Health Statistics, 2018:

- Load DEMO_H.XPT: https://wwwn.cdc.gov/nchs/nhanes/2013-2014/DEMO_H.XPT
- Load DEMO_I.XPT: https://wwwn.cdc.gov/nchs/nhanes/2015-2016/DEMO_I.XPT

```
import pandas as pd
from functools import reduce
import matplotlib.pyplot as plt

## ** Download Demographic (DEMO) Data and Keep Variables Of
Interest **
# Load DEMO_I.XPT
dfDEMO_I = pd.read_sas('DEMO_I.XPT')
#dfDEMO_I.set_index('SEQN') # Respondent sequence number
print('----------------------------------------')
print('dfDEMO_I - keep seqn riagendr ridageyr sdmvstra sdmvpsu
wtmec2yr')
print('----------------------------------------')
print(dfDEMO_I)

print(list(dfDEMO_I.columns))

dfDEMO_IREQ = dfDEMO_I
[['SEQN','RIAGENDR','RIDAGEYR','SDMVSTRA','SDMVPSU',
'WTMEC2YR','RIDRETH3','INDFMIN2','INDHHIN2']]

# keep seqn riagendr ridageyr sdmvstra sdmvpsu wtmec2yr
```

(continued)

```
# https://wwwn.cdc.gov/nchs/data/tutorials/
DB303_Fig1_Stata_LOG.log

# Load DPQ_H.XPT
dfDEMO_H = pd.read_sas('DEMO_H.XPT')
# dfDEMO_I.set_index('SEQN') # Respondent sequence number
print('-------------------------------------')
print('dfDEMO_H - keep seqn riagendr ridageyr sdmvstra sdmvpsu
wtmec2yr')
print('-------------------------------------')
print(dfDEMO_H)

dfDEMO_HREQ = dfDEMO_H
[['SEQN','RIAGENDR','RIDAGEYR','SDMVSTRA',
'SDMVPSU','WTMEC2YR','RIDRETH3','INDFMIN2','INDHHIN2']]

print('-----------------dfDEMO--------------------')

#dfDEMO = pd.merge(dfDEMO_H, dfDEMO_I, left_index=True,
right_index=True)

# ROW BIND
# https://www.datasciencemadesimple.com/append-concatenate-
rows-python-pandas-row-bind/
frames_DEMO = [dfDEMO_HREQ, dfDEMO_IREQ]
dfDEMO = pd.concat(frames_DEMO)

print(dfDEMO)

print('-------------------------------------')
```

Load DEMO_H and DEMO_I file with the following key columns:

- SEQN: Sequence number
- RIAGENDR: Respondent gender
- RIDAGEYR: Respondent age in years
- SDMVSTRA: Masked variance pseudo-stratum[101]
- SDMVPSU: Masked variance unit pseudo-stratum variable for variance estimation
- WTMEC2YR: Full sample 2-year MEC exam weight
- RIDRETH3: Recode of reported race and Hispanic origin information, with non-Hispanic Asian category

[101] DEMO_H.XPT—https://wwwn.cdc.gov/Nchs/Nhanes/2013-2014/DEMO_H.htm#SDMVSTRA

Code or value	Value description
1	Mexican American
2	Other Hispanic
3	Non-Hispanic white
4	Non-Hispanic black
6	Non-Hispanic Asian
7	Other race—including multi-racial
.	Missing

- INDFMIN2: Total family income (reported as a range value in dollars)[102]

Code or value	Value description
1	$ 0 to $ 4,999
2	$ 5,000 to $ 9,999
3	$10,000 to $14,999
4	$15,000 to $19,999
5	$20,000 to $24,999
6	$25,000 to $34,999
7	$35,000 to $44,999
8	$45,000 to $54,999
9	$55,000 to $64,999
10	$65,000 to $74,999
12	$20,000 and Over
13	Under $20,000
14	$75,000 to $99,999
15	$100,000 and Over

- INDHHIN2: Total household income (reported as a range value in dollars)

  Output:

```
[10175 rows x 47 columns]
-----------------dfDEMO--------------------
        SEQN  RIAGENDR  RIDAGEYR  SDMVSTRA  SDMVPSU      WTMEC2YR  RIDRETH3  INDFMIN2  INDHHIN2      INDFMPIR
0    73557.0       1.0      69.0     112.0      1.0  13481.042095       4.0       4.0       4.0  8.400000e-01
1    73558.0       1.0      54.0     108.0      1.0  24471.769625       3.0       7.0       7.0  1.780000e+00
2    73559.0       1.0      72.0     109.0      1.0  57193.285376       3.0      10.0      10.0  4.510000e+00
3    73560.0       1.0       9.0     109.0      2.0  55766.512438       3.0       9.0       9.0  2.520000e+00
4    73561.0       2.0      73.0     116.0      2.0  65541.871229       3.0      15.0      15.0  5.000000e+00
...      ...       ...       ...       ...      ...           ...       ...       ...       ...           ...
9966 93698.0       1.0       2.0     121.0      1.0   9896.007775       1.0       6.0       6.0  5.100000e-01
9967 93699.0       2.0       6.0     129.0      1.0   9875.921047       4.0      15.0      15.0  4.580000e+00
9968 93700.0       1.0      35.0     126.0      1.0  43194.215112       3.0       1.0       1.0  5.397605e-79
9969 93701.0       1.0       8.0     124.0      2.0  48252.504294       3.0      15.0      15.0  4.220000e+00
9970 93702.0       2.0      24.0     119.0      2.0 105080.445194       3.0       7.0      10.0  3.540000e+00

[20146 rows x 10 columns]
```

[102]DEMP_H: INDFMIN2: https://wwwn.cdc.gov/Nchs/Nhanes/2013-2014/DEMO_H.htm#INDFMIN2

5.8.2 Step 2: Mental Health—Depression Screener (DPQ)

- Load DPQ_H.XPT: https://wwwn.cdc.gov/Nchs/Nhanes/2013-2014/DPQ_H.XPT
- Load DPQ_I.XPT: https://wwwn.cdc.gov/Nchs/Nhanes/2015-2016/DPQ_I.XPT

```
# Load DPQ_H.XPT
dfDPQ_H = pd.read_sas('DPQ_H.XPT')
#dfDPQ_H.set_index('SEQN') # Respondent sequence number
#dfDEMO_I.set_index('SEQN') # Respondent sequence number
print('----------------------------------------')
print('dfDPQ_H - keep seqn riagendr ridageyr sdmvstra sdmvpsu
wtmec2yr')
print('----------------------------------------')
print(dfDPQ_H)

# Load DPQ_H.XPT
dfDPQ_I = pd.read_sas('DPQ_I.XPT')
#dfDPQ_I.set_index('SEQN') # Respondent sequence number
#dfDEMO_I.set_index('SEQN') # Respondent sequence number
print('----------------------------------------')
print('dfDPQ_I - keep seqn riagendr ridageyr sdmvstra sdmvpsu
wtmec2yr')
print('----------------------------------------')
print(dfDPQ_I)

print('-----------------dfDPQ--------------------')

# ROW BIND
# https://www.datasciencemadesimple.com/append-concatenate-
rows-python-pandas-row-bind/

frames_DPQ = [dfDPQ_H, dfDPQ_I]
dfDPQ = pd.concat(frames_DPQ)

print(dfDPQ)

print('---------------dfDEMOAndDPQ---------------------')
```

Output:

```
[5735 rows x 11 columns]
-----------------dfDPQ--------------------
        SEQN       DPQ010       DPQ020       DPQ030  ...       DPQ070       DPQ080       DPQ090        DPQ
100
0    73557.0  1.000000e+00 5.397605e-79 5.397605e-79 ...  5.397605e-79 5.397605e-79 5.397605e-79 1.000000e
+00
1    73558.0  2.000000e+00 5.397605e-79 5.397605e-79 ...  5.397605e-79 5.397605e-79 5.397605e-79 5.397605e
-79
2    73559.0  5.397605e-79 5.397605e-79 5.397605e-79 ...  5.397605e-79 5.397605e-79 5.397605e-79
NaN
3    73561.0  2.000000e+00 1.000000e+00 5.397605e-79 ...  5.397605e-79 5.397605e-79 5.397605e-79 1.000000e
+00
4    73562.0  3.000000e+00 3.000000e+00 3.000000e+00 ...  2.000000e+00 1.000000e+00 5.397605e-79 3.000000e
+00
```

5.8.3 Step 3: Merge DEMO and DPQ Data Frame

```
dfDEMOAndDPQ=dfDEMO.merge(dfDPQ, on='SEQN', how='left')
print(dfDEMOAndDPQ)

print('-------------------------------------')
dfDEMOAndDPQ.to_csv
('NHANES_Percentageofpersonsaged20andoverwithdepression.csv')
```

Output:

```
-----------------dfDEMOAndDPQ---------------------
        SEQN  RIAGENDR  RIDAGEYR  SDMVSTRA  ...        DPQ070        DPQ080        DPQ090        DPQ100
0    73557.0       1.0      69.0     112.0  ...  5.397605e-79  5.397605e-79  5.397605e-79  1.000000e+00
1    73558.0       1.0      54.0     108.0  ...  5.397605e-79  5.397605e-79  5.397605e-79  5.397605e-79
2    73559.0       1.0      72.0     109.0  ...  5.397605e-79  5.397605e-79  5.397605e-79           NaN
3    73560.0       1.0       9.0     109.0  ...           NaN           NaN           NaN           NaN
4    73561.0       2.0      73.0     116.0  ...  5.397605e-79  5.397605e-79  5.397605e-79  1.000000e+00
...      ...       ...       ...       ...  ...           ...           ...           ...           ...
20141  93698.0     1.0       2.0     121.0  ...           NaN           NaN           NaN           NaN
20142  93699.0     2.0       6.0     129.0  ...           NaN           NaN           NaN           NaN
20143  93700.0     1.0      35.0     126.0  ...           NaN           NaN           NaN           NaN
20144  93701.0     1.0       8.0     124.0  ...           NaN           NaN           NaN           NaN
20145  93702.0     2.0      24.0     119.0  ...  5.397605e-79  5.397605e-79  5.397605e-79  5.397605e-79

[20146 rows x 20 columns]
----------------------------------
```

5.8.4 Step 4: Create Depression Score[103] (Score Will Be Missing If Any of the Items Are Missing)

Depression _ Score = dpq010 + dpq020 + dpq030 + dpq040 + dpq050 + dpq060 + dpq070 + dpq080 + dpq090;

```
dfNHANES_Depression_DATA['DepressionScore'] =
dfNHANES_Depression_DATA['DPQ010'] +dfNHANES_Depression_DATA
['DPQ020'] +dfNHANES_Depression_DATA['DPQ030'] +
dfNHANES_Depression_DATA['DPQ040'] +dfNHANES_Depression_DATA
['DPQ050'] + dfNHANES_Depression_DATA['DPQ060'] +
dfNHANES_Depression_DATA['DPQ070'] + dfNHANES_Depression_DATA
['DPQ080'] + dfNHANES_Depression_DATA['DPQ090'] +
dfNHANES_Depression_DATA['DPQ100'] #+ dfNHANES_Depression_DATA
['DPQ100'])
```

[103] DB303—https://wwwn.cdc.gov/nchs/data/tutorials/DB303_Fig1_SUDAAN.sas

** Create binary depression indicator as 0/100 variable, to calculate the prevalence of depression **;

- if (0 <= Depression_Score < 10) then Depression_Indicator = 0;
- else if (Depression_Score >= 10) then Depression_Indicator = 100;

```
dfNHANES_Depression_DATA['Depression_Indicator'] =
dfNHANES_Depression_DATA['DepressionScore'].apply(lambda x:
0 if x < 10 else 100)
```

Output:

DHHIN2	...	DPQ030	DPQ040	DPQ050	DPQ060	DPQ070	DPQ080	DPQ090	DPQ100	DepressionScore	Depression_Indicator
4.0	...	5.397605e-79	5.397605e-79	5.397605e-79	5.397605e-79	5.397605e-79	5.397605e-79	5.397605e-79	1.000000e+00	2.0	0
7.0	...	5.397605e-79	5.397605e-79	5.397605e-79	5.397605e-79	5.397605e-79	5.397605e-79	5.397605e-79	5.397605e-79	2.0	0
10.0	...	5.397605e-79	5.397605e-79	5.397605e-79	5.397605e-79	5.397605e-79	5.397605e-79	5.397605e-79	NaN	NaN	100
9.0	...	NaN	NaN	NaN	NaN	NaN	NaN	NaN	NaN	NaN	100
15.0	...	5.397605e-79	3.000000e+00	3.000000e+00	5.397605e-79	5.397605e-79	5.397605e-79	5.397605e-79	1.000000e+00	10.0	100
...	...	...	...	...	...	...	...	...	...	...	...
6.0	...	NaN	NaN	NaN	NaN	NaN	NaN	NaN	NaN	NaN	100
15.0	...	NaN	NaN	NaN	NaN	NaN	NaN	NaN	NaN	NaN	100
1.0	...	NaN	NaN	NaN	NaN	NaN	NaN	NaN	NaN	NaN	100
15.0	...	NaN	NaN	NaN	NaN	NaN	NaN	NaN	NaN	NaN	100
10.0	...	1.000000e+00	1.000000e+00	5.397605e-79	5.397605e-79	5.397605e-79	5.397605e-79	5.397605e-79	5.397605e-79	2.0	0

5.8.5 Step 5: Categorize Age

** Categorize age (apply format, then convert to numeric variable) **;

```
def agecategory_function(x):
    if x > 0 and x < 20:
        return 0
    elif x >= 20 and x < 30:
        return 1
    elif x >= 30 and x < 40:
        return 2
    elif x >= 40 and x < 50:
        return 3
    elif x >= 50 and x < 60:
        return 4
    elif x >= 60 and x <= 69:
        return 5
    else:
     return 6
```

Call the function:

```
dfNHANES_Depression_DATA['ageCategory'] =
dfNHANES_Depression_DATA['RIDAGEYR'].apply(lambda x:
agecategory_function(x))
```

5.8.6 Step 6: Calculate MFC Weight for 4 Years

Calculate MEC weight for 4-year data. Use the MEC exam weights, per the analytic notes in the DPQ documentation file:

```
# https://wwwn.cdc.gov/nchs/data/tutorials/DB303_Fig1_R.R
# https://wwwn.cdc.gov/nchs/nhanes/tutorials/samplecode.aspx
# Generate 4-year MEC weight (Divide weight by 2 because we are
appending 2 survey cycles)
# Note: using the MEC Exam Weights (WTMEC2YR), per the analytic
notes on the
#    Mental Health - Depression Screener (DPQ_H) documentation

dfNHANES_Depression_DATA['WTMEC4YR'] = 1/2 *
dfNHANES_Depression_DATA['WTMEC2YR']
```

Output:

DPQ080	DPQ090	DPQ100	DepressionScore	Depression_Indicator	ageCategory	WTMEC4YR
5.397605e-79	5.397605e-79	1.000000e+00	2.0	0	5	6740.521047
5.397605e-79	5.397605e-79	5.397605e-79	2.0	0	4	12235.884813
5.397605e-79	5.397605e-79	NaN	NaN	100	6	28596.642688
NaN	NaN	NaN	NaN	100	0	27883.256219
5.397605e-79	5.397605e-79	1.000000e+00	10.0	100	6	32770.935614
...	...	...	...	...	...	...
NaN	NaN	NaN	NaN	100	0	4948.003888
NaN	NaN	NaN	NaN	100	0	4937.960523
NaN	NaN	NaN	NaN	100	2	21597.107556
NaN	NaN	NaN	NaN	100	0	24126.252147
5.397605e-79	5.397605e-79	5.397605e-79	2.0	0	1	52540.222597

5.8.7 Step 7: Adults Aged 20 and over with a Valid Depression Score

Select data frame of adults aged 20 and over with a valid depression score:

```
# Define indicator for analysis population of interest: adults aged
20 and over with a valid depression score
#inAnalysis= (RIDAGEYR >= 20 & !is.na(Depression.Score))
dfNHANES_Depression_DATA=dfNHANES_Depression_DATA
[dfNHANES_Depression_DATA['RIDAGEYR']>20]

# Don't drop, just take the rows where DepressionScore is not NA:
# https://stackoverflow.com/questions/13413590/how-to-drop-
rows-of-pandas-dataframe-whose-value-in-a-certain-column-is-
nan
dfNHANES_Depression_DATA = dfNHANES_Depression_DATA
[dfNHANES_Depression_DATA['DepressionScore'].notna()]

dfNHANES_Depression_DATA
```

Output:

IAGENDR	RIDAGEYR	SDMVSTRA	SDMVPSU	WTMEC2YR	RIDRETH3	INDFMIN2	INDHHIN2	DepressionScore	Depression_Indicator	ageCategory	WT
1.0	69.0	112.0	1.0	13481.042095	4.0	4.0	4.0	2.0	0	5	674
1.0	54.0	108.0	1.0	24471.769625	3.0	7.0	7.0	2.0	0	4	1223
2.0	73.0	116.0	2.0	65541.871229	3.0	15.0	15.0	10.0	100	6	3277
1.0	56.0	111.0	1.0	25344.992359	1.0	9.0	9.0	22.0	100	4	1267
2.0	61.0	114.0	1.0	61758.654880	3.0	10.0	10.0	2.0	0	5	3087
...	...	...	...	...	...	...	...	...	...	...	
2.0	72.0	132.0	2.0	20542.102254	4.0	9.0	9.0	3.0	0	6	1027
1.0	34.0	131.0	2.0	31309.362247	4.0	14.0	14.0	3.0	0	2	1565
1.0	53.0	126.0	1.0	22794.734098	1.0	3.0	3.0	2.0	0	4	1139
2.0	76.0	130.0	2.0	66678.487176	3.0	4.0	4.0	3.0	0	6	3333
2.0	24.0	119.0	2.0	105080.445194	3.0	7.0	10.0	2.0	0	1	5254

5.8.8 Step 8: Depression Score Histogram Plot

Plot depression score histogram:

```
import seaborn as sns, numpy as np

sns.histplot(dfNHANES_Depression_DATA['DepressionScore'],
       label="Skewness : %.2f"%(dfNHANES_Depression_MaleDATA
['DepressionScore']\
                  .skew())).legend(loc="best")

plt.show()
```

Plot histogram and correlation values for the following numerical features:

```
numeric_features = ['RIAGENDR', 'RIDAGEYR',
'RIDRETH3','SDMVSTRA', 'SDMVPSU', 'WTMEC2YR','ageCategory',
'WTMEC4YR','INDFMIN2','INDHHIN2', 'INDFMPIR']

dfNHANES_Depression_DATA[numeric_features +
['DepressionScore']].describe()
```

	RIAGENDR	RIDAGEYR	RIDRETH3	SDMVSTRA	SDMVPSU	WTMEC2YR	ageCategory	WTMEC4YR	INDFMIN2	INDHHIN2	INDFMF
count	6155.000000	6155.000000	6155.000000	6155.000000	6155.000000	6155.000000	6155.000000	6155.000000	6155.000000	6155.000000	6.155000e+
mean	1.557271	49.548335	3.225508	118.490820	1.484647	43624.636333	3.484322	21812.318166	7.902356	8.452640	2.401971e+
std	0.496750	17.376597	1.503922	8.813455	0.499805	37073.258810	1.689654	18536.629405	4.399942	5.612659	1.601000e+
min	1.000000	21.000000	1.000000	104.000000	1.000000	5157.019076	1.000000	2578.509538	1.000000	1.000000	5.397605e-
25%	1.000000	34.000000	2.000000	111.000000	1.000000	19507.941629	2.000000	9753.970815	5.000000	5.000000	1.040000e+
50%	2.000000	49.000000	3.000000	118.000000	1.000000	28380.031292	3.000000	14190.015646	7.000000	7.000000	1.950000e+
75%	2.000000	64.000000	4.000000	126.000000	2.000000	54920.080316	5.000000	27460.040158	10.000000	14.000000	3.795000e+
max	2.000000	80.000000	7.000000	133.000000	2.000000	242386.660766	6.000000	121193.330383	15.000000	99.000000	5.000000e+

Plot histogram between numerical features and depression score to witness the influence of the each parameter on depression score:

```
for col in numeric_features:
  fig = plt.figure(figsize=(9, 6))
  ax = fig.gca()
  feature = dfNHANES_Depression_DATA[col]
  feature.hist(bins=100, ax = ax)
  ax.axvline(feature.mean(), color='magenta',
linestyle='dashed', linewidth=2)
  ax.axvline(feature.median(), color='cyan',
linestyle='dashed', linewidth=2)
  ax.set_title(col)
plt.show()
```

Output: During 2013–2016, American adults aged 20 and over had depression in a given 2-week period[104] [43].

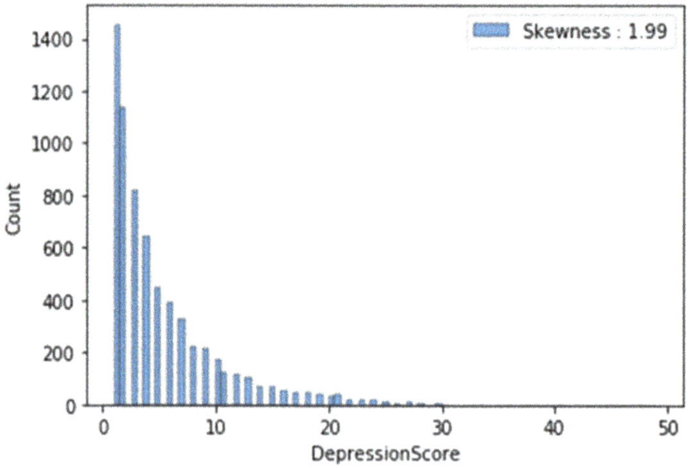

Depression was lower among non-Hispanic Asian adults [43], compared with Hispanic, non-Hispanic black, or non-Hispanic white adults.

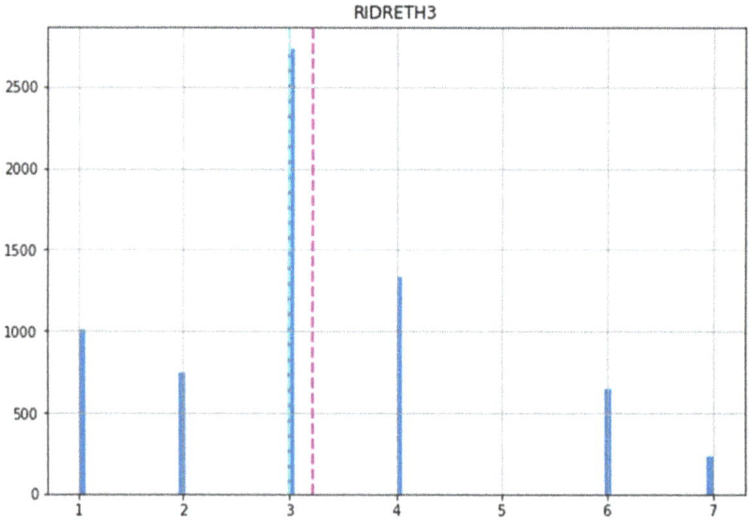

[104] Prevalence of Depression Among Adults Aged 20 and Over: United States—https://www.cdc.gov/nchs/products/databriefs/db303.htm#depression_prevalence_20_and_over

As it can be seen from the RIDRETH3: Depression score graph that the mon-Hispanic Asian (RIDRETH3=6) is lower compared to other race—including multi-racial (RIDRETH3=7).

The prevalence of depression decreased as family income levels increased.

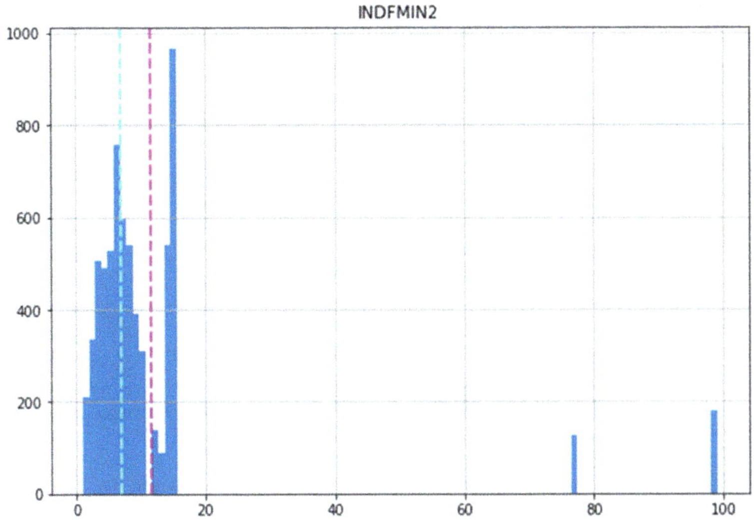

Especially during economic contractions, one tends to see increase in depression scores[105]. For instance, during 2021, real household income per capita fell by 3.8% in the Organisation for Economic Co-operation and Development (OECD) area in the second quarter of 2021 following growth of 5.2% in the first quarter. The fall was driven by a sharp drop in household income in the United States of America, as fiscal support provided by the government during the COVID-19 pandemic began to be withdrawn. The drop in real disposable income exerts a huge pressure on families and could contribute to food insecurity.

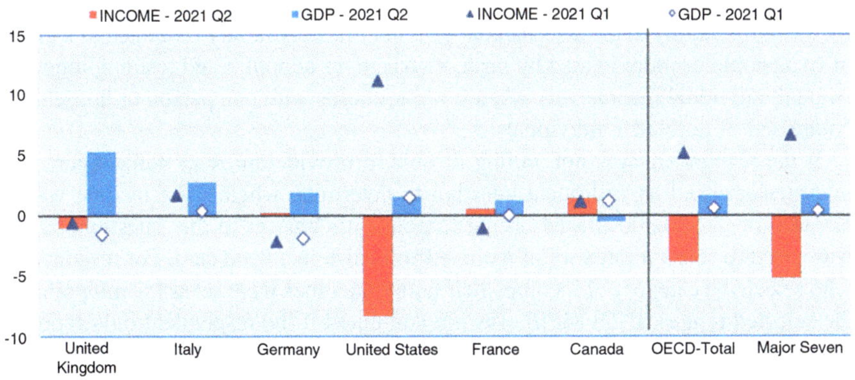

[105] Household income in the OECD area fell sharply in the second quarter of 2021, despite strong growth in GDP—https://www.oecd.org/sdd/na/growth-and-economic-well-being-second-quarter-2021-oecd.htm

As can be seen from the INDFMPIR, the depression scores are pretty much spread across. Ni significant information can be derived except for score 5 that has exhibited higher values of depression score.

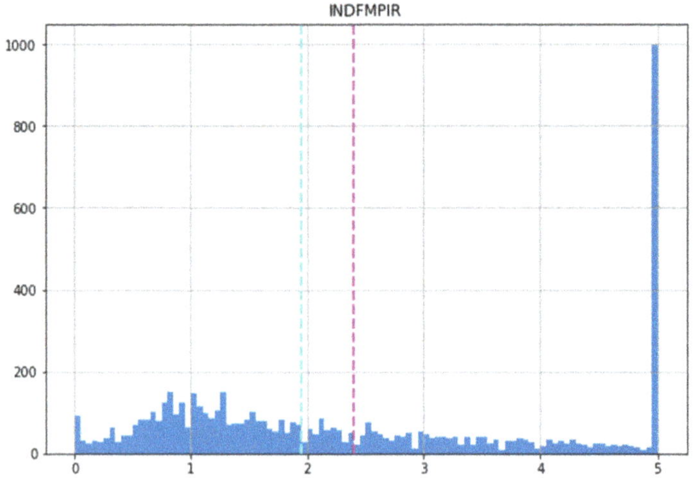

INDFMPIR[106]: This variable is the ratio of family income to poverty. The Department of Health and Human Services (DHHS) poverty guidelines were used as the poverty measure to calculate this ratio. These guidelines are issued each year, in the Federal Register, for determining financial eligibility for certain federal programs, such as Head Start; Supplemental Nutrition Assistance Program (SNAP); Special Supplemental Nutrition Program for Women, Infants, and Children (WIC); and the National School Lunch Program. The poverty guidelines vary by family size and geographic location (with different guidelines for the 48 contiguous states and the District of Columbia, Alaska, and Hawaii).

INDFMPIR was calculated by dividing total annual family (or individual) income by the poverty guidelines specific to the survey year.

During the household interview, the respondent was asked to report total income for the entire family in the last calendar year in dollars. A family is defined as a group of two people or more related by birth, marriage, or adoption and residing together. Annual individual income was asked for households with one person or households comprised of unrelated individuals.

If the respondent was not willing or able to provide an exact dollar figure, the interviewer asked an additional question to determine whether the income was $<$ \$20,000 or $\geq$ \$20,000. Based on the respondent's answer to this question, he/she was asked to select a category of income from a list on a hand card. For respondents who selected a category of income, their family incomes were set as the midpoints of the selected ranges. INDFMPIR was not computed if the respondent only reported income as $<$ \$20,000 or $\geq$ \$20,000, but no additional details provided. INDFMPIR values at or above 5.00 were coded as 5.00 or more because of disclosure concerns. The values were not computed if the income data was missing.

[106]IDNFMPIR—https://wwwn.cdc.gov/Nchs/Nhanes/2017-2018/P_DEMO.htm#INDFMPIR

5.8.9 *Step 9: Correlation Between Key Features and Depression Score*

Let's compute correlation between key attributes and depression score column:

```
for col in numeric_features:
   fig = plt.figure(figsize=(9, 6))
   ax = fig.gca()
   feature = dfNHANES_Depression_DATA[col]
   label = dfNHANES_Depression_DATA['DepressionScore']
   correlation = feature.corr(label)
   plt.scatter(x=feature, y=label)
   plt.xlabel(col)
   plt.ylabel('DepressionScore')
  ax.set_title('DepressionScore vs ' + col + '- correlation: ' + str
(correlation))
plt.show()
```

Output:

Correlation between the depression score and INDFMIN2 ("family income") is negatively correlated with -23.1%. What does it mean is the higher the family income of the respondents, the lower the depression scores. In other words, the prevalence of depression decreased as family income levels increased[107] [43].

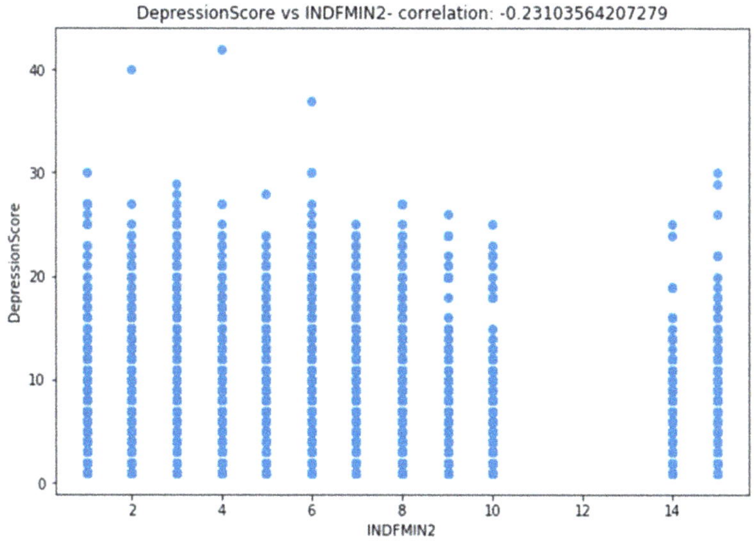

[107]Prevalence of Depression Among Adults Aged 20 and Over: United States of America, 2013–2016—https://www.cdc.gov/nchs/products/databriefs/db303.htm#depression_prevalence_20_and_over

The same can be observed from the correlation between depression score and INDFMPIR. The higher the value of family income over poverty line, the lower the depression scores. In other words, the prevalence of depression decreased as income levels increased.

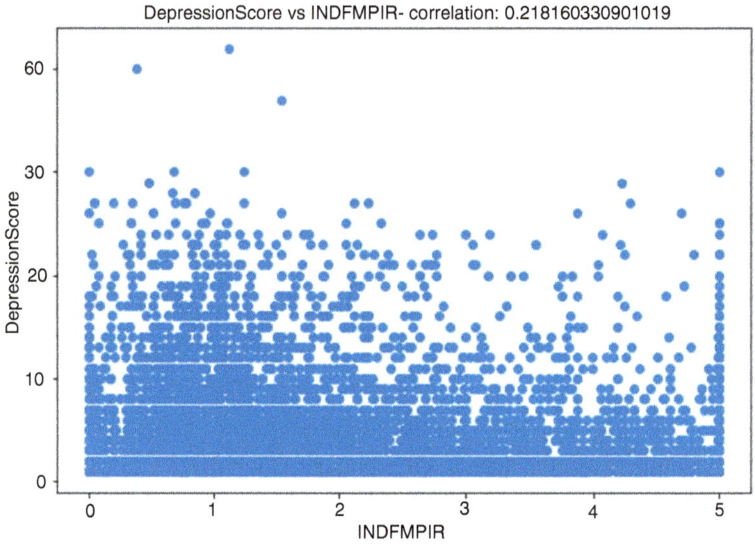

5.8.10 Step 10: Regression Model

Develop regression model. Split the dataset into training and testing data:

```
# Separate features and labels

X, y = dfNHANES_Depression_DATA[['RIDAGEYR', 'SDMVSTRA',
'WTMEC2YR', 'RIDRETH3','WTMEC4YR','INDFMIN2']].values,
dfNHANES_Depression_DATA['DepressionScore'].values
print('Features:',X[:10], '\nLabels:', y[:10], sep='\n')

from sklearn.model_selection import train_test_split

# Split data 70%-30% into training set and test set
X_train, X_test, y_train, y_test = train_test_split(X, y,
test_size=0.30, random_state=0)

print ('Training Set: %d, rows\nTest Set: %d rows' % (X_train.size,
X_test.size))
```

Let's compute the correlation between data frame attributes.
Output:
Training set: 28152 rows
Test set: 12066 rows
Construct regression model:

```
# Train the model
from sklearn.linear_model import LinearRegression

# Fit a linear regression model on the training set
model = LinearRegression(normalize=False).fit(X_train, y_train)
print (model)

print('intercept:', model.intercept_)
print('slope:', model.coef_)
```

Output:
Intercept: 7.698458827541425
Slope: [0.01208841 -0.01792657 -0.00000699 -0.15718569 -0.00000349 -0.00723775]

5.8.11 Step 11: Evaluate the Model

Perform and evaluate the model.

```
import numpy as np

predictions = model.predict(X_test)
np.set_printoptions(suppress=True)
print('Predicted labels: ', np.round(predictions)[:10])
print('Actual labels   : ' ,y_test[:10])
```

Output:
Predicted labels: [5. 4. 6. 5. 6. 5. 5. 6. 6. 5.]
Actual labels: [4. 3. 6. 1. 6. 1. 2. 3. 9. 2.]

```
import matplotlib.pyplot as plt

%matplotlib inline
```

(continued)

```
plt.scatter(y_test, predictions)
plt.xlabel('Actual Labels')
plt.ylabel('Predicted Labels')
plt.title('Depresion Score Predictions')
# overlay the regression line
z = np.polyfit(y_test, predictions, 1)
p = np.poly1d(z)
plt.plot(y_test,p(y_test), color='magenta')
plt.show()
```

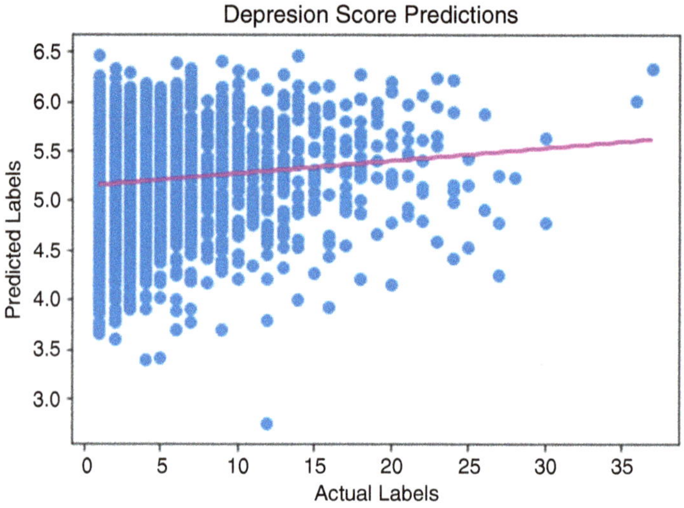


```
from sklearn.metrics import mean_squared_error, r2_score

mse = mean_squared_error(y_test, predictions)
print("MSE:", mse)

rmse = np.sqrt(mse)
print("RMSE:", rmse)

r2 = r2_score(y_test, predictions)
print("R2:", r2)
```

Output:
MSE: 24.77714961903213
RMSE: 4.977665077024782
R2: 0.015155531146714085

5.8.12 Step 12: OLS Method

Using OLS method[108], run the model[109] [43]:

```
#Using OLS method

from statsmodels.formula.api import ols
#'BPXSY1','RIDAGEYR','BMXWT','BMXHT','RIAGENDR',
'RIDRETH3','DMDMARTL','DMDEDUC2'

formula='DepressionScore ~ RIDAGEYR + SDMVSTRA + WTMEC2YR +
RIDRETH3 + INDFMIN2 + WTMEC4YR'
model=ols(formula,dfNHANES_Depression_DATA).fit()
model.summary()
```

Output:

OLS Regression Results

Dep. Variable:	DepressionScore	**R-squared:**	0.012
Model:	OLS	**Adj. R-squared:**	0.011
Method:	Least Squares	**F-statistic:**	15.76
Date:	Sun, 02 Jan 2022	**Prob (F-statistic):**	1.87e-15
Time:	03:34:48	**Log-Likelihood:**	-20377.
No. Observations:	6703	**AIC:**	4.077e+04
Df Residuals:	6697	**BIC:**	4.081e+04
Df Model:	5		
Covariance Type:	nonrobust		

	coef	**std err**	**t**	**P>|t|**	**[0.025**	**0.975]**
Intercept	6.9399	0.867	8.007	0.000	5.241	8.639
RIDAGEYR	0.0159	0.004	4.448	0.000	0.009	0.023
SDMVSTRA	-0.0129	0.007	-1.839	0.066	-0.027	0.001
WTMEC2YR	-7.628e-06	1.36e-06	-5.601	0.000	-1.03e-05	-4.96e-06
RIDRETH3	-0.1554	0.041	-3.798	0.000	-0.236	-0.075
INDFMIN2	-0.0069	0.003	-1.978	0.048	-0.014	-6.26e-05
WTMEC4YR	-3.814e-06	6.81e-07	-5.601	0.000	-5.15e-06	-2.48e-06

[108] Linear Regression - ANOVA, Two Sample testing Added—https://www.kaggle.com/aviskumar/linear-regression-anova-twosampletesting-added

[109] NCHS Data Brief No. 303—Prevalence of Depression Among Adults Aged 20 and Over: United States of America, 2013–2016 - https://wwwn.cdc.gov/nchs/nhanes/tutorials/samplecode.aspx

Omnibus:	2692.764	**Durbin-Watson:**	1.966
Prob(Omnibus):	0.000	**Jarque-Bera (JB):**	11263.656
Skew:	1.976	**Prob(JB):**	0.00
Kurtosis:	7.971	**Cond. No.**	7.42e+15

Warnings:

[1] Standard errors assume that the covariance matrix of the errors is correctly specified.

[2] The smallest eigenvalue is 4.81e-19. This might indicate that there are strong multicollinearity problems or that the design matrix is singular.

5.9 Use Case: NHANES Body Measure Associations Between Body Weight and the Health and Nutritional Status (BPXSY2 vs. BPXSY1)

Systolic blood pressure data is from body measures (BMX_J)[110]. NHANES body measures data are used to monitor trends in infant and child growth; to estimate the prevalence of overweight and obesity in US children, adolescents, and adults; and to examine the associations between body weight and the health and nutritional status of the US population [45].

The body measures data were collected, in the Mobile Examination Center (MEC—image source: Mobile Examination Center (please see Fig. 5.20): 1988[111]), by trained health technicians. The health technician was assisted by a recorder during the body measures examination. The participant's age at the time of the screening interview determined the body measures examination protocol. In some instances, the age at the screening interview and age at the time of the health examination differed by several weeks. The demographics data file includes variables for age in years at screening (RIDAGEYR) for all participants and age in months at examination (RIDEXAGM) for participants who were <240 months of age at the time of examination:

- BPXSY1—Systolic: blood pressure (1st rdg) mm Hg
- BPXSY2—Systolic: blood pressure (2nd rdg) mm Hg

Dataset: BMX_J_DATA[112]

[110] BMX_J: https://wwwn.cdc.gov/Nchs/Nhanes/2017-2018/BMX_J.htm

[111] NHANES Mobile Center—https://collections.nlm.nih.gov/catalog/nlm:nlmuid-101447545-img

[112] BMX_J_DATA—https://wwwn.cdc.gov/Nchs/Nhanes/2017-2018/BMX_J.XPT

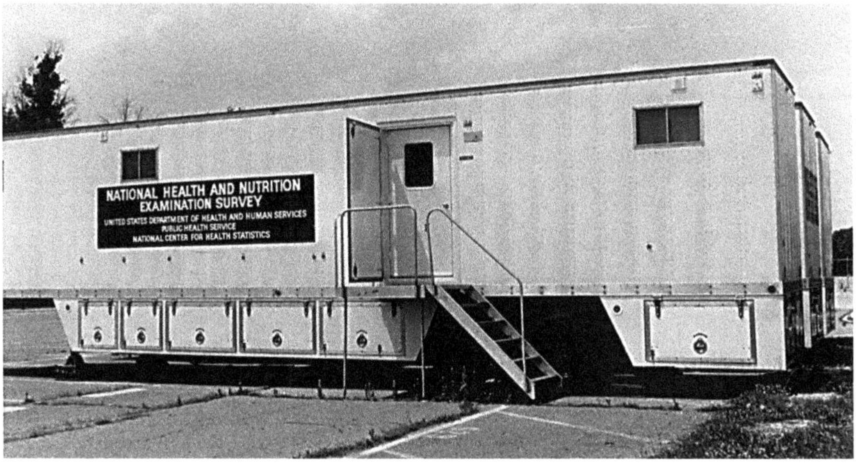

Fig. 5.20 Mobile Examination Center

```
# -*- coding: utf-8 -*-
"""
Created on Wed Dec 29 19:18:19 2021

@author: CHVUPPAL
"""

# Collinearity

import pandas as pd
import matplotlib.pyplot as plt

dfP_BMX = pd.read_sas('P_BMX.XPT')
dfBPX_J = pd.read_sas('BPX_J.XPT')

print(dfP_BMX.columns)

print(dfBPX_J.columns)

print(dfP_BMX)
print(dfBPX_J)

# Create a new data from for BPXSY1 & BPXSY2
dfBPXSY1vsBPXSY2 = dfBPX_J[['BPXSY1', 'BPXSY2']].copy()
```

(continued)

```
print(dfBPXSY1vsBPXSY2)

# Scatter plot to see the distribution of Blood pres (1nd rdg) mm Hg &
Blood pres (2nd rdg) mm Hg
plt.scatter(dfBPXSY1vsBPXSY2['BPXSY1'], dfBPXSY1vsBPXSY2
['BPXSY2'])
plt.title("BPXSY2 vs. BPXSY1")
plt.xlabel("BPXSY1")
plt.ylabel("BPXSY2")

plt.show()
```

Output:

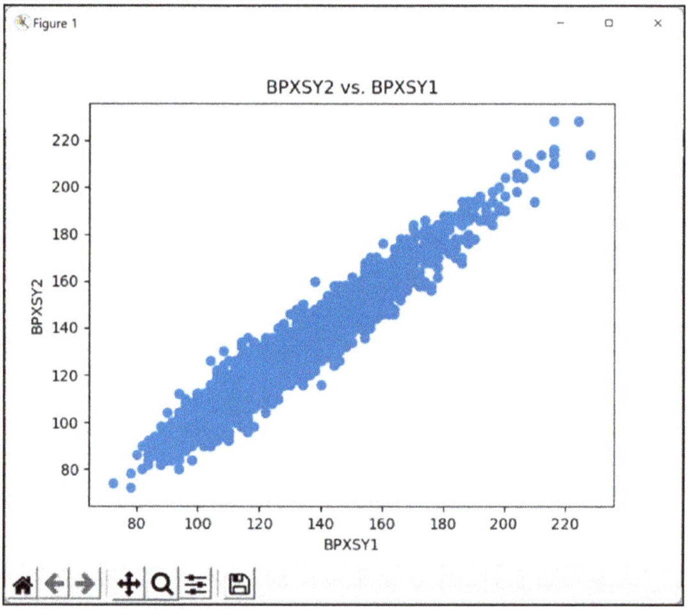

The distribution of systolic: blood pressure first reading vs. blood pressure second reading is very similar.

Using NHANES data tables, one can infer the health parameters of the national United States of America as it captures the vital signs with demographics and other important parameters. For instance, high blood pressure data can provide guidance on the health distribution across the population as high blood pressure continues to be a leading health problem in the US NHANES data tables (NHANES 2018–2020) containing survey data from nearly 10,000 people in the United States in 2018. For this analysis, we will focus on only the following variables to see how high blood pressure reflected across the nation:

- RIAGENDR: Respondent's gender
- RIDAGEYR: Respondent's age in years
  - Age in years, at the time of the screening interview, is reported for survey participants between the ages of 1 and 79 years. Due to disclosure concern, all responses of participants aged 80 years and older are coded as "80." In NHANES 2017–March 2020 pre-pandemic sample, the weighted mean age for participants 80 years and older is 85 years. RIDAGEYR was calculated based on the participant's date of birth. In rare cases, if the actual date of birth was missing but the participant's age in years was provided, the reported age was used.
- RIDRETH3[113]: Respondent's racial or ethnic background
  - This is the race-ethnicity variable included in the demographics file since the 2011–2012 survey cycle to accommodate the oversample of Asian Americans. It was derived from responses to the survey questions on race and Hispanic origin. Respondents who self-identified as "Mexican American" were coded as such (i.e., RIDRETH3=1) regardless of their other race-ethnicity identities. Otherwise, self-identified "Hispanic" ethnicity would result in code "2, Other Hispanic" in the RIDRETH3 variable. All other non-Hispanic participants would then be categorized based on their self-reported races: non-Hispanic white (RIDRETH3=3), non-Hispanic black (RIDRETH3=4), non-Hispanic Asian (RIDRETH3=6), and other non-Hispanic races including non-Hispanic multi-racial (RIDRETH3=7). Code "5" was not used in RIDRETH3.
- BMXWT: Respondent's weight in kilograms
- BPXPLS: Respondent's resting pulse rate
- BPXSY1: Respondent's systolic blood pressure ("top" number in BP)
- BPXD1: Respondent's diastolic blood pressure ("bottom" number in BP)

5.9.1 Step 1: Load P_BMX.XPT, BPX_J.XPT, and P_DEMO.XPT

```
import pandas as pd
import matplotlib.pyplot as plt

# Load P_BMX.XPT
dfP_BMX = pd.read_sas('P_BMX.XPT')
dfP_BMX.set_index('SEQN') # Respondent sequence number
```

(continued)

[113] P_DEMO—https://wwwn.cdc.gov/Nchs/Nhanes/2017-2018/P_DEMO.htm

```
# Load BPX_J.XPT
dfBPX_J = pd.read_sas('BPX_J.XPT')
dfBPX_J.set_index('SEQN') # Respondent sequence number

# Load P_DEMO.XPT

dfP_DEMO = pd.read_sas('P_DEMO.XPT')
dfP_DEMO.set_index('SEQN') # Respondent sequence number
```

5.9.2 Step 2: Inspect Columns of the Dataset

```
print('------------dfP_BMX---------------------')
print(dfP_BMX.columns)
print('-------------dfBPX_J--------------------')

print(dfBPX_J.columns)

print('-------------dfP_DEMO.columns---------------------')
print(dfP_DEMO.columns)

print('---------------------------------')
```

Output:

```
------------dfP_BMX---------------------
Index(['SEQN', 'BMDSTATS', 'BMXWT', 'BMIWT', 'BMXRECUM',
'BMIRECUM', 'BMXHEAD',
    'BMIHEAD', 'BMXHT', 'BMIHT', 'BMXBMI', 'BMDBMIC', 'BMXLEG',
'BMILEG',
    'BMXARML', 'BMIARML', 'BMXARMC', 'BMIARMC', 'BMXWAIST',
'BMIWAIST',
    'BMXHIP', 'BMIHIP'],
   dtype='object')
-------------dfBPX_J--------------------
Index(['SEQN', 'PEASCCT1', 'BPXCHR', 'BPAARM', 'BPACSZ',
'BPXPLS', 'BPXPULS',
    'BPXPTY', 'BPXML1', 'BPXSY1', 'BPXDI1', 'BPAEN1', 'BPXSY2',
'BPXDI2',
    'BPAEN2', 'BPXSY3', 'BPXDI3', 'BPAEN3', 'BPXSY4', 'BPXDI4',
'BPAEN4'],
   dtype='object')
```

(continued)

```
-------------dfP_DEMO.columns----------------------
Index(['SEQN', 'SDDSRVYR', 'RIDSTATR', 'RIAGENDR', 'RIDAGEYR',
'RIDAGEMN',
    'RIDRETH1', 'RIDRETH3', 'RIDEXMON', 'DMDBORN4', 'DMDYRUSZ',
'DMDEDUC2',
    'DMDMARTZ', 'RIDEXPRG', 'SIALANG', 'SIAPROXY', 'SIAINTRP',
'FIALANG',
    'FIAPROXY', 'FIAINTRP', 'MIALANG', 'MIAPROXY', 'MIAINTRP',
'AIALANGA',
    'WTINTPRP', 'WTMECPRP', 'SDMVPSU', 'SDMVSTRA', 'INDFMPIR'],
   dtype='object')
----------------------------------
-------------df_P_DEMO_BPX_J--------------------
Index(['SEQN_x', 'PEASCCT1', 'BPXCHR', 'BPAARM', 'BPACSZ',
'BPXPLS', 'BPXPULS',
    'BPXPTY', 'BPXML1', 'BPXSY1', 'BPXDI1', 'BPAEN1', 'BPXSY2',
'BPXDI2',
    'BPAEN2', 'BPXSY3', 'BPXDI3', 'BPAEN3', 'BPXSY4', 'BPXDI4',
'BPAEN4',
    'SEQN_y', 'SDDSRVYR', 'RIDSTATR', 'RIAGENDR', 'RIDAGEYR',
'RIDAGEMN',
    'RIDRETH1', 'RIDRETH3', 'RIDEXMON', 'DMDBORN4', 'DMDYRUSZ',
'DMDEDUC2',
    'DMDMARTZ', 'RIDEXPRG', 'SIALANG', 'SIAPROXY', 'SIAINTRP',
'FIALANG',
    'FIAPROXY', 'FIAINTRP', 'MIALANG', 'MIAPROXY', 'MIAINTRP',
'AIALANGA',
    'WTINTPRP', 'WTMECPRP', 'SDMVPSU', 'SDMVSTRA', 'INDFMPIR'],
   dtype='object')
```

5.9.3 Step 3: Filter the Data Based on Age, Select Adult, and Scatter Plot Each Important Blood Pressure Items

```
dfBPXSY1vsBPXSY2         =          df_P_DEMO_BPX_J[['RIDAGEYR',
'BPXSY1','BPXDI1','BPXPLS','RIAGENDR','RIDRETH3']].copy()

print(dfBPXSY1vsBPXSY2)

print('------------dfBPXSY1vsBPXSY2
[AGE]---------------------')

# Only Adults
dfBPXSY1vsBPXSY2=dfBPXSY1vsBPXSY2[dfBPXSY1vsBPXSY2
['RIDAGEYR']>18]
print(dfBPXSY1vsBPXSY2)
```

Merge demographic and blood pressure data frames and filter based on the age (respondent age >18).

5.9.4 Step 4: Scatter Plot Blood Pressure (First Reading) mm Hg and Blood Pressure (Second Reading)

```
# Scatter plot to see the distribution of Blood pres (1nd rdg) mm Hg &
Blood pres (2nd rdg) mm Hg
plt.scatter(dfBPXSY1vsBPXSY2['RIDAGEYR'], dfBPXSY1vsBPXSY2
['BPXSY1'])
plt.title("RIDAGEYR vs. BPXSY1")
plt.xlabel("RIDAGEYR")
plt.ylabel("BPXSY1")
```

Output: Scatter plot of respondent's age to systolic blood pressure shows peak outliers can be observed for the age between 40 and 50. And low value can be observed for middle 35.

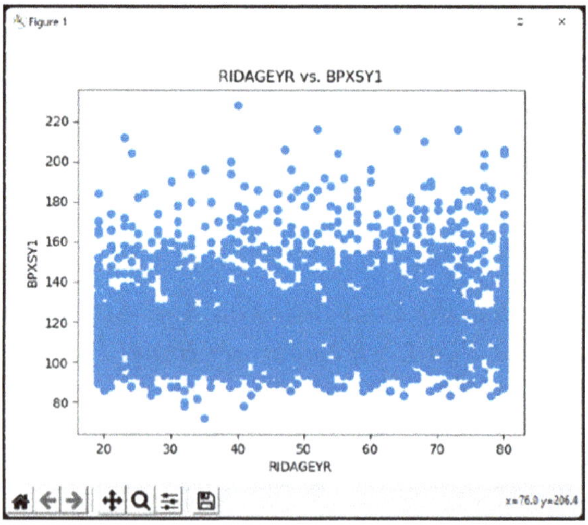

5.9.5 Step 5: Scatter Plot Blood Pressure (First Reading) mm Hg and Gender

```
# Scatter plot to see the distribution of Blood pres (1nd rdg) mm Hg &
Gender
plt.scatter(dfBPXSY1vsBPXSY2['BPXSY1'], dfBPXSY1vsBPXSY2
['RIAGENDR'])
plt.title("BPXSY1 vs. RIAGENDR")
plt.xlabel("BPXSY1")
plt.ylabel("RIAGENDR")

plt.show()
```

Output:

As it can be seen from the scatter plot, male gender has higher number of blood pressure.

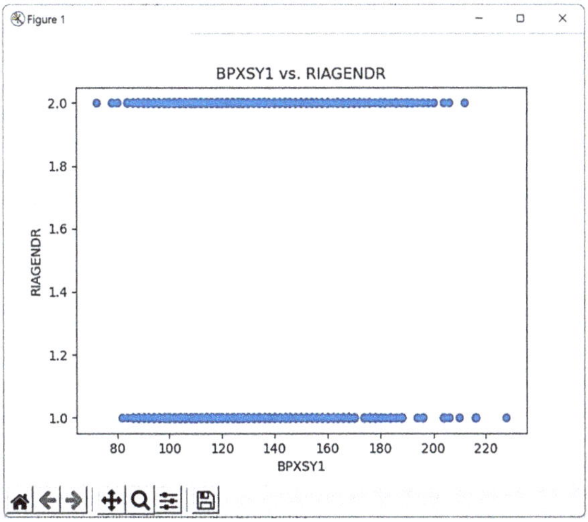

5.9.6 Step 6: Scatter Plot Race and Blood Pressure (First Reading) mm Hg

```
# Scatter plot to see the distribution of RIDRETH3 and Blood pres
(1nd rdg) mm Hg
```

(continued)

```
plt.scatter(dfBPXSY1vsBPXSY2['RIDRETH3'], dfBPXSY1vsBPXSY2
['BPXSY1'])
plt.title("RIDRETH3 vs. BPXSY1")
plt.xlabel("RIDRETH3")
plt.ylabel("BPXSY1")

plt.show()
```

Output:

As per the scatter plot (RIDRETH3 vs. BPXSY1), RIDRETH3=6 exhibited higher values of systolic blood pressure, and the RIDRETH3=7 exhibited lower values. Similarly, RIDRETH3=1 exhibited diverse rage spanning from lower of 80 mm Hg to 190 mm Hg.

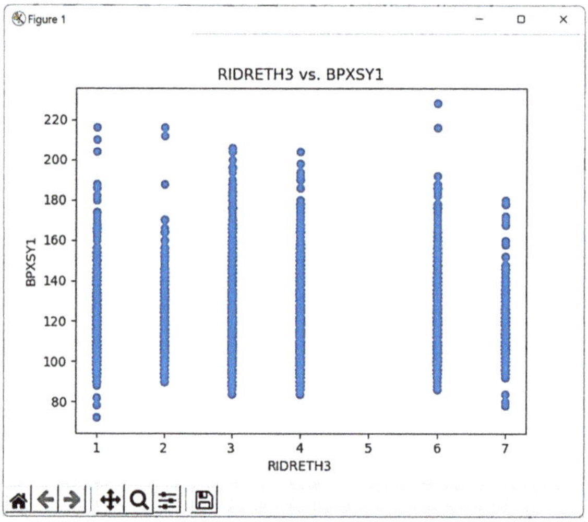

5.9.7 *Step 7: Correlation*

```
correlation_df = dfBPXSY1vsBPXSY2.corr()

print(correlation_df)
```

Output:

	RIDAGEYR	BPXSY1	BPXDI1	BPXPLS	RIAGENDR	RIDRETH3
RIDAGEYR	1.000000	0.055007	0.018333	-0.017127	-0.028697	-0.014725
BPXSY1	0.055007	1.000000	0.435275	-0.187295	0.005979	0.004318
BPXDI1	0.018333	0.435275	1.000000	-0.080569	0.026922	0.007629
BPXPLS	-0.017127	-0.187295	-0.080569	1.000000	0.004844	0.000014
RIAGENDR	-0.028697	0.005979	0.026922	0.004844	1.000000	0.003276
RIDRETH3	-0.014725	0.004318	0.007629	0.000014	0.003276	1.000000

5.9.8 Step 8: Scatter Plot Systolic and Pulse Rate

```
plt.scatter(dfBPXSY1vsBPXSY2['BPXSY1'], dfBPXSY1vsBPXSY2
['BPXPLS'])
plt.title("BPXSY1 vs. BPXPLS")
plt.xlabel("BPXSY1")
plt.ylabel("BPXPLS")
plt.show()
```

Output:

As shown in the scatterplot of systolic blood pressure and pulse rate, higher pulse rates can be observed for the blood pressure that is in the rage of 100 to 140 mm Hg. Please kindly note that higher pulse rate is associated with higher blood pressure.

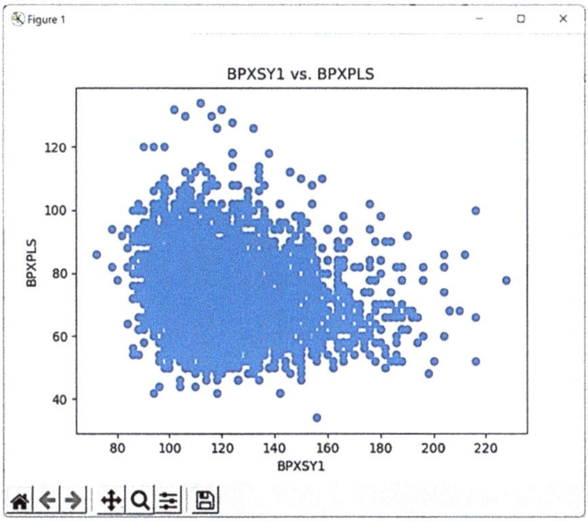

5.10 Use Case: NHANES 2019–2020 Questionnaire Instruments FSQ—Prevalence of Food Security

NHANES 2019–2020 Questionnaire Instruments consist of several questionnaires that are administered to NHANES participants both at home and in the MECs. The questionnaires and a brief description of each section can be referred online[114]. NHANES 2017–2018 Questionnaire Data[115] have various sections spanning from alcohol use to weight history. The comprehensive list (please see Table 5.4) are included in the following table:

The comprehensive list of variables for the 2017–2018 Questionnaire consists of 1372 key indicators[116].

Software code for this model can be found on GitHub Link: NHANES_FSQ_Data.py (Python Code)
NHANES_2017-2018_FSQModel.ipynb (Jupyter Notebook Code)

5.10.1 Step 1: Load DEMO_J.XPT Dataset

The dataset for food security indicator consists of FSO and demographics XPT files.

Load demographic file—https://wwwn.cdc.gov/Nchs/Nhanes/2017-2018/DEMO_J.XPT

```
"""
Created on Sun Jan 2 18:25:00 2022

@author: CHVUPPAL
"""

# https://wwwn.cdc.gov/nchs/nhanes/continuousnhanes/
questionnaires.aspx?Cycle=2019-2020
```

(continued)

[114] NHANES—https://wwwn.cdc.gov/nchs/nhanes/continuousnhanes/questionnaires.aspx?Cycle=2019-2020

[115] Questionnaire Data—https://wwwn.cdc.gov/Nchs/Nhanes/Search/DataPage.aspx?Component=Questionnaire&Cycle=2017-2018

[116] 2017–2018 Questionnaire Variable List—https://wwwn.cdc.gov/Nchs/Nhanes/Search/variablelist.aspx?Component=Questionnaire&Cycle=2017-2018

Table 5.4 NHANES file list

Data file name	Doc file	Data file	Date published
Acculturation	ACQ_J Doc	ACQ_J Data [XPT - 396.3 KB]	February 2020
Alcohol use	ALQ_J Doc	ALQ_J Data [XPT - 434.4 KB]	December 2020
Audiometry	AUQ_J Doc	AUQ_J Data [XPT - 3.9 MB]	July 2020
Blood pressure and cholesterol	BPQ_J Doc	BPQ_J Data [XPT - 531.8 KB]	February 2020
Cardiovascular health	CDQ_J Doc	CDQ_J Data [XPT - 518.7 KB]	February 2020
Consumer behavior	CBQ_J Doc	CBQ_J Data [XPT - 435.4 KB]	April 2021
Consumer behavior phone Follow-up module—adult	CBQPFA_J Doc	CBQPFA_J Data [XPT - 2.7 MB]	June 2021
Consumer behavior phone Follow-up module—child	CBQPFC_J Doc	CBQPFC_J Data [XPT - 1.1 MB]	June 2021
Current health status	HSQ_J Doc	HSQ_J Data [XPT - 590.2 KB]	February 2020
Dermatology	DEQ_J Doc	DEQ_J Data [XPT - 242.4 KB]	February 2020
Diabetes	DIQ_J Doc	DIQ_J Data [XPT - 3.7 MB]	February 2020
Diet behavior and nutrition	DBQ_J Doc	DBQ_J Data [XPT - 3.3 MB]	February 2020
Disability	DLQ_J Doc	DLQ_J Data [XPT - 906.2 KB]	February 2020
Drug use	DUQ_J Doc	DUQ_J Data [XPT - 1.4 MB]	February 2020
Early childhood	ECQ_J Doc	ECQ_J Data [XPT - 243.8 KB]	February 2020
Food security	FSQ_J Doc	FSQ_J Data [XPT - 3.5 MB]	July 2021
Health insurance	HIQ_J Doc	HIQ_J Data [XPT - 1.1 MB]	February 2020
Hepatitis	HEQ_J Doc	HEQ_J Data [XPT - 309.1 KB]	February 2020
Hospital utilization and access to care	HUQ_J Doc	HUQ_J Data [XPT - 725.1 KB]	February 2020
Housing characteristics	HOQ_J Doc	HOQ_J Data [XPT - 218.1 KB]	December 2020
Immunization	IMQ_J Doc	IMQ_J Data [XPT - 797.6 KB]	February 2020
Income	INQ_J Doc	INQ_J Data [XPT - 1.1 MB]	April 2021
Kidney conditions—urology	KIQ_U_J Doc	KIQ_U_J Data [XPT - 699.1 KB]	February 2020

(continued)

Table 5.4 (continued)

Data file name	Doc file	Data file	Date published
Medical conditions	MCQ_J Doc	MCQ_J Data [XPT - 5.2 MB]	February 2020
Mental health—depression Screener	DPQ_J Doc	DPQ_J Data [XPT - 4 77.8 KB]	February 2020
Occupation	OCQ_J Doc	OCQ_J Data [XPT - 4 83.4 KB]	February 2020
Oral Health	OHQ_J Doc	OHQ_J Data [XPT - 3 MB]	February 2020
Osteoporosis	OSQ_J Doc	OSQ_J Data [XPT - 2.2 MB]	March 2020
Pesticide use	PUQMEC_J Doc	PUQMEC_J Data [XPT - 175.5 KB]	June 2020
Physical activity	PAQ_J Doc	PAQ_J Data [XPT - 780.9 KB]	February 2020
Physical activity—youth	PAQY_J Doc	PAQY_J Data [XPT - 88.1 KB]	February 2020
Physical functioning	PFQ_J Doc	PFQ_J Data [XPT - 2.3 MB]	February 2020
Prescription medications	RXQ_RX_J Doc	RXQ_RX_J Data [XPT - 9.2 MB]	March 2020
Prescription medications—drug information	RXQ_ DRUG Doc	RXQ_DRUG Data [XPT - 2.6 KB]	Updated September 2021
Preventive aspirin use	RXQASA_J Doc	RXQASA_J Data [XPT - 244.5 KB]	April 2020
Reproductive health	RHQ_J Doc	RHQ_J Data [XPT - 1.2 MB]	February 2020
Sleep disorders	SLQ_J Doc	SLQ_J Data [XPT - 459. 5 KB]	February 2020
Smoking—cigarette use	SMQ_J Doc	SMQ_J Data [XPT - 2.2 MB]	February 2020
Smoking—household Smokers	SMQFAM_ J Doc	SMQFAM_J Data [XPT - 290.5 KB]	February 2020
Smoking—recent tobacco use	SMQRTU_J Doc	SMQRTU_J Data [XPT - 1.3 MB]	February 2020
Smoking—secondhand Smoke exposure	SMQSHS_J Doc	SMQSHS_J Data [XPT - 1.1 MB]	February 2020
Volatile toxicant	VTQ_J Doc	VTQ_J Data [XPT - 573.9 KB]	December 2020
Weight history	WHQ_J Doc	WHQ_J Data [XPT - 1.7 MB]	June 2020
Weight history—youth	WHQMEC_ J Doc	WHQMEC_J Data [XPT - 42 KB]	June 2020

```
import pandas as pd
from functools import reduce
import matplotlib.pyplot as plt

# https://wwwn.cdc.gov/Nchs/Nhanes/2017-2018/DEMO_J.XPT

dfDEMO_J = pd.read_sas('DEMO_J.XPT')
#dfDEMO_I.set_index('SEQN') # Respondent sequence number
print('---------------------------------------')
print('dfDEMO_J - keep seqn riagendr ridageyr sdmvstra sdmvpsu
wtmec2yr')
print('---------------------------------------')
print(dfDEMO_J)

print(list(dfDEMO_J.columns))

dfDEMO_IREQ = dfDEMO_J
[['SEQN','RIAGENDR','RIDAGEYR','SDMVSTRA',
'SDMVPSU','WTMEC2YR','RIDRETH3','INDFMIN2',
'INDHHIN2','INDFMPIR']]
```

5.10.2 Step 2: Load FSQ_J.XPT Dataset

File name: https://wwwn.cdc.gov/Nchs/Nhanes/2017-2018/FSQ_J.XPT

```
## ** Download Demographic (DEMO) Data and Keep Variables Of
Interest **
# Load DEMO_I.XPT
dfFSQ_J = pd.read_sas('FSQ_J.XPT')
#dfDEMO_I.set_index('SEQN') # Respondent sequence number
print('----------------------------
----------')
print('dfFSQ_J - keep seqn riagendr ridageyr sdmvstra sdmvpsu
wtmec2yr')
print('----------------------
-----------------')
print(dfFSQ_J)
print( 'FSDHH --- ' + str(dfFSQ_J['FSDHH'] ))
print( 'FSDAD --- ' + str(dfFSQ_J['FSDAD'] ))
print( 'FSDCH --- ' + str(dfFSQ_J['FSDCH'] ))

print(list(dfFSQ_J.columns))
```

(continued)

```
dfFSQ_JREQ = dfFSQ_J[['SEQN','FSD032A', 'FSD032B', 'FSD032C',
'FSD041', 'FSD052', 'FSD061', 'FSD071', 'FSD081', 'FSD092',
'FSD102', 'FSD032D', 'FSD032E', 'FSD032F', 'FSD111', 'FSD122',
'FSD132', 'FSD141', 'FSD146', 'FSDHH', 'FSDAD', 'FSDCH']]
```

5.10.3 *Step 3: Merge Demographic and Food Security Data Frames*

```
dfDEMO_J_And_FSQ_J=dfDEMO_IREQ.merge
(dfFSQ_JREQ, on='SEQN', how='left')
print(dfDEMO_J_And_FSQ_J)

print('--------------------
-------------------')
dfDEMO_J_And_FSQ_J.to_csv('NHANES_FSQ2017-2018.csv')
```

Save the file to CSV file to run in Jupyter Notebook.
Output:

```
Name: FSDCH, Length: 9254, dtype: float64
['SEQN', 'FSD032A', 'FSD032B', 'FSD032C', 'FSD041', 'FSD052', 'FSD061', 'FSD071', 'FSD081', 'FSD092', 'FSD102', '
FSD032D', 'FSD032E', 'FSD032F', 'FSD111', 'FSD122', 'FSD132', 'FSD141', 'FSD146', 'FSDHH', 'FSDAD', 'FSDCH', 'FSD
151', 'FSQ165', 'FSD165N', 'FSQ012', 'FSD012N', 'FSD230', 'FSD230N', 'FSD795', 'FSD225', 'FSD235', 'FSD855', 'FSD
860', 'FSQ865', 'FSQ162', 'FSQ760', 'FSD760N', 'FSQ653', 'FSD660ZC', 'FSD675', 'FSD680', 'FSD670ZC', 'FSQ690', 'F
SQ695', 'FSD652ZW', 'FSD672ZW', 'FSD652CW', 'FSD660ZW']
         SEQN  RIAGENDR  RIDAGEYR  SDMVSTRA  SDMVPSU      WTMEC2YR  ...  FSD132  FSD141  FSD146  FSDHH  FSDAD  F
SDCH
0     93703.0       2.0       2.0     145.0      2.0   8539.731348  ...     NaN     NaN     NaN    1.0    1.0
 1.0
1     93704.0       1.0       2.0     143.0      1.0  42566.614750  ...     NaN     NaN     NaN    1.0    1.0
 1.0
2     93705.0       2.0      66.0     145.0      2.0   8338.419786  ...     NaN     NaN     NaN    1.0    1.0
 NaN
3     93706.0       1.0      18.0     134.0      2.0   8723.439814  ...     NaN     NaN     NaN    NaN    NaN
 NaN
4     93707.0       1.0      13.0     138.0      1.0   7064.609730  ...     NaN     2.0     2.0    4.0    4.0
 3.0
...       ...       ...       ...       ...      ...           ...  ...     ...     ...     ...    ...    ...
 ...
9249 102952.0       2.0      70.0     138.0      2.0  18338.711104  ...     NaN     NaN     NaN    3.0    3.0
 NaN
9250 102953.0       1.0      42.0     137.0      2.0  63661.951573  ...     NaN     NaN     NaN    2.0    2.0
 NaN
9251 102954.0       2.0      41.0     144.0      1.0  17694.783346  ...     NaN     NaN     NaN    3.0    3.0
 1.0
```

5.11 Use Case: Predict Systolic Blood Pressure for Adult Men

Let's take NHANES data to build a model to predict systolic blood pressure for adult men. As we have done before, we will start by selecting only the observations of adult males, and let's initially examine a model that uses a man's age (in years), height, weight, and body mass index (BMI) to predict systolic blood pressure. Note that the BMI is computed directly from height and weight, so it is surely collinear with these two variables. We are deliberately specifying a model with substantial collinearity so that we can learn to detect the problem in the output results. Before running the regression, let's look at the correlations among these predictors [45]:

$$\mathrm{BMI} = \frac{\text{Weight (Kg)}}{[\text{Height (m)}]2} \tag{5.5}$$

In other words, BMI is a linear function of weight.

Dataset: body measures—https://wwwn.cdc.gov/Nchs/Nhanes/2017-2018/P_BMX.XPT

- BPXSY1—Systolic: blood pressure (first rdg) mm Hg—2017–2018 (BMX_J_DATA[117])
- Variables: BPXSY1, RIDAGEYR, BMXWT, BMXHT, and BMXBMI.

Software code for this model can be found on GitHub Link: NHANES_Predict_Systolic_BloodPressure_Model.py (Python Code)

5.11.1 *Step 1: Load NHANES Demographic Variables and Sample Weights (P_DEMO)[118], Body Measures (P_BMX)[119], and Blood Pressure (BPX_J)[120] data*

```
import pandas as pd
from functools import reduce
import matplotlib.pyplot as plt

# Load P_BMX.XPT
dfP_BMX = pd.read_sas('P_BMX.XPT')
dfP_BMX.set_index('SEQN') # Respondent sequence number
```

(continued)

[117] BMX_J_DATA—https://wwwn.cdc.gov/Nchs/Nhanes/2017-2018/BMX_J.XPT

[118] P_DEMO—https://wwwn.cdc.gov/Nchs/Nhanes/2017-2018/P_DEMO.htm#RIAGENDR

[119] Body Measures (P_BMX)—https://wwwn.cdc.gov/Nchs/Nhanes/2017-2018/P_BMX.htm

[120] Blood Pressure (BPX_J)—https://wwwn.cdc.gov/Nchs/Nhanes/2017-2018/BPX_J.htm

```
# Load BPX_J.XPT
dfBPX_J = pd.read_sas('BPX_J.XPT')
dfBPX_J.set_index('SEQN') # Respondent sequence number

# Load P_DEMO.XPT

dfP_DEMO = pd.read_sas('P_DEMO.XPT')
dfP_DEMO.set_index('SEQN') # Respondent sequence number
```

5.11.2 Step 2: Combine Data Frames Based on the Respondent Sequence Number (Drop Nulls)

```
df_P_DEMO_BPX_J = pd.merge(dfP_DEMO, dfP_BMX, left_index=True,
right_index=True)
df_P_DEMO_BMX_BPX_J = pd.merge(df_P_DEMO_BPX_J, dfBPX_J,
left_index=True, right_index=True)

# Create a new data from for RIDAGEYR & BPXSY1
dfPDEMOBMXBPXJ = df_P_DEMO_BMX_BPX_J
[['BPXSY1','RIDAGEYR','BMXWT','BMXHT','BMXBMI','RIAGENDR']].
copy() # 'BMXWAIST'

dfPDEMOBMXBPXJ = dfPDEMOBMXBPXJ.dropna()
print(dfPDEMOBMXBPXJ)
```

5.11.3 Step 3: Select Adult and Male Respondents

```
# Only Adults
dfPDEMOBMXBPXJ=dfPDEMOBMXBPXJ[dfPDEMOBMXBPXJ['RIDAGEYR']>18]

# males
dfPDEMOBMXBPXJ=dfPDEMOBMXBPXJ[dfPDEMOBMXBPXJ['RIAGENDR'] == 1]
```

5.11.4 Step 4: Plot BMI Scatter Plot

```
import seaborn as sns, numpy as np

sns.histplot(dfPDEMOBMXBPXJ['BMXBMI'],
        label="Skewness : %.2f"%(dfPDEMOBMXBPXJ['BMXBMI']\
                    .skew())).legend(loc="best")

plt.show()
```

Output:

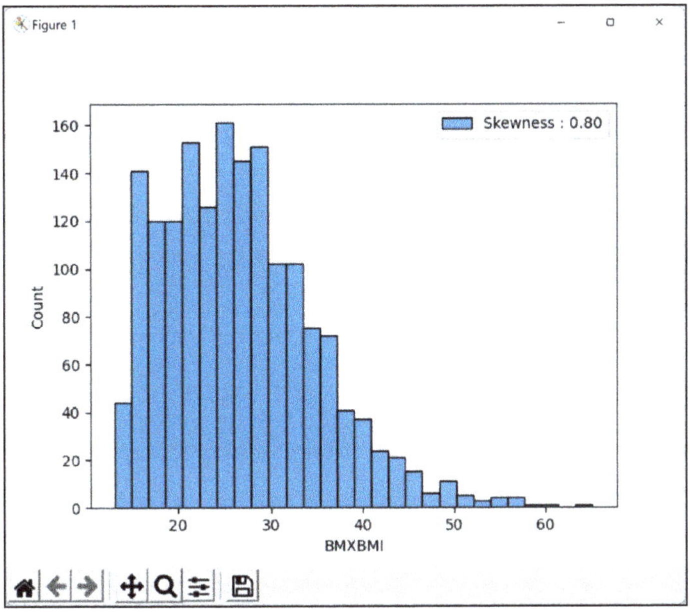

5.11.5 Step 5: Correlation Matrix

```
correlation_df                    =                    dfPDEMOBMXBPXJ
[['BPXSY1','RIDAGEYR','BMXWT','BMXHT','BMXBMI']].corr()
print(correlation_df)
```

Output:

As you can see, the correlation between BMXWT and BMXBMI is 0.9169 (91.69%). That is, BMI is 91.69% depending on weight.

	BPXSY1	RIDAGEYR	BMXWT	BMXHT	BMXBMI
BPXSY1	1.000000	0.092379	0.041613	0.003134	0.056632
RIDAGEYR	0.092379	1.000000	0.001548	-0.020642	0.008648
BMXWT	0.041613	0.001548	1.000000	0.771543	0.916960
BMXHT	0.003134	-0.020642	0.771543	1.000000	0.496362
BMXBMI	0.056632	0.008648	0.916960	0.496362	1.000000

5.11.6 Step 6: Predict Systolic Blood Pressure Model

Predict systolic blood pressure regression model with class variable Y = 'BPXSY1' and feature engineering variables X = { 'RIDAGEYR','BMXWT','BMXHT','BMXBMI' }

```
#separte the predicting attribute into Y for model training
y = dfPDEMOBMXBPXJ['BPXSY1']
print(y)
#separate the other attributes from the predicting attribute

dfPDEMOBMXBPXJ=dfPDEMOBMXBPXJ.drop(['BPXSY1'],axis=1)

x=dfPDEMOBMXBPXJ[['RIDAGEYR','BMXWT','BMXHT','BMXBMI']]

# importing train_test_split from sklearn
from sklearn.model_selection import train_test_split
# splitting the data
x_train, x_test, y_train, y_test = train_test_split(x, y,
test_size = 0.2, random_state = 42)

# importing module
from sklearn.linear_model import LinearRegression
# creating an object of LinearRegression class
LR = LinearRegression()
# fitting the training data
model=LR.fit(x_train,y_train)

y_prediction = LR.predict(x_test)
print(y_prediction)
```

(continued)

```
# importing r2_score module
from sklearn.metrics import r2_score
from sklearn.metrics import mean_squared_error
# predicting the accuracy score
score=r2_score(y_test,y_prediction)

#view model summary
print('intercept:', model.intercept_)
print('slope:', model.coef_)

print('r2 socre is ',score)
print('mean_sqrd_error is==',mean_squared_error(y_test,
y_prediction))
print('root_mean_squared error of is==',np.sqrt
(mean_squared_error(y_test,y_prediction)))
```

Output:

```
intercept: 105.54524951760686
slope: [ 0.12524227 -0.05432815 0.03644214 0.28201776]
r2 socre is -0.012149203927896979
mean_sqrd_error is== 432.64337390075127
root_mean_squared error of is== 20.800081103225324
```

SUMMARY OUTPUT

Regression Statistics	
Multiple R	0.111069
R Square	0.012336
Adjusted R Square	0.009986
Standard Error	20.17537
Observations	1686

ANOVA

	df	*SS*	*MS*	*F*	*ignificance F*
Regression	4	8546.52	2136.63	5.249116	0.000335
Residual	1681	684243.8	407.0457		
Total	1685	692790.4			

	Coefficients	*andard Erro*	*t Stat*	*P-value*	*Lower 95%*	*Upper 95%*	*ower 95.0%*	*Upper 95.0%*
Intercept	115.2546	9.483487	12.15319	1.23E-32	96.65389	133.8553	96.65389	133.8553
RIDAGEYR	0.100897	0.026844	3.758618	0.000177	0.048246	0.153549	0.048246	0.153549
BMXWT	-0.00507	0.098158	-0.05163	0.958832	-0.19759	0.187457	-0.19759	0.187457
BMXHT	-0.02357	0.060198	-0.39153	0.695453	-0.14164	0.094502	-0.14164	0.094502
BMXBMI	0.187136	0.268581	0.696756	0.486052	-0.33965	0.713925	-0.33965	0.713925

The same can be retrieved from running Excel data regression with analysis of variance (ANOVA).

The medical conditions[121] section (prefix MCQ) provides self-reported personal interview data on a broad range of health conditions for both children and adults. The NHANES MCQ questionnaire section is generally modeled on the "Medical Conditions" questionnaire section of the US National Health Interview Survey.

Major topics in the MCQ section includes health conditions/medical history and family history of disease. Under health conditions/medical history, the following items are covered: asthma, childhood and adult, anemia, psoriasis, heart diseases, arthritis, blood transfusions, cancer (multiple varieties), chronic bronchitis, gout, emphysema, liver disease, celiac disease, menses onset, overweight, stroke, thyroid problems, and vision. And under family history of disease, it includes asthma, diabetes, and heart attack/angina.

5.11.7 Micronutrient Availability

Micronutrients[122] include vitamins and minerals and are required [35] in very small (micro) but specific amounts. Vitamins and minerals in foods are necessary for the body to grow, develop, and function properly and are essential for our health and well-being. Our bodies require several different vitamins and minerals, each of which has a specific function in the body and must be supplied in different, sufficient amounts.

5.11.8 The Diet Problem

The "diet problem"[123] (the search of a low-cost diet that would meet the nutritional needs of a US army soldier) is characterized by a long history, whereas most solutions for comparable diet problems were developed in 2000 or later, during which computers with large calculation capacities became widely available and linear programming (LP) tools were developed [46].

---

[121] National Health and Nutrition Examination Survey - https://wwwn.cdc.gov/nchs/nhanes/2009-2010/MCQ_F.htm#Component_Description

[122] The State of Food Security and Nutrition in the World 2021 - https://www.fao.org/3/cb4474en/cb4474en.pdf

[123] A Review of the Use of Linear Programming to Optimize Diets, Nutritiously, Economically and Environmentally - https://www.ncbi.nlm.nih.gov/pmc/articles/PMC6021504/

5.12 An Open-Source Dataset on Dietary Behaviors

The USDA Economic Research Service (ERS) encourages research that makes appropriate use of existing, nationally representative data. Major sources of food-related data are summarized in the table below by type of data[124]:

- Store sales
- Consumer purchases
- Consumption
- Availability
- Food-related

For dietary diversity, let's focus on the consumption data sources (please see Table 5.5):

5.13 Dietary Approaches to Stop Hypertension (DASH)

Hypertension is a very significant health issue in the United States of America. Fifty million or more Americans have high blood pressure that warrants treatment, according to the National Health and Nutrition Examination Survey (NHANES) survey (Joint National Committee on Prevention, Detection, Evaluation, and Treatment of High Blood Pressure 2003). The US Preventive Services Task Force (USPSTF) recommends that clinicians screen adults aged 18 and older for high blood pressure (US Preventive Services Task Force 2007). The most frequent and serious complications of uncontrolled hypertension include coronary heart disease, congestive heart failure, stroke, ruptured aortic aneurysm, renal disease, and retinopathy. The increased risks of hypertension are present in individuals ranging from 40 to 89 years of age. For every 20-mm Hg systolic or 10-mm Hg diastolic increase in blood pressure, there is a doubling of mortality from both ischemic heart disease and stroke (Joint National Committee on Prevention, Detection, Evaluation, and Treatment of High Blood Pressure 2003). Better control of blood pressure has been shown to significantly reduce the probability that these undesirable and costly outcomes will occur. The relationship between the measure (control of hypertension) and the long-term clinical outcomes listed is well established. In clinical trials, antihypertensive therapy has been associated with reductions in stroke incidence (35–40%), myocardial infarction incidence (20–25%), and heart failure incidence (>50%) (Joint National Committee on Prevention, Detection, Evaluation, and Treatment of High Blood Pressure 2003).

---

[124]Food Released Data Sources—https://www.ers.usda.gov/about-ers/partnerships/strengthening-statistics-through-the-icars/food-related-data-sources/#storesales

Table 5.5 Data sources

Name	Source	Description	Time period	Sample	Major uses and content	Others
NHANES	USDA/HHS	Food intake by individuals based on 24-h recall for 2 days; includes personal, economic, health, and demographic characteristics of sampled person only, where food was purchased and eaten	Data collected annually, with 2-year lag between collection and release	Nationally representative. Samples 5,000 individuals yearly	Who eats what in America. Compares intake with dietary guidelines. Analyzes effects of individual characteristics on food consumption. Links food intake with health outcomes	Food intake recall method undercounts calories; no food prices; cross section only
Flexible Consumer Behavior Survey	NCHS for USDA's Economic Research Service	Supplement to NHANES. Questions about diet and health, knowledge, and attitudes. Links diet, health, knowledge, and attitudes with food intake	Collection period: 2005–2014	Nationally representative. NHANES sample, one adult per household	Assess roles of labels, nutrition education, and information programs in food choices, dietary behaviors, and nutrition and weight outcomes	Some questions change in response to research concerns Data may arrive sometime after concerns are stated or relevant
National Eating Trends	NPD Group	Food intake by individuals based on a 2-week diary includes personal, economic, and demographic characteristics of all household members, where food was purchased and eaten	Three-month lag between collection and release	Nationally representative sample of 2,000 households continuously tracked for 30 years	Who eats what in the United States of America. Compares intake with dietary guidelines. Analyzes effects of household characteristics on food consumption. Links food intake with sociodemographic factors	Small sample and self-reported diary reduces reliability. Based on frequency of use rather than amount consumed

National Quality Forum Number 0018 is about controlling high blood pressure. The controlling high blood pressure[125] percentage of patients 18–85 years of age who had a diagnosis of hypertension overlapping the measurement period and whose most recent blood pressure was adequately controlled (<140/90 mm Hg) during the measurement period.

Data-driven approaches in constrained inference models and constrained optimization problems are increasingly gaining traction. Considering their immense impact on applied settings, it deems necessary to provide researchers with open-access and accurate real-world examples and datasets to be utilized in research and methodology evaluation. To this end, we provide a dataset of dietary behavior within the diet recommendation setting. This dataset is gathered and curated from multiple sources and provides researchers with interpretable and accurate (partially self-reporting) data that can be readily used in models and approaches.

The classical diet recommendation optimization problem was first posed by Stigler[126] [47] as finding an optimal set of intakes of different foods with minimum cost while satisfying a given set of minimum nutrient conditions for an average person[127]. We consider this renowned optimization problem in practical settings, where diet plays a vital role in controlling the progression of chronic illnesses such as hypertension and type II diabetes[128] [48].

After reading this chapter, you should be able to answer key drivers of food security quality and safety drivers. Additionally, you should be able to calculate key dietary diversity indicators that include Household Dietary Diversity Score (HDDS) and Food Consumption Score (FCS). Finally, you should be able to develop data-driven food security and nutrition health trends and insights by the application of machine learning and ensemble mathematical models on the National Health and Nutrition Examination Survey (NHANES) data and should be able to link the influence of key economic and demographic consumer behavior and poverty and inequality drivers on the food security.

References

1. Ziad Daoud (Economist) and Felipe Hernandez (Economist), Middle East, North Africa Risk Unrest on High Food Prices, Mon, January 10, 2022, 9:00 PM, https://finance.yahoo.com/news/middle-east-north-africa-risk-050000342.html, Access Date: January 15, 2022

---

[125] Controlling High Blood Pressure—https://ecqi.healthit.gov/ecqm/ep/2020/cms165v8

[126] The Cost of Subsistence—https://math.berkeley.edu/~mgu/MA170F2015/Diet.pdf

[127] The Stigler Diet Problem—https://developers.google.com/optimization/lp/stigler_diet

[128] An Open-Source Dataset on Dietary Behaviors and DASH Eating Plan Optimization Constraints—https://arxiv.org/pdf/2010.07531.pdf

2. Laurel Wamsley (NPR News), Rising food prices have resulted in both food insecurity and improvisation, November 9, 20214:47 PM ET, Heard on All Things Considered, https://www.npr.org/2021/11/09/1054032209/rising-food-prices-have-resulted-in-both-food-insecurity-and-improvisation, Access Date: January 16, 2022
3. Jef L. Leroy, Marie Ruel, Edward A. Frongillo, Jody Harris, and Terri J. Ballard, Measuring the Food Access Dimension of Food Security: A Critical Review and Mapping of Indicators, 2015, https://journals.sagepub.com/doi/pdf/10.1177/0379572115587274, Access Date: January 15, 2022
4. Laura Reiley and Alyssa Fowers, here's why your food prices keep going up, September 15, 2021, at 8:38 a.m. EDT, https://www.washingtonpost.com/business/2021/09/15/food-inflation-faq/, Access Date: January 16, 2022
5. Abdolreza Abbassian, Global food prices rose 'sharply' during 2021, 6 January 2022, https://news.un.org/en/story/2022/01/1109212, Access Date: 17 January 2022, Global food prices rose 'sharply' during 2021 - https://news.un.org/en/story/2022/01/1109212
6. Matthias Kalkuhl (Editor), Joachim von Braun (Editor), Maximo Torero (Editor), Food Price Volatility and Its Implications for Food Security and Policy, Springer; 1st ed. 2016 edition (April 21, 2016), ISBN-13: 978-3319281995
7. Chauncey Alcorn, CNN Business, Global food prices surge to their highest level in a decade, Updated 5:16 PM ET, Fri June 4, 2021, https://www.cnn.com/2021/06/04/business/inflation-food-prices/index.html, Access Date: January 16, 2022
8. The World Food Programme (WFP), COVID-19 will double number of people facing food crises unless swift action is taken, 21 April 2020, https://www.wfp.org/news/covid-19-will-double-number-people-facing-food-crises-unless-swift-action-taken, Access Date: January 03, 2022
9. Rajendra Jadhav, India's sunflower oil imports could jump to record as prices dip below soyoil, August 9, 2021, https://finance.yahoo.com/news/indias-sunflower-oil-imports-could-092259915.html, Access Date: January 15 2022
10. Daniel Grant, Dairy farmers could see $20 milk in 2022, December 15, 2021, https://www.yahoo.com/news/dairy-farmers-could-see-20-120321307.html, Access Date: Dec 16, 2021 8:57 am
11. Mumbi Gitau|Bloomberg, World's record food bill is hitting poorer countries hardest, November 12, 2021 01:54 IST, https://www.business-standard.com/article/international/world-s-record-food-bill-is-hitting-poorer-countries-hardest-121111200065_1.html, Access Date: January 15 2022
12. Sara Heath, Using Quality Measures in Food Security, SDOH Programming, March 2020, https://patientengagementhit.com/news/using-quality-measures-in-food-security-sdoh-programming, Access Date: January 2, 2022
13. Alisha Coleman-Jensen, Matthew P. Rabbitt, Laura Hales, and Christian A. Gregory, Food Security in the U.S. Key Statistics & Graphics, Last updated: Wednesday, September 08, 2021, https://www.ers.usda.gov/topics/food-nutrition-assistance/food-security-in-the-us/key-statistics-graphics.aspx, Access Date: January 09, 2022
14. CHRISTIANNA SILVA, Food Insecurity In The U.S. By The Numbers, Publication Date: September 27, 20204:30 PM ET, https://www.npr.org/2020/09/27/912486921/food-insecurity-in-the-u-s-by-the-numbers, Access Date: January 03, 2022
15. Diane Schanzenbach and Abigail Pitts, How Much Has Food Insecurity Risen? Evidence from the Census Household Pulse Survey, Publication Date: June 2020, https://www.ipr.northwestern.edu/documents/reports/ipr-rapid-research-reports-pulse-hh-data-10-june-2020.pdf, Access Date: January 03, 2022
16. Joseph Daniels, How local organizations are addressing food insecurity in greater Sacramento area, Updated: 6:58 PM PST December 2, 2021, https://www.abc10.com/article/news/local/sacramento/food-insecurity-greater-sacramento-area/103-dcb949b5-8f7f-46d8-85d9-16fc5817a206, Access Date: January 2, 2022

17. Marie T. Ruel, Operationalizing Dietary Diversity: A Review of Measurement Issues and Research Priorities, The Journal of Nutrition, Volume 133, Issue 11, November 2003, Pages 3911S–3926S, https://doi.org/10.1093/jn/133.11.3911S
18. Hoddinott, J. & Yohannes, Y. (2002) Dietary Diversity as a Food Security Indicator. Food Consumption and Nutrition Division Discussion Paper, No. 136. International Food Policy Research Institute, Washington, DC, Access Date: September 18, 2021
19. Mirriam Matita, Ephraim W. Chirwa, Deborah Johnston, Jacob Mazalale, Richard Smith, Helen Walls, does household participation in food markets increase dietary diversity? Evidence from rural Malawi, Global Food Security, Volume 28, 2021, 100486, ISSN 2211-9124, https://doi.org/10.1016/j.gfs.2020.100486.
20. Gina Kennedy, Terri Ballard and MarieClaude Dop, Guidelines for Measuring Household and Individual Dietary Diversity, 2010, https://www.fao.org/3/i1983e/i1983e.pdf, Access Date: January 15, 2022
21. Anne Swindale and Paula Bilinsky, Household Dietary Diversity Score (HDDS) for Measurement of Household Food Access: Indicator Guide, September 2006 , https://www.fantaproject.org/sites/default/files/resources/HDDS_v2_Sep06_0.pdf, Access Date: January 15 2022
22. Marco Stampini, Diana Londoño, Marcos Robles, Pablo Ibarrarán, Effect of Remittances on Food Security in Venezuelan Households, 2022, https://publications.iadb.org/publications/english/document/Effect-of-Remittances-on-Food-Security-in-Venezuelan-Households.pdf, Access Date: January 15, 2022
23. World Health Organization, Indicators for assessing infant and young child feeding practices: definitions and measurement methods,12 April 2021, https://www.who.int/publications/i/item/9789240018389, ISBN: 9789240018389
24. Gina Kennedy, Terri Ballard and Individual Dietary Diversity, ISBN: 20119789251067499, https://www.fao.org/publications/card/en/c/5aacbe39-068f-513b-b17d-1d92959654ea/, Access Date: January 15, 2022
25. Khamis, A.G., Mwanri, A.W., Ntwenya, J.E. et al. The influence of dietary diversity on the nutritional status of children between 6 and 23 months of age in Tanzania. BMC Pediatr 19, 518 (2019). https://doi.org/10.1186/s12887-019-1897-5
26. Isabirye, N., Bukenya, J.N., Nakafeero, M. et al. Dietary diversity and associated factors among adolescents in eastern Uganda: a cross-sectional study. BMC Public Health 20, 534 (2020). https://doi.org/10.1186/s12889-020-08669-7
27. Jenkins RH, Vamos EP, Taylor-Robinson D, Millett C, Laverty AA. Impacts of the 2008 Great Recession on dietary intake: a systematic review and meta-analysis. Int J Behav Nutr Phys Act. 2021;18(1):57. Published 2021 Apr 29. https://doi.org/10.1186/s12966-021-01125-8
28. JAnne Frances Johnson, When the economy goes down, so does the quality of our diets, June 7, 2021, https://nutrition.org/when-the-economy-goes-down-so-does-the-quality-of-our-diets/, Access Date: January 17, 2022
29. Jacqueline Vernarelli, Emma Turchick, Nutritional Impact of Economic Downturn: How the Great Recession Shaped Dietary Behaviors in US Adults, Current Developments in Nutrition, Volume 5, Issue Supplement_2, June 2021, p 191. https://doi.org/10.1093/cdn/nzab035_099
30. Olivier Ecker, Agricultural transformation and food and nutrition security in Ghana: Does farm production diversity (still) matter for household dietary diversity? Food Policy, Volume 79, 2018, Pages 271-282, ISSN 0306-9192, https://doi.org/10.1016/j.foodpol.2018.08.002.
31. Russell Sage Foundation, Understanding Food Insecurity During the Great Recession, 2016, https://web.stanford.edu/group/recessiontrends-dev/cgi-bin/web/resources/research-project/understanding-food-insecurity-during-great-recession, Access Date: January 17, 2022
32. FAO, IFAD, UNICEF, WFP and WHO. 2021. The State of Food Security and Nutrition in the World 2021. Transforming food systems for food security, improved nutrition and affordable healthy diets for all. Rome, FAO. https://doi.org/10.4060/cb4474en
33. Sibhatu, Kibrom T, Krishna, Vijesh V., Qaim, Matin., Production diversity and dietary diversity in smallholder farm households, 2015/08/25, Proceedings of the National Academy of Sciences, https://doi.org/10.1073/pnas.1510982112, Access Date: January 16, 2022

34. R. Maria Saleth, S. Galab and E. Revathi, Issues and Challenges of Inclusive Development: Essays in Honor of Prof. R. Radhakrishna, Publisher: Springer; 1st ed. 2020 edition (June 19, 2020), ISBN-13: 978-9811522284
35. JOHN A. SCHNITTKER, The 1972-73 Food Price Spiral, 1973, https://www.brookings.edu/wp-content/uploads/1973/06/1973b_bpea_schnittker.pdf, Access Date: Access Date: January 17, 2022
36. FSIN, Global Report on Food Crises - 2021, 5 May 2021, https://www.wfp.org/publications/global-report-food-crises-2021, Access Date: January 17, 2022
37. Anthony Faiola, Amid drought, conflict and rocketing prices, a global food crisis could be approaching, top expert warns, December 15, 2021 at 12:01 a.m. EST, https://www.washingtonpost.com/world/2021/12/15/global-food-crisis-pandemic/, Access Date: January 17, 2022
38. National Research Council (US) Committee on Diet and Health. Diet and Health: Implications for Reducing Chronic Disease Risk. Washington (DC): National Academies Press (US); 1989. 3, Dietary Intake and Nutritional Status: Trends and Assessment. Available from: https://www.ncbi.nlm.nih.gov/books/NBK218765/
39. National Research Council, Division on Earth and Life Studies, Commission on Life Sciences, Committee on Diet and Health, Diet and Health: Implications for Reducing Chronic Disease Risk, Publisher: National Academies Press (January 1, 1989), ISBN-13: 978-0309039949
40. Daniela Rubio and Jenny Shapiro, Solving The Need For Up-To-Date Food Consumption Data In Venezuela, 2021, https://www.premise.com/solving-the-need-for-up-to-date-food-consumption-data-in-venezuela/, Access Date: January 15 2022
41. Sara Cincurova, Known unknowns: The challenge of collecting COVID-19 data in Venezuela,4 June 2020, https://www.thenewhumanitarian.org/news-feature/2020/06/04/coronavirus-data-Venezuela, Access Date: January 15 2022
42. Aatish Kumar Sahu, Zhopnu Chüzho, Sanjoy Das, Measuring Household Food Security Index for High Hill Tribal Community of Nagaland, India, 2017, https://pdfs.semanticscholar.org/2c3c/a9e28afc515f710693aba714681e43d32fd4.pdf, Access Date: January 15, 2022
43. Debra J. Brody, M.P.H., Laura A. Pratt, Ph.D., and Jeffery P. Hughes, M.P.H., Prevalence of Depression Among Adults Aged 20 and Over: United States, 2013–2016, https://www.cdc.gov/nchs/products/databriefs/db303.htm#depression_prevalence_20_and_over, Access Date: December 31, 2021
44. Tsz Hin (Stanley) Ng, Blood total mercury levels and depression in the United States: National Health and Nutrition Examination Survey 2005-2008, June 2012, https://idea.library.drexel.edu/islandora/object/idea%3A5374/datastream/OBJ/view, Access Date: January 2, 2022
45. Robert Carver, Practical Data Analysis with JMP, Third Edition 3rd Edition, Publisher: SAS Institute; 3rd edition (October 18, 2019), ISBN-13: 978-1642956108
46. van Dooren C. A Review of the Use of Linear Programming to Optimize Diets, Nutritiously, Economically and Environmentally. Front Nutr. 2018;5:48. Published 2018 Jun 21. https://doi.org/10.3389/fnut.2018.00048
47. Stigler, George J. "The Cost of Subsistence." Journal of Farm Economics, vol. 27, no. 2, [Oxford University Press, Agricultural & Applied Economics Association], 1945, pp. 303–14, https://doi.org/10.2307/1231810
48. Farzin Ahmadi, Fardin Ganjkhanloo, Kimia Ghobadi, An Open-Source Dataset on Dietary Behaviors and DASH Eating Plan Optimization Constraints, Publication Date: 2010, https://arxiv.org/pdf/2010.07531.pdf, Access Date: December 25, 2021

Chapter 6
ML Models: Food Security and Climate Change

This Chapter Covers:

- Agriculture and climate change
- FAO in emergencies countries
- Climate change and most affected countries (2000–2019 and beyond)
- FAO food emergencies and climate change—high-priority list of countries
- Food security and natural resources and resilience
- Climate impacts on agriculture and food supply (CO_2 vs. temperature)
- The Coupled Model Intercomparison Project Phase 6 (CMIP6) climate projections
- Climate model: temperature change (RCP Shared Socioeconomic Pathways (challenges to adaptation vs. mitigation) 6.0)—2006–2100
- Machine learning model: Thailand rice paddy yields and climate change
  - CMIP6 Projections Data and SSP Scenarios 2080–2099
- Machine learning model: Vietnam coffee yields and climate change
  - CMIP6 Projections Data and SSP Scenarios 2080–2099

The chapter introduces the impact of climate change on agriculture productions, develops a framework to identify countries that require immediate action to develop climate-smart agriculture, and establishes a machine learning model to develop such recommendations. As part of identification of countries that are highly sensitive and need immediate action, the chapter studies climate change framework and countries that are under food security emergencies plus the Global Information and Early Warning System on Food and Agriculture (GIEWS). Next, it deep dives on the Coupled Model Intercomparison Project Phase 6 (CMIP6) climate projections and develops Shared Socioeconomic Pathways (challenges to adaptation vs. mitigation)

C. Vuppalapati, *Artificial Intelligence and Heuristics for Enhanced Food Security*, International Series in Operations Research & Management Science 331,
https://doi.org/10.1007/978-3-031-08743-1_6

Food Insecurity Experience Scale (FIES) [8]

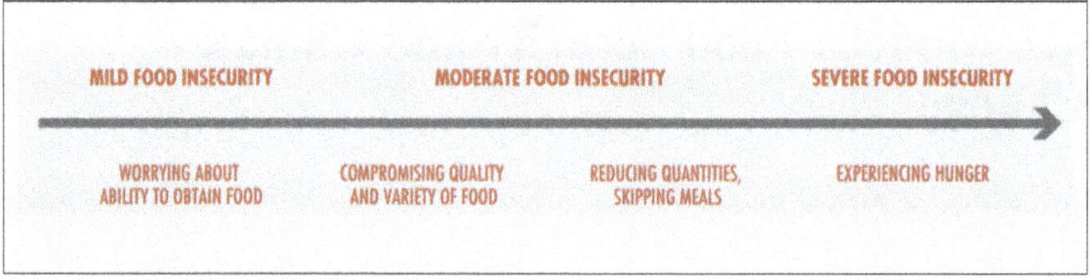

Fig. 6.1 Food insecurity experience scale

framework. Finally, the chapter concludes with the development of machine learning models for Thailand rice paddy yields and Vietnam coffee yields and simulates the food production for SSP 2080–2099.

A household is not "food secure" unless it "feels" food secure[1]. Food security exists when all people, always, have physical and economic access to sufficient, safe, and nutritious food that meets their dietary needs and food preferences for an active and healthy life[2]. Food security is a global need, and every country in the world is actively working to ensure enhancing food security [1]. For instance, the US Department of Agriculture (USDA) Economic Research Service (ERS) monitors the food security of US households through an annual, nationally representative survey. While most US households are food secure, a minority of US households experience food insecurity at times during the year, meaning that their access to adequate food for active, healthy living is limited by lack of money and other resources[3]. Some experience very low food security, a more severe range of food insecurity where food intake of one or more members is reduced and normal eating patterns are disrupted [2]. In 2020, 89.5% of US households were food secure throughout the year. The remaining 10.5% of households were food insecure at least some time during the year, including 3.9% (5.1 million households) that had very low food security[4]. The prevalence[5] of food insecurity (please see Figs. 6.1 and 6.2) is not uniform across the country due to both the characteristics of populations and to state-level policies and economic conditions [2–4].

[1]Issues and Challenges of Inclusive Development: Essays in Honor of Prof. R. Radhakrishna by R. Maria Saleth, S. Galab and E. Revathi, Publisher: Springer; 1st ed. 2020 edition (June 19, 2020), ISBN-13: 978-9811522284

[2]Food Security—https://www.fao.org/fileadmin/templates/faoitaly/documents/pdf/pdf_Food_Security_Cocept_Note.pdf

[3]Food Security and Nutrition Assistance—https://www.ers.usda.gov/data-products/ag-and-food-statistics-charting-the-essentials/food-security-and-nutrition-assistance/

[4]The prevalence of food insecurity in 2020 is unchanged from 2019—https://www.ers.usda.gov/data-products/chart-gallery/gallery/chart-detail/?chartId=58378

[5]Prevalence of food insecurity is not uniform across the country—https://www.ers.usda.gov/data-products/chart-gallery/gallery/chart-detail/?chartId=58392

Fig. 6.2 FAO in emergencies

Food insecurity is one of the most pressing human development challenges faced by many countries. Within the broader context of the ongoing conflicts, climate change, and economic crisis around the world, the combination of a high household dependence on food imports, high food prices, and significantly reduced income is having a devastating impact on people's lives[6]. COVID-19 has changed the terms of food security. Many countries are facing growing levels of food insecurity[7], reversing years of development gains, and threatening the achievement of Sustainable Development Goals by 2030. Even before COVID-19 reduced incomes and disrupted supply chains, chronic and acute hunger were on the rise due to various factors, including conflict, socioeconomic conditions, natural hazards, climate change, and pests. COVID-19 impacts have led to severe and widespread increases in global food insecurity, affecting vulnerable households in almost every country, with impacts expected to continue through 2022.

Food security is a crippling issue at the national level. Numerous countries are experiencing high food price inflation at the retail level, labor shortages, a sharp rise in the price of fertilizer, currency devaluations, shipping costs, supply chain disruptions, extreme weather, and other factors that are increasing food insecurity. Rising food prices have a greater impact on people in low- and middle-income countries since they spend a larger share of their income on food than people in high-income countries.

Experts say extreme weather events are happening with increased frequency and intensity due to climate change; as a result an estimated 13 million people in Kenya, Somalia, and Ethiopia are facing severe hunger as the Horn of Africa experiences its

[6] $127 million from World Bank to shore up food security and rural livelihoods in Yemen—https://www.fao.org/news/story/en/item/1418251/icode/

[7] Food Security and COVID-19—2022, https://www.worldbank.org/en/topic/agriculture/brief/food-security-and-covid-19

worst drought in decades. Nearly 5.5 million children in the four countries are threatened by acute malnutrition, while 1.4 million risked falling into severe acute malnutrition, which can lead to death. Three consecutive rainy seasons have failed as the region has recorded its driest conditions since 1981, the UN's World Food Programme said:

- According to the World Food Program (WFP), some 5.7 million already need food assistance in southern and southeastern Ethiopia, including half a million malnourished children and mothers.
- In Somalia, the number of people classified as seriously hungry is expected to rise from 3.5 million to 4.6 million by May unless urgent interventions are taken.
- Another 2.8 million people need assistance in southeastern and northern Kenya, where a drought emergency was declared in September.

However, Africa contributes the least to global warming, bearing the brunt[8].

> "Harvests are ruined, livestock are dying, and hunger is growing as recurrent droughts affect the Horn of Africa. The situation requires immediate humanitarian action" to avoid a repeat of a crisis like that of Somalia in 2011, when 250,000 died of hunger during a prolonged drought.
> Michael Dunford, WFP's regional director in East Africa

Food insecurity is the global problem and it's catastrophic for children and women. It would be tough to find the solution to food insecurity without addressing climate change and economic issues that are a result of climate change. There is no single solution to address food security (please see the list of countries that are designated under FAO in emergencies[9]). In Bangladesh, an emergency action plan[10], mobilized as part of a Livestock and Dairy Development Project, provided US$87.8 million in cash transfers to 407,000 vulnerable dairy and poultry farmers to support their businesses. Dairy is a cash cow for many small farmers that invest in agriculture and dairy development afloat the small farmer sustainability. Financing also went toward providing personal protection equipment, farm equipment, and enhanced veterinary services through the procurement of 64 mobile veterinary clinics. Recent grants by the World Bank in Yemen targeted to promote the multi-sectoral approach to food insecurity. The approach aimed to increase production of crop, livestock, and fish products, including backyard/garden production; promote climate-smart agriculture; strengthen local-level agri-food systems; and establish national-level agricultural value chains.

[8] 13 million face hunger as Horn of Africa drought worsens: UN—https://mg.co.za/environment/2022-02-09-13-million-face-hunger-as-horn-of-africa-drought-worsens-un/

[9] FAO in emergencies—https://www.fao.org/emergencies/countries/en/

[10] Livestock and Dairy Development Project—https://projects.worldbank.org/en/projects-operations/project-detail/P161246?lang=en

Climate-Smart Dairy Sensors
Small dairy farmer adaptive artificial intelligent sensor technologies can help in two ways: (1) reduce dependency to the impact of climate change on the sustainability of small dairy farmer and (2) improve the financial outcome.

Another innovative way to improve food security is through backyard food production as it is the best deterrent to overcome food insecurity. A recent study conducted by the University of California and Santa Clara University researchers found that people in San Jose who maintained a garden in their yard or a community garden and gardeners consumed more vegetables when they were eating food grown in their gardens[11] [5]. Participants in the pilot study, published in *California Agriculture* journal, reported doubling their vegetable intake to a level that met the number of daily servings recommended by the US Dietary Guidelines. Meals rich in fresh fruits and vegetables are lower in calories and higher in fiber and part of a healthy diet. Growing food saves money and provides a wider variety of fresh produce. For millions of people in war-ridden countries, backyard food production represents a firewall between them and the most severe manifestations of hunger. So, backyard food production is a means to boost local food production to prevent high acute food insecurity from spreading. Agricultural development is one of the best ways to improve food security and financial access to the most venerable communities.

Fertilizer Cost and Food Security
Near-record high fertilizer prices[12]. Most fertilizer prices increased sharply in 2021Q3 and continued rising in early November, reaching levels unseen since the 2008–2009 global financial crisis. Prices have been driven by surging energy costs, supply curtailments, and trade policies. High fertilizer prices could exert inflationary pressures on food prices, compounding food security concerns at a time when the COVID-19 pandemic and climate change are making access to food more difficult [6].

[11] Urban gardens improve food security—https://www.universityofcalifornia.edu/news/urban-gardens-improve-food-security

[12] Soaring fertilizer prices add to inflationary pressures and food security concerns—https://blogs.worldbank.org/opendata/soaring-fertilizer-prices-add-inflationary-pressures-and-food-security-concerns#:~:text=Prices%20have%20been%20driven%20by,Surge%20in%20input%20costs

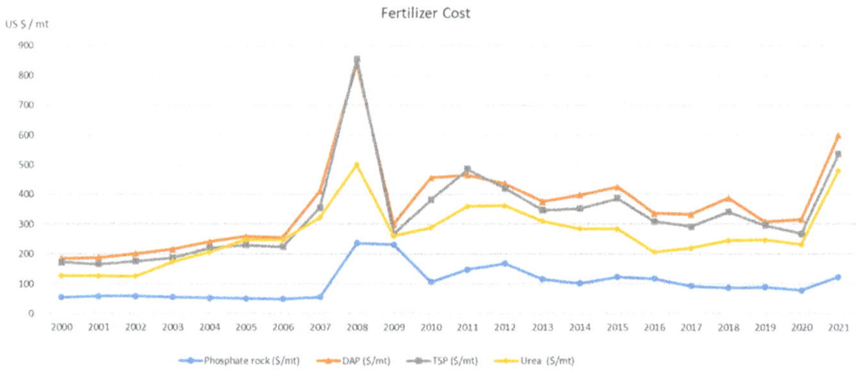

The same increase in prices, of course, could be observed for Producer Price Index by Industry: Nitrogenous Fertilizer Manufacturing: Synthetic Ammonia, Nitric Acid, Ammonium Compounds, and Urea (PCU325311325311A), Index Dec 2014=100[13].

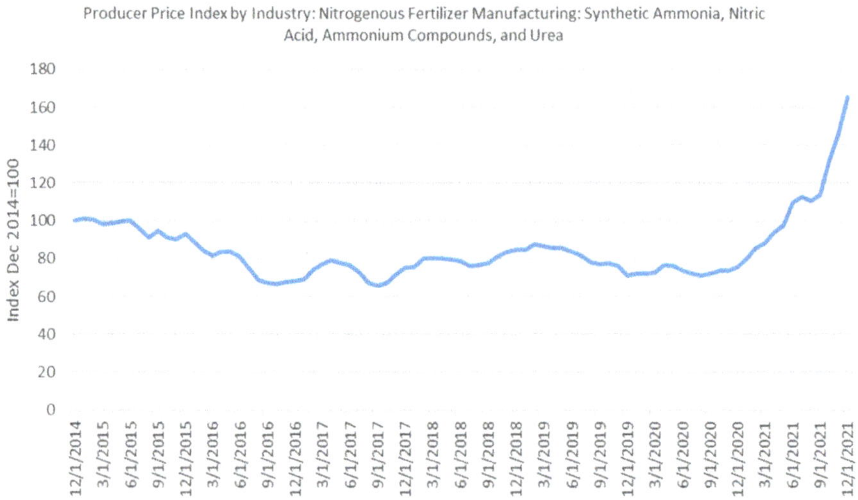

As higher fertilizer prices[14] forces more input agricultural cost and hence more food security concerns [7].

[13] PCU325311325311A—https://fred.stlouisfed.org/series/PCU325311325311A

[14] Fertilizer Prices Continue to Climb, 2022 Planted Acreage Analysis Continues—https://farmpolicynews.illinois.edu/2021/12/fertilizer-prices-continue-to-climb-2022-planted-acreage-analysis-continues/#:~:text=The%20nitrogen%20fertilizer's%20average%20price,%2C%20an%20all%2Dtime%20high

6.1 Agriculture and Climate Change

The threat of climate change and the need to protect natural resources cannot be ignored. For instance, Madagascar[15] is the country that is most exposed to cyclones in Africa (an average of 1.5 cyclones affect Madagascar yearly, the highest number in Africa, and each strong cyclone on average affects 700,000 people[16]) and one of the world's most vulnerable countries to the effects of climate change. Frequent natural disasters and locust threats negatively impact households' livelihoods, pushing thousands of people into poverty and hunger. World hunger is again on the rise driven by conflict and climate change[17]. According to the State of Food Security and Nutrition in the World 2017 report [8], global hunger is on the rise again after steadily declining for over a decade. However, the ambition of a world without hunger and malnutrition by 2030 will be challenging—achieving it will require renewed efforts through new ways of working [8]. Climate change brings social, environmental, and economic impacts.

The report[18] is the first UN global assessment on food security and nutrition to be released following the adoption of the 2030 Agenda for Sustainable Development, which aims to end hunger and all forms of malnutrition by 2030 as a top international policy priority. It singles out conflict—increasingly compounded by climate change [8]—as one of the key drivers behind the resurgence of hunger and many forms of malnutrition. While the impacts of climate change on agriculture and food security are relatively well established, other factors such as migration[19] are contributing to food insecurity (please see Fig. 6.1).

Actions to make agriculture sustainable are among the most effective measures to help nations adapt to and mitigate climate change[20] (climate change renders some of the agricultural practices and arable lands ineffective by 2050) [9]. Coffee production—the altitudinal range for robusta cultivated in the Central Highlands will likely shift from today's 300–900 m to 600–1,000 m by 2050, due to climate change. Suitability below 550 m will decline sharply[21] [9].

To meet demands of food, the world must, by 2050, produce 49% more food than in 2012 as populations grow and diets change. At the same time, almost 80% of the poor live in rural areas where people depend on agriculture, fisheries, or forestry as

---

[15] Madagascar—https://www.fao.org/emergencies/countries/detail/en/c/161541/

[16] Madagascar—https://www.unocha.org/southern-and-eastern-africa-rosea/madagascar

[17] The State of Food Security and Nutrition in the World 2017—https://www.fao.org/3/I7695e/I7695e.pdf

[18] World hunger again on the rise driven by conflict and climate change—https://www.fao.org/emergencies/fao-in-action/stories/stories-detail/en/c/1037511/

[19] Migration, Agriculture and Climate Change—https://www.fao.org/emergencies/resources/documents/resources-detail/en/c/1106745/

[20] FAO'S WORK ON CLIMATE CHANGE—https://www.fao.org/3/i8037e/i8037e.pdf

[21] Coffee Production in the face of Climate Change—https://www.sustaincoffee.org/assets/resources/Vietnam_CountryProfile_Climate_Coffee_6-11.pdf

their main source of income and food. If temperatures continue to rise, then progress toward eradicating hunger and ensuring the sustainability of our natural resource base to achieve the 2030 Agenda for Sustainable Development will be at risk. Climate change will bring further extreme weather events, land degradation and desertification, water scarcity, rising sea levels, and shifting climates—all of which will make the rural poor the first victims, hampering efforts to feed the whole planet. Agricultural economic opportunities shrink due to climate change (the loss of suitable area for robusta will be biggest in Gia Lai and Dak Lak provinces, with about 30% of the currently suitable area[22])—employment days would cut, for instance, agricultural farmers due to shrinkage of crop—cut harvest months from 3 months, normally, to a month, due to cold snap and drought[23,24] [10, 11].

Agriculture and food systems are partly responsible for climate change, but they are also part of the solution. Appropriate actions in agriculture, forestry, and fisheries can mitigate greenhouse gas emission and promote climate adaptation—with efforts to reduce emissions and adjust practices to the new reality often enhancing and supporting one another. For millions of people, especially rural family farmers in developing countries, our actions can make the difference between poverty and prosperity and between hunger and food security and nutrition. Agriculture—where the fight against hunger and climate change comes together—can unlock solutions [9].

We cannot attain zero hunger without tackling climate change.

6.1.1 FAO in Emergencies Countries

The Food and Agriculture Organization (FAO) has identified 46 countries[25] where interventions are required (please see Fig. 6.2). The list was based on hazards[26] that are dangerous phenomena—like floods, tropical storms, or droughts—that can cause loss of life, damage to property and the environment, destruction of livelihoods, and

[22] COFFEE PRODUCTION IN THE FACE OF CLIMATE CHANGE: VIETNAM—https://www.sustaincoffee.org/assets/resources/Vietnam_CountryProfile_Climate_Coffee_6-11.pdf

[23] Coffee Prices Soar After Bad Harvests and Insatiable Demand—https://www.wsj.com/articles/coffee-prices-soar-after-bad-harvests-and-insatiable-demand-11626093703?mod=article_inline

[24] Coffee Prices Jump to Six-Year High as Brazilian Frost Threatens Crop—https://www.wsj.com/articles/coffee-prices-jump-to-six-year-high-as-brazilian-frost-threatens-crop-11627380128?mod=article_inline

[25] FAO in emergencies—https://www.fao.org/emergencies/countries/en/

[26] Hazard and emergency types—https://www.fao.org/emergencies/emergency-types/hazard-and-emergency-types/en/

Table 6.1 Countries most affected in 2019

Ranking 2019 (2018)	Country	Fatalities	Fatalities per 100,000 inhabitants	Absolute losses (in million US$ PPP[a])	Losses per unit GDP in %
1 (54)	Mozambique	700	2.25	4930.08	12.16
2 (132)	Zimbabwe	347	2.33	1836.82	4.26
3 (135)	The Bahamas	56	14.70	4758.21	31.59
4 (1)	Japan	290	0.23	28,899.79	0.53
5 (93)	Malawi	95	0.47	452.14	2.22
6 (24)	Islamic Republic of Afghanistan	191	0.51	548.73	0.67
7 (5)	India	2267	0.17	68,812.35	0.72
8 (133)	South Sudan	185	1.38	85.86	0.74
9 (27)	Niger	117	0.50	219.58	0.74
10 (59)	Bolivia	33	0.29	798.91	0.76

[a]*PPP* purchasing power parities

disruption of services. Hazards can lead to disasters or emergencies, which require urgent action. Such emergencies have a direct impact on food security—floods, storms, tsunamis, and other hazards destroy agricultural infrastructure and assets. In this chapter, we will focus on countries that are part of FAO in emergencies and ranked high in Climate Risk Index (CRI). The reason is simple. We would like to help countries that are facing climate change and that require real help. The learnings from these climate change-sensitive countries include changes in temperature, precipitation on agricultural production, and development of artificial intelligence (AI) models to predict in advance the likelihood of future economic and hunger issues. In essence, we would like to predict the countries that need special care to avert any future food security and hunger issues.

6.1.2 *Climate Change: Countries Most Affected in 2019*

Mozambique, Zimbabwe, and the Bahamas were the most affected countries in 2019 followed by Japan, Malawi, and the Islamic Republic of Afghanistan. Table 6.1 shows the ten most affected countries (Bottom ten) in 2019, with their average weighted ranking (CRI score) and the specific results relating to the four indicators analyzed.

6.1.3 Countries Most Affected in the Period 2000–2019

Puerto Rico, Myanmar, and Haiti have been identified as the most affected countries in this 20-year period. They are followed by the Philippines, Mozambique, and the Bahamas. Table 6.2 shows the ten most affected countries over the last two decades with their average weighted ranking and the specific results relating to the four indicators analyzed.

For the chapter, I have selected countries (please see Table 6.3 and Fig. 6.3) that are part of the FAO in emergencies[27] list and have high risk from the Global Climate Risk Index (CRI)[28] 2021 [12]. Though all countries require total focus to help, the Cartesian joint of FAO emergency country list and high-risk climate index would provide required data to AI models to study the impact of climate change (temperature, precipitation, radiation, vegetation index, and others) on the staple and important crops of these countries, and the learning could be transferred to the world. Herewith are a great gratitude and sheer appreciation to the farmers and people of these countries that I would like to salute.

Additionally, derived the countries from the datasets (please see Fig. 6.4) of FAO GIEWS. The Global Information and Early Warning System[29] on Food and Agriculture (GIEWS) continuously monitors food supply and demand and other key indicators for assessing the overall food security situation in all countries of the world. The information and datasets are analyzed at the sub-national, national, regional, and global level.

The countries include Mozambique, Colombia, Malawi (CRI #5), Bangladesh (CRI #13), Peru (#46), Myanmar (#21), and Pakistan. *Herewith are the highest regard, the greatest gratitude, and the deepest thankful to the all the farmers of these countries that I salute for their contribution to the economies of the world and feeding the food for the planet despite their acute suffering that they endure due to climate and other risks and eventualities.*

The mission of the chapter is to develop machine learning models of crops that are essential to food security of the above countries that simulate the impact of climate change on various Shared Socioeconomic Pathways to assess the impact of climate on these countries and finally to provide recommendation to mitigate or adaptation to save the lives and economies of sensitive list.

---

[27] FAO in emergencies—https://www.fao.org/emergencies/countries/en/

[28] GLOBAL CLIMATE RISK INDEX 2021—https://www.germanwatch.org/sites/default/files/Global%20Climate%20Risk%20Index%202021_2.pdf

[29] Global Information and Early Warning System—https://www.fao.org/giews/background/en/

Table 6.2 Countries most affected in the period 2000–2019

CRI 2000–2019 (1999–2018)	Country	Fatalities	Fatalities per 100,000 inhabitants	Losses in million US$ PPP	Losses per unit GDP in %	Number of events (2000–2019)
1 (1)	Puerto Rico	149.85	4.12	4149.98	3.66	24
2 (2)	Myanmar	7056.45	14.35	1512.11	0.80	57
3 (3)	Haiti	274.05	2.78	392.54	2.30	80
4 (4)	The Philippines	859.35	0.93	3179.12	0.54	317
5 (14)	Mozambique	125.40	0.52	303.03	1.33	57
6 (20)	The Bahamas	5.35	1.56	426.88	3.81	13
7 (7)	Bangladesh	572.50	0.38	1860.04	0.41	185
8 (5)	Pakistan	502.45	0.30	3771.91	0.52	173
9 (8)	Thailand	137.75	0.21	7719.15	0.82	146
10 (9)	Nepal	217.15	0.82	233.06	0.39	191

6.1.4 Colombia

Colombia is a populous country, with an estimated 50.8 million people in 2020 with projections suggesting the country's population could reach nearly 56 million people by 2050. Most of the country's population is concentrated in the Andean highlands and along the Caribbean coast[30]. The expansive eastern and southern Llanos and tropical forests are home to <10% of the country's population. An estimated 81.4% of the country's population live in urban areas, and this is projected to increase to 88.8% by 2050. The Colombian territory is highly vulnerable to extreme events, particularly flooding from "La Niña" phenomena. Vulnerability hotspots include the Caribbean and the Andean regions, with key sectors including housing, transport, energy, agriculture, and health (please see Fig. 6.5—monthly climatology of mean temperature and precipitation in Colombia from 1991 to 2020).

6.1.4.1 Multiple Threats to Food Security

Natural disasters, environmental pollution because of resource exploitation and lack of land tenure, and climate variability are also affecting people's livelihoods. Small-scale farmers are particularly vulnerable to the effects of climate change due to their dependency on rainfed agriculture for food production and income generation, as well as their limited capacity to adapt. Extreme weather events such as drought negatively impact agropastoralists' livelihoods due to the loss of productive assets, severely affecting their food security.

[30] Colombia—https://climateknowledgeportal.worldbank.org/country/colombia

Table 6.3 High-priority countries (both in FAO in emergencies and climate change)

Country	Climate change or FAO emergencies	Ranking (CRI)	Agricultural product	Importance to world economy and food security
 Mozambique	WFP[a]: 1(54)	1(54)	Maize[b] Sorghum Wheat Cashew nuts	Mozambique is one of the largest producer and exporter of cashew nuts
 Colombia	FAO: food security, climate change, and agricultural production	#38 CRI 2019	Maize[c] Rice	Rice and maize are major crops in Colombia
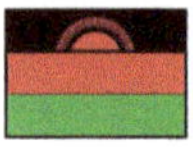 Malawi	FAO[d]: farmers, food prices, floods, and droughts	#5 CRI 2019	Maize Tobacco[e]	Agriculture[f] is the sector in which Malawi competes most successfully in international markets. While maize has been the major food crop in terms of the policy agenda and hectarage planted, tobacco has been, and continues to be, the dominant cash crop in the economy accounting for approximately 58% of the country's total export earnings
 Myanmar		#2 CRI 2000-2019	Rice Dry dean	*Dry beans* are often used to complement a variety of different dishes and foods like rice, salads, tacos, soups, and more. Dry beans can also be seasoned with several different food items, like chicken broth, ham, olive oil, onions, and more. An additional benefit is that dry beans can last for many years if they are stored properly. World's top producing country[g]
 Pakistan	Floods, small-scale farming, hunger[h]	#8 CRI 2000-2019	Cotton Sugarcane Tobacco[i]	Pakistan is one of the leading producers of cotton[j,k]
 Peru	FAO[l]: natural disasters, floods, drought, storms	#45 CRI 2019	Rice Coffee	*Paddy*[m] is another major agricultural product Peru[n] is another big player in the *coffee* industry. Since the 1700s, coffee has been grown in the country across regions in the north, central belt, and south of the country Peruvian coffee comes in two major varieties, divided along with the plantation. The ones

(continued)

Table 6.3 (continued)

Country	Climate change or FAO emergencies	Ranking (CRI)	Agricultural product	Importance to world economy and food security
				grown in the highlands (especially Andes) have a rich floral flavor. Those in the lowlands are usually medium bodied with nutty floral and fruity notes

[a]WFP—https://www.wfp.org/countries/mozambique
[b]Mozambique—https://www.fao.org/giews/countrybrief/country.jsp?lang=en&code=MOZ
[c]Colombia—https://www.fao.org/giews/countrybrief/country.jsp?lang=en&code=COL
[d]Malawi—https://www.fao.org/emergencies/countries/detail/en/c/161513/
[e]Leading tobacco producing countries worldwide in 2020 (in 1,000 metric tons)—https://www.statista.com/statistics/261173/leading-countries-in-tobacco-production/
[f]Malawi—Agricultural Sector—https://www.privacyshield.gov/article?id=Malawi-agricultural-products#:~:text=While%20maize%20has%20been%20the,tea%2C%20cotton%2C%20and%20nuts
[g]The World's Top Dry Bean Producing Countries—https://www.worldatlas.com/articles/the-worlds-top-dry-bean-producing-countries.html
[h]Pakistan—https://www.fao.org/emergencies/countries/detail/en/c/161432/
[i]Leading tobacco producing countries worldwide in 2020 (in 1,000 metric tons)—https://www.statista.com/statistics/261173/leading-countries-in-tobacco-production/
[j]Leading cotton producing countries worldwide in 2020/2021—https://www.statista.com/statistics/263055/cotton-production-worldwide-by-top-countries/
[k]Cotton Exports by Country—https://www.worldstopexports.com/cotton-exports-by-country/
[l]Peru—https://www.fao.org/emergencies/countries/detail/en/c/161514/
[m]GIEWS—Global Information and Early Warning System—https://www.fao.org/giews/countrybrief/country.jsp?code=PER
[n]10 countries that export the best coffee globally—https://seller.alibaba.com/businessblogs/px0897fr-10-countries-that-export-the-best-coffee-globally

Factors[31] such as violence, natural disasters, migration crisis, and climate change and COVID-19[32] have significantly affected agricultural production as well as access and availability of food for communities, thereby undermining their agriculture-based livelihoods [13]. In the immediate aftermath of displacement, people rely mainly on food distribution and solidarity from local communities. During prolonged displacement, people struggle to find livelihood alternatives, and farmers who cannot access their lands risk missing the planting and harvest seasons. As a result, there has been an increase in the adoption of negative coping mechanisms, such as engaging in illegal crop production and illicit mining.

[31] Colombia—https://www.fao.org/emergencies/countries/detail/en/c/168689/

[32] Food insecurity, the other threat of COVID-19 for Colombia's indigenous communities—https://www.unocha.org/story/food-insecurity-other-threat-covid-19-colombia%E2%80%99s-indigenous-communities

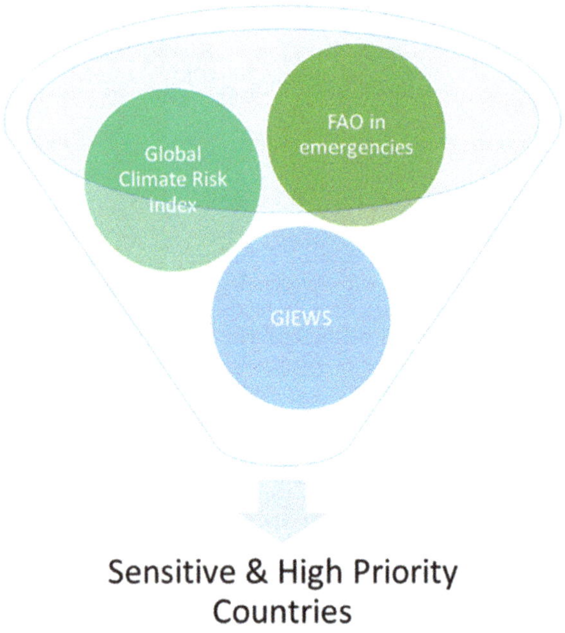

Fig. 6.3 Sensitive and high-priority country list

	8/10 major food crises are driven by conflict[33]. The World Food Programme (WFP)

6.1.5 Malawi

Drought and floods have caused severe food insecurity in Malawi[34]. Floods and droughts pose the most significant and *recurring* risk to Malawi with the highest impacts occurring in the central and south regions[35]. Flooding poses a threat to all low-lying regions around Lake Malawi, with over 100,000 people affected by floods each year, on average (please see Fig. 6.6). Droughts negative effect on average 1.5 million people, the number of higher in dry years. Malawi's small-scale producers—who make up most farmers—are struggling to produce enough to feed themselves and their families. Small landholdings, little access to credit, limited technological

[33]The WFP—https://www.wfp.org/countries

[34]Malawi—https://www.fao.org/emergencies/countries/detail/en/c/161513/

[35]Disaster Risk Profile: Malawi (2019)—https://www.gfdrr.org/en/publication/disaster-risk-profile-malawi-2019

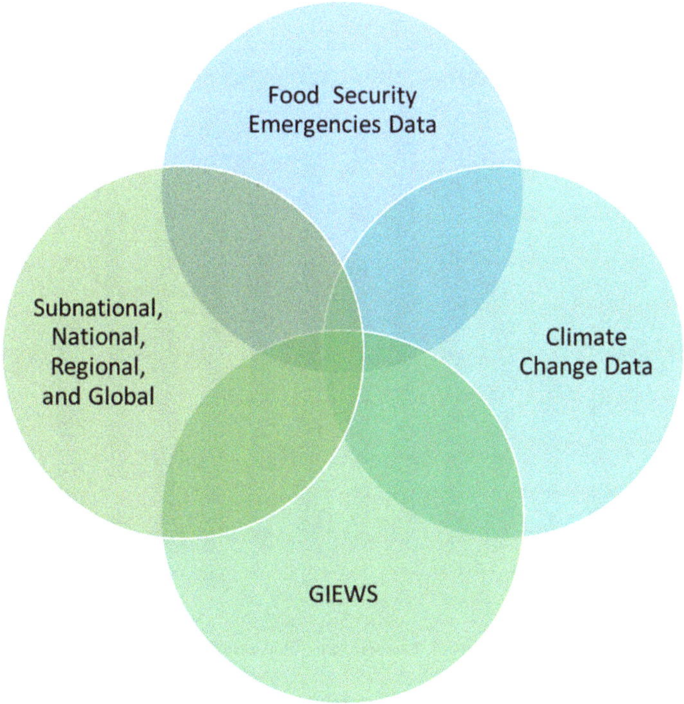

Fig. 6.4 Data view

know-how, and poor market access make it difficult for farmers to move from subsistence to commercial production. And frequent shocks, like dry spells and flooding during the cropping season, outbreaks of crop and livestock diseases, and high food prices, are further undermining their livelihoods. The country's largely rural population depends heavily on crop production for its livelihood, most notably the production of maize, which accounts for three-fifths of daily calorie consumption[36] [14].

Agriculture and downstream agro-processing generate half of gross domestic product (GDP) and four-fifths (almost a third of GDP, occupying nearly 80% of the workforce[37]) of total export earnings and employment[38] [15]. Climate shocks[39]

[36]Olivier Ecker, Economics of Micronutrient Malnutrition: The Demand for Nutrients in Sub-Saharan Africa—ISBN-13: 978-3631595053

[37]Malawi—https://ricepedia.org/malawi & http://irri.org/resources/publications/books/rice-almanac-4th-edition

[38]Agricultural growth and investment options for poverty reduction in Malawi—https://www.ifpri.org/publication/agricultural-growth-and-investment-options-poverty-reduction-malawi-0

[39]Droughts and floods in Malawi Assessing the economywide effects—https://www.ifpri.org/publication/droughts-and-floods-malawi

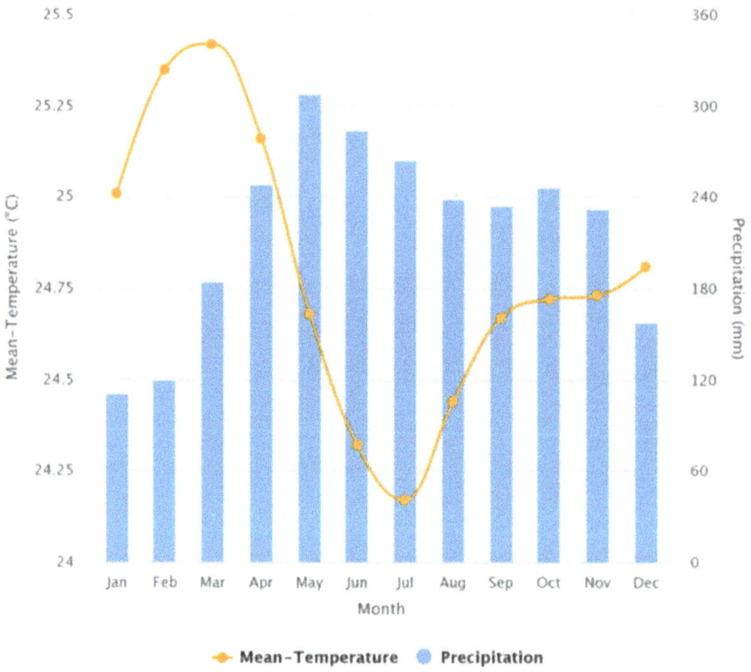

Fig. 6.5 Monthly climatology of mean temperature and precipitation in Colombia from 1991 to 2020

therefore have a potentially profound direct effect on the agricultural sector and farm households while also indirectly affecting other economic sectors and nonfarm households through price and production linkages [16]. Estimating impacts (through computable general equilibrium (CGE) models [15]) from and identifying policy responses to extreme climate events is therefore crucial for designing appropriate agricultural and development strategies. Climate change events, coupled with population and economics, are expected to increase the impacts of droughts and floods. Please see Fig. 6.7 for mean temperature and precipitation for Malawi 1990–2020[40]. And the main production season[41] for cereals like rice (planting November–December and harvesting May–July—please see Fig. 6.8). To assess climate change, it is important to understand crop calendar of Malawi—FAO Country Brief[42]. Please refer to Appendix C for World Data Sources and Appendix D for US Data Sources.

[40]Malawi—https://climateknowledgeportal.worldbank.org/country/malawi

[41]Production Season—https://ricepedia.org/malawi

[42]Country Brief—Malawi—https://www.fao.org/giews/countrybrief/country.jsp?lang=en&code=MWI

Fig. 6.6 Malawi and flood exposure

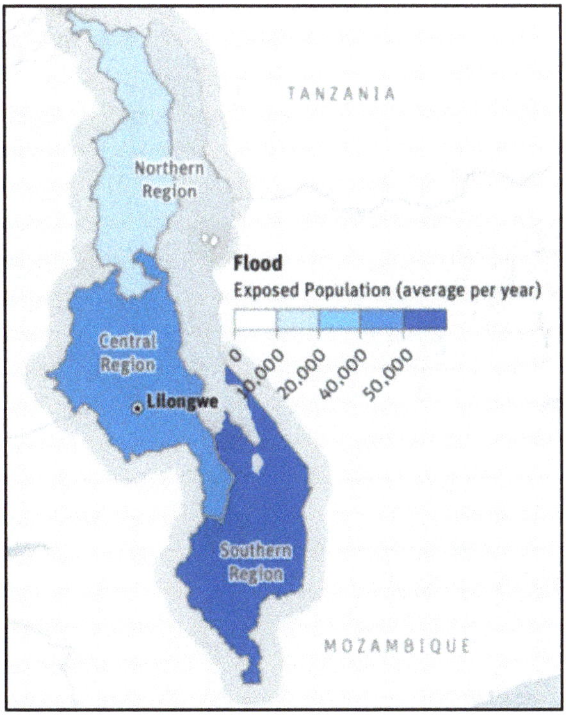

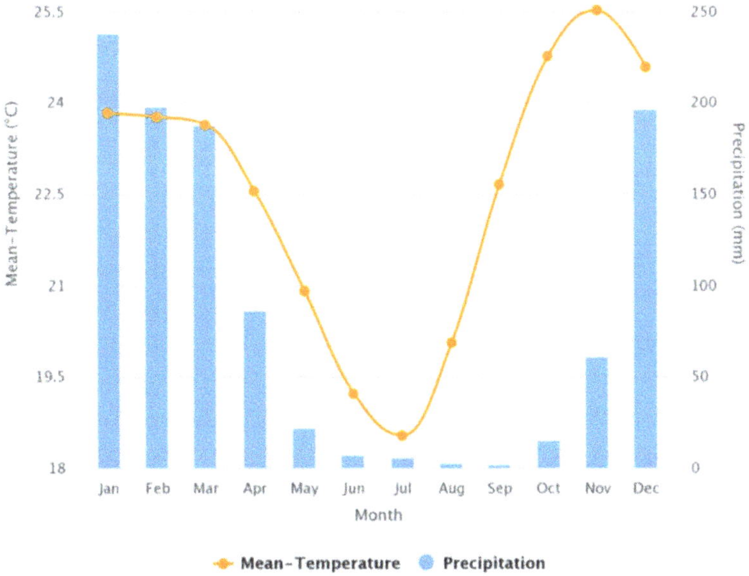

Fig. 6.7 Monthly climatology of mean temperature and precipitation in Malawi from 1991 to 2020

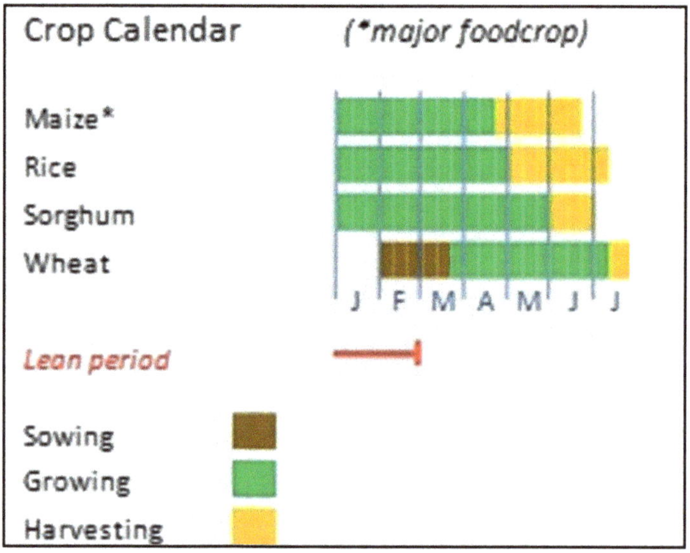

Fig. 6.8 Malawi crop calendar

The Comprehensive Africa Agriculture Development Programme (CAADP)

The Comprehensive Africa Agriculture Development Programme (CAADP) provides an integrated framework of development priorities aimed at restoring agricultural growth, rural development, and food security [52]. The CAADP [43] is an Agenda 2063 continental initiative that aims to help African countries eliminate hunger and reduce poverty by raising economic growth through agriculture-led development.

The principles and values that inform the implementation of CAADP include African ownership and leadership, accountability and transparency, inclusiveness, evidence-based planning and decision-making, and harnessing regional complementarities.

6.1.6 Myanmar

Myanmar[44] is one of the highly affected by climate change events. The observed and projected changes in climate include an increase in temperature, variation in rainfall, and an increased occurrence and severity of extreme weather events such as cyclones, floods, droughts, intense rains, and extreme high temperatures. The country (please see Fig. 6.9) is also experiencing a decrease in the duration of the

[43] CAADP—https://www.fao.org/policy-support/mechanisms/mechanisms-details/en/c/417079/

[44] Myanmar—https://www.fao.org/emergencies/countries/detail/en/c/326208/

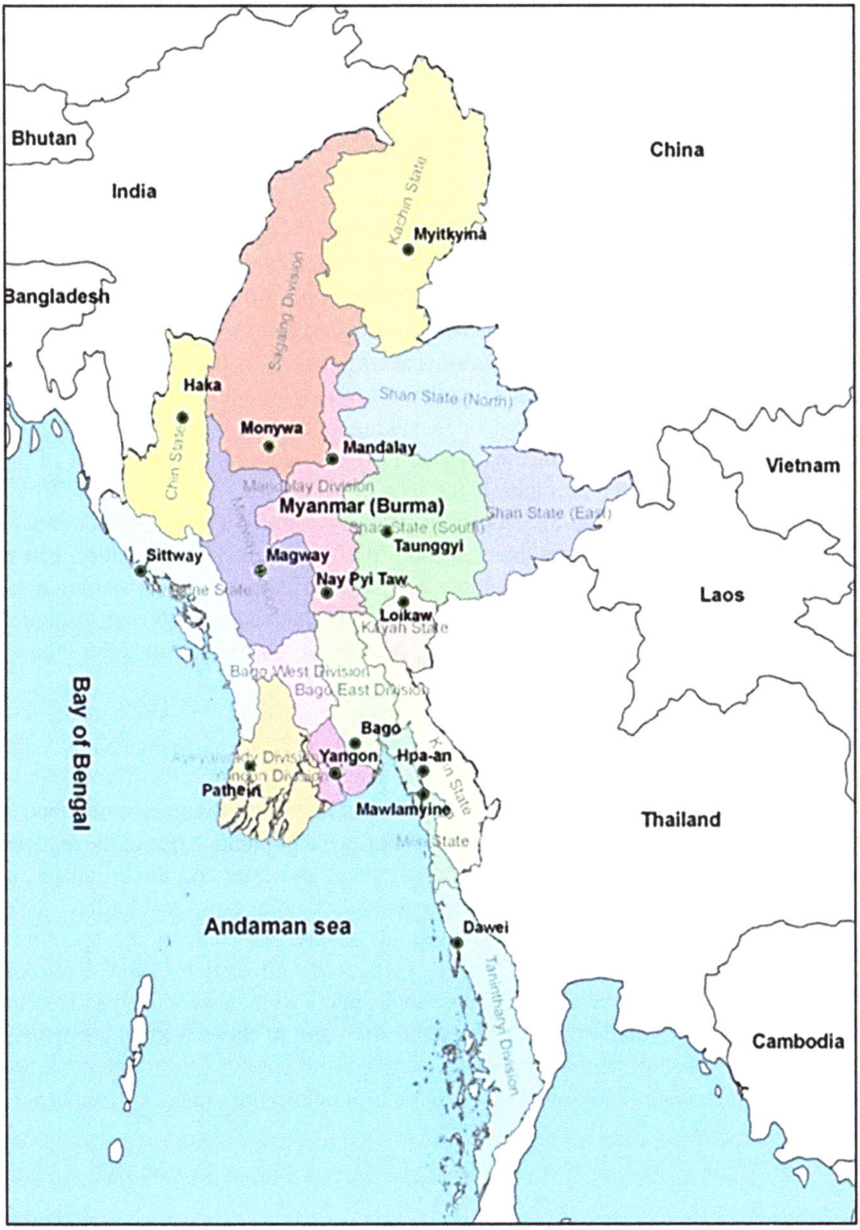

Fig. 6.9 Myanmar

southwest monsoon season due to its late onset and early retreat[45] [17]. The Hazard Profile of Myanmar[46] gives a summary sketch of the nine most recurrent natural hazards that plague the country, which are cyclones, droughts, earthquakes, fires, floods, forest fires, landslides, storm surges, and tsunamis. Myanmar is particularly threatened by cyclones due to its location on the borders of the Bay of Bengal and the Andaman Sea in Southeast Asia, a place of tropical cyclone generation. The country is further exposed because of its long coastline, compounded by the fact that most of its population lives along riverbanks prone to flooding and coastal zones suitable for agricultural production. From 1887 to 2005, 1248 tropical storms formed in the Bay of Bengal, and 80 of them hit the Myanmar coast. In the past 20 years (1995–2014), it has been exposed to 41 extreme weather events resulting in a death toll of 7, 146 (annual average) inhabitants, and an annual average of 0.74% loss per unit in GDP—making it the second-most affected country to extreme weather events[17].

Since 2000 the frequency of cyclones making landfall has gone up significantly from once every 3 years to once each year [18]. Agriculture bore the brunt of the disaster, with the sector accounting for half of all losses. The floods in 2015[47] assessment show more than 527,000 hectares of crops were affected by flooding—mostly paddy rice—and more than 242,000 livestock were killed (poultry, cattle, pigs, and goats). Farmers also lost tools, fertilizer, irrigation systems, fishing nets, traps, and boats. In addition, the assessment found that seasonal job opportunities had been significantly reduced, impacting heavily on the most vulnerable households that often rely on this casual work [18].

The above line chart (Fig. 6.10) provides precipitation data of Rakhine location in Myanmar[48]. The data source was from the World Bank Climate Change Knowledge Portal, with the following parameters (please see Fig. 6.11).

The floods have a severe impact on livestock. The loss of livestock could have a negative impact on the nutritional status of the affected population due to the reduced consumption of animal products, particularly meat and eggs and direct impact on food security. In addition, the loss of large animals such as cows and buffalo could also limit the capacity of affected households to prepare land for the upcoming agricultural winter season. Women were particularly affected by small livestock losses. In the case of female-headed households, many women were unable to rescue livestock since they had no access to boats. As many of these women are largely

[45] Myanmar Climate Change Strategy (2018–2030)—https://myanmar.un.org/sites/default/files/2019-11/MyanmarClimateChangeStrategy_2019.pdf

[46] Hazard profile of Myanmar-http://dpanther.fiu.edu/dpService/dpPurlService/purl/FI13042445/00001

[47] Agriculture and Livelihood Flood Impact Assessment in Myanmar—https://www.fao.org/emergencies/resources/documents/resources-detail/en/c/338553/

[48] Myanmar—https://climateknowledgeportal.worldbank.org/download-data

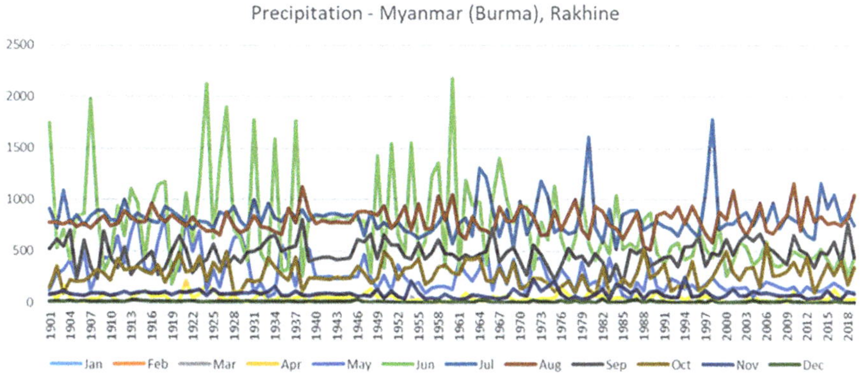

Fig. 6.10 Precipitation 1901–2018

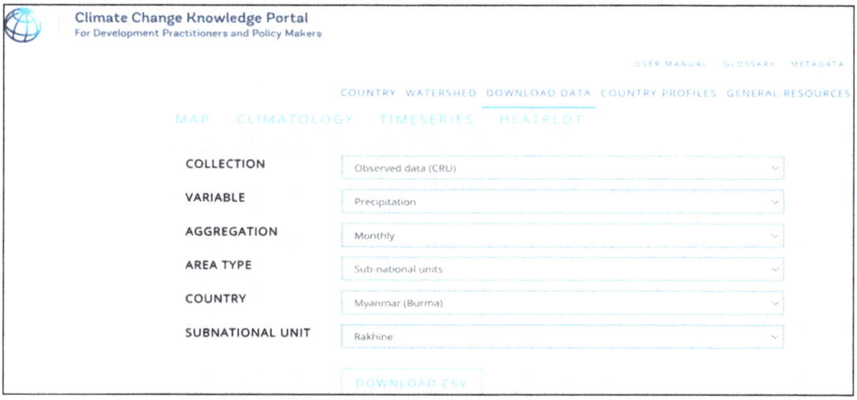

Fig. 6.11 Climate Change Knowledge Portal

dependent on small livestock, these losses will particularly impact on their livelihoods.

The impact of the disaster (please see Fig. 6.12) will most likely result in a reduced yield in all affected areas and no production in areas where paddy fields were entirely washed away. Reduced production will lead to greater food insecurity as well as lower household income which will diminish access to agricultural inputs

Fig. 6.12 Myanmar floods

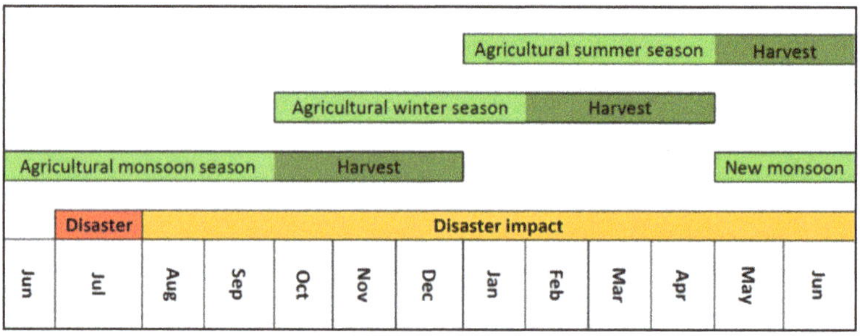

Fig. 6.13 Myanmar seasonal calendar

(mainly seeds and fertilizers) necessary for the upcoming winter and summer agricultural seasons.

The decrease in agricultural activity will also reduce demand for casual agricultural labor, which is another important livelihood source[49], as discussed later in this report. This situation will all have repercussions through to the next monsoon season when farmers will again need cash to invest in agricultural production (please see Fig. 6.13). Other impacts are on availability of labor, equipment, and fisheries [18].

[49] Agriculture and Livelihood Flood Impact Assessment in Myanmar—https://www.fao.org/fileadmin/user_upload/emergencies/docs/Final_Impact_Assessment_Report_final.pdf

6.1.7 *Recent Flooding Events in Myanmar*

Date	Description	Key findings
July 2019–August 2019	Seasonal monsoons[50] have brought strong winds and heavy rains across Myanmar, which further intensified with depressions and low-pressure areas over the Bay of Bengal, causing increased water levels in major rivers and flooding [19]. (1) Chin, (2) Kachin, (3) Mandalay, (4) Rakhine, (5) Sagaing, (6) Magway, and (7) Ayeyarwady 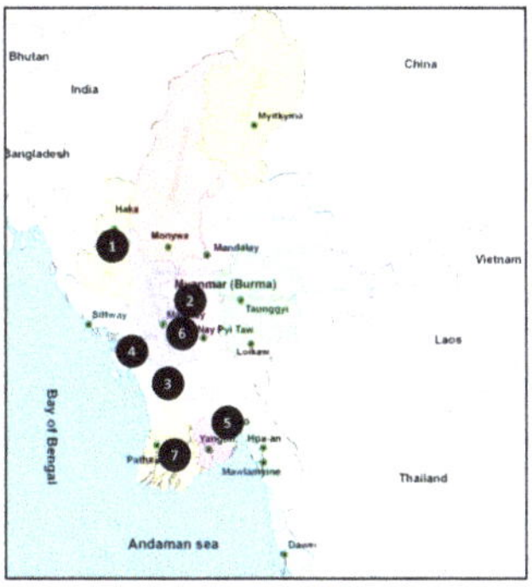	• More than 231,000 people have been affected in various states and regions including Chin, Kachin, Magway, Mandalay, Sagaing, and Rakhine in the first round (July) and in Ayeyarwady, Bago, Kayin, Mon, Tanintharyi, and Yangon in the second round (August to date). • The flooding caused the closure of more than 500 schools, destruction of at least 375 houses, infrastructure, crop harvests, and livestock[51].
July 2015–October 2015	Heavy rains caused floods and landslides in several parts of Myanmar[52] since June 2015. On 30 July, Cyclone Komen made landfall in Bangladesh, bringing strong winds and additional heavy rains to the country, which resulted in widespread flooding across 12 of the country's 14 states and regions (Ayeyarwady, Bago, Chin, Kachin, Kayin, Magway, Mandalay, Mon, Rakhine, Sagaing, Shan, Yangon). On 31 July, the President declared Chin and Rakhine states and Magway and Sagaing regions as natural disaster zones. (1) Chin, (2) Sagaing, (3) Ayeyarwady,	• More than 527,000 hectares of crops affected[53]—mostly paddy rice—with some areas losing the entire crop for the monsoon season. • Almost 2500 hectares of agricultural land affected by landslides, mostly in Chin state. • More than 242,000 livestock killed (poultry, cattle, pigs, and goats). • Almost 30,000 hectares of fish and shrimp ponds lost, as well as fishing equipment such as nets, traps, boats, and engines [20].

(continued)

[50] Myanmar: Floods—Jul 2019—https://reliefweb.int/disaster/fl-2019-000081-mmr

[51] less-crop-damage-floods-year-official—https://www.mmtimes.com/news/less-crop-damage-floods-year-official.html

[52] Myanmar: Floods and Landslides—Jul 2015—https://reliefweb.int/disaster/fl-2015-000080-mmr

[53] Myanmar floods: huge impact on agricultural livelihoods, more international support urgently needed—https://www.fao.org/resilience/resources/resources-detail/en/c/378918/

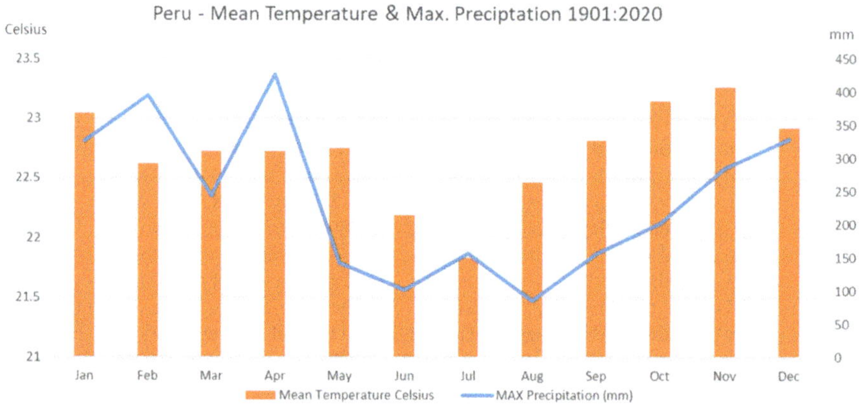

Fig. 6.14 Monthly climatology of mean temperature and precipitation in Peru from 1991 to 2020

Date	Description	Key findings
	(4) Rakhine, (5) Bago, and (6) Magway 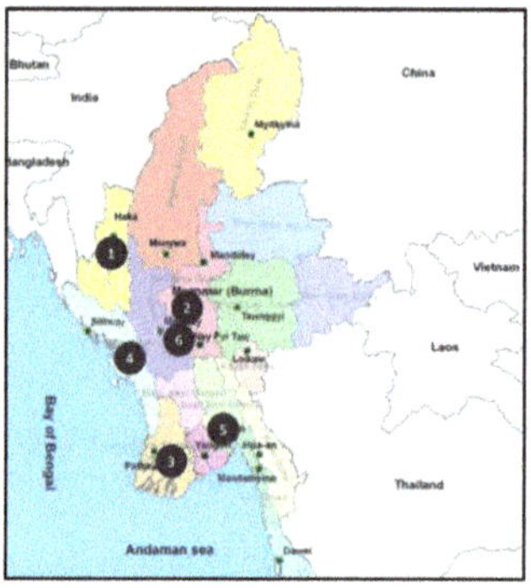	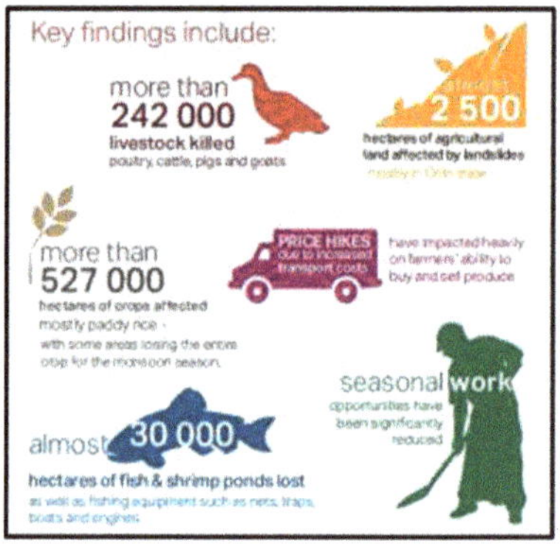

6.1.8 *Peru*

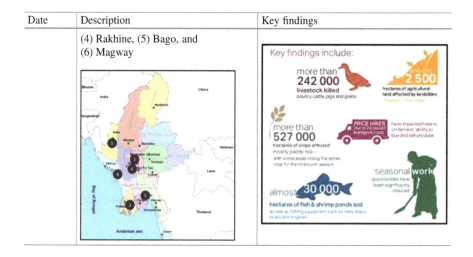

In Peru, climate challenges are a constant reality, making the need for disaster risk reduction and management more pressing[54]. Nearly a third of Peru's 29 million people live below the poverty line. In rural areas, where most people make a living from agriculture, that figure reaches 60%. Frequent natural disasters and the lack of services for agricultural production make it difficult for many to provide for themselves (please see Fig. 6.14). Subsistence farmers and small herders living in the

[54] Peru—https://www.fao.org/emergencies/countries/detail/en/c/161514/

Fig. 6.15 Peru map

Peruvian highlands must contend with extreme weather conditions—from floods and drought to frost and hailstorms—that can chip away at their ability to feed themselves and their families. Most do not have the resources to bounce back each time (please see Fig. 6.15).

Rice is a major agriculture crop in Peru, and irrigated rice accounts for two-thirds of the rice area and most of the supply[55], mainly in the forest rim in the regions of Cajamarca and Amazonas, in the higher forest of San Martín, and in the coastal regions of Tumbes, Piura, Lambayeque, La Libertad, and Arequipa. The above figure is temperature and precipitation of San Martin location from 1900 to 2020 (San Martin—Peru[56]).

Peru's climate zone classifications[57] are derived from the Köppen-Geiger climate classification system (please see Fig. 6.16), which divides climates into five main climate groups divided based on seasonal precipitation and temperature patterns. The five main groups are A (tropical), B (dry), C (temperate), D (continental), and E (polar). All climates except for those in the E group are assigned a seasonal precipitation sub-group (second letter). Climate classifications are identified by hovering your mouse over the legend. A narrative overview of Peru's country context and climate is provided following the visualizations[58].

[55] Peru Rice—https://ricepedia.org/peru

[56] Peru—San Martin—https://www.google.com/maps/place/San+Martin,+Peru/@-6.0807947,-60.0935048,4z/data=!4m5!3m4!1s0x91af8123a4e50da3:0xc0af0e0d9beff07!8m2!3d-7.2444881!4d-76.8259652

[57] Climate Change Knowledge Portal—Peru—https://climateknowledgeportal.worldbank.org/country/peru

[58] Köppen-Geiger Climate Classification, 1991-2020—https://climateknowledgeportal.worldbank.org/country/peru

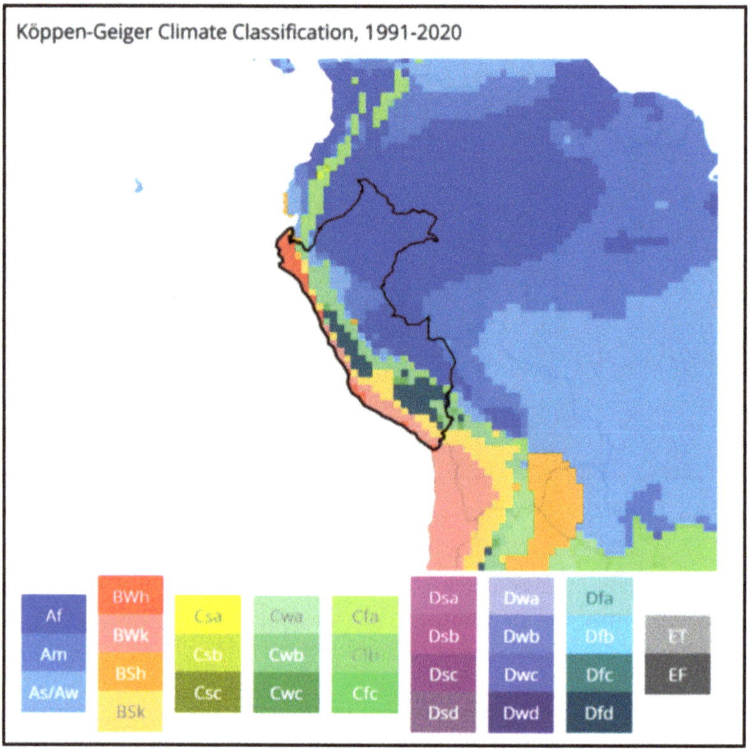

Fig. 6.16 Köppen-Geiger climate classification (1991–2020)

6.1.9 Pakistan

Pakistan[59] ranks as the 25th largest export market for US food and agricultural-related products, with US exports surpassing $1.25 billion in 2019 [21]. The government has introduced "Prime Minister's Agriculture Emergency Programme" for the next 5 years[60]. The Programme aims to increase agricultural productivity, value addition, reduce dependence on imports, and improve lives of farming community. This will also ensure food security in the country and increase the hard-earned income of the farmers. The development of the agriculture sector will further provide stimulus to agro-based industries and overall growth of the economy. Blessed with the fertile land and hardworking farmer community, we can enhance the agricultural productivity to ensure smooth supply of essential food items during COVID-19 [22]. To assess climate change, please see Fig. 6.17 (crop calendar).

[59] Pakistan: Exporter Guide—https://www.fas.usda.gov/data/pakistan-exporter-guide-5

[60] Agriculture Survey Pakistan—https://www.pc.gov.pk/uploads/cpec/PES_2020_21.pdf

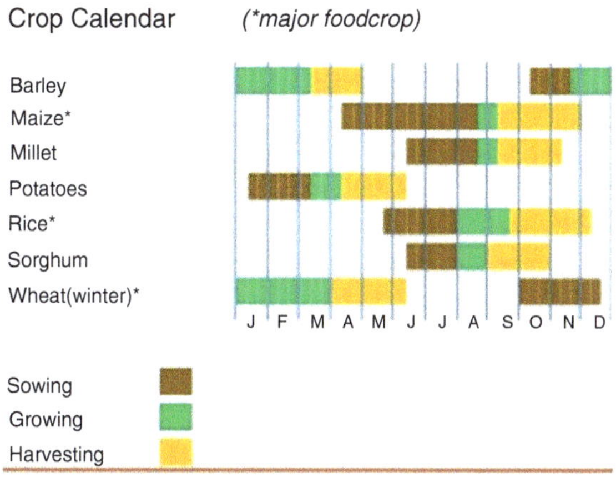

Fig. 6.17 Pakistan calendar

The government has introduced "Prime Minister's Agriculture Emergency Programme" for the next 5 years. The Programme aims to increase agricultural productivity, value addition, reduce dependence on imports, and improve lives of farming community. This will also ensure food security in the country and increase the hard-earned income of the farmers. The development of agriculture sector will further provide stimulus to agro-based industries and overall growth of the economy. Blessed with the fertile land and hardworking farmer community, we can enhance the agricultural productivity to ensure smooth supply of essential food items during COVID-19 [22].

Countries will need to move from ignoring climate change to actively protecting natural resources and building resilience to ensure food security. Having a climate change strategy that covers agricultural adaptation and mitigation will be key to encouraging innovation and driving investment in sustainable agriculture systems. Investment will also need to shore up transport and supply chain infrastructure, from the "first mile" right through to the consumer. This will need to happen for both wealthy countries and in vulnerable regions, to allow all to prepare for these environmental risks. Please refer to Appendix C for World Data Sources and Appendix D for US Data Sources.

The natural resources and resilience category (please see Table 6.4) aims to support building resilience through monitoring countries' adaptations to climate risks and allows policy makers to understand how food security is connected to climate change risks. While it was introduced in the 2017 GFSI as an adjustment factor, it was only in 2020 that it was mainstreamed. The category also allows them

Table 6.4 Natural resources and resilience

Natural resources and resilience[a]
Exposure
Temperature rise
Flooding
Sea-level rise
Water
Agricultural water risk—quantity
Agricultural water risk—quality
Land
Land degradation
Grassland
Forest change
Oceans, rivers, and lakes
Eutrophication
Marine biodiversity
Sensitivity
Food import dependency
Dependence on natural capital
Political commitment to adaptation
Early warning measures/climate-smart agriculture
Commitment to managing exposure
National agricultural adaptation policy
Disaster risk management
Demographic stress
Projected population growth
Urban absorption capacity

[a]The GFSI Methodology—https://my.corteva.com/GFSI?file=ei21_gfsir

to identify opportunities and areas to improve, including building capacity to ensure these risks to food security are addressed[61] [23]. Here are the major indicators[62]:

6.1.10 *Climate Impacts on Agriculture and Food Supply*

Our lifestyle, as we know, would change due to climate change, and it pushes economic opportunities to marginalized communities, especially farmers, to the brink of sustainability issues. Agriculture is an important sector of the US and the

[61]Resilience, natural resources become key as global food security deteriorates slightly—https://www.engineeringnews.co.za/article/resilience-natural-resources-become-key-as-global-food-security-deteriorates-slightly-2021-03-05

[62]Global Food Security Index 2021: The 10-year anniversary October 12, 2021—https://impact.economist.com/sustainability/project/food-security-index/Resources

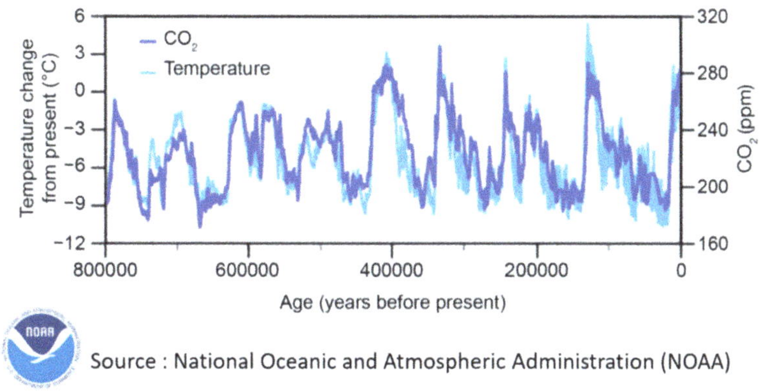

Fig. 6.18 CO2 vs. temperature

world economy. The crops, livestock, and seafood produced in the United States of America contribute more than $300 billion to the economy each year. When food service and other agriculture-related industries are included, the agricultural and food sectors contribute more than $750 billion to the gross domestic product.

Agriculture and fisheries are highly dependent on the climate. Increases in temperature and carbon dioxide (CO_2) can have a significant impact on agriculture (please see Fig. 6.18). Changes in ozone, greenhouse gases, and climate change affect agricultural producers greatly because agriculture and fisheries depend on specific climate conditions. Temperature changes can cause habitat ranges and crop planting dates to shift[63], and droughts and floods due to climate change may hinder farming practices [24]. Climate change may affect the production of maize (corn) and wheat as early as 2030, according to a new NASA study[64]. The net effect is severe for the world as with the interconnectedness of the global food system, impacts in even one region's breadbasket will be felt worldwide. Climate change is a global issue. Finally, paleoclimate data reveal that climate change is not just about temperature. As carbon dioxide has changed in the past, many other aspects of climate have changed too. Temperature change (light blue) and carbon dioxide change (dark blue) measured from the EPICA Dome C[65] ice core in Antarctica [25].

---

[63] Agriculture and Climate—https://www.epa.gov/agriculture/agriculture-and-climate

[64] Global Climate Change Impact on Crops Expected Within 10 Years, NASA Study Finds—https://climate.nasa.gov/news/3124/global-climate-change-impact-on-crops-expected-within-10-years-nasa-study-finds/

[65] Orbital and Millennial Antarctic Climate Variability over the Past 800,000 Years—https://www.ncei.noaa.gov/access/paleo-search/study/6080

6.1.11 CMIP6 Climate Projections

This catalogue entry provides daily and monthly global climate projections data from many experiments, models, and time periods computed in the framework of the sixth phase of the Coupled Model Intercomparison Project (CMIP6)[66]. The CMIP6 will consist of the "runs" from around 100 distinct climate models being produced across 49 different modeling groups. And the purpose of the climate models is to model the climate from historical values, i.e., how the climate has changed in the past and may change in the future. These models [26] simulate the physics, chemistry, and biology of the atmosphere, land, and oceans in detail and require some of the largest supercomputers in the world to generate their climate projections[67].

CMIP6 data underpins the Intergovernmental Panel on Climate Change Sixth Assessment Report, and the data the term "experiments" refers to the three main categories of CMIP6 simulations:

- Historical experiments which cover the period where modern climate observations exist. The period covered is typically 1850–2005.
- Climate projection experiments following the combined pathways of Shared Socioeconomic Pathway (SSP) and Representative Concentration Pathway (RCP). SSPs (please see Fig. 6.19) are outcome of work of an international team of climate scientists, economists, and energy systems modelers that have built a range of new "pathways" that examine how global society, demographics, and economics might change over the next century. In essence, the SSP scenarios provide different pathways of the future climate forcing and provide a toolkit for the climate change research community to carry out integrated, multi-disciplinary analysis[68]. The period covered is typically 2006–2100, and some extended RCP experimental data is available from 2100 to 2300. The main variables that are used for models in the chapter include air temperature (K), capacity of soil to store water (field capacity Kg m^{-2}), daily maximum near-surface air temperature (K), daily minimum near-surface air temperature (K), and precipitation (kg m^{-2} s^{-1}) (please see Fig. 6.20) [27].

In the above figure, a sample CMIP6 annual global average temperature 1850–2100 is provided for illustration purposes. The values for CMIP6 need to be downloaded and contextualized for a location from climatology tab under Climate Change Knowledge Portal[69]. As you can see, the SSP5-8.5, a future to avoid at all

[66] CMIP6 climate projections—https://cds.climate.copernicus.eu/cdsapp#!/dataset/projections-cmip6?tab=overview

[67] CMIP6: the next generation of climate models explained—https://www.carbonbrief.org/cmip6-the-next-generation-of-climate-models-explained

[68] SSP Overview—https://www.unece.org/fileadmin/DAM/energy/se/pdfs/CSE/PATHWAYS/2019/ws_Consult_14_15.May.2019/supp_doc/SSP2_Overview.pdf

[69] Climate Change Knowledge Portal—https://climateknowledgeportal.worldbank.org/download-data

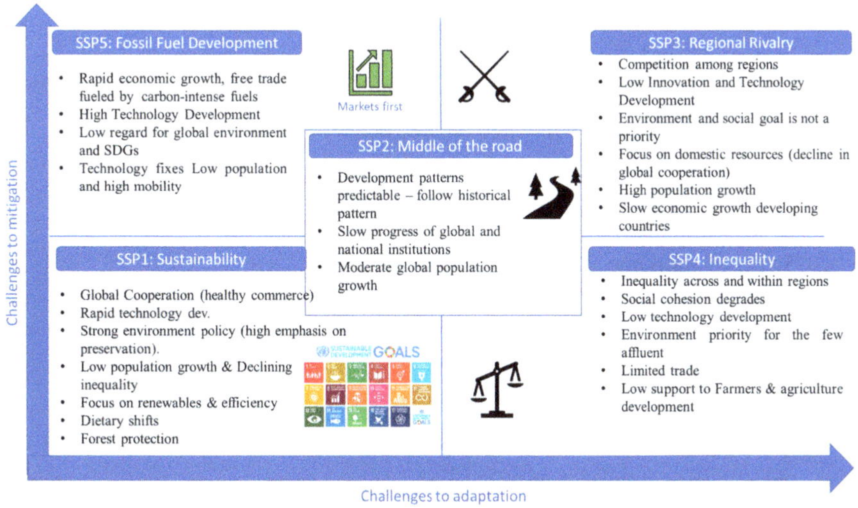

Fig. 6.19 Shared Socioeconomic Pathways (Climate Model—Sea Surface Temperature Change: SSP2 (Middle of the Road)—201–2100—https://sos.noaa.gov/catalog/datasets/724/)

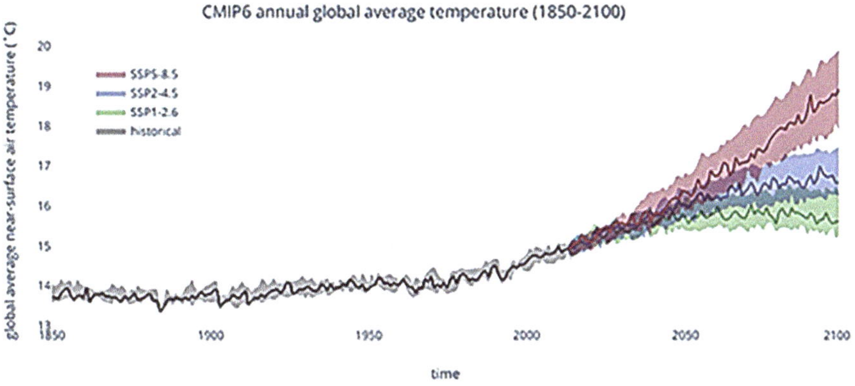

Fig. 6.20 CMIP6 annual global average temperature (1850–2100)

costs[70], projects the current CO_2 emission levels double by 2050. The global economy grows quickly, but this growth is fueled by exploiting fossil fuels and energy-intensive lifestyles. By 2100, the average global temperature is a scorching 4.4 °C higher [28].

Please check Shared Socioeconomic Pathways (please see Fig. 6.19) [29, 30]:

[70] Explainer: The U.N. climate report's five futures—decoded—https://www.reuters.com/business/environment/un-climate-reports-five-futures-decoded-2021-08-09/

- SSP1
  - Mitigation and changes to adaptation are lower. Sustainability is the main theme with the following socioeconomic behavior:
    1. Global cooperation (healthy commerce)
    2. Rapid technology dev.
    3. Strong environment policy (high emphasis on preservation)
    4. Low population growth and declining inequality
    5. Focus on renewables and efficiency
    6. Dietary shifts
    7. Forest protection
- SSP5
  - Market economics is the main aspect of SSP5 with high emphasis on the development and usage of fossil fuel:
    1. Rapid economic growth, free trade fueled by carbon-intense fuels.
    2. High-technology development.
    3. Low regard for global environment and SDGs.
    4. Technology fixes low population and high mobility.
- SSP3
  - Regional rivalry due to competition of resources:
    1. Competition among regions.
    2. Low innovation and technology development.
    3. Environmental and social goal is not a priority.
    4. Focus on domestic resources (decline in global cooperation).
    5. High population growth.
    6. Slow economic growth of developing countries.
- SSP4
  - Inequality of wealth distribution and others:
    1. Inequality across and within regions.
    2. Social cohesion degrades.
    3. Low-technology development.
    4. Environment priority for the few affluent.
    5. Limited trade.
    6. Low support to farmers and agricultural development.
- SSP2
  - Middle of the road with development that is predictable:

1. Development patterns predictable—follow historical pattern.
2. Slow progress of global and national institutions.
3. Moderate global population growth.

Shared Socioeconomic Pathway Narratives for Each Emission Scenario [29–31]

The SSPs are part of a new scenario framework established[71] by the climate change research community that facilitates the integrated analysis of future climate impacts, vulnerabilities, adaptation, and mitigation. The SSPs are based on five narratives describing alternative socioeconomic pathways, including sustainable development, regional rivalry, inequality, fossil-fueled development, and middle-of-the-road development:

- SSP1 Sustainability—Taking the Green Road (low challenges to mitigation and adaptation): The world shifts gradually, but pervasively, toward a more sustainable path, emphasizing more inclusive development that respects perceived environmental boundaries. Management of the global commons slowly improves, educational and health investments accelerate the demographic transition, and the emphasis on economic growth shifts toward a broader emphasis on human well-being. Driven by an increasing commitment to achieving development goals, inequality is reduced both across and within countries. Consumption is oriented toward low material growth and lower resource and energy intensity.
- SSP2
  Middle of the road (medium challenges to mitigation and adaptation): The world follows a path in which social, economic, and technological trends do not shift markedly from historical patterns. Development and income growth proceed unevenly, with some countries making relatively good progress, while others falling short of expectations. Global and national institutions work toward but make slow progress in achieving Sustainable Development Goals. Environmental systems experience degradation, although there are some improvements, and overall, the intensity of resource and energy use declines. Global population growth is moderate and levels off in the second half of the century. Income inequality persists or improves only slowly, and challenges to reducing vulnerability to societal and environmental changes remain[72].

(continued)

[71] Developing detailed Shared Socioeconomic Pathway (SSP) narratives for the Global Forest Sector—https://www.fs.usda.gov/treesearch/pubs/58376

[72] Climate Model—Sea Surface Temperature Change: SSP2 (Middle of the Road)—2015–2100—https://sos.noaa.gov/catalog/datasets/724/

- SSP3
  Regional rivalry—A rocky road (high challenges to mitigation and adaptation): A resurgent nationalism, concerns about competitiveness and security, and regional conflicts push countries to increasingly focus on domestic or, at most, regional issues. Policies shift over time to become increasingly oriented toward national and regional security issues. Countries focus on achieving energy and food security goals within their own regions at the expense of broader-based development. Investments in education and technological development decline. Economic development is slow, consumption is material-intensive, and inequalities persist or worsen over time. Population growth is low in industrialized and high in developing countries. A low international priority for addressing environmental concerns leads to strong environmental degradation in some regions[73,74] [32, 33].
- SSP4 Inequality—A road divided (low challenges to mitigation, high challenges to adaptation): Highly unequal investments in human capital, combined with increasing disparities in economic opportunity and political power, lead to increasing inequalities and stratification both across and within countries. Over time, a gap widens between an internationally connected society that contributes to knowledge- and capital-intensive sectors of the global economy and a fragmented collection of lower-income, poorly educated societies that work in a labor-intensive, low-tech economy. Social cohesion degrades and conflict and unrest become increasingly common. Technology development is high in the high-tech economy and sectors. The globally connected energy sector diversifies, with investments in both carbon-intensive fuels like coal and unconventional oil, but also low-carbon energy sources. Environmental policies focus on local issues around middle- and high-income areas.
- SSP5 Fossil-fueled development[75]—[31] Taking the highway (high challenges to mitigation, low challenges to adaptation): This world places increasing faith in competitive markets, innovation, and participatory

(continued)

[73]Davide Viaggi, The Bioeconomy: Delivering Sustainable Green Growth, Publisher : CABI (January 31, 2019), ISBN-13 : 978-1786392756

[74]Joachim Klement, Geo-Economics: The Interplay between Geopolitics, Economics, and Investments, Publisher : CFA Institute Research Foundation (April 28, 2021), ISBN-13 : 978-1952927065

[75]Climate Model—Sea Surface Temperature Change: SSP2 (Middle of the Road)—2015–2100—https://sos.noaa.gov/catalog/datasets/724/

societies to produce rapid technological progress and development of human capital as the path to sustainable development. Global markets are increasingly integrated. There are also strong investments in health, education, and institutions to enhance human and social capital. At the same time, the push for economic and social development is coupled with the exploitation of abundant fossil fuel resources and the adoption of resource and energy-intensive lifestyles around the world. All these factors lead to rapid growth of the global economy, while global population peaks and declines in the twenty-first century. Local environmental problems like air pollution are successfully managed. There is faith in the ability to effectively manage social and ecological systems, including by geo-engineering if necessary.

6.1.11.1 Projection Models

Table 6.5 provides a short summary[76] of projection model and SSP model [28].

Projection models are powerful and provide data to simulate climatic changes. For instance, consider location Nakhon Sawan in Thailand (please see Fig. 6.21), an agricultural hub that has seen dramatic climate events, October 2021 floods[77], 2018 flooding[78], and 2016 drought[79] [34, 35].

If we would like to assess the impact of climate change on Nakhon Sawan with respect to social economics, we can see the projections by the application of [CMIP6, SSP5-8.5, Multi-Model Ensemble], and see Fig. 6.22 for the increase in precipitation (mm) and amount of water per unit area and time, as follows: an increase of 188% 2091–2100 over the current levels 17.13 2011–2020. The Multi-Model Ensemble[80] is used to estimate the climate change signal, its uncertainty to illustrate the impact of precipitation [36].

Take another climate model projection:

[76] Explainer: The U.N. climate report's five futures—decoded—https://www.reuters.com/business/environment/un-climate-reports-five-futures-decoded-2021-08-09/

[77] PRELIMINARY SATELLITEDERIVED FLOOD ASSESSMENT—https://disasterscharter.org/documents/10180/10624638/vap-844-5-product.pdf

[78] Thailand: Over 20,000 million rai of farming area in Nakhon Sawan affected by flooding—https://reliefweb.int/report/thailand/thailand-over-20000-million-rai-farming-area-nakhon-sawan-affected-flooding

[79] PM in Nakhon Sawan, Chai Nat on drought inspection—https://www.bangkokpost.com/print/835884/

[80] Josep et al., The Mediterranean climate change hotspot in the CMIP5 and CMIP6 projections—https://esd.copernicus.org/preprints/esd-2021-65/esd-2021-65.pdf

Table 6.5 Projection models

Scenario	SSP model	Socioeconomic factors' impact
SSP1-2.6	SSP1	It imagines the same socioeconomic shifts toward sustainability as SSP1-1.9. But temperatures stabilize around 1.8 °C higher by the end of the century
SSP2-4.5	SSP2	Socioeconomic factors follow their historic trends, with no notable shifts. Progress toward sustainability is slow, with development and income growing unevenly. In this scenario, temperatures rise 2.7 °C by the end of the century
SSP3-7.0	SSP3	By the end of the century, average temperatures have risen by 3.6 °C
SSP5-8.5	SSP5	By 2100, the average global temperature is a scorching 4.4 °C higher

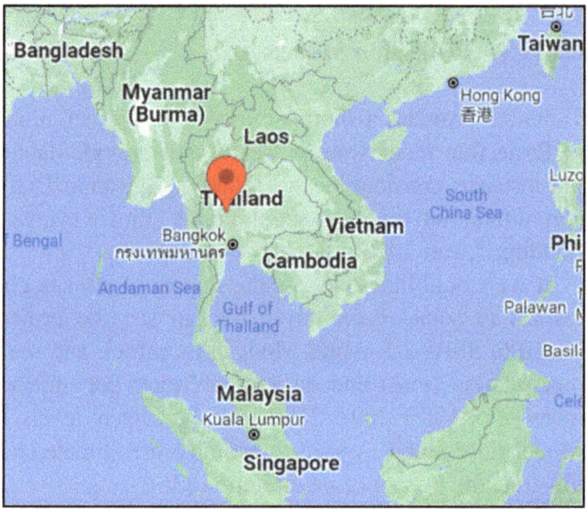

Fig. 6.21 Thailand—Nakhon Sawan

6.1.12 *Climate Model: Temperature Change (RCP 6.0) (2006–2100)*

Climate models are used for a variety of purposes from the study of dynamics of the weather and climate system to projections of future climate[81].

NOAA's Geophysical Fluid Dynamics Laboratory has created several ocean-atmosphere coupled models to predict how greenhouse gas emissions following different population, economic, and energy-use projections may affect the planet.

"Representative Concentration Pathways (RCPs)" are fully integrated scenarios (i.e., they are not a complete package of socioeconomic, emissions and climate projections). They are consistent sets of projections of only the components of

[81] Climate Model: Temperature Change (RCP 6.0)—2006–2100—https://sos.noaa.gov/catalog/datasets/climate-model-temperature-change-rcp-60-2006-2100/

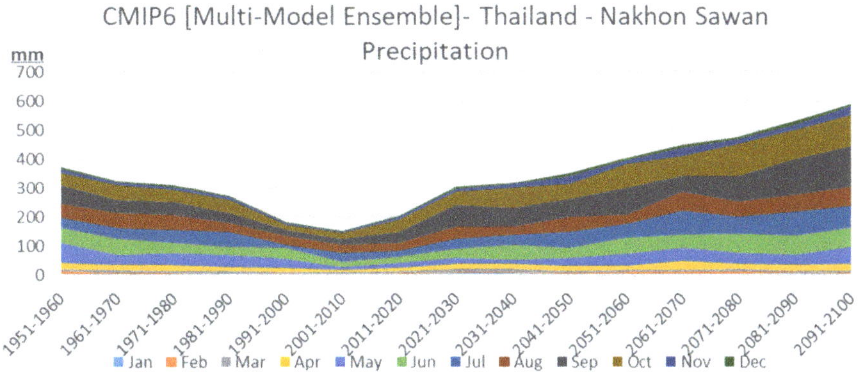

Fig. 6.22 CMIP6 [Multi-Model Ensemble] Thailand—Nakhon Sawan Precipitation

radiative forcing that are meant to serve as input for climate modeling, pattern scaling and atmospheric chemistry modeling, according.

Notable Features

- The earth gets warmer as CO2 increases in the atmosphere.
- The earth doesn't warm uniformly, the oceans warm slower than the continents and arctic.
- Projections for temperature according to RCP 6.0 include continuous global warming through 2100 where CO2 levels rise to 670 ppm by 2100 making the global temperature rise by about 3–4 °C by 2100.

Investing to Stop Climate Change[82] Is Trickier Than It Seems [37]

Given the energy the planet uses, primary world energy consumption (2020), there are four ways the world cuts its use of fossil fuels and so carbon emissions:

1. *Government action*—Government action forces owners of coal, oil, and natural gas to leave the fuel in the ground.
2. *Shareholders force companies* to leave fossil fuels in the ground, replacing executives and closing profitable mines and wells voluntarily. Only shareholders who really care about the environment would do this, and they might indeed make a big difference to the world.
3. *New technology solves the problem.* Think of it as moving beyond the Oil Age: just as the Stone Age didn't end because we ran out of stones, the saying goes, the Oil Age won't end because we run out of oil. Fossil fuels

(continued)

[82] Investing to Stop Climate Change Is Trickier Than It Seems—https://www.wsj.com/articles/investing-to-stop-climate-change-is-trickier-than-it-seems-11643214062?page=1

would be worthless if it were cheaper to use micro-nuclear reactors, solar, wind, green hydrogen, batteries, or fusion.

4. *Consume less*. The final option is to consume less energy, economizing and accepting the hit to growth.

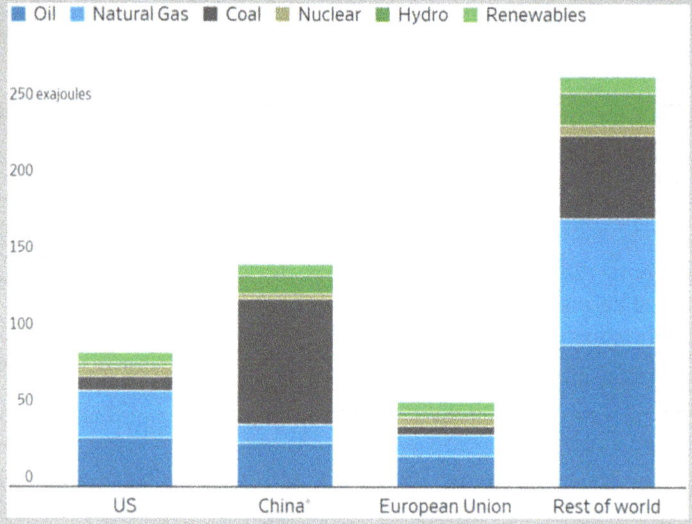

The four options laid out here lead to directly opposing approaches for investors who want to improve the world. The philanthropic approach means buying fossil fuel stocks to leave the carbon in the ground. The new technology approach means buying clean-energy startups to work on fusion, algae, cranes that store power, and the like. When the government gets involved, either with regulation (option 1) or taxes (option 4), investors should expect fossil reserves to be worth a lot less, perhaps zero—although whether fossil fuel stocks are a good investment depends on how much of that government action is already priced in.

6.2 Rice

Rice is a grain that is cultivated on every continent except Antarctica and is the primary food for half the people in the world. Rice cultivation probably originated as early as 10,000 BC in Asia. Rice is grown at varying altitudes (sea level to about 3,000 meters), in varying climates (tropical to temperate), and on dry to flooded land. The growth duration of rice plants is 3–6 months, depending on variety and growing conditions. Rice is harvested by hand in developing countries or by combines in industrialized countries. Asian countries produce about 90% of rice grown worldwide. Rough rice futures and options are traded at the CME Group[83]. World rice production in the 2019–2020 marketing year is expected to fall −0.5% to 741.739 million metric tons, down from the 2018–2019 record high of 745.237. The world's largest rice producers are expected to be China with 28.3% of world production in 2019–2020, India with 23.3%, Indonesia with 7.7%, Bangladesh with 7.3%, Vietnam with 6.1%, and Thailand with 3.8%. US production of rice in 2018–2019 (latest data) is expected to rise +25.8 % year/year to 221.211 million cwt (hundred pounds). On demand, world consumption of rice in 2019–2020 is expected to rise +1.0% to a record high of 486.715 million metric tons. US rice consumption in 2018–2019 (latest data) is expected to rise +0.1% year/year to 135.000 million cwt (hundred pounds) but still below the 2010–2011 record high of 136.921 million cwt (hundred pounds).

Forecasting commodities provide immense help to farmers on better management of supply and demand side[84] (please see Fig. 6.23—Rice futures 1980–2021).

1. 2008, trade restrictions by major rice suppliers

Global rice[85] prices rose to record highs in the spring of 2008, with trading prices tripling from November 2007 to late April 2008. The price increase was not due to crop failure or a particularly tight global rice supply situation [37]. Instead, *trade restrictions by major suppliers, panic buying by several large importers, a weak dollar, and record oil prices* were the immediate causes of the rise in rice prices. Some of the main rice-producing countries have imposed export curbs, and this has combined with low global stocks to drive rice higher[86]. Because rice is critical to the diet of about half the world's population, the rapid increase in global rice prices in late 2007 and early 2008 had a detrimental impact on those rice consumers' well-being. Although rice prices have dropped more than 40% from their April 2008 highs, they remain well above pre-2007 levels.

2. 2002, export controls

In January 2002, prospects for prices were uncertain, although the global supply/demand situation continues to be tight. Indeed, the maintenance of *subsidized export prices* in India had weighed heavily on the market until the end of March and beyond, if the Programme is extended[87]. By March 2002, Vietnam had return on the market with fresh rice supplies for export. The prospects for prices in the next few months will then very much depend on the size of the new rice crops coming onto the market (the 2001 second crops in the Northern Hemisphere and main 2002 crops in the Southern Hemisphere) and on whether China will start buying. In hindsight, the prices become bullish.

(continued)

[83] Rice—https://www.barchart.com/futures/quotes/ZR*0/profile

[84] Union Minister Dr Jitendra Singh says, IMD is continuously working on improving operational forecasting models for various weather and climate extreme events to help farmers and others—https://pib.gov.in/PressReleaseIframePage.aspx?PRID=1779311

[85] Factors Behind the Rise in Global Rice Prices in 2008—https://www.ers.usda.gov/publications/pub-details/?pubid=38490

[86] US Rice Jumps to Record High on Supply Fears—https://www.cnbc.com/2008/04/23/us-rice-jumps-to-record-high-on-supply-fears.html

[87] RICE SITUATION UPDATE AS OF 31 JANUARY 2002—https://www.fao.org/3/ae595e/ae595e.pdf

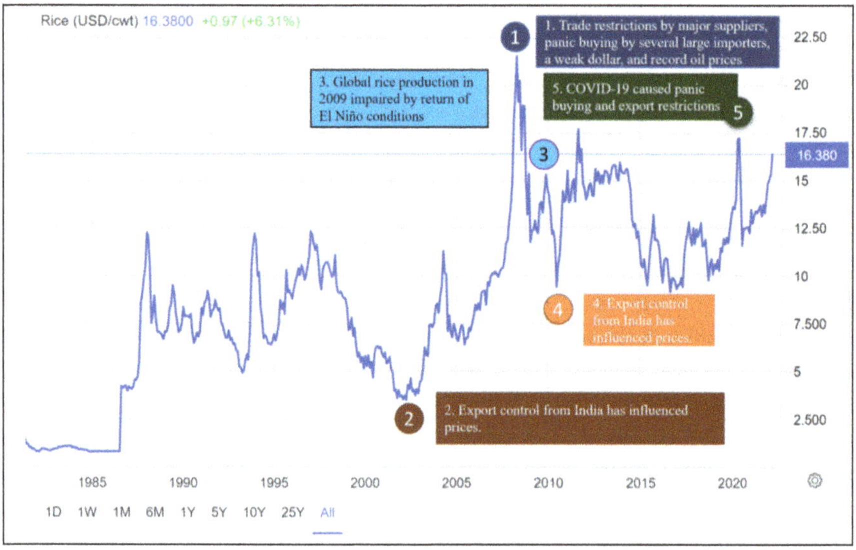

Fig. 6.23 Rice futures 1985–2021

3. 2009, global rice production in 2009 impaired by return of El Niño conditions	
	Global rice production in 2009 impaired by return of El Niño conditions in Asia—resulting in increase in rice prices during the end of 2009 (please see Table 6.6).
4. January 2010, rice prices	
	The rebounding of international rice prices observed at the end of 2009 came to an end by January 2010[88]. This was reflected in the FAO All Rice Price Index, which passed from 251 points in January 2010 to 206 points in April 2010. The weakening has been widespread, with *sluggish world import demand* negatively affecting all the rice market segments[89].
5. January 2020, the COVID-19 and rice prices [38]	
	Despite projections for near-record global rice supplies in 2019–2020[90], trading prices for rice in mid-April 2020 rose to their highest points in more than 7 years after several major exporting countries imposed export bans and restrictions in March 2020 as a response to heightened concerns over domestic food security following the outbreak of COVID-19 [39].

[88] Global rice production in 2009 impaired by return of El Niño conditions—https://www.fao.org/asiapacific/news/detail-events/zh/c/45859/

[89] Global rice production in 2009 impaired by return of El Niño conditions—https://reliefweb.int/report/bangladesh/global-rice-production-2009-impaired-return-el-ni%C3%B1o-conditions

[90] Global trading prices for rice rose to highest level in 7 years due to export restrictions—https://www.ers.usda.gov/data-products/chart-gallery/gallery/chart-detail/?chartId=98504

Table 6.6 El Niño

El Niño[a]
El Niño and La Niña are the warm and cool phases[b] of a recurring climate pattern across the tropical Pacific-the El Niño-Southern Oscillation or "ENSO" for short[c]. The pattern shifts back and forth irregularly every 2 to 7 years, and each phase triggers predictable disruptions of temperature and precipitation. These changes disrupt the large-scale air movements in the tropics, triggering a cascade of global side effects[d]. El Niño is a climatic event that recurs every 2 to 7 years. Please see the below figure to access years ELNO occurred. During 1997–1998, it had strong adverse effects on agricultural production in South America and Southeast Asia. In South America, it was accompanied with floods and storms in Peru, Ecuador, and Chile. In Southeast Asia, in Papua New Guinea, Indonesia, and the Philippines, it was associated with severe drought (La Niña). An El Niño episode is normally preceded in December–January by an unusual warming of the Pacific waters along the Peruvian coast[e].

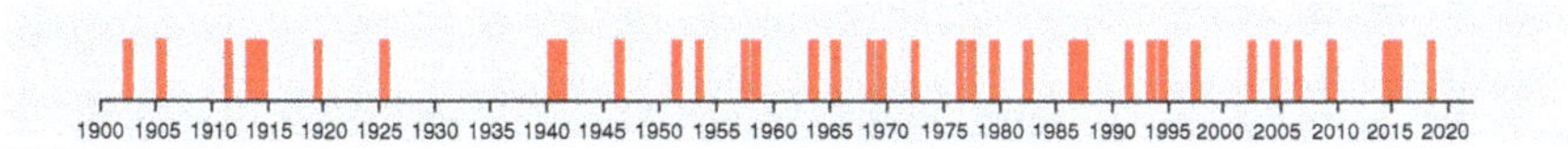

[a]Rice Market Monitor—https://www.fao.org/3/ae595e/ae595e.pdf
[b]Cold & Warm Episodes by Season—https://origin.cpc.ncep.noaa.gov/products/analysis_monitoring/ensostuff/ONI_v5.php
[c]ONI—https://origin.cpc.ncep.noaa.gov/products/analysis_monitoring/ensostuff/ONI_change.shtml
[d]El Niño & La Niña (El Niño-Southern Oscillation)—https://www.climate.gov/enso
[e]El Nino—https://en.wikipedia.org/wiki/El_Ni%C3%B1o

6.3 Mathematical Modeling

Rice demand prediction and rice futures can be analyzed with econometric models' standard ordinary least squares (OLS) regression. In addition, the forecasting performance of the rice futures market is analyzed and compared to out-of-sample forecasts derived from an additive autoregressive integrated moving average (ARIMA)[91] model and the error correction model [40]. In the following section, we have applied ensemble modeling and advanced machine learning to develop Thailand rice production model.

Thailand rice yield production	

[91]Unbiasedness and Market Efficiency Tests of the U.S. Rice Futures Market—https://www.jstor.org/stable/1349773

6.4 Machine Learning Model: Thailand Rice Paddy Yields and Climate Change

Climate change may influence food production via direct and indirect effects on crop growth processes. Direct effects include alterations to carbon dioxide availability, precipitation, and temperatures. Indirect effects include through impacts on water resource availability and seasonality, soil organic matter transformation, soil erosion, changes in pest and disease profiles, the arrival of invasive species, and decline in arable areas due to the submergence of coastal lands and desertification[92]. Projections suggest that Thailand's agriculture sector could be significantly affected from a changing climate, due to its location in the tropics where agricultural productivity is particularly vulnerable.

Let's build a model that depicts the impact of climate change on Thailand rice production:

Software code for this model: FinalThailand_Rice_climateImpact.ipynb (Jupyter Notebook Code)

[92]Thailand—https://climateknowledgeportal.worldbank.org/sites/default/files/2021-08/15853-WB_Thailand%20Country%20Profile-WEB_0.pdf

6.4.1 Data Sources

The data sources for the ML model is from FAO Statistics Data[93], World Bank Climate Change Data specifically Thailand—Nakhon Sawan[94], and World Metrological Organization Data[95] (please see Fig. 6.24). Please refer to Appendix C for World Data Sources and Appendix D for US Data Sources.

Thailand weather calendar data, additionally, were collected from FAO GIEWS[96]. The major sowing, growing, and harvesting months include (please see Fig. 6.25):

- August
- September
- October

Thailand has witnessed climate change and floods during the last decades[97]. The ML model accuracies have improved after only taking sowing, growing, and harvesting months. Floods that occur during the late months of sowing (August), growing (September), and early months of October (harvesting) have a detrimental impact on the production[98]:

- Thailand: Floods and Landslides—August 2019[99]
  - According to media reports, as of 19 September, 33 people are confirmed dead; 23,000 have been evacuated; and 418,000 affected. At least 4,000 houses and 325,000 hectares of crops have been destroyed.

- Flood 2011—The World Bank Supports Thailand'' Post-Floods Recovery Effort[100] [41]
  - The agricultural sector, on the other hand, would lose around THB 40 Bn (US$1.3 Bn) from the loss in agricultural production.
  - In addition, nearly 2.4 million rai of farmland and 20,000 rai of fishery zones are devastated, while 1.6 million animals on farm are affected. The flood damaged 17 roads in eight provinces, including Sukhothai, Phichit,

[93] FAO Statistics data—https://www.fao.org/faostat/en/#data/QCL

[94] Climate Change Data—https://climateknowledgeportal.worldbank.org/download-data

[95] World Metrological Organization Data—https://climatedata-catalogue.wmo.int/explore

[96] FAO GIEWS—Thailand—https://www.fao.org/giews/countrybrief/country.jsp?lang=en&code=THA

[97] Thailand Floods—https://reliefweb.int/disasters?search=Thailand+Floods

[98] List of Floods—https://reliefweb.int/disasters?search=Thailand+Floods

[99] Thailand: Floods and Landslides—Aug 2019—https://reliefweb.int/disaster/fl-2019-000104-tha

[100] The World Bank Supports Thailand's Post-Floods Recovery Effort—https://www.worldbank.org/en/news/feature/2011/12/13/world-bank-supports-thailands-post-floods-recovery-effort

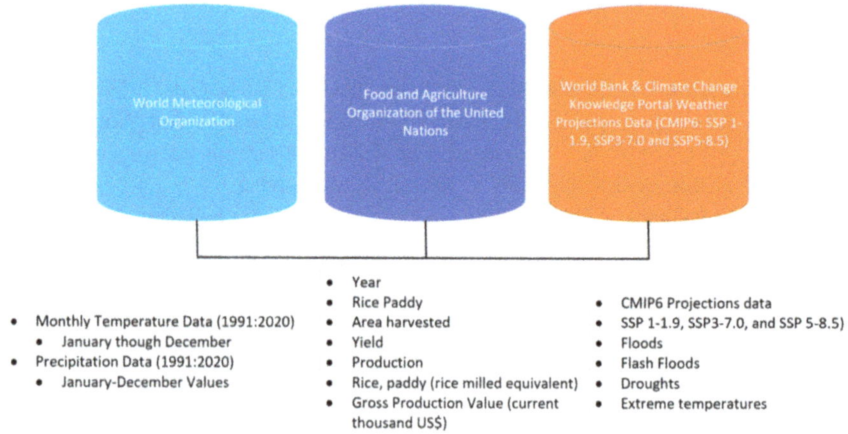

Fig. 6.24 Data sources

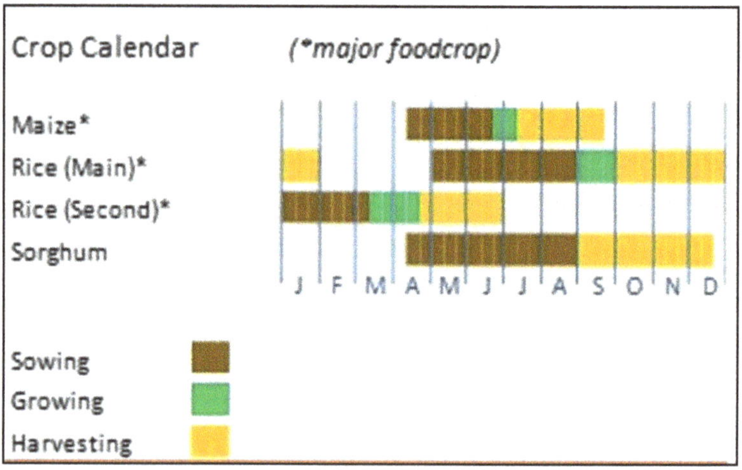

Fig. 6.25 Thailand weather calendar

Phitsanulok, Phetchabun, Nan, Chiang Rai, Nakhon Sawan, and Phra Nakhon Si Ayutthaya[101].

– The 2011 flood affected 69 provinces[102] with a total flood inundation area of 41,381.8 square km (GISTDA). Of these, 19 provinces were most severely

[101] DDPM: 14 provinces still inundated—https://reliefweb.int/report/thailand/ddpm-14-provinces-still-inundated

[102] Impact of the 2011 Floods, and Flood Management in Thailand—https://www.eria.org/Chapter_8.pdf

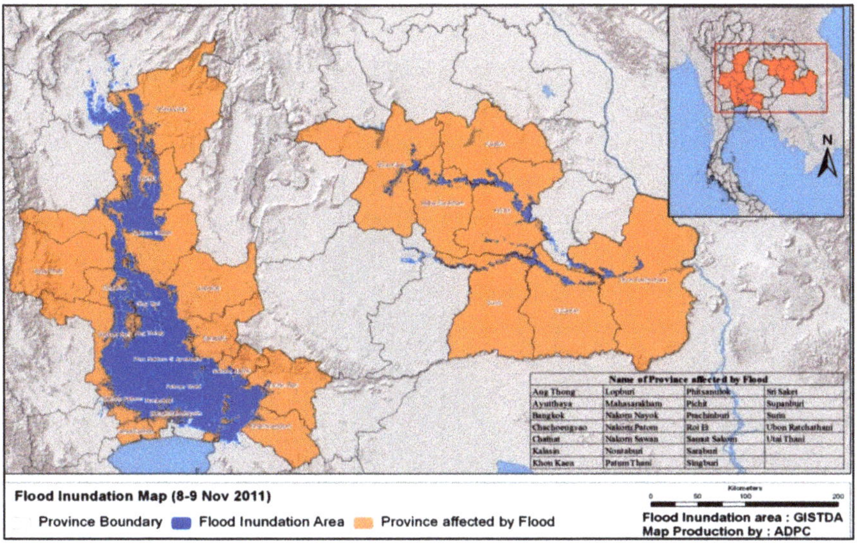

Fig. 6.26 Flood inundation map (8–9 November 2011)

inundated, located in and surrounding provinces. Flooding began around late July 2011 and receded in mid-December 2011 (please see Fig. 6.26) [42].

6.4.2 EDA Framework

Following Exploratory Data Analysis (EDA) framework (please see Figs 6.27 and 6.28) is applied to identify, confirm, and validate data signals:

1. Inspect data frame for nulls and fill with PAD or other methods: In this step, analysis of nulls is performed and fill null values with statistical relevance techniques such as mean or moving averages[103]:
   (a) Univariate (SimpleImputer) vs. Multivariate Imputation (IterativeImputer)
   (b) Simple Imputer: The SimpleImputer class provides basic strategies for imputing missing values. Missing values can be imputed with a provided constant value or using the statistics (mean, median, or most frequent) of each column in which the missing values are located. This class also allows for different missing values encodings[104] [43].

[103] Imputation of missing values—https://scikit-learn.org/stable/modules/impute.html

[104] Roderick J A Little and Donald B Rubin (1986). "Statistical Analysis with Missing Data". John Wiley & Sons, Inc., New York, NY, USA.ISBN-13 : 978-0470526798

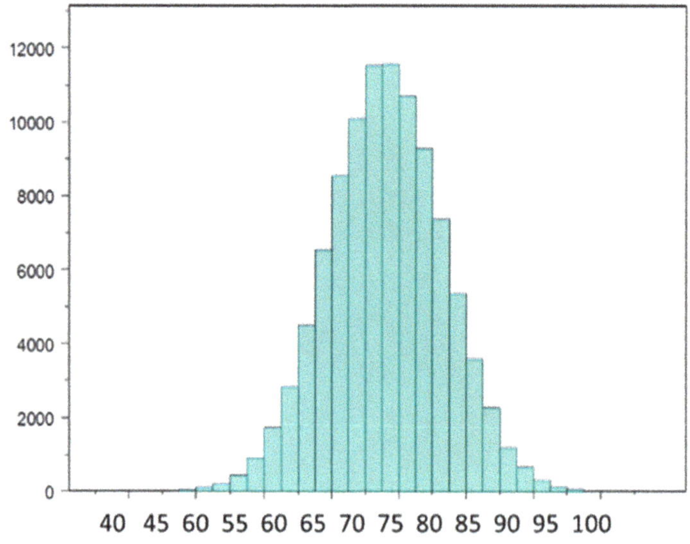

Fig. 6.27 Normal curve

(c) Multivariate Imputer—SCIKIT learn multivariate imputer inspired by MICE R package (Multivariate Imputation by Chained Equations).

2. Describe statistical properties of data frame.
3. Analyze quantile values for numerical items.
4. Evaluate feature variable (X) data distribution.
5. Check for normal or bell curve test.
6. If normal test fails, perform Apply Scalers[105] (MinMaxScaler, StandardScaler[106], and others)[107].
7. Perform class variable (Y) data distribution. Confirm the distribution passes central limit theorem. The central limit theorem[108] states that if you have a population with mean μ and standard deviation σ and take sufficiently large random samples from the population with replacement text annotation indicator, then the distribution of the sample means will be approximately normally distributed. The figure below illustrates a normally distributed characteristic, X, in a population in which the population mean is 75 with a standard deviation of 8 [44].

[105] How to Use StandardScaler and MinMaxScaler Transforms in Python—https://machinelearningmastery.com/standardscaler-and-minmaxscaler-transforms-in-python/

[106] Preprocessing—https://scikit-learn.org/stable/modules/preprocessing.html#preprocessing

[107] Scikit-learn: Machine Learning in Python, Pedregosa et al., JMLR 12, pp. 2825-2830, 2011.

[108] The Role of Probability, Date last modified—https://sphweb.bumc.bu.edu/otlt/mph-modules/bs/bs704_probability/index.html

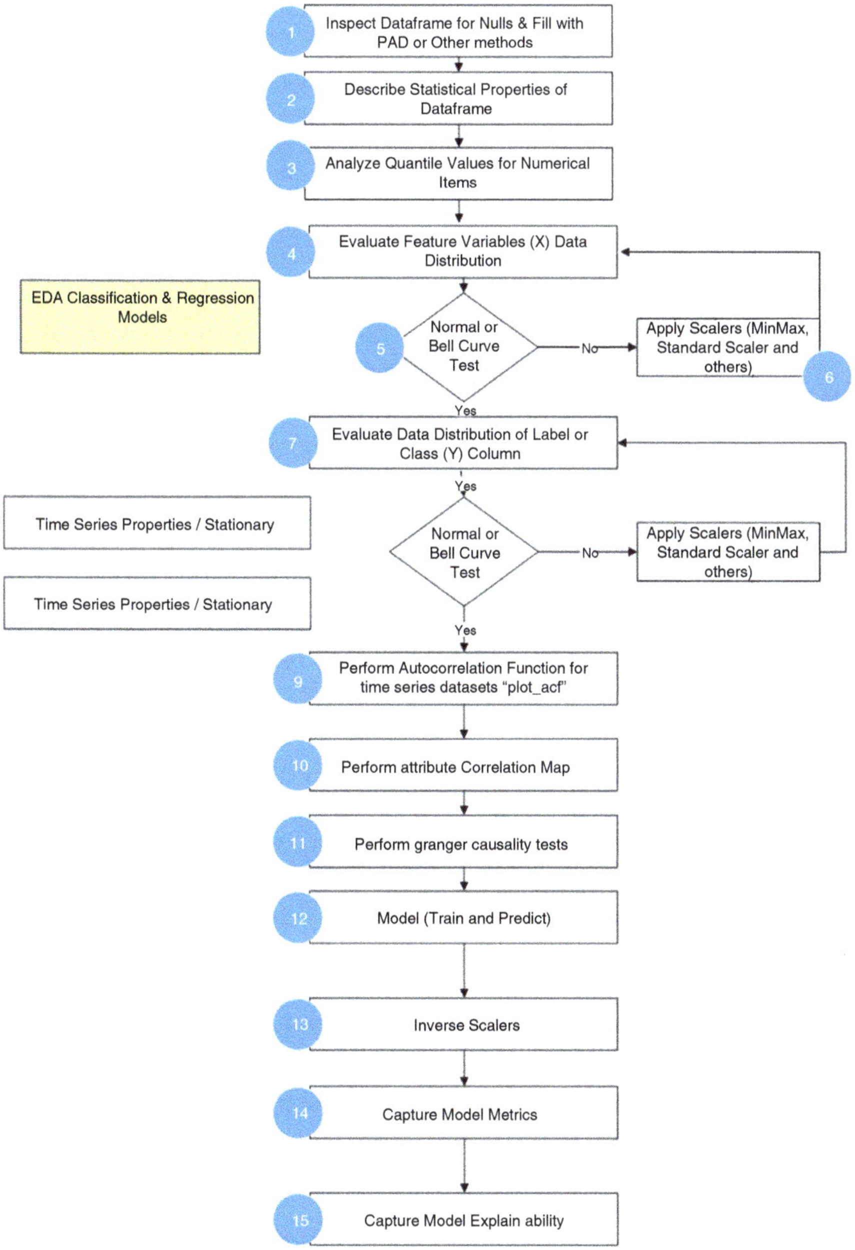

Fig. 6.28 EDA lifecycle

8. If normal test fails, perform Apply Scalers (MinMaxScaler, StandardScaler[109], and others).
9. For time series data, please perform autocorrelation function[110] for time series datasets "plot_acf" [45].
10. For time series data, perform attribute correlation map.
11. Perform granger causality[111] tests for time series [46].
12. Build model.
13. Inverse scalers if scaler operations are performed in steps 6 and 8.
14. Capture model metrics.
15. Capture model explainability.

6.4.3 Step 1: Import Libraries

Import open-source and machine learning libraries to process the data:

```
import numpy as np
import pandas as pd
import functools
from datetime import date
import plotly.graph_objects as go
import seaborn as sns; sns.set()
import statsmodels.api as sm
from statsmodels.tsa.arima_process import ArmaProcess
from scipy import stats
from scipy.stats import pearsonr
import matplotlib.pyplot as plt
from sklearn.preprocessing import MinMaxScaler
from sklearn.model_selection import train_test_split
from sklearn.linear_model import LinearRegression
from sklearn.metrics import r2_score
from sklearn import metrics
from sklearn.impute import SimpleImputer
import eli5
```

[109] Preprocessing—https://scikit-learn.org/stable/modules/preprocessing.html#preprocessing

[110] NIST/SEMATECH e-Handbook of Statistical Methods, http://www.itl.nist.gov/div898/handbook/ , February 12, 2022

[111] A Quick Introduction On Granger Causality Testing For Time Series Analysis—https://towardsdatascience.com/a-quick-introduction-on-granger-causality-testing-for-time-series-analysis-7113dc9420d2

6.4.4 Step 2: Load Thailand FAO Dataset 1991–2020

```
dfFAOThailandRiceProd = pd.read_csv("FAOSTAT_data_Thailand.
csv")
dfFAOThailandRiceProd
```

Output:

Please find output: as you can see, the rice paddy item with production value loaded as part of the data frame:

	Domain Code	Domain	Area Code (FAO)	Area	Element Code	Element	Item Code (FAO)	Item	Year Code	Year	Unit	Value	Flag	Flag Description
0	QV	Value of Agricultural Production	216	Thailand	56	Gross Production Value (current thousand SLC)	27	Rice, paddy	1991	1991	1000 SLC	83415600	Fc	Calculated data
1	QV	Value of Agricultural Production	216	Thailand	56	Gross Production Value (current thousand SLC)	27	Rice, paddy	1992	1992	1000 SLC	76123917	Fc	Calculated data
2	QV	Value of Agricultural Production	216	Thailand	56	Gross Production Value (current thousand SLC)	27	Rice, paddy	1993	1993	1000 SLC	59307941	Fc	Calculated data
3	QV	Value of Agricultural Production	216	Thailand	56	Gross Production Value (current thousand SLC)	27	Rice, paddy	1994	1994	1000 SLC	81360692	Fc	Calculated data
4	QV	Value of Agricultural Production	216	Thailand	56	Gross Production Value (current thousand SLC)	27	Rice, paddy	1995	1995	1000 SLC	90967967	Fc	Calculated data

6.4.5 Step 3: Prepare Data Frame with Gross Production Value

```
dfFAOThailandRiceProd = dfFAOThailandRiceProd
[["Year","Value","Element"]]

dfFAOThailandRiceProd = dfFAOThailandRiceProd.pivot(index =
"Year",columns= "Element" ,values = "Value")
dfFAOThailandRiceProd
```

Output:

Element	Gross Production Value (current thousand SLC)	Gross Production Value (current thousand US$)
Year		
1991	83415600	3269046
1992	76123917	2996993
1993	59307941	2342373
1994	81360692	3235018
1995	90967967	3651103
1996	115878870	4572475
1997	156901852	5002562
1998	132447866	3202364
1999	117376377	3104070
2000	112446713	2803333
2001	140407090	3160051

6.4.6 Step 4: Get Thailand Rice Paddy Yield Data

```
dfFAOThailandRiceYield = pd.read_csv
("FAOSTAT_data_Thailand_yield.csv")
dfFAOThailandRiceYield
```

The above code loads Thailand yield data from FAO.
Output:

	Domain Code	Domain	Area Code (FAO)	Area	Element Code	Element	Item Code (FAO)	Item	Year Code	Year	Unit	Value	Flag	Flag Description
0	QCL	Crops and livestock products	216	Thailand	5312	Area harvested	27	Rice, paddy	1961	1961	ha	6120000	F	FAO estimate
1	QCL	Crops and livestock products	216	Thailand	5312	Area harvested	27	Rice, paddy	1962	1962	ha	6540000	F	FAO estimate
2	QCL	Crops and livestock products	216	Thailand	5312	Area harvested	27	Rice, paddy	1963	1963	ha	6500000	F	FAO estimate
3	QCL	Crops and livestock products	216	Thailand	5312	Area harvested	27	Rice, paddy	1964	1964	ha	6310000	F	FAO estimate
4	QCL	Crops and livestock products	216	Thailand	5312	Area harvested	27	Rice, paddy	1965	1965	ha	6270000	F	FAO estimate
...	...	...	...	...	...	...	...	...	...	...	...	...	...	...
235	QCL	Crops and livestock products	216	Thailand	5510	Production	30	Rice, paddy (rice milled equivalent)	2016	2016	tonnes	21248619	Fc	Calculated data
236	QCL	Crops and livestock products	216	Thailand	5510	Production	30	Rice, paddy (rice milled equivalent)	2017	2017	tonnes	21943568	Fc	Calculated data
237	QCL	Crops and livestock products	216	Thailand	5510	Production	30	Rice, paddy (rice milled equivalent)	2018	2018	tonnes	21576192	Fc	Calculated data

6.4.7 *Step 5: Merge Data [Thailand Production and Yield Data]*

Use data frame inner join to merge the data:

```
dfFAOThailandRiceProdAndYield = dfFAOThailandRiceYield.merge
(dfFAOThailandRiceProd, on = "Year",how = "inner")
```

Output:

	Area harvested RicePaddy	Yield_RicePaddy	ProductionRicePaddy	ProductionMilledRicepaddy	Gross Production Value (current thousand SLC)	Gross Production Value (current thousand US$)
Year						
1991	9052960	22534	20400000	13606800	83415600	3269046
1992	9159680	21745	19917299	13284838	76123917	2996993
1993	9000000	20497	18447260	12304322	59307941	2342373
1994	8975229	23521	21110714	14080846	81360692	3235018
1995	9112951	24158	22015481	14684326	90967967	3651103
1996	9267200	24098	22331638	14895203	115878870	4572475
1997	9912790	23788	23580080	15727913	156901852	5002562
1998	9511520	24180	22998414	15339942	132447866	3202364
1999	9969920	24244	24171412	16122332	117376377	3104070
2000	9891200	26128	25843878	17237867	112446713	2803333
2001	10125424	28739	29099915	19409643	140407090	3160051
2002	9653534	29338	28321137	18890198	143050063	3329835
2003	10163878	29339	29820271	19890121	166069089	4003150
2004	9992868	28895	28873975	19258941	192098556	4775910
2005	10224967	29974	30648248	20442381	212147173	5274656
2006	10165155	29503	29990602	20003732	204895793	5408790

6.4.8 *Step 6: Load Thailand Weather Data*

In this step, load Thailand weather data for SAWAN NAKHON. Dataset includes precipitation, average temperature, minimum temperature, and maximum temperature for SAWAN NAKHON:

```
prcp_NAKHONSAWANThailand = pd.read_excel
("Thailand_NAKHON_SAWAN.xlsx",sheet_name =
"percipitation_Thailand")
prcp_NAKHONSAWANThailand.set_index("Year",inplace = True)
```

(continued)

```
AvgTemp_NAKHONSAWANThailand = pd.read_excel
("Thailand_NAKHON_SAWAN.xlsx",sheet_name = "Thailand_avgtemp")
AvgTemp_NAKHONSAWANThailand.set_index("Year",inplace = True)

minTemp_NAKHONSAWANThailand = pd.read_excel
("Thailand_NAKHON_SAWAN.xlsx",sheet_name = "Thailand_mintemp")
minTemp_NAKHONSAWANThailand.set_index("Year",inplace = True)

maxTemp_NAKHONSAWANThailand = pd.read_excel
("Thailand_NAKHON_SAWAN.xlsx",sheet_name = "Thailand_Maxtemp")
maxTemp_NAKHONSAWANThailand.set_index("Year",inplace = True)
```

6.4.9 Step 7: Create Combined Weather Data Frame

Create combined weather data frame:

```
dfNAKHONSAWANThailandWeather = [prcp_NAKHONSAWANThailand,
AvgTemp_NAKHONSAWANThailand,minTemp_NAKHONSAWANThailand,
maxTemp_NAKHONSAWANThailand]

def agg_df(dfList):
  temp = functools.reduce(lambda left, right: pd.merge(left,
right,
                     on = "Year",
                     how='outer'), dfList)
  return temp

weatherdf = agg_df(dfs)

dfNAKHONSAWANThailandWeather = pd.concat
([prcp_NAKHONSAWANThailand,AvgTemp_NAKHONSAWANThailand,
minTemp_NAKHONSAWANThailand,maxTemp_NAKHONSAWANThailand],axis
= 1)

dfNAKHONSAWANThailandWeather
```

Data frame concat[112] function provides mechanism to combine all data frames.

[112] Pandas Concat—https://pandas.pydata.org/docs/reference/api/pandas.concat.html

Output:

Year	Jan-prcp	feb-prcp	Mar-prcp	Apr-prcp	May-prcp	Jun-prcp	July-prcp	Aug-prcp	Sep-prcp	Oct-prcp	...	Mar-maxTemp	Apr-maxTemp	May-maxTemp	Jun-maxTemp	July-maxTemp	Aug-maxTemp
1939	-999.9	-999.9	-999.9	-999.9	-999.9	-999.9	-999.9	-999.9	-999.9	-999.9	...	34.6	37.2	35.8	33.7	33.4	33.2
1940	0.0	0.0	0.0	-999.9	-999.9	-999.9	-999.9	-999.9	-999.9	-999.9	...	38.0	40.4	37.8	35.3	33.8	33.4
1941	2.0	2.0	17.0	119.0	-999.9	-999.9	97.0	275.0	204.0	234.0	...	-999.9	40.5	-999.9	36.2	35.8	34.4
1942	0.0	13.0	35.0	91.0	188.0	188.0	141.0	234.0	242.0	118.0	...	37.6	37.4	35.9	-999.9	33.7	33.3
1943	0.0	16.0	12.0	-999.9	-999.9	-999.9	-999.9	-999.9	-999.9	-999.9	...	-999.9	-999.9	-999.9	-999.9	-999.9	-999.9
...	...	...	...	...	...	...	...	...	...	...	...	...	...	...	...	...	...
2015	7.0	6.0	59.0	30.0	36.0	35.0	77.0	103.0	258.0	145.0	...	37.4	38.4	39.7	38.3	37.1	35.6
2016	7.0	0.0	0.0	2.0	149.0	-999.9	-999.9	-999.9	-999.9	321.0	...	38.9	41.9	39.5	-999.9	-999.9	-999.9
2017	39.0	9.0	26.0	31.0	383.0	263.0	-999.9	202.0	187.0	199.0	...	37.2	38.1	35.2	34.6	-999.9	33.6
2018	5.0	33.0	9.0	68.0	139.0	103.0	154.0	271.0	55.0	-999.9	...	36.1	-999.9	35.6	34.6	33.3	33.4
2019	NaN	NaN	NaN	NaN	NaN	NaN	NaN	NaN	NaN	NaN	...	38.6	40.5	37.8	35.2	34.8	-999.9

81 rows × 48 columns

Given production and rice paddy yield data that is from 1991, filter data frame from 1991:

```
dfNAKHONSAWANThailandWeather = dfNAKHONSAWANThailandWeather
[dfNAKHONSAWANThailandWeather.index >=1991 ]
dfNAKHONSAWANThailandWeather
```

Output:

Year	Jan-prcp	feb-prcp	Mar-prcp	Apr-prcp	May-prcp	Jun-prcp	July-prcp	Aug-prcp	Sep-prcp	Oct-prcp	...	Mar-maxTemp	Apr-maxTemp	May-maxTemp	Jun-maxTemp	July-maxTemp	Aug-maxTemp
1991	0.0	0.0	19.0	44.0	115.0	25.0	80.0	123.0	66.0	128.0	...	39.0	38.6	38.0	35.0	35.5	33.5
1992	1.0	0.0	0.0	18.0	155.0	136.0	145.0	135.0	207.0	172.0	...	38.1	40.5	38.7	34.8	34.1	33.1
1993	0.0	0.0	64.0	30.0	92.0	54.0	104.0	148.0	322.0	77.0	...	36.5	37.6	37.2	36.6	35.9	33.4
1994	0.0	0.0	175.0	25.0	235.0	186.0	42.0	157.0	100.0	73.0	...	35.7	37.9	35.0	33.7	33.1	32.9
1995	1.0	-999.9	-999.9	-999.9	-999.9	181.0	220.0	-999.9	343.0	90.0	...	38.2	39.6	36.0	35.1	33.4	33.0
1996	0.0	19.0	150.0	131.0	152.0	132.0	63.0	128.0	242.0	155.0	...	37.4	36.8	34.8	34.2	34.1	33.7
1997	0.0	0.0	112.0	68.0	127.0	73.0	110.0	122.0	286.0	116.0	...	-999.9	-999.9	-999.9	37.1	34.3	-999.9
1998	1.0	-888.8	13.0	12.0	199.0	152.0	263.0	174.0	169.0	89.0	...	38.5	39.3	37.8	35.4	34.1	34.1
1999	3.0	22.0	10.0	125.0	345.0	67.0	157.0	124.0	193.0	172.0	...	37.8	35.2	33.4	-999.9	34.1	33.3
2000	0.0	9.0	38.0	181.0	167.0	270.0	117.0	89.0	176.0	210.0	...	36.4	-999.9	34.3	33.5	33.5	-999.9
2001	3.0	30.0	66.0	35.0	217.0	164.0	185.0	116.0	193.0	104.0	...	34.3	39.3	34.6	34.2	34.0	33.3
2002	26.0	0.0	6.0	39.0	174.0	165.0	165.0	216.0	307.0	187.0	...	36.9	-999.9	35.3	34.8	-999.9	33.5
2003	0.0	40.0	49.0	35.0	94.0	205.0	159.0	202.0	258.0	69.0	...	35.6	38.5	37.3	34.3	33.7	33.8
2004	-888.8	72.0	-999.9	13.0	47.0	147.0	228.0	120.0	302.0	0.0	...	37.7	39.8	36.6	34.5	34.1	34.1
2005	-888.8	-888.8	119.0	40.0	44.0	156.0	164.0	269.0	295.0	97.0	...	36.5	37.2	37.4	35.2	34.0	33.9

As a data cleanup process, replace –999 with NANs:

```
dfNAKHONSAWANThailandWeather = dfNAKHONSAWANThailandWeather.
replace(-999.9 , np.nan)
```

Merge both weather and yield data frames:

```
df_merged = dfFAOThailandRiceProdAndYield.merge
(dfNAKHONSAWANThailandWeather,on= "Year", how= "inner")
df_merged
```

Output:

Year	Area harvested RicePaddy	Yield_RicePaddy	ProductionRicePaddy	ProductionMilledRicepaddy	Gross Production Value (current thousand SLC)	Gross Production Value (current thousand US$)	Jan-prcp	feb-prcp	Mar-prcp	Apr-prcp	...	Mar-maxTemp	A
1991	9052960	22534	20400000	13606800	83415600	3269046	0.0	0.0	19.0	44.0	...	39.0	
1992	9159680	21745	19917299	13284838	76123917	2996993	1.0	0.0	0.0	18.0	...	38.1	
1993	9000000	20497	18447260	12304322	59307941	2342373	0.0	0.0	64.0	30.0	...	36.5	
1994	8975229	23521	21110714	14080846	81360692	3235018	0.0	0.0	175.0	25.0	...	35.7	
1995	9112951	24158	22015481	14684326	90967967	3651103	1.0	NaN	NaN	NaN	...	38.2	
1996	9267200	24098	22331638	14895203	115878870	4572475	0.0	19.0	150.0	131.0	...	37.4	
1997	9912790	23788	23580080	15727913	156901852	5002562	0.0	0.0	112.0	68.0	...	NaN	
1998	9511520	24180	22998414	15339942	132447866	3202364	1.0	-888.8	13.0	12.0	...	38.5	
1999	9969920	24244	24171412	16122332	117376377	3104070	3.0	22.0	10.0	125.0	...	37.8	
2000	9891200	26128	25843878	17237867	112446713	2803333	0.0	9.0	38.0	181.0	...	36.4	
2001	10125424	28739	29099915	19409643	140407090	3160051	3.0	30.0	66.0	35.0	...	34.3	
2002	9653534	29338	28321137	18890198	143050063	3329835	26.0	0.0	6.0	39.0	...	36.9	
2003	10163878	29339	29820271	19890121	166069089	4003150	0.0	40.0	49.0	35.0	...	35.6	
2004	9992868	28895	28873975	19258941	192098556	4775910	-888.8	72.0	NaN	13.0	...	37.7	

Merged data frame columns:

```
Index(['Area harvested RicePaddy', 'Yield_RicePaddy',
'ProductionRicePaddy',
    'ProductionMilledRicepaddy',
    'Gross Production Value (current thousand SLC)',
    'Gross Production Value (current thousand US$)', 'Jan-prcp',
'feb-prcp',
    'Mar-prcp', 'Apr-prcp', 'May-prcp', 'Jun-prcp', 'July-prcp',
'Aug-prcp',
    'Sep-prcp', 'Oct-prcp', 'Nov-prcp', 'Dec-prcp', 'Jan-
avgtemp',
    'Feb-avgtemp', 'Mar-avgtemp', 'Apr-avgtemp', 'May-avgtemp',
    'Jun-avgtemp', 'July-avgtemp', 'Aug-avgtemp', 'Sep-avgtemp',
    'Oct-avgtemp', 'Nov-avgtemp', 'Dec-avgtemp', 'Jan-minTemp',
    'Feb-minTemp', 'Mar-minTemp', 'Apr-minTemp', 'May-minTemp',
    'Jun-minTemp', 'July-minTemp', 'Aug-minTemp', 'Sep-minTemp',
    'Oct-minTemp', 'Nov-minTemp', 'Dec-minTemp', 'Jan-maxTemp',
    'Feb-maxTemp', 'Mar-maxTemp', 'Apr-maxTemp', 'May-maxTemp',
    'Jun-maxTemp', 'July-maxTemp', 'Aug-maxTemp', 'Sep-maxTemp',
    'Oct-maxTemp', 'Nov-maxTemp', 'Dec-maxTemp'],
   dtype='object')
```

6.4.10 Step 8: Apply Imputation Strategies to Fill NAN Values (Forward Fill)

In this step, apply data frame forward fill techniques to approximate NAN values to the closest value. Forward fill[113] applies the similarity replacement for temperature. For instance, if a year's precipitation value is NAN, then forward fill applies the next years' value as a NAN replacement:

```
dfNAKHONSAWANThailandWeather = dfNAKHONSAWANThailandWeather.
fillna(method= "ffill")
```

Next, apply imputation strategies to replace the values:

```
dfFAOThailandRiceandWeatherMerged = pd.merge
(dfNAKHONSAWANThailandWeather, dfFAOThailandRiceProdAndYield,
left_index=True, right_index=True)

from sklearn.impute import SimpleImputer
imp = SimpleImputer(missing_values=np.nan,
strategy='most_frequent')
df_impframe = pd.DataFrame(imp.fit_transform
(dfFAOThailandRiceandWeatherMerged),
         columns=dfFAOThailandRiceandWeatherMerged.columns,
         index=dfFAOThailandRiceandWeatherMerged.index)
```

Here we're applying simple imputer, simple imputation transformer[114], for completing missing values.

Imputer data frame has the following columns:

```
Index(['Jan-prcp', 'feb-prcp', 'Mar-prcp', 'Apr-prcp', 'May-
prcp', 'Jun-prcp',
   'July-prcp', 'Aug-prcp', 'Sep-prcp', 'Oct-prcp', 'Nov-prcp',
'Dec-prcp',
   'Jan-avgtemp', 'Feb-avgtemp', 'Mar-avgtemp', 'Apr-avgtemp',
   'May-avgtemp', 'Jun-avgtemp', 'July-avgtemp', 'Aug-avgtemp',
   'Sep-avgtemp', 'Oct-avgtemp', 'Nov-avgtemp', 'Dec-avgtemp',
```

(continued)

[113] pandas.DataFrame.fillna—https://pandas.pydata.org/docs/reference/api/pandas.DataFrame.fillna.html

[114] sklearn.impute.SimpleImputer—https://scikit-learn.org/stable/modules/generated/sklearn.impute.SimpleImputer.html

```
    'Jan-minTemp', 'Feb-minTemp', 'Mar-minTemp', 'Apr-minTemp',
    'May-minTemp', 'Jun-minTemp', 'July-minTemp', 'Aug-minTemp',
    'Sep-minTemp', 'Oct-minTemp', 'Nov-minTemp', 'Dec-minTemp',
    'Jan-maxTemp', 'Feb-maxTemp', 'Mar-maxTemp', 'Apr-maxTemp',
    'May-maxTemp', 'Jun-maxTemp', 'July-maxTemp', 'Aug-maxTemp',
    'Sep-maxTemp', 'Oct-maxTemp', 'Nov-maxTemp', 'Dec-maxTemp',
    'Area harvested RicePaddy', 'Yield_RicePaddy',
'ProductionRicePaddy',
    'ProductionMilledRicepaddy',
    'Gross Production Value (current thousand SLC)',
    'Gross Production Value (current thousand US$)'],
   dtype='object')
```

6.4.11 Step 9: Prepare Final Rice Paddy Data Frame with Weather Details for Regression

Prepare rice paddy data frame with the selected columns as the final ready data frame to be regressed to find the relationships and to predict:

```
Ricedfwithimpframe = df_impframe[['May-prcp', 'Jun-prcp',
    'July-prcp', 'Aug-prcp', 'Sep-prcp', 'Oct-prcp', 'Nov-prcp',
'Dec-prcp',
    'May-avgtemp', 'Jun-avgtemp', 'July-avgtemp', 'Aug-avgtemp',
    'Sep-avgtemp', 'Oct-avgtemp', 'Nov-avgtemp', 'Dec-avgtemp',
    'May-minTemp', 'Jun-minTemp', 'July-minTemp', 'Aug-minTemp',
    'Sep-minTemp', 'Oct-minTemp', 'Nov-minTemp', 'Dec-minTemp',
    'May-maxTemp', 'Jun-maxTemp', 'July-maxTemp', 'Aug-maxTemp',
    'Sep-maxTemp', 'Oct-maxTemp', 'Nov-maxTemp', 'Dec-maxTemp',
    'Area harvested RicePaddy',
'ProductionRicePaddy','Yield_RicePaddy']]
```

Output:

	May-prcp	Jun-prcp	July-prcp	Aug-prcp	Sep-prcp	Oct-prcp	Nov-prcp	Dec-prcp	May-avgtemp	Jun-avgtemp	...	Jun-maxTemp	July-maxTemp	Aug-maxTemp	Sep-maxTemp	Oct-maxTemp	Nov-maxTemp
Year																	
1991	115.0	25.0	80.0	123.0	66.0	128.0	0.0	9.0	31.4	29.8	...	35.0	35.5	33.5	33.5	32.4	32.0
1992	155.0	136.0	145.0	135.0	207.0	172.0	0.0	9.0	31.8	29.6	...	34.8	34.1	33.1	33.3	30.8	30.8
1993	92.0	54.0	104.0	148.0	322.0	77.0	0.0	0.0	31.0	30.6	...	36.6	35.9	33.4	32.9	32.9	33.4
1994	235.0	186.0	42.0	157.0	100.0	73.0	0.0	0.0	29.7	28.9	...	33.7	33.1	32.9	33.3	32.4	32.8
1995	235.0	181.0	220.0	157.0	343.0	90.0	18.0	0.0	30.5	29.6	...	35.1	33.4	33.0	32.8	32.5	31.2
1996	152.0	132.0	63.0	128.0	242.0	155.0	131.0	0.0	29.0	28.7	...	34.2	34.1	33.7	32.6	32.4	32.0
1997	127.0	73.0	110.0	122.0	286.0	116.0	3.0	0.0	30.6	30.6	...	37.1	34.3	33.7	33.2	33.6	33.3
1998	199.0	152.0	263.0	174.0	169.0	89.0	39.0	6.0	31.3	30.0	...	35.4	34.1	34.1	33.6	33.2	32.3
1999	345.0	67.0	157.0	124.0	193.0	172.0	40.0	-888.8	28.4	28.8	...	35.4	34.1	33.3	33.6	32.1	31.8
2000	167.0	270.0	117.0	89.0	176.0	210.0	-888.8	0.0	28.9	28.5	...	33.5	33.5	33.3	32.5	33.1	32.1
2001	217.0	164.0	185.0	116.0	193.0	104.0	6.0	0.0	28.8	28.9	...	34.2	34.0	33.3	32.5	33.1	31.3
2002	174.0	165.0	165.0	216.0	307.0	187.0	14.0	50.0	29.5	29.2	...	34.8	34.0	33.5	32.3	33.3	32.4
2003	94.0	205.0	159.0	202.0	258.0	69.0	1.0	0.0	30.8	28.8	...	34.3	33.7	33.8	33.0	33.9	34.7
2004	47.0	147.0	228.0	120.0	302.0	0.0	-888.8	0.0	30.1	29.0	...	34.5	34.1	34.1	33.4	34.2	34.6
2005	44.0	156.0	164.0	269.0	295.0	97.0	36.0	10.0	31.0	29.8	...	35.2	34.0	33.9	33.0	33.5	32.5
2006	75.0	215.0	105.0	158.0	258.0	218.0	9.0	0.0	29.0	29.1	...	34.3	33.6	33.4	33.0	32.9	34.3

6.4.12 Step 10: Prepare the Final Rice Paddy Data Frame with Weather Details for Regression

Prepare rice paddy data frame with the selected columns as the final ready data frame to be regressed to find the relationships and to predict:

```
numeric_features = ['May-prcp', 'Jun-prcp', 'July-prcp', 'Aug-
prcp', 'Sep-prcp', 'Oct-prcp',
    'Nov-prcp', 'Dec-prcp', 'May-avgtemp', 'Jun-avgtemp', 'July-
avgtemp',
    'Aug-avgtemp', 'Sep-avgtemp', 'Oct-avgtemp', 'Nov-avgtemp',
    'Dec-avgtemp', 'May-minTemp', 'Jun-minTemp', 'July-minTemp',
    'Aug-minTemp', 'Sep-minTemp', 'Oct-minTemp', 'Nov-minTemp',
    'Dec-minTemp', 'May-maxTemp', 'Jun-maxTemp', 'July-maxTemp',
    'Aug-maxTemp', 'Sep-maxTemp', 'Oct-maxTemp', 'Nov-maxTemp',
    'Dec-maxTemp', 'Area harvested RicePaddy']

Ricedfwithimpframe[numeric_features +
['ProductionRicePaddy']].describe()
```

Prepare numeric features for analyzing the data:

	May-prcp	Jun-prcp	July-prcp	Aug-prcp	Sep-prcp	Oct-prcp	Nov-prcp	Dec-prcp	May-avgtemp	Jun-avgtemp	...	May-maxTemp	Jun-maxTemp
count	29.000000	29.000000	29.000000	29.000000	29.000000	29.000000	29.000000	29.000000	29.000000	29.000000	...	29.000000	29.000000
mean	163.344828	139.137931	139.655172	167.793103	230.068966	164.000000	35.000000	9.620690	30.272414	29.548276	...	36.237931	35.055172
std	88.889689	66.280639	54.824575	63.316878	88.641304	104.443765	61.649933	15.218442	1.252397	0.791752	...	1.817458	1.222289
min	36.000000	25.000000	42.000000	64.000000	55.000000	0.000000	0.000000	0.000000	28.400000	28.500000	...	33.400000	33.500000
25%	94.000000	92.000000	104.000000	120.000000	187.000000	97.000000	2.000000	0.000000	29.000000	29.000000	...	34.700000	34.200000
50%	149.000000	136.000000	151.000000	157.000000	242.000000	155.000000	17.000000	0.000000	30.100000	29.300000	...	36.000000	34.800000
75%	217.000000	181.000000	165.000000	217.000000	297.000000	199.000000	40.000000	10.000000	31.100000	29.900000	...	37.400000	35.400000
max	383.000000	270.000000	263.000000	292.000000	380.000000	588.000000	321.000000	50.000000	32.600000	31.500000	...	39.700000	38.300000

8 rows × 34 columns

6.4.13 Step 11: Analyze the Data Distribution Label and Features

```
import pandas as pd
import matplotlib.pyplot as plt

# This ensures plots are displayed inline in the Jupyter notebook
%matplotlib inline

# Get the label column
label = Ricedfwithimpframe['ProductionRicePaddy']

# Create a figure for 2 subplots (2 rows, 1 column)
fig, ax = plt.subplots(2, 1, figsize = (9,12))

# Plot the histogram
ax[0].hist(label, bins=100)
ax[0].set_ylabel('Frequency')

# Add lines for the mean, median, and mode
ax[0].axvline(label.mean(), color='magenta',
linestyle='dashed', linewidth=2)
ax[0].axvline(label.median(), color='cyan',
linestyle='dashed', linewidth=2)

# Plot the boxplot
ax[1].boxplot(label, vert=False)
ax[1].set_xlabel('ProductionRicePaddy')

# Add a title to the Figure
fig.suptitle('ProductionRicePaddy Distribution')

# Show the figure
fig.show()
```

Output:

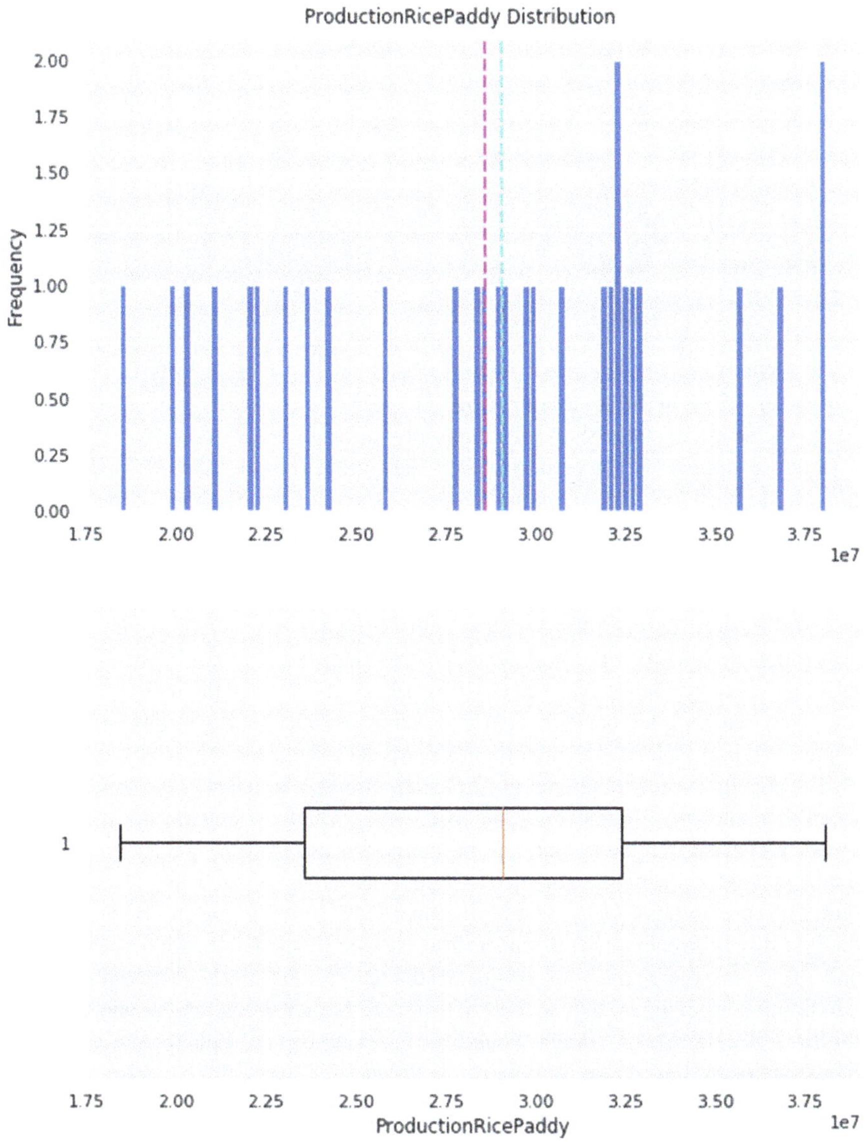

The frequency distribution of rice production with two frequencies: 3.25 and 3.75.

Plot other numerical features:

```
# Plot a histogram for each numeric feature
for col in numeric_features:
  fig = plt.figure(figsize=(9, 6))
  ax = fig.gca()
  feature = Ricedfwithimpframe[col]
  feature.hist(bins=100, ax = ax)
  ax.axvline(feature.mean(), color='magenta',
linestyle='dashed', linewidth=2)
  ax.axvline(feature.median(), color='cyan',
linestyle='dashed', linewidth=2)
  ax.set_title(col)
plt.show()
```

In northern and northeastern regions, the main rice season lasts from May to December, while in the southern region, the main rice crop lasts from September to May. The following table shows the rice-cropping season in the country[115]. Thailand agriculture season[116]: Harvesting of the 2021 main (mostly rainfed) paddy crop, accounting for about 70% of annual output, started in early *October* in the northern region and central plains and will finalize next January in southern parts of the country. Prospects for this crop were boosted by generally favorable weather conditions prevailing between *May* and *August* in the main producing northern and northeastern parts of the country, with adequate water supplies also favoring cultivation in (mostly) irrigated central plains. However, floods across numerous provinces between *September* and *October* negatively affected the main crops just as they were reaching the harvest stage. In essence, weather in the months of August, September, and October are important for overall production.

The average annual temperature of Thailand ranges from 24 °C to 30 °C, and the average annual precipitation reaches 1000 mm, which is suitable for paddy rice growing. Therefore, paddy rice can be cultivated at any time during a year, and double-season or triple-season rice is common. In the data below, October has numerous precipitation days with higher than 1000 mm. In August and September, the intensity of precipitation is higher. The average temperatures are between 24 °C and 30 °C in the months of August and October for 1991–2020 dataset. The maximum temperatures, that is, the intensity, are higher in September and October, the harvest season, a favorable condition for rice crop[117]. Finally, the minimum temperatures in *August and October are between 24 °C and 26 °C, a great condition for rice cultivation* [47].

[115] Thailand—https://www.fao.org/3/Y4347E/y4347e1o.htm#:~:text=In%20northern%20and%20northeastern%20regions,lasts%20from%20September%20to%20May.

[116] Thailand country brief—https://www.fao.org/giews/countrybrief/country.jsp?lang=en&code=THA

[117] FAO Rice Information, Volume 3, December 2002—https://www.fao.org/3/Y4347E/y4347e00.htm#Contents

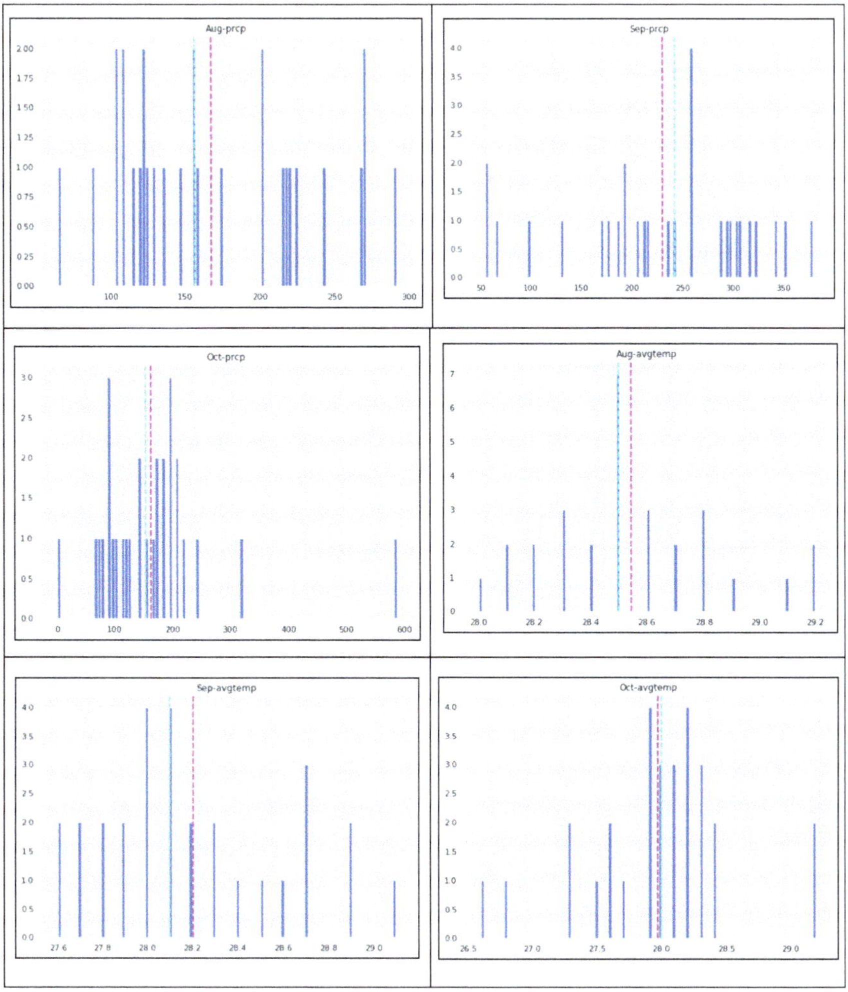
Aug-prcp
Sep-prcp
Oct-prcp
Aug-avgtemp
Sep-avgtemp
Oct-avgtemp

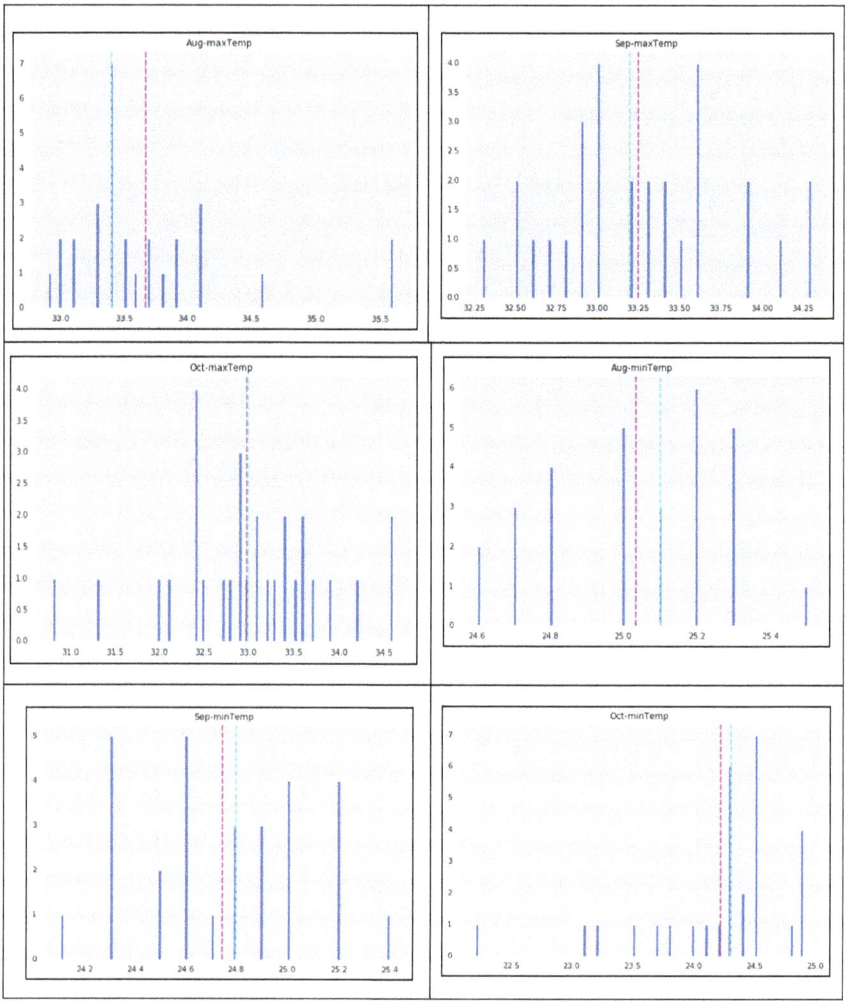

6.4.14 Step 12: Correlation Between Rice Production and Model Feature

The correlation between rice production and model features is depicted in the table below. The dataset exhibits 20% to 40% for precipitation, average temperature, minimum temperature, and maximum temperature feature variable correlated with rice production label. For the month of November, minimum temperature (51.4%) and average temperature (37.51%) are also highly correlated, but given the months are late harvesting season for the main rice production does not influence overall model outcome.

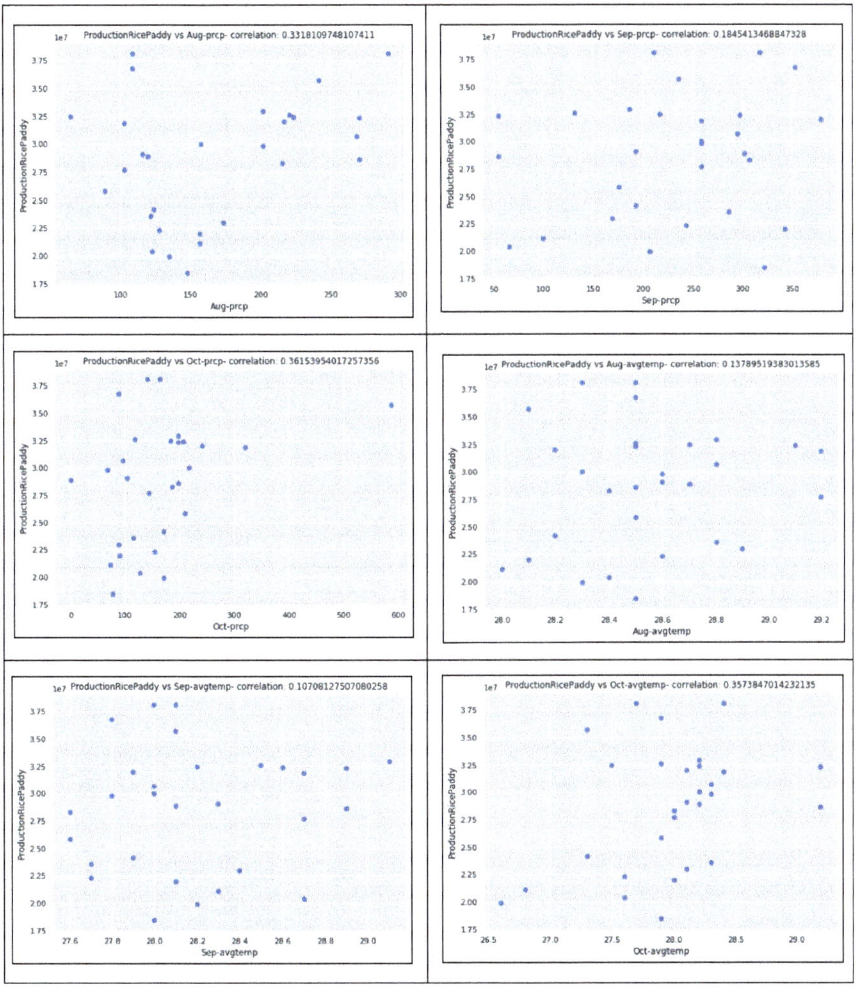

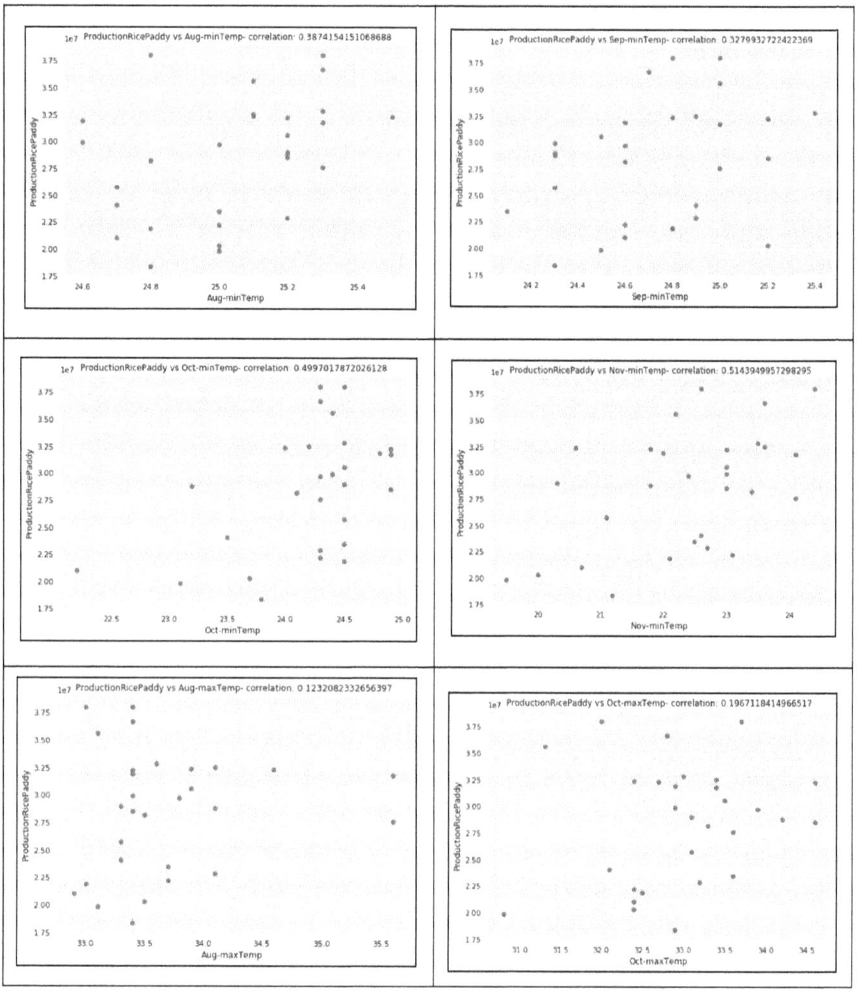

6.4.15 Step 13: Rice Main Season (August to October) Model

Construct the model for all months. But the explainability of model is applicable from August to October for the main rice production months:

```
X5 =df_impframe[[ 'Aug-prcp', 'Sep-prcp', 'Oct-prcp',
    'Aug-avgtemp', 'Sep-avgtemp', 'Oct-avgtemp',
```

(continued)

```
    'Aug-minTemp', 'Sep-minTemp', 'Oct-minTemp',
    'Aug-maxTemp', 'Sep-maxTemp', 'Oct-maxTemp', 'Area harvested
RicePaddy']].values
X5T =df_impframe[[ 'Aug-prcp', 'Sep-prcp', 'Oct-prcp',
    'Aug-avgtemp', 'Sep-avgtemp', 'Oct-avgtemp',
    'Aug-minTemp', 'Sep-minTemp', 'Oct-minTemp',
    'Aug-maxTemp', 'Sep-maxTemp', 'Oct-maxTemp', 'Area harvested
RicePaddy']]
```

Construct regression model:

```
y5 =df_impframe['ProductionRicePaddy'].values

X_train5,X_test5, y_train5 , y_test5 = train_test_split(X5 ,y5,
test_size = 0.2,random_state=0 )
regressor5 = LinearRegression()
regressor5.fit(X_train5,y_train5)
```

Predict the model:

```
# predicting the test
y_pred5 = regressor5.predict(X_test5)
y_pred5
```

Output:

array([21100378.7975116, 41254192.4139603, 33936637.02672544, 27857211.48707293, 25297402.21028363, 32241137.7763059])

```
pred_y_df5 = pd.DataFrame({'Actual Value':y_test5, 'Predicted
value':y_pred5,'Difference': y_test5-y_pred5})
pred_y_df5
```

Output:

	Actual Value	Predicted value	Difference
0	18447260.0	2.110038e+07	-2.653119e+06
1	38102720.0	4.125419e+07	-3.151472e+06
2	32620160.0	3.393664e+07	-1.316477e+06
3	28873975.0	2.785721e+07	1.016764e+06
4	28321137.0	2.529740e+07	3.023735e+06
5	31857000.0	3.224114e+07	-3.841378e+05

Please find regression coefficients and R^2 values:

```
regressor5.coef_
r2_score(y_test5,y_pred5)
```

Output:

```
array([ 7.15486025e+03, -3.87055186e+02, -9.36149073e+03,
-2.24649855e+06,
    4.97484127e+06, -2.71437141e+06, -3.37301811e+06,
-1.37570610e+06,
    1.36151848e+06, 2.24808738e+06, -2.39923333e+06,
1.47605859e+06,
    6.91725140e+00])
0.9321516
```

93.21% of the model explains the rice production.

Let's derive the model governing equation that connects independent and dependent variables:

Y = mx + c

Construct MX part of the equation:

```
mx=""
for ifeature in range(len(X5.columns)):
  if regressor5.coef_[ifeature] <0:
    # format & beautify the equation
   mx += " - " + "{:.2f}".format(abs(regressor5.coef_[ifeature]))
```

(continued)

```
+ " * " + X5.columns[ifeature]
  else:
    if ifeature == 0:
      mx += "{:.2f}".format(regressor5.coef_[ifeature]) + " * " +
X5.columns[ifeature]
    else:
    mx += " + " + "{:.2f}".format(regressor5.coef_[ifeature]) + " *
" + X5.columns[ifeature]

print(mx)
```

Output:

```
7005.45 * Aug-prcp - 2268.66 * Sep-prcp - 4687.05 * Oct-prcp -
1877460.25 * Aug-minTemp + 1082500.91 * Sep-minTemp - 137420.29 *
Oct-minTemp + 1210260.26 * Aug-maxTemp - 1110296.36 * Sep-maxTemp +
749749.23 * Oct-maxTemp + 6.20 * Area harvested RicePaddy
```

Complete the equation:

```
# y=mx+c
if(regressor5.intercept_ <0):
  print("The formula for the " + y5.name + " linear regression line
is = " + " - {:.2f}".format(abs(regressor5.intercept_)) + " + " + mx
)
else:
  print("The formula for the " + y5.name + " linear regression line
is = " + " {:.2f}".format(regressor5.intercept_) + " + " + mx )
```

Output:

$$\begin{aligned}\text{The formula for the ProductionRicePaddy linear regression line is}\\ = -39630641.34 + 7005.45^{*}\ \text{Aug} - \text{prcp} - 2268.66^{*}\ \text{Sep}\\ - \text{prcp} - 4687.05^{*}\ \text{Oct} - \text{prcp} - 1877460.25^{*}\ \text{Aug} - \text{minTemp}\\ + 1082500.91^{*}\ \text{Sep} - \text{minTemp} - 137420.29^{*}\ \text{Oct} - \text{minTemp}\\ + 1210260.26^{*}\ \text{Aug} - \text{maxTemp} - 1110296.36^{*}\ \text{Sep}\\ - \text{maxTemp} + 749749.23^{*}\ \text{Oct} - \text{maxTemp}\\ + 6.20^{*}\ \text{Area harvested RicePaddy}\end{aligned} \tag{6.1}$$

Equation (6.1) provides relationship between dependent variable "ProductionRicePaddy" and independent variables. We will use the equation as part of linkage models.

Finally, let's see explainability of the model:

6.4.16 Step 14: Model Explainability

To identify the important feature attributes of the model, the following code provides explainability—return an explanation of estimator parameters (weights)[118]:

```
eli5.show_weights(regressor5)
```

Output:

y top features

Weight?	Feature
+4974841.271	x4
+2248087.376	x9
+1476058.587	x11
+1361518.479	x8
+7154.860	x0
+6.917	x12
-387.055	x1
-9361.491	x2
-945022.834	<BIAS>
-1375706.100	x7
-2246498.554	x3
-2399233.333	x10
-2714371.408	x5
-3373018.105	x6

Here Y, rice paddy production:

```
eli5.show_weights(regressor5, feature_names =list(X5T.
columns), target_names = "Thailand_rice production" )
```

Output:

Weight?	Feature
+4974841.271	Sep-avgtemp
+2248087.376	Aug-maxTemp
+1476058.587	Oct-maxTemp
+1361518.479	Oct-minTemp
+7154.860	Aug-prcp
+6.917	Area harvested RicePaddy
-387.055	Sep-prcp
-9361.491	Oct-prcp
-945022.834	<BIAS>
-1375706.100	Sep-minTemp
-2246498.554	Aug-avgtemp
-2399233.333	Sep-maxTemp
-2714371.408	Oct-avgtemp
-3373018.105	Aug-minTemp

As the explainability module shows, September average temperature, August maximum temperature, October maximum temperature, and October minimum

[118] ELI5—https://eli5.readthedocs.io/en/latest/autodocs/eli5.html

temperature are some of the top features that have a considerable influence on the model.

6.4.17 Step 14: CMIP6 Projections Data and SSP Scenarios

To model climate projections, see how the model predicts the future projections of rice paddy production for Thailand, Nakhon Sawan province. The goal is to collect for SSP1-2.6, SSP3-7.0, and SSP5-8.5.

Scenario	SSP model	Socioeconomic factors' impact
SSP1-2.6	SSP1	It imagines the same socioeconomic shifts toward sustainability as SSP1-1.9. But temperatures stabilize around 1.8 °C higher by the end of the century
SSP2-4.5	SSP2	Socioeconomic factors follow their historic trends, with no notable shifts. Progress toward sustainability is slow, with development and income growing unevenly. In this scenario, temperatures rise 2.7 °C by the end of the century
SSP3-7.0	SSP3	By the end of the century, average temperatures have risen by 3.6 °C
SSP5-8.5	SSP5	By 2100, the average global temperature is a scorching 4.4 °C higher

Here are the model data simulations:

Collection, CMIP6; aggregation, monthly; calculation, climatology mean; percentile, 90th; and model, Multi-Model Ensemble. Please find the table with all three main variables for scenario SSP5-8.5. The same data are collected for additional two scenarios: SSP1-2.6 and SSP3-7.0.

Minimum temperature	Maximum temperature	Precipitation
Time period: 2020–2039 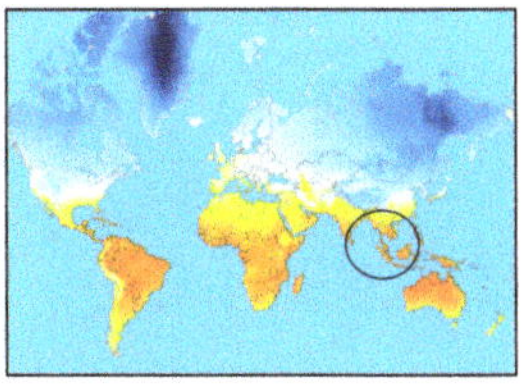	Time period: 2020–2039 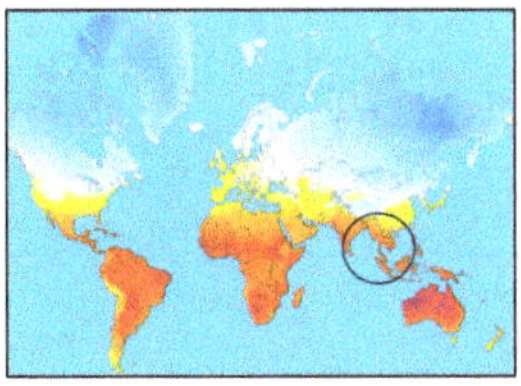	Time period: 2020–2039 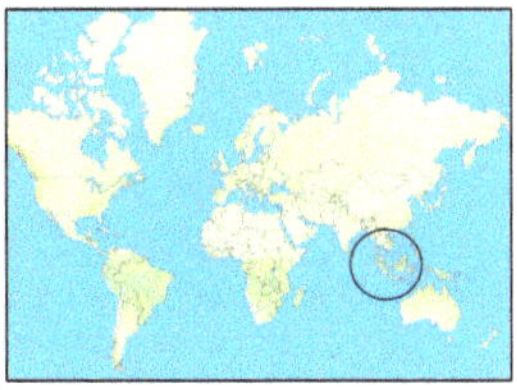
Time period: 2040–2059 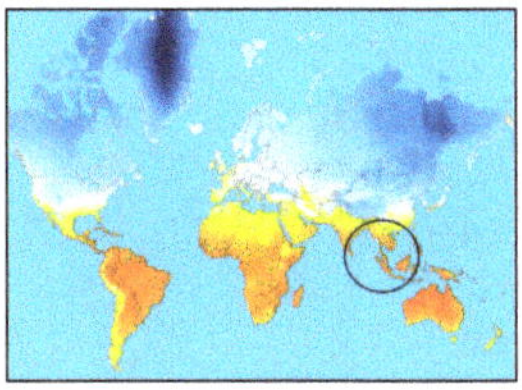	Time period: 2040–2059 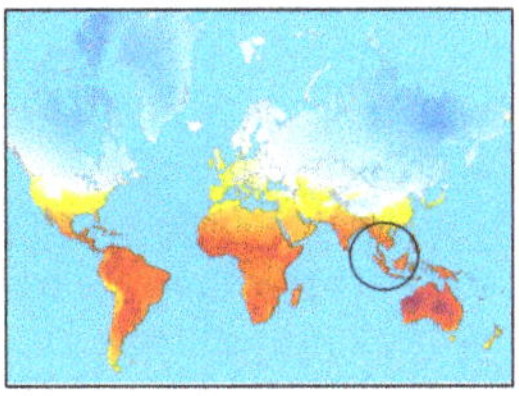	Time period: 2040–2059 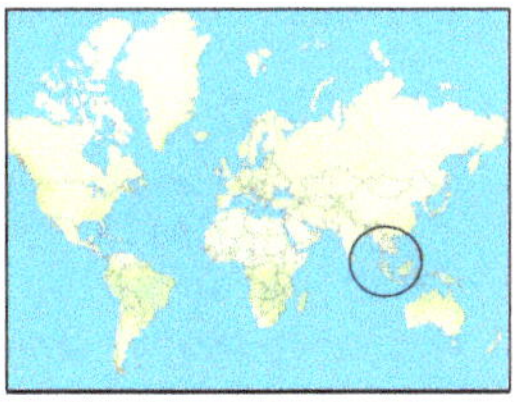

(continued)

Minimum temperature	Maximum temperature	Precipitation
Time period: 2060–2079 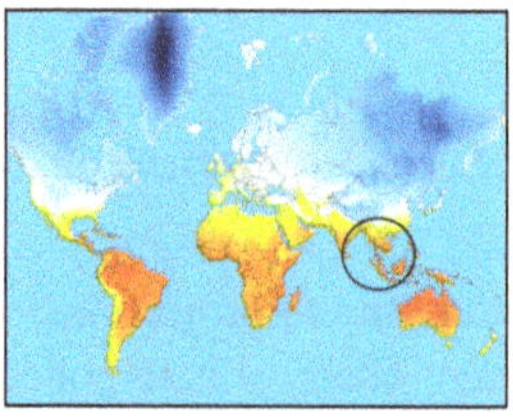	Time period: 2060–2079 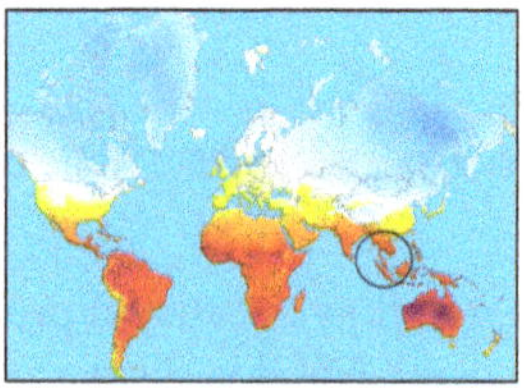	Time period: 2060–2079 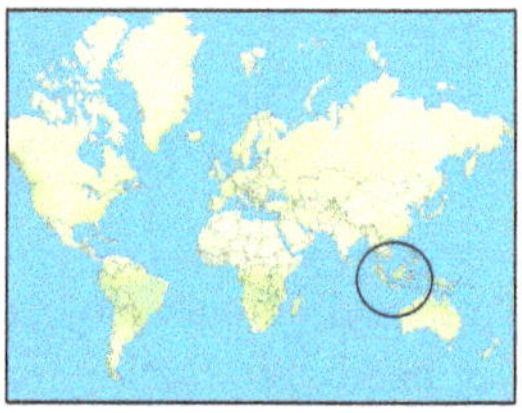
Time period: 2080–2099 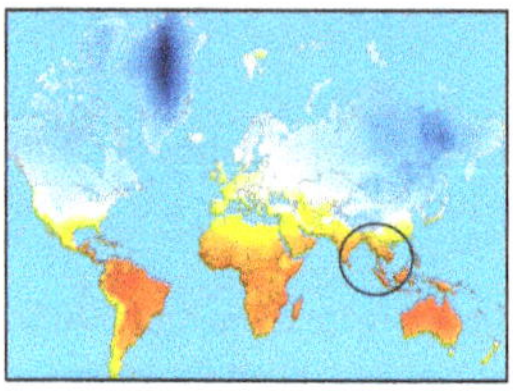	Time period: 2080–2099 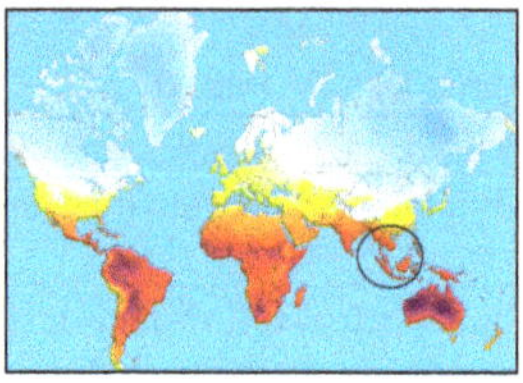	Time period: 2080–2099 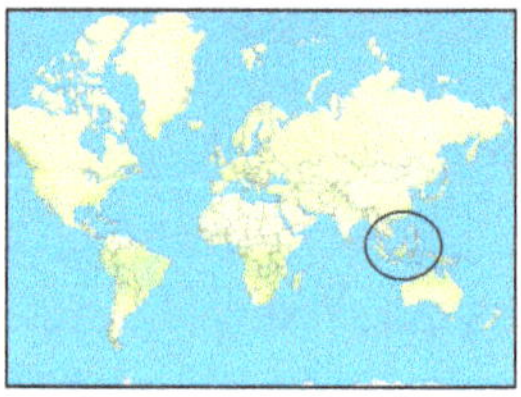

Please find model invocation called scenario SSP5-8.5:

```
cmip6_20202039_climatologymean_90th_ssp5_8p5_
multimodelensembel =np.array
([ 423.24,446.74,247.57,29.27,29.11,28.74,
25.73,25.28,24.28,33.4,
33.42,33.32,9717975.0])
model5.predict([cmip6_20202039_climatologymean_90th_
ssp5_8p5_multimodelensembel])

cmip6_20402059_climatologymean_90th_ssp5_8p5_
multimodelensembel =np.array
([ 415,441.54,253.58,30.15,30.04,29.67,26.47,
26.34,25.32,34.14,34.18,34.23,9717975.0])
model5.predict([cmip6_20402059_climatologymean_90th_
ssp5_8p5_multimodelensembel])

cmip6_20602079_climatologymean_90th_ssp5_8p5_
multimodelensembel =np.array
([ 429.47,466.05,262.2,31.29,31.09,31.05,
27.61,27.51,26.89,35.13,35.2,35.53,9717975.0])
model5.predict([cmip6_20602079_climatologymean_90th_ssp5_
8p5_multimodelensembel])

cmip6_20802099_climatologymean_90th_ssp5_8p5_
multimodelensembel =np.array
([ 479.59,481.84,278.92,32.55,32.56,32.64,28.91,
28.91,28.49,36.57,36.57,37.03,9717975.0])
model5.predict([cmip6_20802099_climatologymean_
90th_ssp5_8p5_multimodelensembel])
```

Output:

Baseline, Thailand rice production for 2015; production, 27,702,191.0 tons			
20,202,039	20,402,059	20,602,079	20,802,099
23,443,996.63	22,101,186.95	19,411,309.34	17,805,754.69
Drop in production from 2015 levels, 27,702,191.0			
−15.37%	−20.21%	−29.92%	−35.72%

SSP5-8.5 is the most outcome we don't want to allow for its climate change implications. A drop of 35% in 2080–2099 can be seen, resulting in a total drop of 9,896,436.4 tons. Soon, 2020–2039, a drop of 4,258,194.4 tons could be due to climate change. Climate change may affect the production of rice as early as 2030, and a similar finding can be observed for maize (corn) due to a high greenhouse gas emission scenario, according to a new NASA study published in the journal[119], *Nature Food*. Maize crop yields are projected to decline 24% [24].

SSP1-1.9:

```
cmip6_20202039_climatologymean_90th_ssp119_8p5_
multimodelensembel =np.array
([ 410.56,416.86,200.63,28.9,28.67,28.22,25.63,
25.11,23.85,32.61,32.45,32.91,9717975.0])
model5.predict([cmip6_20202039_climatologymean_
90th_ssp119_8p5_multimodelensembel])

cmip6_20402059_climatologymean_90th_ssp119_8p5_
multimodelensembel =np.array
([ 462.32,452.93,196.15,29.11,28.82,28.27,25.62,25.41,
23.98,32.6,32.62,33.27,9717975.0])
model5.predict([cmip6_20402059_climatologymean_90th_
ssp119_8p5_multimodelensembel])

cmip6_20602079_climatologymean_90th_ssp119_8p5_
multimodelensembel =np.array
([416.19,419.36,175.27,29.03,28.81,28.2,25.64,25.23,
23.89,32.63,32.57,33.27,9717975.0])
model5.predict([cmip6_20602079_climatologymean_90th_
ssp119_8p5_multimodelensembel])

cmip6_20802099_climatologymean_90th_ssp119_8p5_
multimodelensembel =np.array
([447.55,443.2,174.88,29.02,28.79,28.25,25.58,25.05,
23.73,32.58,32.47,33.19,9717975.0])
model5.predict([cmip6_20802099_climatologymean_
90th_ssp119_8p5_multimodelensembel])
```

[119]Global Climate Change Impact on Crops Expected Within 10 Years, NASA Study Finds—https://climate.nasa.gov/news/3124/global-climate-change-impact-on-crops-expected-within-10-years-nasa-study-finds/

Baseline, Thailand rice production for 2015; production, 27,702,191.0			
20,202,039	20,402,059	20,602,079	20,802,099
23,789,816.21	24,225,919.41	24,669,338.907	24,916,988.005
Drop in production from 2015 levels, 27,702,191.0			
−14.12%	−12.54%	−12.54%	−10.05%

SSP1-1.9 is the desirable outcome that we would like to see to happen. A drop of 14.12% in 2020–2039 can be seen. This would result in a drop of 3,912,374.8 (3.9 million) tons, and by 2080–2099, a drop of 2,785,203 (2.7 million) tons could be due to climate change.

SSP3-7.0:

- Collection: CMIP6
- Aggregation: Monthly
- Calculation: Climatology mean
- Percentile: 90th
- Model: Multi-Model Ensemble

Please find the table with all three main variables for scenario SSP3-7.0.

Minimum temperature	Maximum temperature	Precipitation
Time period: 2020–2039 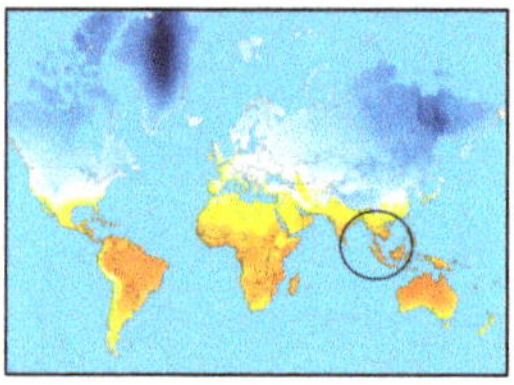	Time period: 2020–2039 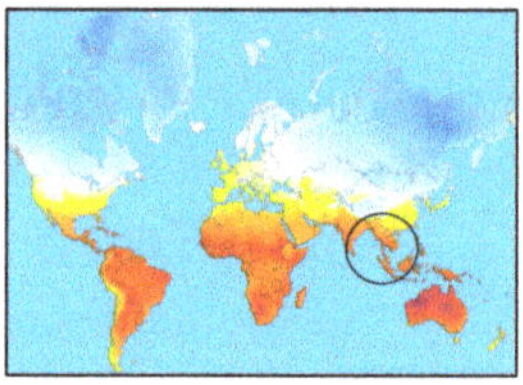	Time period: 2020–2039 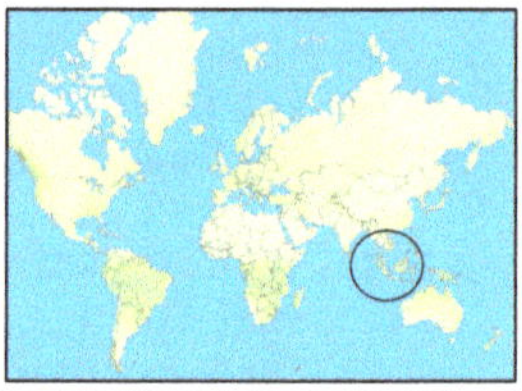
Time period: 2040–2059 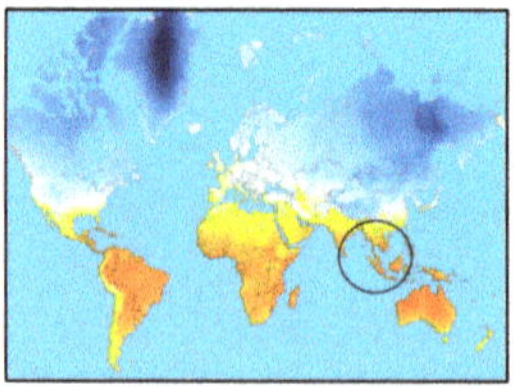	Time period: 2040–2059 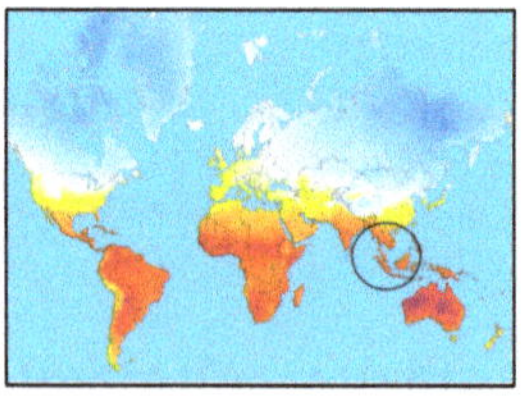	Time period: 2040–2059 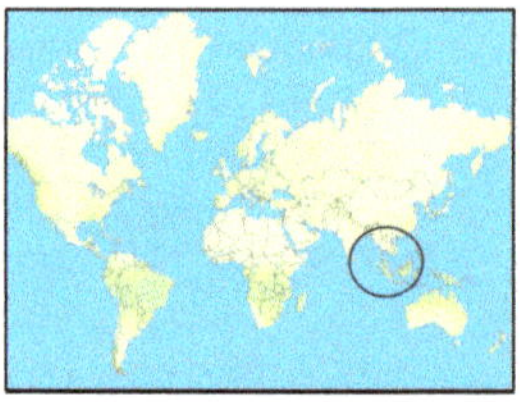

(continued)

Minimum temperature	Maximum temperature	Precipitation
Time period: 2060–2079 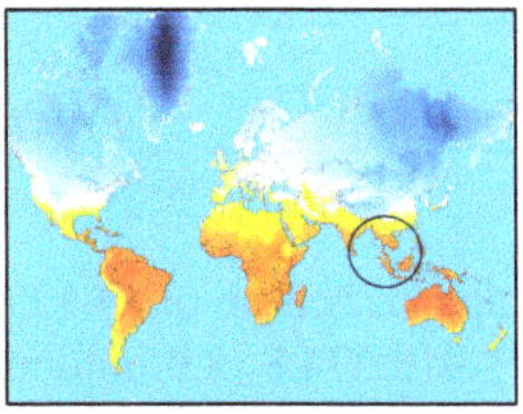	Time period: 2060–2079 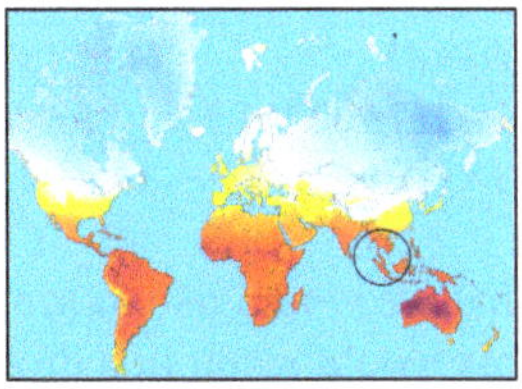	Time period: 2060–2079 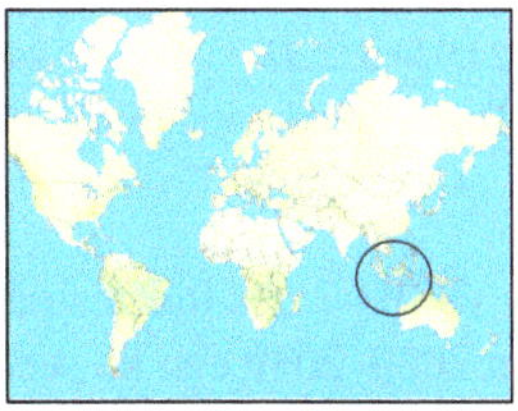
Time period: 2080–2099 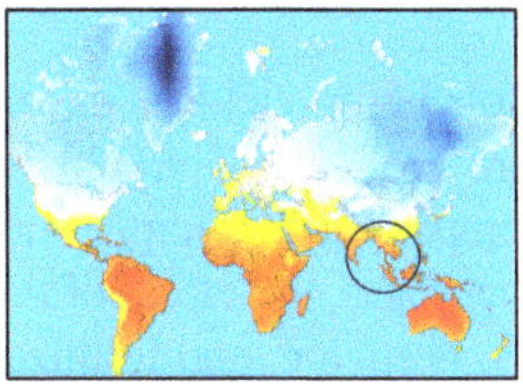	Time period: 2080–2099 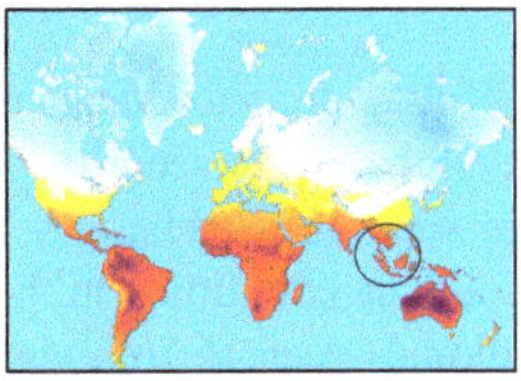	Time period: 2080–2099 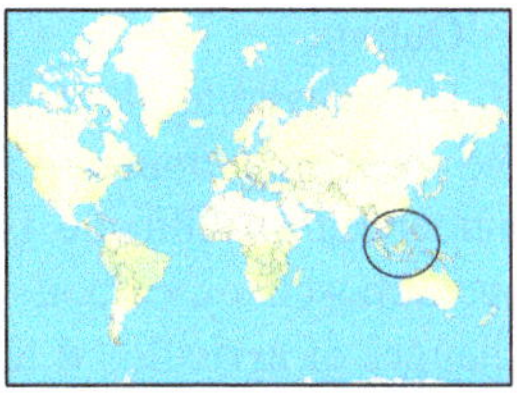

```
cmip6_20202039_climatologymean_90th_ssp3_7p0_
multimodelensembel =np.array
([417.03,441.89,235.49,29.14,29.02,28.55,
25.34,25.17,24.01,32.97,33.05,33.34,9717975.0])
model5.predict([cmip6_20202039_climatologymean_90th_ssp3_
7p0_multimodelensembel])

cmip6_20402059_climatologymean_90th_ssp3_7p0_
multimodelensembel =np.array
([462.32,452.93,196.15,29.11,28.82,28.27,25.62,
25.41,23.98,32.6,32.62,33.27,9717975.0])
model5.predict([cmip6_20402059_climatologymean_90th_
ssp3_7p0_multimodelensembel])

cmip6_20602079_climatologymean_90th_ssp3_7p0_
multimodelensembel =np.array
([416.19,419.36,175.27,29.03,28.81,28.2,25.64,
25.23,23.89,32.63,32.57,33.27,9717975.0])
model5.predict([cmip6_20602079_climatologymean_
90th_ssp3_7p0_multimodelensembel])

cmip6_20802099_climatologymean_90th_ssp3_7p0_
multimodelensembel =np.array
([447.55,443.2,174.88,29.02,28.79,28.25,
25.58,25.05,23.73,32.58,32.47,33.19,9717975.0])
model5.predict([cmip6_20802099_climatologymean_90th_
ssp3_7p0_multimodelensembel])
```

Baseline, Thailand rice production for 2015; production, 27,702,191.0			
20,202,039	20,402,059	20,602,079	20,802,099
24,924,323.32	24,225,919.41	24,669,338.907	24,916,988.00
Drop in production from 2015 levels, 27,702,191.0			
-14.12%	-12.54%	-12.54%	-10.05%

SSP3-7.0 is the desirable that we would like to see to happen. A drop of 14.12% in 2020–2039 can be seen. This would result in a drop of 3,912,374.8 (3.9 million) tons, and by 2080–2099, a drop of 2,785,203 (2.7 million) tons could be due to climate change.

Output:

Understanding the effect of future climate change on global crop yields is one of the most important tasks for global food security. Future crop yields would be influenced by climatic factors such as the changes of temperature, precipitation, and atmospheric carbon dioxide concentration. On the other hand, the effect of the changes of agricultural technologies such as crop varieties, pesticide, and fertilizer input on crop yields has large uncertainty. However, not much is available on the contribution ratio of each factor under the future climate change scenario.

We estimated the future global yields of rice crops under three Shared Socioeconomic Pathways (SSPs) and four Representative Concentration Pathways (RCPs). For this purpose, firstly, we estimated a parameter of a process-based model using a Linear Regression and Ensemble Voting method for each 1.125° spatial grid. The model parameter is relevant to agricultural yield technology. Then, we analyzed the relationship between the values of technological parameter and GDP values (as linkage models, in the next chapters). We found that the estimated values of the agricultural parameters were positively correlated with the GDP. Using the estimated relationship, we predicted future crop yield during 2020 and 2100 under SSP1, SSP2, and SSP3 scenarios and RCP 2.6, 4.5, 6.0, and 8.5. The estimated crop yields were different among SSP scenarios. However, we found that the yield difference attributable to SSPs 1 and 3 was smaller than those attributable to SSP5 and climate change. Particularly, SSP5 has a exhibited substantial decline in overall rice production (please see Fig. 6.29). As it can be seen from the model, climate change would impact the production of rice paddy in Thailand. Client-smart technologies and governmental policies toward climate change are needed to assure global food security[120]. Similar models have conclusions that we've observed in our

[120]Thailand—https://ccpi.org/country/tha/

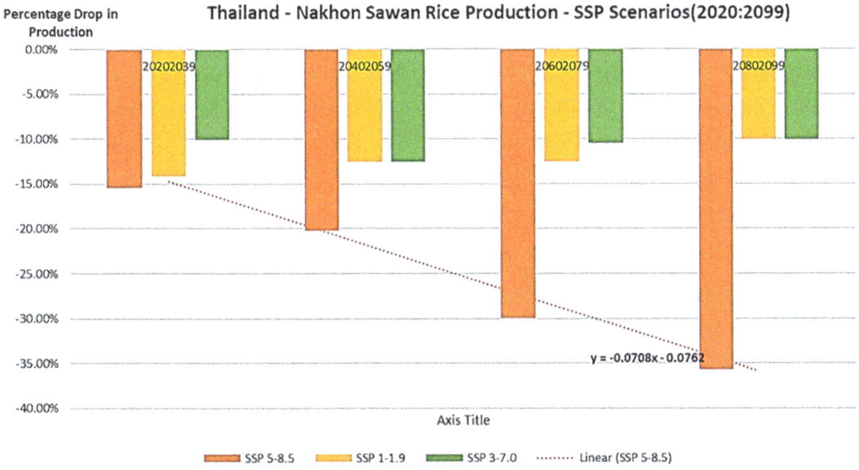

Fig. 6.29 Summary: Thailand Nakhon Sawan—SSPs 2020–2099

model[121]—NASA study[122]: projected decrease in production capacities for rice, corn, and soybeans [24, 48].

6.5 Exposure: Temperature, Floods, and Sea-Level Rise

Food commodities and climate[123] have a special and complicated relationship [49]. The effects of climate change—rising temperatures, less predictable rainfall, wild swings from drought to flooding, new pests, and more—were making it more and more difficult to earn a living from coffee, an experience felt by farmers around the world. Various organizations and companies are seeking solutions to these challenges. They are helping farmers improve production and efficiency, developing new strains of beans, farming wild species, and even growing coffee in labs. Producing coffee makes a significant environmental impact—estimates vary, but about 39 gallons of water are needed for one cup, according to UNESCO's Institute for Water Education.

[121]Future possible crop yield scenarios under multiple SSP and RCP scenarios.—https://ui.adsabs.harvard.edu/abs/2016AGUFMGC54A..07S/abstract

[122]Climate change expected to impact the world's wheat and corn crops by 2030, NASA says—https://www.usatoday.com/story/news/world/2021/11/04/climate-change-impacts-corn-soybeans-rice-crops/6257778001/

[123]Coffee and Climate Have a Complicated Relationship—https://www.nytimes.com/2021/10/31/business/coffee-climate-change.html#:~:text=According%20to%20a%202014%20study,and%20Vietnam%2C%20major%20producing%20countries

Your morning coffee is in danger[124] [50]. To the billions of people around the world who rely on drinking coffee (to put it mildly), that forebodes many difficult mornings and possibly rising prices. To the 100 million or so coffee farmers, to say nothing of the tens of millions more who work in transporting, packaging, distributing, selling, and brewing coffee, the effects of climate change are making an already precarious existence even more so. Climate change[125] is the main driver of lower coffee production and price spikes across the world [51].

Coffee Prices Jump to 6-Year High as Brazilian Frost Threatens Crop[126] [52].

Arabica bean prices have leapt almost a third in July in the latest instance of extreme weather sending jitters through commodity markets.

The worst frost to strike Brazil's coffee-growing region in more than 25 years is set to cut a chunk out of the next year's crop, sending prices of the bean to 6-year highs on global markets.

The cold snap is the second weather shock in recent months to strike farmers in Brazil, the world's biggest coffee producer, threatening to drive up costs at cafes and breakfast tables around the world. Before the frost came, a drought parched the 2021 crop.

Minas Gerais is Brazil's most important coffee-producing state, and the Sul de Minas region produces more coffee than any other part of the state. Temperatures in Sul de Minas fell as low as 23 degrees Fahrenheit in some towns, and the region experienced the coldest weather since 1994, according to Annette Fernandes.

The way the weather has jolted coffee prices demonstrates the risks of the world's growing reliance on Brazil as a source of beans, according to Mr. Copestake. Low-cost Brazilian farms glutted the global market in recent years, pushing less efficient farmers in countries such as Honduras to switch crops or leave their plantations, he said. That leaves buyers in the United States of America and elsewhere vulnerable to events that knock output in Brazil and Vietnam, the biggest source of bitter-tasting robusta beans.

"Everyone in agricultural commodities is a little bit worried about climate change," said Carlos Mera, senior analyst at Rabobank, a major lender to the food and agriculture industries.

---

[124] Your Morning Cup of Coffee Is in Danger. Can the Industry Adapt in Time?—https://time.com/5318245/coffee-industry-climate-change/

[125] A review of evidence of recent climate change in the Central Highlands of Vietnam—https://www.researchgate.net/publication/293109260_A_review_of_evidence_of_recent_climate_change_in_the_Central_Highlands_of_Vietnam

[126] Coffee Prices Jump to Six-Year High as Brazilian Frost Threatens Crop—https://www.wsj.com/articles/coffee-prices-jump-to-six-year-high-as-brazilian-frost-threatens-crop-11627380128?mod=article_inline

Finally, Starbucks, a giant coffee company, has recommended to cut your morning coffee's carbon footprint, skip the Frappuccino, and take a plain black espresso[127]. Adding whipped cream to millions of Starbucks drinks emits 50 times as much greenhouse gas as the company's private jet. Overall, dairy products are the biggest source of carbon dioxide emissions across the coffee giant's operations and supply chain [53].

6.6 Coffee

Coffee originated as a wild crop in Ethiopia and today expanded to become a globally traded[128] and consumed commodity [54]. Nowadays, coffee is cultivated in approximately 70 countries[129] throughout the tropics, in a region referred to as "The Bean Belt." Brazil is leading the pack with 2.64 million metric tons, followed by Vietnam (1.65 million metric tons), Colombia (810K metric tons), Indonesia (660K metric tons), Ethiopia (384K metric tons), and India (348K metric tons). Most coffee (please see Fig. 6.30) is produced on relatively small farms in areas with rugged terrain. Seventy percent of coffee is produced by smallholder farmers, who depend on coffee production for their livelihoods, a same characteristic that would see for milk, rice, wheat, and other essential commodities. Coffee cultivation as such as other cultivation in the tropic area is particularly under threat due to climate change.

Two major commercial varieties, coffee arabica and coffee robusta, dominate the coffee market. Arabica beans comprise approximately 70% of the global market and grow in some specific environmental conditions such as high altitude, mild temperatures, and high rainfall. In fact, the ideal condition for high quality arabica coffee cultivation is:

- Altitude, 900–2000 m
- Rainfall annually, 1500 and 2500 mm
- Temperature, 15–25 °C

Climate change adversely affects coffee production, yield, and affordability/price. A recent spike in coffee prices attributed to unfavorable weather that clipped production in Brazil, the world's largest grower and exporter[130]. Stockpiles of high-

[127] Starbucks has a long way to go to reach environmental goals for 2030—https://fortune.com/2020/01/21/starbucks-carbon-footprint-dairy/

[128] Suitability Study for Coffee Arabica cultivation in relation to climate change.—https://medium.com/@damiano.ciro/suitability-study-for-coffee-arabica-cultivation-in-relation-to-climate-change-fecab562ed6

[129] List of countries by coffee production—https://en.wikipedia.org/wiki/List_of_countries_by_coffee_production

[130] Coffee Reserves Plunge to Lowest in More Than Two Decades—https://finance.yahoo.com/news/world-coffee-reserves-plunge-lowest-193610832.html

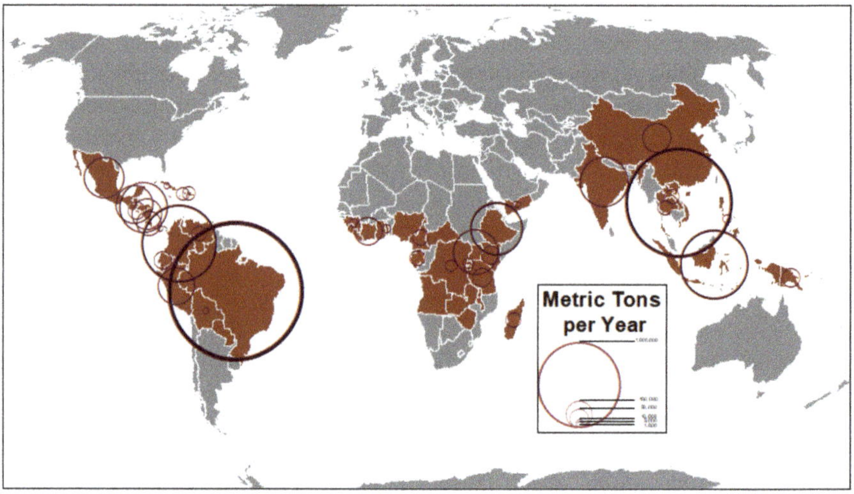

Fig. 6.30 Coffee worldwide

end arabica beans, a favorite of artisan coffee shops and chains like Starbucks Corp., totaled 1.078 million bags or about 143 million pounds, according to data released on Monday, 08 February 2022, by ICE Futures US exchange. That's the lowest level for inventories monitored by the New York exchange since February 2000. Coffee reserves certified by ICE have been falling since September 2021 due to soaring shipping costs and unfavorable weather that clipped production in Brazil, the world's largest grower and exporter. Let's model the impact of climate change on the coffee production in Vietnam [55].

Vietnam coffee production	

6.7 Machine Learning Model: Vietnam Coffee Yields and Climate Change

Coffee is one of the most widely consumed hot beverages all over the world. Brazil, the top coffee-producing country, accounted for 40% of the global coffee supply. Vietnam is a Southeast Asian nation with an extensive coastline and diverse but generally warm climate including temperate and tropical regions. In 2019 Vietnam's

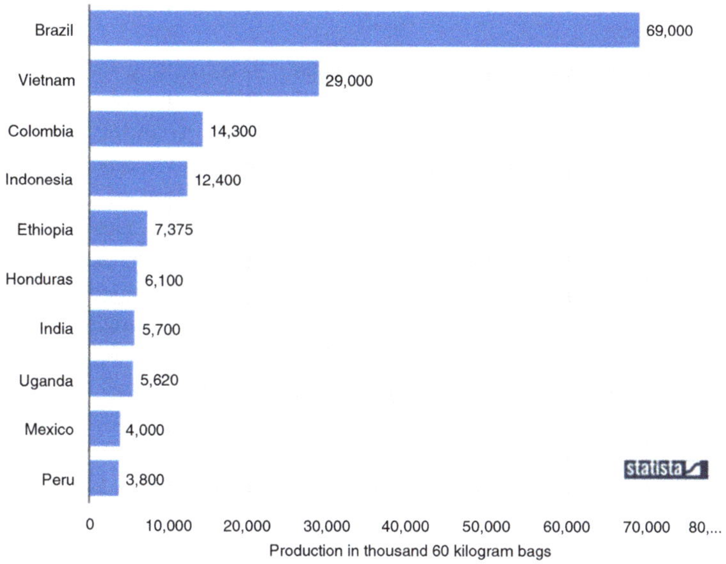

Fig. 6.31 Coffee production

population was estimated at 96.4 million, approximately one-third of whom live in the metropolitan areas of its two mega-cities, Hanoi and Ho Chi Minh City. The relative contribution of agriculture, forestry, and fishing to the country's economy has declined in recent years due to the rapid growth of the industry and service sectors; as of 2017 the agricultural sector contributed 15.3% of gross domestic product, and this is somewhat mismatched against an employment contribution of around 40.3% of the country's labor force.

Rice and coffee productions have a particularly vital role for the country in terms of food security, rural employment, and foreign exchange, employing two-thirds of the rural labor force and positioning Vietnam as consistently one of the world's largest rice exporters (please see Fig. 6.31). Given that a high proportion of the country's population and economic assets (including irrigated agriculture) are in coastal lowlands and deltas and rural areas face issues of poverty and deprivation, Vietnam has been *ranked as among the five countries likely to be most affected by climate change*. It has been estimated that climate change will reduce national income by up to 3.5% by 2050[131]. Climate variability and change are threatening the country's coffee crops. Rising temperatures and extreme weather have subjected Vietnamese coffee farmers to increasing uncertainties: longer droughts, more

[131] Vietnam—https://climateknowledgeportal.worldbank.org/country/vietnam

frequent floods, and severe outbreaks of pests and diseases that result in reduced productivity[132] [56].

Vietnam was the second largest coffee producer[133], accounting for roughly 20% of the world coffee production [57]. Coffee played a major role in Vietnam's stunning economic transformation from the extreme poverty in the late 1980s to the relative prosperity in the 2010s. Vietnam secured as the second most coffee-producing country in the world with annual production [57] of 30,850,000[134] (in 60 kg bags). Coffee production then grew by 20%–30% every year in the 1990s. The industry now employs about 2.6 million people, with beans grown on half a million smallholdings of two to three acres each. This has helped transform the Vietnamese economy. In 1994 some 60% of Vietnamese lived under the poverty line, now <10% do[135] [58].

Common arabica varieties include Catimor, with some Typica and Bourbon. Key regions in Vietnam producing coffee include Central Highlands (*Dak Lak*, *Gia Lai*, *Dak Nong*, *Lam Dong*, and *Kontum*), | North Vietnam (Son La, Dien Bien provinces). The arabica coffee grown in Vietnam is in the regions of Da Lat, Dien Bien, Nghe An, Son La, and Quang Tri, which range in maximum altitude from 1000 to 1400 measured in meters above sea level (m.a.s.l.). Arabica is better suited to the mountainous areas at high altitudes with lower temperatures—*between 20 °C and 22 °C*—and an annual rainfall of *1,300mm to 1,900mm*[136] [59]. Please find the map of Vietnam[137] (see Fig. 6.32).

6.7.1 Key Climate Change Effects

Generally, months of June, July, and August are very important for growing coffee beans. For instance, 2021–2022 coffee harvest in Vietnam suffered a negative impact from *unseasonal dryness* in *June and July* causing reduced yields on trees

[132] Helping Vietnam's Coffee Sector Become More Climate Resilient—https://news.climate.columbia.edu/2020/11/13/vietnam-coffee-climate-resilient/

[133] Coffee production worldwide in 2020, by leading country (in 1,000 60 kilogram bags)*—https://www.statista.com/statistics/277137/world-coffee-production-by-leading-countries/

[134] Vietnam Coffee—https://sucafina.com/na/origins/vietnam#:~:text=Coffee%20Production%20Today,tons%20per%20hectare%20in%20Brazil.

[135] How Vietnam became a coffee giant—https://www.bbc.com/news/magazine-25811724

[136] A breakdown of Vietnamese coffee-producing regions—https://perfectdailygrind.com/2021/12/a-breakdown-of-vietnamese-coffee-producing-regions/#:~:text=As%20mentioned%20previously%2C%20the%20bulk,Lam%20Dong%2C%20and%20Kontum.%E2%80%9D

[137] The CDC—https://wwwnc.cdc.gov/travel/destinations/traveler/none/vietnam

Fig. 6.32 Vietnam

in the key southern producing region known as the Central Highlands[138]. As a result of the weather damage, more severe than expected, the harvest could be 15% smaller [60].

The impact of climate change is severe for Vietnam—for instance, Vietnam Coffee and Cocoa Association (VICOFA) Chairman Luong Van Tu said that the negative impact of *climate change caused disturbing weather patterns for the 2021–2022 Vietnamese coffee cycle which resulted in yield losses as fruit fell to the ground before completing the maturity.*

The month of November plays an important role for Central Highlands. Heavy rains during November could result in huge yield losses such as ripening fruit (please see Fig. 6.33) to fall to the ground and rotten. For instance, during 2021–2022, "Vietnam's coffee harvest starting next November may be reduced by 10–15% due to the heavy rain, which caused ripening fruit to fall to the ground and rotten," said Tu in an interview published on 30 Aug. 2022. When *the bean formation process is*

138 BREAKING: Vietnam's 2021-22 Coffee Harvest Down on Dry Weather- https://stir-tea-coffee.com/tea-coffee-news/breaking-vietnam%E2%80%99s-2021-22-coffee-harvest-down-on-dry-weathe/

Fig. 6.33 Coffee plants

halted by lack of either rain or extreme dryness, this typically also results in producing cherries with a reduced bean size.

6.7.2 Coffee Agronomy

Unusual dry weather was reported across the key southern growing region during the months of June and July which normally is well into the tropical rainy season, according to exporters and traders based on the Vietnamese coffee capital of Buon Ma Thuot in the Central Highlands, where over 70% of Vietnam's coffee is grown. The lack of stable and sufficient rainfall prevented Vietnamese *growers from applying the second round of fertilizer in a timely manner* that in a typical crop cycle will provide trees with a welcome boost to enhance yields, but making matters worse, it occurred at the time of what in coffee agronomy is known as the *crucial bean formation period.* Without sufficient rain and limited fertilizer, this did not allow for beans within cherries to finish their development. The dryness as a "rather strange" weather phenomenon, Vietnam's Simexco Daklak Ltd. coffee exporters said even though "good rains" registered in most parts of the Central Highlands during August, fertilizer prices almost double the rate paid by farmers last year and the need for rainfall to start before any fertilizer is applied led to many farmers to forgo the second round of tree stimulus or not apply it in full. Please see the impact of 2050 on Vietnam coffee production[139] in Fig. 6.34.

[139]Coffee Production in the Face of Climate Change—https://www.sustaincoffee.org/assets/resources/Vietnam_CountryProfile_Climate_Coffee_6-11.pdf

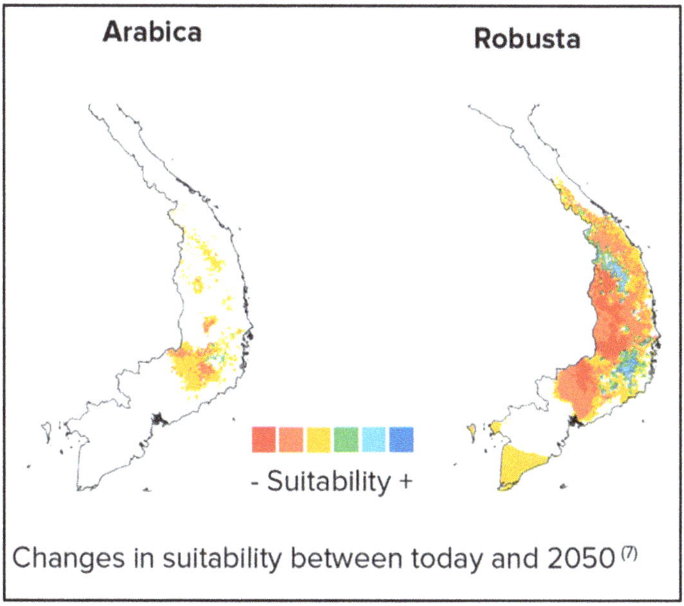

Fig. 6.34 Changes in suitability between 2020 and 2050

Vietnam – Coffee Production – Central Highlands											
Jan	Feb	Mar	Apr	May	Jun	Jul	Aug	Sep	Oct	Nov	Dec
					Crucial bean formation period		Second round of tree stimulus		Harvest		
					Climate Change: Dry weather results in		Climate Change: Heavy Rains or Dry Weather yield losses as fruit fell to the ground before completing the maturity				

Fig. 6.35 Vietnam coffee production—Central Highlands calendar

Harvest months (please see Fig. 6.35) Central Highlands: *November–January* | North Vietnam: *October–January*. Today, with a total production of about 30 million bags, around 95% of coffee grown in Vietnam is robusta. High-end coffee shops mainly buy arabica coffee beans, whereas Vietnam grows the hardier robusta bean. Arabica beans contain between 1% and 1.5% caffeine, while robusta has between 1.6% and 2.7% caffeine, making it taste more bitter. Vietnam has the highest yields globally with an output of 2.8 tons of coffee per hectares. This is a full ton higher than the second-highest yield of 1.4 tons per hectare in Brazil. Privately run farms compose 95% of coffee farms in Vietnam today. Estimates suggest that about 85% of total production area is cultivated by households. Of the area farmed by households, 63%—or approximately 650,000 families—are smallholders with less than one hectare[140]. The remaining 5% are state-run, but that number is gradually shrinking as the land is redistributed to small farmers. Some companies, like Nestle, have processing plants in Vietnam, which roast the beans and pack it [61].

World coffee exports amounted to 9.25 million bags in November 2021, compared with 10.56 million in November 2020. Exports in the first 2 months of coffee year 2021–2022 (October 2021 to November 2021) have decreased by 8.8% to 18.87 million bags compared to 20.69 million bags in the same period in 2020–2021[141]. Please find the map of Vietnam[142].

6.7.3 *Time Series for Terra MODIS-8-Day Vegetation Index*

There are many standard MODIS[143] data products that scientists are using to study global change. These products are being used by scientists from a variety of disciplines, including oceanography, biology, and atmospheric science. This section provides some detail for each product individually, introducing you to the products, explaining the science behind them, and alerting you to known areas of concern with the data products. Additional information about these products can be obtained by going to the appropriate URL's noted below. Please find Vegetation Health (VH) Index for Vietnam[144] and Province-Averaged VH data for CropLand data[145] (please see Table 6.7).

6.7.4 *Data Sources*

The following data sources are used:

- World Metrological Organization: Select a monthly time series (historical observations) under stations data

[140] Production, Consumption, and Food Security in Viet Nam Diagnostic Overview—https://inddex.nutrition.tufts.edu/sites/default/files/Vietnam%20Diagnostic%20Overview%20Sept%2023%5B1%5D.pdf

[141] International Coffee Organization—https://www.ico.org/

[142] Vietnam FAO Profile—https://www.fao.org/3/ap826e/ap826e.pdf

[143] MODIS—https://modis.gsfc.nasa.gov/data/dataprod/index.php#atmosphere

[144] Vietnam—https://www.star.nesdis.noaa.gov/smcd/emb/vci/VH/vh_adminMean.php?type=Province_Weekly_MeanPlot

[145] Province-Averaged VH data for CropLand Data—https://www.star.nesdis.noaa.gov/smcd/emb/vci/VH/get_TS_admin.php?provinceID=8&country=VNM&yearlyTag=Weekly&type=Mean&TagCropland=crop&year1=1982&year2=2022

Table 6.7 Vietnam VH

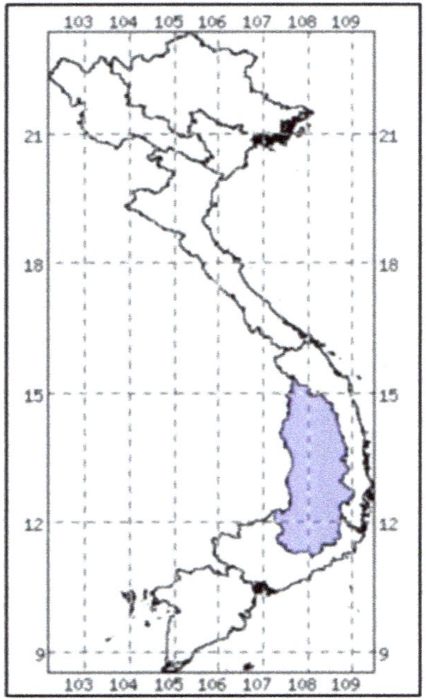

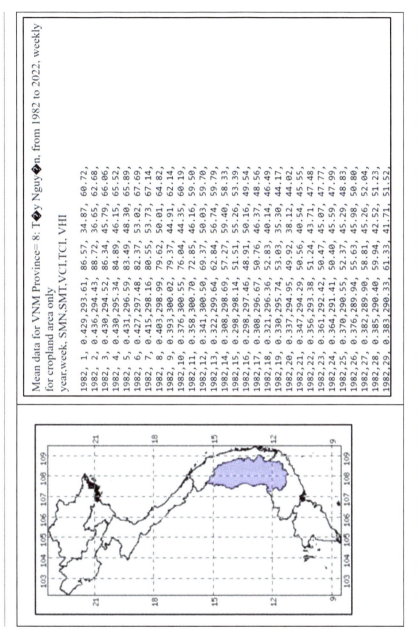

Mean data for VNM Province= 8: T�y Nguy�n, from 1982 to 2022, weekly for cropland area only

year,week, SMN,SMT,VCI,TCI, VHI

```
1982, 1, 0.429,293.61, 86.57, 34.87, 60.72,
1982, 2, 0.436,294.43, 88.72, 36.65, 62.68,
1982, 3, 0.430,294.52, 86.34, 45.79, 66.06,
1982, 4, 0.430,295.34, 84.89, 46.15, 65.52,
1982, 5, 0.431,296.59, 83.49, 48.30, 65.89,
1982, 6, 0.427,297.48, 82.37, 53.02, 67.69,
1982, 7, 0.415,298.16, 80.55, 53.73, 67.14,
1982, 8, 0.403,298.99, 79.62, 50.01, 64.82,
1982, 9, 0.393,300.02, 79.37, 44.91, 62.14,
1982,10, 0.376,300.55, 76.04, 44.35, 60.19,
1982,11, 0.358,300.70, 72.85, 46.16, 59.50,
1982,12, 0.341,300.50, 69.37, 50.03, 59.70,
1982,13, 0.322,299.64, 62.84, 56.74, 59.79,
1982,14, 0.308,298.69, 57.27, 59.40, 58.33,
1982,15, 0.298,298.14, 51.51, 55.26, 53.39,
1982,16, 0.298,297.46, 48.91, 50.16, 49.54,
1982,17, 0.308,296.67, 50.76, 46.37, 48.56,
1982,18, 0.321,296.36, 52.83, 40.14, 46.49,
1982,19, 0.330,295.74, 53.03, 35.30, 44.17,
1982,20, 0.337,294.95, 49.92, 38.12, 44.02,
1982,21, 0.347,294.29, 50.56, 40.54, 45.55,
1982,22, 0.356,293.38, 51.24, 43.71, 47.48,
1982,23, 0.361,292.42, 50.47, 45.07, 47.77,
1982,24, 0.364,291.41, 50.40, 45.59, 47.99,
1982,25, 0.370,290.55, 52.37, 45.29, 48.83,
1982,26, 0.376,289.94, 55.63, 45.98, 50.80,
1982,27, 0.382,289.90, 58.81, 45.26, 52.04,
1982,28, 0.385,290.40, 59.94, 42.52, 51.23,
1982,29, 0.383,290.33, 61.33, 41.71, 51.52,
```

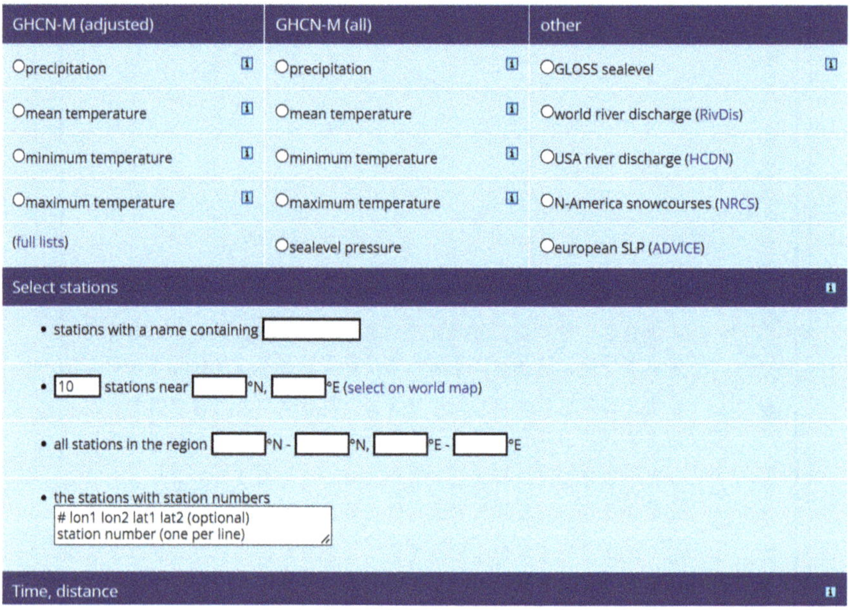

Fig. 6.36 GHCN meteorological Explorer. Source: Meteorological Search Explorer—https://climatedata-catalogue.wmo.int/explore

- Database—GHCN-M(all) stations: Near latitude 13.969221926861373 and longitude 107.9938260890984
- Time distance: At least 10 years of data in the monthly season starting in all months in year 1991–2021

GHCN Meteorological Search Explorer (please see Fig. 6.36) can be accessed from https://climatedata-catalogue.wmo.int/explore.

Search results:

- Looking up 70 stations
- Monthly precipitation_all stations near 13.969221926861373 N 107.9938260890984 E
- Requiring at least 10 years with data in all months

Results:

NHATRANG VIETNAM (VIETNAM) Coordinates: 12.30N, 109.20E, 10m WMO station code: 48877 (get data[146]) Found 19 years with data in 1992–2018	DA-NANG/TOURANE VIETNAM (VIETNAM) Coordinates: 16.00N, 108.20E, 7m WMO station code: 48855 (get data[147]) Found 19 years with data in 1992–2018

(continued)

[146] NHATRANG VIETNAM (VIETNAM)—https://climexp.knmi.nl/getprcpall.cgi?id=someone@somewhere&WMO=48877&STATION=NHATRANG&extraargs=

[147] DA-NANG/TOURANE VIETNAM (VIETNAM)—

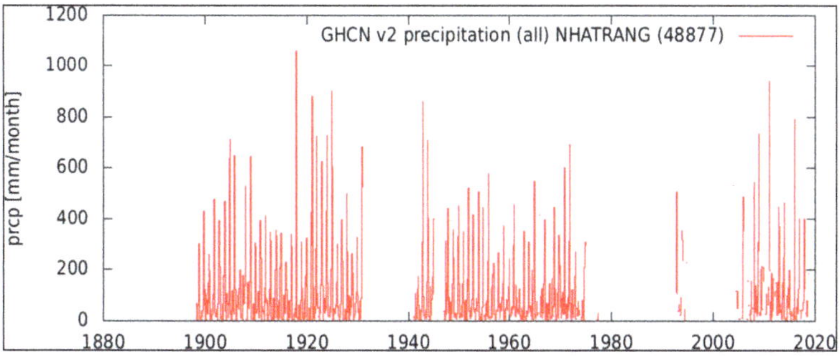

Fig. 6.37 GHCN v2 precipitation

PHU-LIEN VIETNAM (VIETNAM) Coordinates: 20.80N, 106.60E, 116m WMO station code: 48826 (get data[148]) Found 19 years with data in 1992–2018	HA NOI VIETNAM (VIETNAM) Coordinates: 21.02N, 105.80E, 6m WMO station code: 48820 (get data[149]) Found 13 years with data in 1992–2012
QUANGTRI VIETNAM (VIETNAM) coordinates: 16.20N, 107.20E, 1m Near WMO station code: 48852.1 (get data) Found 35 years with data in 1906–1940	HOANG-SA (PATTLE) (VIETNAM) coordinates: 16.50N, 111.60E, 5m WMO station code: 48860 (get data) Found 28 years with data in 1941–1973

Data for NHATRANG Station (please see Fig. 6.37): https://climatedata-catalogue.wmo.int/explore

6.7.4.1 Temperature, Precipitation, and Weather Data

Download time series data from Climate Change Knowledge Portal (CCKP)[150]. The following figure contains precipitation observed values from 1901 to 2020 for Gia Lai (please see Fig. 6.38), a Central Highlands location that is critical to coffee production.

6.7.5 Collected Temperatures for Gia Lai

The following temperatures collected for Central Highlands of Gia Lai (only November to January) displayed the changing temperature trends. As you can see the average temperatures are beyond harvest safe temperatures (October–January) (please see Fig. 6.39).

[148]Vietnam Station data—https://climexp.knmi.nl/getprcpall.cgi?id=someone@somewhere&WMO=48826&STATION=PHU-LIEN&extraargs=

[149]HA NOI VIETNAM (VIETNAM)—https://climexp.knmi.nl/getprcpall.cgi?id=someone@somewhere&WMO=48820&STATION=HA_NOI&extraargs=

[150]Climate Data—https://climateknowledgeportal.worldbank.org/download-data

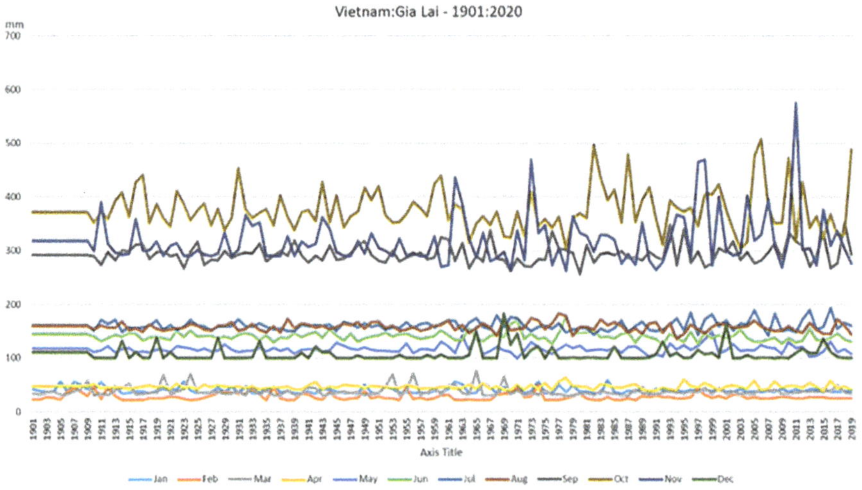

Fig. 6.38 Vietnam: Gia Lai 1901–2020

	Software code for this model: `Final_Release_Vietnam-Coffee-ClimateModeling.ipynb (Jupyter Notebook Code)`

Gia Lai region is experiencing climate change, and the negative effects of climate change can be seen such as increased deforestation[151] and livelihood of small farmers [62]. Additionally, coffee taste and production yield depend on the temperature (please see Fig. 6.40). Higher elevations are associated with sweeter, more complex coffee flavors, but it's correlation, not causation. The real cause of this better-quality, more delicious coffee is temperature. At lower temperatures, coffee trees will grow more slowly. The cherries that contain the seeds we roast and call coffee beans will also ripen more gradually. This means they have more time to develop complex coffee flavors. At these cooler temperatures, it can also be harder for certain pests and diseases to thrive. Take coffee leaf rust, a fungus that attacks the leaves of coffee plants, thereby preventing them from photosynthesizing and getting the energy required to grow healthily. In 2012, it devastated Latin American coffee communities, causing over US $1 billion in just 2 years (USAID). Yet, as Emma Sage, SCA Science Manager, makes clear, leaf rust is not without its own vulnerabilities—and one of these is temperature[152] [63]. Please refer to Appendix C for World Data Sources and Appendix D for US Data Sources.

[151] Assessing land use change in the context of climate change and proposing solutions: Case study in Gia Lai province, Vietnam—http://vnjhm.vn/article/1799

[152] Coffee Quality & M.A.S.L.: How Important Is Elevation REALLY?—https://perfectdailygrind.com/2018/01/coffee-quality-m-a-s-l-how-important-is-altitude-really/

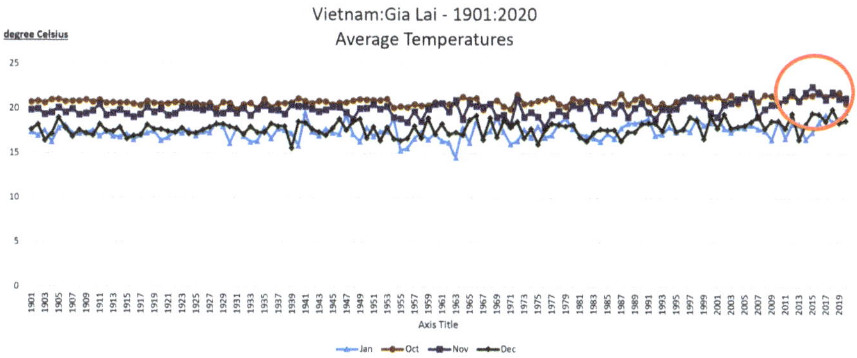

Fig. 6.39 Vietnam Gia Lai 1901–2020

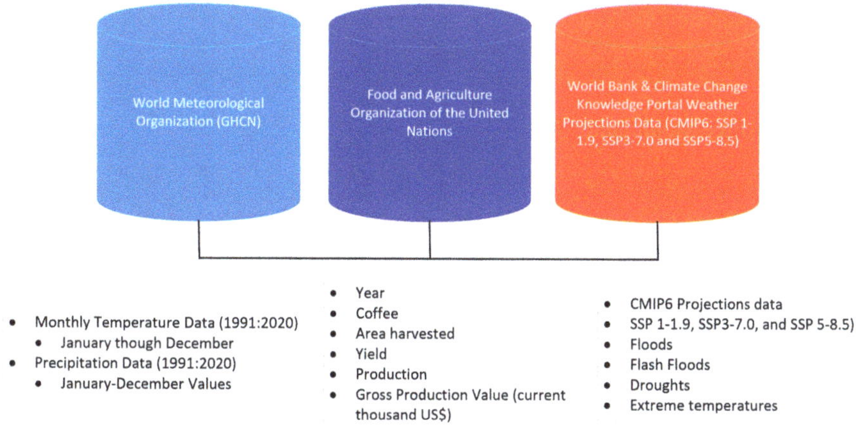

Fig. 6.40 Data sources

In a 2012 Specialty Coffee Association of America (SCAA) blog post[153], the author of the post writes that the optimal temperature for coffee leaf rust is *21–25 °C/ 70–77 °F*, while the disease cannot survive at <15 °C/59 °F [64]. The ideal temperature for growing coffee in is *17–23 °C/63–73 °F*, while it can be grown less effectively at 14–30 °C/57–86 °F. This means that, at lower temperatures, coffee leaf rust should be weaker[154]. Infection with coffee leaf rust not only affects the price of coffee but also has the potential to bankrupt small estates. An ill-timed outbreak can place already vulnerable workers in absolute poverty and destroy communities [65].

---

[153]Some Insights on Coffee Leaf Rust—https://scanews.coffee/2013/02/15/some-insights-on-coffee-leaf-rust-hemileia-vastatrix/

[154]How to Monitor For & Prevent Coffee Leaf Rust—https://perfectdailygrind.com/2019/04/how-to-monitor-for-prevent-coffee-leaf-rust/

6.7.6 Step 1: Load Coffee Dataset

Load libraries and Vietnam Gia Lai coffee dataset:

```
import pandas as pd
import numpy as np
import plotly.graph_objects as go
import seaborn as sns
from sklearn.linear_model import LinearRegression
from sklearn.model_selection import train_test_split
import eli5
from sklearn.preprocessing import MinMaxScaler
import matplotlib.pyplot as plt
import pickle

coffeeAreaHarvested_Original = pd.read_csv
("FAOSTAT_Vietnam_CoffeeareaHarvested.csv")
```

Load coffee yield data:

```
coffeeYield_Original = pd.read_csv
("FAOSTAT_Vietnam_Coffeeyield.csv")
```

Load coffee production quantity data:

```
coffeeProduction_Original  = pd.read_csv
("FAOSTAT_Vietnam_CoffeeProductionQuantity.csv")
coffeeAreaHarvested_Original.head()
```

Output:

	Domain Code	Domain	Area Code (FAO)	Area	Element Code	Element	Item Code (FAO)	Item	Year Code	Year	Unit	Value	Flag	Flag Description
0	QCL	Crops and livestock products	237	Viet Nam	5312	Area harvested	656	Coffee, green	1961	1961	ha	21200	NaN	Official data
1	QCL	Crops and livestock products	237	Viet Nam	5312	Area harvested	656	Coffee, green	1962	1962	ha	24410	NaN	Official data
2	QCL	Crops and livestock products	237	Viet Nam	5312	Area harvested	656	Coffee, green	1963	1963	ha	25000	F	FAO estimate
3	QCL	Crops and livestock products	237	Viet Nam	5312	Area harvested	656	Coffee, green	1964	1964	ha	25000	F	FAO estimate
4	QCL	Crops and livestock products	237	Viet Nam	5312	Area harvested	656	Coffee, green	1965	1965	ha	22800	NaN	Official data

Inspect coffee yield data:

```
coffeeAreaHarvested = coffeeAreaHarvested_Original
[['Year','Value']]
coffeeAreaHarvested.rename(columns={'Value':'AreaHarvested
(ha)'},inplace=True)
coffeeAreaHarvested.head(20)
```

6.7.7 Step 2: Inspect Production and Yield Data

Plot the graph (simple plot[155]):

```
import matplotlib.
pyplot as plt
import numpy as np
# Data for plotting
t =
coffeeAreaHarvested
['Year']
s =
coffeeAreaHarvested
['AreaHarvested(ha)']
fig, ax = plt.subplots()
ax.plot(t, s)
ax.set(xlabel='Year',
ylabel='AreaHarvested
(ha)',

title='AreaHarvested
(ha) over the period
1961:2020')
ax.grid()
plt.show()
```

Output:

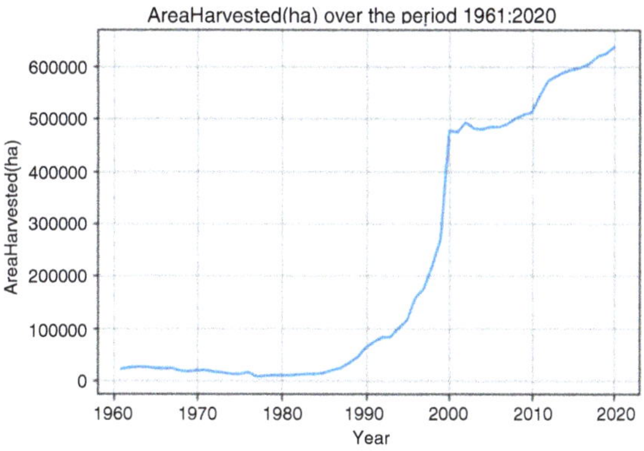

Other plots can be applied for other data[156].
As you can see, a drastic change of area harvested after 2000.
Plot coffee yield data:

[155] Simple Plot—https://matplotlib.org/stable/gallery/lines_bars_and_markers/simple_plot.html#sphx-glr-gallery-lines-bars-and-markers-simple-plot-py
[156] MatPlotLib Gallery—https://matplotlib.org/stable/gallery/

`coffeeYield = coffeeYield_Original [['Year','Value']]`
`coffeeYield.rename (columns= {'Value':'CoffeeYield (hg/ha)'}, inplace=True)`
`coffeeYield.head()`
`import matplotlib. pyplot as plt`
`import numpy as np`
`# Data for plotting`
`t = coffeeYield ['Year']`
`s = coffeeYield ['CoffeeYield (hg/ha)']`
`fig, ax = plt.subplots()`
`ax.plot(t, s)`
`ax.set(xlabel='Year', ylabel='CoffeeYield (hg/ha)',`
`   title='CoffeeYield (hg/ha) over the period 1961:2020')`
`ax.grid()`
`plt.show()`
	Output:
	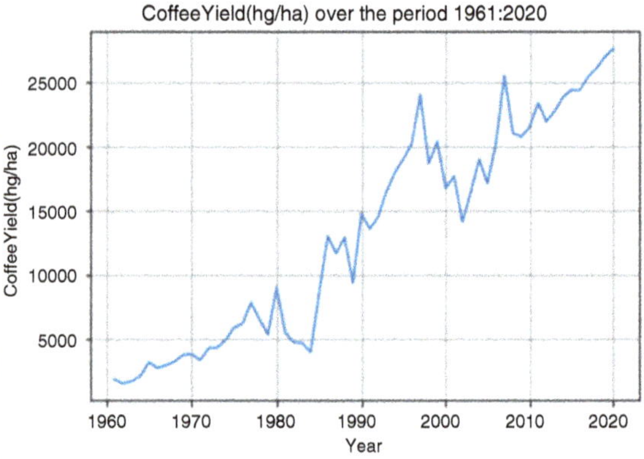

As can be seen after the 1980s, the yield graph skyrocketed.

Plot coffee production data:

`import matplotlib. pyplot as plt`
`import numpy as np`
`# Data for plotting`
`t = coffeeProduction ['Year']`
`s = coffeeProduction ['CoffeeProduction (tonnes)']`
`fig, ax = plt.subplots()`
`ax.plot(t, s)`
`ax.set(xlabel='Year', ylabel= 'CoffeeProduction (tonnes)',`
`   title= 'coffeeProduction over the period 1961:2020')`
`ax.grid()`
`plt.show()`
	Output:
	A truly transformation Vietnam coffee story—lifting small farmers out of poverty and ensuring sustainable income.
	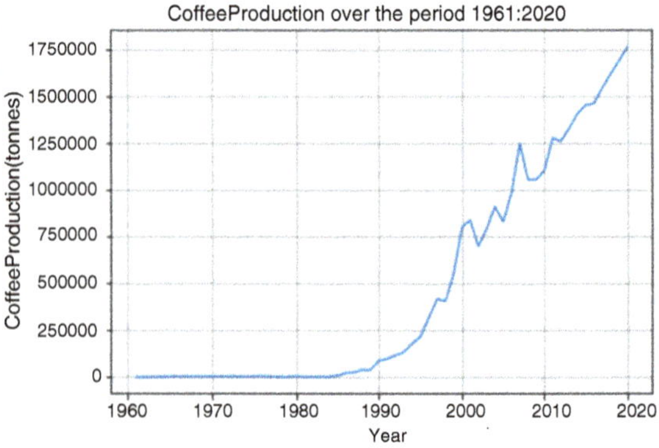

Combine all coffee production data frames (using year) into one final data frame:

```
finaldf = pd.merge(left=coffeeAreaHarvested, right=coffeeYield,
how='outer', left_on='Year', right_on='Year')
finaldf = pd.merge(left=finaldf, right=coffeeProduction,
how='outer', left_on='Year', right_on='Year')
finaldf.head()
```

Output:

	Year	AreaHarvested(ha)	CoffeeYield(hg/ha)	CoffeeProduction(tonnes)
0	1961	21200	1934	4100
1	1962	24410	1647	4020
2	1963	25000	1800	4500
3	1964	25000	2200	5500
4	1965	22800	3289	7500

Plot all three production values into one graph. Of course, apply MinMax transformation:

```
#min max scaler
scaler = MinMaxScaler()
scaled_df[['AreaHarvested(ha)', 'CoffeeYield(hg/ha)',
'CoffeeProduction(tonnes)']] = scaler.fit_transform(finaldf
[['AreaHarvested(ha)', 'CoffeeYield(hg/ha)', 'CoffeeProduction
(tonnes)']])
```

Plot:

```
fig = go.Figure()

fig.add_trace(go.Scatter(x=scaled_df.index,y=scaled_df
['AreaHarvested(ha)'],mode = 'lines',name='AreaHarvested'))

fig.add_trace(go.Scatter(x=scaled_df.index,y=scaled_df
['CoffeeProduction(tonnes)'],mode = 'lines',
name='CoffeeProduction'))

fig.add_trace(go.Scatter(x=scaled_df.index,y=scaled_df
['CoffeeYield(hg/ha)'],mode = 'lines',name='CoffeeYeild'))

fig.show()
```

Output:

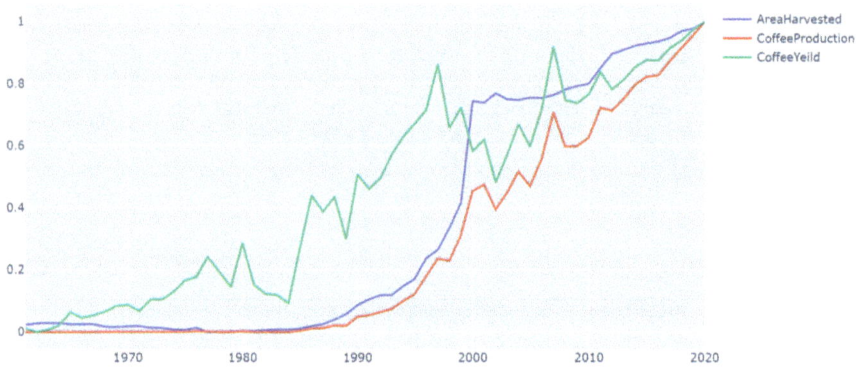

6.7.8 Step 3: Load Weather Data—Gia Lia (Precipitation, Max Temperatures, Min Temperatures, and Average Temperature)

Load weather data:

```
#precipitation data
precipitationDf = pd.read_csv('precipitation_VNM_3344.csv')
precipitationDf.head()

precipitationDf = precipitationDf.add_prefix('Precipitation_')
precipitationDf.rename(columns={'Precipitation_Year':'Year'},
inplace=True)

maxTempDf = pd.read_csv('tmaxGialaiMonthlyData.csv')
maxTempDf.head()

maxTempDf = maxTempDf.add_prefix('maxTemp_')
maxTempDf.head()
```

(continued)

```
maxTempDf.rename(columns={'maxTemp_Year':'Year'},
inplace=True)
maxTempDf.head()

minTempDf = pd.read_csv('tMinGialaiMonthlyData.csv')
minTempDf.head()

minTempDf = minTempDf.add_prefix('minTemp_')
minTempDf.rename(columns={'minTemp_Year':'Year'},
inplace=True)
minTempDf.head()
```

Output:

	Year	minTemp_Jan	minTemp_Feb	minTemp_Mar	minTemp_Apr	minTemp_May	minTemp_Jun	minTemp_Jul	minTemp_Aug	minTemp_Sep
0	1901	17.39	18.30	9.21	21.25	22.00	22.19	22.80	22.65	21.90
1	1902	16.98	17.33	9.28	21.25	22.10	22.61	23.01	22.79	21.83
2	1903	17.59	18.04	9.72	21.68	22.29	21.96	22.81	22.55	21.67
3	1904	16.28	17.59	9.07	20.95	22.00	22.26	22.87	22.86	22.07
4	1905	17.84	18.80	9.46	21.70	22.03	22.29	22.96	23.05	22.12

Add final dataset:

```
finaldf = pd.merge(left=finaldf, right=maxTempDf, how='inner',
left_on='Year', right_on='Year')
finaldf = pd.merge(left=finaldf, right=minTempDf, how='inner',
left_on='Year', right_on='Year')
finaldf = pd.merge(left=finaldf, right=precipitationDf,
how='inner', left_on='Year', right_on='Year')
finaldf.head()
```

	Year	AreaHarvested(ha)	CoffeeYield(hg/ha)	CoffeeProduction(tonnes)	maxTemp_Jan	maxTemp_Feb	maxTemp_Mar	maxTemp_Apr	maxTemp_May
0	1961	21200	1934	4100	24.34	25.87	29.12	29.27	29.46
1	1962	24410	1647	4020	23.59	25.56	28.66	28.77	30.22
2	1963	25000	1800	4500	22.88	25.37	28.46	28.77	30.87
3	1964	25000	2200	5500	25.07	25.48	28.75	29.67	29.22
4	1965	22800	3289	7500	24.15	26.26	28.45	28.85	29.67

6.7.9 Step 4: Correlation (Weather Data and Yield/Production)

The following code outputs the correlation of key attributes to the input feature columns:

```
corr =finaldf.corr()
corr
```

Output:

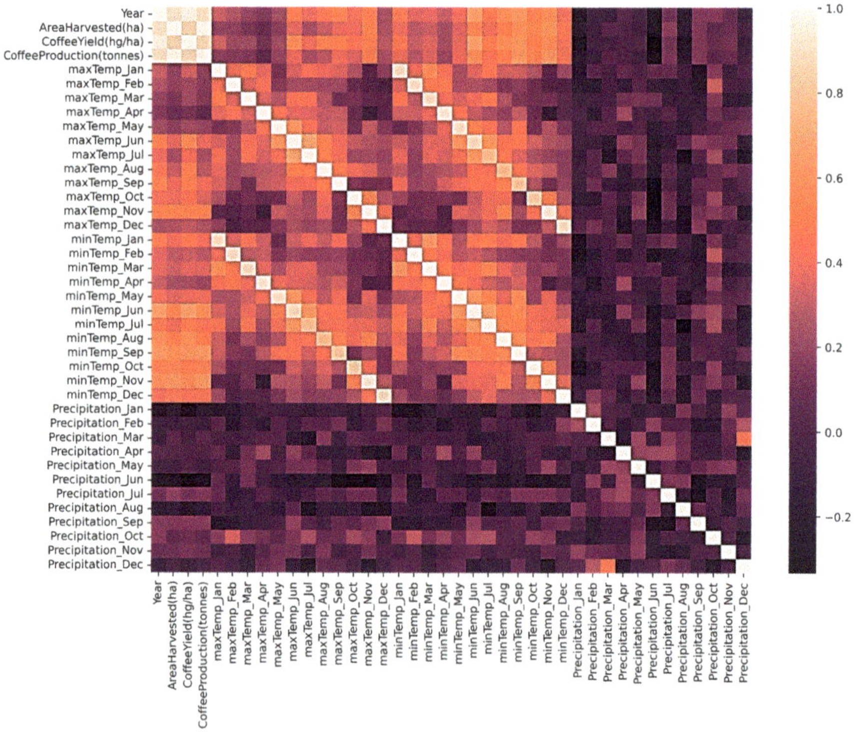

6.7.10 Step 4: Coffee Yield Distribution

To analyze coffee yield distribution, please check the following diagram:

```
import pandas as pd
import matplotlib.
pyplot as plt
# This ensures plots
are displayed inline in
the Jupyter notebook
%matplotlib inline
# Get the label column
label = finaldf
['CoffeeYield
(hg/ha)']
# Create a figure for
2 subplots (2 rows,
1 column)
fig, ax = plt.subplots
(2, 1, figsize = (9,12))
# Plot the histogram
ax[0].hist(label,
bins=100)
ax[0].set_ylabel
('Frequency')
# Add lines for the
mean, median, and mode
ax[0].axvline(label.
mean(),
color='magenta',
linestyle='dashed',
linewidth=2)
ax[0].axvline(label.
median(),
color='cyan',
linestyle='dashed',
linewidth=2)
# Plot the boxplot
ax[1].boxplot(label,
vert=False)
ax[1].set_xlabel
('CoffeeYield
(hg/ha)')
# Add a title to the
Figure
fig.suptitle
('CoffeeYield(hg/ha)
Distribution')
# Show the figure
fig.show()
```

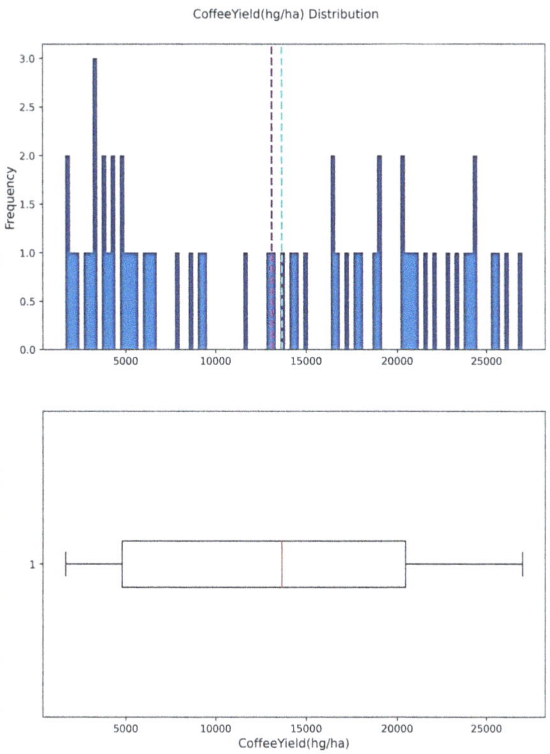

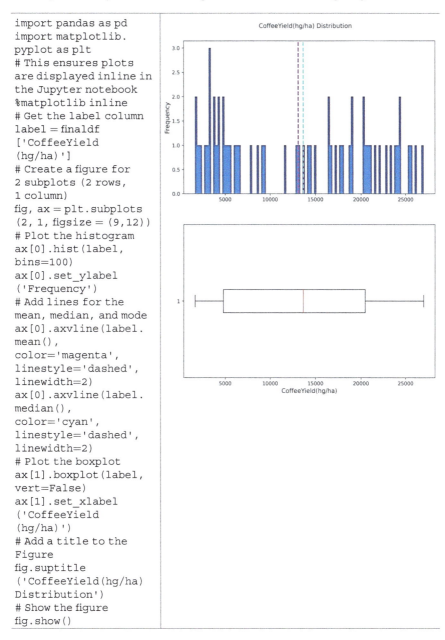

Output: Coffee yield average to 15,000 hg/ha.

6.7.11 Step 5: Feature Distributions—Max. Temperatures

To analyze coffee dataset feature distribution, please check following graphs:

```
# Plot a histogram for each numeric feature
for col in numeric_features:
  fig = plt.figure(figsize=(9, 6))
  ax = fig.gca()
  feature = finaldf[col]
  feature.hist(bins=100, ax = ax)
  ax.axvline(feature.mean(), color='magenta',
linestyle='dashed', linewidth=2)
  ax.axvline(feature.median(), color='cyan',
linestyle='dashed', linewidth=2)
  ax.set_title(col)
plt.show()
```

Output: Temperature plays an important role in coffee production. Especially, June, July, and August temperatures predicate the size of coffee beans/cherries.

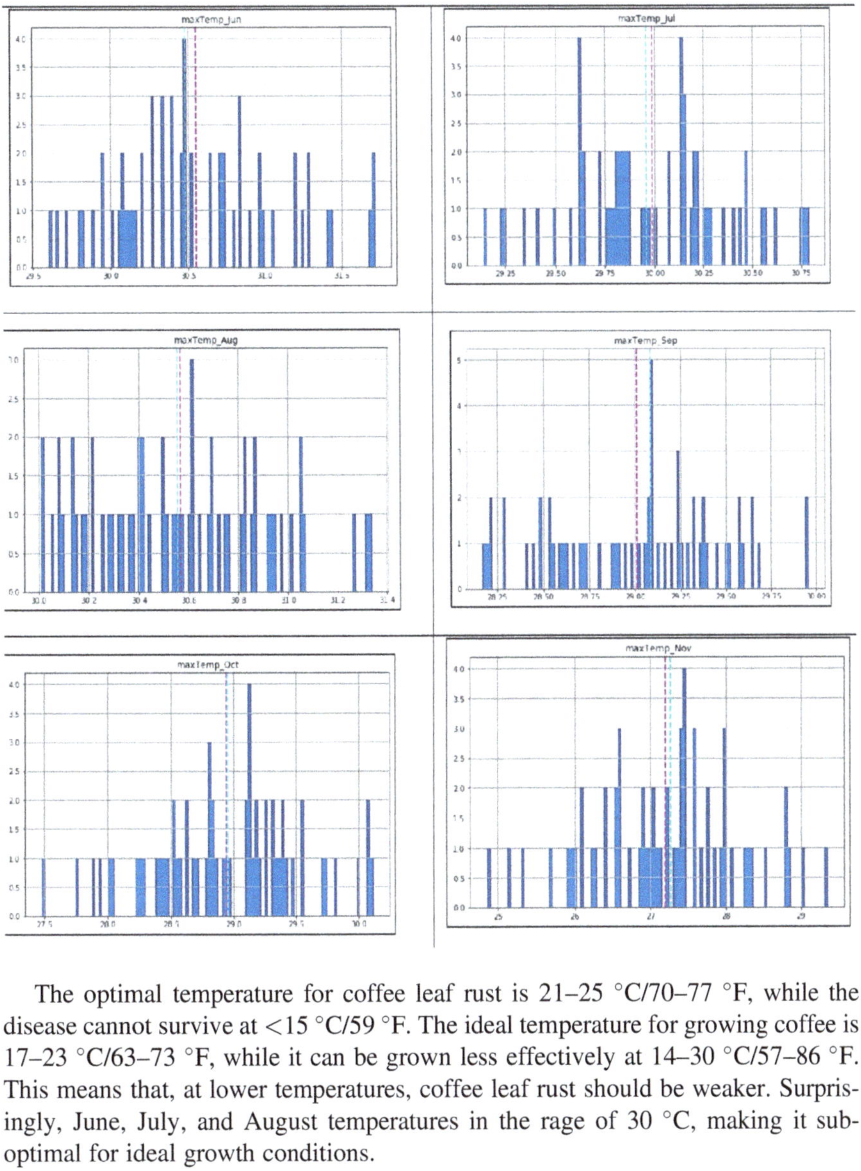

The optimal temperature for coffee leaf rust is 21–25 °C/70–77 °F, while the disease cannot survive at <15 °C/59 °F. The ideal temperature for growing coffee is 17–23 °C/63–73 °F, while it can be grown less effectively at 14–30 °C/57–86 °F. This means that, at lower temperatures, coffee leaf rust should be weaker. Surprisingly, June, July, and August temperatures in the rage of 30 °C, making it sub-optimal for ideal growth conditions.

6.7.12 Step 5: Feature Distributions—Min. Temperatures

To analyze coffee dataset minimum temperature feature distribution, please check the following:

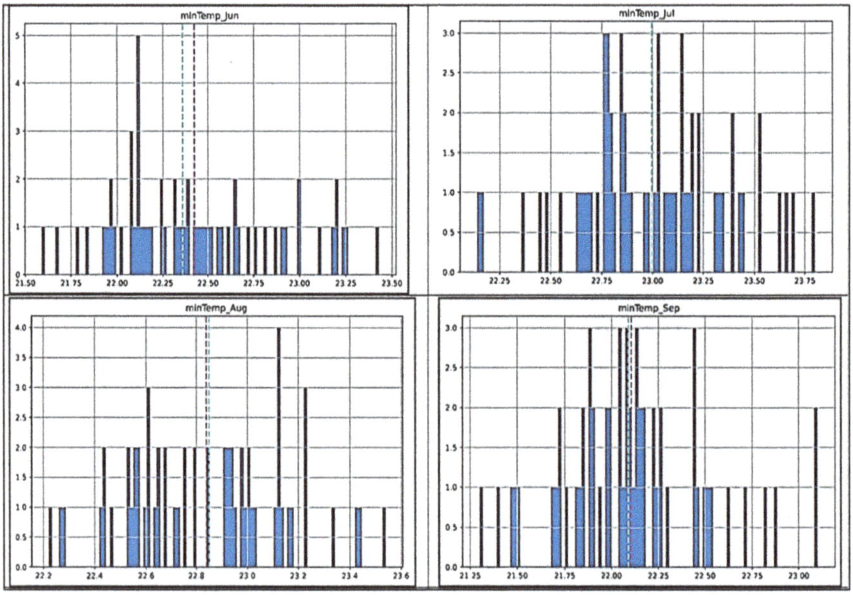

Minimum temperatures in the months of November, December, and January are between 15 °C and 20 °C. Generally, these months are harvest season. Having a higher temperature could be helpful as, at lower temperatures, coffee leaf rust is weaker.

6.7.13 Step 6: Feature Distributions—Precipitations

To analyze coffee dataset precipitation distribution, please check the following:

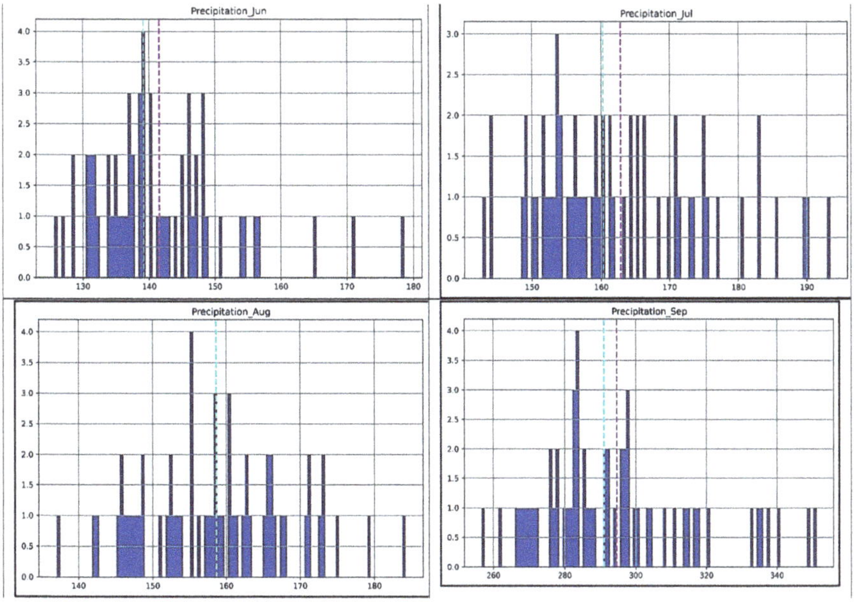

Higher precipitation during September and October could be detrimental as rains during the second round of tree stimulus with heavy rains or dry weather yield losses as fruit fell to the ground before completing the maturity.

Vietnam – Coffee Production – Central Highlands											
Jan	Feb	Mar	Apr	May	Jun	Jul	Aug	Sep	Oct	Nov	Dec
					Crucial bean formation period		Second round of tree stimulus		Harvest		
					Climate Change: Dry weather results in		Climate Change: Heavy Rains or Dry Weather yield losses as fruit fell to the ground before completing the maturity				

6.7.14 *Step 7: Feature Correlations—Label Column—Coffee Production (Tons)/Coffee Yield (hg/ha) vs. Climate Variables*

Application of feature correlations, between "coffee production (tons)" and other parameters would provide inter-relationships. The same can be applied for yield values. October and November maximum temperatures are highly correlated with coffee production. Similarly, June, September, October, and November minimum temperatures are highly influencing coffee production.

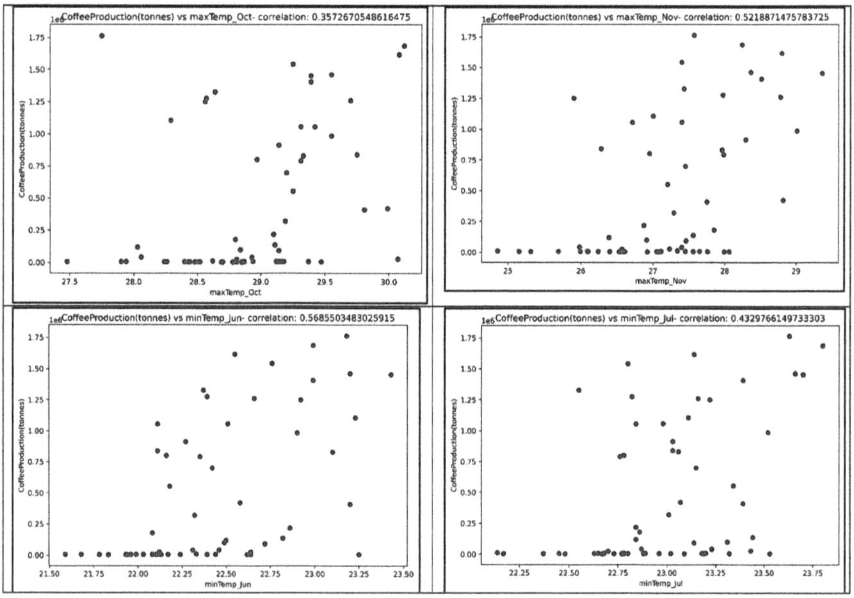

Below-average rainfall[157] and above-average temperatures during the calendar year (CY) 2020 dry season

caused droughts that impacted coffee yields in Vietnam's Central Highlands, the country's main coffee-growing area [70–72]. An average precipitation and above-average temperatures during the dry season caused droughts in some localities in the Central Highlands, Vietnam's main coffee-growing area, which affected yields.

On an interesting note, precipitation has a negative correlation with production, especially, during the months of June (−27.15%), August (−16.66%), and January (−17.2%), a fact that can be observed by the expert statements from Vietnam:

- The lack of stable and sufficient rainfall during June and July prevented Vietnamese growers from applying the second round of fertilizer in a timely manner that in a typical crop cycle will provide trees with a welcome boost to enhance yields, but making matters worse, it occurred at the time of what in coffee agronomy is known as the crucial bean formation period[158].
- The month of November plays an important role for Central Highlands. Heavy rains during November could result in huge yield losses such as ripening fruit to fall to the ground and rotten. For instance, during 2021–2022, "Vietnam's coffee harvest starting next November may be reduced by 10-15% due to the heavy rain,

[157] Coffee Semi-annual—https://apps.fas.usda.gov/newgainapi/api/Report/DownloadReportByFileName?fileName=Coffee%20Semi-annual_Hanoi_Vietnam_11-15-2020

[158] BREAKING: Vietnam's 2021-22 Coffee Harvest Down on Dry Weather—https://stir-tea-coffee.com/tea-coffee-news/breaking-vietnam%E2%80%99s-2021-22-coffee-harvest-down-on-dry-weathe/

which caused ripening fruit to fall to the ground and rotten," said Tu in an interview published on 30 Aug. 2022.

Coffee yield vs. weather variables	
Feature variable	Correlation
Maximum temperature June	53.22%
Maximum temperature August	49.01%
Maximum temperature September	48.17%
Maximum temperature October	52.35%
Maximum temperature November	51.92%
Minimum temperature June	63.28%
Minimum temperature August	54.94%
Minimum temperature September	57.73%
Minimum temperature October	54.07%
Minimum temperature November	54.99%
Precipitation June	−32.52%
Precipitation July	11.98%
Precipitation August	−22.02%
Precipitation September	16.48%
Precipitation January	−28.55%

Yield was influenced by more set of weather feature variables unlike production.

6.7.15 Step 8: Linear Regression Model

Let's perform regression model. Prepare feature and label data frames. X represents features and Y represents label column:

```
X = finaldf[['AreaHarvested(ha)',
       'maxTemp_Jun', 'maxTemp_Jul','maxTemp_Aug',
'maxTemp_Sep','maxTemp_Oct',
'maxTemp_Nov','maxTemp_Dec', 'maxTemp_Jan',
       'minTemp_Jun',
'minTemp_Jul','minTemp_Aug','minTemp_Sep','minTemp_Oct',
'minTemp_Nov',
'minTemp_Dec', 'minTemp_Jan',
       'Precipitation_Jun',
'Precipitation_Jul','Precipitation_Aug','Precipitation_Sep',
'Precipitation_Oct', 'Precipitation_Nov',
       'Precipitation_Dec','Precipitation_Jan'
      ]].values
y = finaldf['CoffeeProduction(tonnes)'].values
```

Model features:

```
model_features = ['AreaHarvested(ha)',
     'maxTemp_Jun', 'maxTemp_Jul','maxTemp_Aug','maxTemp_Sep',
'maxTemp_Oct',
'maxTemp_Nov','maxTemp_Dec', 'maxTemp_Jan',
     'minTemp_Jun', 'minTemp_Jul','minTemp_Aug','minTemp_Sep',
'minTemp_Oct', 'minTemp_Nov','minTemp_Dec', 'minTemp_Jan',
      'Precipitation_Jun',
'Precipitation_Jul','Precipitation_Aug','Precipitation_Sep',
'Precipitation_Oct', 'Precipitation_Nov',
      'Precipitation_Dec','Precipitation_Jan']
```

Prepare train and test data:

```
X_train,X_test, y_train , y_test = train_test_split(X ,y,
test_size = 0.2,random_state = 0)
```

Create a Linear Regression.

```
modelProd = LinearRegression()
```

Predict the test value.

```
# predicting the test
y_pred = modelProd.predict(X_test)

from sklearn.metrics import mean_squared_error, r2_score

mse = mean_squared_error(y_test, y_pred)
print("MSE:", mse)

rmse = np.sqrt(mse)
print("RMSE:", rmse)

r2 = r2_score(y_test, y_pred)
print("R2:", r2)
```

Output:

```
MSE: 18968004071.326603
RMSE: 137724.3771861997
R2: 0.8942914841089066
```

Plot the regressor predicted values:

```
plt.scatter(y_test, y_pred)
plt.xlabel('Actual Value')
plt.ylabel('Predicted value')
plt.title('Vietnam Coffee Prodcution')

z = np.polyfit(y_test, y_pred, 1)
p = np.poly1d(z)
plt.plot(y_test,p(y_test), color='magenta')
plt.show()
```

Output:

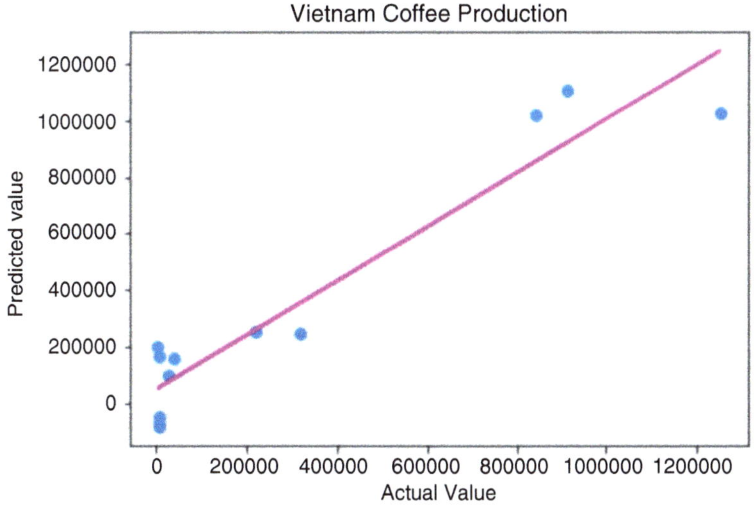

Mode features include:

```
print(list(zip(model_features,modelProd.coef_)))
```

Output:

```
[('AreaHarvested(ha)', 2.0312740651470294), ('maxTemp_Jun',
-20892.1520724203), ('maxTemp_Jul', 5344.5365198533755),
('maxTemp_Aug', -182177.25726630643), ('maxTemp_Sep',
26594.537517568926), ('maxTemp_Oct', 116055.4611482277),
('maxTemp_Nov', -105333.01148693477), ('maxTemp_Dec',
94610.42446407572), ('maxTemp_Jan', -44576.59345736098),
('minTemp_Jun', 19124.071348171365), ('minTemp_Jul',
4103.160545985403), ('minTemp_Aug', 48089.95451450154),
('minTemp_Sep', 191338.8609188045), ('minTemp_Oct',
-55619.37470248132), ('minTemp_Nov', 140201.52609662423),
('minTemp_Dec', -91128.01664478901), ('minTemp_Jan',
24500.117657131577), ('Precipitation_Jun', 94.27173709086583),
('Precipitation_Jul', -565.1831843733653),
('Precipitation_Aug', -91.01843084175016),
('Precipitation_Sep', -296.458971025595),
('Precipitation_Oct', -333.73821194388256),
('Precipitation_Nov', 363.27166784069595),
('Precipitation_Dec', 362.8998454977925),
('Precipitation_Jan', 404.48224175782593)]
```

Let's derive the model governing equation:

```
mx=""
for ifeature in range(len(X.columns)):
  if modelProd.coef_[ifeature] <0:
    # format & beautify the equation
   mx += " - " + "{:.2f}".format(abs(modelProd.coef_[ifeature])) +
" * " + X.columns[ifeature]
  else:
    if ifeature == 0:
     mx += "{:.2f}".format(modelProd.coef_[ifeature]) + " * " + X.
columns[ifeature]
    else:
     mx += " + " + "{:.2f}".format(modelProd.coef_[ifeature]) + " *
" + X.columns[ifeature]

print(mx)
```

Output:

```
2.03 * AreaHarvested(ha) - 20892.15 * maxTemp_Jun + 5344.54 *
maxTemp_Jul - 182177.26 * maxTemp_Aug + 26594.54 * maxTemp_Sep +
116055.46 * maxTemp_Oct - 105333.01 * maxTemp_Nov + 94610.42 *
maxTemp_Dec - 44576.59 * maxTemp_Jan + 19124.07 * minTemp_Jun +
4103.16 * minTemp_Jul + 48089.95 * minTemp_Aug + 191338.86 *
minTemp_Sep - 55619.37 * minTemp_Oct + 140201.53 * minTemp_Nov -
91128.02 * minTemp_Dec + 24500.12 * minTemp_Jan + 94.27 *
Precipitation_Jun - 565.18 * Precipitation_Jul - 91.02 *
Precipitation_Aug - 296.46 * Precipitation_Sep - 333.74 *
Precipitation_Oct + 363.27 * Precipitation_Nov + 362.90 *
Precipitation_Dec + 404.48 * Precipitation_Jan
```

Let's construct Y = mx + C:

```
# y=mx+c
if(modelProd.intercept_ <0):
  print("The formula for the " + y.name + " linear regression line is
= " + " - {:.2f}".format(abs(modelProd.intercept_)) + " + " + mx )
else:
  print("The formula for the " + y.name + " linear regression line is
= " + " {:.2f}".format(modelProd.intercept_) + " + " + mx )
```

Output:

$$
\begin{aligned}
&\text{The formula for the CoffeeProduction(tonnes) linear regression line is}\\
&= -2749432.10 + 2.03^* \text{ AreaHarvested(ha)}\\
&\quad - 20892.15^* \text{ maxTemp\_Jun} + 5344.54^* \text{ maxTemp\_Jul}\\
&\quad - 182177.26^* \text{ maxTemp\_Aug} + 26594.54^* \text{ maxTemp\_Sep}\\
&\quad + 116055.46^* \text{ maxTemp\_Oct} - 105333.01^* \text{ maxTemp\_Nov}\\
&\quad + 94610.42^* \text{ maxTemp\_Dec} - 44576.59^* \text{ maxTemp\_Jan}\\
&\quad + 19124.07^* \text{ minTemp\_Jun} + 4103.16^* \text{ minTemp\_Jul}\\
&\quad + 48089.95^* \text{ minTemp\_Aug} + 191338.86^* \text{ minTemp\_Sep}\\
&\quad - 55619.37^* \text{ minTemp\_Oct} + 140201.53^* \text{ minTemp\_Nov}\\
&\quad - 91128.02^* \text{ minTemp\_Dec} + 24500.12^* \text{ minTemp\_Jan}\\
&\quad + 94.27^* \text{ Precipitation\_Jun} - 565.18^* \text{ Precipitation\_Jul}\\
&\quad - 91.02^* \text{ Precipitation\_Aug} - 296.46^* \text{ Precipitation\_Sep}\\
&\quad - 333.74^* \text{ Precipitation\_Oct} + 363.27^* \text{ Precipitation\_Nov}\\
&\quad + 362.90^* \text{ Precipitation\_Dec} + 404.48^* \text{ Precipitation\_Jan}
\end{aligned}
\tag{6.2}
$$

Equation (6.2) is governing regression model line that connects independent variables to the dependent variable in predicting coffee production (tons). We will revisit the equation in linkage models and linear programming.

6.7.16 Step 9: Pickle the Model to Call Climate Change Values

Prepare the model, and pickle to check the impact of the global climate change:

```
with open('VietnamCoffeeProduction_pklfile.pkl','wb') as f:
  pickle.dump(modelProd,f)

# load the model
with open('VietnamCoffeeProduction_pklfile.pkl','rb') as f:
  modelpklfileVietnamCoffeeProduction = pickle.load(f)
```

6.7.17 Step 10: Model Explainability

To identify the important feature attributes of the model, the following code provides explainability—return an explanation of estimator parameters (weights)[159]:

```
eli5.show_weights(modelProd,feature_names = model_features)
```

Output: As can be seen, minimum temperature of months September and November have huge explainability with the model, followed by October and December. Finally, precipitations in November and January have the influence.

August and November maximum temperatures have negative, although, influence.

[159] ELI5—https://eli5.readthedocs.io/en/latest/autodocs/eli5.html

y top features

Weight?	Feature
+191338.861	minTemp_Sep
+140201.526	minTemp_Nov
+116055.461	maxTemp_Oct
+94610.424	maxTemp_Dec
+48089.955	minTemp_Aug
+26594.538	maxTemp_Sep
+24500.118	minTemp_Jan
+19124.071	minTemp_Jun
+5344.537	maxTemp_Jul
+4103.161	minTemp_Jul
+404.482	Precipitation_Jan
+363.272	Precipitation_Nov
… 3 more positive …	
… 3 more negative …	
-565.183	Precipitation_Jul
-20892.152	maxTemp_Jun
-44576.593	maxTemp_Jan
-55619.375	minTemp_Oct
-91128.017	minTemp_Dec
-105333.011	maxTemp_Nov
-182177.257	maxTemp_Aug
-2749432.103	<BIAS>

6.7.18 Step 11: CMIP6 Projections Data and SSP Scenarios

To model climate projections, see how the model predicts the future projections of coffee production for Vietnam, Gia Lai province. The goal is to collect for SSP1-2.6, SSP3-7.0, and SSP5-8.5.

Scenario	SSP model	Socioeconomic factors' impact
SSP1-2.6	SSP1	It imagines the same socioeconomic shifts toward sustainability as SSP1-1.9. But temperatures stabilize around 1.8 °C higher by the end of the century
SSP2-4.5	SSP2	Socioeconomic factors follow their historic trends, with no notable shifts. Progress toward sustainability is slow, with development and income growing unevenly. In this scenario, temperatures rise 2.7 °C by the end of the century
SSP3-7.0	SSP3	By the end of the century, average temperatures have risen by 3.6 °C
SSP5-8.5	SSP5	By 2100, the average global temperature is a scorching 4.4 °C higher

Here are the model data simulations:

Collection, CMIP6; aggregation, monthly; calculation, climatology mean; percentile, 90th; and model, Multi-Model Ensemble. Please find the table with all three main variables for scenario SSP5-8.5. The same data are collected for additional two scenarios: SSP1-2.6 and SSP3-7.0.

Minimum temperature	Maximum temperature	Precipitation
Time period: 2020–2039 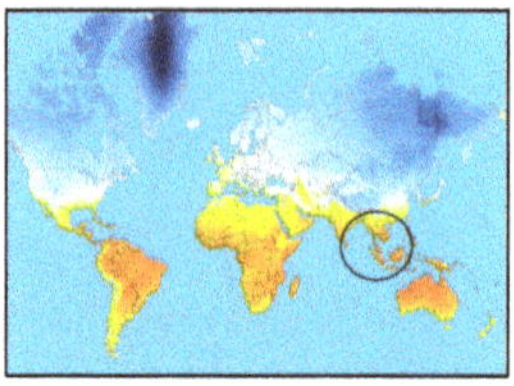	Time period: 2020–2039 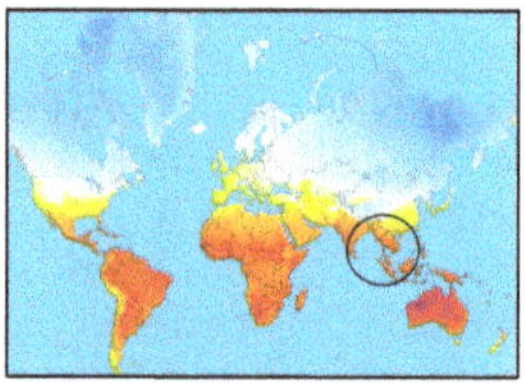	Time period: 2020–2039 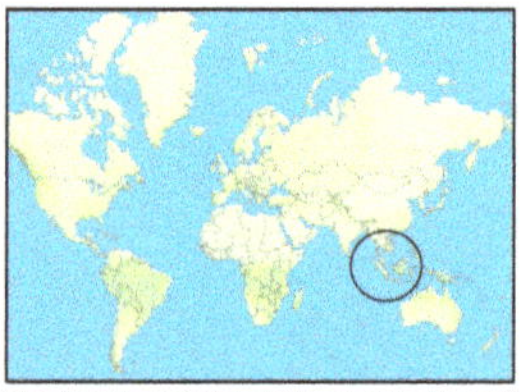
Time period: 2040–2059 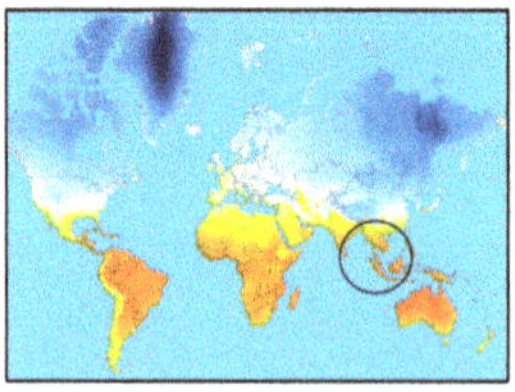	Time period: 2040–2059 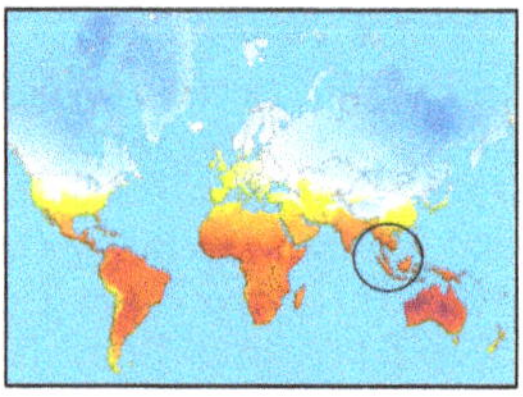	Time period: 2040–2059 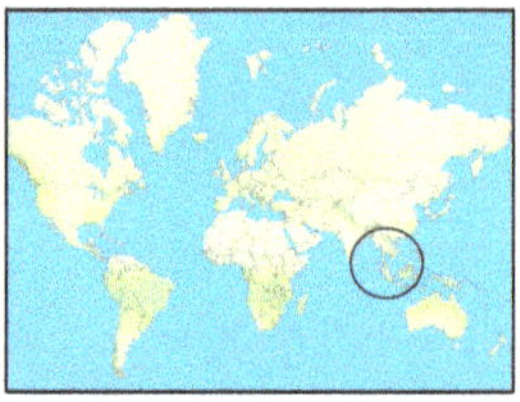
Time period: 2060–2079 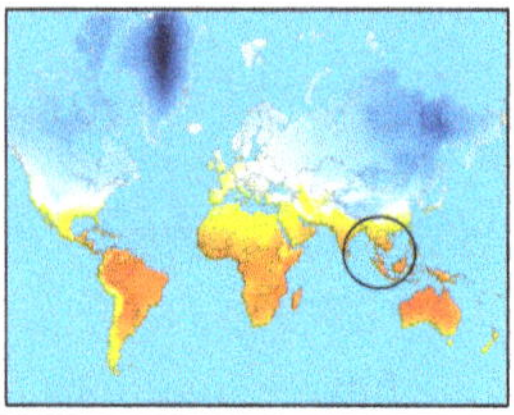	Time period: 2060–2079 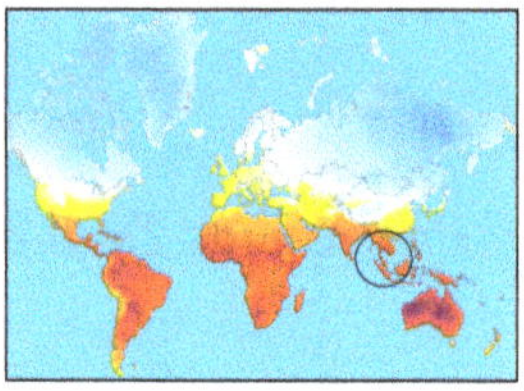	Time period: 2060–2079 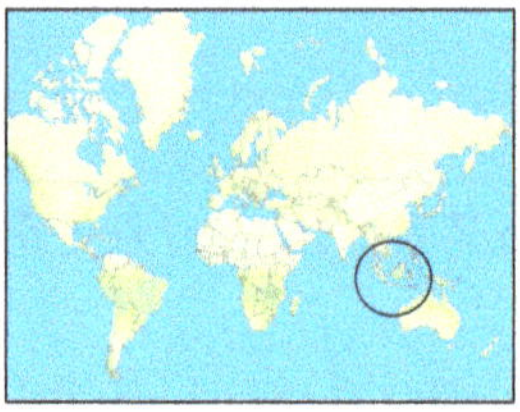
Time period: 2080–2099 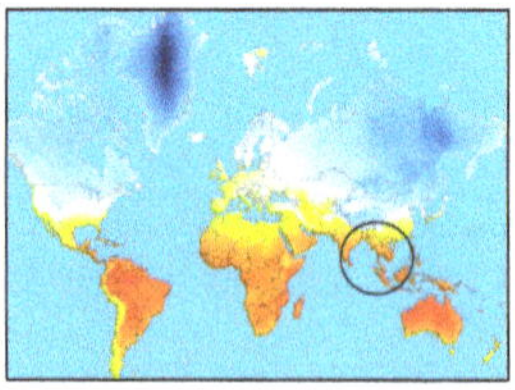	Time period: 2080–2099 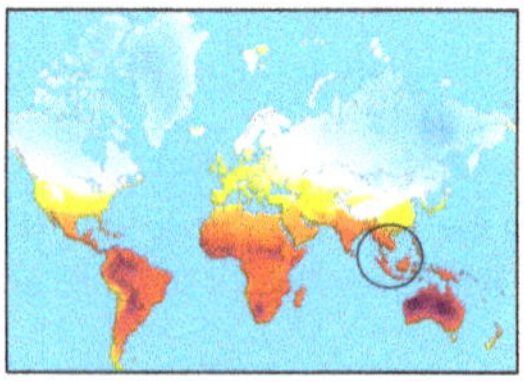	Time period: 2080–2099 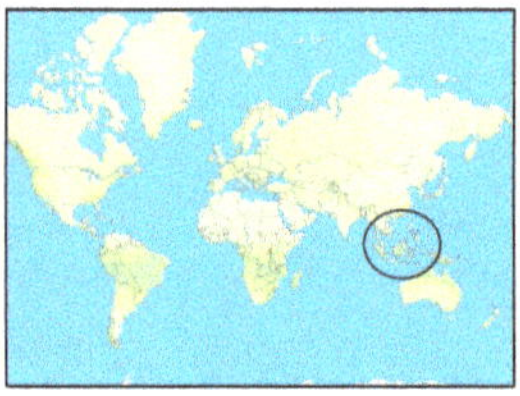

SSP5-8.5—Please find the model invocation:

```
#AreaHarvested(ha)',
    #  'maxTemp_Jun',
'maxTemp_Jul','maxTemp_Aug','maxTemp_Sep',
'maxTemp_Oct','maxTemp_Nov','maxTemp_Dec', 'maxTemp_Jan',
    #  'minTemp_Jun',
'minTemp_Jul','minTemp_Aug','minTemp_Sep','minTemp_Oct',
'minTemp_Nov','minTemp_Dec', 'minTemp_Jan',
    #  'Precipitation_Jun',
'Precipitation_Jul','Precipitation_Aug','Precipitation_Sep',
'Precipitation_Oct', 'Precipitation_Nov',
    #  'Precipitation_Dec','Precipitation_Jan'

cmip6_ssp585_2020_2039_p90_VNM = np.array([[618879,
            30.4,29.65,30.13,29.53,30.08,28.8,27.54,25.16,
22.55,23.14,23.02,22.5,21.72,22.02,20.03,18.62,
135.2,166.49,161.24,351.26,331.97,303.88,100.8,38.83],
             [618879,
             32.42,31.59,31.59,31.17,30.38,29.35,28.25,28.15,
24.51,24.17,24.01,23.6,22.49,21.42,20.15,18.96,
348.81,382,378.74,427.49,389.81,280.07,188.42,84.51]
            ])
modelpklfileVietnamCoffeeProduction.predict(cmip6_ssp585_
2020_2039_p90_VNM[0:2])

cmip6_ssp585_2040_2059_p90_VNM = np.array([[618879,
            30.4,29.65,30.13,29.53,30.08,28.8,27.54,25.16,
22.55,23.14,23.02,22.5,21.72,22.02,20.03,18.62,
135.2,166.49,161.24,351.26,331.97,303.88,100.8,38.83],
             [618879,
             33.39,32.57,32.37,31.87,31.32,30.22,29.24,29.15,
25.73,25.26,25.06,24.73,23.52,22.21,21.05,19.86,
371.18,408.97,413.65,449.53,424.62,339.02,188.76,46.88]
            ])
modelpklfileVietnamCoffeeProduction.predict(cmip6_ssp585_
2040_2059_p90_VNM[0:2])

cmip6_ssp585_2060_2079_p90_VNM = np.array([[618879,
            30.4,29.65,30.13,29.53,30.08,28.8,27.54,25.16,
22.55,23.14,23.02,22.5,21.72,22.02,20.03,18.62,
135.2,166.49,161.24,351.26,331.97,303.88,100.8,38.83],
             [618879,
             35.05,33.82,33.71,33.19,32.45,31.6,30.91,30.89,
27.18,26.66,26.5,26.2,24.76,23.51,22.29,20.98,
398.45,391.61,393.97,447.25,433.55,322.91,179.34,82.85]
            ])
modelpklfileVietnamCoffeeProduction.predict(cmip6_ssp585_
2060_2079_p90_VNM[0:2])
```

(continued)

```
cmip6_ssp585_2080_2099_p90_VNM = np.array([[618879,
            30.4,29.65,30.13,29.53,30.08,28.8,27.54,25.16,
22.55,23.14,23.02,22.5,21.72,22.02,20.03,18.62,
135.2,166.49,161.24,351.26,331.97,303.88,100.8,38.83],
            [618879,
            36.83,35.5,35.46,35.02,34.07,33.06,32.21,32.37,
28.84,28.23,28.07,27.7,26.54,24.92,23.72,22.48,
391.62,408.31,410.28,456.5,448.12,357.53,188.34,73.62]
            ])
modelpklfileVietnamCoffeeProduction.predict(cmip6_ssp585_
2080_2099_p90_VNM[0:2])
```

Output:

Baseline, Vietnam coffee production for 2018; production, 1568899.65 (tons)			
20,202,039	20,402,059	20,602,079	20,802,099
12,73,973.40 (tons)	14,61,875.65 (tons)	17,19,956.86 (tons)	19,11,602.95 (tons)
Drop in production from 2018; production, 15,68,899.65 (tons			
-23.15%	-6.82%	9.62%	21.84%

SSP5-8.5 is the most undesirable climate change outcome. A drop[160,161] of 23.15% in 2020–2039 can be seen. A total drop of 294926.3 tons [24, 66]. For, the 2040–2059 period, a drop of −6.82% could be observed from the model[162] [56]. However, the production sees uptick in 20602079 and 20802099 time periods due to the increase in minimum temperatures (as can be seen from below minimum temperatures) for high elevation of the Central Highlands of Vietnam[163]. As indicated in NASA, climate change effects in terms of the production of coffee in Vietnam can be seen as early as 2030 [67].

[160]Global Climate Change Impact on Crops Expected Within 10 Years, NASA Study Finds—https://climate.nasa.gov/news/3124/global-climate-change-impact-on-crops-expected-within-10-years-nasa-study-finds/

[161]Coffee Semi-annual—https://apps.fas.usda.gov/newgainapi/api/Report/DownloadReportByFileName?fileName=Coffee%20Semi-annual_Hanoi_Vietnam_11-15-2020

[162]Helping Vietnam's Coffee Sector Become More Climate Resilient—https://news.climate.columbia.edu/2020/11/13/vietnam-coffee-climate-resilient/

[163]Expected global suitability of coffee, cashew and avocado due to climate change—https://journals.plos.org/plosone/article?id=10.1371/journal.pone.0261976

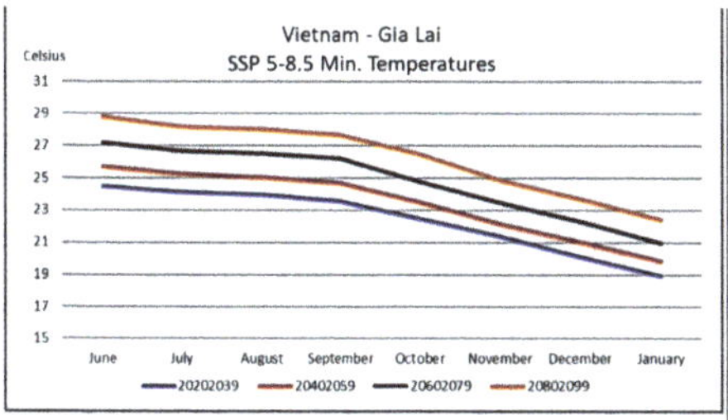

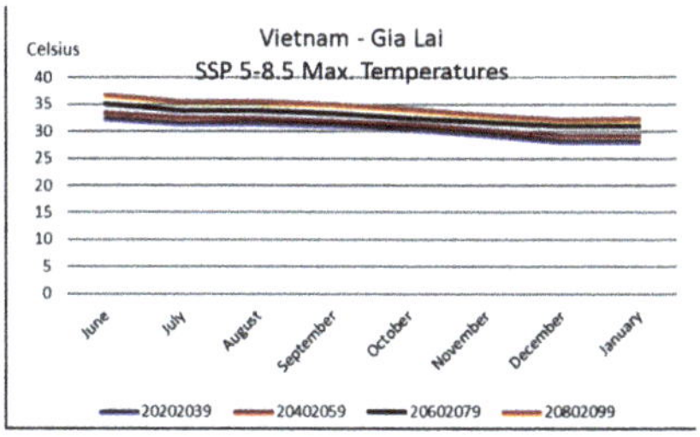

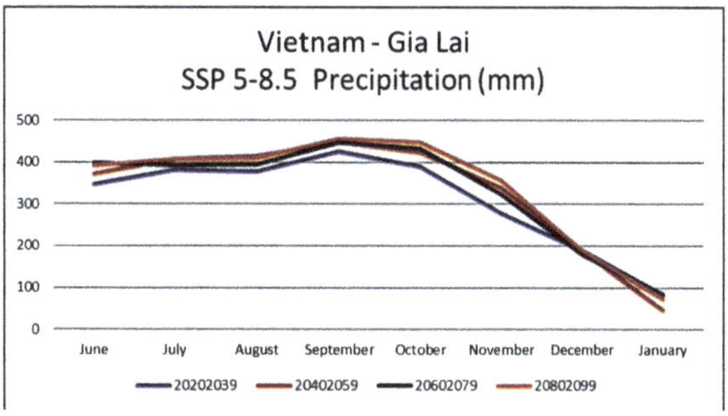

SSP1-1.9—Please find SSP1_1.9 model invocation for Vietnam, Gia Lai. The test data is downloaded from Climate Change Knowledge Portal for Gai Lai Climatology. The output of the model is compared with baseline 2018 production data:

```
cmip6_ssp119_2020_2039_p90_VNM = np.array([[618879,
            30.4,29.65,30.13,29.53,30.08,28.8,27.54,25.16,
22.55,23.14,23.02,22.5,21.72,22.02,20.03,18.62,
135.2,166.49,161.24,351.26,331.97,303.88,100.8,38.83],
             [618879,
             32.32,31.08,30.8,30.41,30.27,29.19,28,27.72,
24.34,23.78,23.74,23.32,22.54,21.48,19.97,18.54,
338.27,368.35,409.17,504.59,417.77,277.06,145.2,52.61]
             ])

modelpklfileVietnamCoffeeProduction.predict
(cmip6_ssp119_2020_2039_p90_VNM[0:2])

cmip6_ssp119_2040_2069_p90_VNM = np.array([[618879,
            30.4,29.65,30.13,29.53,30.08,28.8,27.54,25.16,
22.55,23.14,23.02,22.5,21.72,22.02,20.03,18.62,
135.2,166.49,161.24,351.26,331.97,303.88,100.8,38.83],
             [618879,
             32.32,31.08,30.8,30.41,30.27,29.19,28,27.72,
24.34,23.78,23.74,23.32,22.54,21.48,19.97,18.54,
```

(continued)

```
338.27,368.35,409.17,504.59,417.77,277.06,145.2,52.61]
              ])

modelpklfileVietnamCoffeeProduction.predict
(cmip6_ssp119_2040_2069_p90_VNM[0:2])

cmip6_ssp119_2060_2079_p90_VNM = np.array([[618879,
              30.4,29.65,30.13,29.53,30.08,28.8,27.54,25.16,
22.55,23.14,23.02,22.5,21.72,22.02,20.03,18.62,
135.2,166.49,161.24,351.26,331.97,303.88,100.8,38.83],
              [618879,
              32.2,31.21,30.86,30.52,30.17,29.14,28.03,28.01,
24.33,23.86,23.67,23.45,22.39,21.5,19.97,18.74,
368.07,372.68,390.66,510.66,441.82,263.23,149.3,60.73]
              ])

modelpklfileVietnamCoffeeProduction.predict
(cmip6_ssp119_2060_2079_p90_VNM[0:2])

cmip6_ssp119_2080_2099_p90_VNM = np.array([[618879,
              30.4,29.65,30.13,29.53,30.08,28.8,27.54,25.16,
22.55,23.14,23.02,22.5,21.72,22.02,20.03,18.62,
135.2,166.49,161.24,351.26,331.97,303.88,100.8,38.83],
              [618879,
              32.16,31.26,30.78,30.51,30.17,29.21,28.17,28.2,
24.19,23.82,23.65,23.39,22.37,21.23,19.85,18.71,
354.04,369.7,388.46,541.33,426.91,275.32,142.42,72.05]
              ])

modelpklfileVietnamCoffeeProduction.predict
(cmip6_ssp119_2080_2099_p90_VNM[0:2])
```

Output:

Baseline, Vietnam coffee production for 2018; production, 1,568,899.65 (tons)			
20,202,039	20,402,059	20,602,079	20,802,099
1,279,063.38 (tons)	1,279,909.34 (tons)	1,287,493.06 (tons)	1,261,400.95 (tons)
Drop in production from 2018; production, 1,568,899.65 (tons			
−23.15%	−6.82%	9.62%	21.84%

SSP1-1.9 is the desirable climate change outcome that we would like to see to happen. Nonetheless, a coffee production drop can be observed across all time periods **18.47 (2020–2039), 18.41 (2040–2059), 17.93 (2060–2079)**, and **19.59 (2080–2099)** due to climate change.

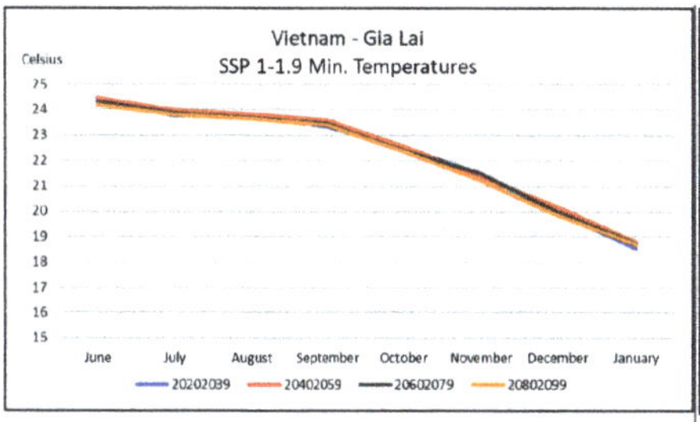

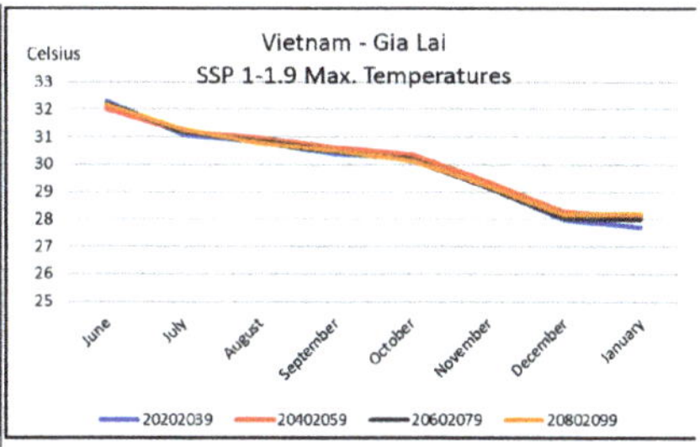

- The month of November (300 mm) plays an important role for Central Highlands. Heavy rains during November could result in huge yield losses such as ripening fruit to fall to the ground and rotten.

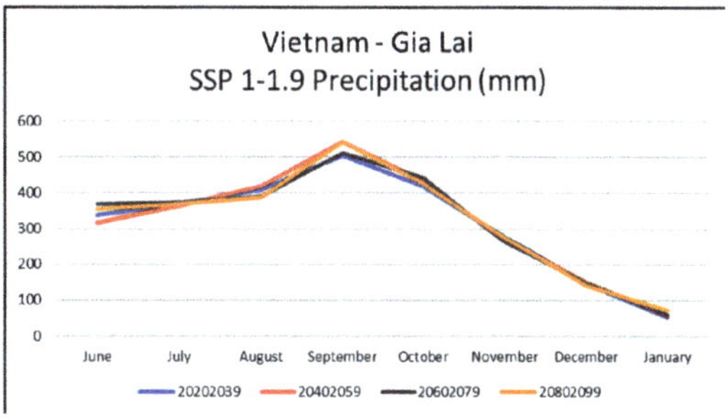

SSP3-7.0:

- Collection: CMIP6
- Aggregation: Monthly
- Calculation: Climatology mean
- Percentile: 90th
- Model: Multi-Model Ensemble

Please find the table with all three main variables for scenario SSP3-7.0.

Minimum temperature	Maximum temperature	Precipitation
Time period: 2020–2039	Time period: 2020–2039	Time period: 2020–2039
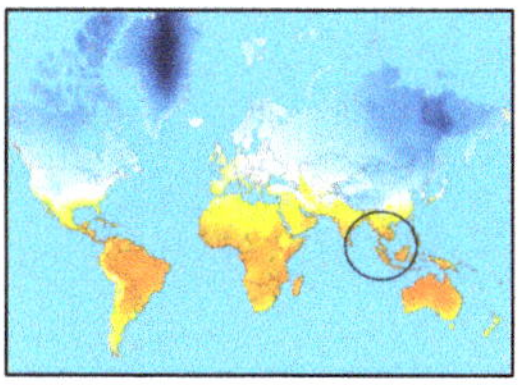	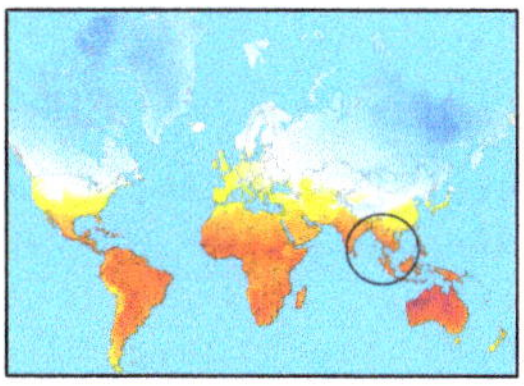	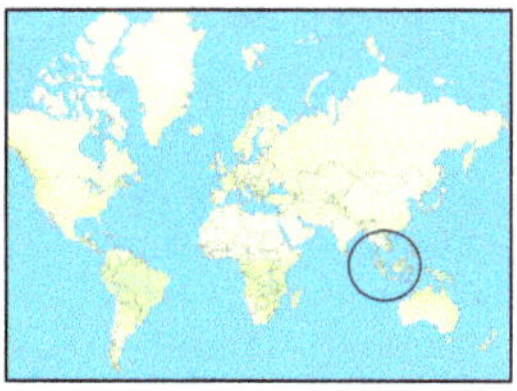

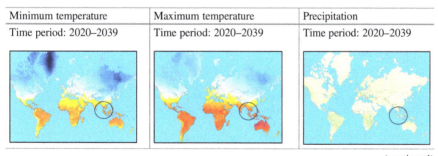

(continued)

Minimum temperature	Maximum temperature	Precipitation
Time period: 2040–2059 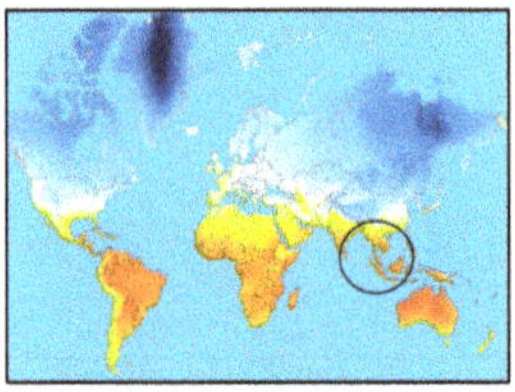	Time period: 2040–2059 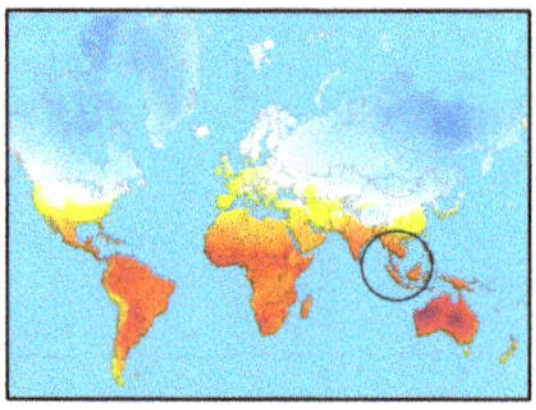	Time period: 2040–2059 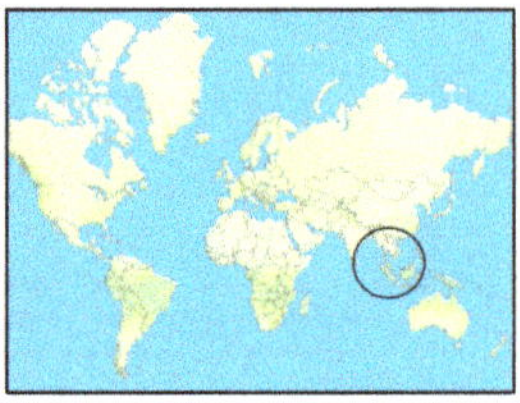
Time period: 2060–2079 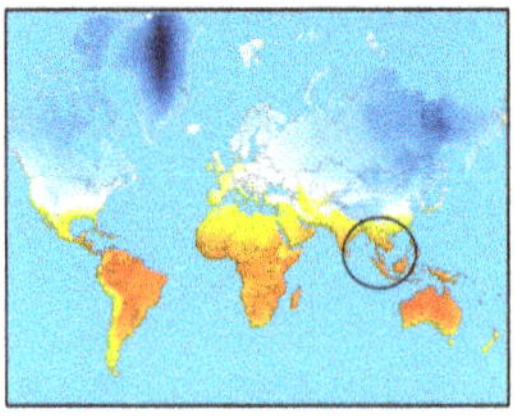	Time period: 2060–2079 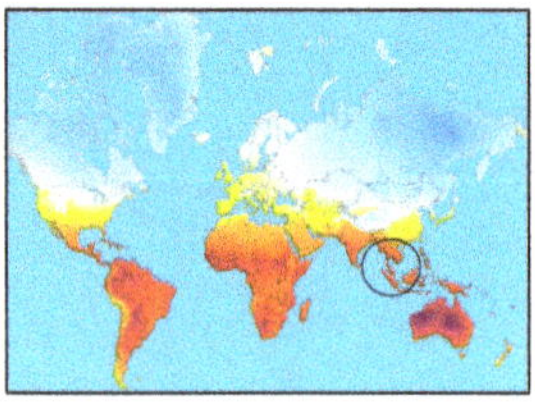	Time period: 2060–2079 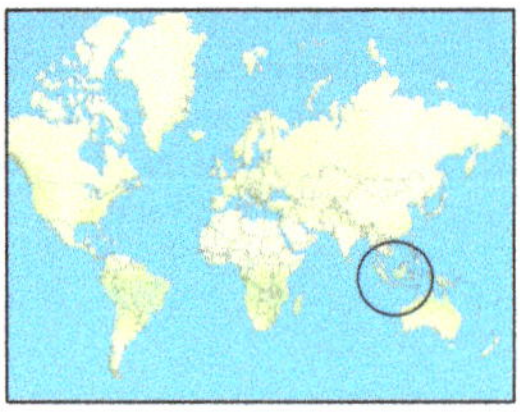
Time period: 2080–2099 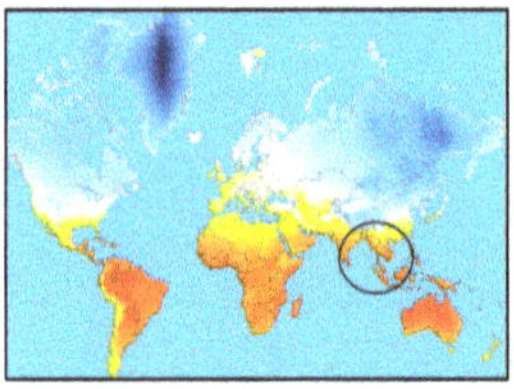	Time period: 2080–2099 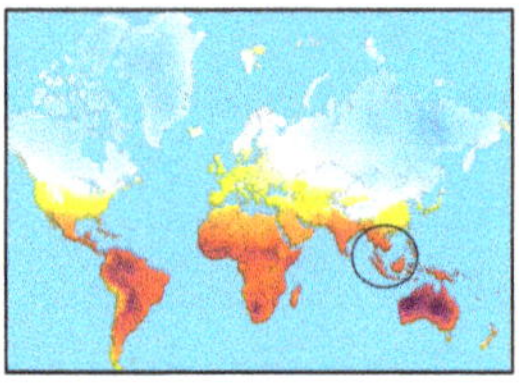	Time period: 2080–2099 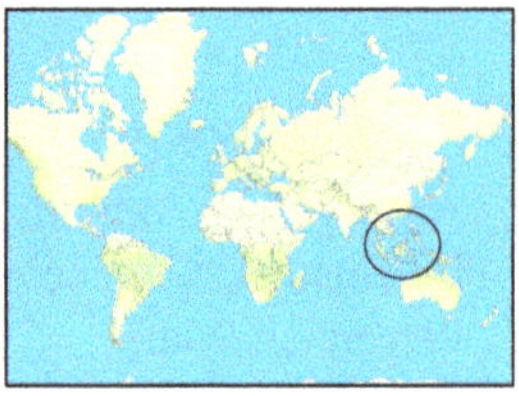

```
cmip6_ssp370_2020_2039_p90_VNM = np.array([[618879,
          30.4,29.65,30.13,29.53,30.08,28.8,27.54,25.16,
22.55,23.14,23.02,22.5,21.72,22.02,20.03,18.62,
135.2,166.49,161.24,351.26,331.97,303.88,100.8,38.83],
          [618879,
          32.35,31.44,31.37,30.93,30.3,29.04,27.97,27.58,
24.39,23.82,23.75,23.43,22.35,21.28,19.82,18.69,
335.91,371.66,375.77,404.14,362.35,301.46,172.17,79.82]
          ])
modelpklfileVietnamCoffeeProduction.predict
(cmip6_ssp370_2020_2039_p90_VNM[0:2])
```

Baseline, Vietnam coffee production for 2018; production, 1,568,899.65 (tons)			
20,202,039	20,402,059	20,602,079	20,802,099
1,314,431.95 (tons)	1,473,165.06 (tons)	1,510,358.35 (tons)	1,707,031.85 (tons)
Drop in production from 2018; production, 1,568,899.65 (tons)			
−16.21%	−6.10%	−3.73%	8.80%

SSP3-7.0 is the desirable climate change outcome that we would like to see to happen. Nonetheless, a coffee production drop can be observed across all time periods 16.21 (2020–2039), 6.10 (2040–2059), and 3.73 (2060–2079), except 8.80 (2080–2099). Higher minimum and maximum temperature (2080–2099) have net growth effect on coffee beans.

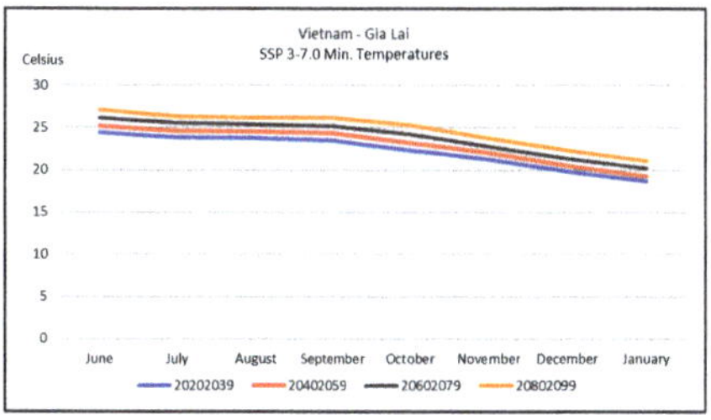

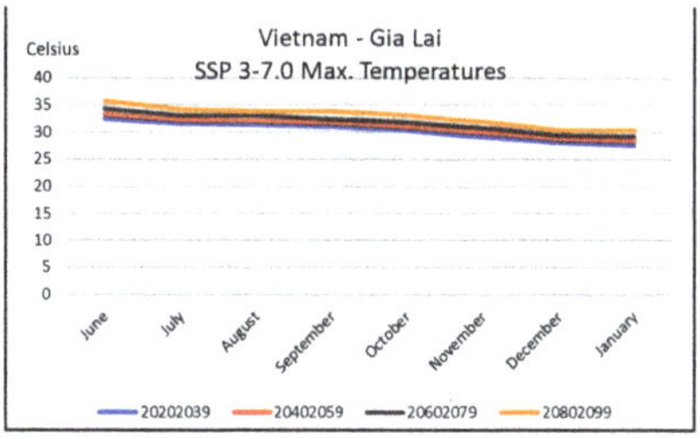

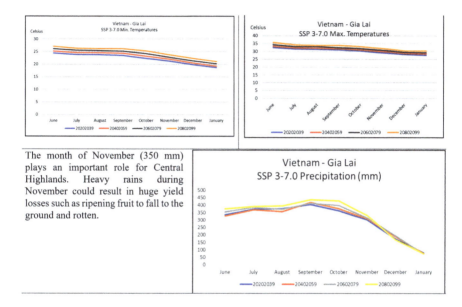

The month of November (350 mm) plays an important role for Central Highlands. Heavy rains during November could result in huge yield losses such as ripening fruit to fall to the ground and rotten.

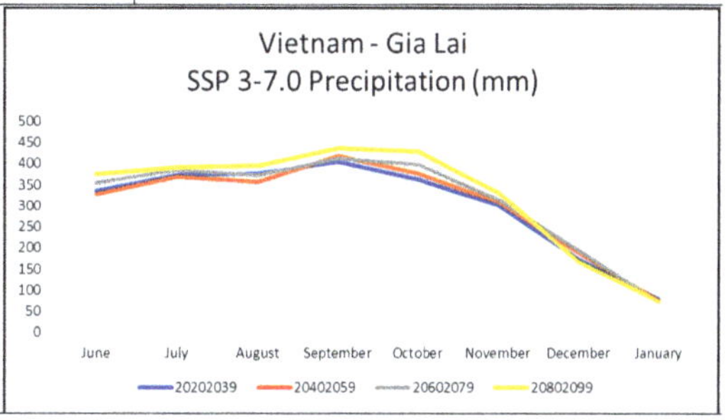

From above SSP5-8.5, SSP3-7.0, and SSP1-1.9, the positive production is highly correlated with higher minimum temperatures (27 °C or higher) for June and July months. As mentioned above, at lower temperatures, coffee leaf rust should be weaker.

Output:

As it can be seen from the model (please see Fig. 6.41), climate change would impact the production of coffee in Vietnam. Client-smart technologies and governmental policies toward climate change are better ways to tackle this future risk[164]. Coffee is of high socioeconomic importance in many tropical smallholder farming systems around the globe. As plantation crops with a long lifespan, coffee cultivation requires long-term planning. The evaluation of climate change impacts on their biophysical suitability is therefore essential for developing adaptation measures and selecting appropriate varieties or crops. In the above model, we modeled the current and future production/yield of coffee arabica on a global scale based on climatic. We have documented climate outputs based on three emission scenarios to model the future (2050) climate change impacts on the crop in the main producing Vietnam. Climatic factors, mainly long dry seasons, mean temperatures (high and

[164] Thailand—https://ccpi.org/country/tha/

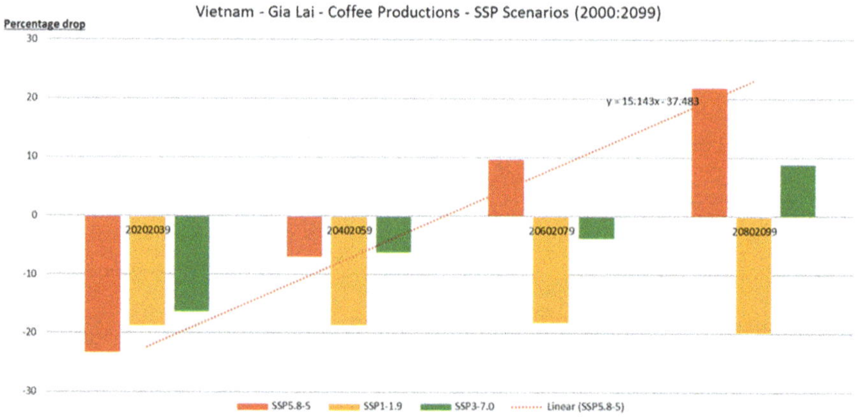

Fig. 6.41 Vietnam Gia Lai coffee production SSPs

low), low minimum temperatures, and annual precipitation (high and low), were more restrictive, and coffee proved to be most vulnerable, with negative climate impacts dominating in all main producing regions. At high latitudes and high altitudes, however, coffee may all profit from increasing minimum temperatures[165] as you can see in 2080 period for SSP5-8.5 [67]. Climate change undoubtedly decreases coffee production and drastically increases the price of coffee. I am certain if proper measures and adaptation policies are not considered, we will wake up to higher coffee prices that, for many, become unaffordable[166,167,168,169] [68, 69].

Brazil and Thailand

While rising temperatures have caught many industries flat-footed, coffee companies have responded in force, bolstering their presence on the ground in coffee-growing countries like Costa Rica, Ethiopia, and Indonesia. Instead of just purchasing coffee, they work with small farms to help them adapt to changing conditions, providing seeds, monitoring production, and suggesting new agricultural practices. "Everybody talks about climate, but the only sector that's actually doing something at scale is the coffee industry," M. Sanjayan, the CEO of Conservation International, tells me as we tour the Starbucks farm in Alajuela.

---

[165] Expected global suitability of coffee, cashew and avocado due to climate change—https://journals.plos.org/plosone/article?id=10.1371/journal.pone.0261976

[166] The daily cup becoming an expensive habit—http://archive.boston.com/business/articles/2011/05/09/coffee_prices_hitting_historic_highs/

[167] Coffee prices expected to rise as a result of poor harvests and growing Demand—https://www.theguardian.com/business/2011/apr/21/commodities-coffee-shortage-price-rise-expected

[168] Coffee is a costly wake-up—https://money.cnn.com/1997/03/06/economy/coffee_prices_pkg/

[169] The cost of a cup of coffee: Then and Now—https://www.mercurynews.com/2016/07/18/the-cost-of-a-cup-of-coffee-then-and-now/

In Costa Rica, farmers say climate change has already made it difficult to predict harvest conditions. "One year it's too short, one year it's too long," says Corrales Cruz of the rainy season. The nation's coffee industry is suffering as a result. Fifteen years ago, the average coffee farm in Costa Rica produced about 14.5 bags of coffee per acre. Today that number is down to fewer than ten bags per acre.

And no matter how much money the industry pumps into research, or how many boots companies put on the ground for retraining, the future of coffee remains at the mercy of a global population that continues to pump greenhouse gases into the atmosphere. Average temperatures are expected to rise by more than 5.5 °F by 2100 even if countries follow through on their commitments to reduce global warming, according to data from the UN Environmental Programme. That's far short of the UN goal of keeping temperature rise below 3.6 °F.

After reading this chapter, you should be able to answer agriculture and climate change, FAO in emergencies countries, climate change and most affected countries, FAO food emergencies and climate change—high-priority list of countries, food security and natural resources and resilience, the Coupled Model Intercomparison Project Phase 6 (CMIP6) climate projections, machine learning model: Thailand rice paddy yields and climate change, and machine learning model: Vietnam coffee yields and climate change.

References

1. Issues and Challenges of Inclusive Development: Essays in Honor of Prof. R. Radhakrishna by R. Maria Saleth, S. Galab and E. Revathi, Publisher: Springer; 1st ed. 2020 edition (June 19, 2020), ISBN-13: 978-9811522284
2. Anikka Martin, Food Security and Nutrition Assistance, Last updated: Monday, November 08, 2021, https://www.ers.usda.gov/data-products/ag-and-food-statistics-charting-the-essentials/food-security-and-nutrition-assistance/, Access Date: December 12, 2021
3. Alisha Coleman-Jensen, Matthew P. Rabbitt, Laura Hales, and Christian A. Gregory, The prevalence of food insecurity in 2020 is unchanged from 2019, https://www.ers.usda.gov/data-products/chart-gallery/gallery/chart-detail/?chartId=58378, Access Date: December 20, 2021
4. lisha Coleman-Jensen, Matthew P. Rabbitt, Laura Hales, and Christian A. Gregory, Prevalence of food insecurity is not uniform across the country, Last updated: Monday, November 08, 2021, https://www.ers.usda.gov/data-products/chart-gallery/gallery/chart-detail/?chartId=58392, Access Date: December 21, 2021
5. Pam Kan-Rice, Urban gardens improve food security, May 31 2016, https://www.universityofcalifornia.edu/news/urban-gardens-improve-food-security, Access Date: January 28, 2022
6. JOHN BAFFES and WEE CHIAN KOH, Soaring fertilizer prices add to inflationary pressures and food security concerns, NOVEMBER 15, 2021, https://blogs.worldbank.org/opendata/soaring-fertilizer-prices-add-inflationary-pressures-and-food-security-concerns, Access Date: February 12, 2022

7. Keith Good, Fertilizer Prices Continue to Climb, 2022 Planted Acreage Analysis Continues, December 5, 2021, https://farmpolicynews.illinois.edu/2021/12/fertilizer-prices-continue-to-climb-2022-planted-acreage-analysis-continues/#:~:text=The%20nitrogen%20fertilizer's%20average%20price,%2C%20an%20all%2Dtime%20high, Access Date: February 12, 2022
8. FAO, IFAD, UNICEF, WFP and WHO. 2017. The State of Food Security and Nutrition in the World 2017. Building resilience for peace and food security. Rome, FAO. ISBN:978-92-5-109888-2
9. FAO, FAO'S WORK ON CLIMATE CHANGE United Nations Climate Change Conference 2017, 2017, https://www.fao.org/3/i8037e/i8037e.pdf, Access Date: January 30, 2022
10. Will Horner and Jeffrey T. Lewis, Coffee Prices Soar After Bad Harvests and Insatiable Demand, July 12, 2021, 8:41 am ET, https://www.wsj.com/articles/coffee-prices-soar-after-bad-harvests-and-insatiable-demand-11626093703?mod=article_inline, Access Date: February 03, 2022
11. Jeffrey T. Lewis and Joe Wallace, Coffee Prices Jump to Six-Year High as Brazilian Frost Threatens Crop, Updated July 27, 2021 4:50 pm ET, https://www.wsj.com/articles/coffee-prices-jump-to-six-year-high-as-brazilian-frost-threatens-crop-11627380128?mod=article_inline, Access Date: February 03, 2022
12. David Eckstein, Vera Künzel, Laura Schäfer, GLOBAL CLIMATE RISK INDEX 2021 - Who Suffers Most from Extreme Weather Events? Weather-Related Loss Events in 2019 and 2000-2019, January 2021, ISBN 978-3-943704-84-6, https://www.germanwatch.org/sites/default/files/Global%20Climate%20Risk%20Index%202021_2.pdf
13. Alexandra Rostiaux, Food insecurity, the other threat of COVID-19 for Colombia's indigenous communities, 26 Feb 2021, https://www.unocha.org/story/food-insecurity-other-threat-covid-19-colombia%E2%80%99s-indigenous-communities, Access Date: February 01, 2022
14. Olivier Ecker, Economics of Micronutrient Malnutrition: The Demand for Nutrients in Sub-Saharan Africa - ISBN-13: 978-3631595053
15. SAMUEL BENIN, JAMES THURLOW, XINSHEN DIAO, CHRISTEN MCCOOL AND FRANKLIN, Agricultural growth and investment options for poverty reduction in Malawi, 2008, https://www.ifpri.org/publication/agricultural-growth-and-investment-options-poverty-reduction-malawi-0, Access Date: January 29, 2022
16. KARL PAUW, JAMES THURLOW AND DIRK VAN SEVENTER, Droughts and floods in Malawi Assessing the economywide effects, 2010, https://www.ifpri.org/publication/droughts-and-floods-malawi, Access Date: January 29, 2022
17. The Ministry of Natural Resources and Environmental Conservation (MONREC), Myanmar Climate Change Strategy (2018 – 2030), 2019, https://myanmar.un.org/sites/default/files/2019-11/MyanmarClimateChangeStrategy_2019.pdf, Access Date: January 29, 2022
18. Jointly led by the Ministry of Agriculture and Irrigation; Ministry of Livestock, Fisheries & Rural Development; FAO and WFP under the framework of the Food Security Sector in partnership with UN Women, World Vision, CESVI, CARE, JICA and LIFT., Agriculture and Livelihood Flood Impact Assessment in Myanmar, Oct 2015, https://www.fao.org/fileadmin/user_upload/emergencies/docs/Final_Impact_Assessment_Report_final.pdf, Access Date: January 29, 2022
19. HTOO THANT, less-crop-damage-floods-year-official, 16 August 2019, https://www.mmtimes.com/news/less-crop-damage-floods-year-official.html, Access Date: January 29, 2022
20. Ms Bui Thi Lan, Myanmar floods: huge impact on agricultural livelihoods, more international support urgently needed, December 2015, https://www.fao.org/fileadmin/user_upload/emergencies/docs/Final%20fact%20sheet%201%20-%20web%20version.pdf, Access Date: March 11, 2022
21. Rashid Yaqoob Raja and Lisa Anderson, Exporter Guide, January 08,2021, https://apps.fas.usda.gov/newgainapi/api/Report/DownloadReportByFileName?fileName=Exporter%20Guide_Islamabad_Pakistan_12-31-2020, Access Date: February 12, 2022

22. Dr. Imtiaz Ahmad, Pakistan Economic Survey 2020-2021, June 2021, https://www.pc.gov.pk/uploads/cpec/PES_2020_21.pdf, Access Date: March 11, 2022
23. SCHALK BURGER, Resilience, natural resources become key as global food security deteriorates slightly, 5TH MARCH 2021, https://www.engineeringnews.co.za/article/resilience-natural-resources-become-key-as-global-food-security-deteriorates-slightly-2021-03-05, Access Date: January 22, 2022
24. Ellen Gray, Global Climate Change Impact on Crops Expected Within 10 Years, NASA Study Finds, November 2, 2021, https://climate.nasa.gov/news/3124/global-climate-change-impact-on-crops-expected-within-10-years-nasa-study-finds/, Access Date: January 28, 2022
25. Jouzel, J., V. Masson-Delmotte, O. Cattani, G. Dreyfus, S. Falourd, G. Hoffmann, B. Minster, J. Nouet, J.M. Barnola, J. Chappellaz, H. Fischer, J.C. Gallet, S. Johnsen, M. Leuenberger, L. Loulergue, D. Luethi, H. Oerter, F. Parrenin, G. Raisbeck, D. Raynaud, A. Schilt, J. Schwander, E. Selmo, R. Souchez, R. Spahni, B. Stauffer, J.P. Steffensen, B. Stenni, T.F. Stocker, J.L. Tison, M. Werner, and E.W. Wolff. 2007. Orbital and Millennial Antarctic Climate Variability over the Past 800,000 Years. Science, Vol. 317, No. 5839, pp.793-797, 10 August 2007.
26. ZEKE HAUSFATHER, CMIP6: the next generation of climate models explained, 02.12.2019 | 8:00am, https://www.carbonbrief.org/cmip6-the-next-generation-of-climate-models-explained, Access Date: February 03, 2022
27. UNECE, SSP Overview, 2019, https://www.unece.org/fileadmin/DAM/energy/se/pdfs/CSE/PATHWAYS/2019/ws_Consult_14_15.May.2019/supp_doc/SSP2_Overview.pdf, Access Date: February 03, 2022
28. Andrea Januta, Explainer: The U.N. climate report's five futures - decoded, August 9, 2021, 1:07 AM PDT, https://www.reuters.com/business/environment/un-climate-reports-five-futures-decoded-2021-08-09/, Access Date: February 03, 2022
29. Keywan Riahi, Detlef P. van Vuuren, Elmar Kriegler, Jae Edmonds, Brian C. O'Neill, Shinichiro Fujimori, Nico Bauer, Katherine Calvin, Rob Dellink, Oliver Fricko, Wolfgang Lutz, Alexander Popp, Jesus Crespo Cuaresma, Samir KC, Marian Leimbach, Leiwen Jiang, Tom Kram, Shilpa Rao, Johannes Emmerling, Kristie Ebi, Tomoko Hasegawa, Petr Havlik, Florian Humpenöder, Lara Aleluia Da Silva, Steve Smith, Elke Stehfest, Valentina Bosetti, Jiyong Eom, David Gernaat, Toshihiko Masui, Joeri Rogelj, Jessica Strefler, Laurent Drouet, Volker Krey, Gunnar Luderer, Mathijs Harmsen, Kiyoshi Takahashi, Lavinia Baumstark, Jonathan C. Doelman, Mikiko Kainuma, Zbigniew Klimont, Giacomo Marangoni, Hermann Lotze-Campen, Michael Obersteiner, Andrzej Tabeau, Massimo Tavoni, The Shared Socioeconomic Pathways and their energy, land use, and greenhouse gas emissions implications: An overview, Global Environmental Change, Volume 42, 2017, Pages 153-168, ISSN 0959-3780, https://doi.org/10.1016/j.gloenvcha.2016.05.009.
30. Daigneault, Adam; Johnston, Craig; Korosuo, Anu; Baker, Justin S.; Forsell, Nicklas; Prestemon, Jeffrey P.; Abt, Robert C. 2019. Developing detailed Shared Socioeconomic Pathway (SSP) narratives for the Global Forest Sector. Journal of Forest Economics. 34(1-2): 7-45. https://doi.org/10.1561/112.00000441.
31. NOAA, Climate Model - Sea Surface Temperature Change: SSP2 (Middle of the Road) - 2015 - 2100, 2022, https://sos.noaa.gov/catalog/datasets/724/, Access Date: March 11, 2022
32. Davide Viaggi, The Bioeconomy: Delivering Sustainable Green Growth, Publisher : CABI (January 31, 2019), ISBN-13 : 978-1786392756, ASIN : B07LCRW5F7, Publication Date: November 30, 2018
33. Joachim Klement, Geo-Economics: The Interplay between Geopolitics, Economics, and Investments, Publisher : CFA Institute Research Foundation (April 28, 2021), ISBN-13: 978-1952927065
34. Govt. Thailand, Thailand: Over 20,000 million rai of farming area in Nakhon Sawan affected by flooding, 30 Sep 2008, https://reliefweb.int/report/thailand/thailand-over-20000-million-rai-farming-area-nakhon-sawan-affected-flooding, Access Date: February 04, 2022

35. Bangkok Post Online Reporters, PM in Nakhon Sawan, Chai Nat on drought inspection, Published: 22/01/2016 at 02:00 PM, https://www.bangkokpost.com/print/835884/, Access Date: February 04, 2022
36. Josep Cos, Francisco Doblas-Reyes1,Martin Jury, Raül Marcos, Pierre-Antoine Bretonnière, and Margarida Samsó, The Mediterranean climate change hotspot in the CMIP5 and CMIP6 projections, 30 July 2021, https://esd.copernicus.org/preprints/esd-2021-65/esd-2021-65.pdf, Access Date: February 04, 2022
37. James Mackintosh, Investing to Stop Climate Change Is Trickier Than It Seems, Updated Jan. 26, 2022, 11:24 am ET, https://www.wsj.com/articles/investing-to-stop-climate-change-is-trickier-than-it-seems-11643214062?page=1, Access Date: February 01, 2022
38. Nathan Childs and James Kiawu, Factors Behind the Rise in Global Rice Prices in 2008, May 2009, https://www.ers.usda.gov/publications/pub-details/?pubid=38490, Access Date: March 04, 2022
39. Nathan Childs, Global trading prices for rice rose to highest level in 7 years due to export restrictions, May 27, 2020, https://www.ers.usda.gov/data-products/chart-gallery/gallery/chart-detail/?chartId=98504, Access Date: March 04, 2022
40. McKenzie, A. M., Jiang, B., Djunaidi, H., Hoffman, L. A., & Wailes, E. J. (2002). Unbiasedness and Market Efficiency Tests of the U.S. Rice Futures Market. Review of Agricultural Economics, 24(2), 474–493. http://www.jstor.org/stable/1349773
41. The World Bank, The World Bank Supports Thailand's Post-Floods Recovery Effort, December 13, 2011, https://www.worldbank.org/en/news/feature/2011/12/13/world-bank-supports-thailands-post-floods-recovery-effort, Access Date: February 04, 2022
42. Poapongsakorn, N. and P. Meethom (2012), 'Impact of the 2011 Floods, and Flood Management in Thailand', in Sawada, Y. and S. Oum (eds.), Economic and Welfare Impacts of Disasters in East Asia and Policy Responses. ERIA Research Project Report 2011-8, Jakarta: ERIA. pp.247-310.
43. Roderick J A Little and Donald B Rubin (1986). "Statistical Analysis with Missing Data". John Wiley & Sons, Inc., New York, NY, USA.ISBN-13: 978-0470526798
44. Lisa Sullivan, PhD, Boston University School of Public Health, The Role of Probability, Date last modified: July 24, 2016, https://sphweb.bumc.bu.edu/otlt/mph-modules/bs/bs704_probability/index.html, Access Date: February 12, 2022
45. NIST/SEMATECH e-Handbook of Statistical Methods, http://www.itl.nist.gov/div898/handbook/, February 12, 2022
46. Susan Li, A Quick Introduction On Granger Causality Testing For Time Series Analysis, Dec 23, 2020, https://towardsdatascience.com/a-quick-introduction-on-granger-causality-testing-for-time-series-analysis-7113dc9420d2, Access Date: February 12, 2022
47. The Secretariat of the International Rice Commission Crop and Grassland Service Plant Production and Protection Division Agriculture Department, FAO Rice Information, Volume 3, December 2002, 2002, https://www.fao.org/3/Y4347E/y4347e00.htm#Contents, Access Date: February 12, 2022
48. Sakurai, G., Yokozawa, M., Nishimori, M., and Okada, M., "Future possible crop yield scenarios under multiple SSP and RCP scenarios.", vol. 2016, 2016.
49. Tatiana Schlossberg, Coffee and Climate Have a Complicated Relationship, Oct. 31, 2021, https://www.nytimes.com/2021/10/31/business/coffee-climate-change.html#:~:text=According%20to%20a%202014%20study,and%20Vietnam%2C%20major%20producing%20countries, Access Date: January 23, 2022
50. JUSTIN WORLAND/ALAJUELA COSTA RICA, Your Morning Cup of Coffee Is in Danger. Can the Industry Adapt in Time, JUNE 21, 2018 6:28 AM EDT, https://time.com/5318245/coffee-industry-climate-change/, Access Date: January 23, 2022
51. Tan Phan-Van and Hiep Van Nguyen, A review of evidence of recent climate change in the Central Highlands of Vietnam, December 2013, https://www.researchgate.net/publication/293109260_A_review_of_evidence_of_recent_climate_change_in_the_Central_Highlands_of_Vietnam, Access Date: February 03, 2022

52. Jeffrey T. Lewis and Joe Wallace, Coffee Prices Jump to Six-Year High as Brazilian Frost Threatens Crop, Updated July 27, 2021, 4:50 pm ET, https://www.wsj.com/articles/coffee-prices-jump-to-six-year-high-as-brazilian-frost-threatens-crop-11627380128?mod=article_inline, Access Date: February 03, 2022
53. ERIC PFANNER AND BLOOMBERG, Starbucks has a long way to go to reach environmental goals for 2030, January 21, 2020, 3:00 PM PST, https://fortune.com/2020/01/21/starbucks-carbon-footprint-dairy/, Access Date: March 11, 2022
54. Damiano Ciro, Suitability Study for Coffee Arabica cultivation in relation to climate change., Dec 1, 2021, https://medium.com/@damiano.ciro/suitability-study-for-coffee-arabica-cultivation-in-relation-to-climate-change-fecab562ed6, Access Date: January 23, 2022
55. Tatiana Freitas, Coffee Reserves Plunge to Lowest in More Than Two Decades,Mon, February 7, 2022, 12:15 PM, https://finance.yahoo.com/news/world-coffee-reserves-plunge-lowest-193610832.html, Access Date: February 09, 2022
56. JOSEPH CONWAY, Helping Vietnam's Coffee Sector Become More Climate Resilient, NOVEMBER 13, 2020, https://news.climate.columbia.edu/2020/11/13/vietnam-coffee-climate-resilient/, Access Date: February 08, 2022
57. M. Shahbandeh, Coffee production worldwide in 2020, by leading country (in 1,000 60 kilogram bags)*, Mar 23, 2021, https://www.statista.com/statistics/277137/world-coffee-production-by-leading-countries/, Access Date: February 06, 2022
58. Chris Summers, How Vietnam became a coffee giant, 25 January 2014, https://www.bbc.com/news/magazine-25811724, Access Date: February 06, 2022
59. Nicholas Castellano, A breakdown of Vietnamese coffee-producing regions,December 1, 2021, https://perfectdailygrind.com/2021/12/a-breakdown-of-vietnamese-coffee-producing-regions/#:~:text=As%20mentioned%20previously%2C%20the%20bulk,Lam%20Dong%2C%20and%20Kontum.%E2%80%9D, Access Date: February 06, 2022
60. Maja Wallengren, BREAKING: Vietnam's 2021-22 Coffee Harvest Down on Dry Weather, September 15, 2021, https://stir-tea-coffee.com/tea-coffee-news/breaking-vietnam%E2%80%99s-2021-22-coffee-harvest-down-on-dry-weathe/, Access Date: February 06, 2022
61. Chungmann Kim, Cristina Alvarez, Abdul Sattar, Arkadeep Bandyopadhyay,Carlo Azzarri, Ana Moltedo, Beliyou Haile, Production, Consumption, and Food Security in Viet Nam Diagnostic Overview, January 7, 2021, https://inddex.nutrition.tufts.edu/sites/default/files/Vietnam%20Diagnostic%20Overview%20Sept%2023%5B1%5D.pdf, Access Date: February 06, 2022
62. Hai, N.N.; Anh, N.T.; Ky, N.M.; Dung, B.Q.; Huong, N.T.N.; Minh, N.H.D.; Ly, N.T. Assessing land use change in the context of climate change and proposing solutions: Case study in Gia Lai province, Vietnam. VN J. Hydrometeorol. 2021, 7, 20-31.
63. Tanya Newton, Coffee Quality & M.A.S.L.: How Important Is Elevation REALLY?, January 16, 2018, https://perfectdailygrind.com/2018/01/coffee-quality-m-a-s-l-how-important-is-altitude-really/, Access Date: February 06, 2022
64. Emma Sage, Some Insights on Coffee Leaf Rust, 2013, https://scanews.coffee/2013/02/15/some-insights-on-coffee-leaf-rust-hemileia-vastatrix/, Access Date: February 06, 2022
65. Neil Soque, How to Monitor For & Prevent Coffee Leaf Rust, April 22, 2019, https://perfectdailygrind.com/2019/04/how-to-monitor-for-prevent-coffee-leaf-rust/, Access Date: February 08, 2022
66. Thanh Vo, Report Name: Coffee Semi-annual, November 17,2020, https://apps.fas.usda.gov/newgainapi/api/Report/DownloadReportByFileName?fileName=Coffee%20Semi-annual_Hanoi_Vietnam_11-15-2020, Access Date: February 07, 2022
67. Grüter R, Trachsel T, Laube P, Jaisli I (2022) Expected global suitability of coffee, cashew and avocado due to climate change. PLoS ONE 17(1): e0261976. https://doi.org/10.1371/journal.pone.0261976
68. Kathleen Pierce, the daily cup becoming an expensive habit, May 9, 2011, http://archive.boston.com/business/articles/2011/05/09/coffee_prices_hitting_historic_highs/, Access Date: February 21, 2022

69. Julia Kollewe, Coffee prices expected to rise as a result of poor harvests and growing Demand, Thu 21 Apr 2011 10.33 EDT, https://www.theguardian.com/business/2011/apr/21/commodities-coffee-shortage-price-rise-expected, Access Date: February 21, 2022
70. Rhonda Schaffler, Coffee is a costly wake-up, March 6, 1997: 7:08 p.m. ET, https://money.cnn.com/1997/03/06/economy/coffee_prices_pkg/, Access Date: February 21, 2022
71. MERCURY NEWS, The cost of a cup of coffee: Then and Now, August 12, 2016 at 12:18 a.m., https://www.mercurynews.com/2016/07/18/the-cost-of-a-cup-of-coffee-then-and-now/, Access Date: February 21, 2022
72. Jiawei Han, Micheline Kamber and Jian Pei, Data Mining: Concepts and Techniques, Publisher: Morgan Kaufmann; 3 edition (June 15, 2011), ISBN-10: 9780123814791

Part III
Linkage Models

Chapter 7
Food Security and Advanced Imaging Radiometer ML Models

This chapter covers:

- Food security and advanced imaging radiometer ML models
- Dairy, food security, and satellite data
- Global vegetation: Cropland and Vegetation Index
- The Normalized Difference Vegetation Index (NDVI)
- Food prices increased driven by depreciation of national currency
- Machine learning model: Mozambique cashew nuts production model
  - CMIP6 Projections Data and SSP Scenarios
- Machine learning model: Mozambique cashew nuts and Normalized Difference Vegetation Index (NDVI) model
  - CMIP6 Projections Data and SSP Scenarios

The chapter introduces food security and advanced imaging radiometer datasets and ML models. As part of the chapter, satellite radiometer, dairy, food security and satellite data, global vegetation—Cropland and Vegetation Index, and the Normalized Difference Vegetation Index (NDVI) are also covered. Next, the chapter also covers Mozambique cashew nuts market, agriculture, and industrialization. Finally, it concludes with two machine learning models that specifically look at Mozambique cashew nuts production model and Mozambique cashew nuts and Normalized Difference Vegetation Index (NDVI) model. The chapter also summarizes cashew nuts production with findings of CMIP6 Projections Data and SSP Scenarios.

C. Vuppalapati, *Artificial Intelligence and Heuristics for Enhanced Food Security*, International Series in Operations Research & Management Science 331,
https://doi.org/10.1007/978-3-031-08743-1_7

Food security is very essential for humanity and for all. A household is not "food secure" unless it "feels" food secure.[1] Food security exists when all people, always, have physical and economic access to sufficient, safe, and nutritious food that meets their dietary needs and food preferences for an active and healthy life.[2] Food security is a global need, and every country in the world is actively working to ensure enhancing food security [1]. For instance, the US Department of Agriculture (USDA) Economic Research Service (ERS) monitors the food security of US households through an annual, nationally representative survey. While most US households are food secure, a minority of US households experience food insecurity at times during the year, meaning that their access to adequate food for active, healthy living is limited by lack of money and other resources.[3] Some experience very low food security and a more severe range of food insecurity where food intake of one or more members is reduced and normal eating patterns are disrupted [2]. In 2020, 89.5% of US households were food secure throughout the year. The remaining 10.5% of households were food insecure at least some time during the year, including 3.9% (5.1 million households) that had very low food security.[4] The prevalence[5] of food insecurity (please see Figs. 7.1 and 7.2) is not uniform across the country due to both the characteristics of populations and state-level policies and economic conditions [2–4].

Food security and climate change are the pressing reality of many low-income countries, and they "can't keep up" with impacts.[6] Historical trends indicate there is a strong relationship between food insecurity and exposure to climate hazards. As 70% of the population depend on climate-sensitive agricultural production for their food and livelihoods, increased frequency and intensity of storms, droughts, and floods are likely to pose pressure on agricultural income undermining 25% of the country's economy (e.g., Mozambique), 70% of livelihoods, as well as the food and nutrition security of the whole country.[7] Many countries need climate data and earth

[1]Issues and Challenges of Inclusive Development: Essays in Honor of Prof. R. Radhakrishna by R. Maria Saleth, S. Galab and E. Revathi, Publisher: Springer; 1st ed. 2020 edition (June 19, 2020), ISBN-13: 978-9811522284.

[2]Food Security—https://www.fao.org/fileadmin/templates/faoitaly/documents/pdf/pdf_Food_Security_Cocept_Note.pdf

[3]Food Security and Nutrition Assistance—https://www.ers.usda.gov/data-products/ag-and-food-statistics-charting-the-essentials/food-security-and-nutrition-assistance/

[4]The prevalence of food insecurity in 2020 is unchanged from 2019—https://www.ers.usda.gov/data-products/chart-gallery/gallery/chart-detail/?chartId=58378

[5]Prevalence of food insecurity is not uniform across the country—https://www.ers.usda.gov/data-products/chart-gallery/gallery/chart-detail/?chartId=58392

[6]Climate change: Low-income countries 'can't keep up' with impacts—https://www.bbc.com/news/world-58080083

[7]Food security and climate change, the pressing reality of Mozambique—https://docs.wfp.org/api/documents/WFP-0000129988/download/?_ga=2.150781622.1610345514.1644972256-1475447051.1638068907

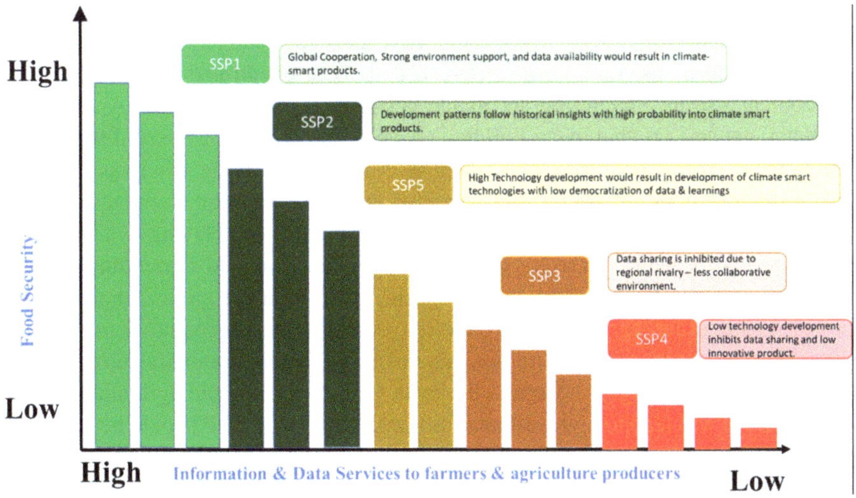

Fig. 7.1 Information and data services vs. food security

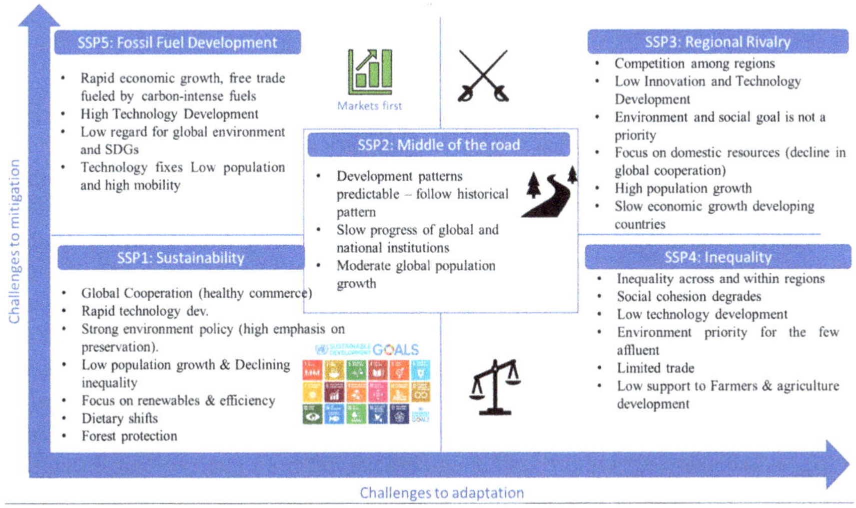

Fig. 7.2 Shared socioeconomic pathways [9] (Climate Model—Sea Surface Temperature Change: SSP2 (Middle of the Road)—2015–2100—https://sos.noaa.gov/catalog/datasets/724/)

observatory data and of course data that is localized to develop climate models.[8] This is where data from weather climate satellite play an important role [5–7].

[8] Meteorologists Can't Keep Up With Climate Change In Mozambique—https://www.npr.org/sections/goatsandsoda/2019/12/11/782918005/meteorologists-cant-keep-up-with-climate-change-in-mozambique

Data play a crucial role in adaptation and mitigating risk. For instance, historically, Caribbean islands used to see category 4 hurricanes and have prepared for with our adaptation plans. Last year, 2020–2021, the Caribbean had a record-breaking 30 tropical storms—including six major hurricanes. On islands like Antigua and Barbuda, experts say that many buildings have been unable to withstand the intense winds these storms have brought. Category 5 hurricanes bring winds as strong as 180 miles per hour which the roofs cannot withstand because it creates stronger pressure inside our houses and don't have adaptation plans. Having availability of authorized information datasets on a timely manner, related agricultural inputs, production, finance, access to market, yield, governmental credit, import/export policies, and other agricultural productivity enhancers (geospatial data, weather data (Manson, temperature, humidity, sun radiation, vegetation index), environmental satellites, data imaging spectroradiometer, market access, and agricultural financial services) will lead to high sustainability with SSP1, and SSP2 could be achieved.

"We used to see category four hurricanes, so that's what we have prepared for with our adaptation plans, but now we are being hit by category five hurricanes. Category five hurricanes bring winds as strong as 180 miles per hour which the roofs cannot withstand because it creates stronger pressure inside our houses" [5].
Chief Climate Negotiator
Alliance of Small Island States
Caribbean

Surprisingly, regions of the world that fall into a long tail, either due to lack of information exchange, regional rivalry, lack of agricultural innovation, or availability of intelligent systems that are specific to sub-national, are forced to SSP3 or SSP4 (please see Figs. 7.1 and 7.2) [8, 9].

Satellite technology plays an important role in ensuring food security. Satellite data acquisition is crucial. For instance, consider NASA Harvest Mission that ensures harvest through agriculture satellite monitoring. Satellite technologies identify crop conditions over the main growing areas for wheat, maize, rice, soybean, and other crops and overlap with datasets that are based on a combination of national and regional crop analyst inputs along with earth observation data. Crops needed to improve health can be intervened by sending real-time insights to farmers. Meteorology is critical to agriculture, and integrated dataset from meteorological satellites is a natural extension of data to generate daily forecasts of the changing weather.

ECOSTRESS
The ECOsystem Spaceborne Thermal Radiometer Experiment on Space Station (ECOSTRESS) will monitor one of the most basic processes in living plants: the loss of water through the tiny pores in leaves.

7.1 Satellite Radiometer

Satellites can be classified by their function since they are launched into space to do a specific job. The type of satellite that is launched to monitor cloud patterns for a weather station will be different than a satellite launched to send television signals across the world. The satellite must be designed specifically to fulfill its role:

- Astronomy satellites—Hubble Space Telescope
- Atmospheric studies satellites—Polar
- Communications satellites—Anik E
- Navigation satellites—Navstar
- Reconnaissance satellites—Kennan, Big Bird, Lacrosse
- Remote sensing satellites—RADARSAT
- Search and rescue satellites—Cospas-Sarsat
- Space exploration satellites—Galileo
- Weather satellites—Meteosat

Our focus in this chapter is on weather satellites and data captured (maps) by the weather satellites for agricultural purposes. Importantly, satellite data provides crop intelligence[9] and helps preserve and protect future security through advance modeling of machine learning with climate change pattern [9].

How Satellite Maps Help Prevent Another "Great Grain Robbery" [10]
The need for crop intelligence dates to 1972. In July of that year, the Soviet Union purchased 15 million tons of wheat, corn, soybeans, and barley from the United States of America at low subsidized prices. Russia was experiencing severe drought and needed foreign grain. But the massive purchase, which took the United States of America by surprise, depleted the country's grain stocks and caused wheat prices to soar, resulting in a domestic food crisis.

Later nicknamed the Great Grain Robbery by congressional leaders, the event highlighted the need for global agricultural monitoring, and it just happened to coincide with a satellite that could provide just that: NASA's Landsat 1.

Landsat 1 was launched into space that same month, allowing for the first-ever view of drought and crop conditions from space. This convergence of events prompted NASA to team up with the US Department of Agriculture (USDA) and NOAA to develop, in 1974, the first satellite crop production forecast: The Large Area Crop Inventory Experiment or LACIE [10].

[9] How Satellite Maps Help Prevent Another 'Great Grain Robbery'—https://www.nasa.gov/feature/how-satellite-maps-help-prevent-another-great-grain-robbery

Coming to radiometers, essentially a telescope with a focal length of 3650 mm, meteorological satellite (Meteosat)'s main piece of equipment is the radiometer,[10] an instrument which is sensitive to some visible and thermal radiation. The radiometer carefully scans the earth's surface by taking data about every piece of land as if it were reading a book line by line. Each "line" that the satellite scans is translated into a series of individual picture elements or pixels. For each pixel, the radiometer measures the energy of the different spectral bands. This measurement is digitally coded and transmitted to the ground station. The coding gives each radioactive intensity level a certain gray value giving Meteosat pictures the look of a black-and-white photograph. Meteosat pictures, however, should not be regarded as true photographs; they are a series of picture elements, and each element comprises a certain digital value according to the measured energy.

7.1.1 Archive: Advanced Very High-Resolution Radiometer

The Polar-orbiting Operational Environmental Satellites (POES) Advanced Very High-Resolution Radiometer[11] (AVHRR) is a cross-track scanning system with five spectral bands having a resolution of 1.1 km and a frequency of earth scans twice per day (0230 and 1430 local solar time). NOAA-12 and NOAA-14 are still in service to produce the data.[12] There are three data types produced from the POES AVHRR [11]:

- The Global Area Coverage (GAC) dataset is a reduced-resolution image data that is processed onboard the satellite taking only one line out of every three and average every four of five adjacent samples along the scan line.
- The Local Area Coverage (LAC) dataset is recorded onboard at original resolution (1.1 km) for part of an orbit and later transmitted to earth.
- The High-Resolution Picture Transmission (HRPT) is real-time downlink data.

The NOAA POES AVHRR's objectives include providing data for weather forecasting and vegetation studies. The data supports many applications, including the following:[13]

- Aerosols
- Albedo and reflectance
- Cloud type, amount, and cloud top temperature

---

[10] Satellite Radiometer—http://satellites.spacesim.org/english/function/weather/radiomet.html

[11] NOAA—https://www.avl.class.noaa.gov/release/data_available/avhrr/index.htm#:~:text=The%20Advanced%20Very%20High%20Resolution,and%201430%20local%20solar%20time).

[12] USGS EROS Archive—Advanced Very High Resolution Radiometer (AVHRR)—Sensor Characteristics—https://www.usgs.gov/centers/eros/science/usgs-eros-archive-advanced-very-high-resolution-radiometer-avhrr-sensor

[13] NOAA POES—https://earth.esa.int/eogateway/missions/noaa

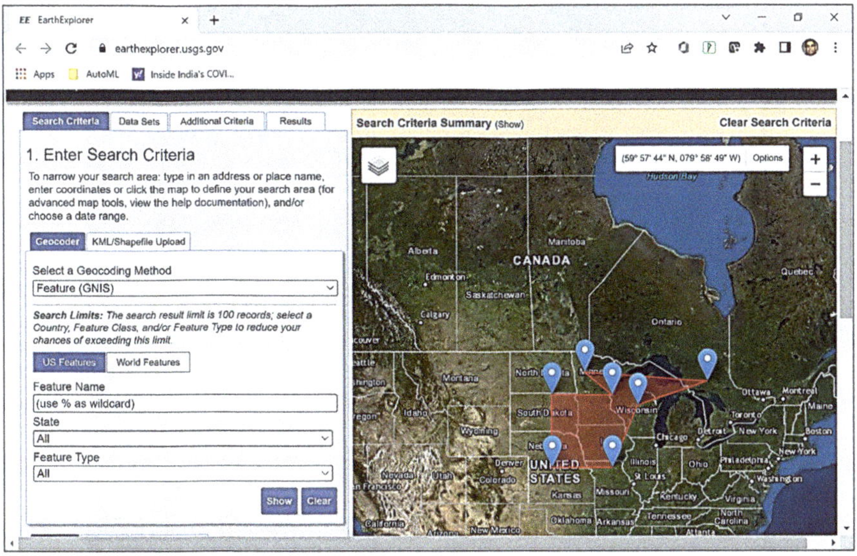

Fig. 7.3 EarthExplorer

- Liquid water and precipitation rate
- Multi-purpose imagery (land and ocean)
- Radiation budget
- Sea ice cover, edge, and thickness
- Snow cover, edge, and depth
- Surface temperature (land and ocean)
- Vegetation

EarthExplorer[14] can be used to search, preview, and download Advanced Very High-Resolution Radiometer (AVHRR). The collection is located under the Advanced Very High-Resolution Radiometer (AVHRR) category (please see Fig. 7.3).

The objective of the AVHRR instrument is to provide radiance data for investigation of clouds, land-water boundaries, snow and ice extent, ice or snow melt inception, day and night cloud distribution, temperatures of radiating surfaces, sea surface temperature, and vegetation classification and greenness, through passively measured visible, near-infrared, and thermal infrared spectral radiation bands.

[14]EarthExplorer—https://earthexplorer.usgs.gov/

Fig. 7.4 NOAA VIIRS—Source: PACE NOAA

7.1.2 Visible Infrared Imaging Radiometer Suite (VIIRS)

The Visible Infrared Imaging Radiometer Suite (VIIRS) (please see Fig. 7.4) instrument collects visible and infrared imagery and global observations of land, atmosphere, cryosphere, and oceans.

Currently flying on the Suomi NPP satellite mission, VIIRS generates many critical environmental products about snow and ice cover, clouds, fog, aerosols, fire, smoke plumes, dust, *vegetation health*, phytoplankton abundance, and chlorophyll. VIIRS will also be on the JPSS-1 and JPSS-2 satellite missions.

VIIRS features daily multi-band imaging capabilities to support the acquisition of high-resolution atmospheric imagery and other instrument products, including visible and infrared imaging of hurricanes and detection of fires, smoke, and atmospheric aerosols.

VIIRS extends and improves upon a series of measurements initiated by the Advanced Very High-Resolution Radiometer (AVHRR), the Moderate Resolution Imaging Spectroradiometer (MODIS), and the Operational Linescan System (OLS). VIIRS image source is from NOAA.[15]

In summary VIIRS provides benefits:[16]

[15]NOAA PACE—https://pace.oceansciences.org/gallery_more.htm?id=148

[16]VIIRS—https://www.jpss.noaa.gov/viirs.html

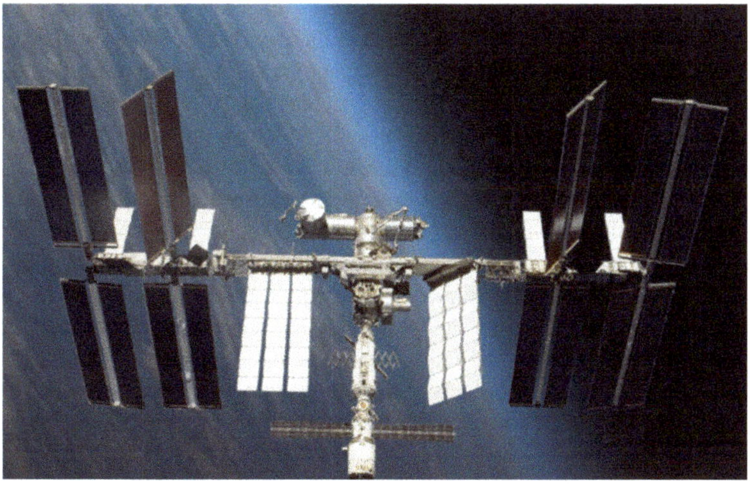

Fig. 7.5 ECOSTRESS (Source: NASA)

- VIIRS generates products for the operational weather community that improves weather, flooding, and storm forecasting abilities, which help protect life and property.
- The maritime forecasting products of sea ice and ocean nutrients from VIIRS also help the maritime and commercial fishing industries.
- Higher-resolution and more accurate measurements of sea surface temperature.
- VIIRS provides global coverage twice a day with 750 m resolution across its entire scan.

7.1.3 ECOSTRESS

The ECOsystem Spaceborne Thermal Radiometer Experiment on Space Station (ECOSTRESS) will monitor one of the most basic processes in living plants: the loss of water through the tiny pores in leaves. When people lose water through their pores, the process is called sweating. The related process in plants is known as transpiration. Because water that evaporates from soil around plants also affects the amount of water that plants can use, ECOSTRESS (please see Fig. 7.5) will measure combined evaporation and transpiration, known as evapotranspiration (ET).[17]

Using a thermal radiometer docked on the International Space Station (ISS), ECOSTRESS measures thermal energy (a.k.a. heat) coming from the earth, which is then used to calculate ET, water stress, and water use efficiency maps. With a

[17] ECOSTRESS—https://www.jpl.nasa.gov/missions/ecosystem-spaceborne-thermal-radiometer-experiment-on-space-station-ecostress

revisit time of every 3–5 days, we can now see how crops are doing at the sub-weekly scale. ECOSTRESS provides a unique dataset because of its orbital path and technology. It gives global coverage at the scale of a farmer's field (~70 m resolution), and because it is on the ISS, it provides data at different hours of the day on each overpass. Many other satellites are in sun-synchronous orbit, meaning they always go over the same geographic location on earth at about the same time of day. With ECOSTRESS, we can now see how plants are responding to different conditions at different times of the day and year.

Although climate change will continue to challenge the global agricultural system throughout the twenty-first century, space technologies such as ECOSTRESS are making it easier for humans to protect food security and continue toward accomplishing the second SDG. Despite being thousands of miles away, ECOSTRESS is connecting farmers to their fields in new and unique ways, helping to ensure that our favorite foods are always in the grocery store and people around the world have enough to eat.

7.1.4 *Defense Meteorological Satellite Program (DMSP)*

Since the mid-1960s, when the Department of Defense (DoD) initiated the Defense Meteorological Satellite Program (DMSP) (please see Fig. 7.6), low, earth-orbiting

Fig. 7.6 DMSP (Source: NOAA)

satellites have provided the military with important environmental information. Each DMSP satellite has a 101-min orbit and provides global coverage twice per day.

The DMSP[18] satellites "see" such environmental features as clouds, bodies of water, snow, fire, and pollution in the visual and infrared spectra. Scanning radiometers record information which can help determine cloud type and height, land and surface water temperatures, water currents, ocean surface features, ice, and snow. Communicated to ground-based terminals, the data is processed, interpreted by meteorologists, and ultimately used in planning and conducting US military operations worldwide.

7.2 Dairy, Food Security, and Satellite Data

Good nutrition is the foundation of health and wellness for adults and children alike, and dairy is a crucial part of a healthy diet now more than ever. In fact, no other type of food or beverage provides the unique combination of nutrients that dairy contributes to the American diet, including high-quality protein, calcium, vitamin D, and potassium, and health benefits including better bone health and lower risk for type 2 diabetes and cardiovascular disease. Dairy companies from coast to coast are making important contributions to the nutrition and good health of people across the country[19] [12].

Heat stress (HS) causes cows to produce less milk with the same nutritional input, which effectively increases farmers' production costs. The economic toll due to higher-temperature, heat stress is a $1 billion annual problem. Not only in the United States o America but also around the globe, heat stress causes an adverse impact on dairy productivity. The opportunities, however, for the dairy industry are to electronically monitor cattle temperature and implement appropriate measures so that the impact of HS can be minimized. The US Department of Agriculture estimates nearly $2.4 billion a year in losses from animal illnesses that lead to death can be prevented by electronically checking on cattle's vital signs[20,21] [13, 14].

Heat stress negatively affects the productivity and health of dairy cattle, and heat stress abatement is common for the lactating herd.[22] However, recent studies

[18] Defense Meteorological Satellite Program (DMSP)—https://www.ospo.noaa.gov/Operations/DMSP/index.html

[19] June is National Dairy Month: Let's Recognize Dairy's Role in Fighting Food Insecurity—https://www.idfa.org/news/june-is-national-dairy-month-lets-recognize-dairys-role-in-fighting-food-insecurity

[20] Keeping cows cool critical to dairy industry as climate warms—https://www.pennlive.com/midstate/2013/04/keeping_cows_cool_critical_to.html

[21] How Cows Are Becoming Smart Connected Products—https://www.forbes.com/sites/ptc/2014/06/18/how-cows-are-becoming-smart-connected-products/?sh=1c94e914419c

[22] FINANCIAL IMPACTS OF LATE-GESTATION HEAT STRESS ON COW AND OFF-SPRING LIFETIME PERFORMANCE—https://edis.ifas.ufl.edu/publication/AN374

indicate that heat stress of dry cows (i.e., late-gestation, non-lactating period between two subsequent lactations) dramatically affects the next lactation and the next generation. In the United States of America, heat stress costs the dairy industry more than $1.5 billion annually due to losses in production and reproductive performance and an increase in morbidity and mortality of lactating dairy cows. That is why heat abatement practices such as shade, fans, soakers, and misters are commonly used by US dairies, especially for lactating cows. During the past decade, numerous studies have shown that the negative effects of heat stress observed during lactation also extend to the dry period. Exposure of dry cows to heat stress negatively affects milk production by reducing milk yield an average of 10 lb/day. Initial estimates of the effect of heat stress exposure during the dry period suggest $810 million in milk losses annually; therefore, cooling dry cows is profitable for 89% of the animals in the US dairies. However, this scenario does not account for the economic impact of the late-gestation heat stress on the performance of the offspring. The effects of in utero exposure to heat stress on survival, milk production, and reproduction across multiple generations have now been quantified, and those impacts persist for at least three subsequent lactations [15].

In the United States of America, climate change is likely to increase average daily temperatures and the frequency of heat waves, which can reduce meat and milk production in animals. Methods that livestock producers use to mitigate thermal stress—including modifications to animal management or housing—tend to increase production costs and capital expenditures. Dairy cows are particularly sensitive to heat stress, and the dairy sector has been estimated to bear over half of the costs of current heat stress to the livestock industry[23] [16].

Satellites provide global coverage to assess health of us, humans, and beyond. For example, the dairy industry, in developing countries, suffers immensely in summers due to extreme heat and heat stress-related events. The production of milk goes down and increases food insecurity. Milk production decreases as the level of heat stress increases. Mild heat stress results in a production decrease of about 2.5 lbs/head/day. Mild to moderate heat stress results in a production decrease of about 6 lbs/head/day. Cows begin to experience heat stress at much lower temperatures than humans. In general, mild heat stress[24] starts around 72 °F with 50% humidity. High-producing cows eat more and generate more heat [17]. They can begin to experience heat stress in well-ventilated barns at air temperatures as low as 65 °F. Keeping cows[25] cool in the summertime is a major concern for dairy farmers, even in the relatively moderate climate of the Northeast [17, 18].

[23] Climate Change, Heat Stress, and U.S. Dairy Production—https://www.ers.usda.gov/webdocs/publications/45279/49164_err175.pdf?v=0

[24] Heat Stress in Dairy Cattle—https://extension.umn.edu/dairy-milking-cows/heat-stress-dairy-cattle#:~:text=Cows%20begin%20to%20experience%20heat,low%20as%2065%C2%B0F.

[25] Managing Dairy Heat stress—https://www.climatehubs.usda.gov/sites/default/files/ARS_heatstress_April2020_508tagged.pdf

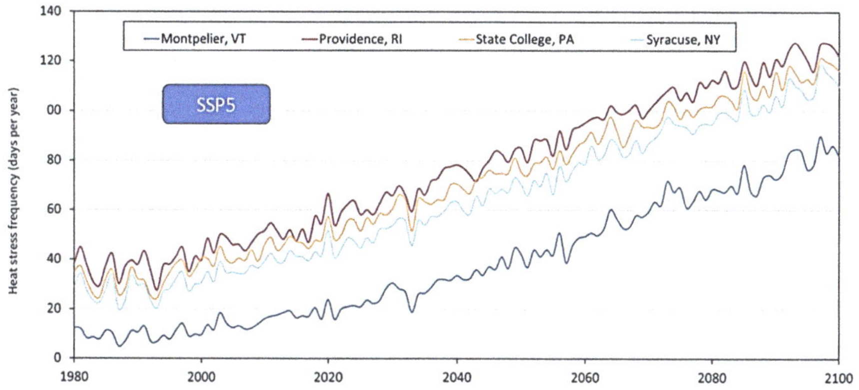

Fig. 7.7 Climate change (SSP5) and heat stress

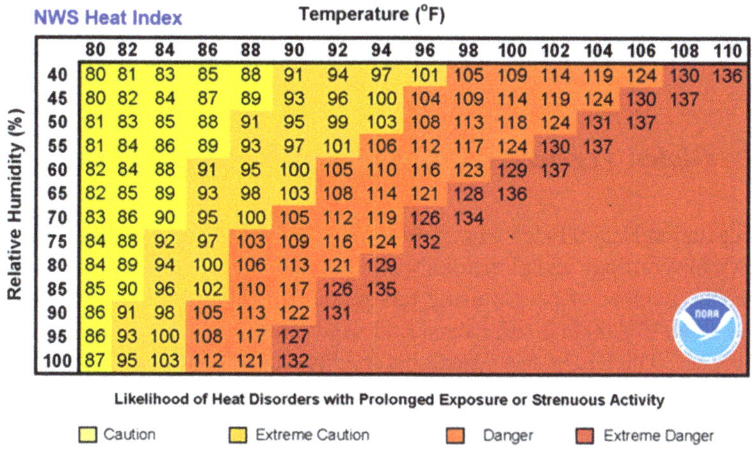

NWS Heat Index
Temperature (°F)
80 82 84 86 88 90 92 94 96 98 100 102 104 106 108 110
40 80 81 83 85 88 91 94 97 101 105 109 114 119 124 130 136
45 80 82 84 87 89 93 96 100 104 109 114 119 124 130 137
50 81 83 85 88 91 95 99 103 108 113 118 124 131 137
55 81 84 86 89 93 97 101 106 112 117 124 130 137
60 82 84 88 91 95 100 105 110 116 123 129 137
65 82 85 89 93 98 103 108 114 121 128 136
70 83 86 90 95 100 105 112 119 126 134
75 84 88 92 97 103 109 116 124 132
80 84 89 94 100 106 113 121 129
85 85 90 96 102 110 117 126 135
90 86 91 98 105 113 122 131
95 86 93 100 108 117 127
100 87 95 103 112 121 132
Relative Humidity (%)
Likelihood of Heat Disorders with Prolonged Exposure or Strenuous Activity
Caution
Extreme Caution
Danger
Extreme Danger

Fig. 7.8 Heat stress

A clear positive trend can be witnessed in heat stress frequency for four representative dairying locations in the Northeast from 1980 to 2100. Predictions of heat stress frequency are based on climate model output that assumes a business-as-usual (a.k.a. high, SSP5) emission scenario for greenhouse gases (please see Fig. 7.7). Heat stress frequency was estimated as the number of days each year when the temperature humidity index exceeded 70. Data is downloaded from Climate Change Portal.

Detecting heat stress (please see Fig. 7.8) event provides dairy farmers to adapt mitigation strategies such as cooling the dairy farms or increase water usage to reduce heat stress. Hanumayamma Innovations and Technologies company,[26] a veterinary apparatus manufacturer of Wearable veterinary sensor, combines field-

[26]Hanumayamma Innovations and Technologies, inc.—https://www.hanuinnotech.com/

collected temperature and humidity with thermal index from satellites to predict the health of cattle. Generally, CLASS 10 classification is reserved to medical apparatus such as surgical, medical, dental, and veterinary apparatus and instruments; artificial limbs, eyes, and teeth; orthopedic articles; and suture materials. Specifically, Hanumayamma' s cow necklace sensors are "Class 10: Wearable veterinary sensor for use in capturing a cow's vital signs, providing data to the farmer to monitor the cow's milk productivity, and improving its overall health." For cattle and bovine, heat stress is one of the leading causes of diseases and low productivity[27] [19–21].

Having data from satellite will provide datasets and coverage for all other dairy farms. Satellite data can help identify hotspots and the processes that drive high wet-bulb temperature in such locations.[28] That's where NASA Earth Observations play an important role. Instruments in space like the Atmospheric Infrared Sounder (AIRS) on NASA's Aqua satellite and the ECOsystem Spaceborne Thermal Radiometer Experiment (ECOSTRESS) on the International Space Station (ISS) provide useful data for studying heat stress. Data from sensors are stitched with data from satellite to develop location-specific heat stress to activity algorithms to facilitate productivity farms across the world [22].

7.2.1 Global Vegetation: Cropland and Vegetation Index

The Vegetation Health Index was created in the mid-1990s by a research scientist in NOAA's meteorology and climatology division named Dr. Felix Kogan. In the past decade, the number of people using this product has jumped from about 2400 in 2010 to about 69,000 in 2020, Kogan said. And it has a bigger role to play in "Crop Intelligence." The USDA has a small but mighty team of meteorologists engaged in a major effort to develop estimates of crop production in more than 120 growing regions, covering 35 countries. They do this for major row crops, such as corn, wheat, sunflowers, soybeans, barley, cotton, and rapeseed—and they rely on the VHI to help create these yield estimates (please see Fig. 7.9).

What's more, the research the VHI informs just might be protecting the price groceries. Knowing how crops are faring in other countries informs decisions about planting, food prices, and foreign market exports, said Mark Brusberg, chief meteorologist for the USDA.[29] For example, said Brusberg, who monitors crops in Brazil, "if an area of the Southern Hemisphere is experiencing drought and there's

[27]Building an IoT Framework for Connected Dairy—https://dl.acm.org/doi/10.1109/BigDataService.2015.39

[28]Too Hot to Handle: How Climate Change May Make Some Places Too Hot to Live—https://climate.nasa.gov/ask-nasa-climate/3151/too-hot-to-handle-how-climate-change-may-make-some-places-too-hot-to-live/

[29]The Great Drain Robbery—https://www.nasa.gov/feature/how-satellite-maps-help-prevent-another-great-grain-robbery

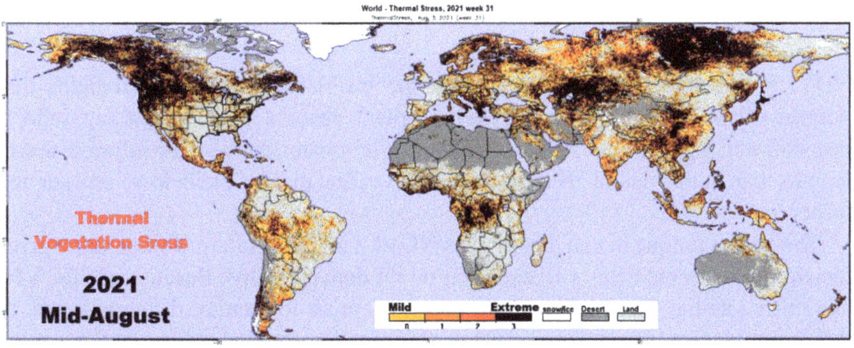

Fig. 7.9 Thermal vegetation stress

going to be a shortage of corn, it might impact prices in the U.S. as well, and farmers might decide to plant more corn here" [22].

There are many standard Moderate Resolution Imaging Spectroradiometer (MODIS) data products that scientists[30] are using to study global change. These products are being used by scientists from a variety of disciplines, including oceanography, biology, and atmospheric science. In this book, I have applied the MODIS dataset for food security[31] [23].

The satellite-based global VH System is designed to monitor, diagnose, and predict long- and short-term land environmental conditions and climate-dependent socioeconomic activities. The System is based on satellite observations of the earth, biophysical theory of vegetation response to the environment, set of algorithms for satellite data processing, interpretation, product development, validation, calibration, and applications.

"We must not lose sight of the urgency of this climate change: protecting food security from the impacts of climate change, while making agriculture more productive and more resilient at the same time."
José Graziano da Silva,[32]
FAO Director-General, 2017 [24]

[30] MODIS Data Products—https://modis.gsfc.nasa.gov/data/dataprod/index.php#atmosphere

[31] An Ongoing Blended Long-Term Vegetation Health Product for Monitoring Global Food Security—https://www.mdpi.com/2073-4395/10/12/1936

[32] FAO'S WORK ON CLIMATE CHANGE—https://www.fao.org/3/i8037e/i8037e.pdf

7.2.2 *The Normalized Difference Vegetation Index (NDVI)*

NDVI is calculated by measuring both the visible and near-infrared lights that bounce off vegetation. Chlorophyll in a plant absorbs visible light but reflects near-infrared light. If the reflected radiation is greater in the near-infrared wavelengths than in the visible, then the vegetation in that pixel is likely to be greener and lusher.

The VHI is unique in that it combines NDVI with temperature. Temperature gives insight into extreme heat or freezes that might damage crops. Because of this, VHI and other satellite-based indices also allow scientists to monitor different stages of the crops as they grow, which makes the resulting crop yield predictions more accurate.

Crops need certain ambient air temperatures for growth. Spring wheat, for example, needs a daily average temperature of at least 41 degrees Fahrenheit to grow, corn needs at least 50 degrees Fahrenheit, and cotton won't grow until the daily average is closer to 60 degrees. Using VHI formula to the crop stage for a particular area, data science teams can model the yield: corn in the silking stage in the Ukraine, for example. It was a game changer!

> "As long as you know when the crop was put in the ground, roughly, you can now estimate with a good degree of accuracy when the crop enters the key stages of development where weather matters the most. and if we didn't have the VHI data, we'd be hosed. It's become an integral part of our operations."
> Eric Luebehusen, a meteorologist at the US Department of Agriculture[33] [10]

Satellite observations are principally represented by the Advanced Very High-Resolution Radiometer (AVHRR) flown on NOAA polar-orbiting satellites. Data are global with the resolution 4 km and 7-day composite. In addition, the MODIS system is using data and products from Geostationary Operational Environmental Satellites (GOES[34]). Meteosat series of satellites are geostationary meteorological satellites operated by EUMETSAT[35] and Defense Meteorological Satellite Program[36] (DMSP). The MODIS system capture contains the following vegetation health indices and products: Vegetation Condition Index (VCI), Temperature Condition Index (TCI), Vegetation Health Index (VHI), Soil Saturation Index (SSI), No noise Normalized Difference Vegetation Index (SMN), No noise Brightness

[33] How Satellite Maps Help Prevent Another 'Great Grain Robbery'—https://www.nasa.gov/feature/how-satellite-maps-help-prevent-another-great-grain-robbery

[34] GOES Image Viewer—https://www.star.nesdis.noaa.gov/goes/index.php

[35] METEOSAT—https://www.eumetsat.int/our-satellites/meteosat-series

[36] Defense Meteorological Satellite Program (DMSP)—https://www.ospo.noaa.gov/Operations/DMSP/index.html

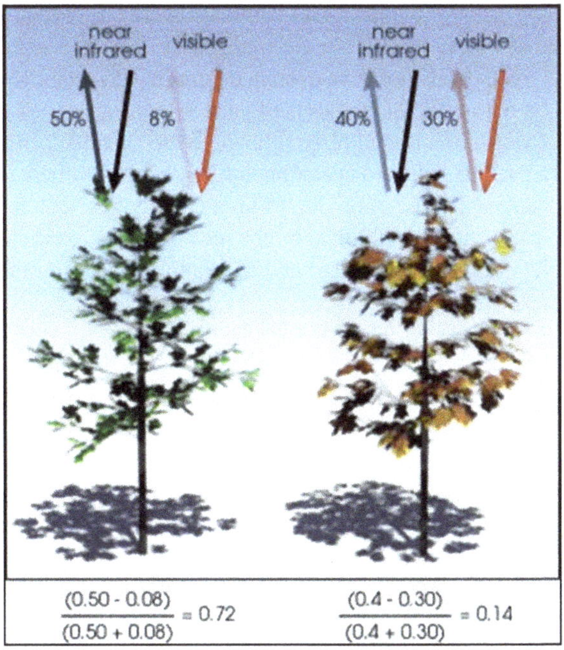

Fig. 7.10 NDVI

Temperature (SMT), Fire Risk Index (FRI)—drought, malaria, vegetation health, ecosystems, land sensitivity to ENSO.[37]

The NDVI (please Fig. 7.10) is a dimensionless index that describes the difference between visible and near-infrared reflectance of vegetation cover and can be used to estimate the density of green on an area of land. Normalized Difference Vegetation Index (NDVI) quantifies vegetation by measuring the difference between near-infrared[38] (which vegetation strongly reflects) and red light (which vegetation absorbs) [25]. Overall, NDVI is a standardized way to measure healthy vegetation. When you have high NDVI values, you have healthier vegetation. When you have low NDVI, you have less or no vegetation. Generally, if you want to see vegetation change over time, then you will have to perform atmospheric correction.[39]

NDVI values range from +1.0 to -1.0. Areas of barren rock, sand, or snow usually show very low NDVI values (e.g., 0.1 or less). Sparse vegetation such as shrubs and grasslands or senescing crops may result in moderate NDVI values (approximately 0.2–0.5). High NDVI values (approximately 0.6–0.9) correspond to dense vegetation such as that found in temperate and tropical forests or *crops at*

[37] STAR—Global Vegetation Health Products: Background and Explanation—https://www.star.nesdis.noaa.gov/smcd/emb/vci/VH/VH-Syst_10ap30.php

[38] NDVI—https://gisgeography.com/ndvi-normalized-difference-vegetation-index/

[39] What is NDVI (Normalized Difference Vegetation Index)?—https://gisgeography.com/ndvi-normalized-difference-vegetation-index/

their peak growth stage.[40] Although there are several vegetation indices, one of the most widely used is the Normalized Difference Vegetation Index (NDVI) [26].

NDVI values can be averaged over time to establish "normal" growing conditions in a region for a given time of year. Further analysis can then characterize the health of vegetation in that place relative to the norm. When analyzed through time, NDVI can reveal where vegetation is thriving and where it is under stress, as well as changes in vegetation due to human activities such as deforestation, natural disturbances such as wildfires, or changes in plants' phenological stage. In the following, NDVI captures of Brazil and Mozambique at two times [27] are shown:

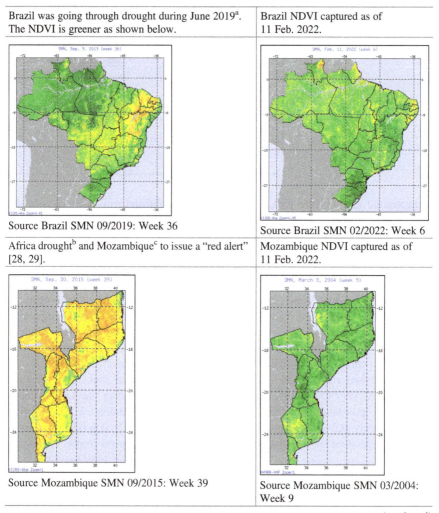

Brazil was going through drought during June 2019[a]. The NDVI is greener as shown below.	Brazil NDVI captured as of 11 Feb. 2022.
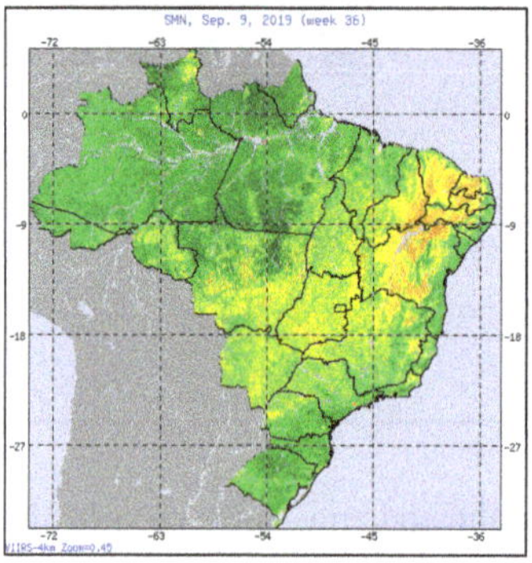 Source Brazil SMN 09/2019: Week 36	 Source Brazil SMN 02/2022: Week 6
Africa drought[b] and Mozambique[c] to issue a "red alert" [28, 29].	Mozambique NDVI captured as of 11 Feb. 2022.
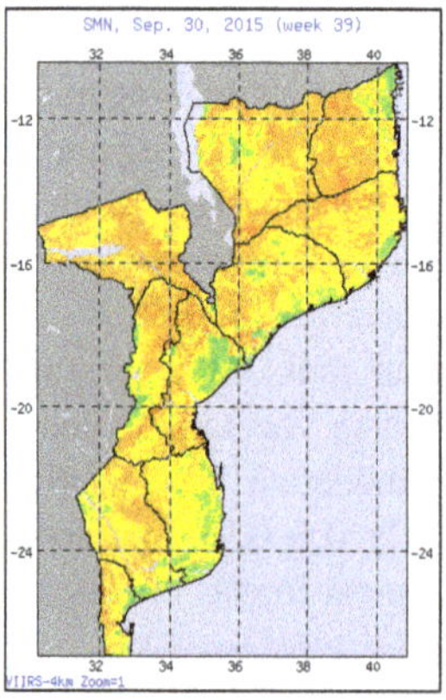 Source Mozambique SMN 09/2015: Week 39	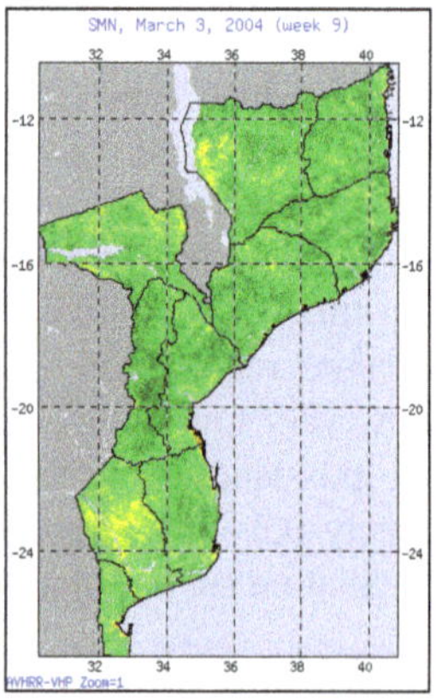 Source Mozambique SMN 03/2004: Week 9

(continued)

[40]NDVI, the Foundation for Remote Sensing Phenology—https://www.usgs.gov/special-topics/remote-sensing-phenology/science/ndvi-foundation-remote-sensing-phenology

[a]Brazil Battered by Drought—https://earthobservatory.nasa.gov/images/148468/brazil-battered-by-drought
[b]Africa drought fears grip Malawi and Mozambique—https://www.bbc.com/news/world-africa-36037414
[c]El Niño: Urgent Need for Support as Drought Affects 14 Million People in Southern Africa—https://www.care.org/news-and-stories/press-releases/el-nino-urgent-need-for-support-as-drought-affects-14-million-people-in-southern-africa/

NOAA Satellite Application Research[41] (STAR)'s web page[42] provides SMN, SMT, VCI, TCI, and VHI indices weekly since 1984 split in provinces:

- SMN= Provincial mean NDVI with noise reduced (greenness)
- SMT=Provincial mean brightness temperature with noise reduced
- VCI = Vegetation Condition Index (VCI <40 indicates moisture stress; VCI >60 favorable condition)
- TCI= Thermal Condition Index (TCI <40 indicates thermal stress; TCI >60 favorable condition)
- VHI =Vegetation Health Index (VHI <40 indicates vegetation stress; VHI >60 favorable condition)

Drought vegetation:

- VHI<15 indicates drought from severe-to-exceptional intensity.
- VHI<35 indicates drought from moderate-to-exceptional intensity.
- VHI>65 indicates good vegetation condition.
- VHI>85 indicates very good vegetation condition.

For example, consider California. As it can be seen, the state is under severe drought (VHI < 35):

```
Mean data for USA Province= 5: California, from 1982 to 2022,
weekly43
for cropland area only
year,week, SMN,SMT,VCI,TCI, VHI
2021,27, 0.337,309.54, 41.18, 36.06, 38.75,
2021,28, 0.336,309.63, 38.41, 34.65, 36.64,
2021,29, 0.334,309.76, 36.15, 32.43, 34.39,
2021,30, 0.330,309.88, 33.48, 28.50, 31.08,
2021,31, 0.325,309.88, 31.49, 25.85, 28.74,
2021,32, 0.320,309.75, 30.06, 24.55, 27.37,
2021,33, 0.312,309.56, 28.76, 23.10, 26.01,
```

(continued)

[41]NOAA—STAR—https://www.star.nesdis.noaa.gov/star/index.php

[42]STAR—Global Vegetation Health Products: Browse Archived Image of selected country—https://www.star.nesdis.noaa.gov/smcd/emb/vci/VH/vh_browseByCountry.php

[43]USA Province= 5: California, from 1982 to 2022, weekly CROP LAND—https://www.star.nesdis.noaa.gov/smcd/emb/vci/VH/get_TS_admin.php?provinceID=5&country=USA&yearlyTag=Weekly&type=Mean&TagCropland=crop&year1=1982&year2=2022

```
2021,34, 0.305,309.31, 28.27, 21.63, 25.02,
2021,35, 0.298,308.80, 28.69, 22.15, 25.53,
2021,36, 0.291,308.11, 29.64, 24.03, 26.96,
2021,37, 0.282,307.21, 30.34, 27.42, 29.02,
2021,38, 0.274,305.95, 31.69, 32.85, 32.42,
2021,39, 0.266,304.39, 33.60, 38.86, 36.36,
2021,45, 0.236,294.02, 55.30, 42.08, 48.71,
2021,46, 0.233,292.69, 58.47, 35.65, 47.09,
2021,47, 0.233,291.52, 61.96, 30.15, 46.07,
2021,48, 0.234,290.26, 64.46, 26.05, 45.26,
2021,49, 0.234,288.87, 65.92, 24.64, 45.26,
2021,50, 0.236,287.90, 66.48, 29.90, 48.18,
2021,51, 0.240,287.30, 67.15, 31.63, 49.38,
2021,52, 0.248,287.27, 68.02, 31.90, 49.97,
2022, 1, 0.256,287.71, 67.64, 32.23, 49.94,
2022, 2, 0.265,288.51, 67.57, 30.69, 49.13,
2022, 3, 0.275,289.39, 67.42, 28.09, 47.75,
2022, 4, 0.285,290.27, 66.43, 25.87, 46.16,
2022, 5, 0.293,290.99, 64.05, 24.46, 44.27,
2022, 6, 0.300,291.92, 61.16, 22.08, 41.64,
```

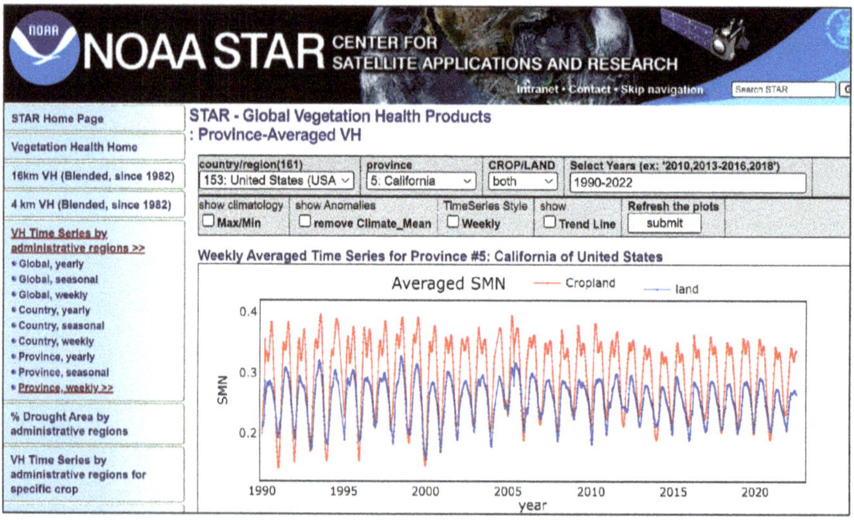

Below, we have applied data from NOAA STAR Vegetation Index (please see Fig. 7.11) to analyze and predict crop yield performance for the countries Pakistan, Thailand, and others. In the prior chapter, we have applied machine learning model for correlating the crop yield to climate change data for PMI6 project data for various SSPs.

As it can be seen, the SMN for California[44] province provides the clear movement of SMN. Week number in Vegetation Health Products "week" defined here is based

[44] Averaged SMN—https://www.star.nesdis.noaa.gov/smcd/emb/vci/VH/vh_adminMean.php?type=Province_Weekly_MeanPlot

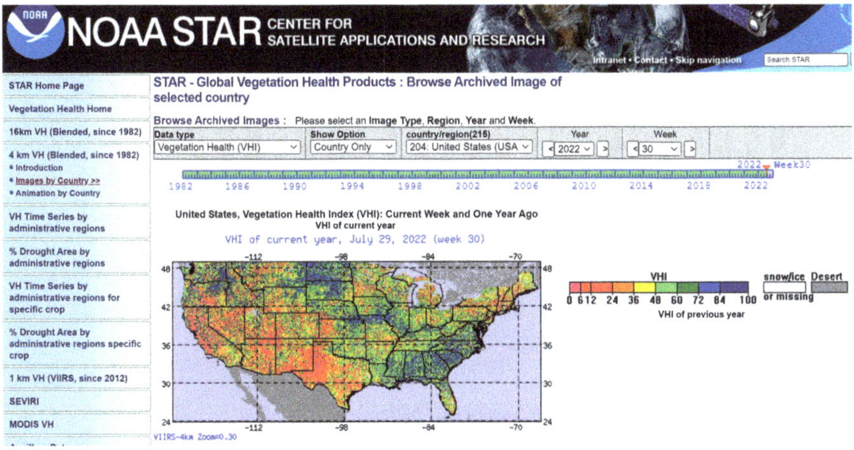

Fig. 7.11 NOAA VHI

on "day of the year," i.e., week 1 covers day of the year 1–7. (For example, in 2016, images will be updated on Friday.[45])

Moderate Resolution Imaging Spectroradiometer (MODIS) vegetation indices,[46] produced on 16-day intervals and at multiple spatial resolutions, provide consistent spatial and temporal comparisons of vegetation canopy greenness, a composite property of leaf area, chlorophyll, and canopy structure. Two vegetation indices are derived from atmospherically corrected reflectance in the red, near-infrared, and blue wavebands: the Normalized Difference Vegetation Index (NDVI), which provides continuity with NOAA's AVHRR NDVI time series record for historical and climate applications and the enhanced vegetation index (EVI), which minimizes canopy-soil variations and improves sensitivity over dense vegetation conditions. The two products more effectively characterize the global range of vegetation states and processes. In Fig. 7.12 please find Weekly Averaged Time Series for Province #29: Nakhon Sawan of Thailand.[47]

NDVI is used to quantify vegetation greenness and is useful in understanding vegetation density and assessing changes in plant health. NDVI is calculated as a ratio between the red (R) and near-infrared (NIR) values in traditional fashion. As vegetation health captures water-related and temperature-related response of crops, it is a good indicator of crop growth and therefore is a good indicator of crop yield due

[45] STAR—Global Vegetation Health Products: Browse Archived Image of selected country—https://www.star.nesdis.noaa.gov/smcd/emb/vci/VH/vh_browseByCountry.php

[46] MODIS—https://modis.gsfc.nasa.gov/data/dataprod/mod13.php

[47] STAR—Global Vegetation Health Products: Province-Averaged VH—https://www.star.nesdis.noaa.gov/smcd/emb/vci/VH/vh_adminMean.php?type=Province_Weekly_MeanPlot

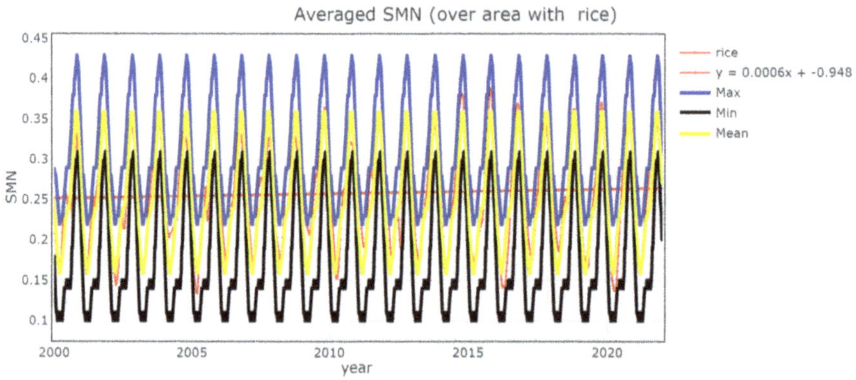

Fig. 7.12 Averaged SMN—Nakhon Sawan of Thailand

to weather impacts. To date, vegetation health products have been widely used in modeling or estimating many kinds of crop yield in many countries or regions. For effective crop yielding modeling, agricultural technology (which may also include agricultural policy, fertilizer usage, and other non-weather factors) and weather are two major factors. As part of the model, we will apply weather factors to deduce the impact of climate/weather on crop yield. Stratification of the NDVI values was required to assess weather impacts on vegetation in non-homogeneous areas. The stratification resulted in the Vegetation Condition Index (VCI), which could be simply illustrated as

$$\mathrm{VCI} = 100 \times (\mathrm{NDVINDVI}_{\min})/(\mathrm{NDVI}_{\max}\,\mathrm{NDVI}_{\min}) \tag{7.1}$$

Where NDVI $_{max}$ and NDVI $_{min}$ are the multi-year absolute maximum and minimum of NDVI.

VCI is more sensitive to rainfall dynamics compared to NDVI, and it is a better indicator of vegetation response to precipitation impact. Please (see Fig. 7.13) find Weekly Averaged Time Series for Province #29: Nakhon Sawan of Thailand.[48]

Although VCI is capable of capturing water-related stress, it could be augmented by the introduction of the Temperature Condition Index (TCI), for determining temperature-related vegetation stress and stress caused by excessive wetness:

$$\mathrm{TCI} = 100 \times (\mathrm{BT}_{\max}\,\mathrm{BT})/(\mathrm{BT}_{\max}\,\mathrm{BT}_{\min}) \tag{7.2}$$

where BT, BT_{max}, and BT_{min} are the brightness temperature, originally derived from AVHRR's fourth channel (10–11.5 μm), its multi-year absolute maximum, and

[48] Weekly Averaged Time Series for Province #29: Nakhon Sawan of Thailand—https://www.star.nesdis.noaa.gov/smcd/emb/vci/VH/vh_adminMeanByCrop.php?type=Province_Weekly_MeanPlot

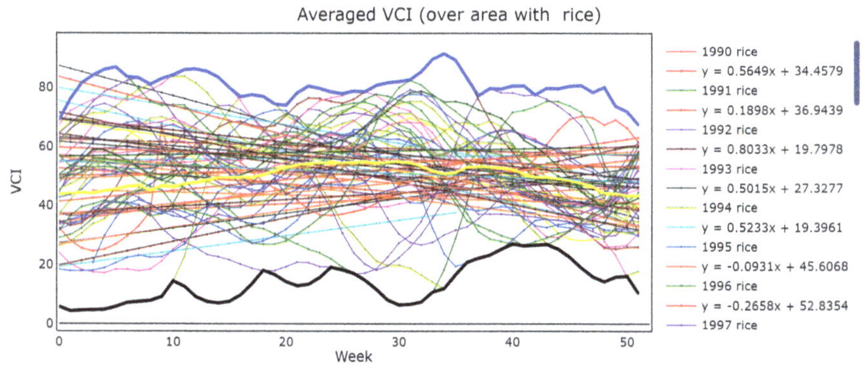

Fig. 7.13 Weekly Averaged Time Series for Province #29—Nakhon Sawan of Thailand

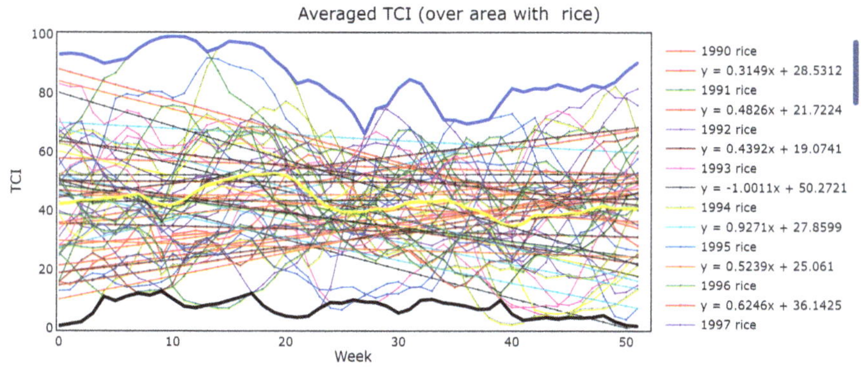

Fig. 7.14 Weekly Averaged Time Series for Province #29—Nakhon Sawan of Thailand

minimum, respectively. Please find TCI (please see Fig. 7.14) for Weekly Averaged Time Series for Province #29: Nakhon Sawan of Thailand.[49]

By combining VCI and TCI through weighted average, one can obtain the Vegetation Health Index (VHI) as

$$\mathrm{VHI} = \alpha \times \mathrm{VCI} + (1\alpha) \times \mathrm{TCI} \tag{7.3}$$

where, in a general sense, the weight α is simply 0.5[50] [30].

The Vegetation Health Index (VHI) illustrates (Fig. 7.15) the severity of drought based on vegetation health and the influence of temperature on plant conditions. The

[49] TCI—https://www.star.nesdis.noaa.gov/smcd/emb/vci/VH/vh_adminMeanByCrop.php?type=Province_Weekly_MeanPlot

[50] Felix Kogan, Remote Sensing for Malaria: Monitoring and Predicting Malaria from Operational Satellites (Springer Remote Sensing/Photogrammetry), Springer; 1st ed. 2020 edition (July 21, 2020), ISBN-13: 978-3030460198

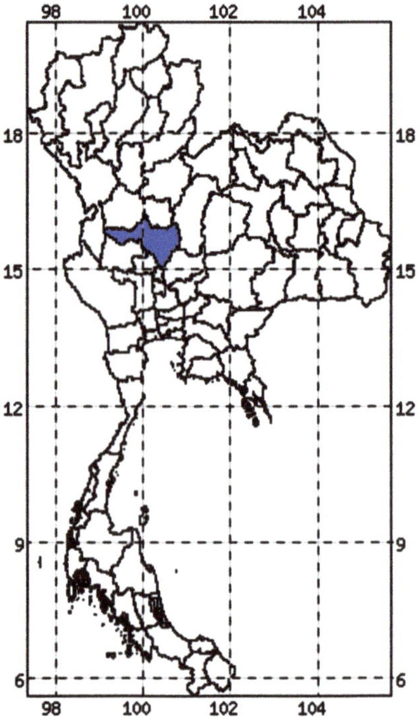

Fig. 7.15 VHI

VHI is a composite index and the elementary indicator used to compute the ASI. It combines both the Vegetation Condition Index (VCI) and the Temperature Condition Index (TCI). The TCI is calculated using a similar equation to the VCI but relates the current temperature to the long-term maximum and minimum, as it is assumed that higher temperatures tend to cause a deterioration in vegetation conditions. *A decrease in the VHI would, for example, indicate relatively poor vegetation conditions and warmer temperatures, signifying stressed vegetation conditions, and over a longer period would be indicative of drought.* The VHI images are computed for the two main seasons and in three modalities: dekadal, monthly, and annual.[51]

The direct application of vegetation health product on drought has been documented throughout the world, e.g., to measure drought's onset time, intensity, duration, and impact on vegetation in the United States of America, Southern Africa, China, India, Mongolia, and elsewhere. Malaria predication can also be detected using VHI. Furthermore, VHI has been selected as one of six key objective drought indicators (the other five selected indicators are Palmer Drought Severity Index (PDSI), Soil Moisture from Climate Prediction Center (CPC/SM), US Geological Survey (USGS) daily streamflow percentiles, percent of normal precipitation, and

[51] Vegetation Indicators—https://www.fao.org/giews/earthobservation/country/index.jsp?lang=en&code=USA

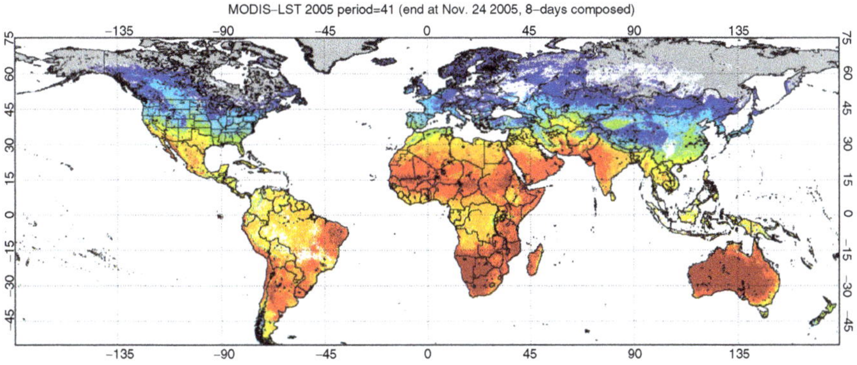

Fig. 7.16 MODIS land surface temperature of 2005 period = 041: composed by MODIS data from 17 November 2005 to 24 November 2005 (each period has 8 days) 8-day composed LST of 2005 (MODIS Land Surface Temperature of 2005 Period=041: composed by MODIS data from 11/17/2005 to 11/24/2005—https://www.star.nesdis.noaa.gov/smcd/emb/vci/VH/modis_browse8daysLST.php)

Standardized Precipitation Index (SPI)) to be used by the US Drought Monitor (USDM)[52,53] [23]. Please find Thailand Data[54] for rice crop for province Nakhon Sawan of Thailand.[55] For worldwide temperature indices, please see Figs. 7.16, 7.17 and 7.18.

Please click under MODIS VH menu Browse MODIS-8-day LST.[56]

7.2.3 Greenhouse Gas Emissions[57]

Total greenhouse gas emissions in kt of CO_2 equivalent are composed of CO_2 totals excluding short-cycle biomass burning (such as agricultural waste burning and Savannah burning) but including other biomass burning (such as forest fires, post-

[52] An Ongoing Blended Long-Term Vegetation Health Product for Monitoring Global Food Security—https://www.mdpi.com/2073-4395/10/12/1936

[53] USING VEGETATION INDICES FROM SATELLITE IMAGES TO ESTIMATE EVAPO-TRANSPIRATION AND VEGETATION WATER USE IN NORTH-CENTRAL PORTUGAL—https://edepot.wur.nl/283768

[54] Thailand—https://www.star.nesdis.noaa.gov/smcd/emb/vci/VH/get_TS_admin.php?provinceID=29&country=THA&yearlyTag=Weekly&type=Mean&TagCropland=RICE&year1=1982&year2=2022

[55] Thailand Region 29—https://www.star.nesdis.noaa.gov/smcd/emb/vci/VH/image_country_G04L01.php?type=PROVINCE&country=THA&provinceID=29

[56] MODIS LST—https://www.star.nesdis.noaa.gov/smcd/emb/vci/VH/modis_browse8daysLST.php

[57] Environment Database—Greenhouse gas emissions—https://stats.oecd.org/Index.aspx?DataSetCode=AIR_GHG

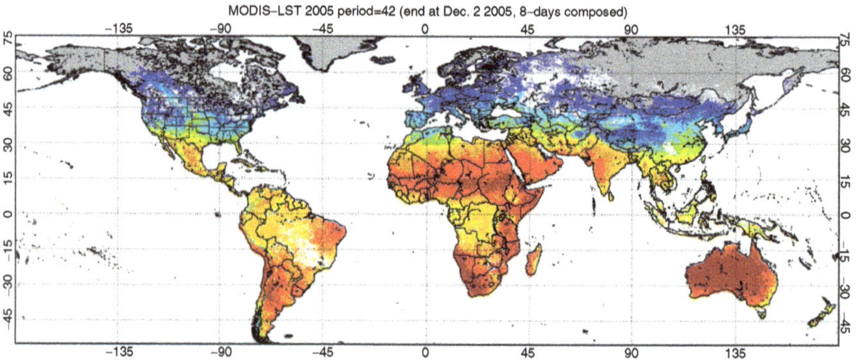

Fig. 7.17 MODIS land surface temperature of 2005 period = 042: composed by MODIS data from 25 November 2005 to 2 December 2005 (MODIS Land Surface Temperature of 2005 Period=042: composed by MODIS data from 11/25/2005 to 12/2/2005—https://www.star.nesdis.noaa.gov/smcd/emb/vci/VH/modis_browse8daysLST.php) (each period has 8 days)

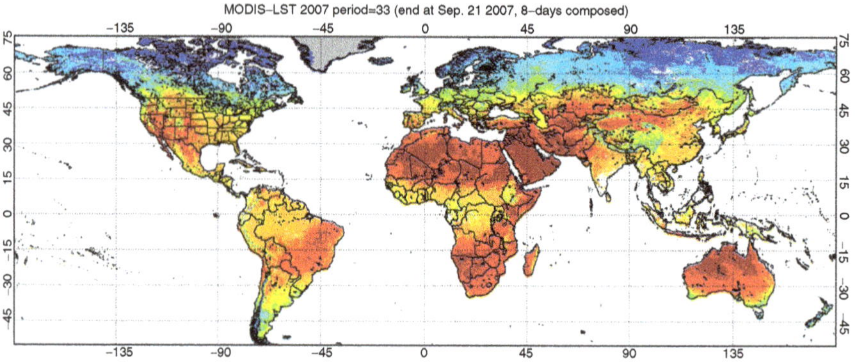

Fig. 7.18 MODIS land surface temperature of 2007 period = 033: composed by MODIS data from 14 September 2007 to 21 September 2007 (each period has 8 days). 8-day composed LST of 2007 (MODIS Land Surface Temperature of 2007 Period=033: composed by MODIS data from 9/14/2007 to 9/21/2007—https://www.star.nesdis.noaa.gov/smcd/emb/vci/VH/modis_browse8daysLST.php)

burn decay, peat fires, and decay of drained peatlands), all anthropogenic CH_4 sources, N_2O sources, and F-gases (HFCs, PFCs, and SF6).[58]

[58]Green gas emissions—https://tradingeconomics.com/greece/total-greenhouse-gas-emissions-kt-of-co2-equivalent-wb-data.html

How Is Hydrogen Produced?[59]

It turns out that the most common way (more than 90% of hydrogen made in the United States of America) of producing industrial amounts of hydrogen is steam-methane reforming (SMR). In other words,[60] you take methane gas (CH_4), and you chuck a load of steam (H_2O) at it under high pressure which results in hydrogen and a large quantity of CO_2.

Steam-Methane Reforming Is a Widely Used Method of Commercial Hydrogen Production

Steam-methane reforming currently accounts for nearly all commercially produced hydrogen in the United States of America. Commercial hydrogen producers and petroleum refineries use steam-methane reforming to separate hydrogen atoms from carbon atoms in methane (CH_4). In steam-methane reforming, high-temperature steam (1300–1800 °F) under 3–25 bar pressure (1 bar = 14.5 pounds per square inch) reacts with methane in the presence of a catalyst to produce hydrogen, carbon monoxide, and a relatively small amount of carbon dioxide (CO_2).

This dataset[61] presents trends in man-made emissions of major greenhouse gases and emissions by gas. Data refers to total emissions of CO_2 (emissions from energy use and industrial processes, e.g., cement production), CH_4 (methane emissions from solid waste, livestock, mining of hard coal and lignite, rice paddies, agriculture, and leaks from natural gas pipelines), nitrous oxide (N_2O), hydrofluorocarbons (HFCs), perfluorocarbons (PFCs), sulfur hexafluoride (SF_6), and nitrogen trifluoride (NF_3). Data excludes indirect CO_2. For UNCCCC Annex I countries, data follow the IPCC 2006 guideline. Territories' coverage is as defined in the Kyoto Protocol. We can download the dataset from the OECD and World Bank.[62]

7.2.3.1 Relationship Between GDP and Greenhouse Gas Emissions

The relationship between economics and the quality of the environment have long been regarded as a very close connection. However, how does the gross domestic

[59] How is hydrogen produced?—https://www.eia.gov/energyexplained/hydrogen/production-of-hydrogen.php

[60] Hydrogen Without CO_2—https://finance.yahoo.com/news/hydrogen-production-without-co2-getting-225107985.html

[61] Greenhouse Gas Emissions—https://stats.oecd.org/Index.aspx?DataSetCode=AIR_GHG

[62] CO_2 Emissions—https://data.worldbank.org/indicator/EN.ATM.CO2E.PP.GD

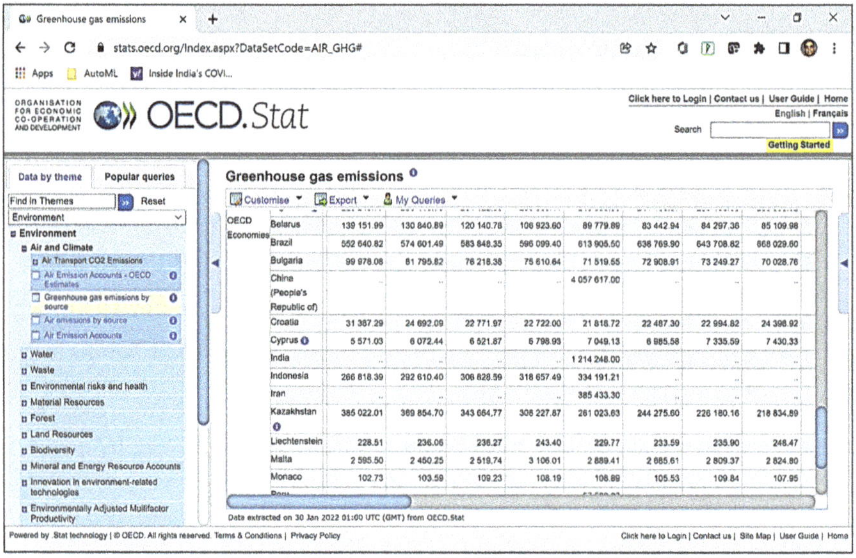

Fig. 7.19 OECD greenhouse gas emissions

product affect how much a country produces in greenhouse gases, particularly carbon dioxide? With data from the World Bank Database, one can observe the positive correlation between GDP pc and carbon dioxide emissions: as a country's GDP pc increases, so does its production of carbon dioxide into the atmosphere. Human activity, which often leads to increased GDP (please see Fig. 7.19) such as goods production and services, frequently produces carbon dioxide emissions. For example, most goods and services involve some use of energy, often in the form of coal or petroleum. Therefore, as the amount of goods produced increases, the amount of fossil fuels spent also increases. Please refer to Appendix E for Economics data sources.

GDP Data for World: Downloaded GDP World data from the World Bank[63] and analyzed CO_2 for the same. As displayed in Fig. 7.20, there exists a positive correlation (15.94%) between these parameters on a series of annual data from 1997 to 2018. Similar results were found when the relationship between real GDP and CO_2 emissions for 17 transitional economies based on a series of annual data from 1997 to 2014 was analyzed.[64] The analysis was conducted using Dynamic Ordinary Least Squares (OLS) (DOLS) and Fully Modified OLS (FMOLS) approaches [31].

The advanced, developing, and developed economies must take the responsibility to take the cost of greenhouse gas emissions and global climate change. Otherwise,

[63] The World Bank GDP Data—https://data.worldbank.org/indicator/NY.GDP.MKTP.KD.ZG

[64] A Cointegration Analysis of Real GDP and CO_2 Emissions in Transitional Countries—https://www.mdpi.com/2071-1050/9/4/568

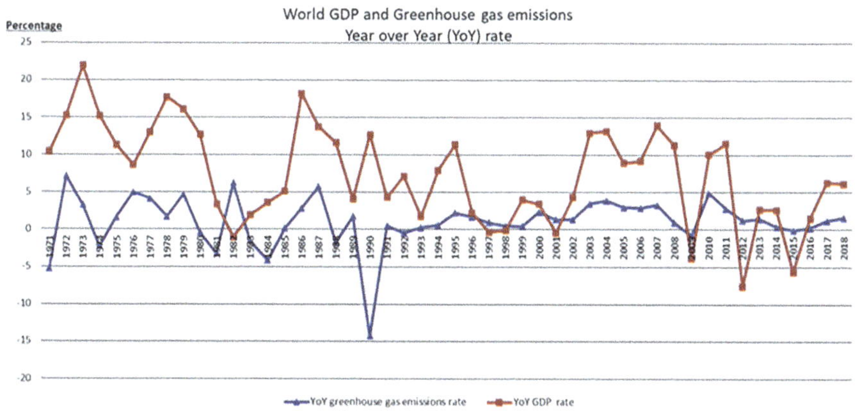

Fig. 7.20 World GDP and greenhouse gas emissions

countries that are the least contributors to global warming, for instance, Africa, are bearing the brunt (please see Fig. 7.21).

7.2.4 Impact of Climate on Staple Cereals—and Crop Yields[65]

7.2.4.1 Wheat

Wheat, which grows best in temperate climates, may see a broader area where it can be grown as temperatures rise, including the Northern United States and Canada, North China Plains, Central Asia, Southern Australia, and East Africa, but these gains may level off mid-century [32].

7.2.4.2 Corn or Maize

Maize, or corn, is grown all over the world, and large quantities are produced in countries nearer the equator. North and Central America, West Africa, Central Asia, Brazil, and China will potentially see their maize yields decline in the coming years and beyond as average temperatures rise across these breadbasket regions, putting more stress on the plants.

[65] Global Climate Change Impact on Crops Expected Within 10 Years, NASA Study Finds—https://climate.nasa.gov/news/3124/global-climate-change-impact-on-crops-expected-within-10-years-nasa-study-finds/

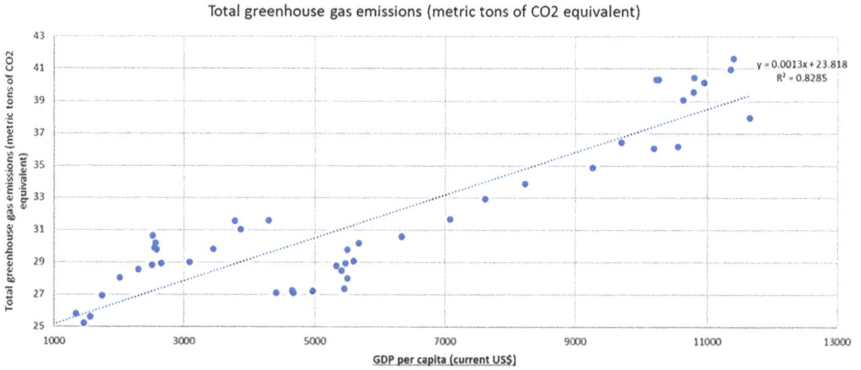

Fig. 7.21 GDP per capita

7.2.4.3 Soybeans

Soybean and rice projections showed a decline in some regions, but at the global scale, the different models still disagree on the overall impacts from climate change. For maize and wheat, the climate effect was much clearer, with most of the model results pointing in the same direction.

7.2.4.4 Climate Models vs. Weather Prediction Models

Unlike weather forecasts,[66] which describe a detailed picture of the expected daily sequence of conditions starting from the present, climate models are probabilistic, indicating areas with higher chances to be warmer or cooler and wetter or drier than usual. Climate models are based on global patterns in the ocean and atmosphere and records of the types of weather that occurred under similar patterns in the past.

Temperature Change and Carbon Dioxide Change
When the carbon dioxide concentration goes up, the temperature goes up. When the carbon dioxide concentration goes down, the temperature goes down.

[66] Climate Models—https://www.climate.gov/maps-data/climate-data-primer/predicting-climate/climate-models

Fig. 7.22 Nampula

7.3 Mozambique

Mozambique[67] borders Tanzania, Malawi, Zambia, Zimbabwe, South Africa, and Eswatini. Its long Indian Ocean coastline of 2500 km faces east to Madagascar. About two-thirds of its population of more than 31 million (2020) live and work in rural areas. It is endowed with ample arable land, water, energy, as well as newly discovered natural gas and mineral resources offshore; three, deep seaports; and a relatively large potential pool of labor. Agriculture remains as one of the most important economic sectors in Mozambique. The country experiences high levels of climate variability and extreme weather events (i.e., droughts, floods, and tropical cyclones).

The sandy soils and temperate climate of northern Mozambique create the perfect growing conditions for cashew trees. Mozambique has more than 32 million cashew trees, and nearly 70% are in the country's "cashew belt" that cuts across the northern provinces of Nampula (please see Fig. 7.22), Zambezia, and Cabo Delgado. With the proper attention and care of the trees, average yields can be 11 kg of raw cashew nut per year, and the productive lifespan of a tree can reach upward of 50 years. At these rates, cashew production is a business opportunity for smallholder farmers that lasts for generations.[68] Mozambique has reported the biggest cashew nuts production in

[67] Climate Change Portal—https://climateknowledgeportal.worldbank.org/country/mozambique

[68] Cashew Production—http://www.mozambicancashew.com/from-tree-to-trade#:~:text=Cashew%20Production,-The%20sandy%20soils&text=During%20the%20harvest%2C%20which%20occurs,sale%20to%20nearby%20processing%20facilities

3 years, 2018–2021. The once world cashew export leader sees its industry recovering slowly.[69]

Mozambique was a top global producer of cashew nuts in the 1970s and had a thriving processing sector that exported primary processed cashew nuts. The cashew nuts industry is one of the major employers in Mozambique. The importance of cashew in Mozambique was shown by a question on the national census, which asked "do you have any cashew trees?". In the 1960s, Mozambique produced half of the world's cashew nuts. Cashew remained Mozambique's largest export until 1982. It involved millions of peasant growers, after railways, sugar, and textiles.[70] Mozambique was a top global producer of cashew nuts in the 1970s and had a thriving processing sector that exported primary processed cashew nuts. However, following a prolonged civil war from the 1970s to the mid-1990s, and a cyclone in 1994 that destroyed 40% of productive cashew trees, Mozambique's cashew production levels fell dramatically [33].

Cashew production is the main source of income for close to 1.4 million rural producers in Mozambique. Not only Mozambique, but it is also a cash crop for many other countries.[71] As one of only a few reliable cash crops that farmers can grow, cashew production is the economic backbone of thousands of communities throughout the country [34]. Smallholder cashew producers typically manage small farms with 10–20 cashew trees mixed with other crops. During the harvest, which occurs from *October to February* in Mozambique, the average cashew farmer produces about 100 kg of raw nut for sale to nearby processing facilities. In 2013, total production of raw cashew nut was 64,000 metric tons, making Mozambique the second largest cashew producer in East and Southern Africa and the eighth largest producer globally.

In Mozambique today, nearly 1 million households grow cashew—though much of it is grown relatively passively. Smallholder farmers[72] are responsible for more than 95% of the country's cashew production. In Nampula province alone, cashews account for nearly one-fifth of the total household income and approximately two-thirds of the total cash income (in cashew-producing areas) [35].

As per the Global Climate Risk Index (CRI)[73] 2021, Mozambique, Zimbabwe, and the Bahamas were the countries most affected by the impacts of extreme weather events in 2019.

---

[69]Mozambique records highest cashew crop in 3 years—http://www.afrol.com/articles/35980

[70]Power without Responsibility: The World Bank & Mozambican Cashew Nuts.—https://www.jstor.org/stable/4006634

[71]Guinea-Bissau's cashew farmers survive tough times—https://www.bbc.com/news/world-africa-57286241

[72]MozaCajú Impact Report—https://www.technoserve.org/wp-content/uploads/2018/01/mozacaju-impact-report.pdf

[73]GLOBAL CLIMATE RISK INDEX 2021—https://www.germanwatch.org/sites/default/files/Global%20Climate%20Risk%20Index%202021_2.pdf

African cashew nut industries contribute over half of the world's supply of raw cashew nut (RCN), a product that is growing in demand globally as incomes rise and diets change. However, African countries still only process around 7% of the world's cashews. Most of the African RCN production is exported, primarily as an intermediate good to countries such as India and Vietnam which are the top two importers of RCN with shell. These countries then process the cashew nut for export to markets in North America and Europe, which house 40% of the global cashew demand. African countries are trying to climb the industrial ladder, and the processing of agricultural commodities seems a natural first step. By roasting coffee and spinning cotton, they hope to boost export earnings and create jobs. For example, a fifth of the retail price of cashews goes to primary processors. By reviving its industry, Mozambique[74] has captured some of that value [36].

Industrialization[75] of Cashew [37]

The cashew tree, *Anacardium occidentale* L., is native to northeast Brazil from where it has spread to other parts of South and Central America. It was introduced into southeast Africa and India by the Portuguese to help control

(continued)

[74] Mozambique's nut factories have made a cracking comeback—https://www.economist.com/middle-east-and-africa/2019/09/12/mozambiques-nut-factories-have-made-a-cracking-comeback

[75] Industrialization of Cashew in the state of Ceara, Brazil—https://repository.arizona.edu/bitstream/handle/10150/300508/pa-26-02-13-16.pdf?sequence=1&isAllowed=y

erosion and has since been planted extensively in these areas. Because of its sensitivity to cold, especially when young, its growth is limited to tropical areas, and the tree is being cultivated in the United States of America only in southernmost Florida.

The tree grows well in sandy coastal soils, needs very little attention, and does well with 30–40 in. of rainfall per year. Rainfall exceeding this amount, especially during the time of blossoming, promotes the growth of anthracnose which results in a fall of leaves and blossoms and a decay of forming fruit. All these factors contribute to a considerable reduction in yield.[76]

7.3.1 Food Prices Increased Driven by Depreciation of National Currency

The annual food inflation rate was estimated at 11% in April 2021, partly driven by the depreciation of the national currency throughout 2020 and early 2021. In February 2021, the metical reached its lowest value in the previous 4 years and, as of March, had lost over 10% of its value against the US dollar on a yearly basis.

A significant inflow of US dollars into the country due to higher revenues from exports of primary commodities spurred an appreciation of the domestic currency in April and May 2021, which is likely to have helped contain import inflationary pressure on food prices. The metical has, however, begun to depreciate again in June.[77]

7.3.2 Near-Average Cereal Import Requirements in 2021–2022

Harvesting of the 2021 cereal crops, mainly maize, is nearly complete. Cereal outputs in central and southern provinces are anticipated at average to above-average levels, while outputs in northern provinces, particularly in Cabo Delgado, are expected at a reduced level. FAO's forecast of the national cereal output stands at a near-average level of 2.8 million tons in 2021, 5% below the high level in 2020, reflecting the negative effects of weather hazards and the conflict on plantings and crop yields.

[76] Model Profile for 1.0 ha Cashew Cultivation—https://agricoop.nic.in/sites/default/files/Cashewnut%20Cultivation%20%281%29.pdf

[77] Mozambique—https://www.fao.org/giews/countrybrief/country.jsp?lang=en&code=MOZ

Mozambique has a low average yield[78] of raw cashew nut (RCN) of 3 kg/tree [38]. The latest census of agriculture in 2015 estimated that 1.33 million households owned cashew trees. Another 30,000 households were involved in post-harvest. One-half of RCN production sold was processed in 2015, up from one-third in 2008. A large share of cashew exports are raw nuts, mostly "informal" (no tax). In 2017, national production was only two-thirds of 1972, when Mozambique was the world leader in cashew exports. An export tax was imposed on RCN exports in 2001, currently 18% of the FOB price, to promote domestic processing. Key challenges for production include replacing aging trees with improved root-stock and stepped-up anti-fungal spraying. Industrial processing now comprises 15 factories employing 17,000 workers, 57% of whom are women. The main recommendations are a multi-stakeholder platform to periodically review cashew developments, smallholder participation in producer organizations, privatization of seedling distribution and tree-spraying without subsidies, public and private commercial infrastructure (warehouses, transportation, access roads), accessible international market and technical information, using cashew shells to generate energy, using cashew apple to produce packaged fermented beverages, and a cross-ministry push on food safety protocols for cashew.

Agroforestry

Agroforestry[79] is the intentional integration of trees and shrubs into crop and animal farming systems to create environmental, economic, and social benefits. It has been practiced in the United States of America and around the world for centuries. For a management practice to be called agroforestry, it typically must satisfy the four "i"s:

- Intentional
- Intensive
- Integrated
- Interactive

There are five widely recognized categories of agroforestry in the United States of America:[80]

- *Alley cropping* means planting crops between rows of trees to provide income while the trees mature. The system can be designed to produce fruits, vegetables, grains, flowers, herbs, bioenergy feedstocks, and more.

(continued)

[78] The Cashew Value Chain in Mozambique—https://openknowledge.worldbank.org/handle/10986/31863

[79] Agroforestry—https://www.usda.gov/topics/forestry/agroforestry#:~:text=Agroforestry%20is%20the%20intentional%20integration,around%20the%20world%20for%20centuries

[80] The USDA Agroforestry Strategic Framework—https://www.usda.gov/sites/default/files/documents/usda-agroforestry-strategic-framework.pdf

- *Forest farming* operations grow food, herbal, botanical, or decorative crops under a forest canopy that is managed to provide ideal shade levels as well as other products. Forest farming is also called multi-story cropping.
- *Silvopasture* combines trees with livestock and their forages on one piece of land. The trees provide timber, fruit, or nuts as well as shade and shelter for livestock and their forages, reducing stress on the animals from the hot summer sun, cold winter winds, or a downpour.

7.3.3 Climate Change

Mozambique's long coastline, sprawling river delta, and changing weather patterns make it susceptible to multiple hazards as the climate changes. Flooding, heat waves, cyclones, and drought are all getting more[81] frequent and severe [7] as the earth gets hotter.[82] Adapting to drought here will likely require comparatively big, expensive infrastructure projects to bring water out to the fields. Putting the brakes on climate change in general will require a global shift away from fossil fuels that has been slow at best [39].

In 2019, Mozambique was the most affected country worldwide by the impacts of extreme weather events. It scored fifth over the period 2000–2019 (Global Climate Risk Index 2021). While the country only contributes 0.1–0.2% to global emission, Mozambique is the 38th most vulnerable and the 13th least ready country to address the effects of climate change[83] [6].

Mozambique is one of many countries around the world where weather forecasting is lagging even as climate change drives more extreme and variable weather. In Bangladesh, tropical storm warnings are not adequately reliable. In Peru, the national government is trying to get weather and climate information to residents who are experiencing more extreme weather. The World Meteorological Organization is working on upgrading flash flood warning systems in more than 50 countries around the world, including Mozambique [7]. Everyday weather is also becoming more extreme and harder to predict. As is true in much of the world, as the earth gets hotter, climate change is also making extreme rain more likely. That means that a

---

[81] Meteorologists Can't Keep Up With Climate Change in Mozambique—https://www.npr.org/sections/goatsandsoda/2019/12/11/782918005/meteorologists-cant-keep-up-with-climate-change-in-mozambique

[82] Mozambique Is Racing to Adapt to Climate Change. The Weather Is Winning—https://www.npr.org/sections/goatsandsoda/2019/12/27/788552728/mozambique-is-racing-to-adapt-to-climate-change-the-weather-is-winning

[83] Food security and climate change, the pressing reality of Mozambique—https://docs.wfp.org/api/documents/WFP-0000129988/download/?_ga=2.150781622.1610345514.1644972256-1475447051.1638068907

larger share of the precipitation that falls on the country is forecast to come in big dumps as opposed to more moderate rainstorms.

> "Nowadays, we have more intense phenomena. More cases of strong rains, the heat is stronger, the winds are stronger. It's different from the past. Mozambique is already suffering longer droughts and more frequent and severe storms than it was earlier this century because the climate is changing."
> Acacio Tembe, Mozambique's lead weather forecaster

Climate change has disproportionate effects on women and girls in Mozambique, since they are more dependent on natural resources for household and agricultural tasks. Women are normally responsible for crop production (men oversee livestock) and availability of food and water for the household. Women's rights and control over natural resources are less than men's, and they are often underrepresented in decision-making bodies. Women's burdens are aggravated if they are left alone by men who migrate to larger cities or even abroad (which is according to some an increasingly common coping strategy to climate-related hazards, while other studies report reduced male migration in recent years). As a result, in many areas, over 50% of households is female-headed, and women and girls need to cope with the burdens of reduced water availability and food security[84] [40] (please see 1991–2020 mean temperature values in Mozambique) (Fig. 7.23).

7.3.3.1 NOAA: Advanced Very High-Resolution Radiometer (AVHRR) and Climate Change Parameters

All three vegetation indicators are based on 10-day (dekadal) vegetation data from the METOP-AVHRR sensor at 1 km resolution (2007 and after). Data at 1 km resolution for the period 1984–2006 are derived from the NOAA-AVHRR dataset at 16 km resolution. Precipitation estimates for all African countries (except Cabo Verde and Mauritius) are taken from NOAA/FEWSNet, while for the remaining countries data is obtained from ECMWF.

Vegetation Health Index:[85] The Vegetation Health Index (VHI) illustrates the severity of drought based on vegetation health and the influence of temperature on plant conditions. The VHI is a composite index and the elementary indicator used to compute the ASI. It combines both the Vegetation Condition Index (VCI) and the Temperature Condition Index (TCI). The TCI is calculated using a similar equation to the VCI but relates the current temperature to the long-term maximum and minimum, as it is assumed that higher temperatures tend to cause a deterioration in vegetation conditions. A decrease in the VHI would, for example, indicate relatively

[84] Climate Change Profile MOZAMBIQUE—https://ees.kuleuven.be/klimos/toolkit/documents/689_CC_moz.pdf

[85] Mozambique—https://www.fao.org/giews/earthobservation/country/index.jsp?lang=en&code=MOZ

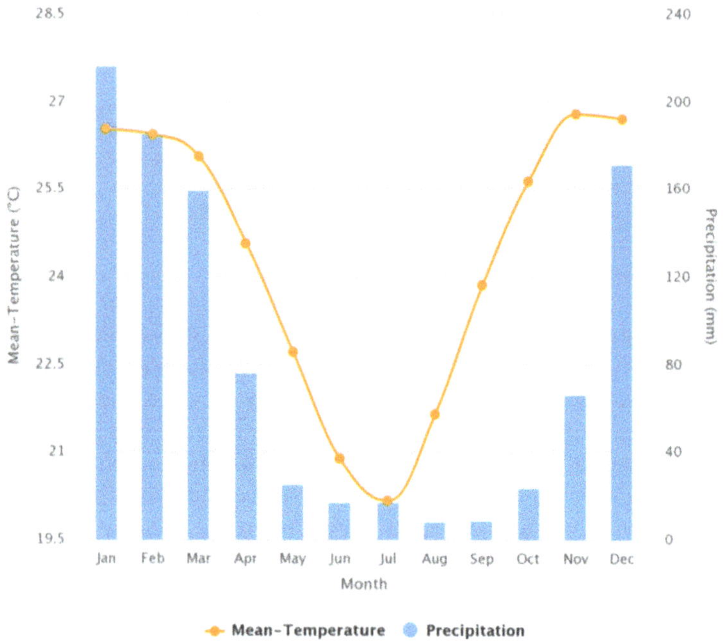

Fig. 7.23 Monthly climatology of mean-temperature and precipitation in Mozambique from 1991 to 2020

poor vegetation conditions and warmer temperatures, signifying stressed vegetation conditions, and over a longer period would be indicative of drought. The VHI images are computed for the two main seasons and in three modalities: dekadal, monthly, and annual.

Here is the 2020 VHI for Mozambique[86] (please see below figures):

[86]Mozambique—https://www.fao.org/giews/earthobservation/country/index.jsp?lang=en&code=MOZ

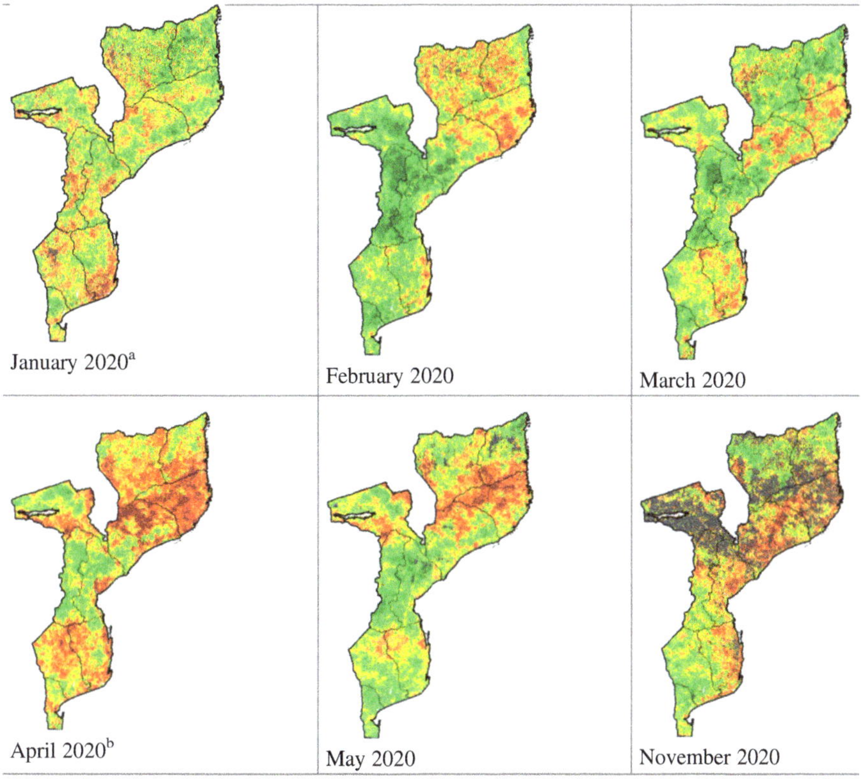

[a]Mozambique January 2020—https://www.fao.org/giews/earthobservation/asis/data/country/MOZ/MAP_NDVI_ANOMALY/HR/om2001h.png

[b]Mozambique April 2020—https://www.fao.org/giews/earthobservation/asis/data/country/MOZ/MAP_NDVI_ANOMALY/HR/om2004h.png

7.3.3.2 Province Averaged VH

Please find Vegetation Health Data 1990–2021 of Mozambique, Nampula.[87] For cashew nuts, select crop type and other pulses. Other parameters include VCI/TCI/VHI, VHI; Select Years, 1990–2021 (please see Fig. 7.24).

Please find Weekly Averaged Time Series for Province #6: Nampula of Mozambique—Averaged VHI (Fig. 7.25):[88]

[87]VH Score—https://www.star.nesdis.noaa.gov/smcd/emb/vci/VH/vh_adminMeanByCrop.php?type=Province_Weekly_PAreaPlot

[88]Nampula of Mozambique: Averaged VHI—https://www.star.nesdis.noaa.gov/smcd/emb/vci/VH/vh_adminMeanByCrop.php?type=Province_Weekly_MeanPlot

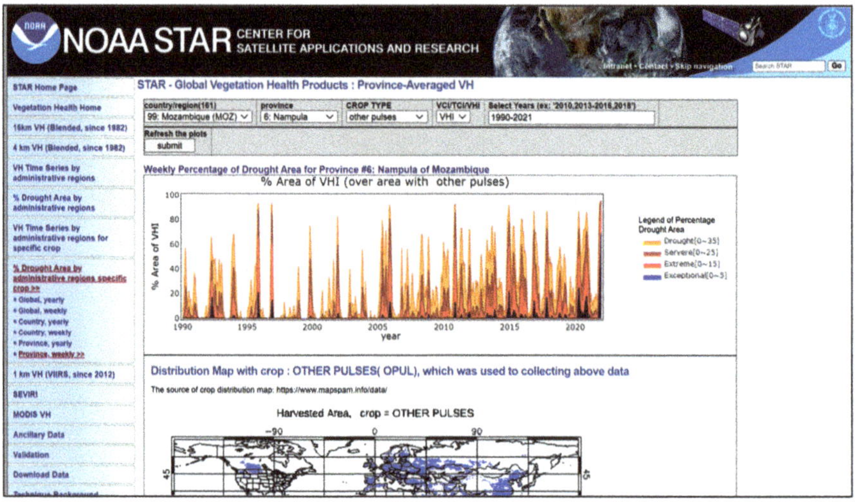

Fig. 7.24 VHI; select years, 1990–2021

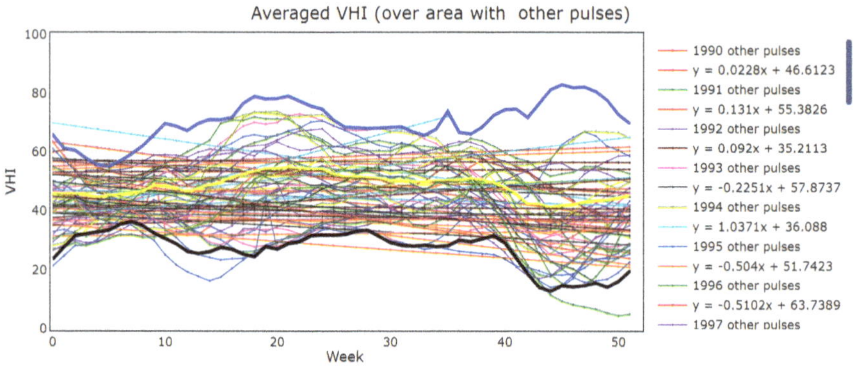

Fig. 7.25 Averaged VHI

As it can be seen (please see Fig. 7.26), the VHI (y) score is negatively correlated with x (years). What it tells us is that the Vegetation Health Index (y) is decaying, either due to climate change or extreme weather factors.

$$y = 0.56.3x + 1173.8016 \tag{7.4}$$

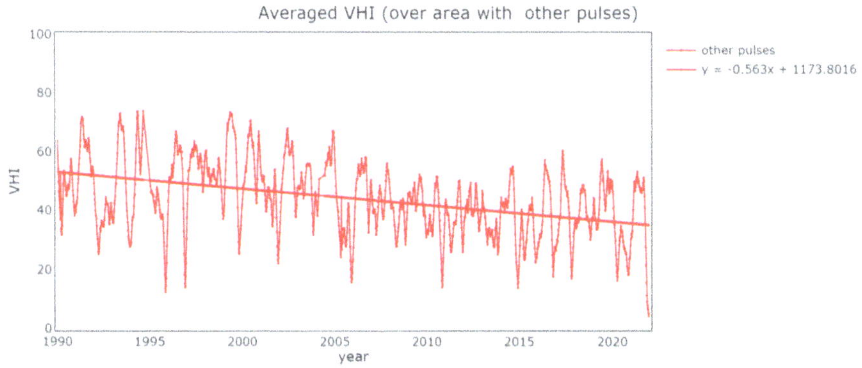

Fig. 7.26 Averaged VHI

7.3.3.3 No Noise (Smoothed) Normalized Difference Vegetation Index (SMN)

Global, 4 km, 7-day composite, validated. The SMN is derived from no noise NDVI, which components were pre- and post-launch calibrated. SMN (please see Fig. 7.27) can be used to estimate the start and senescence of vegetation, start of the growing season, and phenological phases.

$$y = 0.0007x + 1.7027 \tag{7.5}$$

Averaged SMN and greenness,[89] no noise NDVI are trending negative (-.0007x). Over the years, as per the above graph, the greenness is reduced, albeit a slower slope.

Mozambique cashew nuts	

[89] Mozambique, Greenness (No Noise NDVI)—https://www.star.nesdis.noaa.gov/smcd/emb/vci/VH/vh_browseByCountry.php

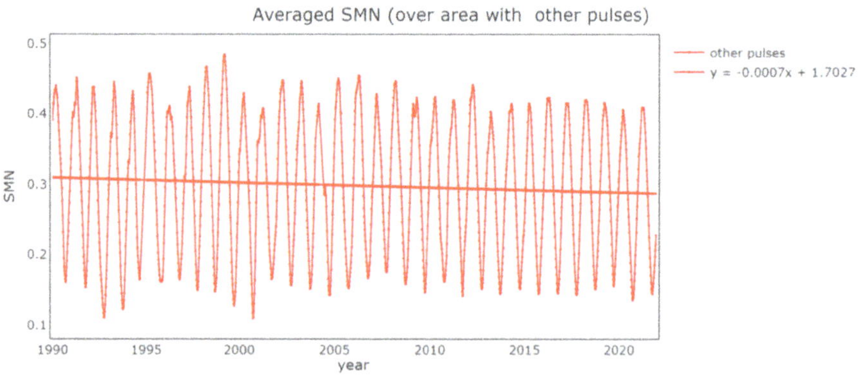

Fig. 7.27 Averaged SMN

7.4 Machine Learning Model: Mozambique Cashew Nuts Production Model

Let's build a model that depicts the impact of climate change on Thailand rice production:

	Software code for this model: Mozambique_Namapula_CashewNuts_code.ipynb (Jupyter Notebook Code)

7.4.1 Data Sources

The data sources for the ML model is from FAO Statistics Data,[90] World Bank Climate Change Data specifically Mozambique—Nampula,[91] World Metrological Organization Data,[92] FAO Earth Observation,[93] and NOAA STAR—STAR—Global Vegetation Health Products: Province-Averaged VH[94] (please see

[90] FAO Statistics data—https://www.fao.org/faostat/en/#data/QCL

[91] Climate Change Data—https://climateknowledgeportal.worldbank.org/download-data

[92] World Metrological Organization Data—https://climatedata-catalogue.wmo.int/explore

[93] Earth Observation Mozambique—https://www.fao.org/giews/earthobservation/country/index.jsp?lang=en&code=MOZ

[94] NOAA STAR—VHI—https://www.star.nesdis.noaa.gov/smcd/emb/vci/VH/vh_adminMeanByCrop.php?type=Province_Weekly_MeanPlot

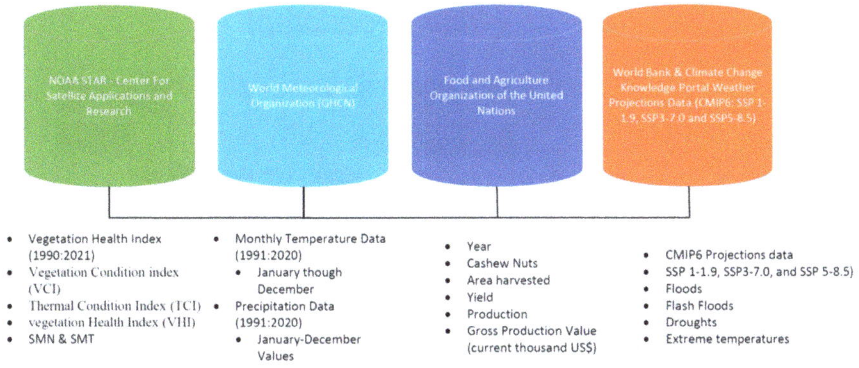

Fig. 7.28 Data sources

Fig. 7.28). Please refer to Appendix C for World Data Sources, Appendix D for United States of America, and Appendix E for Economics data sources.

Mozambique weather calendar data is, additionally, collected from FAO GIEWS.[95] The major sowing, growing, and harvesting months include (please see Fig. 7.29):

- January–May
- October–December

Mozambique has witnessed climate change and floods during the last decades.[96] Additionally, recently, 2022, food prices increased amid depreciation of national currency, and worsening conflict and effects of the COVID-19 pandemic deteriorate food security in 2021. The annual food inflation rate was estimated at 11% in April 2021, partly driven by the depreciation of the national currency throughout 2020 and early 2021. In February 2021, the metical reached its lowest value in the previous 4 years and, as of March, had lost over 10% of its value against the US dollar on a yearly basis. Finally, as per the latest IPC analysis released in January 2021, the number of people facing acute food insecurity in the April–September 2021 period was projected at 1.7 million. Although, at the national level, the prevalence of food insecurity has decreased compared to the January–March 2021 period, primarily reflecting an improvement in food supplies from the main harvest, the situation in the northern province of Cabo Delgado has worsened. The number of food-insecure people in Cabo Delgado is estimated at 770,000 in the April–September 2021 period, an increase of about 15% over the previous period because of the persisting conflict and the impact of rainfall deficits on agricultural production.

---

[95]FAO GIEWS—Mozambique—https://www.fao.org/giews/countrybrief/country.jsp?lang=en&code=MOZ

[96]Mozambique—https://reliefweb.int/disasters?search=Mozambique

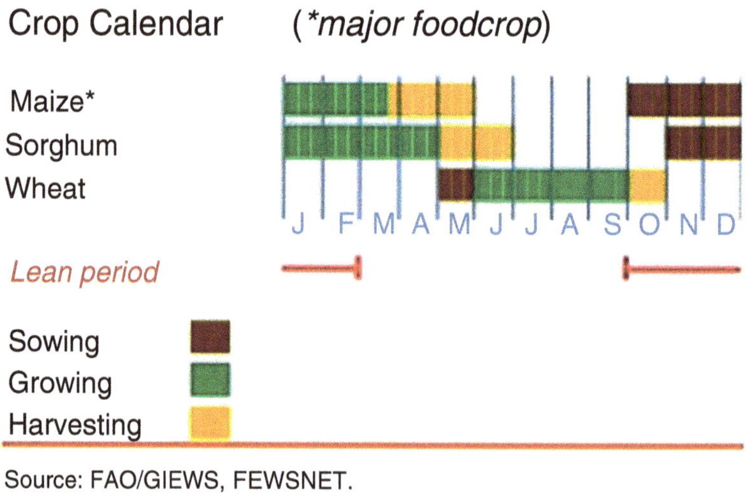

Fig. 7.29 Mozambique calendar

Cashew is a tropical plant and can thrive even at high temperatures.[97] *Young plants are sensitive to frost.* The distribution of cashew is restricted to altitudes up to 700 m *above mean sea level where the temperature does not fall below 20 °C for prolonged period.* Areas where the temperatures range from 20 to 30 °C with an annual precipitation of 1000–2000 mm are ideal for cashew growing. However, temperatures above 36 °C between the flowering and fruiting periods could adversely affect the fruit setting and retention. Heavy rainfall, evenly distributed throughout the year, is not favorable though the trees may grow and sometimes set fruit. Cashew needs a climate with a well-defined dry season of at least 4 months to produce the best yields. Coincidence of excessive rainfall and high relative humidity with flowering may result in flower/fruit drop and heavy incidence of fungal diseases. The Biologische Bundesanstalt, Bundessortenamt und Chemische Industrie (BBCH)[98] scale has been standardized for certain fruit crops like mango, stone fruits, pome fruits, etc. However, for cashew no such scale is available. Different phenological growth stages identified in cashew and their codification according to the modified BBCH scale are listed.[99]

---

[97] Model Profile for 1.0 ha Cashew Cultivation—https://agricoop.nic.in/sites/default/files/Cashewnut%20Cultivation%20%281%29.pdf

[98] New BBCH growth stage keys—https://gd.eppo.int/reporting/article-6991

[99] BBCH Cashew Scale—https://cashew.icar.gov.in/phenology/wp-content/uploads/2018/11/table-phenology-stages.pdf

Growth state (not all captured)—for complete comprehensive—please refer to:[a]	
Principal growth stage 8: Nut and apple maturity stages	
Principal growth stage 9: Senescence stages	

(continued)

[a]PHENOLOGICAL GROWTH STAGES IDENTIFIED IN CASHEW—https://cashew.icar.gov.in/phenology/wp-content/uploads/2018/11/figures-cashew-phenology.pdf

By *mid-October*, the cashew harvest season will begin in this northernmost district of Nangade, the major cashew-producing region of Cabo Delgado province. By *November*, the trees in the rest of the cashew-producing areas of the country will be ready for harvest, which typically lasts through *December*, when commercialization begins.[100]

So, October–December is a growing season; January–March is a harvesting season.

The ML model accuracies have improved after only taking sowing, growing, and harvesting months. Floods occur during the late months of sowing (august), growing (September), and early months of October (harvesting) which have a detrimental impact on the production:[101]

- Southern Africa: Drought (2018–2022)[102]
  - Crisis (IPC Phase 3) outcomes are ongoing across rainfall-deficit areas and areas of northern Mozambique and eastern DRC impacted by continued conflict. Between December 2021 and March 2022, around 336,000 people (29% of the population in Eswatini) are estimated to be facing high acute food insecurity (IPC Phase 3 crisis or above) and require urgent humanitarian assistance. Of this population, 286,000 people are classified in IPC Phase 3 (crisis) and 50,000 in IPC Phase 4 (emergency). An additional 376,000 are classified in IPC Phase 2 (stressed) and require livelihood support.
- Mozambique: Floods (December 2019[103])
  - Mozambique was severely affected by the rising level of the water basin, which led to floods. These floods in the region caused the isolation of communities in floodplains; community members took refuge in treetops, houses, and ravines waiting for rescue.

Mozambique is one of the highly vulnerable climate change countries and requires every measure of help to overcome.

[100]Cashew Farmers—https://static1.squarespace.com/static/53971e4fe4b0240065924592/t/57ced3e803596e075fcb3958/1473172637648/Cashew+farmers+prepare+for+a+busy+harvest+season+9.06.2016-small.pdf

[101]List of Floods—https://reliefweb.int/disasters?search=Thailand+Floods

[102]Southern Africa: Drought—2018-2022—https://reliefweb.int/disaster/dr-2018-000429-zwe

[103]Mozambique: Floods—Dec 2019—https://reliefweb.int/disaster/fl-2020-000011-moz

7.4.2 Step 1: Import Libraries

Import open-source and machine learning libraries to process the data. For graphics, Matplot Lib is used and ELI5 for explainability. ELI5[104] is a Python library which allows you to visualize and debug various machine learning models using unified API:

```
import pandas as pd
import numpy as np
from sklearn.linear_model import LinearRegression
from sklearn.model_selection import StratifiedKFold
from sklearn.metrics import mean_squared_error
from math import sqrt
import random
from sklearn.impute import SimpleImputer
from sklearn.experimental import enable_iterative_imputer
from sklearn.impute import IterativeImputer
import seaborn as sns
import matplotlib.pyplot as plt
import seaborn as sn

import plotly.graph_objects as go
import seaborn as sns; sns.set()
import matplotlib.pyplot as plt
from scipy import stats
from scipy.stats import pearsonr
import matplotlib.pyplot as plt
from sklearn.preprocessing import MinMaxScaler
from sklearn.model_selection import train_test_split
from sklearn.linear_model import LinearRegression
from sklearn.ensemble import RandomForestClassifier
from sklearn.metrics import r2_score
from sklearn import metrics
import eli5
from statsmodels.tsa.stattools import adfuller
from statsmodels.graphics.tsaplots import plot_pacf
from statsmodels.graphics.tsaplots import plot_acf
from sklearn.metrics import mean_squared_error
from sklearn.model_selection import train_test_split
from statsmodels.tsa.arima_model import ARIMA
from statsmodels.tsa.stattools import grangercausalitytests
from statsmodels.tsa.api import VAR
import lime
import lime.lime_tabular
```

[104] ELI5—https://eli5.readthedocs.io/en/latest/

7.4.3 Step 2: Load Mozambique Cashew Nuts FAO Dataset 1991–2020

```
dfMozambiqueNamaoulsaCashewProductionData = pd.read_csv
('namapula_cashewdata.csv')
dfMozambiqueNamaoulsaCashewProductionData.tail()
```

Output:
Please find output: as you can see, the cashew nuts with shell item:

	Area	Element	Item	Year	Area_harvested_Unit	Area_harvested_Value	Yield(unit)	Yield_value	Production(units)	Production_value
55	Mozambique	Area harvested	Cashew nuts, with shell	2016	ha	124160	hg/ha	8391	tonnes	104179
56	Mozambique	Area harvested	Cashew nuts, with shell	2017	ha	165764	hg/ha	8391	tonnes	139088
57	Mozambique	Area harvested	Cashew nuts, with shell	2018	ha	157114	hg/ha	8274	tonnes	130000
58	Mozambique	Area harvested	Cashew nuts, with shell	2019	ha	169139	hg/ha	8277	tonnes	140000
59	Mozambique	Area harvested	Cashew nuts, with shell	2020	ha	154858	hg/ha	8280	tonnes	128225

As it can be seen, Mozambique had a peak area harvested during the 1970s.

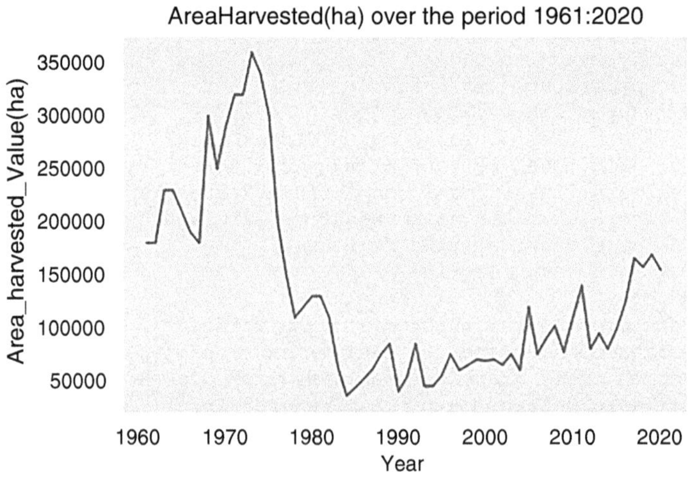


```
import matplotlib.pyplot as plt
import numpy as np

# Data for plotting
t = dfMozambiqueNamaoulsaCashewProductionData['Year']
```

(continued)

```
s = dfMozambiqueNamaoulsaCashewProductionData
['Production_value']

fig, ax = plt.subplots()
ax.plot(t, s)

ax.set(xlabel='Year', ylabel='Production_value(tonnes)',
    title='Cashew Nuts with Shell Production_value (tonnes) over
the period 1961:2020')
ax.grid()

plt.show()
```

And can be seen in the production of cashew nuts:

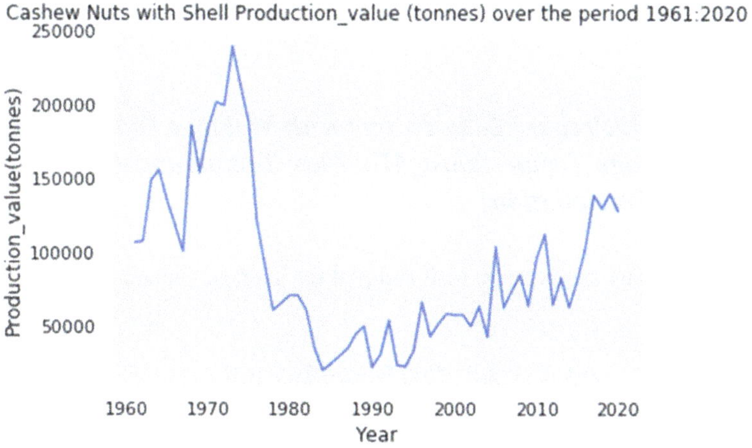

One fact we can see is that yield rates are low[105] during the 1970s but improved recently [38]:

[105]Mozambique has a low average yield of raw cashew nut (RCN) of 3 kg/tree—https://openknowledge.worldbank.org/handle/10986/31863

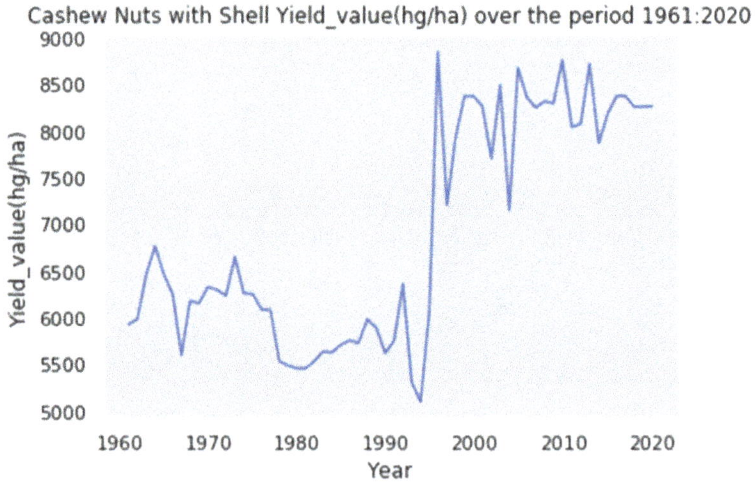

7.4.4 Step 3: Prepare Data Frame with Weather Data (Average Temperature, Min/Max Temperatures, and Precipitation)

Please find average temperature load and plot the average temperatures:

```
dfMozambiqueNamaoulsaCashewProductionPrecp = pd.read_csv
('Nampula_pr.csv')
dfMozambiqueNamaoulsaCashewProductionPrecp.head()


fig, axes = plt.subplots(nrows=6, ncols=2, dpi=120, figsize=(10,6))
for i, ax in enumerate(axes.flatten()):
  data = dfMozambiqueNamaoulsaCashewProductionTemp
[dfMozambiqueNamaoulsaCashewProductionTemp.columns[i+1]]
  ax.plot(data, color='red', linewidth=1)
  ax.set_title(dfMozambiqueNamaoulsaCashewProductionTemp.
columns[i+1])
  ax.xaxis.set_ticks_position('none')
  ax.yaxis.set_ticks_position('none')
  ax.spines["top"].set_alpha(0)
  ax.tick_params(labelsize=6)
plt.tight_layout();
```

Output:
As can be seen, March and April have maximum average temperatures.

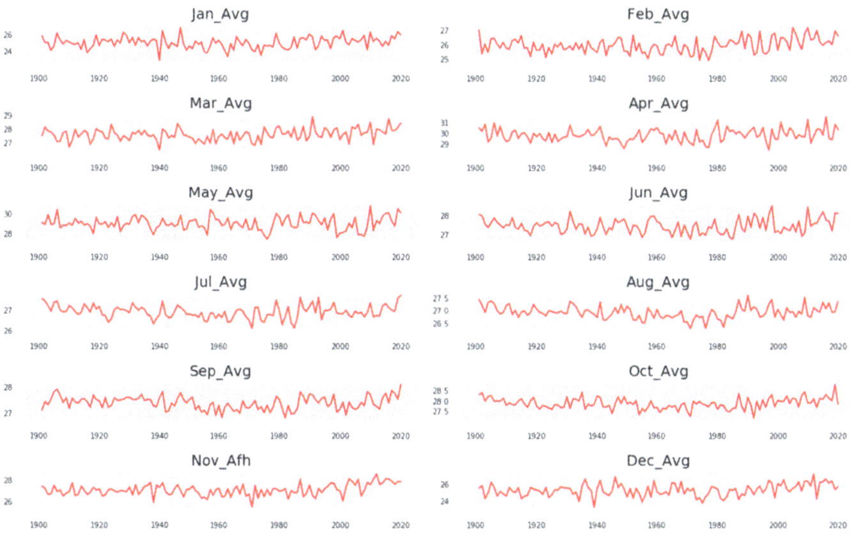

Please find minimum temperature details:

```
dfMozambiqueNamaoulsaCashewProductionTempMin = pd.read_csv
('tasmin_timeseries_monthly_cru_1901-2020_MMR_2126.csv')
dfMozambiqueNamaoulsaCashewProductionTempMin.head()
```

Output: (plotting minimum temperature data frame) December (17.5 °C) and January (15 °C) have, on average, minimum temperatures.

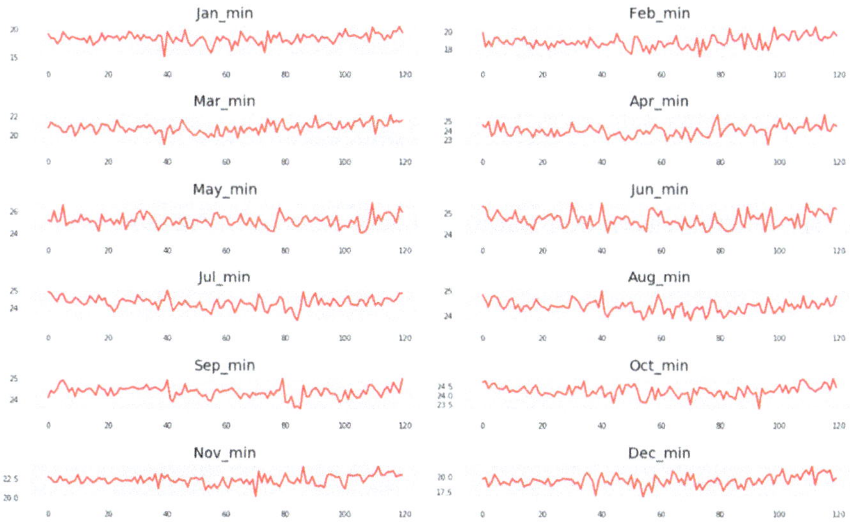

Please find maximum temperature values:

```
dfMozambiqueNamaoulsaCashewProductionTempMax = pd.read_csv
('tasmax_timeseries_monthly_cru_1901-2020_MMR_2126.csv')
dfMozambiqueNamaoulsaCashewProductionTempMax.head()
```

Output plot: Historically, months of March (35 °C), April (36 °C), and May (35 °C) have exhibited higher temperatures.

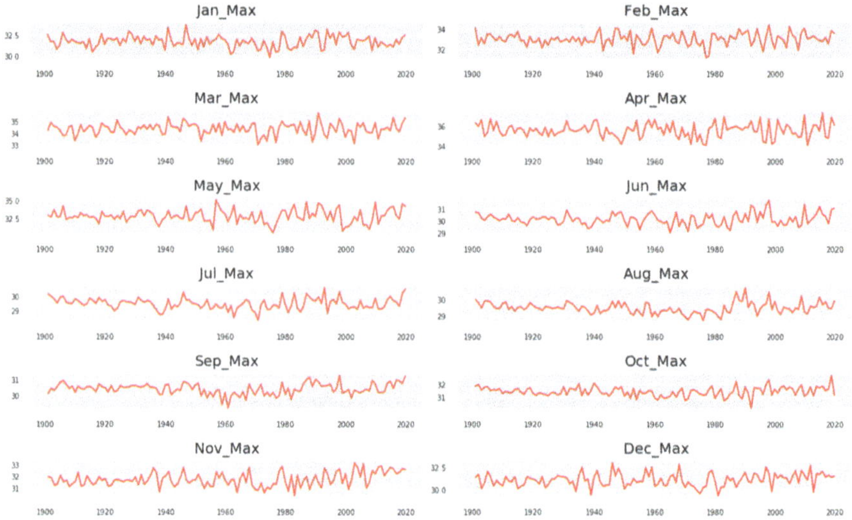

Please find precipitation values:

```
dfMozambiqueNamaoulsaCashewProductionPrecp = pd.read_csv
('Nampula_pr.csv')
dfMozambiqueNamaoulsaCashewProductionPrecp.head()
```

Output: Generally, cashew nuts need annual precipitation of 1000–2000 mm; January–May and November and December are the favorable months.[106]

[106]Cashew Cultivation—https://agricoop.nic.in/sites/default/files/Cashewnut%20Cultivation%20%281%29.pdf

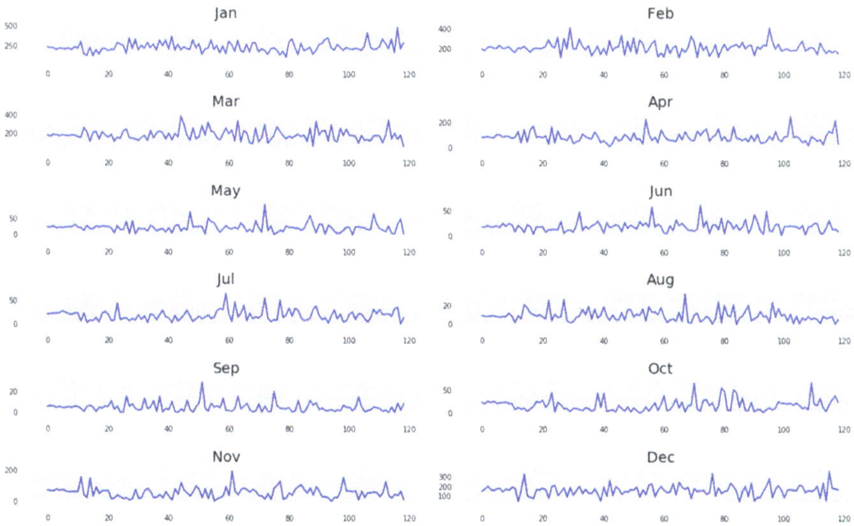

7.4.5 Step 4: Concat Weather Data into a Frame to Prepare the Data for Modeling

```
weatherdf = pd.concat
([dfMozambiqueNamaoulsaCashewProductionPrecp,
         dfMozambiqueNamaoulsaCashewProductionTempMin,
         dfMozambiqueNamaoulsaCashewProductionTempMax],axis =
1)
```

The combined weather data frame columns include:

```
Index(['Jan_Precp', 'Feb_Precp', 'Mar_Precp', 'Apr_Precp',
'May_Precp',
   'Jun_Precp', 'Jul_Precp', 'Aug_Precp', 'Sep_Precp',
'Oct_Precp',
   'Nov_Precp', 'Dec_Precp', 'Year_precp', 'Jan_min', 'Feb_min',
'Mar_min',
   'Apr_min', 'May_min', 'Jun_min', 'Jul_min', 'Aug_min',
'Sep_min',
   'Oct_min', 'Nov_min', 'Dec_min', 'dupYear', 'Jan_Max',
'Feb_Max',
   'Mar_Max', 'Apr_Max', 'May_Max', 'Jun_Max', 'Jul_Max',
'Aug_Max',
   'Sep_Max', 'Oct_Max', 'Nov_Max', 'Dec_Max', 'dupYear'],
  dtype='object')
```

7.4.6 Step 5: Merge Production and Weather Data Frames to Prepare the Data for Modeling

```
df = dfMozambiqueNamaoulsaCashewProductionData[['dupYear' ,
'Area_harvested_Unit', 'Area_harvested_Value' , 'Yield(unit)',
'Yield_value' , 'Production(units)', 'Production_value']]

dfMozambiqueNamaoulsaCashewProduction = weatherdf[weatherdf.
index >=1961 ]
dfMozambiqueNamaoulsaCashewProduction
df_final= df
df_final

df_merged = df_final.merge
(dfMozambiqueNamaoulsaCashewProduction,on= "Year", how=
"inner")
df_merged
```

The combined data frame columns include:

```
Index(['dupYear_x', 'Area_harvested_Unit',
'Area_harvested_Value',
    'Yield(unit)', 'Yield_value', 'Production(units)',
'Production_value',
    'Jan_Precp', 'Feb_Precp', 'Mar_Precp', 'Apr_Precp',
'May_Precp',
    'Jun_Precp', 'Jul_Precp', 'Aug_Precp', 'Sep_Precp',
'Oct_Precp',
   'Nov_Precp', 'Dec_Precp', 'Year_precp', 'Jan_min', 'Feb_min',
'Mar_min',
    'Apr_min', 'May_min', 'Jun_min', 'Jul_min', 'Aug_min',
'Sep_min',
    'Oct_min', 'Nov_min', 'Dec_min', 'dupYear_y', 'Jan_Max',
'Feb_Max',
    'Mar_Max', 'Apr_Max', 'May_Max', 'Jun_Max', 'Jul_Max',
'Aug_Max',
    'Sep_Max', 'Oct_Max', 'Nov_Max', 'Dec_Max', 'dupYear_y']

dfFinal = df_merged[['dupYear_x', 'Area_harvested_Value',
     'Nov_Precp', 'Dec_Precp', 'Jan_Precp',
    'Feb_Precp', 'Mar_Precp', 'Apr_Precp', 'Jan_Max_temp',
'Feb_Max_temp', 'Mar_Max_temp', 'Apr_Max_temp',
    'May_Max_temp', 'Nov_Max_temp', 'Dec_Max_temp' ,
    'Jan_min_temp', 'Feb_min_temp', 'Mar_min_temp',
'Apr_min_temp',
    'May_min_temp', 'Nov_min_temp', 'Dec_min_temp',
'Production_value']]
```

7.4.7 Step 6: Merge Data—The Final Data Frame

Based on the cashew nuts growing and harvesting calendar, prepare the final data frame:

```
import pandas as pd
import matplotlib.pyplot as plt
# This ensures plots are displayed inline in the Jupyter notebook
%matplotlib inline
# Get the label column
label = dfFinal['Production_value']
# Create a figure for 2 subplots (2 rows, 1 column)
fig, ax = plt.subplots(2, 1, figsize = (9,12))
# Plot the histogram
ax[0].hist(label, bins=100)
ax[0].set_ylabel('Frequency')
# Add lines for the mean, median, and mode
ax[0].axvline(label.mean(), color='magenta', linestyle='dashed', linewidth=2)
ax[0].axvline(label.median(), color='cyan', linestyle='dashed', linewidth=2)
# Plot the boxplot
ax[1].boxplot(label, vert=False)
ax[1].set_xlabel('Production_value')
# Add a title to the Figure
fig.suptitle('Production_value tonnes')
# Show the figure
fig.show()
```

The cashew nuts production value distribution is as follows:

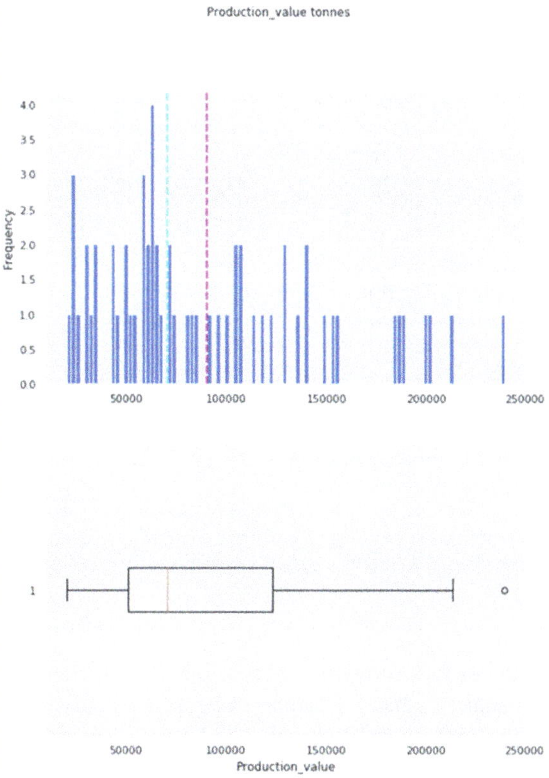

7.4.8 Step 7: Correlation Values

The following code provides production to feature attribute correlation ship values:

```
numeric_features=[
    'Area_harvested_Value',
    'Nov_Precp', 'Dec_Precp', 'Jan_Precp',
    'Feb_Precp', 'Mar_Precp', 'Apr_Precp', 'Jan_Max_temp',
'Feb_Max_temp', 'Mar_Max_temp', 'Apr_Max_temp',
    'May_Max_temp', 'Nov_Max_temp', 'Dec_Max_temp' ,
    'Jan_min_temp', 'Feb_min_temp', 'Mar_min_temp',
'Apr_min_temp',
    'May_min_temp', 'Nov_min_temp', 'Dec_min_temp',
'Production_value']

for col in numeric_features:
  fig = plt.figure(figsize=(9, 6))
  ax = fig.gca()
  feature = dfFinal[col]

  label = dfFinal['Production_value']
  correlation = feature.corr(label)
  plt.scatter(x=feature, y=label)
  plt.xlabel(col)
  plt.ylabel('Production Value (tonnes)')
  ax.set_title('Production Value (tonnes) vs ' + col + '-
correlation: ' + str(correlation))
plt.show()
```

Please find correlation of precipitation, and temperature for months of November, December, January, February, March, April, and May vs. production of cashew nuts is positively correlated with April precipitation (19.28%); in other months precipitation is negatively correlated with higher precipitation, or rainfall could have a detrimental impact on the production. For instance, Brazil (during 2013–2017[107]) has suffered considerably from adverse climate conditions [38]. Consecutive years of drought and then very heavy rainfall when it came severely decreased cashew production in Ceará and São Paulo provinces, together accounting for over 80% of surfaces cultivated, preventing the country from maintaining its level of cashew production[108] [41].

[107] The Cashew Value Chain in Mozambique—https://documents1.worldbank.org/curated/en/397581559633461087/pdf/The-Cashew-Value-Chain-in-Mozambique.pdf

[108] Effects of moisture conditions and management on production of Cashew—https://webapps.itc.utwente.nl/librarywww/papers_2004/msc/nrm/teshome_demissie_tolla.pdf

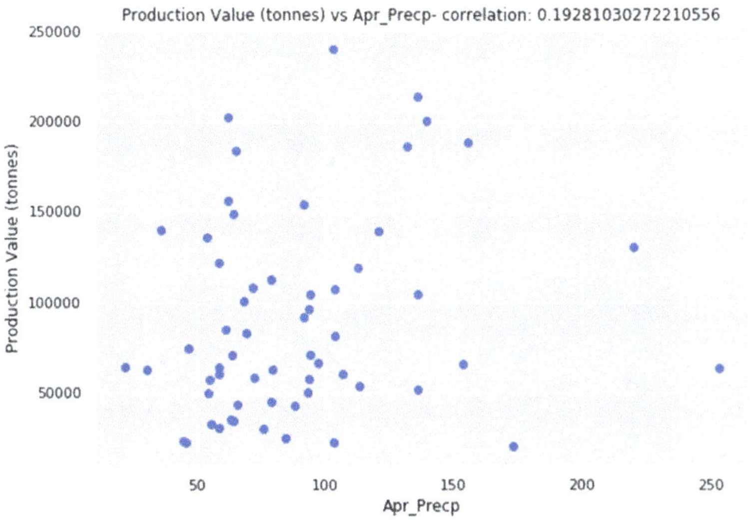

7.4.9 Step 8: Correlation Matrix

A correlation matrix would help understand the influence of feature variables:

```
import seaborn as sn
import matplotlib.pyplot as plt

import plotly.graph_objects as go
import seaborn as sn; sn.set()

fig = plt.figure(figsize=(30, 20))
corrMatrix = dfFinal.corr()
sn.heatmap(corrMatrix, annot=True)
plt.show()
```

Output:

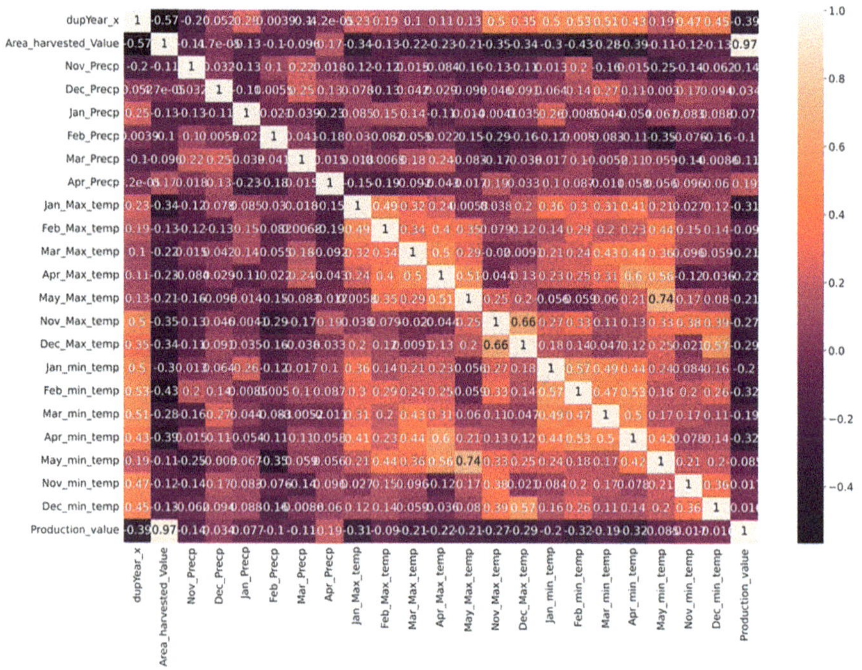

7.4.10 Step 9: Prepare the Final Cashew Nut Data Frame with Weather Details for Regression

Prepare cashew nuts data frame with the selected columns as the final ready data frame to be regressed to find the relationships and to predict:

```
X = dfFinal[['Area_harvested_Value', 'Nov_precp', 'Dec_precp',
'Jan_precp',
    'Feb_precp', 'Mar_precp', 'Apr_precp', 'Jan_Max_temp',
'Feb_Max_temp',
    'Mar_Max_temp', 'Apr_Max_temp', 'May_Max_temp',
'Nov_Max_temp',
    'Dec_Max_temp', 'Jan_min_temp', 'Feb_min_temp',
'Mar_min_temp',
    'Apr_min_temp', 'May_min_temp', 'Nov_min_temp',
'Dec_min_temp' ]]
y = dfFinal['Production_value']

from sklearn.model_selection import train_test_split
X_train, X_test, y_train, y_test = train_test_split(X, y,
test_size = 0.3, random_state = 0)
```

Create a Linear Regression model:

```
# Train the model
from sklearn.linear_model import LinearRegression

# Fit a linear regression model on the training set
model = LinearRegression(normalize=False).fit(X_train, y_train)
print (model)
```

Plot the regression:

```
import matplotlib.pyplot as plt

%matplotlib inline

plt.scatter(y_test, predictions)
plt.xlabel('Actual Labels')
plt.ylabel('Predicted Labels')
plt.title('Predictions')
# overlay the regression line
z = np.polyfit(y_test, predictions, 1)
p = np.poly1d(z)
plt.plot(y_test,p(y_test), color='magenta')
plt.show()
```

Output:

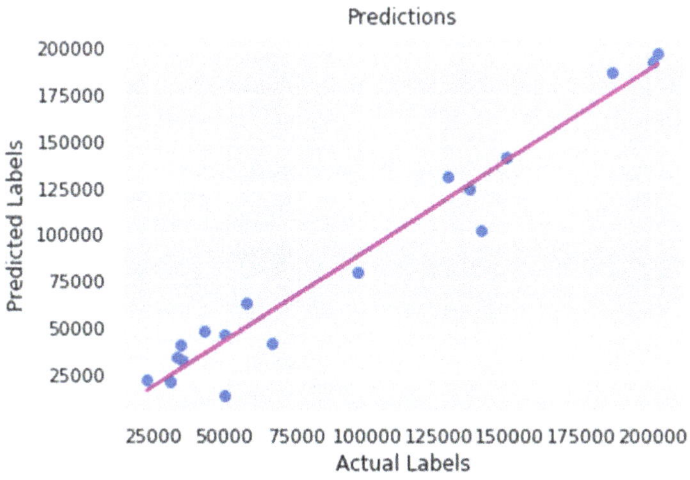

7.4.11 Step 10: Regression Metrics

Model coefficients:

```
model.coef_
```

Output:

```
array([  0.73007022,  -18.29348097,  -33.63937156,
56.18265129,
     -70.69066602,  34.72855531,  -26.26085163, 1413.00243657,
     86.37193125, -3676.68748831, -1472.0506279 , 10316.282367 ,
    1316.0962829 ,  19.29377086, -259.69726528, 12095.64297155,
     3348.30256476,  3843.3152941 , -18145.15189753,
4834.61209436,
     4452.29376214])
```

Feature variable	Coefficients	Summary
Area_harvested_Value	0.73007022	Area-harvested variables have a statistically substantial effect on cashew productivity in Mozambique
Nov_precp Dec_precp Feb_precp Apr_precp	-18.29348097 -33.63937156 -70.69066602 -26.26085163	Rainfall during harvesting season is detrimental to production. Hence, rainfall negatively correlated. Heavy rainfall, evenly distributed throughout the year, is not favorable though the trees may grow and sometimes set fruit
Jan_precp Mar_precp	56.18265129 34.72855531	January and March precipitation variables have a statistically substantial effect on cashew productivity in Mozambique
Jan_Max_temp Feb_Max_temp Nov_Max_temp May_Max_temp Dec_Max_temp	1413.00243657 86.37193125 1316.0962829 10,316.282367 19.29377086	January, February, May, November, and December maximum temperature variables have a statistically substantial effect on cashew productivity in Mozambique Areas where the temperatures range from 20 to 30 °C are ideal for cashew growing. Hence, we're seeing the influence
Mar_Max_temp Apr_Max_temp	-3676.68748831 -1472.0506279	Maximum temperatures of March and April ranged from 32 to 36 °C temperatures in the range, and above 36 °C between the flowering and fruiting periods could adversely affect the fruit setting and retention
Nov_Min_temp Feb_Min_temp Mar_Min_temp Apr_Min_temp May_Min_temp		November, February, March, April, and May minimum temperatures have a statistically substantial positive effect on cashew productivity in Mozambique as temperatures range from 20 to 30 °C

(continued)

Feature variable	Coefficients	Summary
Dec_Min_temp Jan_Min_temp	- 18,145.15189753 -259.69726528	The temperature below 20 °C for prolonged period is not good for cashew productions. December and January temperatures below 20 °C have a negative influence on the production

```
from sklearn.metrics import mean_squared_error, r2_score

mse = mean_squared_error(y_test, predictions)
print("MSE:", mse)
rmse = np.sqrt(mse)
print("RMSE:", rmse)
r2 = r2_score(y_test, predictions)
print("R2:", r2)
```

Output:

```
MSE: 254629997.466921
RMSE: 15957.129988407094
R2: 0.9287767076636368
```

Model could explain 92.87% of data signals and a good candidate for prediction. Let's derive the model governing mathematical equation:

```
mx=""
for ifeature in range(len(X.columns)):
  if model.coef_[ifeature] <0:
    # format & beautify the equation
   mx += " - " + "{:.2f}".format(abs(model.coef_[ifeature])) + " * "
+ X.columns[ifeature]
  else:
    if ifeature == 0:
      mx += "{:.2f}".format(model.coef_[ifeature]) + " * " + X.
columns[ifeature]
    else:
      mx += " + " + "{:.2f}".format(model.coef_[ifeature]) + " * " + X.
columns[ifeature]

print(mx)
```

Output:

```
0.73 * Area_harvested_Value - 18.29 * Nov_Precp - 33.64 * Dec_Precp
+ 56.18 * Jan_Precp - 70.69 * Feb_Precp + 34.73 * Mar_Precp - 26.26 *
Apr_Precp + 1413.00 * Jan_Max_temp + 86.37 * Feb_Max_temp - 3676.69
* Mar_Max_temp - 1472.05 * Apr_Max_temp + 10316.28 * May_Max_temp +
1316.10 * Nov_Max_temp + 19.29 * Dec_Max_temp - 259.70 *
Jan_min_temp + 12095.64 * Feb_min_temp + 3348.30 * Mar_min_temp +
3843.32 * Apr_min_temp - 18145.15 * May_min_temp + 4834.61 *
Nov_min_temp + 4452.29 * Dec_min_temp
```

Derive Y=MX + C complete equation:

```
# y=mx+c
if(model.intercept_ <0):
  print("The formula for the " + y.name + " linear regression line is
= " + " - {:.2f}".format(abs(model.intercept_)) + " + " + mx )
else:
  print("The formula for the " + y.name + " linear regression line is
= " + " + {:.2f}".format(model.intercept_) + " + " + mx)
```

Output:

$$
\begin{aligned}
&\text{The formula for the Production\_value linear regression line is} \\
&= -381942.82 + 0.73^{*}\ \text{Area\_harvested\_Value} - 18.29^{*}\ \text{Nov\_Precp} \\
&\quad - 33.64^{*}\ \text{Dec\_Precp} + 56.18^{*}\ \text{Jan\_Precp} - 70.69^{*}\ \text{Feb\_Precp} \\
&\quad + 34.73^{*}\ \text{Mar\_Precp} - 26.26^{*}\ \text{Apr\_Precp} + 1413.00^{*}\ \text{Jan\_Max\_temp} \\
&\quad + 86.37^{*}\ \text{Feb\_Max\_temp} - 3676.69^{*}\ \text{Mar\_Max\_temp} \\
&\quad - 1472.05^{*}\ \text{Apr\_Max\_temp} + 10316.28^{*}\ \text{May\_Max\_temp} \\
&\quad + 1316.10^{*}\ \text{Nov\_Max\_temp} + 19.29^{*}\ \text{Dec\_Max\_temp} \\
&\quad - 259.70^{*}\ \text{Jan\_min\_temp} + 12095.64^{*}\ \text{Feb\_min\_temp} \\
&\quad + 3348.30^{*}\ \text{Mar\_min\_temp} + 3843.32^{*}\ \text{Apr\_min\_temp} \\
&\quad - 18145.15^{*}\ \text{May\_min\_temp} + 4834.61^{*}\ \text{Nov\_min\_temp} \\
&\quad + 4452.29^{*}\ \text{Dec\_min\_temp}
\end{aligned} \tag{7.6}
$$

The above equation connects independent and dependent variables. We will use the equation to develop heuristics model.

7.4.12 Step 11: Model Explainability

Model explainability can be achieved using ELI5 library:

```
eli5.show_weights(model, feature_names=[
'Area_harvested_Value', 'Nov_precp',
   'Dec_precp', 'Jan_precp', 'Feb_precp', 'Mar_precp',
'Apr_precp',
   'Jan_Max_temp', 'Feb_Max_temp', 'Mar_Max_temp',
'Apr_Max_temp',
   'May_Max_temp', 'Nov_Max_temp', 'Dec_Max_temp',
'Jan_min_temp',
   'Feb_min_temp', 'Mar_min_temp', 'Apr_min_temp',
'May_min_temp',
   'Nov_min_temp', 'Dec_min_temp'])
```

Output:
Intercept (bias) feature is shown as <BIAS> in the above table.[109]

Weight	Feature
+12095.643	Feb_min_temp
+10316.282	May_Max_temp
+4834.612	Nov_min_temp
+4452.294	Dec_min_temp
+3843.315	Apr_min_temp
+3348.303	Mar_min_temp
+1413.002	Jan_Max_temp
+1316.096	Nov_Max_temp
+86.372	Feb_Max_temp
+56.183	Jan_precp
+34.729	Mar_precp
+19.294	Dec_Max_temp
… 1 more positive …	
… 1 more negative …	
-26.261	Apr_precp
-33.639	Dec_precp
-70.691	Feb_precp
-259.697	Jan_min_temp
-1472.051	Apr_Max_temp
-3676.687	Mar_Max_temp
-18145.152	May_min_temp
-381942.817	<BIAS>

[109] ELI5—https://eli5.readthedocs.io/en/latest/tutorials/sklearn-text.html#baseline-model

7.5 Machine Learning Model: Mozambique Cashew Nuts and Climate Projections Model

To build the Climate Projections Model, we have used two sets of models. First is the simulation from the international Coupled Model Intercomparison Project Phase 6 (CMIP6). Each of the five CMIP6 climate models is used to model the impact of climate on Mozambique—Nampula and its own unique response of the earth's atmosphere to greenhouse gas emission scenarios through 2100. Second is the cashew[110] nuts crop production and yield [32].

For modeling climate projections, the data is downloaded from Climate Change Portal (CCP) with feature variables that include minimum temperature, maximum temperature, and precipitation variables. Calculations are performed on anomaly from reference period 1995–2004, for Mozambique national and Nampula subunit with 90th percentile, SSP1-1.9. SSP3-7.0, and SSP5-8.5, Multi-Model Ensemble (please see Fig. 7.30). The maximum temperature heat plot is as shown below: as it can be seen from the above graph, 6.7 °C difference of temperature from baseline of 1995–2004.

Parameter summary (please see Fig. 7.31):

- Collection: Projections data (CMIP6)
- Variable: Max temperature
- Calculation: Anomaly (from reference period, 1995–2004)
- Area type: Sub-national

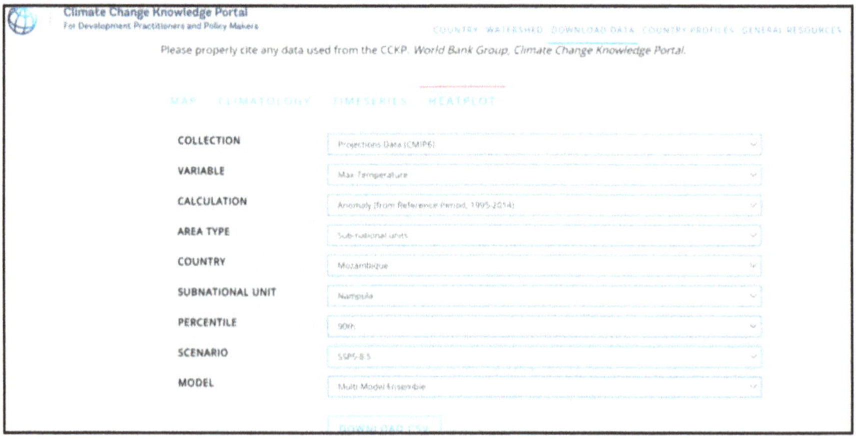

Fig. 7.30 Mozambique—Nampula 1951:2100

[110]Global Climate Change Impact on Crops Expected Within 10 Years, NASA Study Finds—https://climate.nasa.gov/news/3124/global-climate-change-impact-on-crops-expected-within-10-years-nasa-study-finds/

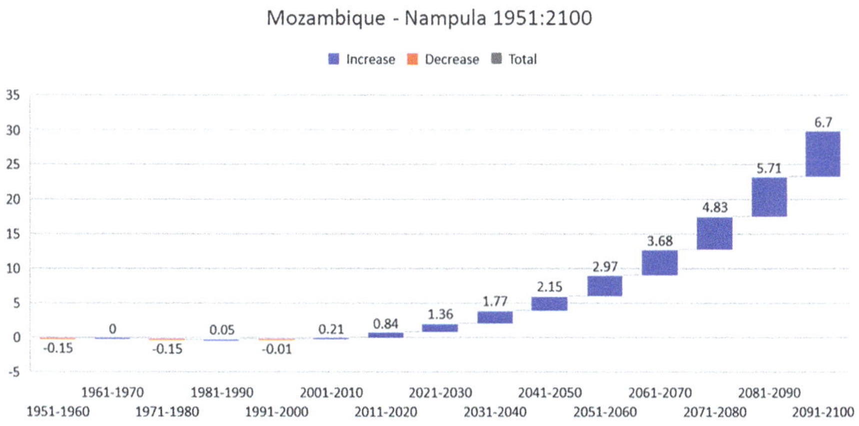

Fig. 7.31 Climate change portal

- Country: Mozambique
- Sub-national unit: Nampula
- Percentile: 90th
- Scenario: SSP5-8.5
- Model: Multi-Model Ensemble

7.5.1 CMIP6 Projections Data and SSP Scenarios

To model climate projections and assess the impact of climate change, the cashew nuts production machine learning (ML) for Mozambique is subjected to temperatures and precipitation inputs as per CMIP6 projections and for scenarios as described in the below table. The goal was to collect cashew production for SSP1-2.6, SSP3-7.0, and SSP5-8.5 scenarios and compare with baseline (2019 production—with area harvested 169,139 hectares (HA) and production of 140,000 tons with cashew nuts with shell (please see Table 7.1).

Table 7.1 SSP scenarios

Scenario	SSP model	Socioeconomic factors' impact
SSP1-2.6	SSP1	It imagines the same socioeconomic shifts toward sustainability as SSP1-1.9. But temperatures stabilize around 1.8 °C higher by the end of the century
SSP2-4.5	SSP2	Socioeconomic factors follow their historic trends, with no notable shifts. Progress toward sustainability is slow, with development and income growing unevenly. In this scenario, temperatures rise 2.7 °C by the end of the century
SSP3-7.0	SSP3	By the end of the century, average temperatures have risen by 3.6 °C
SSP5-8.5	SSP5	By 2100, the average global temperature is a scorching 4.4 °C higher

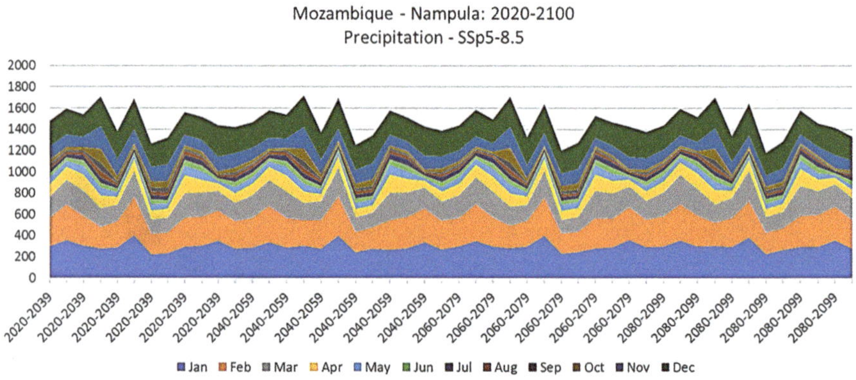

Fig. 7.32 Mozambique precipitation SSP5-8.5

The projection models are powerful and provide data to simulate climatic changes. Mozambique has seen its share climate change events such as December 2019 floods,[111] 2013 flash flooding,[112] and 2016 drought,[113] and ML model quantifies the future impact [42].

By the way Mozambique is one of the most climate-impacted countries, and understanding climate projects would help device adaptation and government policies to protect small farmers and marginalized communities. For instance, [CMIP6, SSP5-8.5, Multi-Model Ensemble] projections predict a 700% increase in precipitation (mm) over 1995–2014 reference period. The Multi-Model Ensemble[114] is used to estimate the climate change signal and its uncertainty to illustrate (please see Fig. 7.32) the impact of precipitation [43].

7.5.2 Climate Model: Temperature Change (RCP 6.0) (2006–2100)

Climate models are used for a variety of purposes from the study of dynamics[115] of the weather and climate system to projections of future climate [44].

[111] Mozambique: Floods—Dec 2019—https://reliefweb.int/disaster/fl-2020-000011-moz

[112] Mozambique: Floods—Jan 2013—https://reliefweb.int/disaster/fl-2013-000008-moz

[113] PM in Nakhon Sawan, Chai Nat on drought inspection—https://www.bangkokpost.com/print/835884/

[114] Josep et al., The Mediterranean climate change hotspot in the CMIP5 and CMIP6 projections—https://esd.copernicus.org/preprints/esd-2021-65/esd-2021-65.pdf

[115] Climate Model: Temperature Change (RCP 6.0)—2006–2100—https://sos.noaa.gov/catalog/datasets/climate-model-temperature-change-rcp-60-2006-2100/

NOAA's Geophysical Fluid Dynamics Laboratory has created several ocean-atmosphere coupled models to predict how greenhouse gas emissions following different population, economic, and energy-use projections may affect the planet.

"Representative Concentration Pathways (RCPs) are not new, fully integrated scenarios (i.e., they are not a complete package of socioeconomic, emissions and climate projections). They are consistent sets of projections of only the components of radiative forcing that are meant to serve as input for climate modeling, pattern scaling and atmospheric chemistry modeling," according.

Notable Features

- The earth gets warmer as CO_2 increases in the atmosphere.
- The earth doesn't warm uniformly; the oceans warm slower than the continents and arctic.
- Projections for temperature according to RCP 6.0 include continuous global warming through 2100 where CO_2 levels rise to 670 ppm by 2100 making the global temperature rise by about 3–4 °C by 2100.

Here are the model data simulations:

Collection: [*CMIP6—aggregation, monthly; calculation, climatology mean; Percentile, 90th; and model, Multi-Model Ensemble*]. Please find the table with all three main variables for scenario SSP5-8.5. Similar data was collected for additional two scenarios: SSP1-2.6 and SSP3-7.0.

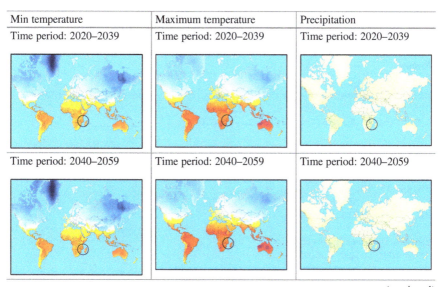

Min temperature	Maximum temperature	Precipitation
Time period: 2020–2039 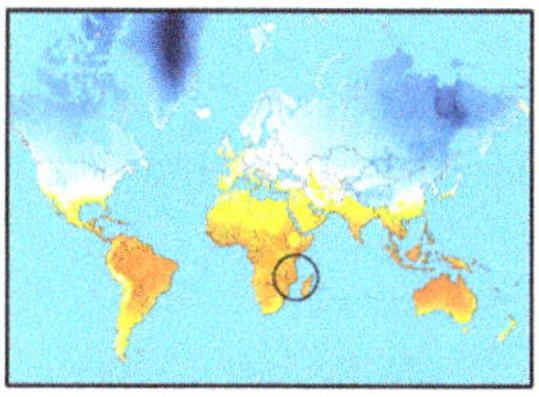	Time period: 2020–2039 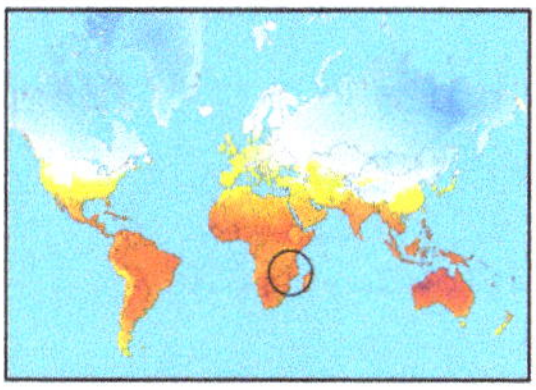	Time period: 2020–2039 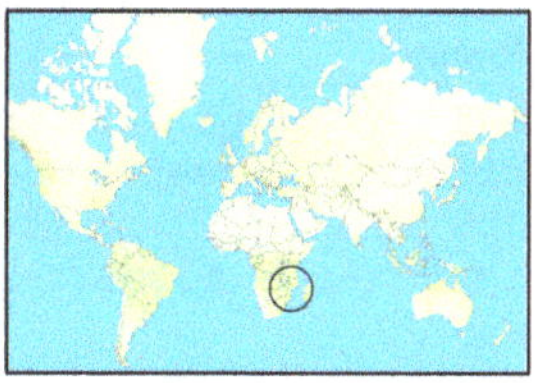
Time period: 2040–2059 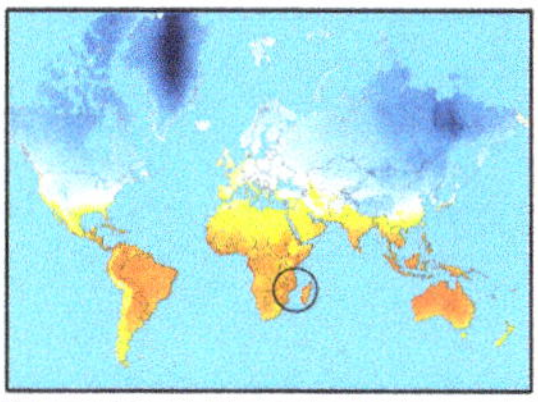	Time period: 2040–2059 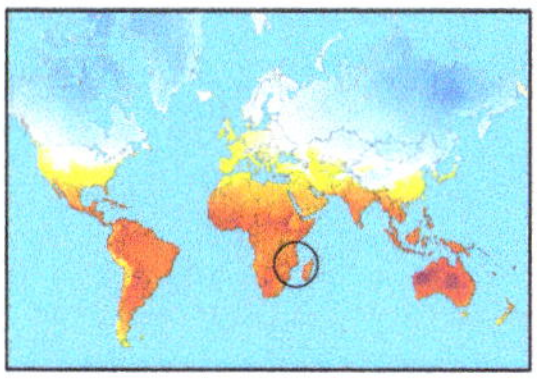	Time period: 2040–2059 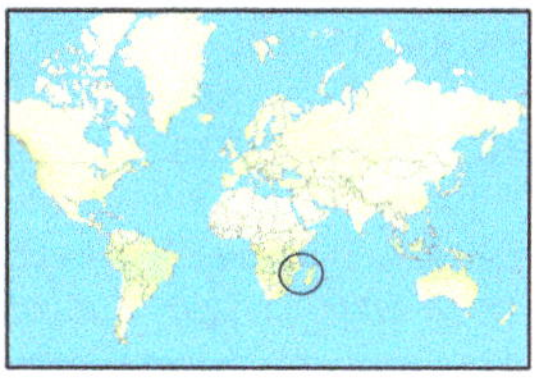

(continued)

Min temperature	Maximum temperature	Precipitation
Time period: 2060–2079 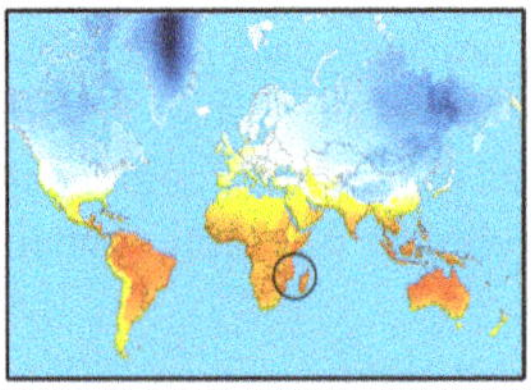	Time period: 2060–2079 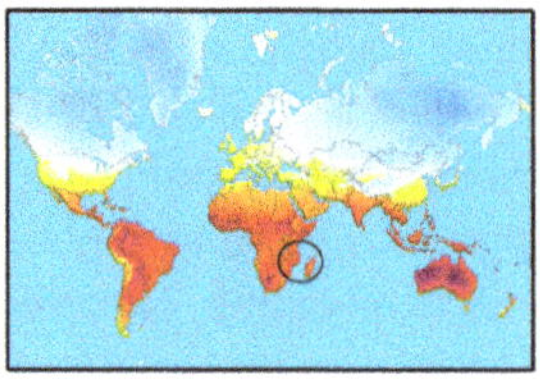	Time period: 2060–2079 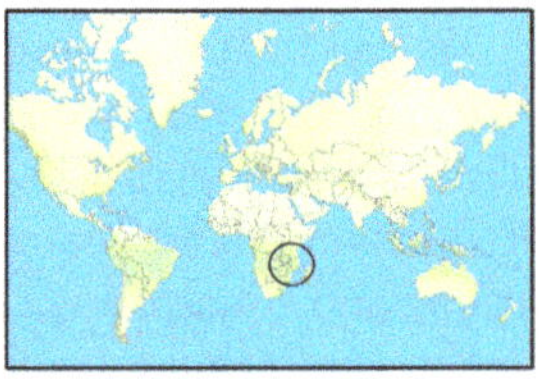
Time period: 2080–2099 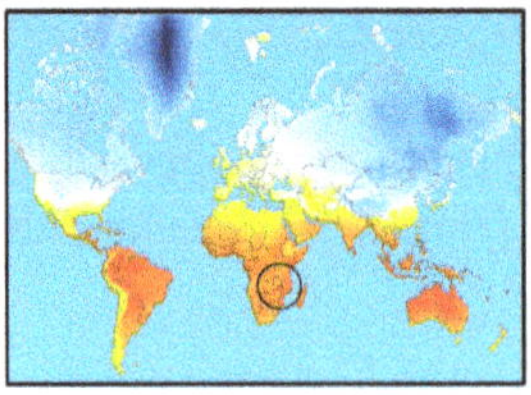	Time period: 2080–2099 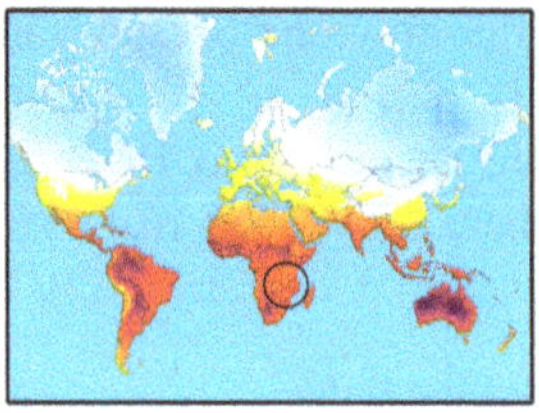	Time period: 2080–2099 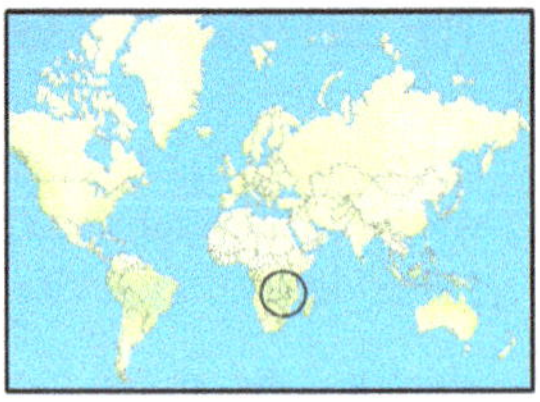

7.5.2.1 SSP5-8.5

Please find model invocation called scenario SSP5-8.5:

CMIP6, 20202039_climatologymean 90th Percentile SSP5_8.5_multimodelensembel—Results

```
MOZ_NAM_cmip6_20202039_
climatologymean_90th_ssp5_
8p5_multimodelensembel = [[ 169139   ,
           103.57   ,  266.38,  340.05   ,
309.27   ,
177.16   , 89.42   ,

32.65,32.19,31.73,31.04,29.32,
35.44,33.75,
           23.88,23.91,23.13,22.00,
20.25,23.32,23.67
  ]]
result = model.predict
(MOZ_NAM_cmip6_20202039_
climatologymean_
90th_ssp5_8p5_multimodelensembel)
```

Cashew nuts production, 316,353.675 tons

y (score **316353.675**) top features

Contribution?	Feature
+329502.059	May_Max_temp
+318236.367	Feb_min_temp
+124684.646	Nov_min_temp
+123483.347	Area_harvested_Value
+117362.464	Dec_min_temp
+93776.893	Apr_min_temp
+85683.961	Mar_min_temp
+60314.361	Nov_Max_temp
+50105.066	Jan_Max_temp
+19074.572	Jan_precp
+10686.360	Mar_precp
+2990.196	Feb_Max_temp
+707.503	Dec_Max_temp
-1789.285	Nov_precp
-4552.581	Apr_precp
-6840.426	Jan_min_temp
-8118.862	Dec_precp
-23455.163	Feb_precp
-48724.876	Apr_Max_temp
-124766.773	Mar_Max_temp
-381942.817	<BIAS>
-410443.336	May_min_temp

(continued)

CMIP6, 20402059_climatologymean 90th Percentile SSP5_8.5_multimodelensembel—Results

```
MOZ_NAM_cmip6_20402059_climatologymean_
90th_ssp5_8p5_
multimodelensembel = [[ 169139,
   255.94 , 340.85, 340.05 ,
326.76 , 308.8 ,
199.26,
   33.72,33.14,32.64,32.03,
30.42,36.82,35.35,
   24.94,24.98,24.4,23.08,21.3,24.44,25
]]
result = model.predict
(MOZ_NAM_cmip6_20402059_
climatologymean_90th_ssp5_8p5_
multimodelensembel)
print(result)
```

Cashew nuts production, 256,902.32701409 tons

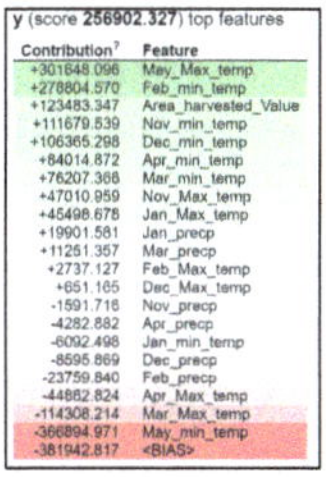

y (score **256902.327**) top features

Contribution?	Feature
+301648.096	May_Max_temp
+278804.570	Feb_min_temp
+123483.347	Area_harvested_Value
+111679.539	Nov_min_temp
+106365.298	Dec_min_temp
+84014.872	Apr_min_temp
+76207.368	Mar_min_temp
+47010.959	Nov_Max_temp
+45498.678	Jan_Max_temp
+19901.581	Jan_precp
+11251.357	Mar_precp
+2737.127	Feb_Max_temp
+651.165	Dec_Max_temp
-1591.716	Nov_precp
-4282.882	Apr_precp
-6092.498	Jan_min_temp
-8595.869	Dec_precp
-23759.840	Feb_precp
-44862.824	Apr_Max_temp
-114308.214	Mar_Max_temp
-366894.971	May_min_temp
-381942.817	<BIAS>

CMIP6, 20602079_climatologymean 90th Percentile SSP5_8.5_multimodelensembel—Results

```
MOZ_NAM_cmip6_20602079_climatologymean_
90th_ssp5_8p5_
multimodelensembel = [[ 169139,

97.81,241.35,339.51,331.8,313.47,173.36,

35.46,34.62,33.94,33.1,31.94,38.23,36.67,

26.34,26.31,25.65,24.4,22.62,25.79,26.36
]]
result = model.predict(MOZ_NAM_cmip6_
20602079_climatologymean_
90th_ssp5_8p5_multimodelensembel)
print(result)
```

Cashew nuts production, 260,635.17 tons

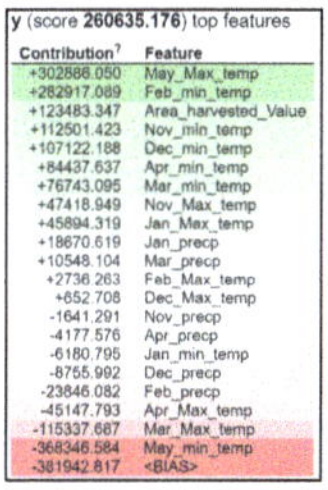

y (score **260635.176**) top features

Contribution?	Feature
+302886.050	May_Max_temp
+282917.089	Feb_min_temp
+123483.347	Area_harvested_Value
+112501.423	Nov_min_temp
+107122.188	Dec_min_temp
+84437.637	Apr_min_temp
+76743.095	Mar_min_temp
+47418.949	Nov_Max_temp
+45894.319	Jan_Max_temp
+18670.619	Jan_precp
+10548.104	Mar_precp
+2736.263	Feb_Max_temp
+652.708	Dec_Max_temp
-1641.291	Nov_precp
-4177.576	Apr_precp
-6180.795	Jan_min_temp
-8755.992	Dec_precp
-23846.082	Feb_precp
-45147.793	Apr_Max_temp
-115337.687	Mar_Max_temp
-368346.584	May_min_temp
-381942.817	<BIAS>

CMIP6, 20602079_climatologymean 90th Percentile SSP5_8.5_multimodelensembel—Results

```
MOZ_NAM_cmip6_20802099_climatologymean_
90th_ssp5_8p5_multimodelensembel =
[[ 169139,
   80.03,252.85,331.31,
344.33,352.08,205.63,
   37.06,36.08,35.49,
34.79,33.6,40.13,38.13,

28.05,27.7,27.17,26.12,24.4,27.65,27.96
]]
result = model.predict
(MOZ_NAM_cmip6_20802099_
climatologymean_
90th_ssp5_8p5_multimodelensembel)
print(result)
```

Cashew nuts production, 262,719.60510445 tons

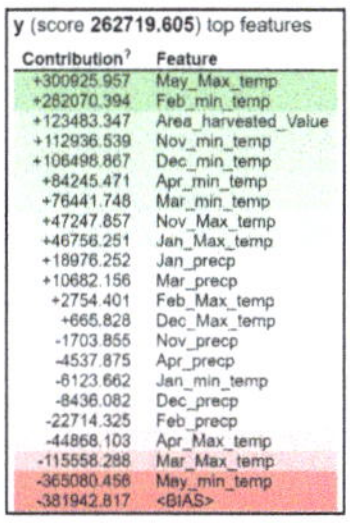

y (score **262719.605**) top features

Contribution?	Feature
+300925.957	May_Max_temp
+282070.394	Feb_min_temp
+123483.347	Area_harvested_Value
+112936.539	Nov_min_temp
+106498.867	Dec_min_temp
+84245.471	Apr_min_temp
+76441.748	Mar_min_temp
+47247.857	Nov_Max_temp
+46756.251	Jan_Max_temp
+18976.252	Jan_precp
+10682.156	Mar_precp
+2754.401	Feb_Max_temp
+665.828	Dec_Max_temp
-1703.855	Nov_precp
-4537.875	Apr_precp
-6123.662	Jan_min_temp
-8436.082	Dec_precp
-22714.325	Feb_precp
-44868.103	Apr_Max_temp
-115558.288	Mar_Max_temp
-365080.456	May_min_temp
-381942.817	<BIAS>

Output (Table 7.2):

SSP5-8.5 is the most climatic severe outcome that could potentially increase in 4.4 °C. Nevertheless, it had production increase with respect to cashew nuts

Table 7.2 SSP5-8.5 summary

Baseline: Mozambique cashew nuts 2019 production with area harvested of 169,139 hectares[a] (HA) and production of 140,000 tons with cashew nuts with shell			
20,202,039	20,402,059	20,602,079	20,802,099
263,753.85	282,252.081	316,353.67	341,179.57
Change in production from 2019 levels, 140,000 tons			
+88.39%	+101.60%	+125.96%	+143.69%

[a]Model Profile for 1.0 ha Cashew Cultivation—https://agricoop.nic.in/sites/default/files/Cashewnut%20Cultivation%20%281%29.pdf

production in Mozambique. An increase of 88.39% can be seen with respect to 2020–2039 time. As it turns out, cashew crop is cash crop with climate resilience to the Nampula region of Mozambique and will become more suitable for cashew production under climate change. The adoption of cashew as a cash crop brings valuable income to farmers. "*If you plant cashew an acre, it's better than farming three acres of maize because there is more profit in cashew than even in cocoa.*"[116] Similar findings from NASA team have shown wheat crops could increase under climate change.[117] The models showed that the global yield of wheat would increase about 17%. Wheat crops typically grow best in temperate climates, but rising temperatures could mean the crops could expand into other areas, such as the northern parts of the United States of America. But projections showed the increase would stop around 2050 [45, 46].

7.5.2.2 SSP1-1.9

Please find model invocation called scenario SSP1-1.9:

CMIP6, 20202039_climatologymean 90th Percentile SSP1_1.9_multimodelensembel—Results

```
MOZ_NAM_cmip6_20202039_climatologymean_90th_
ssp1_1p9_multimodelensembel = [[169139,
    87.01,255.53,354.23,336.11,323.98,163.09,
    32.2,31.69,31.09,30.49,29.24,35.72,33.75,
    23.46,23.05,22.76,21.86,20.22,23.1,23.89
]]
result = model.predict
(MOZ_NAM_cmip6_20202039_climatologymean_90th_
ssp1_1p9_multimodelensembel)
print(result)
```

Cashew nuts production, 256,902.32 tons

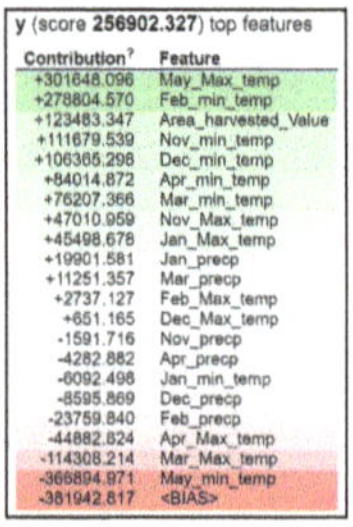
y (score **256902.327**) top features

Contribution?	Feature
+301648.096	May_Max_temp
+278804.570	Feb_min_temp
+123483.347	Area_harvested_Value
+111679.539	Nov_min_temp
+106365.298	Dec_min_temp
+84014.872	Apr_min_temp
+76207.366	Mar_min_temp
+47010.959	Nov_Max_temp
+45498.678	Jan_Max_temp
+19901.581	Jan_precp
+11251.357	Mar_precp
+2737.127	Feb_Max_temp
+651.165	Dec_Max_temp
-1591.716	Nov_precp
-4282.882	Apr_precp
-6092.498	Jan_min_temp
-8595.869	Dec_precp
-23759.840	Feb_precp
-44882.824	Apr_Max_temp
-114308.214	Mar_Max_temp
-366894.971	May_min_temp
-381942.817	<BIAS>

(continued)

[116]Predicting the Impact of Climate Change on Cashew Growing Regions in Ghana and Cote d'Ivoire—https://www.africancashewalliance.com/sites/default/files/documents/ghana_ivory_coast_climate_change_and_cashew.pdf

[117]Climate change expected to impact the world's wheat and corn crops by 2030, NASA says—https://www.usatoday.com/story/news/world/2021/11/04/climate-change-impacts-corn-soybeans-rice-crops/6257778001/

CMIP6, 20402059_climatologymean 90th Percentile SSP1_1.9_multimodelensembel—Results

```
MOZ_NAM_cmip6_20402059_climatologymean_90th_
ssp1_1p9_multimodelensembel = [[ 169139,

102.78,247.59,327.11,321.95,311.92,160.82,
    32.6,31.94,31.39,30.66,29.16,36.08,34.33,
    23.85,23.33,22.9,22.11,20.18,23.38,23.9 ]]
result = model.predict
(MOZ_NAM_cmip6_20402059_climatologymean_90th_
ssp1_1p9_multimodelensembel)
print(result)
```

Cashew nuts production, 261,731.35 tons

y (score **261731.359**) top features

Contribution?	Feature
+300822.794	May_Max_temp
+282191.351	Feb_min_temp
+123483.347	Area_harvested_Value
+113033.231	Nov_min_temp
+106409.821	Dec_min_temp
+84975.701	Apr_min_temp
+76676.129	Mar_min_temp
+47484.754	Nov_Max_temp
+46063.879	Jan_Max_temp
+18377.907	Jan_precp
+10832.531	Mar_precp
+2758.719	Feb_Max_temp
+662.355	Dec_Max_temp
-1880.204	Nov_precp
-4223.270	Apr_precp
-6193.780	Jan_min_temp
-8328.772	Dec_precp
-22758.860	Feb_precp
-45133.072	Apr_Max_temp
-115411.220	Mar_Max_temp
-366169.165	May_min_temp
-381942.817	<BIAS>

CMIP6, 20602079_climatologymean 90th Percentile SSP1_1.9_multimodelensembel—Results

```
MOZ_NAM_cmip6_20602079_climatologymean_
90th_ssp1_1p9_multimodelensembel = [[ 169139,
    89.72,260.29,332.32,337.33,303.73,159.08,

32.48,31.68,31.37,30.67,29.36,36.03,33.83,
    23.8,23.39,22.92,21.97,20.3,23.27,24.06 ]]
result = model.predict(MOZ_NAM_cmip6_20602079_
climatologymean_90th_
ssp1_1p9_multimodelensembel)
print(result)
```

Cashew nuts production, 260,635.17 tons

y (score **260635.176**) top features

Contribution?	Feature
+302886.050	May_Max_temp
+282917.089	Feb_min_temp
+123483.347	Area_harvested_Value
+112501.423	Nov_min_temp
+107122.188	Dec_min_temp
+84437.637	Apr_min_temp
+76743.095	Mar_min_temp
+47418.949	Nov_Max_temp
+45894.319	Jan_Max_temp
+18670.619	Jan_precp
+10548.104	Mar_precp
+2736.263	Feb_Max_temp
+652.708	Dec_Max_temp
-1641.291	Nov_precp
-4177.576	Apr_precp
-6180.795	Jan_min_temp
-8755.992	Dec_precp
-23846.082	Feb_precp
-45147.793	Apr_Max_temp
-115337.687	Mar_Max_temp
-368346.584	May_min_temp
-381942.817	<BIAS>

CMIP6, 20802099_climatologymean 90th Percentile SSP1_1.9_multimodelensembel—Results

```
MOZ_NAM_cmip6_20802099_climatologymean_90th_
ssp1_1p9_multimodelensembel = [[ 169139,
    93.14,250.78,337.76,321.32,307.59,172.8,
    33.09,31.89,31.43,30.48,29.17,35.9,34.51,
    23.58,23.32,22.83,21.92,20.12,23.36,23.92
]]
result = model.predict(MOZ_NAM_cmip6_20802099_
climatologymean_90th_ssp5_
8p5_multimodelensembel)
print(result)
```

Cashew nuts production, 262,719.60510445 tons

y (score **262719.605**) top features

Contribution?	Feature
+300925.957	May_Max_temp
+282070.394	Feb_min_temp
+123483.347	Area_harvested_Value
+112936.539	Nov_min_temp
+106498.867	Dec_min_temp
+84245.471	Apr_min_temp
+76441.748	Mar_min_temp
+47247.857	Nov_Max_temp
+46756.251	Jan_Max_temp
+18976.252	Jan_precp
+10682.156	Mar_precp
+2754.401	Feb_Max_temp
+665.828	Dec_Max_temp
-1703.855	Nov_precp
-4537.875	Apr_precp
-6123.662	Jan_min_temp
-8436.082	Dec_precp
-22714.325	Feb_precp
-44868.103	Apr_Max_temp
-115558.288	Mar_Max_temp
-365080.456	May_min_temp
-381942.817	<BIAS>

Output:

The SSP1-1.9 has a positive impact (please see Table 7.3) on the cashew nuts production in Mozambique. An increase of 83.50% can be seen with respect to 2020–2039 time. As it turns out, cashew crop is a cash crop with climate resilience to the Nampula region of Mozambique and will become more suitable for cashew production under climate change.

Table 7.3 SSP1-1.9 summary

Baseline: Mozambique cashew nuts 2019 production with area harvested of 169,139 hectares[a] (HA) and production of 140,000 tons with cashew nuts with shell			
20,202,039	20,402,059	20,602,079	20,802,099
263,753.85	282,252.081	316,353.67	341,179.57
Change in production from 2019 levels, 140,000 tons			
+83.50%	+86.95%	+86.16%	+87.65%

[a]Model Profile for 1.0 ha Cashew Cultivation—https://agricoop.nic.in/sites/default/files/Cashewnut%20Cultivation%20%281%29.pdf

7.5.2.3 SSP3-7.0

Please find model invocation called scenario SSP1-1.9:

CMIP6, 20202039_climatologymean 90th Percentile SSP3_7.0_multimodelensembel—Results

```
MOZ_NAM_cmip6_20202039_climatologymean_
90th_ssp3_7p0_multimodelensembel = [[169139,
    96.22,239.52,337.45,310.01,297.75,175.16,
    32.39,32.02,31.49,30.78,29.43,35.06,33.56,
    23.46,23.47,23.11,21.84,20.16,23.09,23.44]]
result = model.predict(MOZ_NAM_cmip6_20202039_
climatologymean_
90th_ssp3_7p0_multimodelensembel)
print(result)
```

Cashew nuts production, 261,646.04 tons

y (score **261646.042**) top features

Contribution?	Feature
+303608.190	May_Max_temp
+283884.741	Feb_min_temp
+123483.347	Area_harvested_Value
+111631.193	Nov_min_temp
+104361.766	Dec_min_temp
+83938.006	Apr_min_temp
+77379.272	Mar_min_temp
+46142.336	Nov_Max_temp
+45767.149	Jan_Max_temp
+18958.836	Jan_precp
+10340.427	Mar_precp
+2765.629	Feb_Max_temp
+647.499	Dec_Max_temp
-1760.199	Nov_precp
-4599.851	Apr_precp
-6092.498	Jan_min_temp
-8057.302	Dec_precp
-21914.813	Feb_precp
-45309.718	Apr_Max_temp
-115778.889	Mar_Max_temp
-365806.262	May_min_temp
-381942.817	<BIAS>

CMIP6, 20402059_climatologymean 90th Percentile SSP3_7.0_multimodelensembel—Results

```
MOZ_NAM_cmip6_20402059_climatologymean_
90th_ssp3_7p0_multimodelensembel = [[169139,
    94.55,259.42,337.08,324.18,293.17,170.09,
33.41,33.03,32.33,31.51,30.18,36.24,34.64,
24.12,24.16,23.71,22.64,21.02,23.91,24.2]]
result = model.predict(MOZ_NAM_cmip6_20402059_
climatologymean_90th_ssp3_7p0_multimodelensembel)
print(result)
```

Cashew nuts production, 271,636.82 tons

y (score **271636.828**) top features

Contribution?	Feature
+311345.402	May_Max_temp
+292230.734	Feb_min_temp
+123483.347	Area_harvested_Value
+115595.575	Nov_min_temp
+107745.509	Dec_min_temp
+87012.658	Apr_min_temp
+79388.254	Mar_min_temp
+47695.329	Nov_Max_temp
+47208.411	Jan_Max_temp
+18938.048	Jan_precp
+10181.371	Mar_precp
+2852.865	Feb_Max_temp
+668.336	Dec_Max_temp
-1729.649	Nov_precp
-4466.708	Apr_precp
-6263.898	Jan_min_temp
-8726.726	Dec_precp
-22916.500	Feb_precp
-46384.315	Apr_Max_temp
-118867.306	Mar_Max_temp
-381411.093	May_min_temp
-381942.817	<BIAS>

(continued)

CMIP6, 20602079_climatologymean 90th Percentile SSP3_7.0_multimodelensembel—Results

```
MOZ_NAM_cmip6_20602079_climatologymean_90th_
ssp3_7p0_multimodelensembel = [[169139,
    89.57,227.51,330.96,313.57,297.61,152.74,
34.45,33.84,33.24,32.37,31.11,37.62,35.83,
25.29,25.01,24.53,23.75,21.93,25.08,25.47]]
result = model.predict(MOZ_NAM_cmip6_20602079_
climatologymean_90th_ssp3_
7p0_multimodelensembel)
print(result)
```

Cashew nuts production, 293,966.38 tons

y (score **293966.382**) top features

Contribution?	Feature
+320939.544	May_Max_temp
+302512.031	Feb_min_temp
+123483.347	Area_harvested_Value
+121252.071	Nov_min_temp
+113399.922	Dec_min_temp
+91278.738	Apr_min_temp
+82133.862	Mar_min_temp
+49511.542	Nov_Max_temp
+48677.934	Jan_Max_temp
+18594.210	Jan_precp
+10335.565	Mar_precp
+2922.826	Feb_Max_temp
+691.296	Dec_Max_temp
-1638.547	Nov_precp
-4011.082	Apr_precp
-6567.744	Jan_min_temp
-7653.293	Dec_precp
-22166.472	Feb_precp
-47650.279	Apr_Max_temp
-122213.092	Mar_Max_temp
-381942.817	<BIAS>
-397923.181	May_min_temp

CMIP6, 20802099_climatologymean 90th Percentile SSP1_1.9_multimodelensembel—Results

```
MOZ_NAM_cmip6_20802099_climatologymean_
90th_ssp3_7p0_multimodelensembel = [[169139,
    81.45,226.71,337.84,331.24,306.26,177.17,
35.61,35.06,34.6,33.75,32.33,38.75,37.29,
26.45,26.29,25.66,24.72,23.11,26.17,26.56]]
result = model.predict(MOZ_NAM_cmip6_20802099_
climatologymean_
90th_ssp3_7p0_multimodelensembel)
print(result)
```

Cashew nuts production, 313,156.29 tons

y (score **313156.292**) top features

Contribution?	Feature
+333525.409	May_Max_temp
+317994.454	Feb_min_temp
+126521.799	Nov_min_temp
+123483.347	Area_harvested_Value
+118252.922	Dec_min_temp
+95006.754	Apr_min_temp
+85917.444	Mar_min_temp
+50998.731	Nov_Max_temp
+50317.017	Jan_Max_temp
+18980.747	Jan_precp
+10635.967	Mar_precp
+3028.200	Feb_Max_temp
+719.465	Dec_Max_temp
-1490.004	Nov_precp
-4652.635	Apr_precp
-6868.993	Jan_min_temp
-7626.382	Dec_precp
-23415.576	Feb_precp
-49681.709	Apr_Max_temp
-127213.387	Mar_Max_temp
-381942.817	<BIAS>
-419334.460	May_min_temp

Baseline: Mozambique cashew nuts 2019 production with area harvested of 169,139 hectares[a] (HA) and production of 140,000 tons with cashew nuts with shell			
20,202,039	20,402,059	20,602,079	20,802,099
261,646.04	271,636.82	293,966.38	313,156.29
Change in production from 2019 levels, 140,000 tons			
+86.89%	+94.02%	+109.97%	+123.68%

[a]Model Profile for 1.0 ha Cashew Cultivation—https://agricoop.nic.in/sites/default/files/Cashewnut%20Cultivation%20%281%29.pdf

As it can be seen that climate change has a positive impact on cashew nuts production in Mozambique, Nampula. An increase of 86.89% can be seen with respect to 2020–2039 time. Please find the overall summary (please see Fig. 7.33):

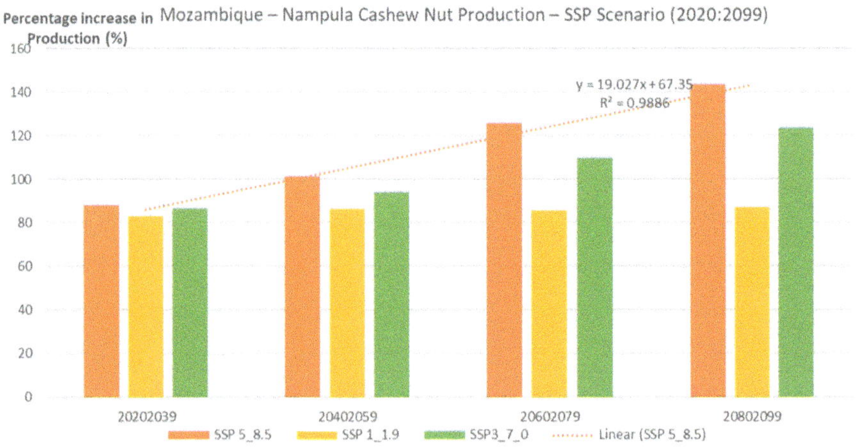

Fig. 7.33 SSP summary

As demonstrated by machine learning model, the suitability to grow cashew is projected to remain stable[118] **or to increase under future climate change conditions in the Mozambique—Nampula area** [47]**.**

Taking into consideration that cashew production allows for the commercialization of several by-products like wood, cashew fruit, and products from beekeeping, cashew plantations become even more profitable. Cashew plantations can bring valuable income to local farmers and allow for intercropping of legumes.

Alley cropping is a farming practice where shrubs or trees are grown in alternate rows with food crops. It is one subcategory of agroforestry. Under good management practices, this can result in higher yields and better soil conditions, while the farmer can additionally profit from the produce of the trees. Alley cropping in many parts of the world is most done with cashew or mango trees [47]. Please refer to Appendix—Food Security and Nutrition: The 17 Sustainable Development Goals (SDGs).

	As a valuable cash crop, cashew brings additional income to households. Income from cashew plantation is highest shortly before the rainy season when it is most needed for farmers. Selling by-products (e.g., cashew apple juice, fruit as feed for livestock) can create additional income. Opportunities to engage in beekeeping are growing under cashew production, which comes with a favorable impact of further income from honey, beeswax, and propolis [47].

(continued)

[118] Climate Risk Analysis for Identifying and Weighing Adaptation Strategies for the Agricultural Sector in Northern Ghana—https://www.adaptationcommunity.net/wp-content/uploads/2021/09/Climate-Risk-Analysis-for-Identifying-and-Weighing-Adaptation-Strategies-for-the-Agricultural-Sector-in-Northern-Ghana.pdf

	Stable annual income from cashew can contribute to no hunger in the communities. The intercropping of groundnuts can provide rich nutrients and proteins already in the first years of planting.
	Stable annual income from cashew can contribute to increased living standards and better health conditions. Cashew is rich in fiber, heart-healthy fats, and plant protein and is a good source of copper, magnesium, and manganese. Using cashew for own consumption can help diversify food sources (also through by-products like honey and cashew apple) and thus contribute to SDG-3.
	According to Evans et al. (2014), increased cashew production has enabled farmers to pay for their children's education.
	Higher income opportunities due to cashew plantations can enhance living conditions in rural communities and can decrease rural exodus.

Mozambique cashew nuts and Normalized Difference Vegetation Index (NDVI)	

7.6 Machine Learning Model: Mozambique Cashew Nuts and Normalized Difference Vegetation Index (NDVI) Model

The Vegetation Health Index (VHI) illustrates the severity of drought based on vegetation health and the influence of temperature on plant conditions. The VHI is a composite index and the elementary indicator used to compute the ASI. It combines both the Vegetation Condition Index (VCI) and the Temperature Condition Index (TCI). The TCI is calculated using a similar equation to the VCI but relates the current temperature to the long-term maximum and minimum, as it is assumed that higher temperatures tend to cause a deterioration in vegetation conditions. A

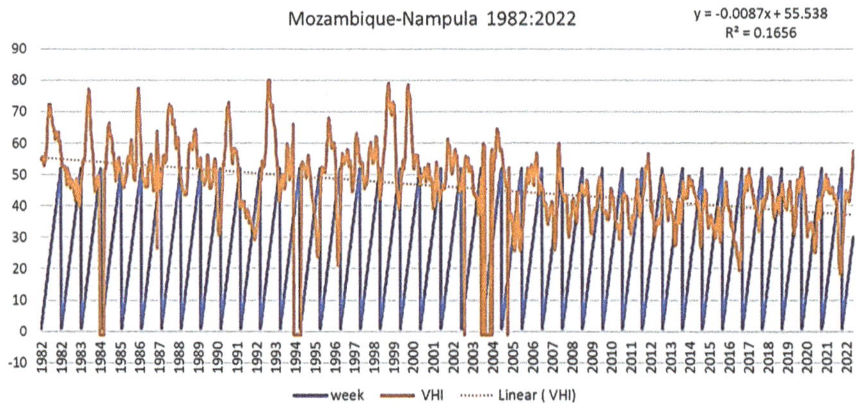

Fig. 7.34 NDVI

decrease in the VHI would, for example, indicate relatively poor vegetation conditions and warmer temperatures, signifying stressed vegetation conditions, and over a longer period would be indicative of drought (2018–2022 drought, 2015–2017 drought, and 1985 Mozambique drought). A downward trend (slope of negative—equation) as seen in the figure is indicative of the long-term stress on the crops (please see Fig. 7.34).

$$y = 6E^{06}x + 0.5912 \tag{7.7}$$

Software code for this model:
Mozambique_Namapula_CashewNuts_code_MODIS_LSAT_NDVI.ipynb (Jupyter Notebook Code)

Landsat Normalized Difference Vegetation Index[119] (NDVI) is used to quantify vegetation greenness and is useful in understanding vegetation density and assessing changes in plant health. Global and regional *vegetation health* (VH) is a NOAA/NESDIS system estimating vegetation health, moisture condition, thermal condition, and their products. It contains Vegetation Health Indices (VHI) derived from the radiance observed by the Advanced Very High-Resolution Radiometer[120] (AVHRR) onboard afternoon polar-orbiting satellites: the NOAA-7, NOAA-9,

[119] NDVI—https://www.usgs.gov/landsat-missions/landsat-normalized-difference-vegetation-index

[120] Advanced Very High-Resolution Radiometer—https://www.usgs.gov/centers/eros/science/usgs-eros-archive-advanced-very-high-resolution-radiometer-avhrr

NOAA-11, NOAA-14, NOAA-16, NOAA-18, and NOAA-19 and Visible Infrared Imaging Radiometer Suite[121] (VIIRS) from Suomi-NPP satellite.[122]

VH products from VIIRS were also processed from 2012 to present (1 km resolution, 7-day composite). The VH products can be used *as proxy data for monitoring vegetation health*, drought, soil saturation, moisture and thermal conditions, fire risk, greenness of vegetation cover, vegetation fraction, leave area index, start/end of the growing season, crop and pasture productivity, teleconnection with ENSO, desertification, mosquito-borne diseases, invasive species, ecological resources, land degradation, etc.

The following indices and products are available:

- Vegetation Health Index (VHI)
- Vegetation Condition Index (VCI)
- Temperature Condition Index (TCI)
- Soil Saturation Index (SSI)
- No noise Normalized Difference Vegetation Index (SMN)
- No noise Brightness Temperature (SMT)

In the following model, we have used VHI to model the impact of climate change-induced thermal and moisture on production of cashew nuts production in Mozambique. We have downloaded Vegetation Health Index from Earth Observation.[123] Three indicators are included as part of the Earth Observation:

- Seasonal indicators
  - Cropland
  - Grassland
- Vegetation indicators
  - NDVI anomaly
- Precipitation indicators

All three vegetation indicators are based on 10-day (dekadal) vegetation data from the METOP-AVHRR sensor at 1 km resolution (2007 and after). Data at 1 km resolution for the period 1984–2006 are derived from the NOAA-AVHRR dataset at 16 km resolution. Precipitation estimates for all African countries (except Cabo Verde and Mauritius) are taken from Famine Early Warning Systems Network[124]

[121]Visible Infrared Imaging Radiometer Suite (VIIRS)—https://www.jpss.noaa.gov/viirs.html

[122]STAR—Global Vegetation Health Products: Introduction—https://www.star.nesdis.noaa.gov/smcd/emb/vci/VH/index.php

[123]Earth Observation—https://www.fao.org/giews/earthobservation/country/index.jsp?lang=en&code=MOZ

[124]Famine Early Warning Systems Network—https://earlywarning.usgs.gov/fews

Indicat(	Country	ADM1_	Province	Land_T	Date	Data	Data_lc	Year	Month	Dekad	Unit	Source
Normalize	Mozambique	2118	Nampula	Crop Area	1/1/1984	0.658	0.616	1984	1	1		METOP
Normalize	Mozambique	2118	Nampula	Crop Area	1/11/1984	0.7	0.643	1984	1	2		METOP
Normalize	Mozambique	2118	Nampula	Crop Area	1/21/1984	0.73	0.666	1984	1	3		METOP
Normalize	Mozambique	2118	Nampula	Crop Area	2/1/1984	0.718	0.684	1984	2	1		METOP
Normalize	Mozambique	2118	Nampula	Crop Area	2/11/1984	0.741	0.695	1984	2	2		METOP
Normalize	Mozambique	2118	Nampula	Crop Area	2/21/1984	0.755	0.71	1984	2	3		METOP
Normalize	Mozambique	2118	Nampula	Crop Area	3/1/1984	0.756	0.72	1984	3	1		METOP
Normalize	Mozambique	2118	Nampula	Crop Area	3/11/1984	0.773	0.721	1984	3	2		METOP
Normalize	Mozambique	2118	Nampula	Crop Area	3/21/1984	0.785	0.738	1984	3	3		METOP
Normalize	Mozambique	2118	Nampula	Crop Area	4/1/1984	0.803	0.746	1984	4	1		METOP
Normalize	Mozambique	2118	Nampula	Crop Area	4/11/1984	0.842	0.752	1984	4	2		METOP
Normalize	Mozambique	2118	Nampula	Crop Area	4/21/1984	0.844	0.756	1984	4	3		METOP
Normalize	Mozambique	2118	Nampula	Crop Area	5/1/1984	0.839	0.754	1984	5	1		METOP
Normalize	Mozambique	2118	Nampula	Crop Area	5/11/1984	0.81	0.73	1984	5	2		METOP

Fig. 7.35 NDVI anomaly

(NOAA/FEWSNet), while for the remaining countries data is obtained from European Centre for Medium-Range Weather Forecasts[125] (ECMWF).

7.6.1 NDVI Anomaly

The Normalized Difference Vegetation Index (NDVI) measures the "greenness" of ground cover and is used as a proxy to indicate the density and health of vegetation. NDVI values range from +1 to -1, with high positive values corresponding to dense and healthy vegetation and low and/or negative NDVI values indicating poor vegetation conditions or sparse vegetative cover. The NDVI anomaly indicates the variation of the current dekad to the long-term average, where a positive value (e.g., 20%) would signify enhanced vegetation conditions compared to the average, while a negative value (for instance, -40%) would indicate comparatively poor vegetation conditions. The following data was downloaded spatial aggregated time series data for Mozambique[126]—Source Meteorological Operational satellite (METOP). The data is from 1984 and would look as follows (please see Fig. 7.35).

7.6.2 Step 1: Load Mozambique Vegetation Health Index (VHI) DEKAD Data

```
#  https://www.fao.org/giews/earthobservation/country/index.
jsp?lang=en&code=MOZ
```

(continued)

[125] European Centre for Medium-Range Weather Forecasts—https://www.ecmwf.int/

[126] Spatial NDVI Data—https://www.fao.org/giews/earthobservation/asis/data/country/MOZ/GRAPH_NDVI_AGRI/ndvi_adm1_data.csv

```
# Vegetation Index
# Vegetation Health Index
data_MODIS_VHI = pd.read_csv
('Mozambique_vhi_adm_dekad_data_ML.csv')
data_MODIS_VHI.head()
```

Output:

	Indicator	Country	ADM1_CODE	Province	Date	Data	Year	Month	Dekad	Unit	Source	monthlabel
0	Vegetation Health Index (VHI)	Mozambique	2118	Nampula	1/1/1984	0.552	1984	1	1		FAO-ASIS	month_1_wk_dekad_1
1	Vegetation Health Index (VHI)	Mozambique	2118	Nampula	1/11/1984	0.568	1984	1	2		FAO-ASIS	month_1_wk_dekad_2
2	Vegetation Health Index (VHI)	Mozambique	2118	Nampula	1/21/1984	0.570	1984	1	3		FAO-ASIS	month_1_wk_dekad_3
3	Vegetation Health Index (VHI)	Mozambique	2118	Nampula	2/1/1984	0.578	1984	2	1		FAO-ASIS	month_2_wk_dekad_1
4	Vegetation Health Index (VHI)	Mozambique	2118	Nampula	2/11/1984	0.596	1984	2	2		FAO-ASIS	month_2_wk_dekad_2

```
data_MODIS_VHI = data_MODIS_VHI.drop(['Indicator','Country',
'ADM1_CODE','Province','Date','Month','Dekad','Unit','Source'],
axis=1)
data_MODIS_VHI
```

Output:

	Data	Year	monthlabel
0	0.552	1984	month_1_wk_dekad_1
1	0.568	1984	month_1_wk_dekad_2
2	0.570	1984	month_1_wk_dekad_3
3	0.578	1984	month_2_wk_dekad_1
4	0.596	1984	month_2_wk_dekad_2
...	...	...	...
1367	0.354	2021	month_12_wk_dekad_3
1368	0.324	2022	month_1_wk_dekad_1
1369	0.324	2022	month_1_wk_dekad_2
1370	0.254	2022	month_1_wk_dekad_3
1371	0.219	2022	month_2_wk_dekad_1

Need to pivot table to arrange columns as year based:

```
data_MODIS_VHI = data_MODIS_VHI.pivot_table('Data', ['Year'],
'monthlabel')
data_MODIS_VHI
```

Pivot tables offer a ton of flexibility for me as a data scientist. pivot_table requires a data and an index parameter:

- Data is the Pandas DataFrame you pass to the function.
- Index is the feature that allows you to group your data.

In our case, we use data as the column and year as the pivot.
Output:

monthlabel	month_10_wk_dekad_1	month_10_wk_dekad_2	month_10_wk_dekad_3	month_11_wk_dekad_1	month_11_wk_dekad_2	month_11_wk_dekad_3
Year						
1984	0.659	0.625	0.597	0.592	0.621	0.694
1985	0.572	0.590	0.603	0.597	0.653	0.713
1986	0.549	0.628	0.709	0.737	0.749	0.751
1987	0.600	0.639	0.594	0.499	0.421	0.343
1988	0.510	0.556	0.603	0.565	0.527	0.495
1989	0.532	0.506	0.450	0.452	0.552	0.688
1990	0.452	0.453	0.396	0.324	0.293	0.324
1991	0.562	0.595	0.586	0.556	0.565	0.590
1992	0.509	0.544	0.477	0.494	0.551	0.654
1993	0.616	0.613	0.618	0.579	0.535	0.454
1994	0.569	0.601	0.588	0.563	0.570	0.598
1995	0.415	0.402	0.322	0.278	0.318	0.405
1996	0.573	0.512	0.387	0.245	0.239	0.354

where week 1 represents January 1st through January 7th.

7.6.3 *Step 2: Prepare Production Data Frame*

```
dfdfMozambiqueNamaoulsaCashewDataProduction = data[['Year' ,
'Area_harvested_Unit', 'Area_harvested_Value' , 'Yield(unit)',
'Yield_value' , 'Production(units)', 'Production_value']]
dfdfMozambiqueNamaoulsaCashewDataProduction
```

Output: The cashew production data frame:

	Year	Area_harvested_Unit	Area_harvested_Value	Yield(unit)	Yield_value	Production(units)	Production_value
0	1961	ha	180000	hg/ha	5944	tonnes	107000
1	1962	ha	180000	hg/ha	6000	tonnes	108000
2	1963	ha	230000	hg/ha	6478	tonnes	149000
3	1964	ha	230000	hg/ha	6783	tonnes	156000
4	1965	ha	210000	hg/ha	6476	tonnes	136000
5	1966	ha	190000	hg/ha	6263	tonnes	119000
6	1967	ha	180000	hg/ha	5611	tonnes	101000
7	1968	ha	300000	hg/ha	6197	tonnes	185900
8	1969	ha	250000	hg/ha	6168	tonnes	154200
9	1970	ha	290000	hg/ha	6345	tonnes	184000
10	1971	ha	320000	hg/ha	6313	tonnes	202000
11	1972	ha	320000	hg/ha	6250	tonnes	200000

7.6.4 Step 3: Merge Cashew Nuts and MODIS VHI

```
#https://www.fao.org/giews/earthobservation/country/index.
jsp?lang=en&code=MOZ
dfdfMozambiqueNamaoulsaCashewProductionMODISLSNDVI = pd.merge
(data_MODIS_VHI, df, left_index=True, right_index=True,how =
"inner")
dfdfMozambiqueNamaoulsaCashewProductionMODISLSNDVI

# Check Columns of merged data frame:
dfdfMozambiqueNamaoulsaCashewProductionMODISLSNDVI.columns
```

Output:

```
Index(['month_10_wk_dekad_1', 'month_10_wk_dekad_2',
'month_10_wk_dekad_3',
    'month_11_wk_dekad_1', 'month_11_wk_dekad_2',
'month_11_wk_dekad_3',
    'month_12_wk_dekad_1', 'month_12_wk_dekad_2',
'month_12_wk_dekad_3',
    'month_1_wk_dekad_1', 'month_1_wk_dekad_2',
'month_1_wk_dekad_3',
    'month_2_wk_dekad_1', 'month_2_wk_dekad_2',
'month_2_wk_dekad_3',
    'month_3_wk_dekad_1', 'month_3_wk_dekad_2',
'month_3_wk_dekad_3',
    'month_4_wk_dekad_1', 'month_4_wk_dekad_2',
```

(continued)

```
'month_4_wk_dekad_3',
    'month_5_wk_dekad_1', 'month_5_wk_dekad_2',
'month_5_wk_dekad_3',
    'month_6_wk_dekad_1', 'month_6_wk_dekad_2',
'month_6_wk_dekad_3',
    'month_7_wk_dekad_1', 'month_7_wk_dekad_2',
'month_7_wk_dekad_3',
    'month_8_wk_dekad_1', 'month_8_wk_dekad_2',
'month_8_wk_dekad_3',
    'month_9_wk_dekad_1', 'month_9_wk_dekad_2',
'month_9_wk_dekad_3',
    'Area_harvested_Unit', 'Area_harvested_Value', 'Yield
(unit)',
    'Yield_value', 'Production(units)', 'Production_value'],
   dtype='object')
```

7.6.5 Step 4: Merged Data Frame Correlation Matrix

```
fig = plt.figure(figsize=(30, 20))
corrMatrix =
dfdfMozambiqueNamaoulsaCashewProductionMODISLSNDVI.corr()
sn.heatmap(corrMatrix, annot=True)
plt.show()
```

Output: most of all the Vegetation Health Index values are negatively correlated. Since the Vegetation Health Index (VHI) illustrates the severity of drought based on the vegetation health and the influence of temperature on plant conditions, the production of cashew nuts, of course, negatively correlated.

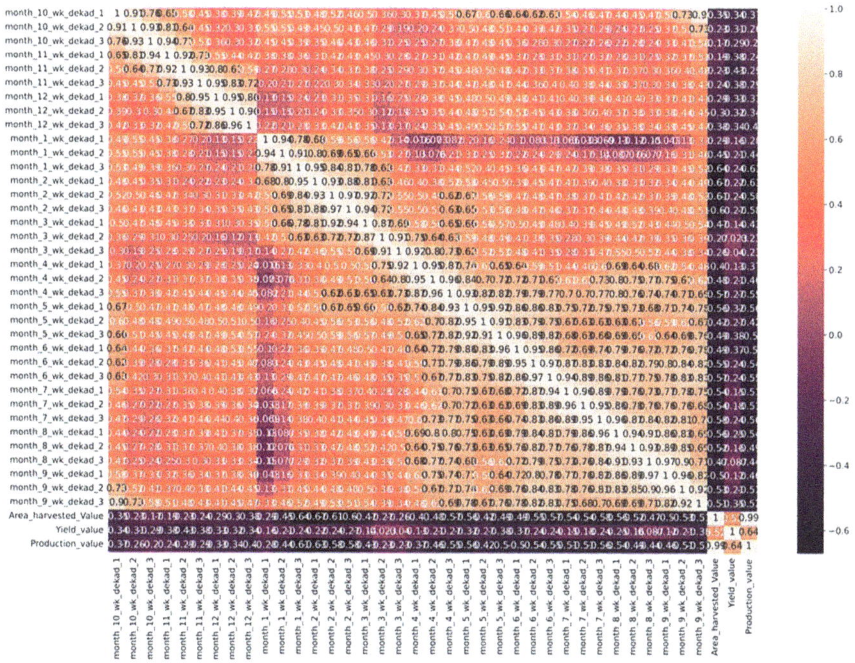

7.6.6 *Step 5: Prepare Data for Regression Model*

The feature columns represent peak growing and harvesting season (January–May and October–December):

```
X = result[['Area_harvested_Value', 'month_1_wk_dekad_1',
'month_1_wk_dekad_2', 'month_1_wk_dekad_3',
    'month_2_wk_dekad_1', 'month_2_wk_dekad_2',
'month_2_wk_dekad_3',
    'month_3_wk_dekad_1', 'month_3_wk_dekad_2',
'month_3_wk_dekad_3',
    'month_4_wk_dekad_1', 'month_4_wk_dekad_2',
'month_4_wk_dekad_3',
    'month_5_wk_dekad_1', 'month_5_wk_dekad_2',
'month_5_wk_dekad_3', 'month_11_wk_dekad_1',
'month_11_wk_dekad_2', 'month_11_wk_dekad_3',
    'month_12_wk_dekad_1', 'month_12_wk_dekad_2',
'month_12_wk_dekad_3']]
y = result['Production_value']

from sklearn.model_selection import train_test_split
X_train, X_test, y_train, y_test = train_test_split(X, y,
test_size = 0.3, random_state = 0)
```

That is:

- January through May (Month 1–Month 5)
- November and December (Month 11–Month 12)

```
# Train the model
from sklearn.linear_model import LinearRegression

# Fit a linear regression model on the training set
model = LinearRegression(normalize=False).fit(X_train, y_train)
print (model)
```

Output:

```
LinearRegression()
```

7.6.7 Step 6: Regression Metrics

```
import matplotlib.pyplot as plt

%matplotlib inline

plt.scatter(y_test, predictions)
plt.xlabel('Actual Labels')
plt.ylabel('Predicted Labels')
plt.title('Predictions')
# overlay the regression line
z = np.polyfit(y_test, predictions, 1)
p = np.poly1d(z)
plt.plot(y_test,p(y_test), color='magenta')
plt.show()
```

Output:

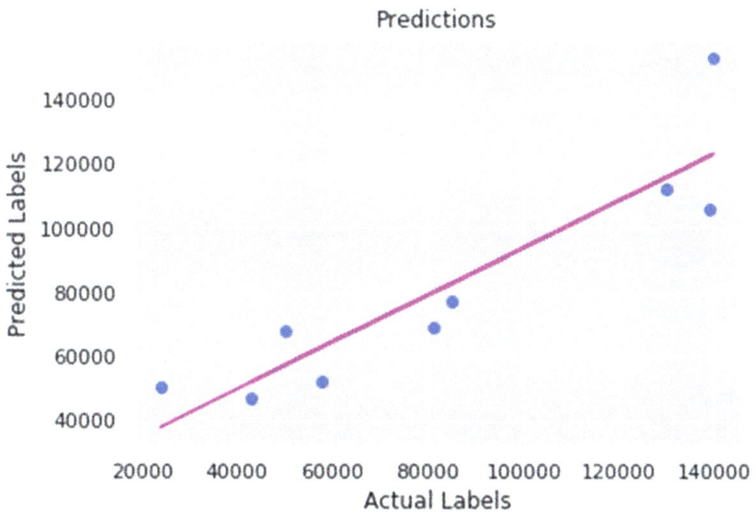

Regression metrics:

```
from sklearn.metrics import mean_squared_error, r2_score

mse = mean_squared_error(y_test, predictions)
print("MSE:", mse)

rmse = np.sqrt(mse)
print("RMSE:", rmse)

r2 = r2_score(y_test, predictions)
print("R2:", r2)
```

Output:

```
MSE: 320846336.63368446
RMSE: 17912.18402746255
R²: 0.8125123590310175
```

Let's derive the model governing equation: construct MX part of the Y = MX + C:

```
mx=""
for ifeature in range(len(X.columns)):
  if model.coef_[ifeature] <0:
    # format & beautify the equation
```

(continued)

```
    mx += " - " + "{:.2f}".format(abs(model.coef_[ifeature])) + " * "
+ X.columns[ifeature]
  else:
    if ifeature == 0:
      mx += "{:.2f}".format(model.coef_[ifeature]) + " * " + X.
columns[ifeature]
    else:
     mx += " + " + "{:.2f}".format(model.coef_[ifeature]) + " * " + X.
columns[ifeature]

print(mx)
```

Output:

```
0.83 * Area_harvested_Value - 199211.44 * month_1_wk_dekad_1 +
359427.05 * month_1_wk_dekad_2 + 39776.52 * month_1_wk_dekad_3 -
303306.09 * month_2_wk_dekad_1 + 215089.23 * month_2_wk_dekad_2 -
355906.50 * month_2_wk_dekad_3 + 363689.73 * month_3_wk_dekad_1 -
305043.25 * month_3_wk_dekad_2 + 184213.55 * month_3_wk_dekad_3 +
94688.84 * month_4_wk_dekad_1 - 445018.16 * month_4_wk_dekad_2 +
359763.91 * month_4_wk_dekad_3 + 160432.46 * month_5_wk_dekad_1 -
215951.80 * month_5_wk_dekad_2 + 25001.98 * month_5_wk_dekad_3 +
136083.31 * month_11_wk_dekad_1 - 294193.39 * month_11_wk_dekad_2
+ 112777.57 * month_11_wk_dekad_3 + 10569.97 *
month_12_wk_dekad_1 + 39316.55 * month_12_wk_dekad_2 - 56331.15 *
month_12_wk_dekad_3
```

Construct model equation:

```
# y=mx+c
if(model.intercept_ <0):
  print("The formula for the " + y.name + " linear regression line is
= " + " - {:.2f}".format(abs(model.intercept_)) + mx )
else:
  print("The formula for the " + y.name + " linear regression line is
= " + " + {:.2f}".format(model.intercept_) + mx )
```

Output:

The formula for the Production_value linear regression line is

$$
\begin{aligned}
= \; & +45568.460.83^{*}\ \text{Area\_harvested\_Value} \\
& - 199211.44^{*}\ \text{month\_1\_wk\_dekad\_1} \\
& + 359427.05^{*}\ \text{month\_1\_wk\_dekad\_2} \\
& + 39776.52^{*}\ \text{month\_1\_wk\_dekad\_3} \\
& - 303306.09^{*}\ \text{month\_2\_wk\_dekad\_1} \\
& + 215089.23^{*}\ \text{month\_2\_wk\_dekad\_2} \\
& - 355906.50^{*}\ \text{month\_2\_wk\_dekad\_3} \\
& + 363689.73^{*}\ \text{month\_3\_wk\_dekad\_1} \\
& - 305043.25^{*}\ \text{month\_3\_wk\_dekad\_2} \\
& + 184213.55^{*}\ \text{month\_3\_wk\_dekad\_3} \\
& + 94688.84^{*}\ \text{month\_4\_wk\_dekad\_1} \\
& - 445018.16^{*}\ \text{month\_4\_wk\_dekad\_2} \\
& + 359763.91^{*}\ \text{month\_4\_wk\_dekad\_3} \\
& + 160432.46^{*}\ \text{month\_5\_wk\_dekad\_1} \\
& - 215951.80^{*}\ \text{month\_5\_wk\_dekad\_2} \\
& + 25001.98^{*}\ \text{month\_5\_wk\_dekad\_3} \\
& + 136083.31^{*}\ \text{month\_11\_wk\_dekad\_1} \\
& - 294193.39^{*}\ \text{month\_11\_wk\_dekad\_2} \\
& + 112777.57^{*}\ \text{month\_11\_wk\_dekad\_3} \\
& + 10569.97^{*}\ \text{month\_12\_wk\_dekad\_1} \\
& + 39316.55^{*}\ \text{month\_12\_wk\_dekad\_2} \\
& - 56331.15^{*}\ \text{month\_12\_wk\_dekad\_3}
\end{aligned}
\tag{7.8}
$$

The above equation provides cashew nuts production that would enable to assess the near end of the century cashes nuts production.

7.6.8 Step 7: Other Regression Ensemble Models

I have used Voting Ensemble.[127] The idea behind the Voting Regressor[128] is to combine conceptually different machine learning regressors and return the average predicted values. Such a regressor can be useful for a set of equally well-performing models in order to balance out their individual weaknesses:

```
import numpy as np
from sklearn.linear_model import LinearRegression
from sklearn.ensemble import RandomForestRegressor
```

(continued)

[127] Ensemble Methods—https://scikit-learn.org/stable/modules/classes.html#module-sklearn.ensemble

[128] Voting Regressor—https://scikit-learn.org/stable/modules/ensemble.html#voting-regressor

```
from sklearn.ensemble import GradientBoostingRegressor
from sklearn.ensemble import VotingRegressor
r1 = LinearRegression()
r2 = RandomForestRegressor(n_estimators=10,random_state=1)
r3 = GradientBoostingRegressor()
erT = VotingRegressor([('lr', r1), ('rf', r2) ,('gb',r3)])
erT.fit(X_train, y_train)
```

Output:

```
VotingRegressor(estimators=[('lr', LinearRegression()),
                ('rf',
                 RandomForestRegressor(n_estimators=10,
                                       random_state=1)),
                ('gb', GradientBoostingRegressor())])
```

Call Predict to predict values.

```
yT1_pred = erT.predict(X_test)
yT1_pred

pred_yT1_df = pd.DataFrame({'Actual Value':y_test, 'Predicted
value':yT1_pred,'Difference': y_test-yT1_pred})
pred_yT1_df
```

Output:

Year	Actual Value	Predicted value	Difference
1993	23935	39500.013136	-15565.013136
2019	140000	121045.957141	18954.042859
2004	42988	41964.596907	1023.403093
2001	58000	61638.074590	-3638.074590
2017	139088	102241.823677	36846.176323
2015	81240	82996.249671	-1756.249671
2018	130000	108459.630169	21540.369831
2002	50177	61875.438480	-11698.438480
2008	85000	85036.482179	-36.482179

7.6.9 Step 8: Model R^2

```
import matplotlib.pyplot as plt

%matplotlib inline

plt.scatter(y_test,yT1_pred)
plt.xlabel('Actual Labels')
plt.ylabel('Predicted Labels')
plt.title('Predictions')
# overlay the regression line
z = np.polyfit(y_test, yT1_pred, 1)
p = np.poly1d(z)
plt.plot(y_test,p(y_test), color='magenta')
plt.show()
```

Output:

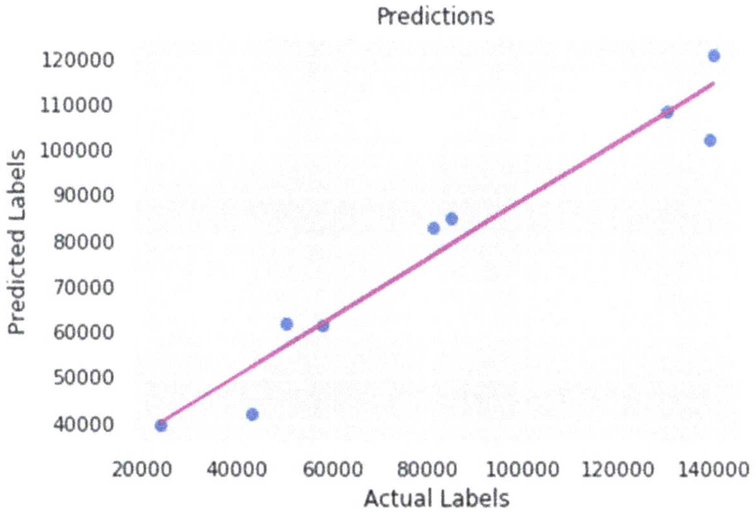


```
r2_score(y_test,yT1_pred)
```

Output:
0.8326557514653319 (83.26%)

7.6.10 Step 9: Explainability of the Model

Please find explainability:

```
eli5.show_weights(model, feature_names=[
'Area_harvested_Value', 'month_1_wk_dekad_1',
'month_1_wk_dekad_2', 'month_1_wk_dekad_3',
    'month_2_wk_dekad_1', 'month_2_wk_dekad_2',
'month_2_wk_dekad_3',
    'month_3_wk_dekad_1', 'month_3_wk_dekad_2',
'month_3_wk_dekad_3',
    'month_4_wk_dekad_1', 'month_4_wk_dekad_2',
'month_4_wk_dekad_3',
    'month_5_wk_dekad_1', 'month_5_wk_dekad_2',
'month_5_wk_dekad_3', 'month_11_wk_dekad_1',
'month_11_wk_dekad_2', 'month_11_wk_dekad_3',
    'month_12_wk_dekad_1', 'month_12_wk_dekad_2',
'month_12_wk_dekad_3'])
```

Output:

y top features

Weight?	Feature
+363689.728	month_3_wk_dekad_1
+359763.911	month_4_wk_dekad_3
+359427.052	month_1_wk_dekad_2
+215089.227	month_2_wk_dekad_2
+184213.552	month_3_wk_dekad_3
+160432.461	month_5_wk_dekad_1
+136083.314	month_11_wk_dekad_1
+112777.570	month_11_wk_dekad_3
+94688.845	month_4_wk_dekad_1
+45568.458	<BIAS>
+39776.520	month_1_wk_dekad_3
+39316.547	month_12_wk_dekad_2
… 3 more positive …	
-56331.151	month_12_wk_dekad_3
-199211.443	month_1_wk_dekad_1
-215951.804	month_5_wk_dekad_2
-294193.387	month_11_wk_dekad_2
-303306.095	month_2_wk_dekad_1
-305043.246	month_3_wk_dekad_2
-355906.500	month_2_wk_dekad_3
-445018.163	month_4_wk_dekad_2

After reading this chapter, you should be able to answer food security and advanced imaging radiometer datasets and ML models and to construct ML models with satellite radiometer, food security and satellite data, and global

(continued)

vegetation—Cropland and Vegetation Index. You should also use the Normalized Difference Vegetation Index (NDVI) as part of advanced ML modeling. Finally, you should extend machine learning models that specifically look at Mozambique cashew nuts production model and Mozambique cashew nuts and Normalized Difference Vegetation Index (NDVI) model. You should also be able to summarize cashew nuts production with findings of CMIP6 Projections Data and SSP Scenarios.

References

1. Issues and Challenges of Inclusive Development: Essays in Honor of Prof. R. Radhakrishna by R. Maria Saleth, S. Galab and E. Revathi, Publisher: Springer; 1st ed. 2020 edition (June 19, 2020), ISBN-13: 978-9811522284
2. Anikka Martin, Food Security and Nutrition Assistance, Last updated: Monday, November 08, 2021, https://www.ers.usda.gov/data-products/ag-and-food-statistics-charting-the-essentials/food-security-and-nutrition-assistance/, Access Date: December 12, 2021
3. Alisha Coleman-Jensen, Matthew P. Rabbitt, Laura Hales, and Christian A. Gregory, The prevalence of food insecurity in 2020 is unchanged from 2019, https://www.ers.usda.gov/data-products/chart-gallery/gallery/chart-detail/?chartId=58378, Access Date: December 20, 2021
4. Alisha Coleman-Jensen, Matthew P. Rabbitt, Laura Hales, and Christian A. Gregory, Prevalence of food insecurity is not uniform across the country, Last updated: Monday, November 08, 2021, https://www.ers.usda.gov/data-products/chart-gallery/gallery/chart-detail/?chartId=58392, Access Date: December 21, 2021
5. Navin Singh Khadka, Climate change: Low-income countries 'can't keep up' with impacts, 8 August 2021, https://www.bbc.com/news/world-58080083, Access Date: March 12, 2022
6. WFP, Food security and climate change, the pressing reality of Mozambique, 7 July 2021, https://docs.wfp.org/api/documents/WFP-0000129988/download/?_ga=2.150781622.1610345514.1644972256-1475447051.1638068907, Access Date: February 15, 2022
7. Rebecca Hersher, Meteorologists Can't Keep Up With Climate Change In Mozambique, December 11, 20195:09 AM ET, https://www.npr.org/sections/goatsandsoda/2019/12/11/782918005/meteorologists-cant-keep-up-with-climate-change-in-mozambique, Access Date: February 15, 2022
8. NOAA, Climate Model – Sea Surface Temperature Change: SSP2 (Middle of the Road) – 2015–2100, 2022, https://sos.noaa.gov/catalog/datasets/724/, Access Date: March 11, 2022
9. ZEKE HAUSFATHER, Explainer: How "Shared Socioeconomic Pathways" explore future climate change, 19 April 2018, https://www.carbonbrief.org/explainer-how-shared-socioeconomic-pathways-explore-future-climate-change, Access Date: February 03, 2022
10. Jennifer Fadoul, How Satellite Maps Help Prevent Another 'Great Grain Robbery', Aug 20 2021, https://www.nasa.gov/feature/how-satellite-maps-help-prevent-another-great-grain-robbery, Access Date: February 19, 2022
11. Earth Resources Observation and Science (EROS) Center, USGS EROS Archive - Advanced Very High Resolution Radiometer (AVHRR) – Sensor Characteristics, July 9, 2018, https://www.usgs.gov/centers/eros/science/usgs-eros-archive-advanced-very-high-resolution-radiometer-avhrr-sensor, Access Date: March 12, 2022

12. Michelle Albee Matto, June is National Dairy Month: Let's Recognize Dairy's Role in Fighting Food Insecurity, June 1, 2020, https://www.idfa.org/news/june-is-national-dairy-month-lets-recognize-dairys-role-in-fighting-food-insecurity, Date: March 12, 2022
13. Dave Warner, keeping cows cool critical to dairy industry as climate warms, Updated: Jan. 05, 2019, 7:56 p.m. | Published: Apr. 24, 2013, 2:15 a.m., https://www.pennlive.com/midstate/2013/04/keeping_cows_cool_critical_to.html, Date: March 12, 2022
14. Jon Marcus, How Cows Are Becoming Smart Connected Products, Jun 18, 2014, 09:31 am EDT, https://www.forbes.com/sites/ptc/2014/06/18/how-cows-are-becoming-smart-connected-products/?sh=1c94e914419c, Access Date: March 12, 2022
15. L. T. Casarotto, J. Laporta, and G. E. Dahl, FINANCIAL IMPACTS OF LATE-GESTATION HEAT STRESS ON COW AND OFFSPRING LIFETIME PERFORMANCE, https://edis.ifas.ufl.edu/publication/AN374, Date: 11/11/2021, Access Date: March 12, 2022
16. Nigel Key, Stacy Sneeringer, and David Marquardt, Climate Change, Heat Stress, and U.-S. Dairy Production, September 2014, https://www.ers.usda.gov/webdocs/publications/45279/49164_err175.pdf?v=0, Access Date: March 12, 2022
17. Joe Armstrong, DVM, Extension cattle production systems educator and Kevin Janni, Extension bioproducts and biosystems engineer, Heat stress in dairy cattle, Reviewed in 2020, https://extension.umn.edu/dairy-milking-cows/heat-stress-dairy-cattle#:~:text=Milk%20production%20decreases%20as%20the,lbs.%2Fhead%2Fday, Access Date: March 12, 2022
18. Climate Hubs USDA, Managing Dairy Heat Stress in the Northeast's Rapidly Changing Climate, https://www.climatehubs.usda.gov/hubs/northeast/topic/managing-dairy-heat-stress-northeasts-rapidly-changing-climate, Access Date: March 12, 2022
19. A. Ilapakurti and C. Vuppalapati, "Building an IoT Framework for Connected Dairy," 2015 IEEE First International Conference on Big Data Computing Service and Applications, 2015, pp. 275–285, https://doi.org/10.1109/BigDataService.2015.39.
20. C. Vuppalapati, A. Ilapakurti, S. Kedari, J. Vuppalapati, S. Kedari and R. Vuppalapati, "Democratization of AI, Albeit Constrained IoT Devices & Tiny ML, for Creating a Sustainable Food Future," 2020 3rd International Conference on Information and Computer Technologies (ICICT), 2020, pp. 525-530, doi: https://doi.org/10.1109/ICICT50521.2020.00089.
21. J. S. Vuppalapati, S. Kedari, A. Ilapakurthy, A. Ilapakurti and C. Vuppalapati, "Smart Dairies—Enablement of Smart City at Gross Root Level," 2017 IEEE Third International Conference on Big Data Computing Service and Applications (BigDataService), 2017, pp. 118-123, doi: https://doi.org/10.1109/BigDataService.2017.35.
22. Alan Buis, Too Hot to Handle: How Climate Change May Make Some Places Too Hot to Live, March 9, 2022, 11:41 PST, https://climate.nasa.gov/ask-nasa-climate/3151/too-hot-to-handle-how-climate-change-may-make-some-places-too-hot-to-live/, Access Date: March 12, 2022
23. Yang W, Kogan F, Guo W. An Ongoing Blended Long-Term Vegetation Health Product for Monitoring Global Food Security. Agronomy. 2020; 10(12):1936. https://doi.org/10.3390/agronomy10121936
24. FAO, FAO'S WORK ON CLIMATE CHANGE United Nations Climate Change Conference 2017, 2017, https://www.fao.org/3/i8037e/i8037e.pdf, Access Date: January 30, 2022
25. GISGeography, What is NDVI (Normalized Difference Vegetation Index)? Last Updated: October 29, 2021, https://gisgeography.com/ndvi-normalized-difference-vegetation-index/, Access Date: March 12, 2022
26. Jesslyn Brown, NDVI, the Foundation for Remote Sensing Phenology, November 27, 2018, https://www.usgs.gov/special-topics/remote-sensing-phenology/science/ndvi-foundation-remote-sensing-phenology, Access Date: March 12, 2022
27. Kasha Patel, Brazil Battered by Drought, June 17, 2021, https://earthobservatory.nasa.gov/images/148468/brazil-battered-by-drought, Access Date: February 13, 2022
28. BBC, Africa drought fears grip Malawi and Mozambique, 13 April 2016, https://www.bbc.com/news/world-africa-36037414, Access Date: February 13, 2022
29. Holly Frew, El Niño: Urgent Need for Support as Drought Affects 14 million People in Southern Africa, Jan. 28, 2016, https://www.care.org/news-and-stories/press-releases/el-nino-

urgent-need-for-support-as-drought-affects-14-million-people-in-southern-africa/, Access Date: February 13, 2022
30. Felix Kogan, Remote Sensing for Malaria: Monitoring and Predicting Malaria from Operational Satellites (Springer Remote Sensing/Photogrammetry), Springer; 1st ed. 2020 edition (July 21, 2020), ISBN-13: 978-3030460198
31. Mitić P, Munitlak Ivanović O, Zdravković A. A Cointegration Analysis of Real GDP and CO_2 Emissions in Transitional Countries. Sustainability. 2017; 9(4):568. https://doi.org/10.3390/su9040568
32. Ellen Gray, Global Climate Change Impact on Crops Expected Within 10 Years, NASA Study Finds, November 2, 2021, https://climate.nasa.gov/news/3124/global-climate-change-impact-on-crops-expected-within-10-years-nasa-study-finds/, Access Date: January 28, 2022
33. Hanlon, Joseph. "Power without Responsibility: The World Bank & Mozambican Cashew Nuts." Review of African Political Economy, vol. 27, no. 83, [ROAPE Publications Ltd, Taylor & Francis, Ltd.], 2000, pp. 29–45, http://www.jstor.org/stable/4006634.
34. Ricci Shryock, Guinea-Bissau's cashew farmers survive tough times, 13 June 2021, https://www.bbc.com/news/world-africa-57286241, Access Date: February 16, 2022
35. Jane Grob and Daria Gage, MozaCajú Impact Report, DECEMBER 2017, https://www.technoserve.org/wp-content/uploads/2018/01/mozacaju-impact-report.pdf, Access Date: February 15, 2022
36. The Economist, Mozambique's nut factories have made a cracking comeback, Sep 12th 2019, https://www.economist.com/middle-east-and-africa/2019/09/12/mozambiques-nut-factories-have-made-a-cracking-comeback, Access Date: February 15, 2022
37. Ralph L. Price, Industrialization of Cashew in the state of Ceara, Brazil, 2013, https://repository.arizona.edu/bitstream/handle/10150/300508/pa-26-02-13-16.pdf?sequence=1&isAllowed=y, Access Date: February 18, 2022
38. Carlos Costa, The Cashew Value Chain in Mozambique, 2019, https://documents1.worldbank.org/curated/en/397581559633461087/pdf/The-Cashew-Value-Chain-in-Mozambique.pdf, Access Date: February 18, 2022
39. Rebecca Hersher, Mozambique Is Racing To Adapt To Climate Change. The Weather Is Winning, December 27, 20194:27 PM ET, https://www.npr.org/sections/goatsandsoda/2019/12/27/788552728/mozambique-is-racing-to-adapt-to-climate-change-the-weather-is-winning, Access Date: February 15, 2022
40. Dutch Sustainability, Climate Change Profile MOZAMBIQUE, 2015, https://ees.kuleuven.be/klimos/toolkit/documents/689_CC_moz.pdf, Access Date: February 15, 2022
41. Teshome Demissie Tolla, Effects of moisture conditions and management on production of Cashew, 2004, https://webapps.itc.utwente.nl/librarywww/papers_2004/msc/nrm/teshome_demissie_tolla.pdf, Access Date: February 18, 2022
42. Bangkok Post Online Reporters, PM in Nakhon Sawan, Chai Nat on drought inspection, Published: 22/01/2016 at 02:00 PM, https://www.bangkokpost.com/print/835884/, Access Date: February 04, 2022
43. Josep Cos, Francisco Doblas-Reyes1, Martin Jury, Raül Marcos, Pierre-Antoine Bretonnière, and Margarida Samsó, The Mediterranean climate change hotspot in the CMIP5 and CMIP6 projections, 30 July 2021, https://esd.copernicus.org/preprints/esd-2021-65/esd-2021-65.pdf, Access Date: February 04, 2022
44. Climate Model: Temperature Change (RCP 6.0) – 2006–2100, https://sos.noaa.gov/catalog/datasets/climate-model-temperature-change-rcp-60-2006-2100/, Access Date: March 12. 2022
45. Anton Eitzinger, Armando Martinez and Narioski Castro, "Predicting the Impact of Climate Change on Cashew Growing Regions in Ghana and Cote d'Ivoire", 2011, https://www.africancashewalliance.com/sites/default/files/documents/ghana_ivory_coast_climate_change_and_cashew.pdf, Access Date: February 18, 2022
46. Jordan Mendoza, Climate change expected to impact the world's wheat and corn crops by 2030, NASA says, November 4, 2021, https://www.usatoday.com/story/news/world/2021/11/04/climate-change-impacts-corn-soybeans-rice-crops/6257778001/, Access Date: February 19, 2022

47. Aschenbrenner, P., Chemura, A., Jarawura, F., Habtemariam, L., Lüttringhaus, S., Murken, L., Roehrig, F., Tomalka, J., & Gornott, C., (2021). Climate Risk Analysis for Identifying and Weighing Adaptation Strategies for the Agricultural Sector in Northern Ghana – A Study at District Level in the Upper West Region. A report prepared by the Potsdam Institute for Climate Impact Research (PIK) for the Deutsche Gesellschaft für Internationale Zusammenarbeit (GIZ) GmbH on behalf of the German Federal Ministry for Economic Cooperation and Development (BMZ), 118 pp. https://doi.org/10.48485/pik.2021.001.

Chapter 8
Composite Models: Food Security and Natural Resources

This chapter covers the following:

- Food security
- Linkages of agriculture (A), food (F), and nutrition (N)
- Key drivers of food security
- Food security indicators and drivers
- The Global Food Security Index (GFSI)
- Machine learning models
  - Economic access and food security predictive model
  - Commodity Price Prediction and Linear Programming
  - Food affordability predictive model
- Food security and technological innovation
- Small farm sustainability
- Data-driven food security models

The chapter introduces food security, key drivers of food insecurity, and food security indicators and drivers. Additionally, it introduces the Global Food Security Index (GFSI) framework and the United Nations Suite of Food Security Indicators. Next, machine learning and linear programming are applied to develop commodity price prediction, economic access and food security, and food affordability models. Finally, the chapter concludes with the role of technological innovations, data-driven models, and small farm sustainability to enhance food security.

The world is at a critical juncture. In 2020, nearly one in three people did not have access to adequate food—between 720 and 811 million people faced hunger.

C. Vuppalapati, *Artificial Intelligence and Heuristics for Enhanced Food Security*,
International Series in Operations Research & Management Science 331,
https://doi.org/10.1007/978-3-031-08743-1_8

Compared with that in 2019[1] [1], 46 million more people in Africa, almost 57 million more in Asia, and about 14 million more in Latin America and the Caribbean were affected by hunger. Based on forecasts of global population growth, current deficit to feed people around the world, decline of agricultural production due to conflicts,[2,3] and increased demand for greener fuel and biodiesel, food security will remain an important economic development issue over the next several decades. As food-versus-fuel tension becomes more intense,[4] the day will come when more agricultural products will be used for energy than food. And it seems the day has arrived [2–4].

Many governments that have high dependencies on imported oil to drive their economies, now, clearly moving towards biofuels. For instance, an ambitious ethanol plan, announced by the most populous country, has spurred food security fears.[5,6] Ethanol production using food crops is a double-edged sword. Sugarcane is a major source of ethanol production. However, the center has plans to push for greater production from non-sugar sources and has offered credit-linked subsidy schemes to set up distillation units and biorefineries that utilize molasses and surplus and damaged food grains. But the adverse impact of grain diversion on ethanol production cannot be ignored. In years of deficit food grain production, the diversion of food grains and sugarcane to distilleries and biorefineries can create pressure on the stocks earmarked for human consumption through the Public Distribution System (PDS), resulting in food inflation. Sugarcane and paddy consume almost 70% of the total water used in irrigation. Despite being a water-intensive crop and expensive raw material for ethanol production, government has considered sugarcane as a lucrative food crop. On the other hand, maize, millet, and barley require less water and are low-cost crops and farmers can grow them instead of water-intensive sugarcane and rice. To solve the food–fuel conundrum, we need to strike a balance between promises and pitfalls[7] [4–6].

---

[1]Food Security—https://www.fao.org/state-of-food-security-nutrition

[2]U.S. economy appeared ready to surge, but Russia's invasion of Ukraine could send shockwaves—https://www.washingtonpost.com/business/2022/02/25/economy-us-russia-ukraine-gas/

[3]The war in Ukraine is likely to slow global growth, the I.M.F. warns—https://www.nytimes.com/2022/03/10/business/economy/imf-global-outlook-ukraine-war.html

[4]Cargill CEO Says Global Food Prices to Stay High on Labor Crunch—https://finance.yahoo.com/news/cargill-ceo-says-global-food-125939772.html

[5]An Ambitious Ethanol Plan Spurs Food Security Fears in India—https://www.bloomberg.com/news/articles/2021-10-05/an-ambitious-ethanol-plan-sparks-food-security-fears-in-india#:~:text=India's%20ambitious%20plan%20to%20cut,world's%20second%2Dmost%20populous%20country

[6]Government assures ethanol blending plan will not affect India's food security—https://www.businesstoday.in/latest/economy/story/govt-assures-ethanol-blending-plan-will-not-affect-indias-food-security-308747-2021-10-07

[7]The food-fuel conundrum—https://www.thehindubusinessline.com/opinion/the-food-fuel-conundrum/article35348867.ece

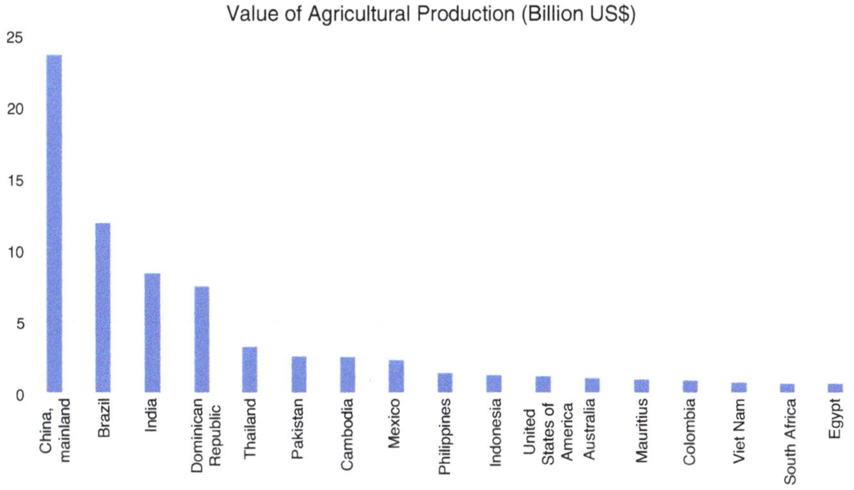

Fig. 8.1 Value of agricultural production (billion USD)

Sugarcane is an important commercial crop worldwide and one of the principal sources of sugar, ethanol, and jaggery (a semi-refined sugar product used in the Indian subcontinent) globally. Its by-products are also used as a fodder to feed livestock in many countries. In 2018, the worldwide gross production value of sugarcane was 75.6 billion USD.[8] Most of the rainfed and irrigated commercial sugarcane is grown between 35°N and S of the equator. The crop flourishes under a long, warm growing season with a high incidence of radiation and adequate moisture, followed by a dry, sunny, and cool but frost-free ripening and harvesting period.

The present area of sugarcane (*Saccharum officinarum*) is about 26.46 million hectares (ha) with a total commercial world production of about 1729.6 million tons of cane/year (FAOSTAT, 2020). Sugarcane[9] originated in Asia; though its cultivation started around 327 BC in the Indian subcontinents, it gradually found its way to the rest of the world via trade routes through the Middle East. Later, it arrived and flourished as an industry in the New World. In India, sugarcane is still used in many religious rituals. Today, sugarcane is cultivated in most countries with warm climates. Peru, Senegal, and Guatemala are the leading countries with the highest yield (1,237,593, 1,141,924, and 1,129,402 hg/ha, respectively). Moreover, China, Brazil, and India are the leading countries in terms of agricultural production value (US$23.49 billion, US$11.81 billion, and US$8.32 billion, respectively) (please see Fig. 8.1).

Brazil tops the list of sugarcane producers, with an annual production of 686,844 thousand metric tons. The South-Central region of Brazil is accountable for more than 90% of this national production output. Sugar is the main product sourced from

[8]FAO Stat—Sugarcane Gross Production—https://www.fao.org/faostat/en/#data/QV/visualize

[9]Sugarcane—https://www.fao.org/land-water/databases-and-software/crop-information/sugarcane/en/

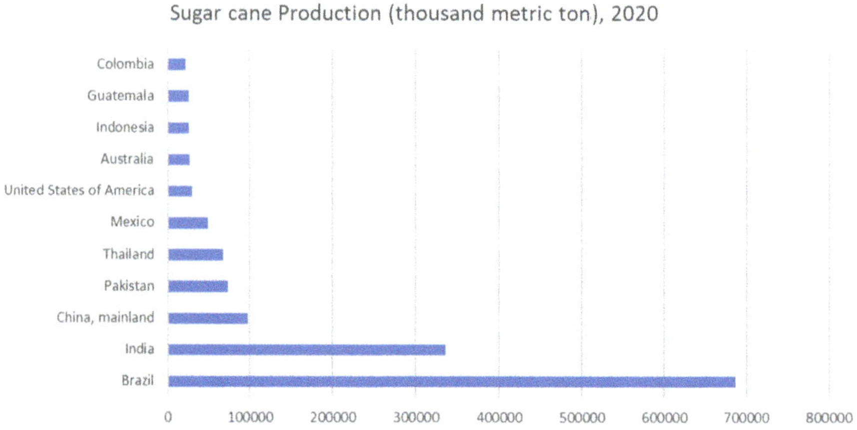

Fig. 8.2 Sugarcane production, 2020

the sugarcane cultivated in Brazil (The state of the world's land and water resources for food and agriculture: Systems at breaking point [SOLAW 2021]).[10] However, production of ethanol from residual molasses is now growing more popular, as ethanol is used as a fuel to power automobiles, which Brazilians are owning in larger numbers.

In India, the states of Maharashtra, Uttar Pradesh, Punjab, and Bihar produce the maximum quantities of sugarcane. A total of 370,500 thousand metric tons (TMT) of annual produce was recorded in the year 2020. It is not, however, any wonder that India is one of the largest exporters of sugar worldwide. In India, sugarcane is cultivated to produce crystal sugar, jaggery (gur), and numerous alcoholic beverages. It is estimated that the nation's sugarcane industry provides employment for more than 6 million Indians. The country exports sugar to Sri Lanka, Bangladesh, Somalia, Sudan, Indonesia, and the United Arab Emirates. India's sugar export to Iran is declining, however, especially over the past few years (please see Fig. 8.2).

An estimated 80% of sugar production in China comes from the sugarcane grown in its south and southwest regions. In fact, the country has a long history of sugarcane cultivation. The earliest records of such suggest that in the fourth century BC, Chinese people were already knowledgeable regarding sugarcane cultivation. However, they did not become proficiently familiar with sugar refining techniques until much later, around 645 AD. Despite being one of the largest producers, China, to meet its high domestic demand, imports sugar from other countries as well. Namely, the chief among these are Brazil, Thailand, Australia, Myanmar, Vietnam, and Cuba. Currently, the country is the largest market for sugar sourced from Myanmar. The country has also plunged into the production of ethanol from

[10]Top Sugarcane-Producing Countries—https://www.worldatlas.com/articles/top-sugarcane-producing-countries.html

sugarcane as a full-fledged affair to help meet the rising fuel demand of its vast population.[11]

8.1 Sugarcane Cultivation

Sugarcane, a tall perennial grass reaching 3–4 m in height, comprised of jointed, fibrous stalks, is grown in tropical and semitropical climates. After the planting of cane stalk cuttings, the plant matures in 1–2 years. Two to four crops are harvested from the original plantings, unless the plants are impaired or destroyed by frost, disease, or other causes. Once harvested, sugarcane must be processed quickly before its sucrose deteriorates. Sugarcane and sugar beet yields can vary widely from year to year due to weather but yields for both have tended to increase over time.

In the United States, sugarcane is produced in Florida, Louisiana, and Texas. Acreage of sugarcane for sugar rose from an average 704,000 acres in the first half of the 1980s to 903,400 acres in FY 2020/2021. Over the same period, sugar produced from sugarcane grew from 2.910 million STRV to 4.251 million STRV in 2020/2021.

Florida's sugarcane production has expanded significantly since the United States ceased importing sugar from Cuba in 1960. Florida is the largest cane-producing region in the United States. Most of the sugarcane is produced in organic soils along the southern and southeastern shore of Lake Okeechobee in southern Florida, where the growing season is long, and winters are generally warm. Florida produced an average 2.06 million short tons, raw value (STRV) of sugar between FY 2017 and FY 2021.

Annual Cash Receipts

Cash receipts for US sugar growers vary with sugar yields and prices. Cash receipts for sugar beets were $1.184 billion in the 2018/2019 crop year and $1.098 billion in the 2019/2020 crop year. Sugarcane cash receipts were $1.000 billion in the 2018/2019 crop year and $1.160 billion in the 2019/2020 crop year. On average, the sugar crops account for less than 1% of the cash receipts received by US farmers for all agricultural commodities.

The optimum temperature for sprouting (germination) of stem cuttings is *32–38 °C*. The optimum growth is achieved with mean daily *temperatures between 22 and 30 °C*. The minimum temperature for active growth is *approximately 20 °C*. For *ripening*, however, relatively lower temperatures in the range of *20–10 °C* are desirable, since this has a noticeable influence on the reduction of vegetative growth rate and the enrichment of sucrose in the cane (please see Fig. 8.3).

A *long growing season* is essential for high yields. This prerequisite essential would have very negative yield curves due to climate change. The normal length of

[11] Top Sugarcane-Producing Countries—https://www.worldatlas.com/articles/top-sugarcane-producing-countries.html

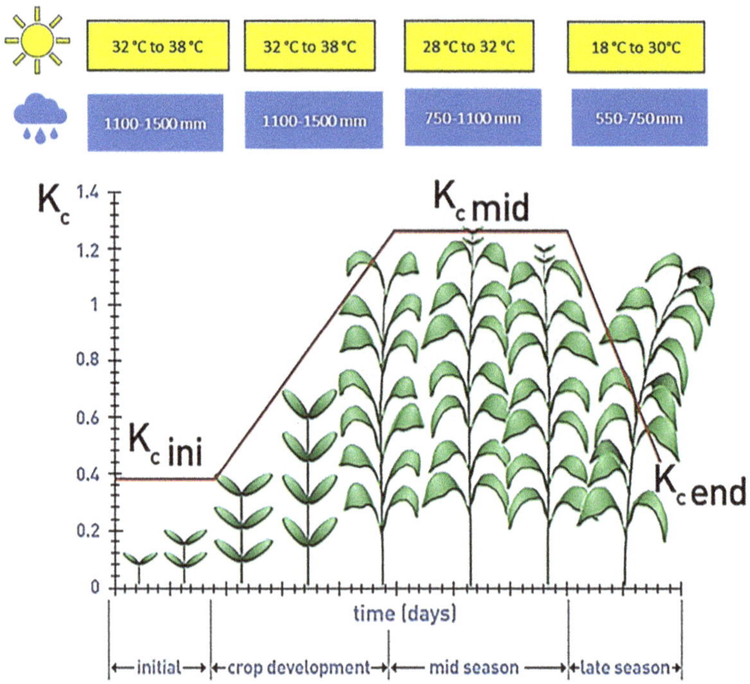

Fig. 8.3 Sugarcane phenology: temperature and precipitation

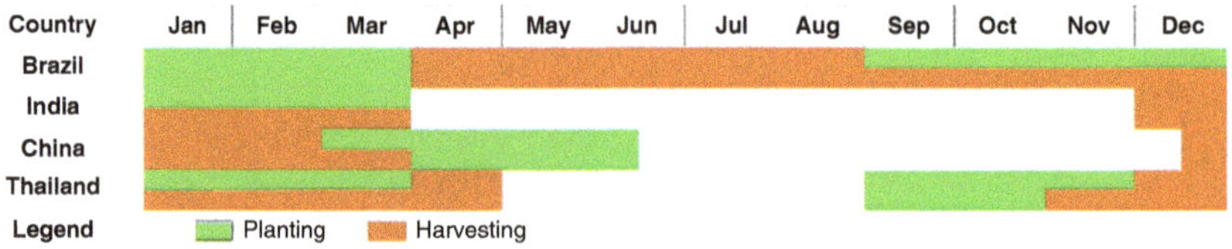

Fig. 8.4 Sugar calendar

the total growing period *varies between 9 months with harvest before winter frost and 24 months* in Hawaii, but it is *generally 15–16 months*. In Queensland, Australia, this is generally a norm of time to cultivate. Plant (first) crop is normally followed by two to four ratoon crops, and in certain cases up to a maximum of eight crops are taken, each taking about 1 year to mature. Growth of the stool is slow at first, gradually increasing until the maximum growth rate is reached after which growth slows down as the cane begins to ripen and mature (please see Fig. 8.4). The flowering of sugarcane is controlled by day length, but it is also influenced by water and nitrogen supply. Flowering has a progressive deleterious effect on sucrose content. Normally, therefore, flowering is prevented, or nonflowering varieties are used[12] [7].

Within the four main development stages of sugarcane, the required optimal climatic conditions are as follows [7]:

- The *germination stage* requires rainfall from *1100 to 1500 mm, 32–38 °C* average temperature, solar energy 18–36 MJ/m^2, and high relative humidity (80–85%). The optimal temperature is a mandatory requirement for sprouting of the stem cuttings.
- For the *tillering stage*, the climatic conditions required are like the first phase; however, water supply must be controlled to maximize growth.
- The grand growth stage needs rainfall between *750 and 1100 mm, 28–32 °C* average temperature, sunlight at 10–18 MJ/m^2, and high relative humidity of 80–87%. This stage requires high humidity for rapid cane elongation, while temperature above 38 °C and high light intensity are critical to increase the rate of photosynthesis and respiration.
- Moderate relative humidity values (40–65%) and deficiency of water supply are desirable. As the day length (photoperiod) (10–14 h) is important for sucrose, accumulation enough solar radiation (31–36 MJ/m^2) is necessary, while low temperatures of *18–30 °C led to ripening* (please see Fig. 8.5).

For soil and fertilizer needs, sugarcane does not require a special type of soil. The best soils are those that are more than 1 m deep but deep rooting to a depth of up to 5 m is possible. The soil should preferably be well-aerated (after heavy rain the pore space fills with air to >10–12%) and have a total available water content of 15% or more. When there is a groundwater table, it should be more than 1.5–2.0 m below the surface. The optimum soil pH is about 6.5 but sugarcane will grow in soils with pH in the range of 5–8.5. Sugarcane has high nitrogen and potassium needs and relatively low phosphate requirements, or 100–200 kg/ha N, 20–90 kg/ha P, and 125–160 kg/ha K for a yield of 100 ton/ha cane, but application rates are sometimes higher. At maturity, the nitrogen content of the soil must be as low as possible for a good sugar recovery, particularly where the ripening period is moist and warm.

Time Course for Growth of Stalks at Constant Temperature

Time courses for the development of the primary stalk at temperatures from 18 to 30 °C are shown in the figure below. The shape of these curves is mainly determined by the effect of temperature on the rate of increase of leaf area and not on the rate of photosynthesis, since direct measurements made in full sunlight show that photosynthesis per unit leaf area is virtually independent of temperature over the range from 9 to 34 °C. During the juvenile stage, the

(continued)

[12]Remote Sensing Applications in Sugarcane Cultivation: A Review—https://www.mdpi.com/2072-4292/13/20/4040/htm

optimum temperature for plant and stalk growth is about 30 °C. As plants approach the early adult stage, the leaf area per stalk tends to a constant value, the rate of production of new leaves being balanced by the senescence of old leaves. At this stage, the rate of stalk development also tends to a constant value which is independent of temperature, but as the stalk matures further, the rate decreases. Thus, stalks maintained at 22 °C enter early adult stage at about 200 days, attaining a growth rate between 200 and 300 days which is approximately equal to the maximum rate attained at 30 °C.

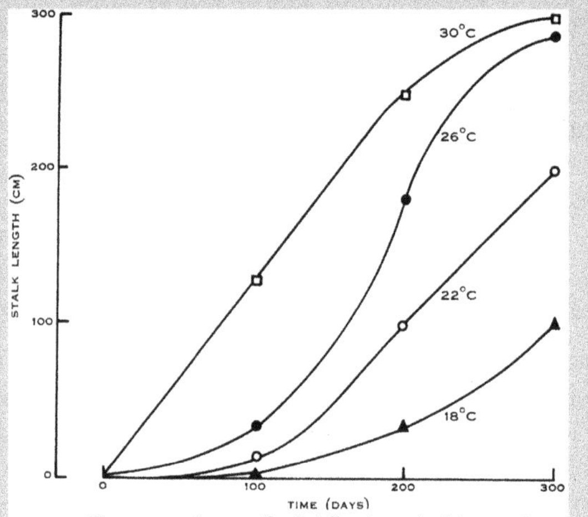

Time course for growth of stalks at constant temperatures.

8.2 Energy Production

Economic interest in sugarcane has increased significantly in recent years due to the increased worldwide demand for sustainable energy production. The Brazilian experience in sugarcane ethanol production has paved the way for the establishment of a consolidated world supply to meet the demand of a proposed ethanol addition of approximately 10% to gasoline worldwide. It is estimated that the Brazilian production of sugarcane must double in the next decade to meet this goal.

Estimated ethanol conversion[13] factors for sugar per unit of feedstock are shown in the table below with comparisons to corn. In sugarcane or sugar beet factories, cane or beet juice could be used to make ethanol rather than sugar and molasses. The 2003–2005 average raw sugar recovery factor from sugarcane in the United States was 12.26% (ERS, USDA). Based on actual raw sugar and molasses production

[13]Ethanol from Sugar—https://www.fsa.usda.gov/Internet/FSA_File/ethanol_fromsugar_july06.pdf

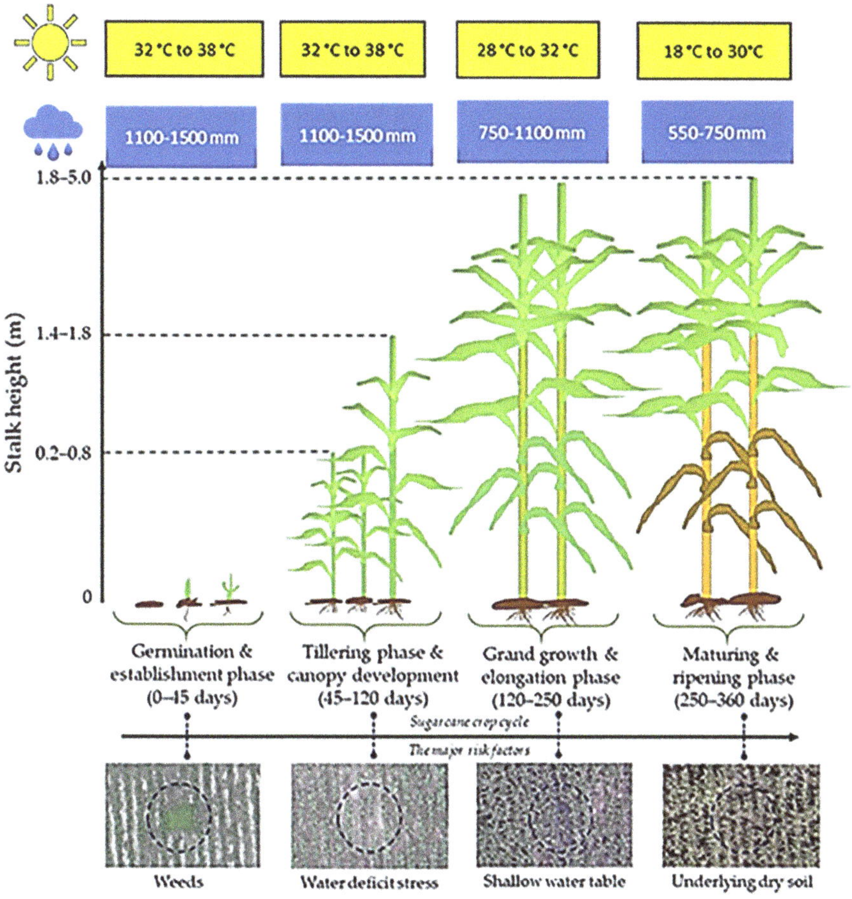

Fig. 8.5 Sugarcane phenology: temperature and precipitation

values from sugarcane, approximately 3.0 gallons of sugarcane molasses is produced as a by-product for every 100 pounds of raw sugar produced (American Sugar Cane League; Louisiana State University Agricultural Center). Sugarcane molasses have been estimated to be 49.2% sucrose (Rein). Using these recovery factors, 1 ton of sugarcane would yield approximately 277 pounds of sucrose. This would be sufficient to produce 19.5 gallons of ethanol (Table 8.1) [8].

- 1/ Based on the 2003–2005 US average raw sugar recovery rate of 12.26% per ton of cane and sucrose recovery from cane molasses at 41.6 pounds per ton of sugarcane
- 2/ Based on the 2003–2005 US average refined sugar recovery rate of 15.5% per ton of beets and sucrose recovery from beet molasses at 40.0 pounds per ton of sugar beets
- 3/ Based on an average sucrose recovery of 49.2% per gallon of cane molasses

Table 8.1 Commodity and ethanol conversion factor

Commodity	Ethanol conversion factor
Corn	98.21 gallons per ton (2.75 gallons per bushel)
Sugarcane 1/	19.50 gallons per ton
Sugar beets 2/	24.80 gallons per ton
Molasses 3/	69.40 gallons per ton
Raw sugar	135.40 gallons per ton
Refined sugar	141.0 gallons per ton

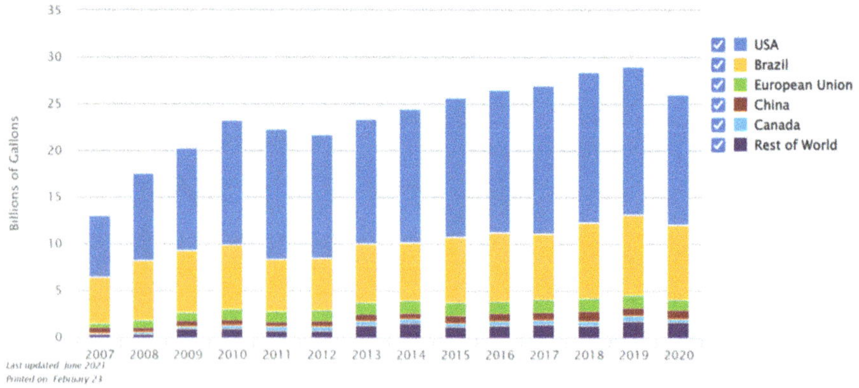

Fig. 8.6 Ethanol production

Globally, the United States and Brazil are the leading countries of ethanol production. The global ethanol production by country or region, from 2007 to 2020, is shown below (source: Energy.gov[14]/annual ethanol production[15]). Overall, global production continues to increase, but production fell worldwide in 2020 due to the COVID-19 pandemic. The United States is the world's largest producer of ethanol, having produced over 13.9 billion gallons in 2020. Together, the United States and Brazil produce 84% of the world's ethanol. The vast majority of US ethanol is produced from corn, while Brazil primarily uses sugarcane (please see Fig. 8.6).

In the future, biotechnological advances might help to reduce the environmental impacts of increased sugarcane production by developing solutions that would produce more sugarcane with decreased requirements for fertilizers and water. Among these solutions, GM sugarcane will play a key role in providing growers with more productive and resistant varieties. In Brazil, the National Biosafety Technical Commission (CTNBio) has already approved field trials with genetically modified sugarcane, incorporating traits such as increased yield, drought tolerance,

[14]Global Ethanol Production by Country or Region—https://afdc.energy.gov/data/10331

[15]Annual Ethanol Production—https://ethanolrfa.org/markets-and-statistics/annual-ethanol-production

insect resistance, and herbicide tolerance. It is expected that the enormous research and development efforts being conducted by government and private institutions will in the medium term result in the commercial release in Brazil of genetically modified sugarcane[16] [9].

8.3 Australia

Australia[17] is the sixth largest country in the world, with a land area of 7.7 million km^2, and the third largest marine jurisdiction. Australia's population is 25.7 million people (2020). Most people live in urban areas along the eastern and western coastal regions. Over 85% of Australians live within 50 km of the coastline. Australia has a strong and open economy that has grown continuously for over 25 consecutive years. Its GDP per capita was approximately $51,812 in 2020. Australia has a service-based economy with services accounting for 66% GDP (2020). The country is an exporter of natural resources, energy, and food.

The Australian sugar industry[18] produces raw and refined sugar from sugarcane. Around 95% of sugar produced in Australia is grown in Queensland and about 5% in northern New South Wales, along 2100 km of coastline between Mossman in far north Queensland and Grafton in northern New South Wales. A sugar industry was established in Western Australia in the Ord River Irrigation Area in the mid-1990s but ceased operations in 2007. More than 80% of all sugar produced in Australia is exported as bulk raw sugar, making Australia the second largest raw sugar exporter in the world. In recent years, Asia has become a major focus with key export markets including South Korea, Indonesia, Japan, and Malaysia which are our most important markets.

Queensland really encapsulates the imagery of the east coast with so many popular tourist destinations: the Whitsundays, Great Barrier Reef, Fraser Island, Surfers Paradise, and the list goes on! Getting a campervan and experiencing Queensland is a great idea, you can get a tan, do some cash fruit picking, and really explore the sunshine state.

Harvest dates	Main crops	Locations
May to Dec	Sugarcane	Ayr, Innisfail, Ingham

The Australian continent covers a large range of climate zones, from the tropics in the north to the arid interior and temperate regions in the south. Overall, Australia is

[16] Sugarcane (*Saccharum X officinarum*): A Reference Study for the Regulation of Genetically Modified Cultivars in Brazil—https://www.ncbi.nlm.nih.gov/pmc/articles/PMC3075403/

[17] Climate Change Knowledge Portal—https://climateknowledgeportal.worldbank.org/country/australia

[18] Sugar—https://www.awe.gov.au/agriculture-land/farm-food-drought/crops/sugar#:~:text=Overview,in%20northern%20New%20South%20Wales

the driest of all inhabited continents, with considerable rainfall and temperature variability both across the country and from year to year. Australia's geography, coastal population concentrations, and biodiversity render it particularly vulnerable to small variations in climate.

8.3.1 Sugarcane and Australian Economy

Australia is 13th largest producer of sugarcane (27,472.72 metric tons) and it is worth around A$1.75 billion to the Australian economy. It is estimated that the Australian sugar industry directly employs approximately 16,000 people across the growing, harvesting, milling, and transport sectors.[19] The Australian sugar industry produces both raw and refined sugar from sugarcane [10]. The industry's major product is raw crystal sugar which is sold domestically and exported. The industry is largely concentrated along Australia's eastern coastline—between Mossman in far north Queensland and Grafton in northern New South Wales. Approximately 95% of the sugar produced in Australia is grown in Queensland with the balance being grown in northern New South Wales.

The sugarcane industry is one of Australia's largest and most important rural industries and sugar has been identified as Queensland's most important rural crop. Approximately 35 million tons of sugarcane grown annually can produce up to 4.5 million tons of raw sugar, 1 million tons of molasses, and 10 million tons of bagasse. Approximately 80% of Australia's sugar production is exported as bulk raw sugar, making Australia the second largest sugar exporter in the world. Over recent years, Asian exports have become a major focus, with markets such as South Korea, Indonesia, Japan, and Malaysia becoming some of the most important. Around 85% of the raw sugar produced in Queensland is exported and generates up to $2 billion in export earnings. The majority of Australia's domestic market is supplied by sugarcane grown in New South Wales.

With over 20 different varieties of cane growing in the area, the Mossman Sugar Mill production area spans over 8500 ha all the way from the Daintree Rainforest up to Atherton Tablelands. The sugarcane flowers from early *May through June and July to November* is typically the cane harvesting season for our local sugar industry. Originally, the cane was burnt before harvesting, but nowadays in the tropical north, it is generally cut "green" (please see Table 8.2). The remaining roots then produce new shoots, and several crops may be grown from the same stock before ploughing and replanting is necessary.[20]

[19]Chapter 2: The Australian sugar industry—https://www.aph.gov.au/Parliamentary_Business/Committees/Senate/Rural_and_Regional_Affairs_and_Transport/Sugar/Report/c02

[20]Sugar Cane Harvesting—Tropical North Queensland—https://www.tropicaltours.com.au/sugar-cane-harvesting-tropical-north-queensland/#:~:text=The%20sugar%20cane%20flowers%20early,is%20generally%20cut%20'green'

Table 8.2 Australia Queensland sugar production calendar

	Flowering	Harvesting
Australia / Queensland Sugar Season	May through June	July to November

Most of Australia's sugarcane is grown in high-rainfall areas along the coastal plains and river valleys on 2100 km of the eastern coastline between Mossman in Far *North Queensland* and Grafton, NSW. Sugarcane grown in Queensland accounts for about 94% of Australia's raw sugar production, while nearly 5% is produced in Northern New South Wales and the remainder in Western Australia's Ord River Irrigation Area[21] [11].

8.3.2 World Sugar Prices

The global sugar market is less integrated (compared to the United States) than other major agricultural commodity markets, however, due to diverse and complex domestic policies of most major sugar-producing and trading countries. As a result, wholesale and retail sugar prices around the globe are influenced by local agricultural and trade policies and vary greatly from market to market, rather than simply reflecting the world futures price.

The world raw sugar price is represented by the nearby futures contract listed by the Intercontinental Exchange (ICE) (commonly referred to as the Number 11 contract). The above chart (please see Fig. 8.7) is for ICE 11 sugar prices.[22] As you can see, sugar prices soared[23] in 1974 due to adverse weather conditions [12]. Please note US domestic raw sugar price is based on the price of the nearby futures contract settlement price listed by the Intercontinental Exchange (ICE) (often referred to as the Number 16 contract).

[21] Crops ready for a different future climate—https://research.csiro.au/climate/themes/agriculture/crops-ready-different-future-climate/

[22] ICE Contract 11—Sugar—https://tradingeconomics.com/commodity/sugar

[23] Soaring Sugar Cost Arouses Consumers And U.S. Inquiries—https://www.nytimes.com/1974/11/15/archives/soaring-sugar-cost-arouses-consumers-and-u-s-inquiries-what-sent.html

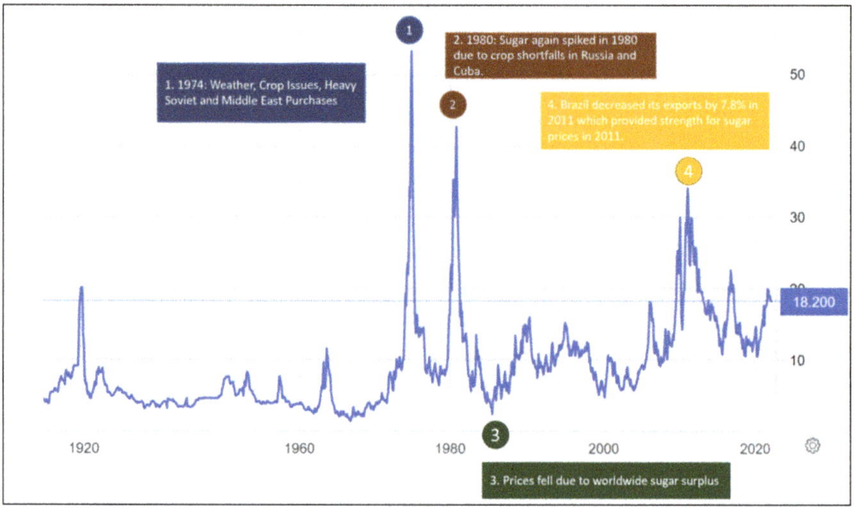

Fig. 8.7 ICE sugar futures

(1) 1974 Soared Prices	
	*Adverse weather conditions* in 1974 have sharply reduced the beet-sugar crop in Europe. Additionally, mild, dry conditions and an infectious plant disease, the "yellow virus," had cut the latest harvest in Europe by 30%, virtually assuring it will be insufficient to meet the year's demand. Finally, the surge was due to the supply gap of sugar production that would be unable to catch up before October 1975, when next year's crop is harvested.
	Commodity speculation, mostly by Middle Eastern countries, faced with a need to invest the revenues resulting from their *quadrupled oil* prices, has also driven up prices. Algeria, for example, has bought about 3 million tons R sugar and Kuwait about 2 million within the last month—equivalent to about one-third of the American consumption and considerably more than their own domestic needs.
(2) 1980 Sugar Prices[24] [13]	
	Weather caused a price spike during 1980; the Soviet Union, normally the largest sugar producer, has not received *enough rain* in its sugar beet area in the Ukraine, where lack of topsoil moisture has led to delayed plantings.[25] Late planting means that the new crop might be damaged by frosts in September before harvest. At the same time, Western Europe, another major sugar-producing area, has been afflicted with too much rain. Wet weather has also imperiled Cuba's cane crop, "where

(continued)

[24] FAO—Cuba Conference, Cuba—SUGAR MARKETS IN DISARRAY, https://www.fao.org/3/X4988E/x4988e05.htm

[25] Commodities: Price Rises For Sugar Continue—https://www.nytimes.com/1983/06/06/business/commodities-price-rises-for-sugar-continue.html

	torrential rains have delayed the harvest by 6–8 weeks." The crop in South Africa, which has endured one of the worst droughts in its history, has been cut sharply. As a producer, it found itself overcommitted by some 600,000 tons. It has gone into the market to buy sugar for its customers or has asked buying nations like Japan, Canada, and South Korea to seek sugar elsewhere. Drought also has hampered crops in Australia, the Philippines, and Thailand [14].
(3) 1985 Sugar Prices—Fell	
	The price of sugar futures fell to new lows in April 21, 1985, as the worldwide *surplus* weighed on the market.[26] Sugar prices fell again, bottoming in May 1985 at less than US$0.03 per pound and averaging US$0.04 per pound for the year. In early 1990, prices had recovered to US$0.14 cents per pound and then declined to approximately US$0.08 to US$0.09 per pound[27] [15, 16].
(4) 1984 Spike in Sugar Prices [17]	
	Brazil is the largest sugar-producing country in the world. The production of sugar has increased 352% since 1990. About 34% of Brazilian sugar consumed domestically is converted into ethanol for fuel. Exports have risen from 1.2 million metric tons in 1990 to 26.9 million metric tons in 2011. Sugar that is converted into ethanol is subsidized at prices higher than the world price. Recent increases in the world oil price have increased the price of ethanol[28] which in turn increased Brazil's conversion of sugar into ethanol, reducing potential sugar exports from Brazil. That reduction in the growth of sugar exports could be one of the main forces for world sugar price increases. Brazil decreased its exports by 7.8% in 2011 which provided strength for sugar prices in 2011 [17].

In essence, the above weather and market signals have a direct impact on food security. In summary, weather conditions, oil and gas prices, weather-induced spikes, insufficient rain, and biofuel demand could induce higher costs to food commodities and thus impact food security.

8.3.3 Sugar Demand

Since 1971, the worldwide sugar consumption has exceeded worldwide sugar production. No narrowing of the gap is expected for at least a year, and consequently, prices are expected to continue to increase. The monthly world price averaged about 19.23 cents a pound between 2009/2010 and 2016/2017 but has ranged between 13.42 cents per pound in 2014/2015 and 28.42 cents per pound in

[26] FUTURES/OPTIONS; Sugar Surplus Sends Prices Plummeting—https://www.nytimes.com/1985/04/24/business/futures-options-sugar-surplus-sends-prices-plummeting.html

[27] Philippine—Sugar—http://countrystudies.us/philippines/64.htm

[28] 2012 Outlook of the U.S. and World Sugar Markets, 2011–2021—https://ageconsearch.umn.edu/record/128037/files/AAE692.pdf

2010/2011. The world refined sugar price is also measured by the nearby futures contract settlement prices in the ICE—often referred to as the Number 5 contract. Refined prices are higher than raw sugar prices, reflecting the cost of refining and storing sugar to a higher polarity for human consumption. World refined sugar prices have averaged about 23.75 cents a pound from 2009/2010 to 2015, ranging between 17.07 cents per pound in 2014/2015 and 32.63 cents per pound in 2010/2011.

8.3.4 Climate Change

Australia is experiencing higher temperatures, more extreme droughts, fire seasons, floods, and more extreme weather due to climate change. Rising sea levels add to the intensity of high-sea-level events and threaten housing and infrastructure. Drought is a frequent visitor in Australia.[29] The Australian Bureau of Meteorology describes the typical rainfall over much of the continent as "not only low, but highly erratic." Over the last decade, drought has dominated Australian weather. Between 2007 and 2009, the southeastern states of Victoria and New South Wales endured 3 years of drought, which broke only when La Niña rains drenched the region in late 2010.

The number of days that breaks heat records has doubled in the past 50 years. Heatwaves are of particular concern: they are occurring more often and are more intense than in the past. In recent decades more people have died in Australia in heat waves than all other natural disasters combined. Some parts of Australia—inland areas particularly—are expected to warm faster than along the coasts.[30]

Higher temperatures create a range of extreme weather and climate events (please see Fig. 8.8): longer droughts in some areas of the continent and, in others, heavier rainstorms due to greater evaporation. Marine heatwaves are on the rise devastating Australia's kelp forests, seagrass meadows, coral reefs, and all the underwater creatures that depend on them. These impacts also affect humans—creatures and habitats are sources of food and income. Coral bleaching has increased in frequency and severity on the Great Barrier Reef. It is now occurring so frequently that large areas are unlikely to ever recover.

As the oceans absorb not just heat but also excess carbon from the atmosphere, oceans and seas have become more acidic. This acidity reduces the capacity of crustaceans, hard corals, and coralline algae to draw out calcium carbonate from the water, to grow and strengthen their skeletons.

---

[29] Drought is a frequent visitor in Australia—https://earthobservatory.nasa.gov/world-of-change/AustraliaNDVI

[30] Climate Change—https://australian.museum/learn/climate-change/climate-change-impacts/#:~:text=Australia%20is%20experiencing%20higher%20temperatures,in%20the%20past%2050%20years

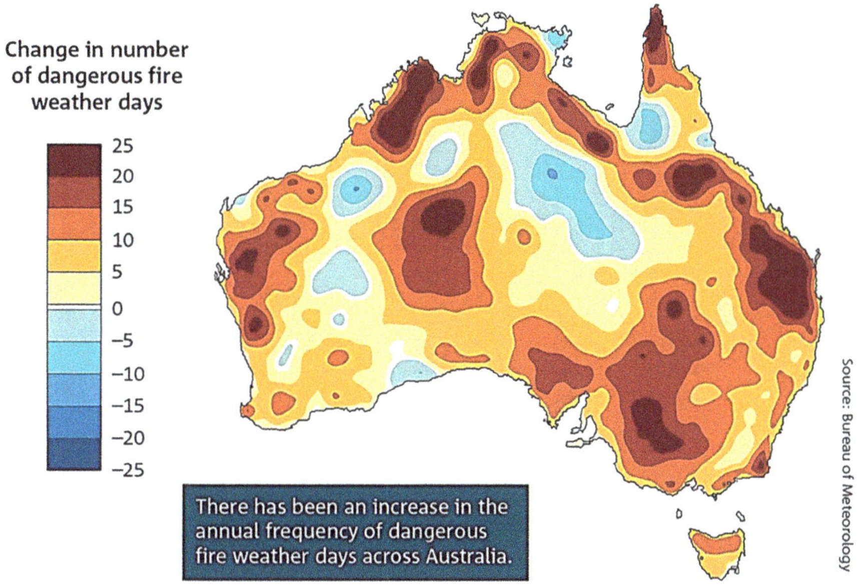

Fig. 8.8 Australia climate events

Australia's changing climate represents a significant challenge to individuals, communities, governments, businesses, industry, and the environment. Australia has already experienced increases in average temperatures over the past 60 years, with more frequent hot weather, fewer cold days, shifting rainfall patterns, and rising sea levels. More of the same is expected in the future.[31]

The projections are presented for eight distinct regions of Australia, each of which will be affected differently by climate change. Climate change brings social, environmental, and economic impacts.

The projections are based on data from up to 40 global climate models, developed by institutions around the world, that were driven by four greenhouse gas and aerosol emission scenarios. Results have been prepared for 21 climate variables (both on the land and in the ocean) and for four 20-year time periods (centered on 2030, 2050, 2070, and 2090).

[31] Climate Change—https://www.csiro.au/en/research/environmental-impacts/climate-change/climate-change-information

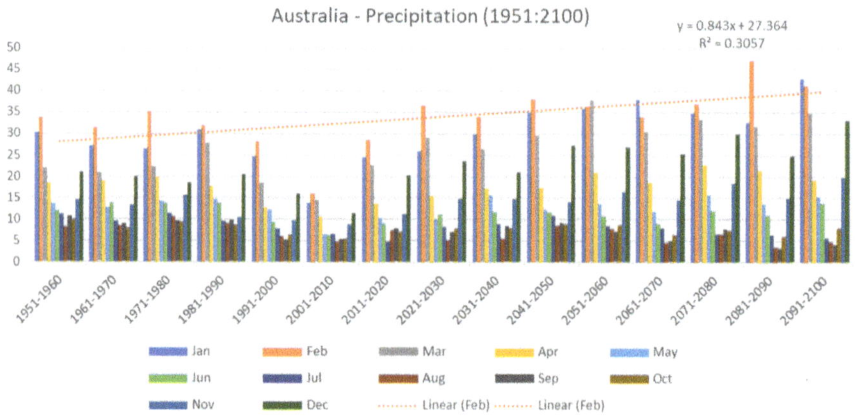

Fig. 8.9 Australia precipitation 1951–2020

> "Our new extremes of heat and other severe weather mean we now need to re-imagine how our towns and city's function, ensure we provide essential climate safety services, and rethink how we go about our daily lives and care for others."
> Australian Climate Council

8.3.5 *Precipitation and Rainfall*

Seasonal-average rainfall changes will vary across Australia (please see Fig. 8.9).

- Later in the century (2090), the precipitation increases. In the above graph, for the month of February, it is highly likely for precipitation to have a positive slope.
- Except for the months of June, July, August, September, and October, all other months would see an increase in precipitation, late in the century (2090).

8.3.6 *Maximum Temperature*

As a nation, Australia will be hotter by 6 °C by the end of the century. Seasonal temperatures will trend higher and will result in vegetation stress (please see Fig. 8.10).

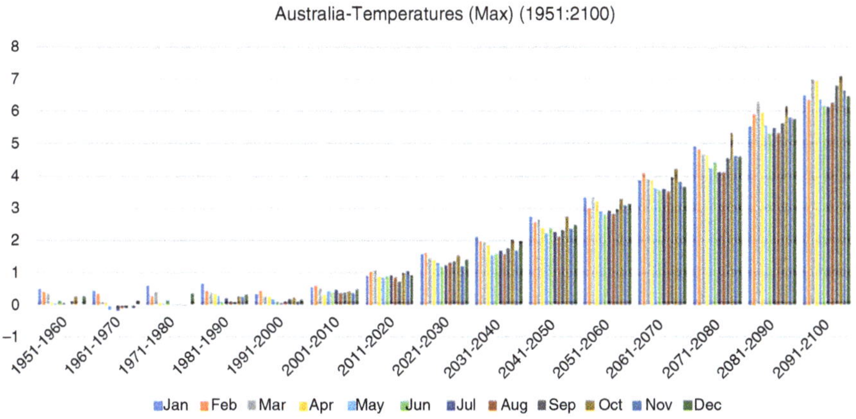

Fig. 8.10 Australia max temperatures (1951–2020)

The above temperature change was captured for Projections Data CMIP6, Scenario SSP 5-8.5, and multi-model ensembles. Higher temperatures[32] are not good for the country and to the rest of the world [18].

Month	Trend line
January	$y = 0.9803x + 22.491R^2 = 0.3942$
February	$y = 0.843x + 27.364R^2 = 0.3057$
March	$y = 1.1165x + 17.889R^2 = 0.5887$
April	$y = 0.2885x + 15.676R^2 = 0.148$
May	$y = 0.1187x + 12.034R^2 = 0.0465$
June	$y = -0.0577x + 11.921R^2 = 0.0143$
July	$y = -0.2615x + 10.438R^2 = 0.3369$
August	$y = -0.2852x + 9.121R^2 = 0.4008$
September	$y = -0.345x + 10.086R^2 = 0.4642$
October	$y = -0.1066x + 8.6962R^2 = 0.1398$
November	$y = 0.3975x + 11.057R^2 = 0.3379$
December	$y = 0.9123x + 15.398R^2 = 0.5476$

[32] Climate change: Australia pledges net zero emissions by 2050—https://www.bbc.com/news/world-australia-59046032

8.3.6.1 NOAA: Advanced Very-High-Resolution Radiometer (AVHRR) and Climate Change Parameters

The Vegetation Condition Index (VCI)[33] evaluates the current vegetation health in comparison to the historical trends. The VCI relates the current dekadal Normalized Difference Vegetation Index (NDVI) to its long-term minimum and maximum, normalized by the historical range of NDVI values for the same dekad. The VCI was designed to separate the weather-related component of the NDVI from the ecological element.

Here is 2021 VCI for Australia: Queensland:[34] Especially for the weeks that have the model explainability from the sugarcane production point of view. Consider VCI for May through October 2021 (please see Fig. 8.11).

$$VCI = 100 \times (NDVI - NDVI_{min})/(NDVI_{max} - NDVI_{min}) \qquad (8.1)$$

where $NDVI_{max}$ and $NDVI_{min}$ are the multiyear absolute maximum and minimum of NDVI.

8.3.7 *Vegetation Index*

Vegetation Indices[35] (VIs) are used to monitor and characterize terrestrial landscapes; Vegetation Indices are related to the absorption of *photosynthetically*[36] active radiation by vegetation and correlate with biomass or primary productivity.

- Spectral Vegetation Indices (VIs) are optical measures of vegetation canopy "greenness," a composite property of the following:
  - Leaf chlorophyll
  - Leaf area
  - Canopy cover
  - Canopy architecture
- VIs are widely used in studies involving vegetation dynamics:
  - Land surface phenology
  - Climate–vegetation interactions
- VIs can be used to produce estimates of:

---

[33] Australia—https://www.fao.org/giews/earthobservation/country/index.jsp?lang=en&code=AUS

[34] Australia—https://www.fao.org/giews/earthobservation/country/index.jsp?lang=en&code=AUS

[35] Vegetation Index—https://www.star.nesdis.noaa.gov/jpss/vi.php

[36] MODIS Photosynthesis—https://www.ntsg.umt.edu/files/modis/ATBD_MOD17_v21.pdf

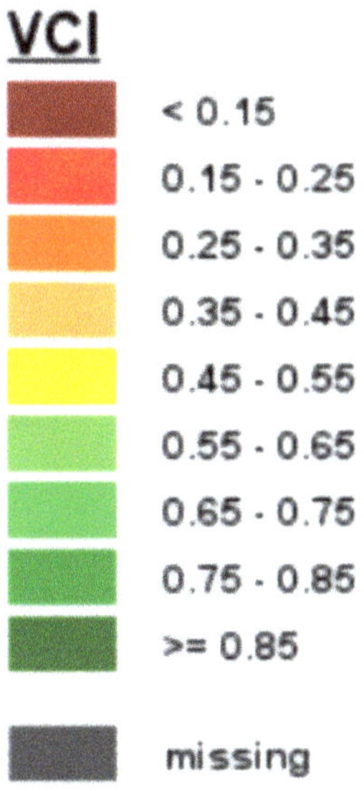

Fig. 8.11 VCI legend

- Leaf Area Index (LAI)
- Fraction of photosynthetically active radiation (fPAR)
- Net photosynthesis (PSN)
- Annual net primary production (NPP)
- Green vegetation fraction (GVF)

For the purpose of identifying photosynthesis, we have used VI (please see table below).

May 2021[37] 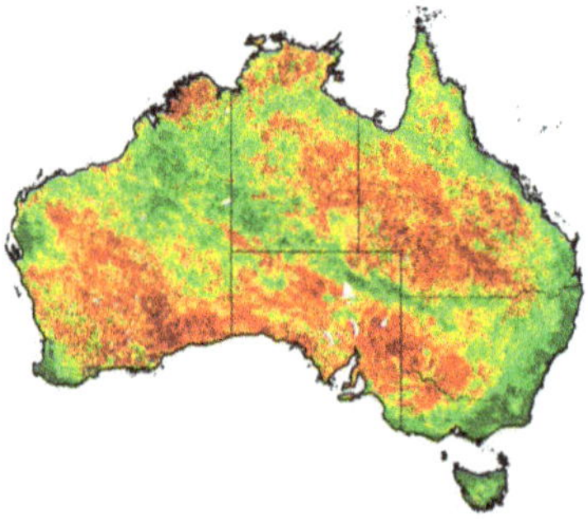	June 2021[38] 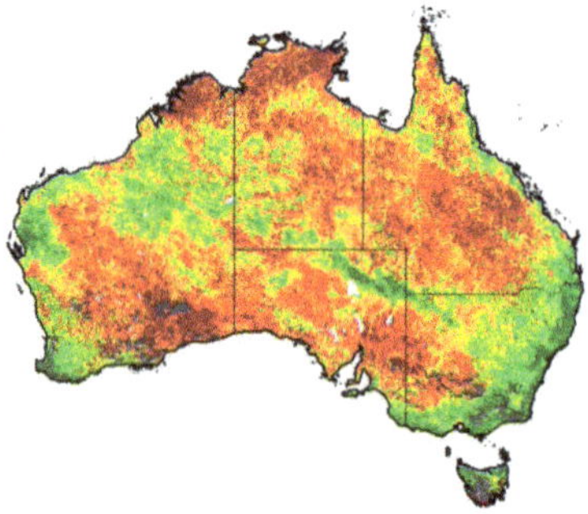
July 2021[39] 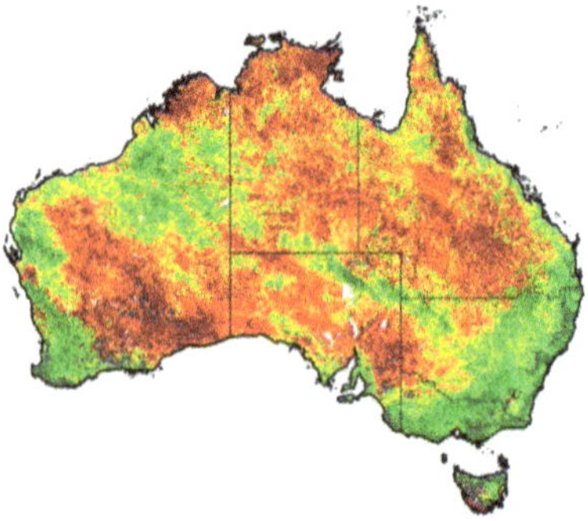	August 2021[40] 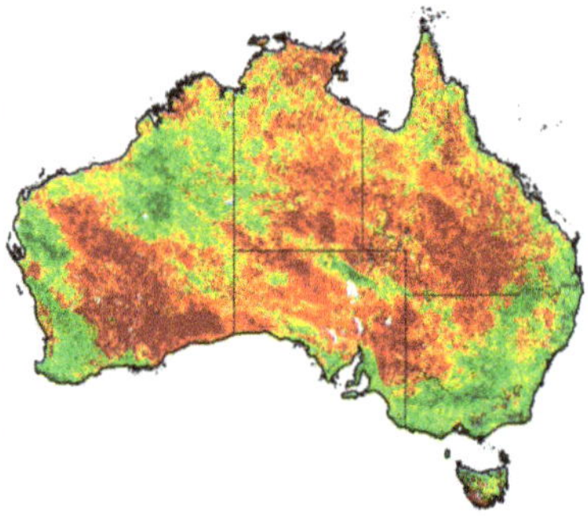

(continued)

[37] Australia, May 2021 VCI—https://www.fao.org/giews/earthobservation/asis/data/country/AUS/MAP_NDVI_ANOMALY/HR/om2105c.png

[38] Australia, June 2021—https://www.fao.org/giews/earthobservation/asis/data/country/AUS/MAP_NDVI_ANOMALY/HR/om2106c.png

[39] Australia, July 2021—https://www.fao.org/giews/earthobservation/asis/data/country/AUS/MAP_NDVI_ANOMALY/HR/om2107c.png

[40] Australia, August 2021—https://www.fao.org/giews/earthobservation/asis/data/country/AUS/MAP_NDVI_ANOMALY/HR/om2108c.png

Setptember 2021[41]	October 2021[42]
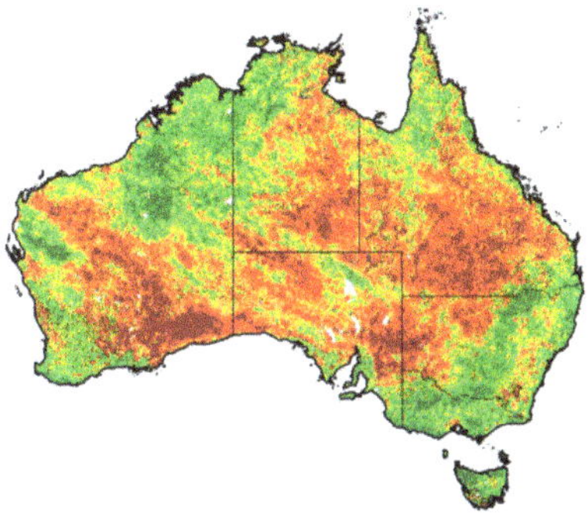	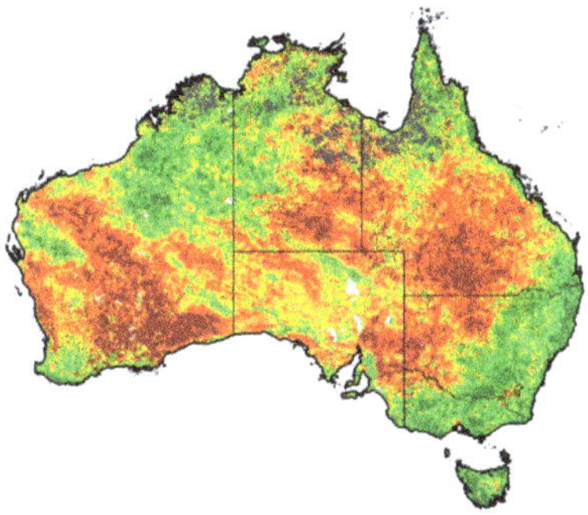

8.3.7.1 Province Averaged VH

Please find Australia, Queensland, Vegetation Health Data 1990:2021.[43] Select sugarcane crop. Other parameters include VCI/TCI/VHI: VHI, select years: 1990–2021 (please see Fig. 8.12).

Please find Weekly Averaged Time Series for Province: Queensland of Australia: Averaged VHI:[44]

As can be seen (please see Fig. 8.13), the VHI (y) score is negatively correlated with x (years), especially for drought years. For instance, the 1982–1983 drought;[45] the Millennium drought, 1997 to 2009 (acute dry 2002–2003); and 2017–2019.

$$Y_{1982} = 0.6786x + 20.2478 \tag{8.2}$$

$$Y_{2002} = 0.2796x + 10.0921 \tag{8.3}$$

$$Y_{2018} = 0.1965x + 4.6112 \tag{8.4}$$

$$Y_{2019} = 0.3165x + 11.3283 \tag{8.5}$$

[41] Australia, September 2021—https://www.fao.org/giews/earthobservation/asis/data/country/AUS/MAP_NDVI_ANOMALY/HR/om2109c.png

[42] Australia, October 2021—https://www.fao.org/giews/earthobservation/asis/data/country/AUS/MAP_NDVI_ANOMALY/HR/om2110c.png

[43] VH Score—https://www.star.nesdis.noaa.gov/smcd/emb/vci/VH/vh_adminMeanByCrop.php?type=Province_Weekly_PAreaPlot

[44] Nampula of Mozambique: Averaged VHI—https://www.star.nesdis.noaa.gov/smcd/emb/vci/VH/vh_adminMeanByCrop.php?type=Province_Weekly_MeanPlot

[45] Australia Previous Droughts—http://www.bom.gov.au/climate/drought/knowledge-centre/previous-droughts.shtml

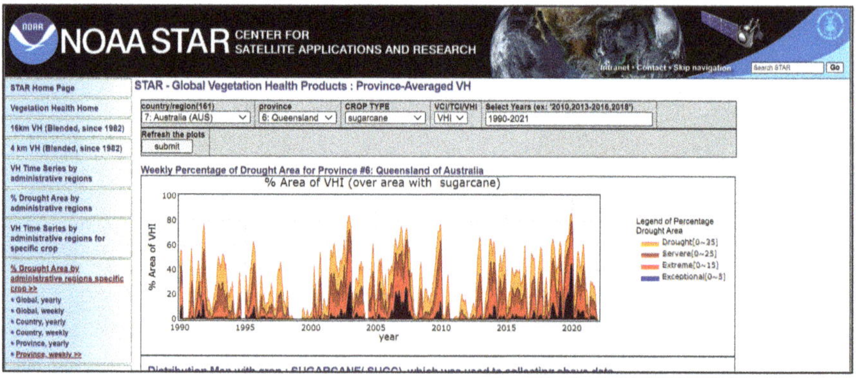

Fig. 8.12 NOAA

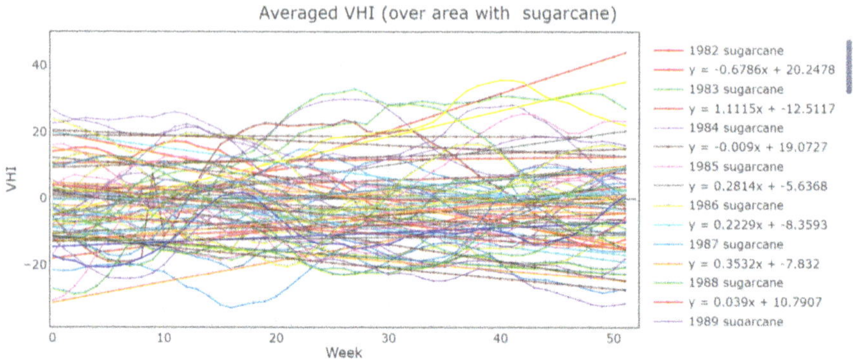

Fig. 8.13 Weekly averaged time series for province #6 Queensland

Drought is a frequent visitor in Australia.[46] In the 2017–2019 drought, after a particularly wet winter and spring in 2016 over much of Australia, conditions turned dry in 2017. Rainfall in most of the Murray–Darling Basin was substantially below average in 2017, 2018, and 2019. Though these years saw the most widespread dry conditions, in some parts of the country, such as western Victoria and western Queensland, dry conditions had been observed during 2012–2015 (please see Fig. 8.14).

The VHI is trending downward with a negative slope, indicative of lower scores of VHI (please see Fig. 8.15).

$$Y_{\mathrm{VHI}} = 0.4273x + 902.2338 \tag{8.6}$$

[46]Droughts in Australia—https://earthobservatory.nasa.gov/world-of-change/AustraliaNDVI

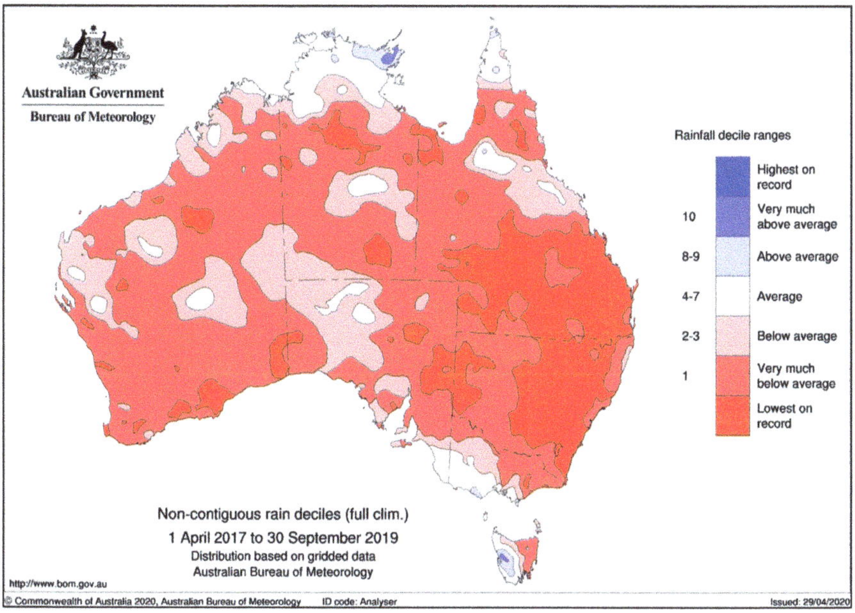

Fig. 8.14 Australia rain

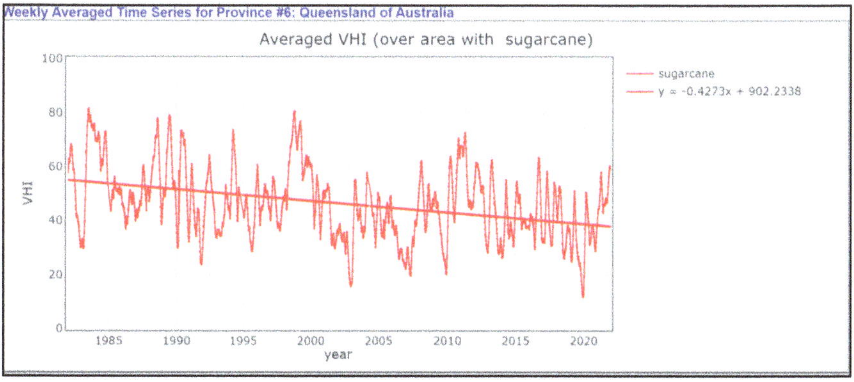

Fig. 8.15 Queensland VHI

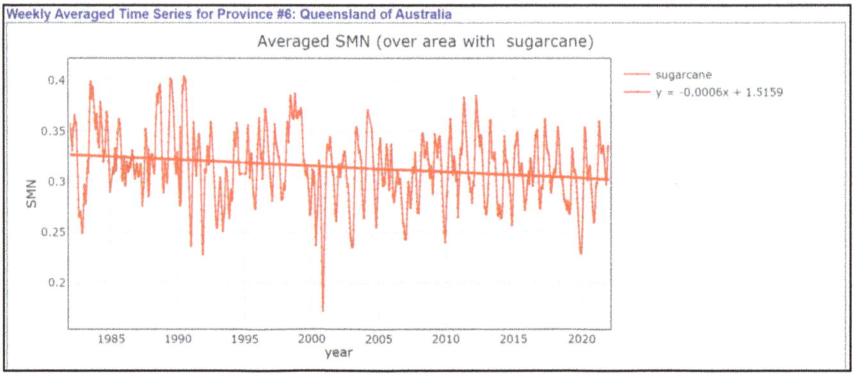

Fig. 8.16 Queensland SMN

8.3.7.2 No Noise (Smoothed) Normalized Difference Vegetation Index (SMN)

The global, 4-km, 7-day composite is validated. The SMN is derived from no noise NDVI, in which components were pre- and postlaunch calibrated. SMN can be used to estimate the start and senescence of vegetation, start of the growing season, and phenological phases (please see Fig. 8.16).

$$Y_{\mathrm{SMN}} = 0.0006x + 1.5159 \tag{8.7}$$

The average SMN, greenness[47] (no noise NDVI), is trending negative (-0.0006x). Over the years, as per the above graph, the greenness is reduced, albeit at a slower slope (Eq. 8.7).

The overall global harvested crop area for sugarcane.[48] Australia, Queensland, is a dominant area; additional countries include Brazil, India, middle-and-central Africa, and China (please see Fig. 8.17).

[47] Mozambique, Greenness (No Noise NDVI)—https://www.star.nesdis.noaa.gov/smcd/emb/vci/VH/vh_browseByCountry.php

[48] Global Sugarcane Harvested Areas—https://www.star.nesdis.noaa.gov/smcd/emb/vci/crop/SPAM_2010/HavestedArea_GT0/global_HarvestedArea__SUGC.png

Australia sugarcane	

8.4 Machine Learning Model: Australia Sugarcane Production Model

Let us build the Queensland, Australia, sugarcane production model:

	Software code for this model: Australia_Sugarcane_code.ipynb (Jupyter Notebook code)

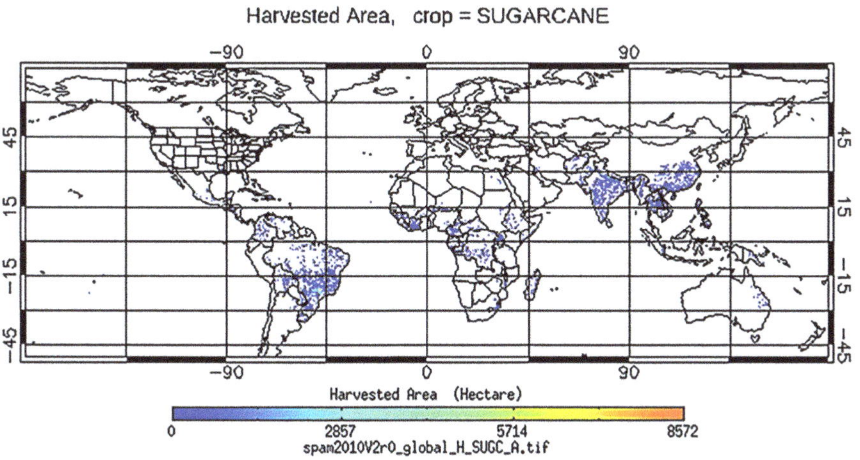

Fig. 8.17 Global sugarcane production

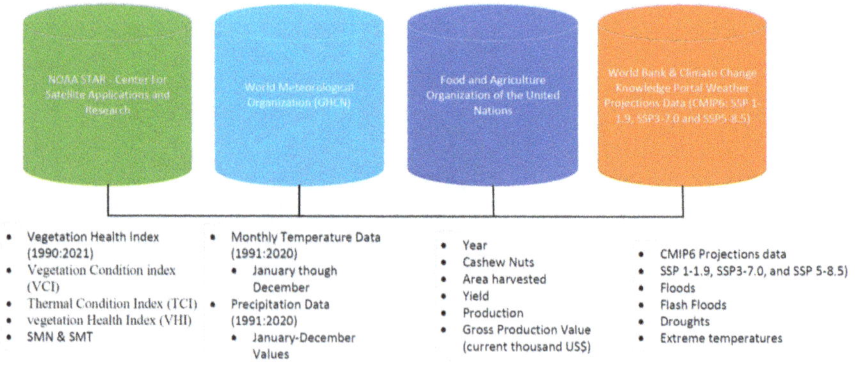

Fig. 8.18 Data sources

8.4.1 Data Sources

The data sources for the ML model is from FAO Statistics Data,[49] World Bank Climate Change Data specifically Australia–Queensland,[50] World Metrological Organization Data,[51] FAO Earth Observation,[52] and NOAA STAR–STAR–Global Vegetation Health Products: Province-Averaged VH[53] (please see Fig. 8.18). Please refer to Appendix C for world data sources, Appendix D for the US data sources, and Appendix E for economics data sources.

The Australia weather calendar data, additionally, were collected from FAO GIEWS. The major sowing, growing, and harvesting months include (as described above):

	Flowering	Harvesting
Australia/Queensland sugar season[54]	May through June	July to November

[49]FAO Statistics data—https://www.fao.org/faostat/en/#data/QCL

[50]Climate Change Data—https://climateknowledgeportal.worldbank.org/download-data

[51]World Metrological Organization Data—https://climatedata-catalogue.wmo.int/explore

[52]Earth Observation Mozambique—https://www.fao.org/giews/earthobservation/country/index.jsp?lang=en&code=AUS

[53]NOAA STAR—VHI—https://www.star.nesdis.noaa.gov/smcd/emb/vci/VH/vh_adminMeanByCrop.php?type=Province_Weekly_MeanPlot

[54]Sugar Cane Harvesting—Tropical North Queensland—https://www.tropicaltours.com.au/sugar-cane-harvesting-tropical-north-queensland/#:~:text=The%20sugar%20cane%20flowers%20early,is%20generally%20cut%20'green'

Australia has witnessed climate change and floods during the last decades[55,56] (as described above).

8.4.2 Step 1: Import Libraries

Import open source and machine learning libraries to process the data. For graphics, Matplot Lib is used and ELI5 for Explainability. ELI5[57] is a Python library which allows you to visualize and debug various machine learning models using unified API.

```
import pandas as pd
import numpy as np
from sklearn.linear_model import LinearRegression
from sklearn.model_selection import StratifiedKFold
from sklearn.metrics import mean_squared_error
from math import sqrt
import random
from sklearn.impute import SimpleImputer
from sklearn.experimental import enable_iterative_imputer
from sklearn.impute import IterativeImputer
import seaborn as sns
import matplotlib.pyplot as plt
import seaborn as sn

import plotly.graph_objects as go
import seaborn as sns; sns.set()
import matplotlib.pyplot as plt
from scipy import stats
from scipy.stats import pearsonr
import matplotlib.pyplot as plt
from sklearn.preprocessing import MinMaxScaler
from sklearn.model_selection import train_test_split
from sklearn.linear_model import LinearRegression
from sklearn.ensemble import RandomForestClassifier
from sklearn.metrics import r2_score
from sklearn import metrics
import eli5
from statsmodels.tsa.stattools import adfuller
from statsmodels.graphics.tsaplots import plot_pacf
```

(continued)

[55] Australia—https://reliefweb.int/disasters?search=Australia

[56] Australia Previous Droughts—http://www.bom.gov.au/climate/drought/knowledge-centre/previous-droughts.shtml

[57] ELI5—https://eli5.readthedocs.io/en/latest/

```
from statsmodels.graphics.tsaplots import plot_acf
from sklearn.metrics import mean_squared_error
from sklearn.model_selection import train_test_split
from statsmodels.tsa.arima_model import ARIMA
from statsmodels.tsa.stattools import grangercausalitytests
from statsmodels.tsa.api import VAR
import lime
import lime.lime_tabular
```

8.4.3 Step 2: Load Australia Sugarcane Production: FAO Dataset 1991–2020

```
dfAustraliaSugarcaneProductionData              =          pd.read_csv
('AustraliaSugarcaneProduction.csv')
dfAustraliaSugarcaneProductionData.tail()
```

Output:
Please find output: as you can see, it is sugarcane production.

	Year	Areaharvested(ha)	SugarcaneProduction(tonnes)	Yield(hg/ha)
55	2016.0	447204	34403004	769291
56	2017.0	453470	36561497	806261
57	2018.0	442958	33506830	756434
58	2019.0	395399	32415352	819814
59	NaN	366426	30283457	826455

As it can be seen, Queensland, Australia, had a peak production in 1997—39,938,000 tons with 401,000 ha area harvested.

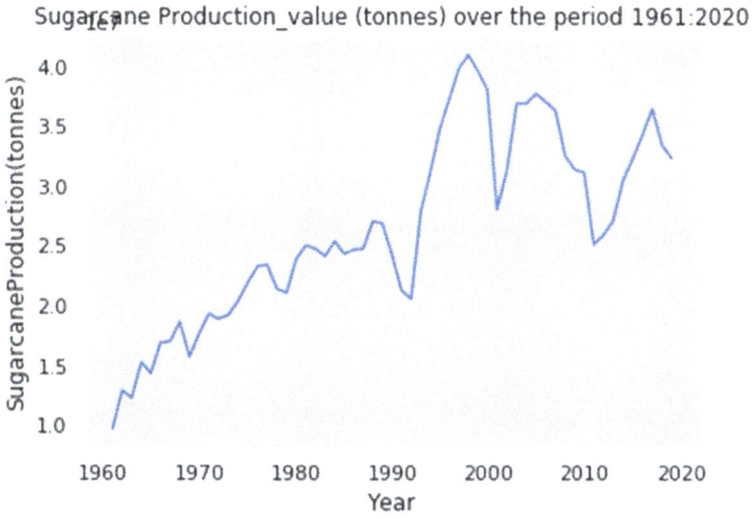


```
import matplotlib.pyplot as plt
import numpy as np

# Data for plotting
t = dfAustraliaSugarcaneProductionData['dupYear']
s = dfAustraliaSugarcaneProductionData['SugarcaneProduction
(tonnes)']

fig, ax = plt.subplots()
ax.plot(t, s)

ax.set(xlabel='Year', ylabel='SugarcaneProduction(tonnes)',
   title='Sugarcane Production_value (tonnes) over the period
1961:2020')
ax.grid()

plt.show()
```

As can be seen in the sugarcane production:

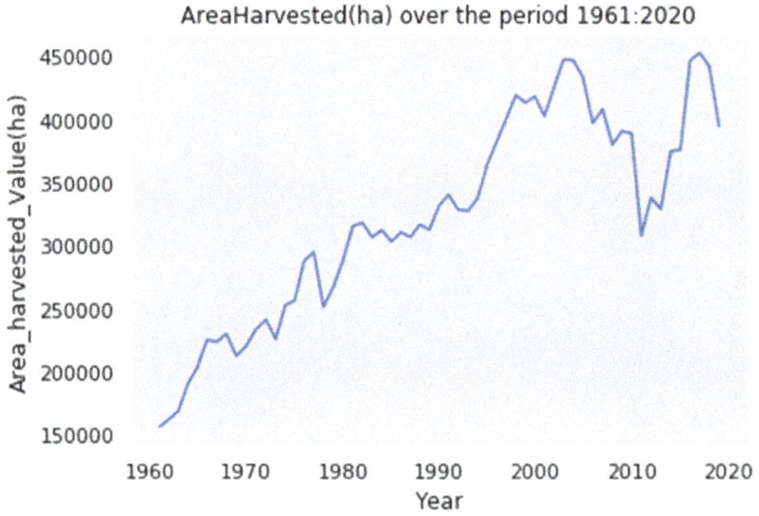

And finally, yield rates peaked in the late 1990s and early 2000. The profitability of cane farming is directly linked to production, so farmers know better than anyone the impact of declining yields. At the same time, average yields have either plateaued or fallen,[58] and this is a concerning trend that must be overcome if the viability of the industry is to be assured [19].

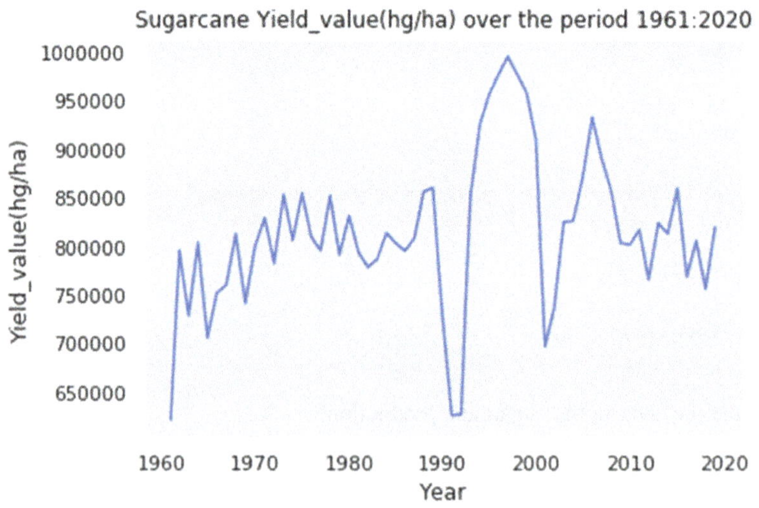

[58] Sugar production figures for 2019 down—https://www.weeklytimesnow.com.au/news/rural-weekly/sugar-production-figures-for-2019-down/news-story/4a301a678336cdaaa11842933e528c6b

Queensland sugarcane crop for 2019 was 28.4 million tons, down to 6.7% from 2018s 30.4 million tons and some 6 million tons less than the 2016 crop of 34.4 million tons (-17.4%). The crop area harvested was also down to 12,000 ha across the state's sugar regions last year, 2018. There were 350,082 ha under cane in 2019 compared to 362,414 ha in 2018. The figures also showed cane yield in terms of tons per hectare in 2019 was lower than the average result for the past 8 years (2012–2018) across every Queensland sugarcane production region.

8.4.4 *Step 3: Prepare DataFrame with Weather Data (Average Temperature, Min/Max Temperatures, and Precipitation)*

Please find average temperature load and plot the average temperatures:

```
dfAustraliaQueenslandSugarcaneCashewProductionMeanTemp = pd.
read_csv('tas_timeseries_monthly_cru_1901-2020_AUS_473.csv')
dfAustraliaQueenslandSugarcaneCashewProductionMeanTemp.head()
```

Plot average temperatures for Queensland for 1900–2020.

```
fig, axes = plt.subplots(nrows=6, ncols=2, dpi=120, figsize=(10,6))
for i, ax in enumerate(axes.flatten()):
  fig, axes = plt.subplots(nrows=6, ncols=2, dpi=120, figsize=
(10,6))
for i, ax in enumerate(axes.flatten()):
  data = dfAustraliaQueenslandSugarcaneCashewProductionMeanTemp
[dfAustraliaQueenslandSugarcaneCashewProductionMeanTemp.
columns[i]]
  ax.plot(data, color='red', linewidth=1)
  ax.set_title
(dfAustraliaQueenslandSugarcaneCashewProductionMeanTemp.
columns[i])
  ax.xaxis.set_ticks_position('none')
  ax.yaxis.set_ticks_position('none')
  ax.spines["top"].set_alpha(0)
  ax.tick_params(labelsize=6)
plt.tight_layout();
```

Output:

On average, in1900–2020, as can be seen, January and February have higher average temperatures compared to the other months of the year.

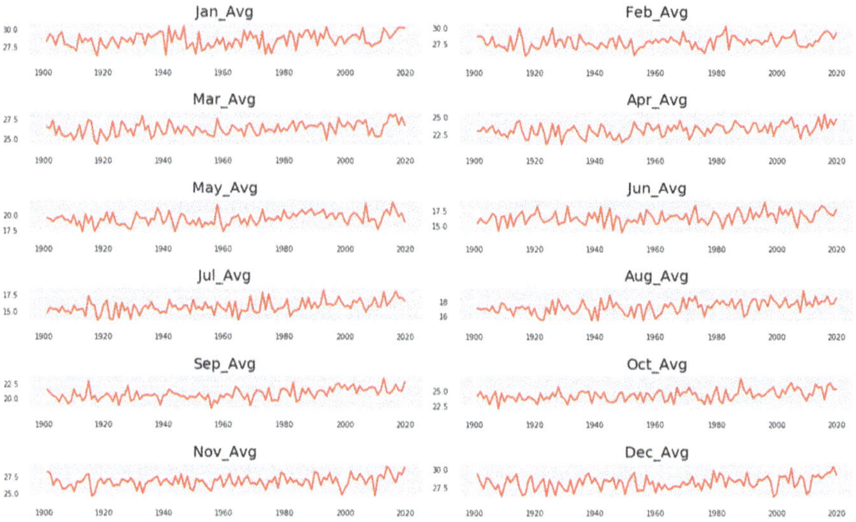

Please find minimum temperature details:

```
dfAustraliaQueenslandSugarcaneCashewProductionTeamMin = pd.
read_csv('tasmin_timeseries_monthly_cru_1901-2020_AUS_473.
csv')
dfAustraliaQueenslandSugarcaneCashewProductionTeamMin.head()
```

Output: (plotting the minimum temperature in DataFrame) July[59] (10 °C) and August (12.5 °C) have, on an average, minimum temperatures. The minimum threshold temperature[60] for cane growth is 16 °C.

[59] Australia: Queensland—Coldest Month—https://www.timeanddate.com/weather/australia/brisbane/climate

[60] Weather Conditions Suitable for Sugarcane Crop—http://indiaagronet.com/community/specificfarmers/Weather%20conditions.htm

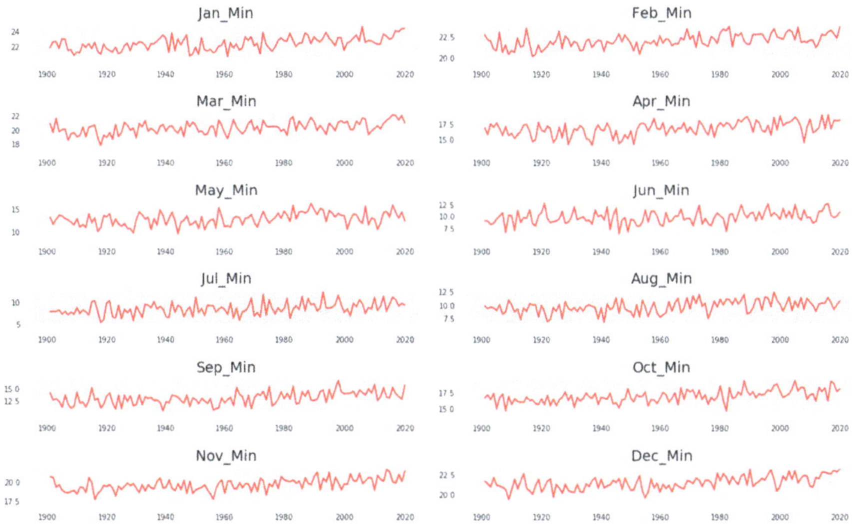

Sugarcane is a warmth-loving crop, and restrictions in growing season can arise due to insufficient soil warmth for germination and checking of growth of the cane stalk by low air temperature. The minimum soil temperature for the germination of sugarcane sets is in the range of 19–21 °C; the optimum range is from 27 to 38 °C, while temperatures above 38 °C are not conducive. Though temperatures up to 43 °C are tolerated under high soil moisture status, the weather situation is conducive for borer pest attack and sprouts suffer extensive damage under these conditions. The effects of low root temperatures are operative even with warm air temperatures. Warm air temperatures when root temperatures are low lead to nitrogen starvation of leaves and poor juice quality. Root temperatures exert modifying influence on the effect of air temperatures on the rate of node formation. The minimum threshold temperature for cane growth is 16 °C. A check in vegetative growth occurs when the mean daily air temperature drops below 21 °C. Thus, low air temperatures are most effective in inducing ripening and lead to better juice quality. For flower development, higher night temperatures may be required. The optimum temperature for sucrose synthesis in leaves is 30 °C.

Please find the maximum temperature values:

```
dfAustraliaQueenslandSugarcaneCashewProductionTeamMax = pd.
read_csv('tasmax_timeseries_monthly_cru_1901-2020_AUS_473.
csv')
dfAustraliaQueenslandSugarcaneCashewProductionTeamMax.head()
```

Output plot: Historically, December and January are the hottest months[61] (37.5°C).

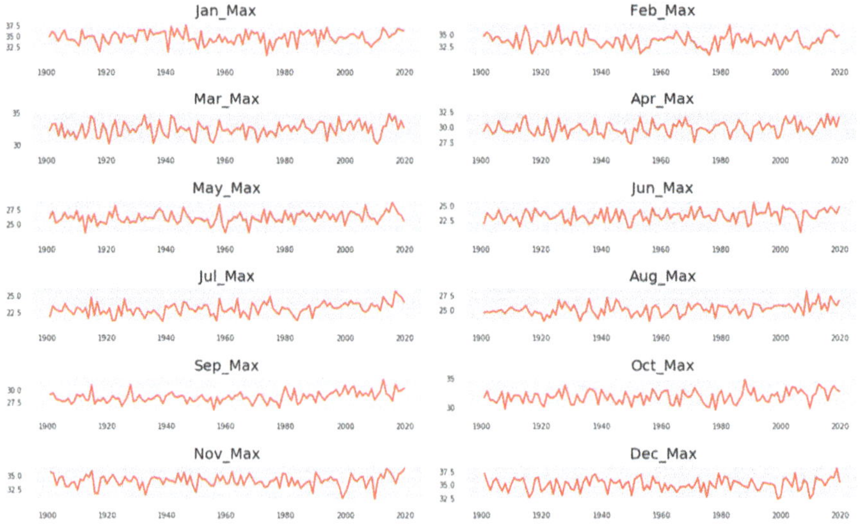

Please find precipitation values:

```
dfAustraliaQueenslandSugarcaneCashewProductionPrecp = pd.
read_csv('pr_timeseries_monthly_cru_1901-2020_AUS_473.csv')
dfAustraliaQueenslandSugarcaneCashewProductionPrecp.head()
```

Output:

Drier than normal conditions prevailed in many inland and southern parts of Queensland over the period of 2013–2019. In 2016, rainfall was above average in western Queensland and below average in some northern and southeastern areas. Although the long-term (1900–2015) statewide average annual rainfall for Queensland is 616 mm, this is highly variable on an annual and a decadal basis. From mid-2015 to early 2017, more than 80% of the state was affected by drought declared under state government processes, and more than 65% of the state remained affected by drought declared at the end of 2019 (please see Fig. 8.19). Generally, sugarcane needs annual precipitation of 1000–1500 mm; during early development, we can witness precipitation stress.[62]

---

[61] Queensland Islands Weather and Climate—https://www.queenslandislands.com/weather.html

[62] Cashew Cultivation—https://agricoop.nic.in/sites/default/files/Cashewnut%20Cultivation%20%281%29.pdf

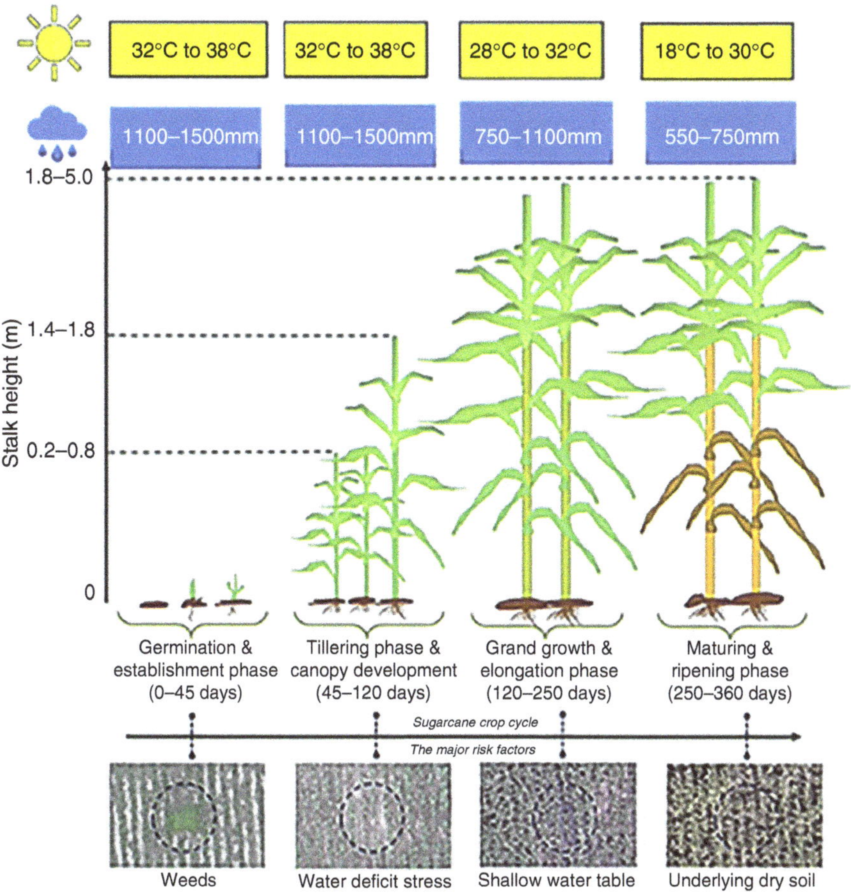

Fig. 8.19 Sugarcane phenology: temperature and precipitation

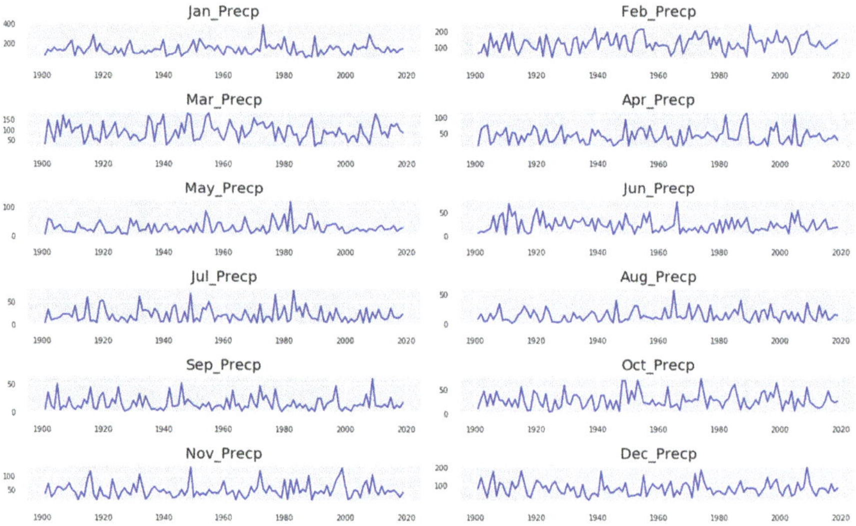

Sugarcane can grow over a prolonged season. Under warm humid conditions, it can continue its growth, unless terminated by flowering. However, its height is strongly influenced by the age of the crop and season. Temperatures above 50 °C arrest its growth; those below 20 °C slow it down markedly and severe frost proves fatal. The crop does best in the tropical regions receiving a rainfall of 750–1200 mm. It can also be grown in subtropical areas, but where the climate is subject to extremes, and the dry season is long, the growth period of the crop is restricted to a bare 4 months in the year, and the yields tend to be distinctly lower than those in the tropics.[63]

8.4.5 *Step 4: Concat Weather Data to Prepare the Data for Modeling*

```
weatherdf                          =                    pd.concat
([dfAustraliaQueenslandSugarcaneCashewProductionPrecp,

dfAustraliaQueenslandSugarcaneCashewProductionTeamMax,

dfAustraliaQueenslandSugarcaneCashewProductionTeamMin,

dfAustraliaQueenslandSugarcaneCashewProductionMeanTemp],axis
= 1)
```

[63]Weather Conditions Suitable for Sugarcane Crop—http://indiaagronet.com/community/specificfarmers/Weather%20conditions.htm

The combined weather DataFrame columns include:

```
Index(['Jan_Precp', 'Feb_Precp', 'Mar_Precp', 'Apr_Precp',
'May_Precp',
    'Jun_Precp', 'Jul_Precp', 'Aug_Precp', 'Sep_Precp',
'Oct_Precp',
   'Nov_Precp', 'Dec_Precp', 'Year_precp', 'Jan_Max', 'Feb_Max',
'Mar_Max',
    'Apr_Max', 'May_Max', 'Jun_Max', 'Jul_Max', 'Aug_Max',
'Sep_Max',
    'Oct_Max', 'Nov_Max', 'Dec_Max', 'dupYear', 'Jan_Min',
'Feb_Min',
    'Mar_Min', 'Apr_Min', 'May_Min', 'Jun_Min', 'Jul_Min',
'Aug_Min',
    'Sep_Min', 'Oct_Min', 'Nov_Min', 'Dec_Min', 'dupYear',
'Jan_Avg',
    'Feb_Avg', 'Mar_Avg', 'Apr_Avg', 'May_Avg', 'Jun_Avg',
'Jul_Avg',
    'Aug_Avg', 'Sep_Avg', 'Oct_Avg', 'Nov_Avg', 'Dec_Avg',
'dupYear'],
   dtype='object')
```

8.4.6 Step 5: Merge Production and Weather DataFrames to Prepare the Data for Modeling

```
dfAustraliaQueenslandSugarcaneWeather = weatherdf[weatherdf.
index >=1961 ]
dfAustraliaQueenslandSugarcaneWeather

df_merged = df_final.merge
(dfAustraliaQueenslandSugarcaneWeather,on= "Year", how=
"inner")
df_merged
```

The combined DataFrame columns include:

```
Index(['dupYear_x', 'Areaharvested(ha)', 'SugarcaneProduction
(tonnes)',
    'Yield(hg/ha)', 'Jan_Precp', 'Feb_Precp', 'Mar_Precp',
'Apr_Precp',
    'May_Precp', 'Jun_Precp', 'Jul_Precp', 'Aug_Precp',
'Sep_Precp',
```

(continued)

```
    'Oct_Precp', 'Nov_Precp', 'Dec_Precp', 'Year_precp',
'Jan_Max',
    'Feb_Max', 'Mar_Max', 'Apr_Max', 'May_Max', 'Jun_Max',
'Jul_Max',
    'Aug_Max', 'Sep_Max', 'Oct_Max', 'Nov_Max', 'Dec_Max',
'dupYear_y',
    'Jan_Min', 'Feb_Min', 'Mar_Min', 'Apr_Min', 'May_Min',
'Jun_Min',
    'Jul_Min', 'Aug_Min', 'Sep_Min', 'Oct_Min', 'Nov_Min',
'Dec_Min',
    'dupYear_y', 'Jan_Avg', 'Feb_Avg', 'Mar_Avg', 'Apr_Avg',
'May_Avg',
    'Jun_Avg', 'Jul_Avg', 'Aug_Avg', 'Sep_Avg', 'Oct_Avg',
'Nov_Avg',
    'Dec_Avg', 'dupYear_y'],
   dtype='object')

dfFinal = df_merged[['dupYear_x', 'Jan_Precp', 'Feb_Precp',
'Mar_Precp', 'Apr_Precp',
    'May_Precp', 'Jun_Precp', 'Jul_Precp', 'Aug_Precp',
'Sep_Precp',
    'Oct_Precp', 'Nov_Precp', 'Dec_Precp', 'Year_precp',
'Jan_Max',
    'Feb_Max', 'Mar_Max', 'Apr_Max', 'May_Max', 'Jun_Max',
'Jul_Max',
    'Aug_Max', 'Sep_Max', 'Oct_Max', 'Nov_Max', 'Dec_Max',
'dupYear_y',
    'Jan_Min', 'Feb_Min', 'Mar_Min', 'Apr_Min', 'May_Min',
'Jun_Min',
    'Jul_Min', 'Aug_Min', 'Sep_Min', 'Oct_Min', 'Nov_Min',
'Dec_Min',
    'dupYear_y', 'Jan_Avg', 'Feb_Avg', 'Mar_Avg', 'Apr_Avg',
'May_Avg',
    'Jun_Avg', 'Jul_Avg', 'Aug_Avg', 'Sep_Avg', 'Oct_Avg',
'Nov_Avg',
    'Dec_Avg','Areaharvested(ha)', 'SugarcaneProduction
(tonnes)',
    'Yield(hg/ha)']]
```

8.4.7 Step 6: Merge Data: The Final DataFrame

Based on the sugarcane growing and harvesting calendar, prepare the final DataFrame.

```
dfFinal = dfFinal[dfFinal.index < 2020 ]
  dfFinal
```

`import pandas as pd`
`import matplotlib.pyplot as plt`
`# This ensures plots are displayed inline in the Jupyter notebook`
`%matplotlib inline`
`# Get the label column`
`label = dfFinal ['Production_value']`
`import pandas as pd`
`import matplotlib.pyplot as plt`
`# This ensures plots are displayed inline in the Jupyter notebook`
`%matplotlib inline`
`# Get the label column`
`label = dfFinal ['SugarcaneProduction (tonnes)']`
`# Create a figure for 2 sub-plots (2 rows, 1 column)`
`fig, ax = plt.subplots(2, 1, figsize = (9,12))`
`# Plot the histogram`
`ax[0].hist(label, bins=100)`
`ax[0].set_ylabel('Frequency')`
`# Add lines for the mean, median, and mode`
`ax[0].axvline(label.mean(), color='magenta', linestyle='dashed', linewidth=2)`
`ax[0].axvline(label.median(), color='cyan', linestyle='dashed', linewidth=2)`
`# Plot the boxplot`
`ax[1].boxplot(label, vert=False)`
`ax[1].set_xlabel ('SugarcaneProduction (tonnes)')`
`# Add a title to the Figure`
	The cashew nut production value distribution is as follows:
	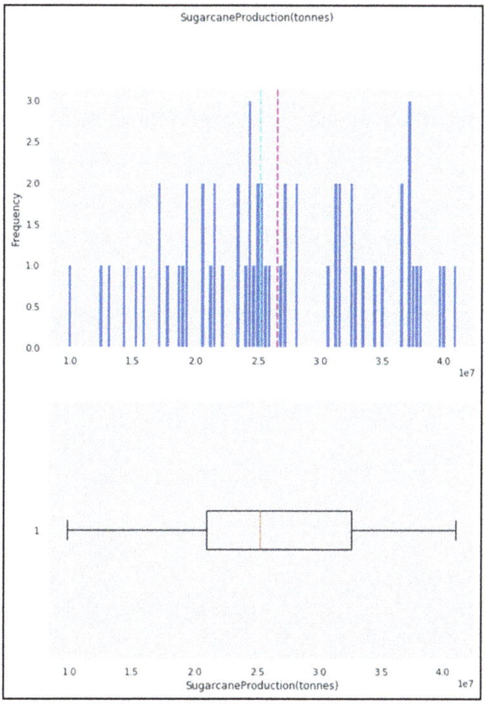

(continued)

fig.suptitle
('SugarcaneProduction
(tonnes)')
# Show the figure 	
fig.show()	

8.4.8 Step 7: Correlation Values

The following code provides production to feature attribute correlation ship values.

```
numeric_features=[
    'Jan_Precp', 'Feb_Precp', 'Mar_Precp', 'Apr_Precp',
    'May_Precp', 'Jun_Precp', 'Jul_Precp', 'Aug_Precp',
'Sep_Precp',
    'Oct_Precp', 'Nov_Precp', 'Dec_Precp', 'Jan_Max',
    'Feb_Max', 'Mar_Max', 'Apr_Max', 'May_Max', 'Jun_Max',
'Jul_Max',
    'Aug_Max', 'Sep_Max', 'Oct_Max', 'Nov_Max', 'Dec_Max',
    'Jan_Min', 'Feb_Min', 'Mar_Min', 'Apr_Min', 'May_Min',
'Jun_Min',
    'Jul_Min', 'Aug_Min', 'Sep_Min', 'Oct_Min', 'Nov_Min',
'Dec_Min',
    'Jan_Avg', 'Feb_Avg', 'Mar_Avg', 'Apr_Avg', 'May_Avg',
    'Jun_Avg', 'Jul_Avg', 'Aug_Avg', 'Sep_Avg', 'Oct_Avg',
'Nov_Avg',
    'Dec_Avg','Areaharvested(ha)', 'SugarcaneProduction
(tonnes)',
    'Yield(hg/ha)']

for col in numeric_features:
  fig = plt.figure(figsize=(9, 6))
  ax = fig.gca()
  feature = dfFinal[col]

  label = dfFinal['SugarcaneProduction(tonnes)']
  correlation = feature.corr(label)
  plt.scatter(x=feature, y=label)
  plt.xlabel(col)
  plt.ylabel('Sugarcane Production Value (tonnes)')
  ax.set_title('Sugarcane Production Value (tonnes) vs ' + col + '-
correlation: ' + str(correlation))
plt.show()
```

Please find the correlation of precipitation, and temperature for months of January to December. Average temperatures play an important role and it is proven through data—positively correlated with sugarcane production. A fact[64] that can be deduced from that crop need certain ambient air temperatures for growth. Spring wheat, for example, needs a daily average temperature of at least 41 °F to grow, corn needs at least 50 °F, and cotton will not grow until the daily average is closer to 60 °F. For sugarcane, check when vegetative growth[65] occurs when the mean daily air temperature drops below 21 °C [20].

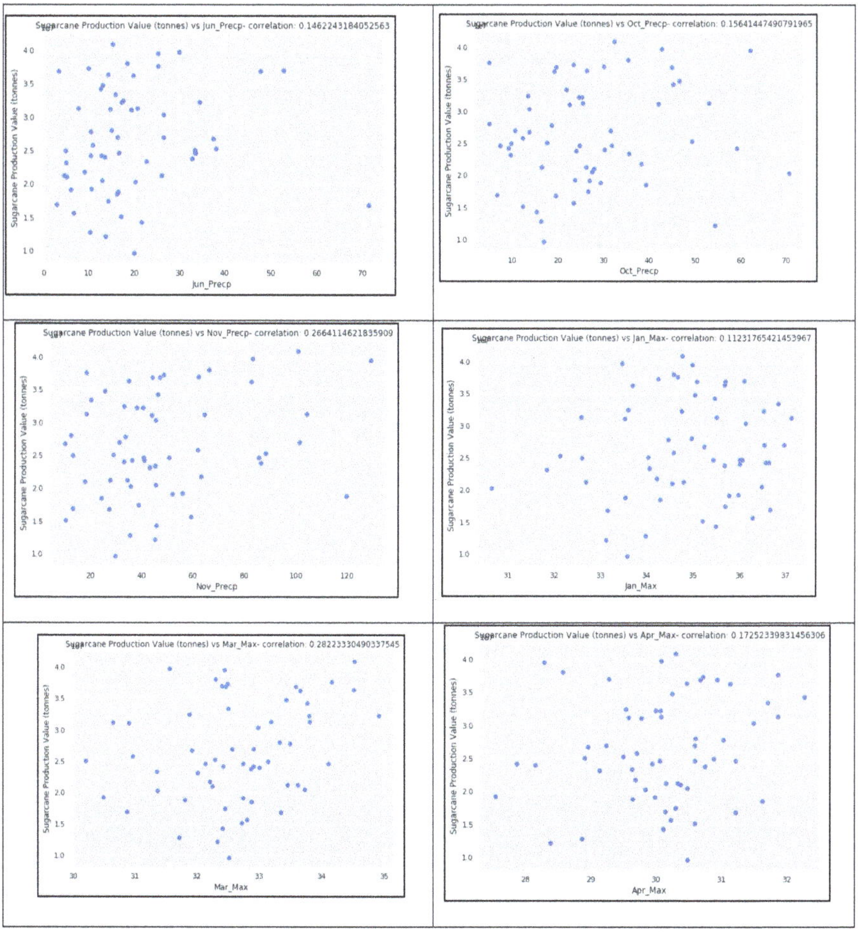

64 How Satellite Maps Help Prevent Another 'Great Grain Robbery'—https://www.nasa.gov/feature/how-satellite-maps-help-prevent-another-great-grain-robbery

65 Weather Conditions Suitable for Sugarcane Crop—http://indiaagronet.com/community/specificfarmers/Weather%20conditions.htm

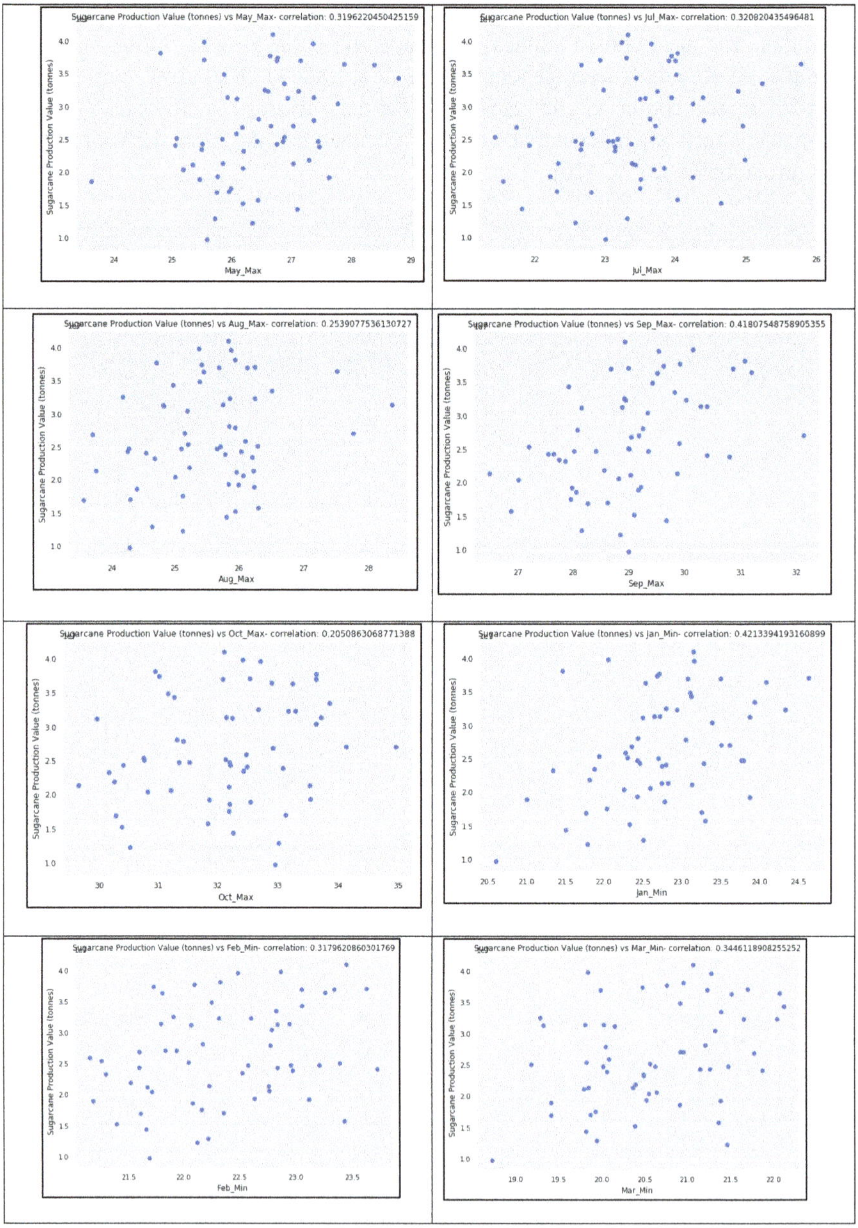
Sugarcane Production Value (tonnes) vs May_Max- correlation: 0.3196220450425159
May_Max
Sugarcane Production Value (tonnes)
Sugarcane Production Value (tonnes) vs Jul_Max- correlation: 0.320820435496481
Jul_Max
Sugarcane Production Value (tonnes)
Sugarcane Production Value (tonnes) vs Aug_Max- correlation: 0.2539077536130727
Aug_Max
Sugarcane Production Value (tonnes)
Sugarcane Production Value (tonnes) vs Sep_Max- correlation: 0.41807548758905355
Sep_Max
Sugarcane Production Value (tonnes)
Sugarcane Production Value (tonnes) vs Oct_Max- correlation: 0.2050863068771388
Oct_Max
Sugarcane Production Value (tonnes)
Sugarcane Production Value (tonnes) vs Jan_Min- correlation: 0.4213394193160899
Jan_Min
Sugarcane Production Value (tonnes)
Sugarcane Production Value (tonnes) vs Feb_Min- correlation: 0.3179620860301769
Feb_Min
Sugarcane Production Value (tonnes)
Sugarcane Production Value (tonnes) vs Mar_Min- correlation: 0.3446118908255252
Mar_Min
Sugarcane Production Value (tonnes)

8.4.9 Step 8: Correlation Matrix

A correlation matrix would help to understand the influence of feature variables.

```
import seaborn as sn
import matplotlib.pyplot as plt

import plotly.graph_objects as go
import seaborn as sn; sn.set()

fig = plt.figure(figsize=(30, 20))
corrMatrix = dfFinal.corr()
sn.heatmap(corrMatrix, annot=True)
plt.show()
```

Output:

As can be seen from the correlation matrix, precipitation values have lower influence on sugarcane production. On the other hand, temperatures, min/max/average, have a positive correlation value with sugarcane production. Net temperatures will have considerable influence on the yield and production of sugarcane.

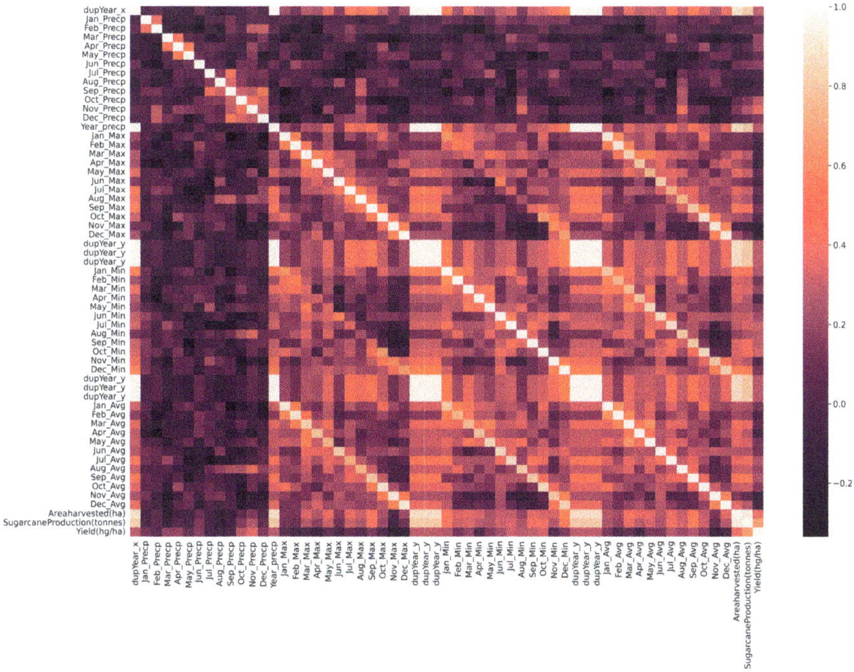

8.4.10 Step 9: Prepare Final Sugarcane DataFrame with Weather Details for Regression

Prepare sugarcane DataFrame with the selected columns as final ready DataFrame to be regressed to find the relationships and to predict. The selected columns are based on the phenological stages and corresponding growing and harvesting seasons.

```
X = dfFinal[[ 'May_Precp','Jun_Precp', 'Jul_Precp', 'Aug_Precp',
'Sep_Precp',
    'Oct_Precp', 'Nov_Precp', 'May_Max', 'Jun_Max', 'Jul_Max',
    'Aug_Max', 'Sep_Max', 'Oct_Max', 'Nov_Max', 'May_Min',
'Jun_Min',
    'Jul_Min', 'Aug_Min', 'Sep_Min', 'Oct_Min', 'Nov_Min',
'May_Avg',
    'Jun_Avg', 'Jul_Avg', 'Aug_Avg', 'Sep_Avg', 'Oct_Avg',
'Nov_Avg',
    'Areaharvested(ha)']]
y = dfFinal['SugarcaneProduction(tonnes)']

from sklearn.model_selection import train_test_split
X_train, X_test, y_train, y_test = train_test_split(X, y,
test_size = 0.3, random_state = 0)
```

Create a linear regression model.

```
# Train the model
from sklearn.linear_model import LinearRegression

# Fit a linear regression model on the training set
model = LinearRegression(normalize=False).fit(X_train, y_train)
print (model)
```

Plot the regression.

```
import matplotlib.pyplot as plt

%matplotlib inline

plt.scatter(y_test, predictions)
plt.xlabel('Actual Labels')
plt.ylabel('Predicted Labels')
plt.title('Predictions')
```

(continued)

```
# overlay the regression line
z = np.polyfit(y_test, predictions, 1)
p = np.poly1d(z)
plt.plot(y_test,p(y_test), color='magenta')
plt.show()
```

Output:

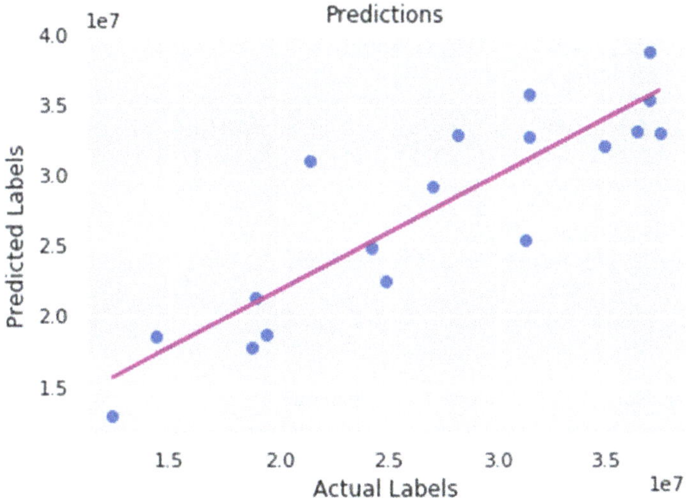

The actual and predicted labels follow the line of regression, although some data points are far from the line of intercepts.

8.4.11 Step 10: Regression Metrics

The model coefficients are as follows:

```
model.coef_
```

Output:

```
array([  4.28779758e+04, 6.10781563e+04, -7.61514630e+04,
-4.76527898e+04,
    6.19163311e+04, 6.11762010e+04, 2.52564209e+04,
-1.18228147e+08,
   -6.37092952e+07, 3.27330577e+07, 5.77095878e+07,
-6.23722992e+07,
    5.00718426e+07, 2.39208995e+07, -1.17305421e+08,
-6.26152877e+07,
    3.31204173e+07, 5.60991474e+07, -6.19087713e+07,
5.18483100e+07,
    2.64589383e+07, 2.34905601e+08, 1.26960057e+08,
-6.67525812e+07,
   -1.13842306e+08, 1.25134355e+08, -1.01110930e+08,
-4.97440967e+07,
    7.32063856e+01])
```

Feature variable	Coefficients	Summary
Area_harvested_Value	42,877.9758	The area harvested has a statistically substantial effect on sugarcane production. If all other feature variables are held constant, a unit increase in area harvested will have 73.2 times sugarcane production outcomes.
May_Precp	42,877.9758	The germination stage requires rainfall from 1100 to 1500 mm. May and June precipitation would have a positive influence on sugarcane production.
Jun_Precp	61,078.1563	
Jul_Precp	-76,151.46	For the tillering stage, the climatic conditions required are like the first phase; however, water supply must be controlled to maximize growth. Any higher rainfall could have a negative impact on the overall sugarcane production.
Aug_Precp	-47,652.79	
Sep_Precp Oct_Precp Nov_Precp	61,916.33 61,176.20 25,256.42	The grand growth stage needs rainfall between 750 and 1100 mm and high relative humidity of 80–87%. This stage requires high humidity for rapid cane elongation and high light intensity is critical to increase the rate of photosynthesis and respiration. Hence, September, October, and November precipitation would have a positive influence on sugarcane production. Moderate relative humidity values (40–65%) led to ripening! This is particularly true for Queensland. Given values are lower during the September, October, and November, hence, we see positive coefficients.

(continued)

Feature variable	Coefficients	Summary
May_Max Jun_Max Sep_Max	-118,228,147.00 -63,709,295.20 -62,372,299.20	Given May and June maximum temperatures, as per the dataset, ranged from 22.5 to 27 °C, nonetheless, the germination stage requires 32–38 °C average temperature. The optimal temperature is a mandatory requirement for sprouting of the stem cuttings. Given temperatures are below the optimal threshold, a negative influence is observed on sugarcane production.
Jul_Max Aug_Max Oct_Max Nov_Max	32,733,057.70 57,709,587.80 50,071,842.60 23,920,899.50	The grand growth stage needs an average temperature of 28–32 °C, sunlight at 10–18 MJ/m^2, and high relative humidity of 80–87%. This stage requires high humidity for rapid cane elongation, while temperatures *above 38 °C and high light intensity are critical to increase* the rate of photosynthesis and respiration. Maximum temperature is very important!
Dec_Min_temp Jan_Min_temp	- 18,145.15189753 -259.69726528	A temperature below 20 °C for a prolonged period is not good for cashew production. December and January temperatures below 20 °C have a negative influence on the production.
May_Avg Jun_Avg	234,905,601.00 126,960,057.00	The optimal average temperature is a mandatory requirement for sprouting of the stem cuttings. Hence, May and June average temperatures play an important role.

```
from sklearn.metrics import mean_squared_error, r2_score

mse = mean_squared_error(y_test, predictions)
print("MSE:", mse)
rmse = np.sqrt(mse)
print("RMSE:", rmse)
r2 = r2_score(y_test, predictions)
print("R2:", r2)
```

Output:

```
MSE: 13995963886926.055
RMSE: 3741117.999599325
R²: 0.7771396116114835
```

The model could explain 77% of the data signals, which is a good candidate for prediction.

Let us derive the model equation: $Y = \mathrm{m}x + \mathrm{C}$.

```
mx=""
for ifeature in range(len(X.columns)):
  if model.coef_[ifeature] <0:
    # format & beautify the equation
   mx += " - " + "{:.2f}".format(abs(model.coef_[ifeature])) + " * "
+ X.columns[ifeature]
  else:
    if ifeature == 0:
      mx += "{:.2f}".format(model.coef_[ifeature]) + " * " + X.
columns[ifeature]
    else:
     mx += " + " + "{:.2f}".format(model.coef_[ifeature]) + " * " + X.
columns[ifeature]

print(mx)
```

Output:

```
42877.98 * May_Precp + 61078.16 * Jun_Precp - 76151.46 * Jul_Precp -
47652.79 * Aug_Precp + 61916.33 * Sep_Precp + 61176.20 * Oct_Precp +
25256.42 * Nov_Precp - 118228147.47 * May_Max - 63709295.24 *
Jun_Max + 32733057.68 * Jul_Max + 57709587.78 * Aug_Max -
62372299.23 * Sep_Max + 50071842.59 * Oct_Max + 23920899.51 *
Nov_Max - 117305421.33 * May_Min - 62615287.74 * Jun_Min +
33120417.25 * Jul_Min + 56099147.37 * Aug_Min - 61908771.28 *
Sep_Min + 51848310.01 * Oct_Min + 26458938.31 * Nov_Min +
234905601.28 * May_Avg + 126960056.93 * Jun_Avg - 66752581.20 *
Jul_Avg - 113842306.34 * Aug_Avg + 125134354.74 * Sep_Avg -
101110930.07 * Oct_Avg - 49744096.72 * Nov_Avg + 73.21 *
Areaharvested(ha)
```

Model equation:

```
# y=mx+c
if(model.intercept_ <0):
  print("The formula for the " + y.name + " linear regression line is
= " + " - {:.2f}".format(abs(model.intercept_)) + " + " + mx )
else:
  print("The formula for the " + y.name + " linear regression line is
= " + " {:.2f}".format(model.intercept_) + " + " + mx )
```

Output:

The following equation provides the governing rule that connects climate-independent variables to sugarcane production–dependent variable.

The formula for the SugarcaneProduction(tonnes) linear regression line is

$$\begin{aligned} = \; & 3036621.00 + 42877.98^{*}\text{ May\_Precp} + 61078.16^{*}\text{ Jun\_Precp} \\ & - 76151.46^{*}\text{ Jul\_Precp} - 47652.79^{*}\text{ Aug\_Precp} \\ & + 61916.33^{*}\text{ Sep\_Precp} + 61176.20^{*}\text{ Oct\_Precp} \\ & + 25256.42^{*}\text{ Nov\_Precp} - 118228147.47^{*}\text{ May\_Max} \\ & - 63709295.24^{*}\text{ Jun\_Max} + 32733057.68^{*}\text{ Jul\_Max} \\ & + 57709587.78^{*}\text{ Aug\_Max} - 62372299.23^{*}\text{ Sep\_Max} \\ & + 50071842.59^{*}\text{ Oct\_Max} + 23920899.51^{*}\text{ Nov\_Max} \\ & - 117305421.33^{*}\text{ May\_Min} - 62615287.74^{*}\text{ Jun\_Min} \\ & + 33120417.25^{*}\text{ Jul\_Min} + 56099147.37^{*}\text{ Aug\_Min} \\ & - 61908771.28^{*}\text{ Sep\_Min} + 51848310.01^{*}\text{ Oct\_Min} \\ & + 26458938.31^{*}\text{ Nov\_Min} + 234905601.28^{*}\text{ May\_Avg} \\ & + 126960056.93^{*}\text{ Jun\_Avg} - 66752581.20^{*}\text{ Jul\_Avg} \\ & - 113842306.34^{*}\text{ Aug\_Avg} + 125134354.74^{*}\text{ Sep\_Avg} \\ & - 101110930.07^{*}\text{ Oct\_Avg} - 49744096.72^{*}\text{ Nov\_Avg} \\ & + 73.21^{*}\text{ Areaharvested(ha)} \end{aligned} \tag{8.8}$$

X_test, y_test, predictions

8.4.12 Step 11: Other Regressors (Ensemble)

Apply the ensemble model with the voting regressor. Here, we are voting using linear regression, random forest regressor, gradient boost regressor, and voting regressor.

```
import numpy as np
from sklearn.linear_model import LinearRegression
from sklearn.ensemble import RandomForestRegressor
from sklearn.ensemble import GradientBoostingRegressor
from sklearn.ensemble import VotingRegressor
r1 = LinearRegression()
r2 = RandomForestRegressor(n_estimators=10,random_state=1)
r3 = GradientBoostingRegressor()
erT = VotingRegressor([('lr', r1), ('rf', r2) ,('gb',r3)])
erT.fit(X_train, y_train)
```

Output:

```
VotingRegressor(estimators=[('lr', LinearRegression()),
                            ('rf',
                             RandomForestRegressor(n_estimators=10,
                                                   random_state=1)),
                            ('gb', GradientBoostingRegressor())])
```

Predict the sugarcane output:

```
yT1_pred = erT.predict(X_test)
yT1_pred
```

Output:

```
array([24007559.79962087, 32430003.53121917,
36658269.01432661,
    26192629.61135609, 21007236.00276946, 13662689.61060354,
    32106371.21641877, 36327875.91038544, 34281962.15876026,
    23866940.86545394, 16828176.03024909, 17687004.0131907 ,
    27084681.71496051, 34743219.90651048, 25593451.99028583,
    35236016.39273275, 34651245.00703724, 16818958.27382595])
```

Plot the prediction:

```
plt.scatter(y_test,yT1_pred)
plt.xlabel('Actual Labels')
plt.ylabel('Predicted Labels')
plt.title('Predictions')
# overlay the regression line
z = np.polyfit(y_test, yT1_pred, 1)
p = np.poly1d(z)
plt.plot(y_test,p(y_test), color='magenta')
plt.show()
```

Output:

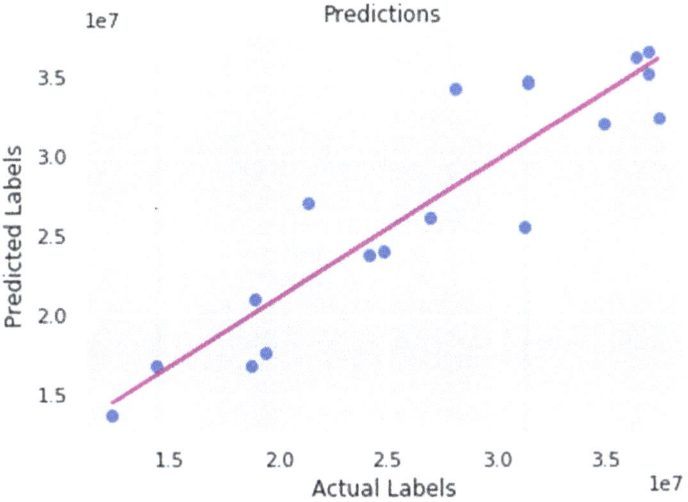

Score the model.

```
r2_score(y_test,yT1_pred)
```

Output: *83.99%*

8.4.13 Step 12: Model Explainability

Model explainability can be achieved using the ELI5 library.

```
eli5.show_weights(model, feature_names=[
'May_Precp','Jun_Precp', 'Jul_Precp', 'Aug_Precp', 'Sep_Precp',
   'Oct_Precp', 'Nov_Precp', 'May_Max', 'Jun_Max', 'Jul_Max',
'Aug_Max', 'Sep_Max', 'Oct_Max', 'Nov_Max', 'May_Min',
'Jun_Min', 'Jul_Min', 'Aug_Min', 'Sep_Min', 'Oct_Min',
'Nov_Min', 'May_Avg', 'Jun_Avg', 'Jul_Avg', 'Aug_Avg',
'Sep_Avg', 'Oct_Avg', 'Nov_Avg', 'Areaharvested(ha)'])
```

Output: The intercept (bias) feature is shown as <BIAS> in the above table.[66]

[66]ELI5—https://eli5.readthedocs.io/en/latest/tutorials/sklearn-text.html#baseline-model

Weight?	Feature
+234905601.277	May_Avg
+126960056.927	Jun_Avg
+125134354.739	Sep_Avg
+57709587.785	Aug_Max
+56099147.372	Aug_Min
+51848310.007	Oct_Min
+50071842.591	Oct_Max
+33120417.254	Jul_Min
+32733057.678	Jul_Max
+26458938.310	Nov_Min
… 8 more positive …	
… 2 more negative …	
-49744096.723	Nov_Avg
-61908771.282	Sep_Min
-62372299.226	Sep_Max
-62615287.741	Jun_Min
-63709295.243	Jun_Max
-66752581.203	Jul_Avg
-101110930.071	Oct_Avg
-113842306.335	Aug_Avg
-117305421.327	May_Min
-118228147.474	May_Max

8.5 Machine Learning Model: Queensland, Australia, Sugar Production with Climate Projections (CMPI6) and Shared Socioeconomic Pathways Scenarios—Time Period 2020–2099

Let us build a model that depicts the impact of climate change on sugarcane production in Queensland, Australia:

Software code for this model: Final_-Australia_Queensland_Sugarcane_code_MODIS_LSAT_NDVI_Release.ipynb (Jupyter Notebook Code)

To build the climate projections model, we have used two sets of models. First, scenario simulations based on the Climate Model Intercomparison Project-Phase 6 (CMIP6) for each of the five CMIP6 climate models.[67]

[67] Global Climate Change Impact on Crops Expected Within 10 Years, NASA Study Finds—https://climate.nasa.gov/news/3124/global-climate-change-impact-on-crops-expected-within-10-years-nasa-study-finds/

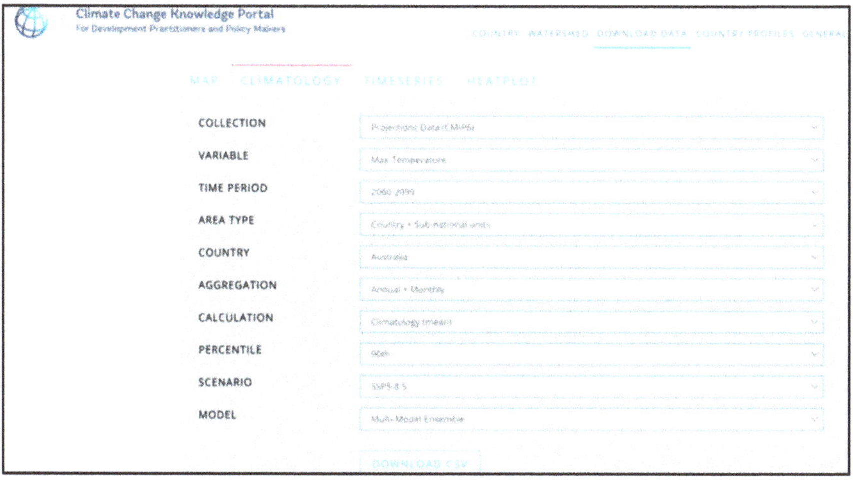

Fig. 8.20 Climate Change Knowledge Portal

For climate modeling, the data is downloaded from the Climate Change Knowledge Portal (CCKP) with feature variables that include minimum temperature, maximum temperature, average temperature, and precipitation. The data is downloaded for climatology (mean) and time periods include 2020–2039, 2040–2059, 2060–2070, and 2080–2099, specifically the Australia–Queensland cropland with 90th percentile. The model predictions are captured for SSP 1-1.9. SSP 3-7.0, and SSP 5-8.5, based on the multi-model ensemble.

As compared to CMIP5, a higher end of the range of temperature projections in CMIP6, with a corresponding rise in heat extremes, has many implications for the ecological and socioeconomic impact of projected climate change under a high-emission scenario. Greater impacts on terrestrial and marine ecosystems, human health, the economy, and society are projected at the high end in CMIP6 compared to CMIP5 (please see Fig. 8.20). However, as framed differently, CMIP6 indicates a lower magnitude of total greenhouse gas emissions is required for all models to be below a given warming threshold (e.g., 2 °C) than the one produced by CMIP5. With this data, the projection models would provide prescriptive guidance to lawmakers and overall agricultural and governmental agencies to increase food security adaptation measures for the future[68] [21].

Parameter Summary:

- Collection: projections data (CMIP6)
- Variable: max temperature
- Calculation: anomaly (from reference period, 1995–2014)
- Area type: subnational

[68] Insights from CMIP6 for Australia's future climate—https://agupubs.onlinelibrary.wiley.com/doi/full/10.1029/2019EF001469

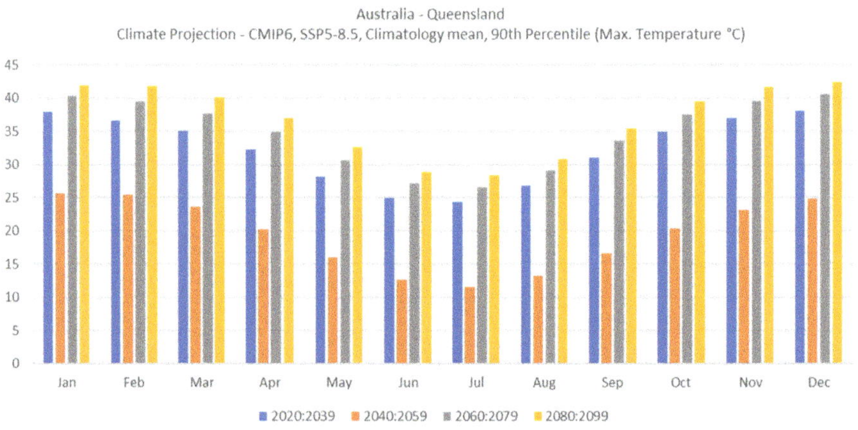

Fig. 8.21 Climate projections

- Country: Australia
- Subnational unit: Queensland
- Percentile: 90th
- Scenario: SSP 5-8.5
- Model: multi-model ensemble

As can be seen from the climate projections data (please see Fig. 8.21), SSP 5-8.5 has increased temperatures to the end of the century (2080–2099).

"The U.S. farming industry is already battling the effects of climate change, including increased drought and flooding. We don't have time to waste."
U.S. Secretary of Agriculture Thomas Vilsack [22][69]

8.5.1 CMIP6 Projections Data and SSP Scenarios

To model climate projections and assess the impact of climate change, the sugarcane production machine learning (ML) for Queensland, Australia, is updated to apply temperatures and precipitation inputs per CMIP6 projections and for scenarios as described in Table 8.3. The goal was to collect sugarcane production for SSP 1-2.6, SSP 3-7.0, and SSP 5-8.5 scenarios and compare with baseline (2019 production—

[69]Exclusive-U.S.-UAE push for another $4bln in farming climate change investment—https://finance.yahoo.com/news/exclusive-u-uae-push-another-084920821.html

Table 8.3 SSP scenarios

Scenario	SSP model	Impact of socioeconomic factors
SSP 1-2.6	SSP 1	It imagines the same socioeconomic shifts towards sustainability as SSP 1-1.9. But temperatures stabilize around 1.8 °C higher by the end of the century.
SSP 2-4.5	SSP2	Socioeconomic factors follow their historic trends, with no notable shifts. Progress towards sustainability is slow, with development and income growing unevenly. In this scenario, temperatures rise to 2.7 °C by the end of the century.
SSP 3-7.0	SSP3	By the end of the century, average temperatures have risen by 3.6 °C.
SSP 5-8.5	SSP5	By 2100, the average global temperature is a scorching 4.4 °C higher.

with area harvested at 169,139 hectares (HA) and production of 140,000 tons of sugarcane).

The projection models are powerful and provide data to simulate climatic changes. Australia[70] is experiencing higher temperatures, more extreme droughts, fire seasons, floods,[71] and more extreme weather due to climate change. Rising sea levels add to the intensity of high-sea-level events and threaten housing and infrastructure. The number of days that break heat records has doubled in the past 50 years.

After a particularly wet winter and spring in 2016 over much of Australia, conditions turned dry in 2017. Rainfall (please see Fig. 8.22) in most of the Murray–Darling Basin was substantially below average in 2017, 2018, and 2019. Though these years saw the most widespread dry conditions, in some parts of the country, such as western Victoria and western Queensland, dry conditions had been observed during 2012–2015.

Following precipitation levels derived from Climate Projections (CMIP6, SSP5-8.5, multi-model ensemble), projections predict an 700% increase in precipitation (mm) over the 1995–2014 reference period. The multi-model ensemble[72] is used to estimate the climate change signal and its uncertainty to illustrate the impact of precipitation [23].

Near the end of the century, the precipitation levels are higher during the months of January, February, March, and December. Let us see its impact on sugarcane production in Queensland (please see Fig. 8.23).

[70] Climate Change—https://australian.museum/learn/climate-change/climate-change-impacts/#:~:text=Australia%20is%20experiencing%20higher%20temperatures,in%20the%20past%2050%20years

[71] Australia previous droughts—http://www.bom.gov.au/climate/drought/knowledge-centre/previous-droughts.shtml

[72] Josep et al., The Mediterranean climate change hotspot in the CMIP5 and CMIP6 projections—https://esd.copernicus.org/preprints/esd-2021-65/esd-2021-65.pdf

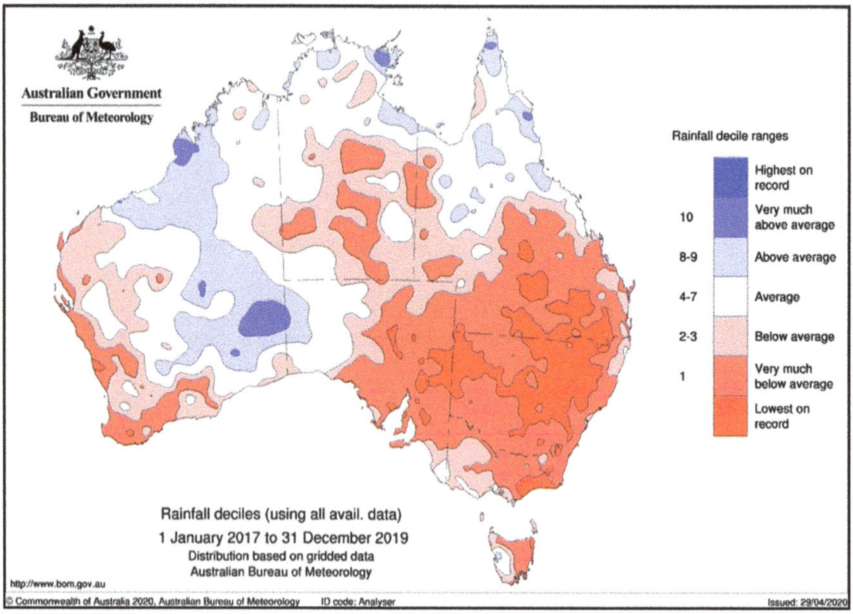

Fig. 8.22 Australia rainfall

8.5.2 Climate Model: Temperature Change (RCP 6.0): 2006–2100

Climate models are used for a variety of purposes from the study of dynamics of the weather and climate system to projections of future climate.[73]

NOAA's Geophysical Fluid Dynamics Laboratory has created several ocean–atmosphere-coupled models to predict how greenhouse gas emissions following different population, economic, and energy-use projections may affect the planet.

"Representative Concentration Pathways (RCPs)" are fully integrated scenarios (i.e., they are not a complete package of socioeconomic, emissions and climate projections). They are consistent sets of projections of only the components of radiative forcing that are meant to serve as input for climate modeling, pattern scaling and atmospheric chemistry modeling," according.

Notable features:

- The Earth gets warmer as CO_2 increases in the atmosphere.
- The Earth does not warm uniformly; the oceans warm slower than the continents and arctic.

[73]Climate Model: Temperature Change (RCP 6.0)—2006–2100, https://sos.noaa.gov/catalog/datasets/climate-model-temperature-change-rcp-60-2006-2100/

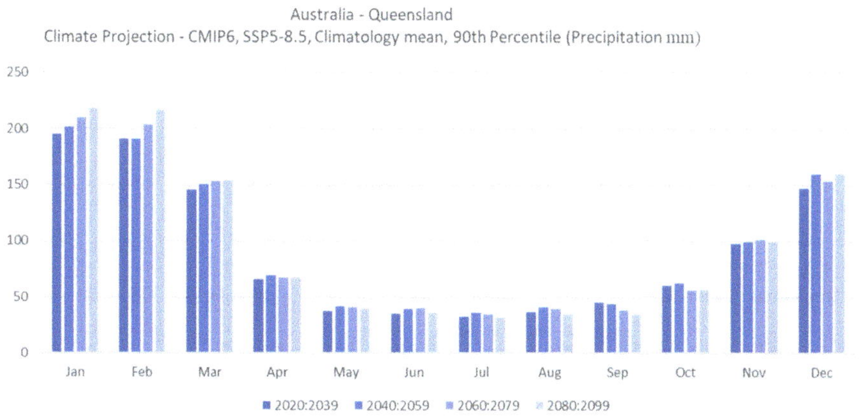

Fig. 8.23 Queensland, Australia, climate projections

- Projections for temperature according to RCP 6.0 include continuous global warming through 2100 where CO_2 levels rise to 670 ppm by 2100 making the global temperature rise by about 3–4 °C by 2100.

Here are the model data simulations:

Collection: [*CMIP6—Aggregation: Monthly—Calculation: Climatology Mean—Percentile: 90th, and Model: Multi-Model Ensemble*]. Please find the table with all three main variables for scenario SSP 5-8.5. Similar data was collected for the additional two scenarios: SSP 1-2.6 and SSP 3-7.0.

Min temperature	Maximum temperature	Precipitation
Time period: 2020–2039 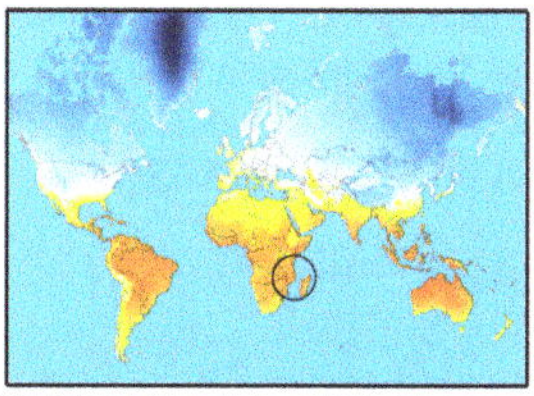	Time period: 2020–2039 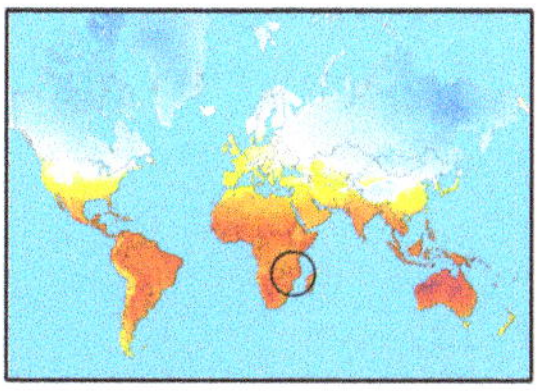	Time period: 2020–2039 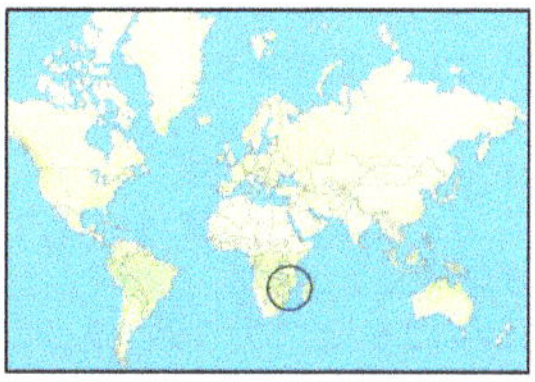
Time period: 2040–2059 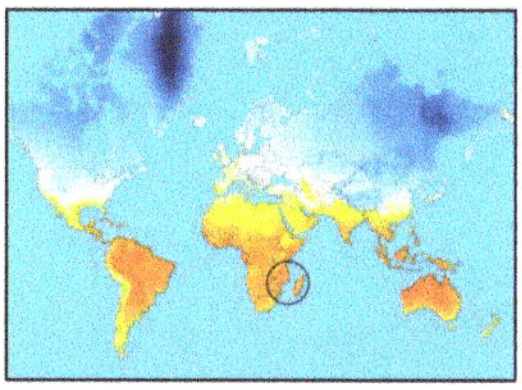	Time period: 2040–2059 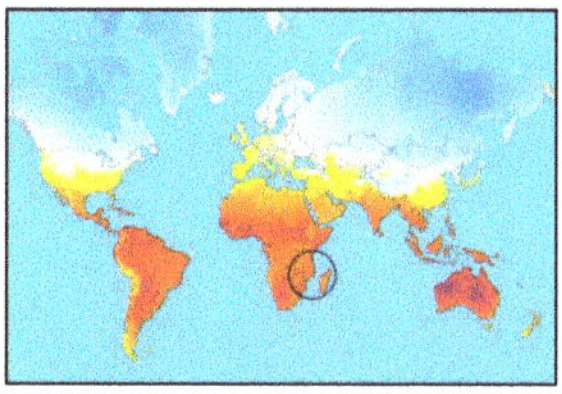	Time period: 2040–2059 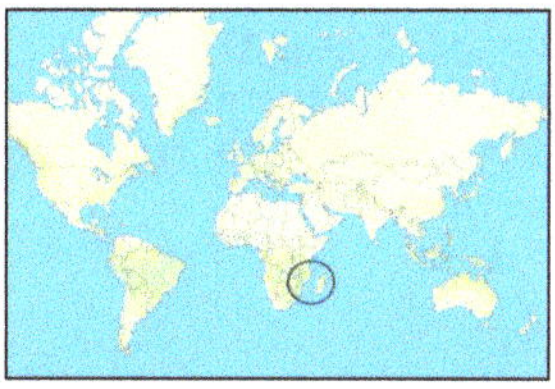

(continued)

Min temperature	Maximum temperature	Precipitation
Time period: 2060–2079 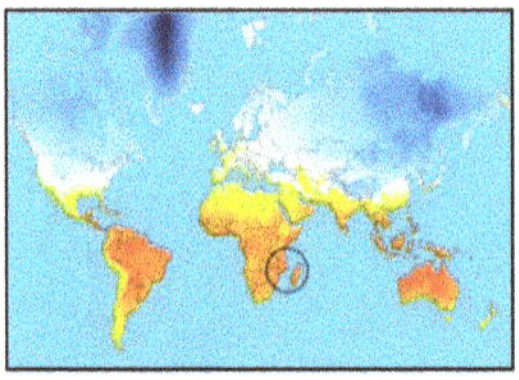	Time period: 2060–2079 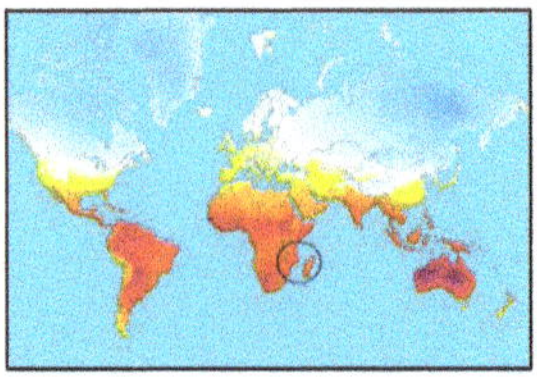	Time period: 2060–2079 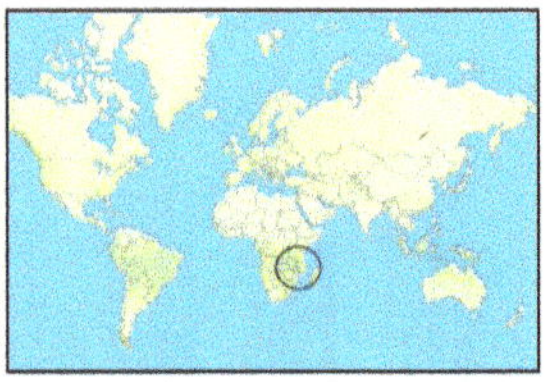
Time period: 2080–2099 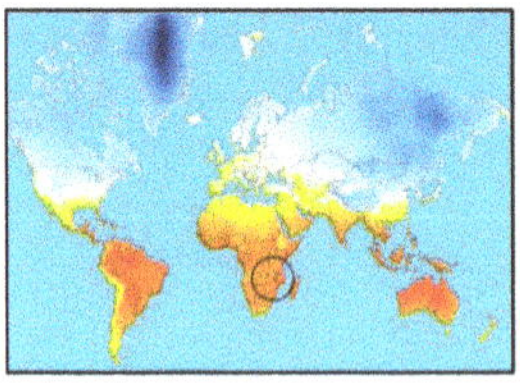	Time period: 2080–2099 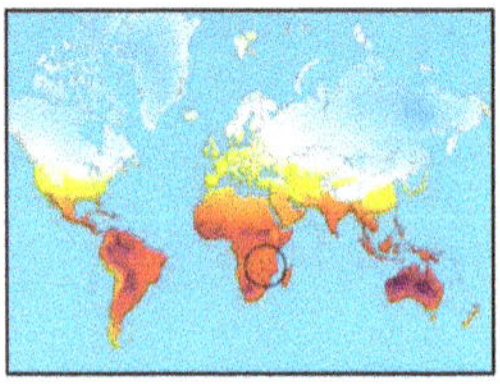	Time period: 2080–2099 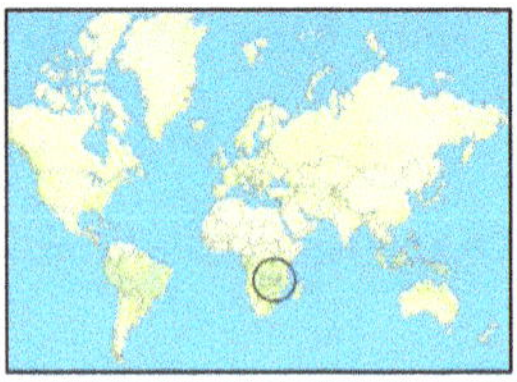

8.5.2.1 SSP 5-8.5

Please find model invocation called Scenario SSP5-8.5:

CMIP 6, 20202039_climatologymean 90th Percentile SSP 5_8.5_multimodelensembel—Results

```
AUS_QNS_cmip6_20202039_
climatologymean_90th_
ssp5_8p5_multimodelensembel = [[
                38,35.88,32.4,36.82,
45.11,60.57,97.91,
28.13,24.91,24.26,26.81,
31,34.87,36.95,
14.83,11.53,10.4,12,15.52,
19.06,21.64,
21.3,18.1,17.12,19.21,23.12,
26.86,29.29,395399
]]
result = model.predict(AUS_QNS_
cmip6_20202039_
climatologymean_
90th_ssp5_8p5_multimodelensembel)
print(result)
```

For area under acreage 395,399 ha, SSP5_8.5 2020: 2039 CMIP6 projections had generated 13,473,543.24 tons of sugarcane against 2019 baseline of 32,415,352 tons. A decrease in **-58.43%**.

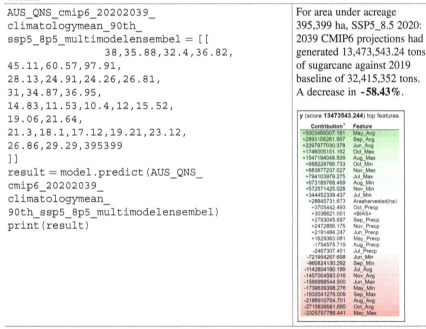

y (score **13473543.244**) top features

Contribution?	Feature
+5003489307.191	May_Avg
+2893106281.567	Sep_Avg
+2297977030.378	Jun_Avg
+1746005151.162	Oct_Max
+1547194048.509	Aug_Max
+988228788.733	Oct_Min
+883877237.027	Nov_Max
+794103979.275	Jul_Max
+673189768.469	Aug_Min
+572571425.028	Nov_Min
+344452339.437	Jul_Min
+28945731.673	Areaharvested(ha)
+3705442.493	Oct_Precp
+3036621.001	<BIAS>
+2793045.697	Sep_Precp
+2472856.175	Nov_Precp
+2191484.247	Jun_Precp
+1629363.081	May_Precp
-1754575.719	Aug_Precp
-2467307.401	Jul_Precp
-721954267.658	Jun_Min
-960824130.292	Sep_Min
-1142804190.189	Jul_Avg
-1457004593.016	Nov_Avg
-1586998544.500	Jun_Max
-1739639398.276	May_Min
-1933541276.009	Sep_Max
-2186910704.701	Aug_Avg
-2715839581.695	Oct_Avg
-3325757788.441	May_Max

(continued)

CMIP 6, 20402059_climatologymean 90th Percentile SSP 5_8.5_multimodelensembel—Results

```
AUS_QNS_cmip6_20402059_climatologymean_
90th_ssp5_8p5_multimodelensembel = [[
                    41.59,40.09,36.42,
41.21,44.37,
62.21,99.24,
29.19,25.77,25.23,27.81,32.03,
35.9,38.02,
15.97,12.58,11.51,13.24,16.57,
20.31,23.06,
22.47,19.04,18.13,20.36,24.27,
28.19,30.51,395399 ]]
result = model.predict(AUS_QNS_cmip6_
20402059_
climatologymean_
90th_ssp5_8p5_multimodelensembel)
print(result)
```

For area under acreage 395,399 ha, SSP5_8.5 2040:2059 CMIP6 projections had generated 24,544,771.33 tons of sugarcane against 2019 baseline of 32,415,352 tons.
A decrease in -24.28%.

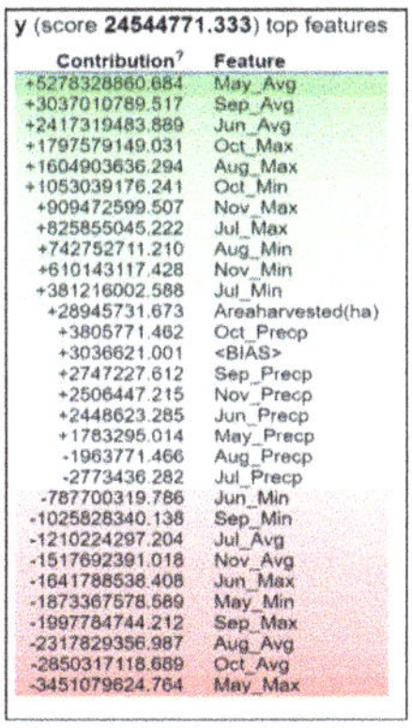

y (score **24544771.333**) top features

Contribution?	Feature
+5278328860.684	May_Avg
+3037010789.517	Sep_Avg
+2417319483.889	Jun_Avg
+1797579149.031	Oct_Max
+1604903636.294	Aug_Max
+1053039176.241	Oct_Min
+909472599.507	Nov_Max
+825855045.222	Jul_Max
+742752711.210	Aug_Min
+610143117.428	Nov_Min
+381216002.588	Jul_Min
+28945731.673	Areaharvested(ha)
+3805771.462	Oct_Precp
+3036621.001	<BIAS>
+2747227.612	Sep_Precp
+2506447.215	Nov_Precp
+2448623.285	Jun_Precp
+1783295.014	May_Precp
-1963771.466	Aug_Precp
-2773436.282	Jul_Precp
-787700319.786	Jun_Min
-1025828340.138	Sep_Min
-1210224297.204	Jul_Avg
-1517692391.018	Nov_Avg
-1641788538.408	Jun_Max
-1873367578.589	May_Min
-1997784744.212	Sep_Max
-2317829356.987	Aug_Avg
-2850317118.689	Oct_Avg
-3451079624.764	May_Max

CMIP 6, 20602079_climatologymean 90th Percentile SSP 5_8.5_multimodelensembel—Results

```
AUS_QNS_cmip6_20602079_climatologymean_
90th_ssp5_8p5_multimodelensembel = [[
      41.51,40.47,34.9,39.62,38.5,
56.27,101.09,
30.71,27.12,26.57,29.15,33.6,
37.54,39.56,
17.62,14.06,13.02,14.53,17.82,
21.65,24.33,
24.01,20.41,19.64,21.67,25.66,29.54,31.89,
      395399]]
result = model.predict
(AUS_QNS_cmip6_20602079_
climatologymean_
90th_ssp5_8p5_multimodelensembel)
print(result)
```

For area under acreage 395,399 ha, SSP5_8.5 2060:2079 CMIP6 projections had generated 16,985,278.46 tons of sugarcane against the 2019 baseline of 32,415,352 tons.
A decrease in -47.60%.

y (score **16985278.465**) top features

Contribution?	Feature
+5640083466.650	May_Avg
+3210947542.604	Sep_Avg
+2591254761.879	Jun_Avg
+1879696970.881	Oct_Max
+1682234483.925	Aug_Max
+1122515911.651	Oct_Min
+946310784.758	Nov_Max
+869717342.511	Jul_Max
+815120611.321	Aug_Min
+643745969.082	Nov_Min
+431227832.641	Jul_Min
+28945731.673	Areaharvested(ha)
+3442384.829	Oct_Precp
+3036621.001	<BIAS>
+2553171.593	Nov_Precp
+2471832.984	Jun_Precp
+2383778.748	Sep_Precp
+1779864.776	May_Precp
-1888003.530	Aug_Precp
-2657686.058	Jul_Precp
-880370945.643	Jun_Min
-1103214304.240	Sep_Min
-1311020694.820	Jul_Avg
-1586339244.495	Nov_Avg
-1727798086.986	Jun_Max
-2066921523.778	May_Min
-2095709253.997	Sep_Max
[illegible]	Aug_Avg

(continued)

CMIP 6, 20802099_climatologymean 90th Percentile SSP 5_8.5_multimodelensembel—Results

```
AUS_QNS_cmip6_20802099_climatologymean_
90th_ssp5_8p5_multimodelensembel = [[

39.63,36.12,31.8,34.82,34.85,56.46,99.11,
32.55,28.89,28.4,30.8,35.33,39.45,41.64,
19.87,15.97,14.78,16.28,19.75,23.52,26.36,
26.11,22.36,21.49,23.38,27.43,31.39,33.91,
    395399
]]
result = model.predict
(AUS_QNS_cmip6_20802099_
climatologymean_
90th_ssp5_8p5_multimodelensembel)
print(result)
```

For area under acreage 395,399 ha, SSP5_8.5 2080: 2099 CMIP6 projections had generated 39,922,953.22 tons of sugarcane against 2019 baseline of 32,415,352 tons. An increase in 23.16%.

y (score 39922953.227) top features

Contribution?	Feature
+6133385249.331	May_Avg
+3432435350.492	Sep_Avg
+2838826872.886	Jun_Avg
+1975334190.231	Oct_Max
+1777455303.770	Aug_Max
+1219472251.364	Oct_Min
+996066255.746	Nov_Max
+929618838.063	Jul_Max
+913294119.222	Aug_Min
+697457613.851	Nov_Min
+489519767.008	Jul_Min
+28945731.673	Areaharvested(ha)
+3454008.307	Oct_Precp
+3036621.001	<BIAS>
+2503163.880	Nov_Precp
+2206143.004	Jun_Precp
+2157784.140	Sep_Precp
+1699254.182	May_Precp
-1659270.139	Aug_Precp
-2421616.523	Jul_Precp
-999966145.229	Jun_Min

SSP 5-8.5 is the most climatic severe outcome scenario (SSP 5-8.5) that could potentially increase atmospheric temperatures by 4.4 °C from baseline (2019). Increase in temperatures and precipitation levels had a negative effect on sugarcane production. All climatic scenarios under SSP 5_8.5 (2020:2039, 2040:2059, and 2060:2079) had a negative impact on sugarcane production (-58.43, -24.28, and -47.60), except 2080:2099 that had a positive impact (23.16%).

Similar model runs for SSP 1_1.9 and SSP 3_7.0 have generated the following results.

8.5.2.2 SSP 1_1.9

Sugarcane production will drastically reduce under SSP 1: 2020:2039 (please see Table 8.5) where production drops by **154.42%**. Except 2080:2099 that has minimal reduction (**-37.18%**), other time periods have a triple-digit drop (please see Table 8.4).

Output:

Table 8.4 SSP 5-8.5

Baseline: 2019 Queensland, Australia, sugarcane production with area harvested at 395,399 hectares (HA) and production of 32,415,352 tons			
20,202,039	20,402,059	20,602,079	20,802,099
13,473,543.24	24,544,771.33	16,985,278.46	39,922,953.22
Change in production from 2019 levels: 140,000 tons			
-58.43%	-24.28%	-47.60%	23.16%

Table 8.5 SSP 1_1.9

Baseline: 2019 Queensland, Australia, sugarcane production with area harvested at 395,399 hectares (HA) and production of 32,415,352 tons			
20,202,039	20,402,059	20,602,079	20,802,099
-17,642,679.19	**-23,537,262.06**	**-9,699,014.14**	20,363,297.05
Change in production from 2019 levels: 32,415,352 sugarcane tons			
-154.42%	**-172.61%**	**-129.92%**	**-37.18%**

8.5.2.3 SSP 3_7.0

Sugarcane production will drastically reduce under SSP 3: 2080:2099 where production drops by **257.05%** (Fig. 8.24). Except 2050:2059 that has minimal reduction (**-36.17%**), other time periods have a triple-digit drop (please see Table 8.6).

As demonstrated by the machine learning model, sugarcane production will decrease in Queensland, Australia, near the end of the century. The adaption plan would consider options such as growing and harvest calendar shift and more climate-smart technologies.

8.6 Machine Learning Model: Queensland, Australia, Sugar Production with the Normalized Difference Vegetation Index (NDVI) Model

The Vegetation Health Index[74] (VHI) illustrates the severity of drought based on vegetation health and the influence of temperature on plant conditions. The VHI is a composite index and the elementary indicator used to compute the Agricultural Stress Index(ASI). It combines both the Vegetation Condition Index (VCI) and the Temperature Condition Index (TCI). The TCI is calculated using a similar equation to the VCI but relates the current temperature to the long-term maximum and minimum, as it is assumed that higher temperatures tend to cause a deterioration in vegetation conditions. A decrease in the VHI would, for example, indicate relatively poor vegetation conditions and warmer temperatures, signifying stressed vegetation conditions, and over a longer period would be indicative of drought (drought is a

[74]Landsat Spectral Indices products courtesy of the U.S. Geological Survey Earth Resources Observation and Science Center—https://www.usgs.gov/landsat-missions/landsat-surface-reflectance-derived-spectral-indices

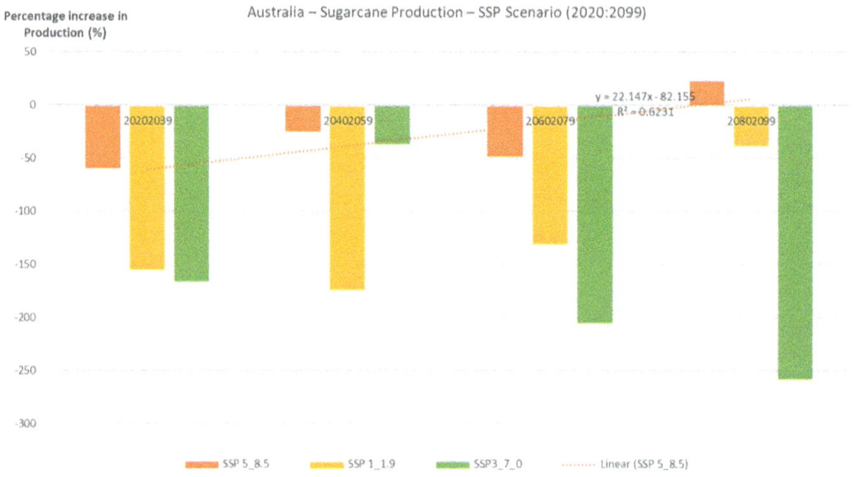

Fig. 8.24 Australia sugarcane production SSP scenarios

Table 8.6 SSP 3_7.0

Baseline: 2019 Queensland, Australia, sugarcane production with area harvested at 395,399 hectares (HA) and production of 32,415,352 tons			
20,202,039	20,402,059	20,602,079	20,802,099
-21,232,850.69	20,688,826.03	**-33,811,418.99**	**-50,910,245.21**
Change in production from 2019 levels: 32,415,352 sugarcane tons			
-165.50%	**-36.17%**	**-204.30%**	**-257.05%**

frequent visitor in Australia[75] and historically Australia has experienced droughts[76] in 1901–1902, 1914–1915, 1937–1945, 1965–1968, 1982–1983, 1997–2009, and 2017–2019). A downward trend (slope of negative—equation) as seen in the figure is indicative of long-term stress on the crops. Finally, for sugarcane, we would like to factor in photosynthesis as the production and yield depend upon the amount of sunlight received. For that, we consider VHI as an indicator (please see Fig. 8.25).

Our goal is straightforward: Knowledge of the key atmospheric variables (rainfall, solar radiation, temperature) and time of year influencing cane yields may help refine yield forecasting techniques and improve decision-making capabilities throughout the sugar industry. A review of productivity trends in the Wet Tropics over a 35-year period identified excessive wetness, especially early in the growing season, and low solar radiation adversely affected sugarcane productivity. In other words, excessive rainfall coincides with low solar radiation and extreme waterlogging which adversely affect crop growth, increase nutrient losses

[75] Droughts in Australia—https://earthobservatory.nasa.gov/world-of-change/AustraliaNDVI

[76] Historical droughts—http://www.bom.gov.au/climate/drought/knowledge-centre/previous-droughts.shtml

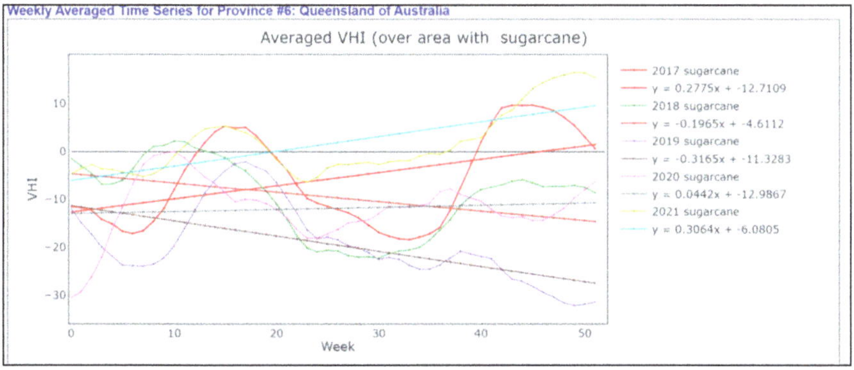

Fig. 8.25 Averaged VHI

(especially nitrogen), and may prevent crop production practices being completed in a timely manner (which may increase weed competition, delay fertilizer application[77] or hilling up of plant cane) [24].

$$Y_{2018} = 0.1965x + 4.6112 \tag{8.9}$$

Queensland, Australia, sugarcane NDVI	

	Software code for this model: Final_-Australia_Queensland_Sugarcane_code_MODIS_LSAT_NDVI_Release.ipynb (Jupyter Notebook Code)

The Landsat Normalized Difference Vegetation Index[78] (NDVI) is used to quantify vegetation greenness and is useful in understanding vegetation density and assessing changes in plant health. Global and Regional *Vegetation Health* (VH) is a NOAA/NESDIS system estimating vegetation health, moisture condition, thermal condition, and their products. It contains Vegetation Health Indices (VHIs)

[77] Improving the drought monitoring capability of VHI at the global scale via ensemble indices for various vegetation types from 2001 to 2018, Weather and Climate Extremes—https://www.sciencedirect.com/science/article/pii/S2212094722000068

[78] NDVI—https://www.usgs.gov/landsat-missions/landsat-normalized-difference-vegetation-index

derived from the radiance observed by the Advanced Very-High-Resolution Radiometer[79] (AVHRR) onboard afternoon polar-orbiting satellites—the NOAA-7, 9, 11, 14, 16, 18, and 19—and Visible Infrared Imaging Radiometer Suite[80] (VIIRS) from Soumi-NPP satellite.[81]

VH products from VIIRS were also processed from 2012 to present (1-km resolution, 7-day composite). The VH products can be used *as proxy data for monitoring vegetation health*, drought, soil saturation, moisture and thermal conditions, fire risk, greenness of vegetation cover, vegetation fraction, Leaf Area Index, start/end of the growing season, crop and pasture productivity, teleconnection with ENSO, desertification, mosquito-borne diseases, invasive species, ecological resources, land degradation, etc.

The following indices and products are available:

- Vegetation Health Index (VHI)
- Vegetation Condition Index (VCI)
- Temperature Condition Index (TCI)
- Soil Saturation Index (SSI)
- No noise Normalized Difference Vegetation Index (SMN)
- No noise brightness temperature (SMT)

In the following model, we have used VHI to model the impact of climate change–induced thermal and moisture on the production of cashew nuts in Mozambique. We have downloaded the Vegetation Index from Earth Observation.[82] Three indicators are included as part of the Earth Observation:

- Seasonal indicators
  - Cropland
  - Grassland
- Vegetation indicators
  - NDVI anomaly
- Precipitation indicators

All three vegetation indicators are based on 10-day (dekadal) vegetation data from the METOP-AVHRR sensor at 1-km resolution (2007 and after). Data at 1-km resolution for the 1984–2006 period are derived from the NOAA-AVHRR dataset at 16-km resolution. Precipitation estimates for all African countries (except Cabo

[79] Advanced Very High-Resolution Radiometer—https://www.usgs.gov/centers/eros/science/usgs-eros-archive-advanced-very-high-resolution-radiometer-avhrr

[80] Visible Infrared Imaging Radiometer Suite (VIIRS)—https://www.jpss.noaa.gov/viirs.html

[81] STAR—Global Vegetation Health Products : Introduction—https://www.star.nesdis.noaa.gov/smcd/emb/vci/VH/index.php

[82] Earth Observation—https://www.fao.org/giews/earthobservation/country/index.jsp?lang=en&code=MOZ

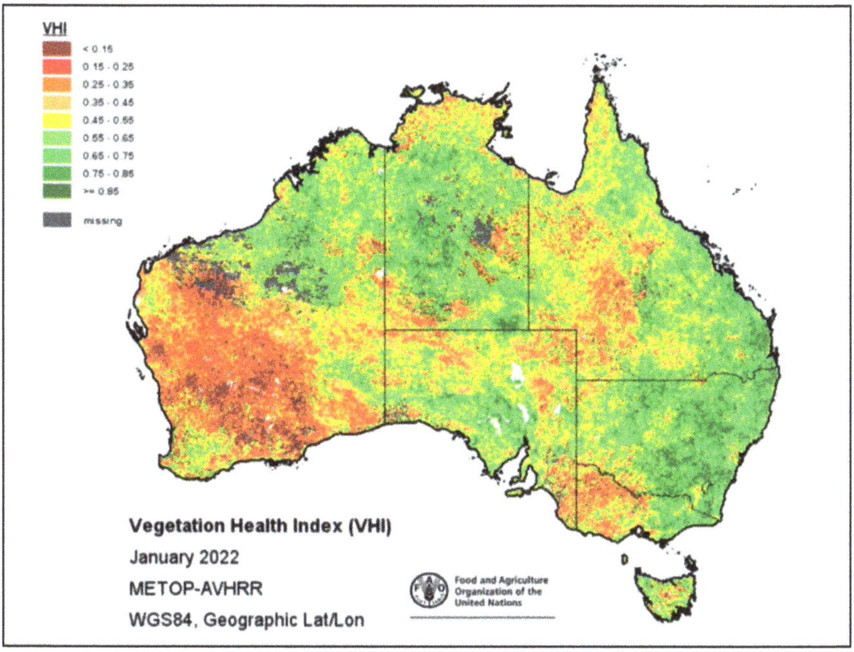

Fig. 8.26 VHI

Verde and Mauritius) are taken from the Famine Early Warning Systems Network[83] (NOAA/FEWSNet), while for the remaining countries, data are obtained from the European Centre for Medium-Range Weather Forecasts[84] (ECMWF).

8.6.1 NDVI Anomaly

The Normalized Difference Vegetation Index (NDVI) measures the "greenness" of ground cover and is used as a proxy to indicate the density and health of vegetation. NDVI values range from +1 to - 1, with high positive values corresponding to dense and healthy vegetation and low and/or negative NDVI values indicating poor vegetation conditions or sparse vegetative cover (please see Fig. 8.26). The NDVI anomaly indicates the variation of the current dekad to the long-term average, where a positive value (for example, 20%) would signify enhanced vegetation conditions compared to the average, while a negative value (for instance, -40%) would indicate comparatively poor vegetation conditions. The following data was downloaded from

[83] Famine Early Warning Systems Network—https://earlywarning.usgs.gov/fews

[84] European Centre for Medium-Range Weather Forecasts—https://www.ecmwf.int/

the spatial aggregated time series data for Australia[85] (source Meteorological Operational Satellite (METOP): the data is from 2022 and would look as follows[86]).

To determine the density of green on a patch of land, researchers must observe the distinct colors (wavelengths) of visible and near-infrared sunlight reflected by the plants. As can be seen through a prism, many different wavelengths make up the spectrum of sunlight. When sunlight strikes objects, certain wavelengths of this spectrum are absorbed, and other wavelengths are reflected. The pigment in plant leaves, chlorophyll, strongly absorbs visible light (from 0.4 to 0.7 μm) for use in photosynthesis. The cell structure of the leaves, on the other hand, strongly reflects near-infrared light (from 0.7 to 1.1 μm). The more leaves a plant has, the more these wavelengths of light are affected.[87]

8.6.2 Step 1: Load Mozambique Vegetation Health Index (VHI) DEKAD Data

```
#   https://www.fao.org/giews/earthobservation/country/index.
jsp?lang=en&code=MOZ
# Vegetation Index
# Vegetation Health Index
data_MODIS_VHI = pd.read_csv
('Mozambique_vhi_adm_dekad_data_ML.csv')
data_MODIS_VHI.head()
```

Output:

	Indicator	Country	ADM1_CODE	Province	Date	Data	Year	Month	Dekad	Unit	Source	monthlabel
0	Vegetation Health Index (VHI)	Mozambique	2118	Nampula	1/1/1984	0.552	1984	1	1		FAO-ASIS	month_1_wk_dekad_1
1	Vegetation Health Index (VHI)	Mozambique	2118	Nampula	1/11/1984	0.568	1984	1	2		FAO-ASIS	month_1_wk_dekad_2
2	Vegetation Health Index (VHI)	Mozambique	2118	Nampula	1/21/1984	0.570	1984	1	3		FAO-ASIS	month_1_wk_dekad_3
3	Vegetation Health Index (VHI)	Mozambique	2118	Nampula	2/1/1984	0.578	1984	2	1		FAO-ASIS	month_2_wk_dekad_1
4	Vegetation Health Index (VHI)	Mozambique	2118	Nampula	2/11/1984	0.596	1984	2	2		FAO-ASIS	month_2_wk_dekad_2

85 Spatial NDVI Data—https://www.fao.org/giews/earthobservation/asis/data/country/AUS/MAP_NDVI_ANOMALY/DATA/vhi_adm1_dekad_data.csv

86 VHI January 2022—https://www.fao.org/giews/earthobservation/asis/data/country/AUS/MAP_NDVI_ANOMALY/HR/om2201h.png

87 Photosynthesis and NDVI—https://earthobservatory.nasa.gov/features/MeasuringVegetation/measuring_vegetation_2.php

```
data_MODIS_VHI = data_MODIS_VHI.drop(['Indicator',
'Country','ADM1_CODE','Province','Date','Month','Dekad',
'Unit','Source'], axis=1)
data_MODIS_VHI
```

Output:

	Data	Year	monthlabel
0	0.489	1984	month_1_wk_dekad_1
1	0.558	1984	month_1_wk_dekad_2
2	0.660	1984	month_1_wk_dekad_3
3	0.661	1984	month_2_wk_dekad_1
4	0.581	1984	month_2_wk_dekad_2
...	...	...	...
1367	0.484	2021	month_12_wk_dekad_3
1368	0.489	2022	month_1_wk_dekad_1
1369	0.473	2022	month_1_wk_dekad_2
1370	0.398	2022	month_1_wk_dekad_3
1371	0.467	2022	month_2_wk_dekad_1

1372 rows × 3 columns

Need to pivot table to arrange columns as year based.

```
data_MODIS_VHI = data_MODIS_
VHI.pivot_table('Data', ['Year'], 'monthlabel')
data_MODIS_VHI
```

Pivot tables offer a ton of flexibility for me as a data scientist. pivot_table requires a data and an index parameter:

- Data is the Pandas DataFrame you pass to the function.
- Index is the feature that allows you to group your data.

In our case, we use Data as the column and Year as pivot.

Output:

monthlabel	month_10_wk_dekad_1	month_10_wk_dekad_2	month_10_wk_dekad_3	month_11_wk_dekad_1	month_11_wk_dekad_2	month_11_wk_dekad_3	r
Year							
1984	0.598	0.568	0.581	0.527	0.461	0.420	
1985	0.368	0.314	0.336	0.363	0.436	0.525	
1986	0.308	0.333	0.405	0.418	0.434	0.419	
1987	0.379	0.405	0.396	0.396	0.400	0.367	
1988	0.312	0.281	0.313	0.429	0.400	0.396	
1989	0.527	0.507	0.521	0.482	0.409	0.472	
1990	0.676	0.689	0.660	0.579	0.478	0.402	
1991	0.454	0.471	0.424	0.339	0.310	0.348	

where week 1 represents January 1st through January 7th.

8.6.3 Step 2: Prepare Production DataFrame

```
dfAustraliaSugarcaneProductionData          =          pd.read_csv
('AustraliaSugarcaneProduction.csv')

dfAustraliaSugarcaneProductionData.tail()
```

Output: The sugarcane production data frame:

	Year	Areaharvested(ha)	SugarcaneProduction(tonnes)	Yield(hg/ha)
0	1961.0	156521	9730296	621661
1	1962.0	162514	12940301	796258
2	1963.0	168981	12312569	728636
3	1964.0	190355	15312357	804411
4	1965.0	203635	14382227	706275
5	1966.0	225407	16952688	752092
6	1967.0	223829	17025232	760636
7	1968.0	230052	18708304	813221
8	1969.0	212786	15784354	741795
9	1970.0	220521	17644816	800142
10	1971.0	233737	19390512	829587
11	1972.0	241699	18928304	783135
12	1973.0	225854	19278000	853560
13	1974.0	253142	20417808	806575
14	1975.0	256800	21959008	855102

8.6.4 Step 3: Sugarcane and MODIS VHI

```
#https://www.fao.org/giews/earthobservation/country/index.
jsp?lang=en&code=AUS
dfAustraliaSugarcaneProductionDataMODISLSNDVI = pd.merge
(data_MODIS_VHI, dfAustraliaSugarcaneProductionData,
left_index=True, right_index=True,how = "inner")
dfAustraliaSugarcaneProductionDataMODISLSNDVI
```

Output:

```
Index(['month_10_wk_dekad_1', 'month_10_wk_dekad_2',
'month_10_wk_dekad_3',
    'month_11_wk_dekad_1', 'month_11_wk_dekad_2',
'month_11_wk_dekad_3',
    'month_12_wk_dekad_1', 'month_12_wk_dekad_2',
'month_12_wk_dekad_3',
    'month_1_wk_dekad_1', 'month_1_wk_dekad_2',
'month_1_wk_dekad_3',
    'month_2_wk_dekad_1', 'month_2_wk_dekad_2',
'month_2_wk_dekad_3',
    'month_3_wk_dekad_1', 'month_3_wk_dekad_2',
'month_3_wk_dekad_3',
    'month_4_wk_dekad_1', 'month_4_wk_dekad_2',
'month_4_wk_dekad_3',
    'month_5_wk_dekad_1', 'month_5_wk_dekad_2',
'month_5_wk_dekad_3',
    'month_6_wk_dekad_1', 'month_6_wk_dekad_2',
'month_6_wk_dekad_3',
    'month_7_wk_dekad_1', 'month_7_wk_dekad_2',
'month_7_wk_dekad_3',
    'month_8_wk_dekad_1', 'month_8_wk_dekad_2',
'month_8_wk_dekad_3',
    'month_9_wk_dekad_1', 'month_9_wk_dekad_2',
'month_9_wk_dekad_3',
    'Areaharvested(ha)', 'SugarcaneProduction(tonnes)', 'Yield
(hg/ha)],
   dtype='object')
```

8.6.5 Step 4: Merged DataFrame Correlation Matrix

```
import seaborn as sn
import matplotlib.pyplot as plt

import plotly.graph_objects as go
import seaborn as sn; sn.set()

fig = plt.figure(figsize=(30, 20))
corrMatrix = dfAustraliaSugarcaneProductionDataMODISLSNDVI.
corr()
sn.heatmap(corrMatrix, annot=True)
plt.show()
```

Output: Almost all the Vegetation Health Index values are negatively correlated. Since the Vegetation Health Index (VHI) illustrates the severity of drought based on the vegetation health and the influence of temperature on plant conditions, the production of cashew nuts, of course, is negatively correlated.

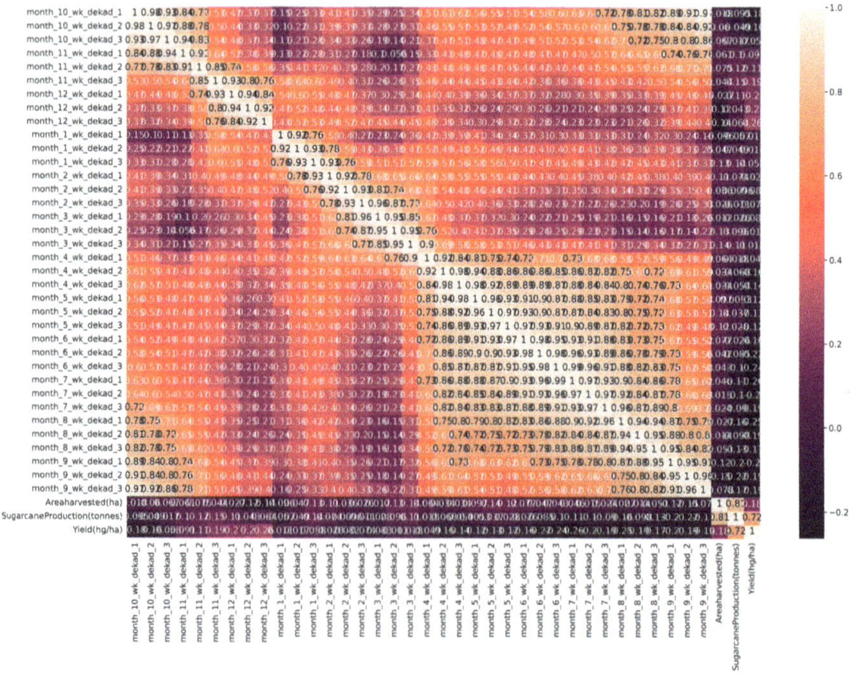

8.6.6 Step 5: Prepare Data for Regression Model

The feature columns represent the peak growing and harvesting season (June to July and August to September).

```
X = dfAustraliaSugarcaneProductionDataMODISLSNDVI
[['Areaharvested(ha)',

    'month_6_wk_dekad_1', 'month_5_wk_dekad_2',
'month_5_wk_dekad_3',
 'month_7_wk_dekad_3',
    'month_8_wk_dekad_1', 'month_8_wk_dekad_2',
'month_8_wk_dekad_3',
    'month_9_wk_dekad_2', 'month_9_wk_dekad_3'
                                         ]]
y = dfAustraliaSugarcaneProductionDataMODISLSNDVI
['SugarcaneProduction(tonnes)']

from sklearn.model_selection import train_test_split
X_train, X_test, y_train, y_test = train_test_split(X, y,
test_size = 0.3, random_state = 0)
```

That is,

- June through July (month 6 to month 7)
- August and September (month 8 to month 9)

```
# Train the model
from sklearn.linear_model import LinearRegression

# Fit a linear regression model on the training set
model = LinearRegression(normalize=True).fit(X_train, y_train)
print (model)
```

Output:

```
LinearRegression(normalize=True)
```

8.6.7 Step 6: Regression Metrics

```
import numpy as np

predictions = model.predict(X_test)
np.set_printoptions(suppress=True)
print('Predicted labels: ', np.round(predictions))
print('Actual labels  : ' ,y_test)

import matplotlib.pyplot as plt

%matplotlib inline

plt.scatter(y_test, predictions)
plt.xlabel('Actual Labels')
plt.ylabel('Predicted Labels')
plt.title('Predictions')
# overlay the regression line
z = np.polyfit(y_test, predictions, 1)
p = np.poly1d(z)
plt.plot(y_test,p(y_test), color='magenta')
plt.show()
```

Output:

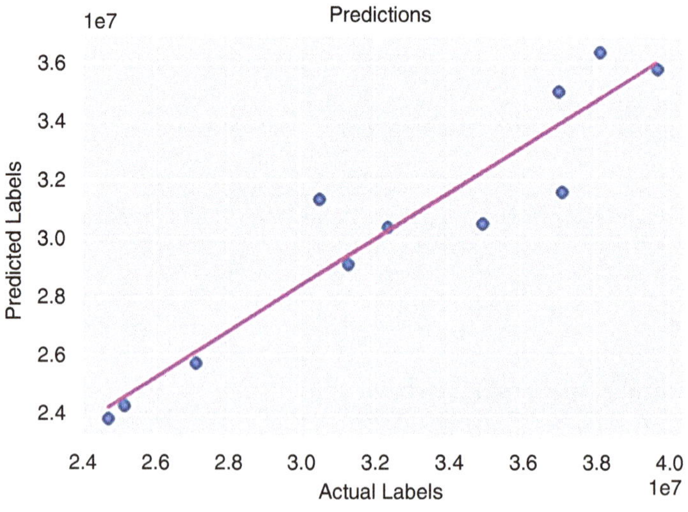

Regression metrics:

```
from sklearn.metrics import mean_squared_error, r2_score

mse = mean_squared_error(y_test, predictions)
print("MSE:", mse)

rmse = np.sqrt(mse)
print("RMSE:", rmse)

r2 = r2_score(y_test, predictions)
print("R2:", r2)
```

Output:

```
MSE: 7734470517784.285
RMSE: 2781091.6054283944
R²: 0.6954045154186335
```

Let us derive the climate change model governing the equation:

```
mx=""
for ifeature in range(len(X.columns)):
  if model.coef_[ifeature] <0:
    # format & beautify the equation
   mx += " - " + "{:.2f}".format(abs(model.coef_[ifeature])) + " * "
+ X.columns[ifeature]
  else:
    if ifeature == 0:
      mx += "{:.2f}".format(model.coef_[ifeature]) + " * " + X.
columns[ifeature]
    else:
      mx += " + " + "{:.2f}".format(model.coef_[ifeature]) + " * " + X.
columns[ifeature]

print(mx)
```

Output:

```
87.45 * Areaharvested(ha) - 16880409.73 * month_6_wk_dekad_1 -
15183064.41 * month_5_wk_dekad_2 + 33223480.62 *
month_5_wk_dekad_3 + 10494115.85 * month_7_wk_dekad_3 -
47461048.87 * month_8_wk_dekad_1 + 14843232.23 *
month_8_wk_dekad_2 + 17939509.52 * month_8_wk_dekad_3 -
3220717.83 * month_9_wk_dekad_2 + 4768851.46 * month_9_wk_dekad_3
```

The model equation:

```
# y=mx+c
if(model.intercept_ <0):
  print("The formula for the " + y.name + " linear regression line is
= " + " - {:.2f}".format(abs(model.intercept_)) + " + " + mx )
else:
  print("The formula for the " + y.name + " linear regression line is
= " + " {:.2f}".format(model.intercept_) + " + " + mx)
```

Output: The below equation provides the governing rule that connects VDSI dekad to sugarcane production.

$$\begin{aligned}
&\text{The formula for the SugarcaneProduction(tonnes) linear regression line is} \\
&= -1075604.43 + 87.45^{*}\ \text{Areaharvested(ha)} \\
&\quad - 16880409.73^{*}\ \text{month_6_wk_dekad_1} \\
&\quad - 15183064.41^{*}\ \text{month_5_wk_dekad_2} \\
&\quad + 33223480.62^{*}\ \text{month_5_wk_dekad_3} \\
&\quad + 10494115.85^{*}\ \text{month_7_wk_dekad_3} \\
&\quad - 47461048.87^{*}\ \text{month_8_wk_dekad_1} \\
&\quad + 14843232.23^{*}\ \text{month_8_wk_dekad_2} \\
&\quad + 17939509.52^{*}\ \text{month_8_wk_dekad_3} \\
&\quad - 3220717.83^{*}\ \text{month_9_wk_dekad_2} \\
&\quad + 4768851.46^{*}\ \text{month_9_wk_dekad_3}
\end{aligned} \tag{8.10}$$

8.6.8 *Step 7: Other Regression Ensemble Model*

I have used the voting ensemble.[88] The idea behind the voting regressor[89] is to combine conceptually different machine learning regressors and return the average predicted values. Such a regressor can be useful for a set of equally well-performing models in order to balance out their individual weaknesses.

[88] Ensemble Methods—https://scikit-learn.org/stable/modules/classes.html#module-sklearn.ensemble

[89] Voting Regressor—https://scikit-learn.org/stable/modules/ensemble.html#voting-regressor

```
import numpy as np
from sklearn.linear_model import LinearRegression
from sklearn.ensemble import RandomForestRegressor
from sklearn.ensemble import GradientBoostingRegressor
from sklearn.ensemble import VotingRegressor
from sklearn.model_selection import RepeatedStratifiedKFold
r1 = LinearRegression()
r2 = RandomForestRegressor(n_estimators=10,random_state=1)
r3 = GradientBoostingRegressor()
cv = RepeatedStratifiedKFold(n_splits=10,n_repeats=3,
random_state=1)
erT = VotingRegressor([('lr', r1), ('rf', r2) ,('gb',r3)],
verbose=True)
erT.fit(X_train, y_train)
```

Output:

```
[Voting] ....................... (1 of 3) Processing lr, total=
0.0s
[Voting] ....................... (2 of 3) Processing rf, total=
0.0s
[Voting] ....................... (3 of 3) Processing gb, total=
0.0s
VotingRegressor(estimators=[('lr', LinearRegression()),
                ('rf',
                RandomForestRegressor(n_estimators=10,
                                random_state=1)),
                ('gb', GradientBoostingRegressor())],
        verbose=True)
```

Call predict to predict values.

```
pred_yT1_df = pd.DataFrame({'Actual Value':y_test, 'Predicted
value':yT1_pred,'Difference': y_test-yT1_pred})
pred_yT1_df
```

Output:

Year	Actual Value	Predicted value	Difference
2015	32379153	3.358331e+07	-1.204154e+06
2004	36993454	3.734138e+07	-3.479277e+05
2000	38164688	3.702789e+07	1.136802e+06
2014	30517650	3.147040e+07	-9.527508e+05
2006	37128107	3.526028e+07	1.867827e+06
1999	39699000	3.784643e+07	1.852571e+06
1994	31312000	2.533079e+07	5.981215e+06
1986	24742000	2.406524e+07	6.767639e+05
1995	34943000	3.338101e+07	1.561993e+06
2013	27136082	2.376308e+07	3.373004e+06
2011	25181814	2.524547e+07	-6.365751e+04

8.6.9 *Step 8: Model* $\mathbf{R}^2$

```
import matplotlib.pyplot as plt

%matplotlib inline

plt.scatter(y_test,yT1_pred)
plt.xlabel('Actual Labels')
plt.ylabel('Predicted Labels')
plt.title('Predictions')
# overlay the regression line
z = np.polyfit(y_test, yT1_pred, 1)
p = np.poly1d(z)
plt.plot(y_test,p(y_test), color='magenta')
plt.show()
```

Output:

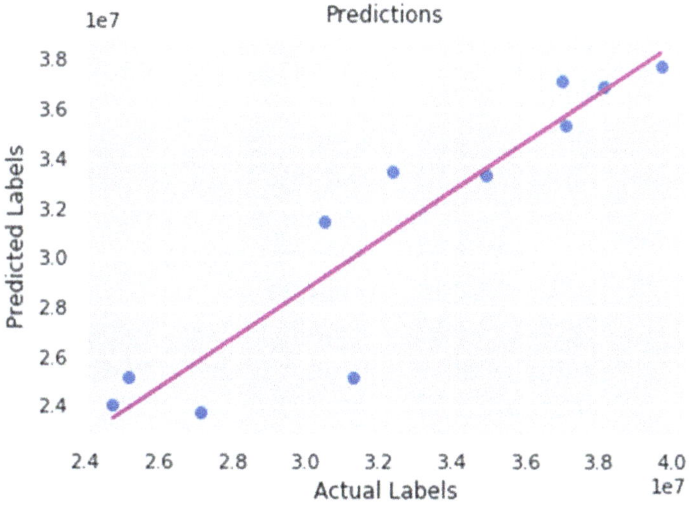


```
r2_score(y_test,yT1_pred)
```

Output:
0.7768759630672404 (77.68%)

8.6.10 Step 9: Explainability of the Model

Please find explainability:

```
eli5.show_weights(model, feature_names=[
'Areaharvested(ha)',

    'month_6_wk_dekad_1', 'month_5_wk_dekad_2',
'month_5_wk_dekad_3',
 'month_7_wk_dekad_3',
    'month_8_wk_dekad_1', 'month_8_wk_dekad_2',
'month_8_wk_dekad_3',
    'month_9_wk_dekad_2', 'month_9_wk_dekad_3' ])
```

Output:

Weight?	Feature
+33223480.616	month_5_wk_dekad_3
+17939509.522	month_8_wk_dekad_3
+14843232.226	month_8_wk_dekad_2
+10494115.850	month_7_wk_dekad_3
+4768851.461	month_9_wk_dekad_3
+87.449	Areaharvested(ha)
-1075604.431	<BIAS>
-3220717.828	month_9_wk_dekad_2
-15183064.409	month_5_wk_dekad_2
-16880409.728	month_6_wk_dekad_1
-47461048.874	month_8_wk_dekad_1

8.7 Conclusion: Sugarcane Yield and Climate Change Impact

Under future climate scenarios,[90] the annual mean rainfall for Queensland is projected to decrease by 1–14% by the year 2030 and between 2 and 42% by 2070 [7]. Annual mean temperatures are projected to increase by 0.5–1.2 °C by 2030 and 1.0–3.7 °C by 2070. The biophysical effects of climate change on sugarcane yield is not certain, but preliminary modelling shows that increased temperature will speed up crop development but will also increase water stress and so have both positive and negative effects. The potential for growing sugarcane in the dry tropics is constrained by limited irrigation as well as rainfall and the current average cost of water stress to the industry is estimated at more than A$200 million a year. In favorable regions, sugarcane production is economically profitable. However, sugarcane management is labor-intensive and requires an adequate water supply. Productivity is highly weather-dependent. Recent climate change has increased the frequency and severity of droughts and floods, negatively affecting growing conditions. Although government policies, related agencies, and sugarcane mills have tried to address these problems, sugarcane crop production has suffered in recent years.

[90] Remote Sensing Applications in Sugarcane Cultivation: A Review—https://www.mdpi.com/2072-4292/13/20/4040/htm

Earth observation (EO) can provide sugarcane as well as other crop-related information over large areas in a timely and cost-efficient manner and thereby address at least a part of the information requirements of the global sugar industry. Since the 1980s, satellite remote sensing has become a relevant data source to detect, map, and monitor crop growth and to support health management and crop productivity.

Weather conditions present the most important challenge for sugarcane yield. Both production and productivity are significantly affected by climate factors including rainfall, temperature, solar radiation, and relative humidity. Sustained sugarcane growth during the different phases requires specific climatic conditions.

Thermal time controls the phenological development of sugarcane. The generally accepted model of thermal time is based on the accumulation of the daily (starting from the first day of planting) average temperature (simplified as the average of maximum and minimum temperature values), from which the baseline temperature for growth is subtracted. This calculation [7] is expressed by growing degree days (GDD) as in Eqs. (8.1) and (8.2).

8.7.1 Yield Estimation and Prediction

Most approaches for sugarcane yield estimation utilize regression analysis between a predictor (e.g., NDVI) and actual yield measurements and they were mostly developed within research studies with no operational solution yet. Beyond challenges related to data availability (both from satellite observations and from actual yield measurements used for calibration and validation), the adoption of a new algorithm has seen a trend towards the application of machine learning approaches. Supervised machine learning algorithms such as RFR have shown highly satisfactory results. However, the challenge remains regarding the implementation of robust predictive models that can be calibrated and then transferred for application to different regions or cropping seasons.

After reading this chapter, you should be able to answer key drivers of food insecurity and food security indicators and drivers. Additionally, you should be able to apply the Global Food Security Index (GFSI) framework to develop machine learning and mathematical optimization models. You should be able to mathematically develop food affordability vs. median income and yearly food cost regressive model. You should be able to answer the role of technological innovations, data-driven models, and small farm sustainability to enhance food security.

References

1. Abha Bhattarai, Tony Romm and Rachel Siegel, U.S. economy appeared ready to surge, but Russia's invasion of Ukraine could send shockwaves, February 25, 2022 at 6:00 a.m. EST, https://www.washingtonpost.com/business/2022/02/25/economy-us-russia-ukraine-gas/, Access Date: March 13, 2022
2. Alan Rappeport, The war in Ukraine is likely to slow global growth, the I.M.F. warns, March 10, 2022, https://www.nytimes.com/2022/03/10/business/economy/imf-global-outlook-ukraine-war.html, Access Date: March 13, 2022
3. Alfred Cang, Cargill CEO Says Global Food Prices to Stay High on Labor Crunch, November 17, 2021, 10:54 PM PST, https://www.bloomberg.com/news/articles/2021-11-17/cargill-ceo-says-global-food-prices-to-stay-high-on-labor-crunch?sref=fyhEsXfZ, Access Date: December 25, 2021
4. Bibhudatta Pradhan & Pratik Parija, An Ambitious Ethanol Plan Spurs Food Security Fears in India, October 2021, https://www.bloombergquint.com/business/an-ambitious-ethanol-plan-sparks-food-security-fears-in-india?msclkid=68fc2699a2f611ec8fd74348a99a8ed7, Access Date: March 13, 2022
5. Business Today, Govt assures ethanol blending plan will not affect India's food security, Oct 07, 2021, 4:49 PM IST, https://www.businesstoday.in/latest/economy/story/govt-assures-ethanol-blending-plan-will-not-affect-indias-food-security-308747-2021-10-07, Access Date: March 13, 2022
6. Rauniyar, The food-fuel conundrum, Jul 15, 2021, https://www.thehindubusinessline.com/opinion/the-food-fuel-conundrum/article35348867.ece , Access Date: March 13, 2022
7. Som-ard J, Atzberger C, Izquierdo-Verdiguier E, Vuolo F, Immitzer M. Remote Sensing Applications in Sugarcane Cultivation: A Review. Remote Sensing. 2021; 13(20):4040. https://doi.org/10.3390/rs13204040
8. Dr. Hossein Shapouri, OEPNU/OCE, USDA and Dr. Michael Salassi, J. Nelson Fairbanks Professor of Agricultural Economics, Department of Agricultural Economics and Agribusiness, LSU Agricultural Center, THE ECONOMIC FEASIBILITY OF ETHANOL PRODUCTION FROM SUGAR IN THE UNITED STATES, July 2006, https://www.fsa.usda.gov/Internet/FSA_File/ethanol_fromsugar_july06.pdf, Access Date: February 19 , 2022
9. Cheavegatti-Gianotto A, de Abreu HM, Arruda P, et al. Sugarcane (Saccharum X officinarum): A Reference Study for the Regulation of Genetically Modified Cultivars in Brazil. Trop Plant Biol. 2011;4(1):62-89. doi:https://doi.org/10.1007/s12042-011-9068-3
10. Commonwealth of Australia 2015, Current and future arrangements for the marketing of Australian sugar, 24 June 2015, https://www.aph.gov.au/Parliamentary_Business/Committees/Senate/Rural_and_Regional_Affairs_and_Transport/Sugar/Report, Access Date: February 19, 2022
11. Dr Mark Stafford Smith, Crops ready for a different future climate, 2020, https://research.csiro.au/climate/themes/agriculture/crops-ready-different-future-climate/, Access Date: February 19, 2022
12. Isadore Barmash, Soaring Sugar Cost Arouses Consumers and U.S. Inquiries, Nov. 15, 1974, https://www.nytimes.com/1974/11/15/archives/soaring-sugar-cost-arouses-consumers-and-u-s-inquiries-what-sent.html , Access Date: February 21, 2022
13. Mr. Helmut Ahlfeld,FAO – Cuba Conference, Cuba – SUGAR MARKETS IN DISARRAY,7–9 December 1999, https://www.fao.org/3/X4988E/x4988e05.htm, Access Date: February 21, 2022
14. Elizabeth M. Fowler, Commodities; Price Rises for Sugar Continue, June 6, 1983, https://www.nytimes.com/1983/06/06/business/commodities-price-rises-for-sugar-continue.html , Access Date: February 21, 2022
15. The Associated Press, FUTURES/OPTIONS; Sugar Surplus Sends Prices Plummeting, April 24, 1985, https://www.nytimes.com/1985/04/24/business/futures-options-sugar-surplus-sends-prices-plummeting.html, Access Date: February 21, 2022

16. Ronald E. Dolan, ed. Philippines: A Country Study. Washington: GPO for the Library of Congress, 1991, http://countrystudies.us/philippines/, Access Date: February 21, 2022
17. Won W. Koo and Richard D. Taylor, 2012 Outlook of the U.S. and World Sugar Markets, 2011–2021, April 2012, https://ageconsearch.umn.edu/record/128037/files/AAE692.pdf , Access Date: February 21, 2022
18. Shaimaa Khalil, Climate change: Australia pledges net zero emissions by 2050, 26 October 2021, https://www.bbc.com/news/world-australia-59046032, Access Date: February 20, 2022
19. Kirili Lamb, Sugar production figures for 2019 down, February 17, 2020 – 5:20PM, https://www.weeklytimesnow.com.au/news/rural-weekly/sugar-production-figures-for-2019-down/news-story/4a301a678336cdaaa11842933e528c6b, Access Date: February 21, 2022
20. Jennifer Fadoul, How Satellite Maps Help Prevent Another 'Great Grain Robbery', Aug 20, 2021, https://www.nasa.gov/feature/how-satellite-maps-help-prevent-another-great-grain-robbery, Access Date: February 19, 2022
21. Ellen Gray, Global Climate Change Impact on Crops Expected Within 10 Years, NASA Study Finds, November 2, 2021, https://climate.nasa.gov/news/3124/global-climate-change-impact-on-crops-expected-within-10-years-nasa-study-finds/, Access Date: January 28, 2022
22. Lisa Barrington, Exclusive-U.S.-UAE push for another $4bln in farming climate change investment, Sun, February 20, 2022, 12:49 AM, https://finance.yahoo.com/news/exclusive-u-uae-push-another-084920821.html, Access Date: February 22, 2022
23. Josep Cos, Francisco Doblas-Reyes, Martin Jury, Raül Marcos, Pierre-Antoine Bretonnière, and Margarida Samsó, The Mediterranean climate change hotspot in the CMIP5 and CMIP6 projections, 30 July 2021, https://esd.copernicus.org/preprints/esd-2021-65/esd-2021-65.pdf, Access Date: February 04, 2022
24. Jingyu Zeng, Rongrong Zhang, Yanping Qu, Virgílio A. Bento, Tao Zhou, Yuehuan Lin, Xiaoping Wu, Junyu Qi, Wei Shui, Qianfeng Wang, Improving the drought monitoring capability of VHI at the global scale via ensemble indices for various vegetation types from 2001 to 2018, Weather and Climate Extremes, Volume 35, 2022, 100412, ISSN 2212-0947, https://doi.org/10.1016/j.wace.2022.100412

Chapter 9
Linkage Models: Economic Key Drivers and Agricultural Production

This chapter covers the following:

- Linkage models
  - Agricultural production, macroeconomic variables, co-movement variables, and farm inputs
- Food-versus-fuel conundrum
- Machine learning linkage models
  - Australia Macroeconomic Drivers and Sugarcane Production Predictive Model
  - Myanmar's Macroeconomic Drivers and Rice Production Predictive Model

This chapter introduces the linkage models that explain the relationship among agricultural production, macroeconomic variables, co-movement variables, and farm inputs such as labor and fertilizer costs. The chapter deep dives on each of the important linkage model variables and explains its relation to machine learning models developed in the chapter. Additionally, it explains the food-versus-fuel conundrum and the role it plays in food security. Next, the chapter introduces the linkage models for two use cases: the Australia Macroeconomic Drivers and Sugarcane Production Predictive Model and Myanmar's Macroeconomic Drivers and Rice Production Predictive Model. Finally, the chapter concludes with highly influential exogenous weather variable on multifaceted weather variable and implications of severe weather on global commerce and food security.

C. Vuppalapati, *Artificial Intelligence and Heuristics for Enhanced Food Security*, International Series in Operations Research & Management Science 331,
https://doi.org/10.1007/978-3-031-08743-1_9

Food Security, Inflation, and Linkage Models[1]

There is a direct correlation between food security and energy inflation. For instance, energy prices in Turkey and other countries have soared in the past year due to recovering demand and geopolitical tensions. But Turks have also seen their overall purchasing power dwindle dramatically amid a currency crisis and two-decade high inflation that reached almost 50% last month.

As a result of high energy prices and the falling currency, several families have been forced to drastically cut the amount they spend on food. Breakfast foods like eggs, cheese, and olives have become luxuries.

Around 160,000 children and young people have dropped out of school in 2021. Some feel obliged to contribute to the family income and leave school to work. Others leave school because they cannot pay for transport or other expenses. With sharp increases of prices in food including basic goods and baby formula, which has risen by 55.6%, malnutrition is now a serious risk for children. Some give their babies dehydrated soups.

Food security is an economic and access argument. The key drivers of economics are the enablers of food security. Conversely, the drivers that adversely impact economics worsen food insecurity. For instance, economies facing an exceptional shortfall in aggregate food production/supplies because of crop failure, natural disasters, weather extremes, interruption of imports, disruption of distribution, excessive postharvest losses, or other supply bottlenecks are at the highest risk of food insecurity. Additionally, countries with widespread lack of access, where most of the population is unable to procure food from local markets, due to very low incomes, exceptionally high food prices, or the inability to circulate within the country, experience severe food insecurity. Countries with severe localized food insecurity due to the influx of refugees, a concentration of internally displaced persons, or areas with combinations of crop failure and deep poverty also experience food insecurity.

The key drivers of food insecurity are high food prices, food price instability, oil prices, deficient grain stock levels, higher exchange rate volatility, rising demand for food due to population growth, commodity futures speculations, climate change, political instability/conflicts,[2] and abrupt change in consumption patterns and pressure on food production rate from climate change, scarcity of natural resource availability due to land degradation and water scarcity, food tradeoff biofuel production, and a lack public and private investment in infrastructure [1]. The degree of importance of each key driver varies between countries and regions according to their unique set of physical, economic, and social circumstances. Weather and

[1]High energy prices put more pressure on Turks, Tue—https://www.yahoo.com/news/high-energy-prices-put-more-071920590.html

[2]Macron tells French farmers: Ukraine war will weigh on you, and it will last—https://www.yahoo.com/news/macron-tells-french-farmers-ukraine-075753530.html

monsoons will have a direct impact on the food security as the pressure on price spikes could result due to bad monsoons.[3] Weather in main production countries[4] could provide a guidance on the staple food shortfall or consequential price increases [2, 3].

A special emphasis must be drawn to conflicts, people displacement, and social unrest. They are the main catalyst and drivers of food insecurity and result in a huge toll on humanitarian and economic aspects of life. The main victims of conflicts and wars are children, women, senior citizens, and marginal communities. As seen recently, political conflicts, sadly, are recurrent and destabilizing the world, resulting in more human suffering! The conflicts and wars not only drastically increase food insecurity in the affected areas but also induces unprecedented price and food instability across many other parts of the world. For example, during the Ukraine 2022 war, French president, Macron, tells French farmers that war will weigh heavily on agriculture and prices: trade restrictions resulting from European Union sanctions on Russia will weigh on French exports such as wine and grains while a further rise in energy prices will hit livestock farming. Measures need to be taken to protect farmers from cost pressures and compensate for lost revenues. During conflict times, a surge in commodity prices can be witnessed. Paris wheat futures[5] rose to new record peaks, while Chicago wheat consolidated near a 9-year high, as conflict weighs on uncertainties, quality risks in Australia's harvest, and a deterioration in US growing conditions kept attention on tightening global supplies [4]. Increase in wheat prices has certainly benefited grain producers but squeezed livestock farmers for whom grain feed is a major cost. The crisis in Ukraine has increased volatility in agricultural markets (hard red winter wheat historical pricing[6]), with Paris wheat futures hitting a record high. Farmers are also worried that the crisis could exacerbate supply tensions in fertilizers and disrupt the spring growing season for crops (please see Fig. 9.1).

Spike in Hard Red Winter Wheat: Due to Ukraine Conflict

Chicago wheat futures eased to below $9 per bushel but remained at over a 9-year high as the Russian invasion of Ukraine brought the possibility of supply disruptions from two of the world's largest producers. With Russia

(continued)

[3]Prices of basmati rice, edible oils and pulses ride high going into festive season—https://economictimes.indiatimes.com/news/economy/agriculture/prices-of-basmati-rice-edible-oils-and-pulses-ride-high-going-into-festive-season/articleshow/86782579.cms?from=mdr

[4]Low monsoon rains in India means less rice for the world in 2012—UN agency—https://news.un.org/en/story/2012/08/417072

[5]Paris wheat sets new record highs as traders FRET on world supplies—https://www.agriculture.com/markets/newswire/grains-paris-wheat-sets-new-record-highs-as-traders-fret-on-world-supplies

[6]Hard Red Winter Wheat—https://www.investing.com/commodities/hard-red-winter-wheat-historical-data

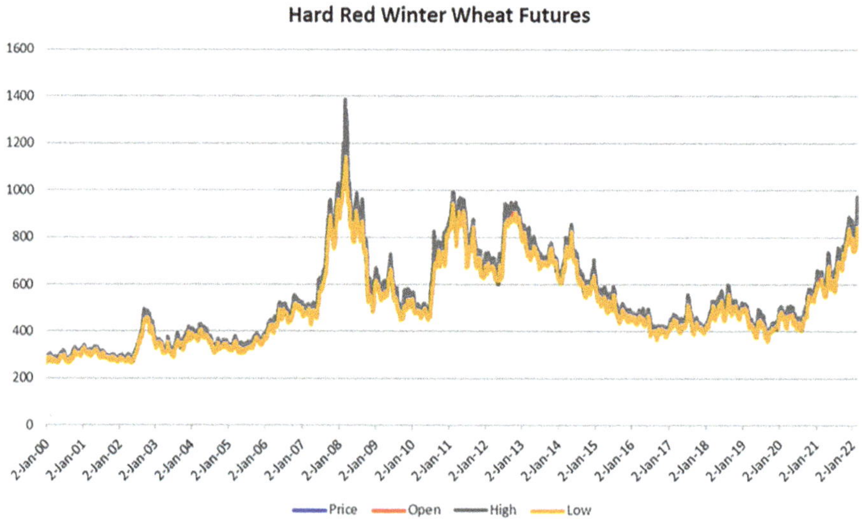

Fig. 9.1 Hard red winter wheat futures

and Ukraine accounting for roughly 30% of the world's wheat exports, conflict in the region jeopardizes crucial supply from an already tight market. Meanwhile, wheat stocks in major exporting countries are already at low levels, and ongoing droughts are reducing maize availability in South America, the United States, and Canada.[7]

Wheat futures are available for trading in the Chicago Board of Trade (CBOT), Euronext, Kansas City Board of Trade (KCBT), and the Minneapolis Grain Exchange (MGEX). The standard contract unit is 5,000 bushels. The United States is the biggest exporter of wheat followed by the European Union, Australia, and Canada. Wheat prices displayed in Trading Economics are based on over-the-counter (OTC) and contract for difference (CFD) financial instruments. Our market prices are intended to provide you with a reference only, rather than as a basis for making trading decisions.[8]

Unit: USD/BU

Frequency: daily

Peace is essential for ensuring sustained food security. Peace and prosperity, as the old saying goes, go together and importantly, reinforce each other. We must preserve peace through good governance and administrative efforts and must always resort to diplomatic pursuits in ensuring peaceful resolution of any conflicts. To

[7] Wheat—https://tradingeconomics.com/commodity/wheat

[8] Wheat —https://tradingeconomics.com/commodity/wheat

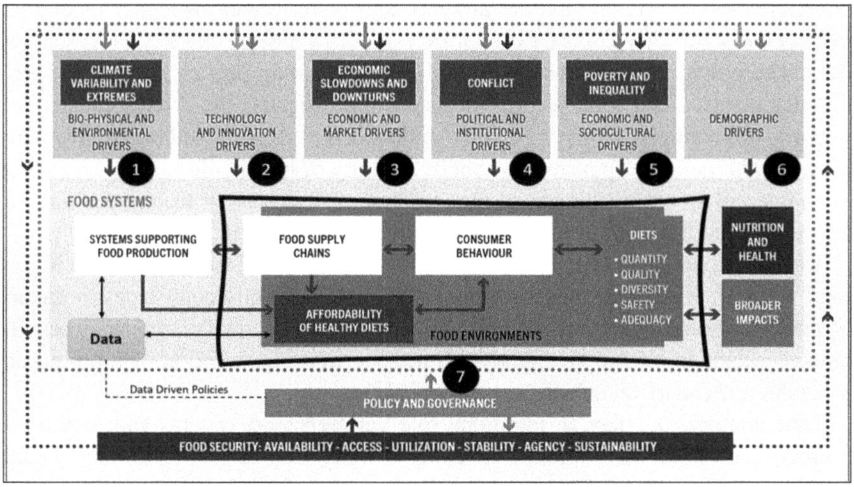

Fig. 9.2 Food security influencer. Source: Adapted from HLPE. 2020. Food security and nutrition: building a global narrative towards 2030. A report by the High-Level Panel of Experts on Food Security and Nutrition of the Committee on World Food Security. Rome [5] (The State of Food Security and Nutrition in the World—https://www.fao.org/3/cb4474en/cb4474en.pdf)

assure food security for current and future generations, it is prudent to investigate the abovementioned factors and model the agricultural and social distributions to avert any future risks to food insecurity.

To view the overall influence of other drivers on food security, please see Fig. 9.2. It presents a food system diagram to illustrate how the drivers behind food security and nutrition trends specifically create multiple impacts throughout food systems (food systems, including food environments), leading to impacts on the four dimensions of food security (availability, access, utilization, and stability), as well as the two additional dimensions of agency and sustainability. These drivers have impacts on attributes of diets (quantity, quality, diversity, safety, and adequacy) and nutrition and health outcomes (nutrition and health). The major drivers are indicated in the dark boxes that are behind the recent rise in hunger and slowdown in progress in reducing all forms of malnutrition:

1. Climate variability and extremes
2. Technology and innovation drivers⁺
3. Economic slowdown and downturns
4. Conflict
5. Poverty and inequality
6. Demographic drivers⁺
7. Policy and governance

⁺Positive Drivers

In Fig. 9.11, conflict (political and institutional drivers), climate variability and extremes (biophysical and environmental drivers), economic slowdowns and downturns (economic and market drivers), and poverty and inequality (economic and sociocultural drivers) are external drivers that act upon food systems (curved box). These drivers tend to create multiple, compounding impacts on food systems that negatively affect food security and nutrition. Because the drivers coexist and interact, this complexity must be fully understood and addressed when designing program and policy responses. The unaffordability of healthy diets is regarded here as an internal driver resulting from the effect of other drivers or factors that directly affect the cost of nutritious foods throughout the food system.

Efficient food production systems and food supply chains contribute to higher food security. Higher farm production diversity significantly contributes to dietary diversity in some situations, but not in all. Improving small farmers' access to markets seems to be a more effective strategy to improve nutrition than promoting production diversity on subsistence farms [6].[9]

Data components play an important role between food systems and food environments. It collects and utilizes information from food drivers (1 through 7) and facilitates optimization by feeding systems supporting food prediction. It feeds back food environment data (supply chain, consumer behavior, and affordability of healthy foods) to improve overall food system effectiveness to drive higher food security. Finally, it enables analytics and heuristics to provide a forecasting capability that improves overall food system efficiencies plus feeds into food policies and governance to have effective policies to overcome food security issues. Data-driven policies, derived from food systems, help to improve governance and policy. Linkage models try to uncover these complex interactions and provide a predictive capability to identify and recommend a future potential food insecurity event.

Agriculture and downstream agro-processing generate half of the gross domestic product (GDP) and four-fifths (almost a third of the GDP, occupying nearly 80% of the workforce[10]) of total export earnings and employment [7].[11] Climate shocks[12] therefore have a potentially profound direct effect on the agricultural sector and farm households while also indirectly affecting other economic sectors and nonfarm households through price and production linkages [8].

[9]Production diversity and dietary diversity in smallholder farm households—https://www.pnas.org/content/112/34/10657

[10]Malawi—https://ricepedia.org/malawi & http://irri.org/resources/publications/books/rice-almanac-4th-edition

[11]Agricultural growth and investment options for poverty reduction in Malawi—https://www.ifpri.org/publication/agricultural-growth-and-investment-options-poverty-reduction-malawi-0

[12]Droughts and floods in Malawi Assessing the economywide effects—https://www.ifpri.org/publication/droughts-and-floods-malawi

9.1 Linkage Models

The purpose of the linkage models is to provide actionable insights by weaving in dynamics of commodity markets on the farm side inputs plus co-movement/causation drivers, thereby holistically looking both supply and demand sides. Put it differently, linkage models are forecast models that account supply–demand aspects of the commodity markets that could manifest in price dynamics, agriculture yields, and harvest pattern. The governing rules for linkage models are derived from data signals that are based on the dynamics of weather pattern, consumption pattern, agriculture transportation, and supply and demand movements that potentially could exert pressure on agricultural crop prices. In other words, linkage models closely observe agricultural market movements to develop forecast models using advanced machine learning and data science. On a broad scale, linkage models are applicable to any commodities that are exposed to market perturbations. Nevertheless, a special emphasis on agriculture commodities is that they are not only affected by global and macro-level movements but also greatly influenced by local markets and national government mandates and policies. In a way, linkage models must be fine-tuned to each local agricultural market that is being modeled. Table 9.1 shows the contributing factors to develop linkage models.

The goal of the linkage model is to develop enhanced food security through the application of artificial intelligence and heuristics. The outcome of the models is considered to enhance food security if the models forestall any potential food crisis issues by combing through the agricultural yield, macroeconomic, and food dependency ontologies or relationships. Before we investigate the linkage model framework, please find below a table that has captured some of the historical events that had resulted in food insecurity. In addition to the factors listed below, many countries across the world have been affected by the COVID-19 pandemic, and as a result, the impact of the pandemic is considered as a key factor that has worsened food insecurity and increased the need for humanitarian assistance in all countries, although it may not be mentioned specifically.[13]

	Nature of food insecurity	Main reason	Country, year
	Exceptional shortfall in aggregate food production/supplies	Poor seasonal rains	In Kenya, about 2.4 million people are estimated to be severely food insecure between November 2021 and January 2022, reflecting consecutive poor rainy seasons since late 2020 that affected crop and livestock production, mainly in northern and eastern pastoral, agropastoral, and

(continued)

[13] Crop Prospects and Food Situation—https://www.fao.org/giews/reports/crop-prospects/en/

Table 9.1 Linkage model contributing factors

Linkage model: contributing factors
Demand–supply dynamics
Long run
Demand
Export demand growth
Due to food demand growth
Due to population growth
New use/innovation: biofuels
IGC's Grains and Oilseeds (GOI) Price Index
Supply
Slow production growth
Declining R&D investment in agricultural innovation
Land retirement (heavily influenced by government policies)
Slower adaptation of data innovations and multimodel data
Artificial intelligence[a] and machine learning readiness [9]
Short run
Demand
Government food policies
Supply
Government food policies
Weather-induced crop losses/failure
Macroeconomic policies
Economic growth
Depreciation of US dollar
Rising oil prices
Accumulation of petrodollars/foreign reserves
Futures market/speculation
Inflation
FAO Food Price Index
Financial crisis
Farm inputs
Employment in agriculture (% of total employment)
Rural population
Area under harvest
Outside farm employment opportunities
Food security
Average value of food production
Domestic food price level index
Depth of the food deficit
Family-level income
Cereal import dependency ratio
Value of food imports over total merchandise exports
Domestic food price volatility (FAO Food Price Index[b]) [10]

(continued)

Table 9.1 (continued)

Per capita food production variability
Per capita food supply variability

[a]McKinsey Global Institute (MGI)—Notes from the AI frontier Modeling the impact of AI on the world economy—https://www.mckinsey.com/~/media/McKinsey/Featured%20Insights/Artificial%20Intelligence/Notes%20from%20the%20frontier%20Modeling%20the%20impact%20of%20AI%20on%20the%20world%20economy/MGI-Notes-from-the-AI-frontier-Modeling-the-impact-of-AI-on-the-world-economy-September-2018.ashx
[b]Food Price Index—https://www.fao.org/worldfoodsituation/foodpricesindex/en/

	Nature of food insecurity	Main reason	Country, year
			marginal agricultural areas [11].[14]
			Somalia, due to poor rainy seasons since late 2020, which severely affected crop and livestock production, about 3.5 million people are estimated to food insecure.
	Exceptional shortfall in aggregate food production/ supplies	Shortfall in cereal production, localized shortfalls in staple food production, tight supplies	In Nigeria, between June and August 2022, 3.64 million people are projected to face severe food insecurity due to shortfall in cereal production caused by the effects of adverse weather.
			In Mozambique, an estimated 1.7 million people required humanitarian assistance until September 2021, and this number is expected to rise moderately until the next harvest period in March 2022, as households exhaust supplies of food from their harvests.
			In Burkina Faso,[15] in 2018, tight food supplies and high prices had resulted in higher food insecurity [11].
			In Pakistan,[16] in 2016, population displacement and localized cereal production shortfalls had contributed to food insecurity [12].

(continued)

[14]Crop prospects and quarterly global report food situation—https://www.fao.org/3/cb7877en/cb7877en.pdf

[15]Crop prospects and food situation, June 2018—https://www.fao.org/3/I9666EN/i9666en.pdf

[16]Crops and food—https://www.fao.org/3/i6558e/i6558e.pdf

	Nature of food insecurity	Main reason	Country, year
	Exceptional shortfall in aggregate food production/supplies	Weather extremes, floods, weather shock, monsoon floods	In Chad, domestic cereal production is anticipated to fall to a below-average level in 2021, due to the effects of adverse weather and the civil conflict. As a result, between June and August 2022, 1.74 million people are projected to face severe food insecurity.
			In Ethiopia, about 7.4 million people were estimated to be severely food insecure between July and September 2021 due to high food prices and floods.
			In Kenya, adverse weather on crop production during the first semester of 2016 had resulted in food insecurity [12].
			In Somalia, late and erratic October-to-December rains severely affected prospects for 2016 "deyr" season crops in most southern and central areas
			Economic constraints, monsoon floods, and high prices of the main staple food—prices of rice, the country's main staple, reached near-record levels in most markets in January 2021, constraining access to food [13].[17]
			In Mozambique,[18] food security conditions are worsened in 2018 in southern and some central provinces due to the unfavorable weather conditions that had reduced the 2018 cereal harvest [12].
	Widespread lack of access	Macroeconomic challenges	In 2021, in Eritrea, macroeconomic challenges had caused the population's vulnerability to food insecurity.

(continued)

[17] Crops and food—https://www.fao.org/3/cb3672en/cb3672en.pdf

[18] Crop prospects and food situation, June 2018—https://www.fao.org/3/I9666EN/i9666en.pdf

	Nature of food insecurity	Main reason	Country, year
	Widespread lack of access	High food prices of the main food staple, soaring food prices due to conflict	In Ethiopia, about 7.4 million people were severely food insecure between July and September 2021 due to high food prices and floods.
			In Sierra Leone, about 1.1 million people were food insecure between October and December 2021 due to high food prices and low purchasing power, resulting in acute constraints on households' economic access to food.
			In Sudan, soaring food prices between October 2021 and February 2022 had caused severe food insecurity in around 6 million people.
			In Pakistan, poverty levels have increased due to losses of income-generating opportunities owing to the effects of the COVID-19 pandemic on the economy. Prices of wheat flour, the country's main staple, were at record or near-record levels in most markets in October 2021, constraining access.
			In Myanmar,[19] the impact of floods for a second consecutive year and renewed conflict in northern parts of Rakhine State affected around 500,000 people in 2016 [14].
	Exceptional shortfall in aggregate food production/ supplies	Economic downturn, severe economic crisis, reduced incomes, economic constraints	In Eswatini, an estimated 316 000 people were food insecure between October 2021 and March 2022 due to food access constraints, largely underpinned by the impacts of the COVID-19 pandemic on the economy.
			In Lesotho, the slow economic recovery continues to impose constraints on households'

(continued)

[19]Crop prospects and food situation, December 2016—https://www.fao.org/3/i6558e/i6558e.pdf

	Nature of food insecurity	Main reason	Country, year
			incomes, impinging on their economic capacity to access food—as a result, the number of people projected to be food insecure between January and March 2022 was estimated at 312 000.
			In Myanmar, in 2021–2022, income losses due to the impact of the COVID-19 pandemic have affected the food security situation of vulnerable households.
	Widespread lack of access	High fuel prices	In Yemen, conflict, poverty, floods, high food, and fuel prices had resulted in high food insecurity [15].[20]
	Severe localized food insecurity	Deterioration of pastoral conditions	In Chad, about 990,000 are projected to be food insecure during June to August due to the serious deterioration of pastoral conditions in the Sahel [12].
			In Djibouti, due to the impact of consecutive unfavorable rainy seasons on pastoral livelihoods, about 197,000 people were severely food insecure, mainly concentrated in pastoral areas [12].
			In Cabo Verde, poor performance of the 2017 agro-pastoral cropping season causing significant loss of livelihoods caused food insecurity [12].
			In Mauritania, 2018, declines in agricultural and pastoral production resulted in unfavorable food security outcomes [12].
			In Senegal, poor pastoral conditions have led to an unfavorable food security situation in northern parts of the country.

(continued)

[20]The critical role of escalating food prices in Yemen's food security crisis—https://blogs.worldbank.org/arabvoices/critical-role-escalating-food-prices-yemens-food-security-crisis

	Nature of food insecurity	Main reason	Country, year
	Exceptional shortfall in aggregate food production/supplies	Drought	In Mozambique [14], drought affected the 2016 production and higher food prices—drought conditions resulted in lower cereal output
	Exceptional shortfall in aggregate food production/supplies	Economic downturn and steep depreciation of the local currency	In Nigeria, due to the economic downturn, steep depreciation of the local currency, population displacements, and severe insecurity in northern areas, more than 8 million people were estimated to be food insecure [14].

In summary, the linkage model goal is to analyze macroeconomic and farm yield data and recommend the following:

- Proactive insights on shortfall in food production/supplies because of crop failure, natural disasters (weather), interruption of imports, disruption of distribution, excessive postharvest losses, or other supply bottlenecks [16]:
  - Linear regression, vector autoregression (VAR), and ensemble models enable artificial intelligence to capture data signals and provide a proactive recommendation.
- Food insecurity warnings stem from lack of access due to very low incomes, exceptionally high food prices, or the inability to circulate within the country:
  - Econometric and machine learning pricing models that factor in organic versus speculative trends
- Impact of adverse weather conditions on crops and a potential shortfall of crops because of the area planted, which indicate a need for close monitoring of the crop for the remainder of the growing season:
  - Climate change models with the Vegetation Index to provide advanced trends and warnings about a shortfall of essential commodity yield to forestall future food insecurity events

The Food security monitoring system (FSMS) framework[21] also underscores key macroeconomic drivers, as illustrated in the abovementioned main reasons, covers 35 key indicators that are classified along the four dimensions of food security[22]: availability (5 indicators), access (10 indicators), utilization (13 indicators), and stability (7 indicators) [17]. The indicators are classified along availability, access,

[21] Sudan—Food Security Monitoring System, http://fsis.sd/SD/EN/FoodSecurity/Monitoring/

[22] Handbook for Defining and Setting up a Food Security Information and Early Warning System (FSIEWS)—https://www.fao.org/3/X8622E/x8622e00.htm#TopOfPage

utilization, and stability[23] [18]. Please see the Food Security Monitoring System (FSMS)/Appendix—Food Security and Nutrition for all lists of food security indicators. To enhance food security, it is important to pay attention to the FSMS key indicators and be able to predict the occurrence of such through the framework of economic-and-food security linkage. Additionally, the derived key food security indicators are synthesized from the Integrated Food Security Phase Classification (IPC) recommendations.[24] Some of the influencing parameters include the following:

		Recommendation
	Pakistan[25]: acute food insecurity situation due to crop failures Time period: October 2021 to March/April 2022 and projection for April/May to June 2022	Training on *climate-smart crop* and fodder production.
	Yemen[26]: Acute food insecurity situation due to money access Time period: October to December 2020 and projection for January to June 2021	Conflict, high food prices, depreciation of local currency, and disrupted livelihoods are the major drivers of acute food insecurity. Injection of *foreign currency reserves*.
	Afghanistan[27]: acute food insecurity situation due to crop failures Time period: September to October 2021 and projection for November 2021 to March 2022	Scale-up of livelihood assistance for the winter wheat season, the spring season crops, and vulnerable herding households is essential to prevent further deterioration of household food production capacity in rural areas.
	Honduras[28]: acute food insecurity situation due to economic opportunity loss caused by the COVID-19 pandemic Time period: December 2020 to March 2021 and projections for April to June 2021 and July to September 2021	Economic reactivation measures: economic opportunity creations for localities where a higher proportion of households have depleted their reserves and are employing crisis or emergency strategies.

(continued)

[23] Food Security Information and Knowledge Sharing System—Sudan—https://www.fao.org/family-farming/detail/en/c/450039/

[24] IPC Food Security—https://www.ipcinfo.org/ipc-country-analysis/en/

[25] Pakistan Acute Food Insecurity Situation—https://www.ipcinfo.org/ipc-country-analysis/details-map/en/c/1155374/?iso3=PAK

[26] Yemen: Acute Food Insecurity Situation October–December 2020 and Projection for January–June 2021—https://www.ipcinfo.org/ipc-country-analysis/details-map/en/c/1152947/

[27] Afghanistan: Acute Food Insecurity Situation September–October 2021 and Projection for November 2021–March 2022—https://www.ipcinfo.org/ipc-country-analysis/details-map/en/c/1155210/?iso3=AFG

[28] Honduras: Acute Food Insecurity Situation December 2020–March 2021 and Projections for April–June 2021 and July–September 2021—https://www.ipcinfo.org/ipc-country-analysis/details-map/en/c/1153046/?iso3=HND

		Recommendation
	Guatemala[29]: acute food insecurity situation due to storm damage of essential commodities, spike in essential food commodities, and COVID-19-induced employment issues. Time period: projection update for November 2020 to March 2021	Employment: through the economic reactivation plan, identify population that got affected—a behavior of the demand for agricultural and nonagricultural employment; temporary and permanent employment, loss of employment, and reduction of wages should be monitored. The price of maize, beans, and other foods should be monitored in the most affected departments, as they could increase due to storm damage or if the restrictions on mobilization due to the COVID-19 pandemic become strict again.
	Haiti[30]: acute food insecurity situation due to economic decline and poor harvest. *Economic decline*: inflation, exchange rate deterioration, reduction in remittances *Poor harvests*: Poor agricultural harvests due to below-normal rainfall. Time period: August 2020 to February 2021 and projection for March to June 2021	Support for livelihoods: to overcome economic challenges due to recurrent shocks in recent years, 2020 (drought, cyclones, price hikes), assistance (credit) to be made available. Linkage between emergency and development.
	Tajikistan[31]: acute food insecurity situation due to low purchasing capacity, fewer harvest, and low livestock asset holding. Time period: September to December 2012	Recommendations: improvement in economic conditions to improve purchasing power and climate-smart technologies to reduce animal diseases.

As mentioned in the above table, some of the root causes of food insecurity include poor harvest, price hikes, access to financial resources, high food prices, high oil prices, lack of drinking and irrigation water, economic decline, unavailability of high-cost fertilizers, animal diseases,[32] and extreme weather. As part of the linkage model, I have derived key economic parameters that could provide venue to develop forecast models to address food insecurity. Contributing factors related to food

[29] Guatemala: Acute Food Insecurity Situation Projection Update for November 2020–March 2021—https://www.ipcinfo.org/ipc-country-analysis/details-map/en/c/1152979/?iso3=GTM

[30] Haiti: Acute Food Insecurity Situation August 2020–February 2021 and Projection for March–June 2021—https://www.ipcinfo.org/ipc-country-analysis/details-map/en/c/1152816/?iso3=HTI

[31] Tajikistan: Acute Food Insecurity Situation September–December 2012—https://www.ipcinfo.org/ipc-country-analysis/details-map/en/c/459517/?iso3=TJK

[32] Tajikistan: Acute Food Insecurity Situation September–December 2012—https://www.ipcinfo.org/ipc-country-analysis/details-map/en/c/459517/?iso3=TJK

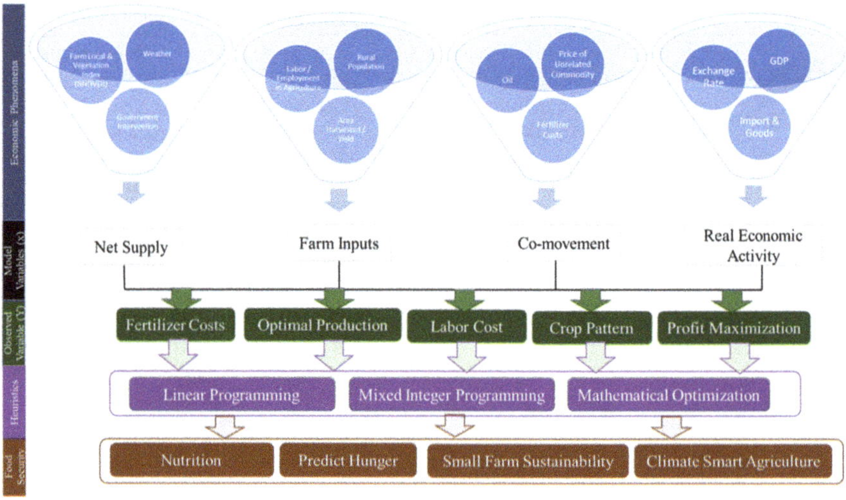

Fig. 9.3 Linkage model

safety are maintained at favorable or normal levels: no adverse weather events, grain prices below the historical average, and employment opportunities.[33]

> Climate-Smart Dairy Edge Device
>
> Dairy milk is a cash crop for many small farmers. Farmers tend to lock in their investment on farm inputs such as high-cost fertilizers, crop seeds, and labor, could see returns after harvest season. Dairy milk and dairy animal husbandry are survival that keep them afloat.
>
> Dairy cattle productivity goes down due to weather, temperature, and humidity stresses. Climate-smart agriculture and dairy products could help farmers to counterbalance risks.

As part of the linkage modeling, the following visual representation illustrates (please see Fig. 9.3) key economic parameters. This study identified four key drivers of wheat futures prices in each futures market: real economic drivers and the corresponding outcome observed variables from the machine learning point of view. Each of these drivers encapsulates numerous economic phenomena, as discussed earlier. Linear regression, ensemble modeling, and voting ensemble consisting of linear regression are applied on the dataset to generate a forecasting model [19].[34]

[33]Honduras: Acute Food Insecurity Situation in Southern Honduras December 2012–January 2013—https://www.ipcinfo.org/ipc-country-analysis/details-map/en/c/459515/?iso3=HND

[34]Deconstructing Wheat Price Spikes: A Model of Supply and Demand, Financial Speculation, and Commodity Price Comovement—https://www.ers.usda.gov/webdocs/publications/45199/46439_err165.pdf?v=0

9.2 Farm Drivers (Exogenous)

Weather, government interventions, local farm laws, trade (import/export) restrictions, Vegetation Health Index, and agricultural grain transportation cost indicators are major farm economic drivers, albeit small farmers and farms have no control over it. Nonetheless, they could influence the global commodity prices.

9.2.1 Composite Index: IGC's Grains and Oilseeds (GOI) Price Index and GOFI

The International Grains Council (IGC) circulates two price indices which are the Grain and Oilseeds Index (GOI) and the Grain and Oilseeds Freight Market Index (GOFI). These two indices indicate the respective market prices [20].[35] The GOI markets are affected by various factors such as supply and demand, weather, and freight markets. A good proxy for export and import demand as the price index rises when importers buy greater quantities from exporters amid rising consumption is the component of the equation which helps to explain the rise in price.[36] From a machine learning point of view, it has data from 2000. The International Grains Council (IGC) is an intergovernmental organization that seeks to further international cooperation in grains trade; promote expansion, openness, and fairness in the grains sector; and contribute to grain market stability and enhance world food security.[37] The demand for essential staple commodities: GOI data[38] can be downloaded in the IGC website (please see Figs. 9.4 and 9.5).

The GOI Price Index is a proxy variable to understand the agriculture export and import market as it exhibits a very close correlation (82.04%) with wheat prices (please see Table 9.2 and Fig. 9.6). In classical statistics, a proxy or proxy variable is a variable that is not in itself directly relevant, but that serves in place of an unobservable or immeasurable variable. For a variable to be a good proxy, it must have a close correlation, not necessarily linear, with the variable of interest.

One can induce COVID-19 shocks and bounce back in the graph. As COVID-19 stunned the world in early March 2020, you can see a drop in the GOFI. As COVID-19 vaccination and boosters became available, the market opened with high-demand imports, a fact that can be observed in the overlapping of the index with wheat price, one of the most staple and necessary grain (please see Fig. 9.7).

---

[35] Analyzing Volatility spillovers between grain and freight markets—https://www.researchgate.net/publication/347557656_Analysing_Volatility_spillovers_between_grain_and_freight_markets

[36] The Grains and Oil Seeds Index—https://www.igc.int/en/markets/marketinfo-goi.aspx

[37] International Grains Council (IGC)—https://www.igc.int/en/about/aboutus.aspx

[38] GOI—http://www.igc.int/en/_csv/igc__goi.xlsb

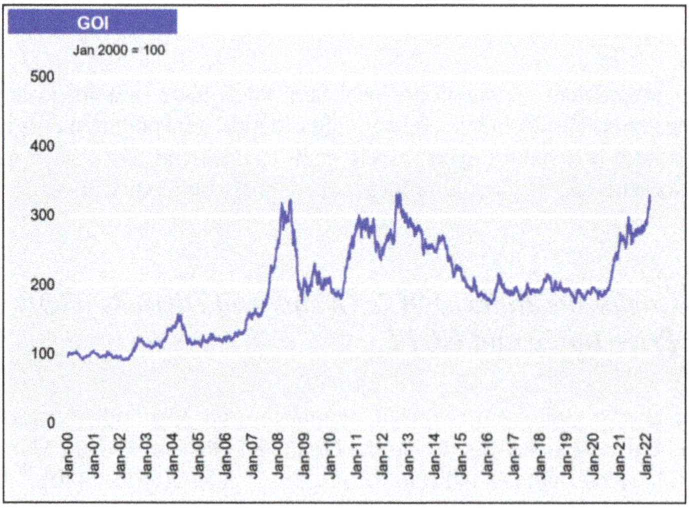

Fig. 9.4 GOI

The dry bulk freight complex, which serves many global grain and oilseed trade, appears to have been resilient in the face of the COVID-19[39] pandemic [21]. No major disruption has been witnessed so far, although COVID-related measures at ports have generally slowed cargo movement. After receding in early 2020, voyage costs on major grain/oilseed routes have posted sharp increases, especially in 2021. While gains were underpinned by optimism about improving economic conditions and robust demand for raw materials amid government stimulus in many countries, additional support stemmed from recovering marine fuel prices and spillover demand from the container market. Additionally, China's new port restrictions, linked to its zero-tolerance approach to the pandemic, have worsened congestion at local ports, thereby contributing to overall tonnage supply tightness. Reflecting market strength, the IGC Grains and Oilseeds Freight Index (GOFI)—a composite trade-weighted measure of total costs on major grains/oilseed routes – has increased by 78% since the start of 2020 and touched[40] its highest on record (since Jan 2013) in early July 2021 [22].

[39]Covid 19 crisis and the Grains/oilseeds and rice trade Agriculture Committee of WTO—https://www.wto.org/english/tratop_e/agric_e/petit_28jul20_e.pdf

[40]COVID-19 and agriculture—https://docs.wto.org/dol2fe/Pages/SS/directdoc.aspx?filename=q:/G/AG/GEN191.pdf&Open=True

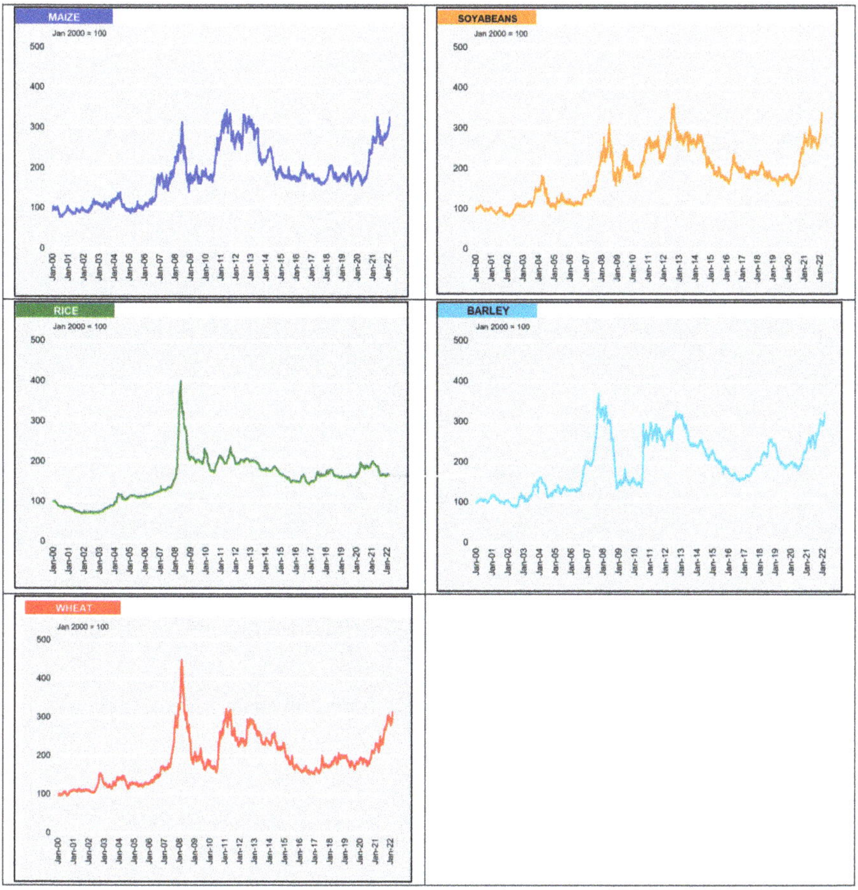

Fig. 9.5 Commodity prices

Table 9.2 Wheat prices and GOFI correlation

	Wheat price[a]	GOFI[b]
Wheat price	1	
GOFI	0.820309303	1

[a]Wheat Prices—https://www.investing.com/commodities/us-wheat-historical-data
[b]GOFI—https://www.igc.int/en/_csv/igc_gofi.xlsb

Revamped IGC Grains and Oilseeds Freight Index

As part of efforts to further promote transparency in international markets, the IGC Secretariat compiles weekly data on nominal ocean freight rates for a wide range of grain- and oilseed-carrying routes (mainly focusing on heavy

(continued)

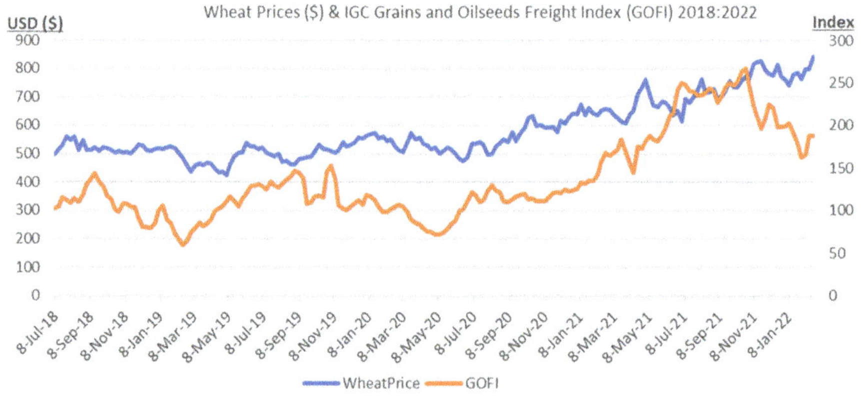

Fig. 9.6 Wheat price and GOFI

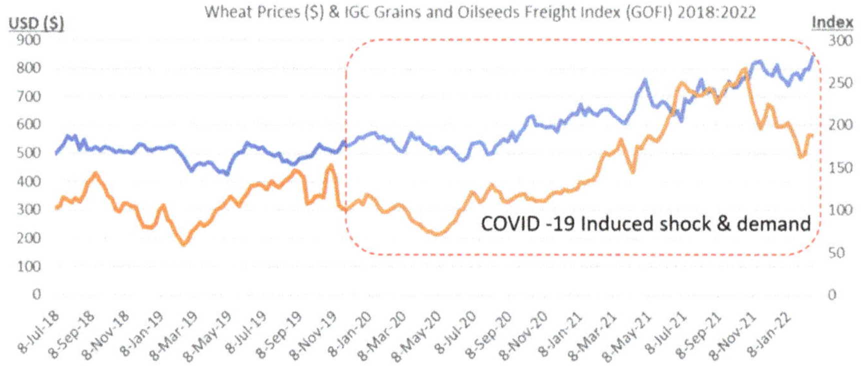

Fig. 9.7 Wheat prices and freight index

grains [wheat, durum], sorghum, and soybeans). The coverage has expanded significantly over the years, to a large extent reflecting the organization's improving expertise in the dry bulk freight market as well as its expanded commodity coverage. In May 2007, the secretariat introduced a weekly IGC Grain Freight Index (GFI), presenting an overall measure of movements in freight prices across key selected grain and oilseed routes.[41]

To provide a better reflection of freight costs in grain and oilseed trade, the secretariat proposes the introduction of a revamped weekly index, namely, the IGC Grains and Oilseeds Freight Index (GOFI). The new index builds

(continued)

[41] Revamped IGC grains and oilseeds Freight Index—https://www.igc.int/en/grainsupdate/gen1718info6rev1.pdf

significantly on the prior measure and covers 68 routes from key exporting origins, including Argentina, Australia, Brazil, Black Sea, Canada, the EU, and the United States (compared to 15 routes covered previously).

9.2.2 *Grain Transportation Cost Indicators*

Grain transportation indicator plays an important role in assessing demand/supply flows. In other words, it is a great leading indicator that pulses the state of agriculture demand. A higher index implies a higher demand, and the converse is true. One note of caution, the index reflects how transportation is in demand to carry agriculture products across the nation and the world, s definite indicator variable for long-run phenomena.

The data shows the weekly cost indices of transporting grain by each mode: truck, rail, barge, and ocean-going vessels. The base of each index (set to 100) is the average of monthly costs in the year 2000. For trucks, the base rate is $1.49 per gallon. For unit trains, the base rate is $1,815.15 per railcar, including the tariff and fuel surcharges (weekly changes reflect the monthly tariff rate plus fuel surcharge and weekly secondary railcar market bids). For the shuttle train, the base rate is $2,338.28 per railcar, including the tariff and fuel surcharges (weekly changes reflect the monthly tariff rate plus fuel surcharge and weekly secondary railcar market bids). For barge, the base rate is 180 and is based on the Illinois River barge rates (see the Downbound Grain Barge Rates dataset for more information). For the Gulf-to-Japan ocean route, the base rate is $22.36/metric ton (please see Fig. 9.8). For the Pacific Northwest-to-Japan ocean route, the base rate is $14.10/metric ton.[42]

Please kindly note that the index peaked in 2008, the great recession. The grain transportation indicator is another important variable of the demand. It is a great proxy variable as it exhibits a very strong correlation with staple wheat price (please see Table 9.3).

9.2.3 *Government Interventions and Credit to Agriculture*

There is an ever-increasing need to invest in agriculture due to a drastic rise in global population and changing dietary preferences of the growing middle class in emerging markets towards higher-value agricultural products. In addition, climate risks increase the need for investments to make agriculture more resilient to such risks. Estimates suggest that demand for food will increase by 70% by 2050 and at least $80 billion annual investments will be needed to meet this demand, most of which

[42] Grains Transportation Cost Indicators—https://agtransport.usda.gov/Transportation-Costs/Grain-Transportation-Cost-Indicators/8uye-ieij

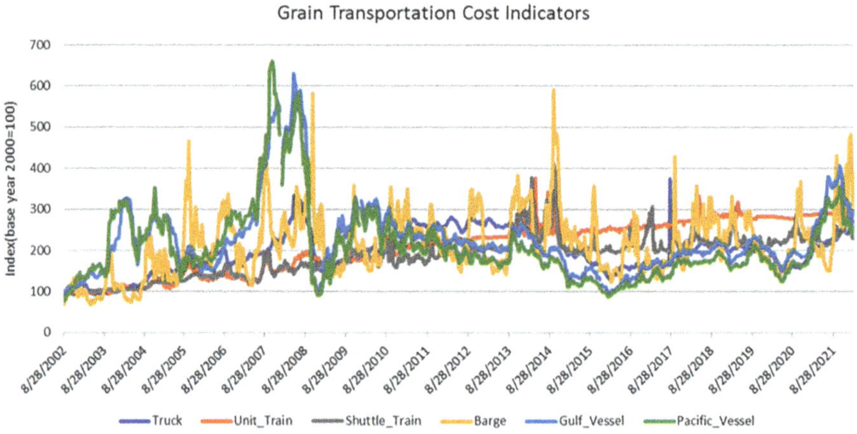

Fig. 9.8 Grain transportation cost indicators

Table 9.3 Wheat price, GOFI, and Vessel Index

	Wheat price[a]	GOFI[b]	Gulf vessel[c]	Pacific vessel
Wheat price	1			
GOFI	0.808325407	1		
Gulf Vessel	0.788814165	0.952294	1	
Pacific Vessel	0.780479927	0.949026	0.993340121	1

[a]Wheat Prices—https://www.investing.com/commodities/us-wheat-historical-data
[b]GOFI—https://www.igc.int/en/_csv/igc_gofi.xlsb
[c]Transportation Data—https://agtransport.usda.gov/api/views/8uye-ieij/rows.csv?accessType=DOWNLOAD

needs to come from the private sector. Financial sector institutions in developing countries lend a disproportionately lower share of their loan portfolios to agriculture compared to the agriculture sector's share of GDP. Agriculture finance empowers poor farmers to increase their wealth and food production to be able to feed 9 billion people by 2050. Finally, agricultural innovation serves to improve farmers' productivity, reduce their environmental impact, and address the challenges associated with ever-changing soil, weather, and market conditions. Since high costs and low yields may undermine the profitability of innovation,[43] policy makers often employ policy instruments to encourage producers to experiment with new production methods [23].

In addition to the supply of digital, farm equipment, capital availability to farmers is very critical to have a sustained and healthy agriculture. Countries across the world have invested in digital banking and announced credit to spur infrastructure improvement and investment. Here are the top 10 2019 countries that have extended

[43]Government Interventions to Promote Agricultural Innovation, https://web.stanford.edu/~kostasb/publications/agricultural_innovation.pdf

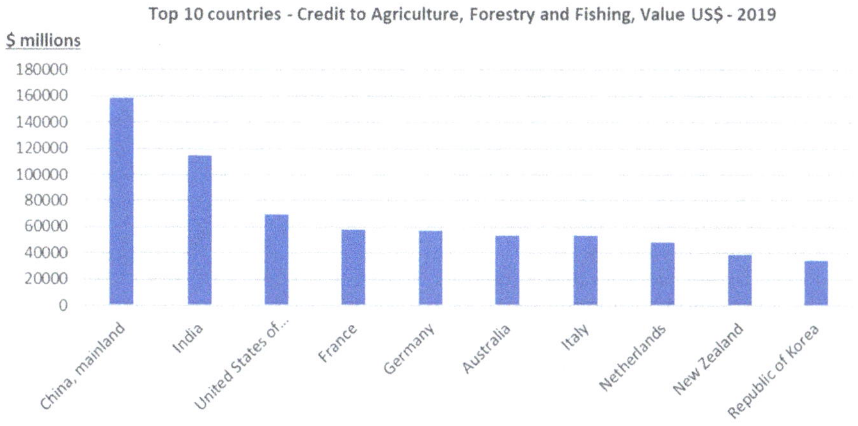

Fig. 9.9 Top 10 countries—credit to agriculture

credit to agriculture[44]: China, India, the United States, France, Germany, Australia, Italy, Netherlands, New Zealand, and Republic of Korea (please see Fig. 9.9).

9.2.3.1 The United States of America

The US Federal government offers a farm loan program to help farmers and ranchers get the financing they need to start, expand, or maintain a family farm.[45] The examples of farm loans include operation loans, farm ownership loans, microloans, youth loans, Native American tribal loans, emergency loans, and targeted loan funding. The ISDA announces the farmers rate on a quarterly.[46]

9.2.3.2 China

China is perhaps the most prominent example of a developing country that has transitioned from taxing to supporting agriculture.[47] In 2013, Chinese price supports and subsidies have risen at an accelerating pace after they were linked to rising production costs [24].

[44]Top 10 Countries—2019 http://www.fao.org/faostat/en/#data/IC/visualize

[45]Farm Loan Program—https://www.fsa.usda.gov/programs-and-services/farm-loan-programs/

[46]USDA Announces March 2021 Lending Rates for Agricultural Producers—https://www.fsa.usda.gov/news-room/news-releases/2021/usda-announces-march-2021-lending-rates-for-agricultural-producers

[47]Growth and Evolution in China's agricultural support policies— https://www.ers.usda.gov/webdocs/publications/45115/39368_err153.pdf?v=41491

Table 9.4 Chinese agricultural support programs

Year	Policy measure	Commodities
2000	Pilot reforms of rural taxes and fees	
	Soybean seed subsidy and pilot grain subsidy programs in several regions	Soybeans, rice, wheat, corn
2004–2006	Direct payment to grain producers General-input subsidy Improved seed subsidy Machinery subsidy Transfer payments to grain counties Reform of grain marketing system Eliminated agricultural, specialty crop, and animal slaughter Taxes Rice and wheat price supports	Rice, wheat, corn, soybeans
2007	Package of pork industry subsidies introduced and expanded Seed subsidy for cotton and rapeseed Transfer payments to oilseed and pork counties	Pork, cotton, rapeseed
2008	General-input subsidy linked to input prices Support prices for corn, soybeans, rapeseed Strategy of raising price supports annually adopted	Soybeans, rice, wheat, corn, rapeseed
2009	Hog price intervention program	Pork
2011	Cotton price support Grassland protection program	Cotton Cattle and sheep

Please find the timeline of Chinese agricultural support programs (please see Table 9.4).

9.2.3.3 India

India is one of the fastest-growing G20 economies, largely reflecting an ambitious reform agenda under implementation since 2014. Against this background, agriculture is a key sector in terms of its contribution to both employment and GDP. Sustained by improved access to inputs such as fertilizers and seeds, as well as better irrigation and credit coverage, production has been increasing on average at about 3.6% annually since 2011. The sector has also been diversifying from grains towards pulses, fruit, vegetables, and livestock products, largely driven by evolving demographics, urbanization, and changing demand patterns. India has achieved a significant fall in the proportion of the population that is undernourished, from around 24% in 1990–1992 to 15% in 2014–2016. Moreover, it has also emerged as a major agricultural exporter of several key commodities, currently being the largest exporter of rice globally and the second largest of cotton.

India (sets agricultural loan target at $226 billion in 2020–2021) to boost farm credit and infrastructure to spur output [25][48]. India also proposed to impose extra taxes on imports of some commodities, including edible oils, pulses, gold, silver, and some alcoholic beverages to finance more infrastructure facilities in rural areas. The impact of agricultural credit is applicable to the entire economy. A multiplier effect can be seen in the investment of agriculture [26].[49] The following are the positive effects on the macroeconomic indicators:

- Real GDP growth
- Unemployment rate (% of labor force)
- Overall budgetary surplus/deficit (% of GDP)
- Exports/imports

As farmers go, so goes the nation.
How did agriculture help the United States recover from the Great Depression in the spring of 1933 [26]?

9.2.3.4 France

The French agriculture sector is a dominant force within the EU 28, accounting for 18% of the EU 28 agricultural output in 2018. With 456 500 farms, the French agriculture sector also accounted for 3.9% of the total employment in the EU 28 in 2018. The value of French agricultural production reached EUR 76.6 billion in 2018. The major agricultural produce that places France among the top producers in the EU are cereals (21.7% of the EU 28), wine (46%), industrial crops (19.7%), cattle (23%), and dairy (16%). Nearly 60% of the agricultural output in France is generated through crop production, with wine representing 17.4% and cereals 14%. The livestock sector generates 34.5% of agricultural output,9 with dairy accounting for 12.5% of this subsector and cattle 9.9%. 10 With a value of EUR 63.5 billion in 2018, agriculture exports accounted for nearly 13% of France's total exports (please see Fig. 9.10), with the cattle sector at 9.9%. 10 With a value of EUR 63.5 billion in 2018, agriculture exports accounted for nearly 13% of France's total exports. France has provided $60 million agriculture credit to farmers. The major purposes of bank loans include working capital, investment in new machinery[50] and equipment,

[48] India to Boost Farm Credit and Infrastructure to Spur Output—https://www.bloomberg.com/news/articles/2021-02-01/india-plans-to-enhance-farm-credit-to-boost-agricultural-output

[49] As farmers go, so goes the nation—https://www.aeaweb.org/research/farmer-recovery-great-depression

[50] Financial needs in the agriculture and agri-food sectors in France—https://www.fi-compass.eu/sites/default/files/publications/financial_needs_agriculture_agrifood_sectors_France_0.pdf

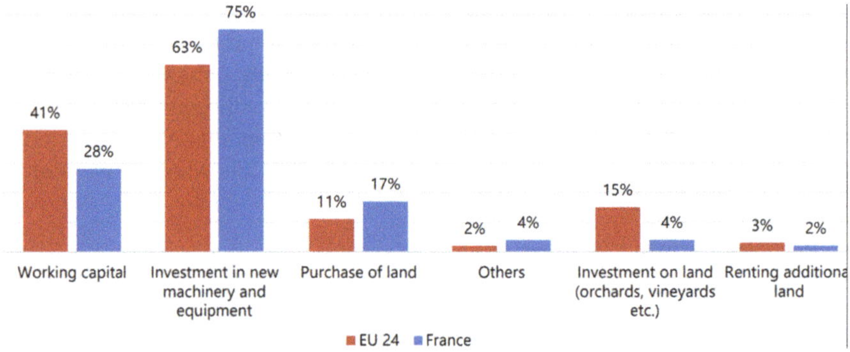

Fig. 9.10 France credit to agriculture

purchase of land, investment on land (orchards, vineyards), and renting additional land [27].

Agriculture finance and agricultural insurance are strategically important for eradicating extreme poverty and boosting shared prosperity. Globally, there are an estimated 500 million smallholder farming households—representing 2.5 billion people—relying, to varying degrees, on agricultural production for their livelihoods. The benefits of our work include the following: growing income of farmers and agricultural SMEs through commercialization and access to better technologies, increasing resilience through climate-smart production, risk diversification and access to financial tools, and smoothing the transition of noncommercial farmers out of agriculture and facilitating the consolidation of farms, assets, and production (financing structural change).

9.2.3.5 History of Credit to Agriculture

2010

During the 2010s, the top countries that have budgets and offered credit to agriculture include China, India, the United States, Netherlands, Italy, Australia, Germany, Republic of Korea, New Zealand, and Spain (please see Fig. 9.11).

2000

During the 2000s, the top countries that have budgets and offered credit to agriculture include the United States, Germany, Australia, Germany, UK, Spain, Canada, India, Argentina, New Zealand, and Iran (please see Fig. 9.15).

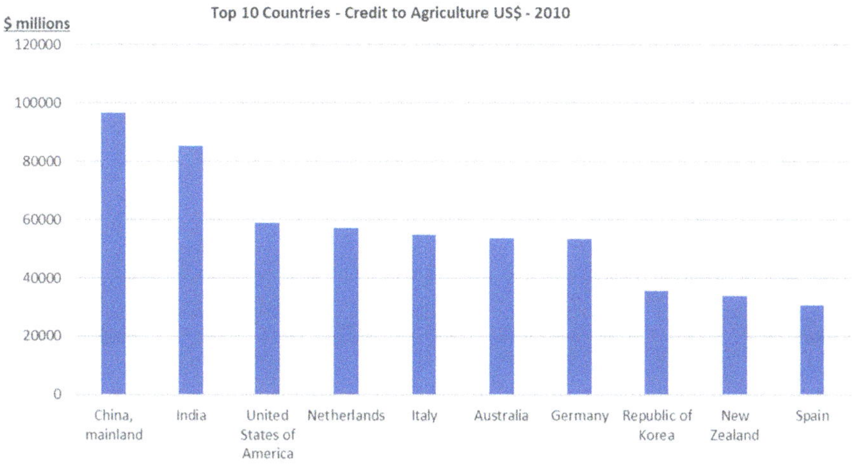

Fig. 9.11 Credit to agriculture—2010

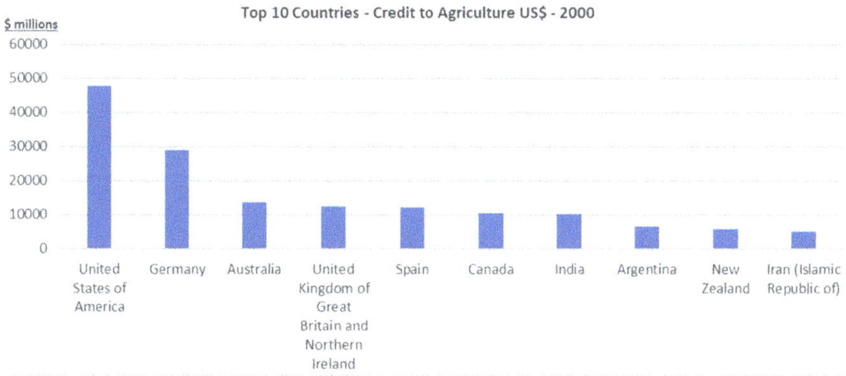

1991

During 1990s, the top countries that have budgets and offered credit to agriculture include Germany, the United States, Canada, Iran, Thailand, Israel, New Zealand, Egypt, Tunisia, and Nigeria (please see Fig. 9.12).

Please note: Iran was not there in the 2010 list.

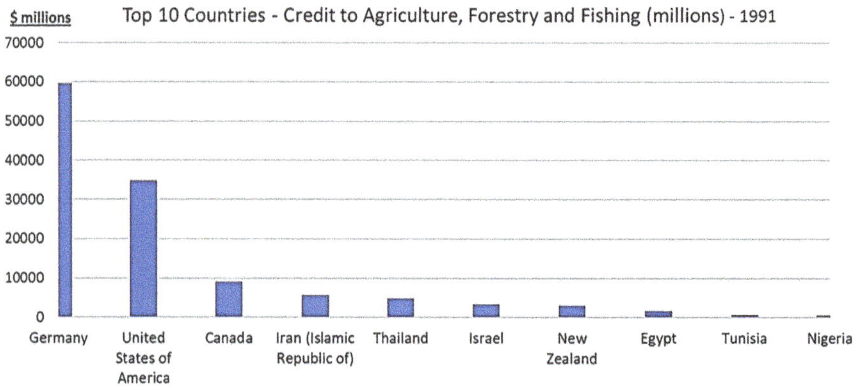

Fig. 9.12 Top 10 countries—credit to agriculture

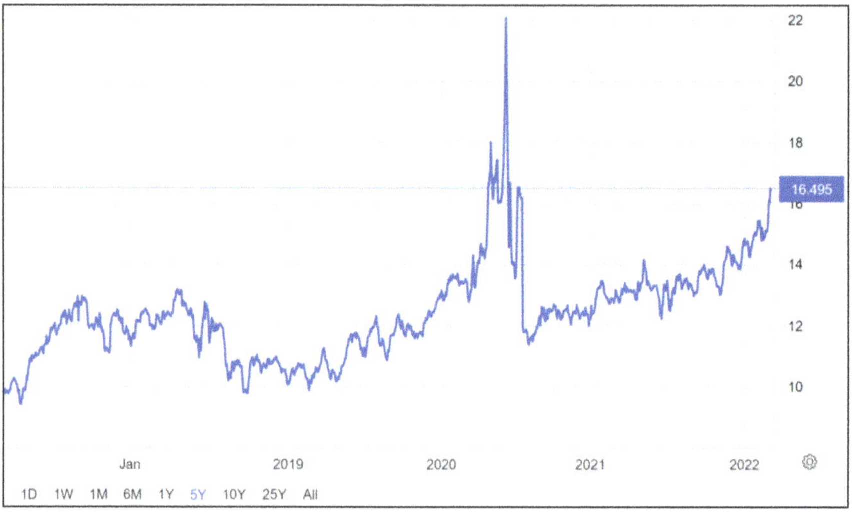

Fig. 9.13 Rice futures

9.2.4 Trade (Import/Export) Restrictions

Agriculture commodity prices are very sensitive to trade restrictions. The reason for trade restrictions could be due to either political events, conflict/war events, or pandemic outbreaks. For instance, the trading prices for rice (please see Fig. 9.13) in mid-April 2020 rose to their highest points in more than 7 years after several major exporting countries imposed trade restrictions in March 2020 as a response to heightened concerns over domestic food security following the outbreak of COVID-19. The rapid increase in global prices occurred as trade restrictions first announced in March in Southeast Asia sharply reduced global exportable supplies.

Please note, despite near global rice supplies in 2019/2020, the prices spiked. Global rice price movements typically have major impacts on import levels in price-sensitive markets, particularly in sub-Saharan Africa, the world's largest rice importing region, where rice is becoming increasingly important to the diet of many lower-income households.[51]

Wars and conflicts tend to increase input costs to agriculture. Government intervention[52] could stabilize the market from price spikes and any potential price gouging [28].

9.2.5 Location and Weather

Weather is a significant influencer of agricultural prices. Bad weather, monsoon, and lower than expected rains have a huge influence on agriculture yield and production. Weather has severe implication on energy, agriculture, and food security. This is especially true in countries whose GDP depends on the agriculture and countries whose food imports drive the food security of the country. Severe weather conditions such as extreme heat, rains, drought, and heat index are determent to agriculture. Please see below table on whose country's energy depends on hydropower.

Drought in these countries, especially food producers, would have (a) a rippling effect on food prices internationally, (b) higher energy bills (given drought negatively affects hydropower generation), and (c) food insecurity caused by a prospective increase in energy imports due to weather-hit food imports. Additionally, depletion of foreign reserves would increase exchange rate and hence impacts purchasing power that has direct implications on food security. In summary, extreme weather events are the food insecurity drivers.

9.2.6 Vegetation Index: Drought and Other Extreme

Vegetation Indices[53] (VIs) are used to monitor and characterize terrestrial landscapes. Vegetation Indices are related to absorption of *photosynthetically*[54] active radiation by vegetation and correlate with biomass or primary productivity:

---

[51] Global trading prices for rice rose to highest level in 7 years due to export restrictions—https://www.ers.usda.gov/data-products/chart-gallery/gallery/chart-detail/?chartId=98504

[52] USDA Warns Fertilizer Firms on Price Gouging Over Ukraine—https://www.bloomberg.com/news/articles/2022-02-24/usda-chief-warns-fertilizer-firms-on-price-gouging-over-ukraine

[53] Vegetation Index—https://www.star.nesdis.noaa.gov/jpss/vi.php

[54] MODIS Photosynthesis—https://www.ntsg.umt.edu/files/modis/ATBD_MOD17_v21.pdf

- Spectral Vegetation Indices (VIs) are optical measures of vegetation canopy "greenness," a composite property of:
  - Leaf chlorophyll
  - Leaf area
  - Canopy cover
  - Canopy architecture
- VIs are widely used in studies involving vegetation dynamics:
  - Land surface phenology
  - Climate–vegetation interactions
- VIs can be used to produce estimates of:
  - Leaf Area Index (LAI)
  - Fraction of photosynthetically active radiation (fPAR)
  - Net photosynthesis (PSN)
  - Annual net primary production (NPP)
  - Green vegetation fraction (GVF)

9.3 Real Economic Activity

Real economic activity variables represent mostly demand of the agriculture. Some of the critical parameters include macroeconomic drivers, Food Price Index, the Dairy Price Index (part of food price index), exchange rate, population growth, rural population (% of total population), Consumer Price Index, EXIM, employment in agriculture, GDP, and total reserves.

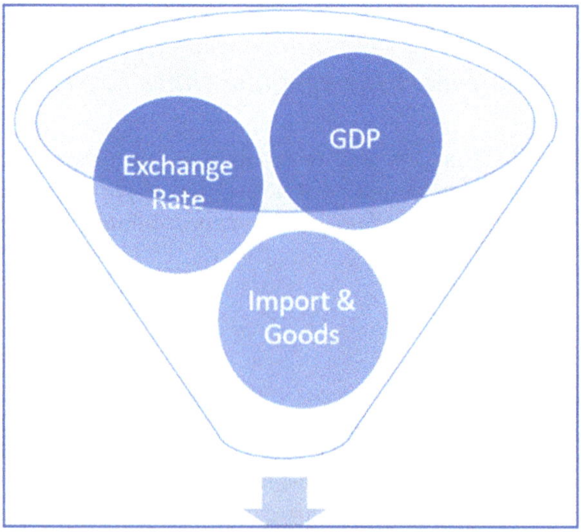

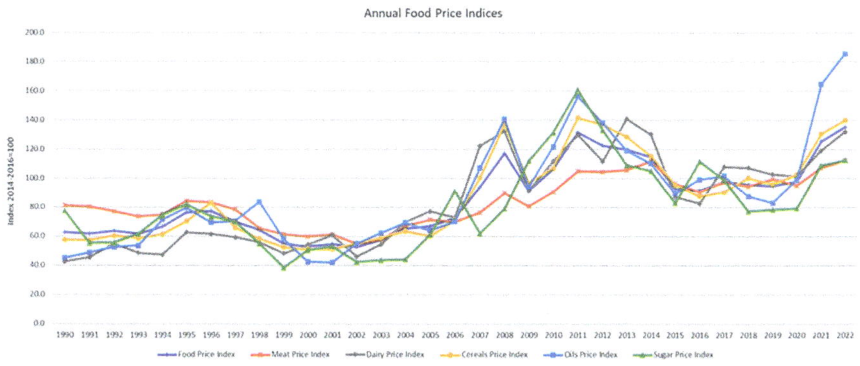

Fig. 9.14 Food Price Index (1960–2022)

9.3.1 Food Price Index

The FAO Food Price Index (FFPI) is a measure of the monthly change in international prices of a basket of food commodities. It consists of the average of five commodity group price indices (Cereal Price Index, Vegetable Oil Price Index, Dairy Price Index, Meta Price Index, and Sugar Price Index) weighted by the average export shares of each of the groups over 2014–2016.[55] The complete dataset consists of 1960–2022[56] (please see Fig. 9.14).

9.3.2 The Dairy Price Index

Availability of dairy products is very important for food security. In food nutrition, dairy product weights are as high as other important food items such as vegetables and protein. The higher price index could be due to (a) increased consumer demand; (b) supply constraint; (c) a tightening in global markets, reflecting a reduction in export availabilities, especially from major dairy-producing countries such as Western Europe; (d) steep increase in commodity prices; and (e) reduction in milk deliveries in some large milk-producing countries and lower stock level–supported prices. Another important factor is milk processing and transportation delays due to COVID-19-related labor shortages further contributing to higher dairy prices (please see Fig. 9.15).

---

[55] The Food Price Index—https://www.fao.org/worldfoodsituation/foodpricesindex/en/

[56] The FAO Food Price Index Dataset—https://www.fao.org/fileadmin/templates/worldfood/Reports_and_docs/food_price_index_nominal_real_feb295.xls

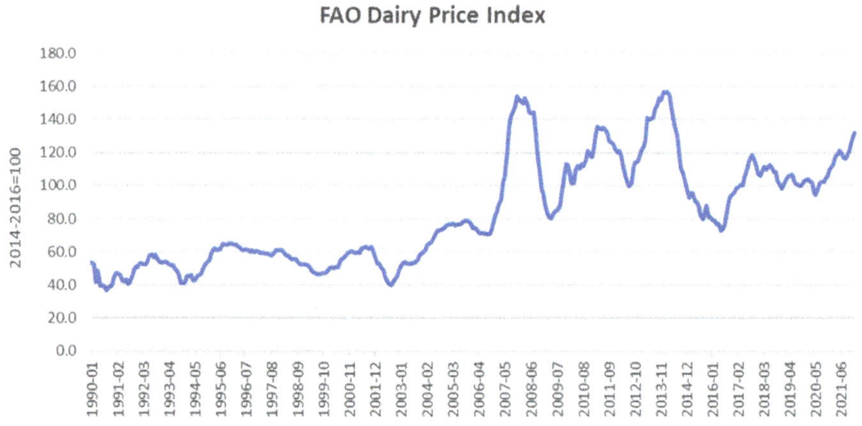

Fig. 9.15 FAO Dairy Price Index

9.3.2.1 World Sugar Prices: High Food Price Time Series

The global sugar market is less integrated (compared to the United States) than other major agricultural commodity markets, however, due to diverse and complex domestic policies of most major sugar-producing and trading countries. As a result, wholesale and retail sugar prices around the globe are influenced by local agricultural and trade policies and vary greatly from market to market, rather than simply reflecting the world futures price (please see Fig. 9.16).

The world raw sugar price is represented by the nearby futures contract listed by the Intercontinental Exchange (ICE) (commonly referred to as the Number 11 contract). The above chart is for ICE 11 sugar prices.[57] As you can see, sugar prices soared in 1974 due to adverse[58] weather conditions [29]. Please note the US domestic raw sugar price is based on the price of the nearby futures contract settlement price listed by the Intercontinental Exchange (ICE) (often referred to as the Number 16 contract).

(1) 1974 soared prices	
	*Adverse weather conditions* in 1974 have sharply reduced the beet sugar crop in Europe. Additionally, mild, dry conditions and an infectious plant disease, the "yellow virus," had cut the latest harvest in Europe by 30%, virtually assuring it will be insufficient to meet the year's demand. Finally, the surge was due to the supply gap of sugar production that would be unable to catch up before October 1975, when the next year's crop is harvested.
	Commodity speculation, mostly by Middle Eastern countries, faced with a need to invest the revenues resulting from their *quadrupled oil* prices, has also driven up

(continued)

[57] ICE Contract 11—Sugar—https://tradingeconomics.com/commodity/sugar

[58] Soaring Sugar Cost Arouses Consumers And U.S. Inquiries—https://www.nytimes.com/1974/11/15/archives/soaring-sugar-cost-arouses-consumers-and-u-s-inquiries-what-sent.html

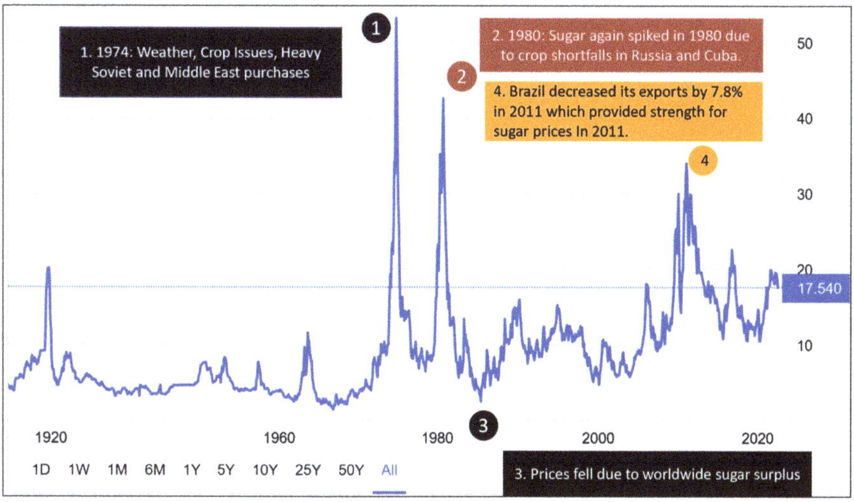

Fig. 9.16 Sugar prices

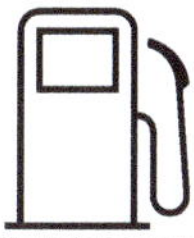	prices. Algeria, for example, has bought about 3 million tons R sugar and Kuwait about 2 million within the last month—equivalent to about one-third of the American consumption and considerably more than their own domestic needs.
(2) 1980 sugar[a] prices [30]	
	Weather caused price spike during 1980. The Soviet Union, normally the largest sugar producer, has not received *enough rain* in its sugar beet area in the Ukraine, where lack of topsoil[b] moisture has led to delayed plantings [31]. Late planting means that the new crop might be damaged by frosts in September before harvest. At the same time, Western Europe, another major sugar-producing area, has been afflicted with too much rain. Wet weather has also imperiled Cuba's cane crop, "where torrential rains have delayed the harvest by six to eight weeks." The crop in South Africa, which has endured one of the worst droughts in its history, has been cut sharply. As a producer, it found itself overcommitted by some 600,000 tons. It has gone into the market to buy sugar for its customers or has asked buying nations like Japan, Canada, and South Korea to seek sugar elsewhere. Drought also has hampered crops in Australia, the Philippines, and Thailand.
(3) 1985 sugar prices—fell	
	The price of sugar futures fell[c] to new lows in April 21, 1985, as the worldwide *surplus* weighed on the market [32]. Sugar prices fell again,[d] bottoming in May 1985 at less than US$0.03 per pound and averaging US$0.04 per pound for the year. In early 1990, prices had recovered to US$0.14 cents per pound and then declined to approximately US$0.08 to US$0.09 per pound [33].
(4) 1984 spike in sugar prices	
	Brazil is the largest sugar-producing country in the world. The production of sugar has increased to 352% since 1990. About 34% of Brazilian sugar consumed domestically is converted into ethanol for fuel. Exports have risen from 1.2 million metric tons in 1990 to 26.9 million metric tons in 2011. Sugar that is converted into

(continued)

ethanol is subsidized at prices higher than the world price. Recent increases in the world oil price have increased the price of ethanol[e] which in turn increased Brazil's conversion of sugar into ethanol, reducing potential sugar exports from Brazil [34]. That reduction in the growth of sugar exports could be one of the main forces for world sugar price increases. Brazil decreased its exports by 7.8% in 2011 which provided strength for sugar prices in 2011.

[a]FAO—Cuba Conference, Cuba—sugar markets in disarray, https://www.fao.org/3/X4988E/x4988e05.htm
[b]Commodities: price rises for sugar continue-https://www.nytimes.com/1983/06/06/business/commodities-price-rises-for-sugar-continue.html
[c]Futures/options; Sugar Surplus Sends Prices Plummeting—https://www.nytimes.com/1985/04/24/business/futures-options-sugar-surplus-sends-prices-plummeting.html
[d]Philippine—Sugar—http://countrystudies.us/philippines/64.htm
[e]2012 Outlook of the U.S. and World Sugar Markets, 2011-2021—https://ageconsearch.umn.edu/record/128037/files/AAE692.pdf

In essence, the above weather and market signals have a direct impact on food security. In summary, weather conditions, oil and gas prices, weather-induced spikes, insufficient rain, and biofuel demand could induce higher costs to food commodities and thus impact food security.

Do Not Expect Grocery Store Prices to Come Down Anytime Soon
Goldman Sachs (GS) projected that the food-at-home category of the Consumer Price Index will increase by another 5% to 6% this year. The Wall Street bank cited a 6% jump in food commodity prices so far this year and "soaring costs" of some farming inputs, including a quintupling of some fertilizer prices. The stage has been set for further substantial increases in retail food prices this year.

A perfect storm of bad weather, poor crop yields, tight inventories,[59] and strong demand has lifted food commodities to nearly 40% over the past 2 years, Goldman Sachs said [35].

9.3.3 *The USD Exchange Rate*

Trade in many agricultural commodities is denominated in USD. A depreciating USD causes dollar-denominated international commodity prices to rise, although not to the full extent of the depreciation. The opposite occurs when the dollar appreciates. For example, consider below the sugar futures (US Sugar #11 Futures[60]) graph. The Sugar No. 11 contract is the world benchmark contract for raw sugar trading and is available on the Intercontinental Exchange[61] (ICE). The biggest producer and

[59]Don't expect grocery store prices to come down anytime soon—https://www.cnn.com/2022/02/08/business/grocery-store-prices-goldman/index.html
[60]US Sugar #11 Futures—https://www.investing.com/commodities/us-sugar-no11-historical-data
[61]The ICE Sugar Contract 11—https://tradingeconomics.com/commodity/sugar

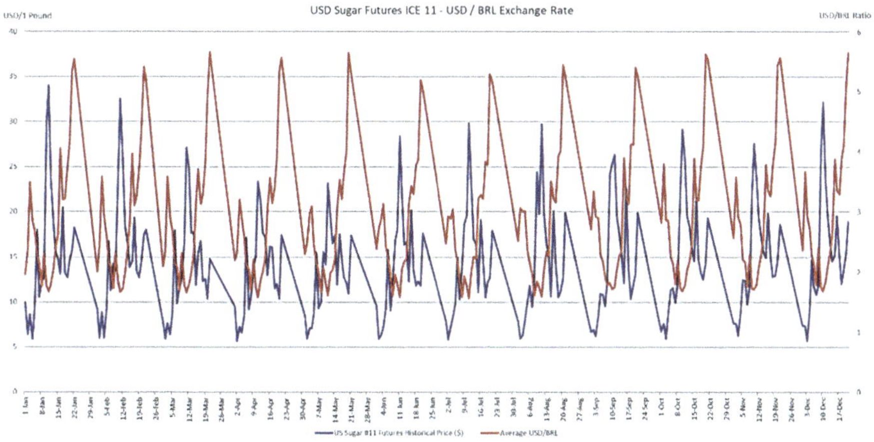

Fig. 9.17 Sugar futures vs. BRL exchange rate

exporter of sugar in the world is Brazil (21% of total production and 45% of total exports). Now, let us analyze the signal between USD/Brazilian Real[62] exchange rate and ICE 11 futures. As indicated in the graph, higher USD/BRL values (i.e., appreciate in USD) coincided with a drop in ICE 11 futures sugar price, and vice versa. That is, a drop in USD/BRL (depreciate in USD) shows in increase in sugar futures. These currency movements added to the amplitude of the price changes observed. (They also help to explain why demand remained strong in countries where the currency was appreciating against the dollar and why falling prices were not fully felt in the same countries once the dollar began to appreciate again.) Exchange rate volatility per se is beyond the scope of this book, but if the future is marked by increased exchange rate volatility, this will also have repercussions for the volatility of international prices of commodities (please see Fig. 9.17).

9.3.4 *Population Growth*

The average annual percent changes with the population, resulting from a surplus (or deficit) of births over deaths and the balance of migrants entering and leaving a country.[63] The rate may be positive or negative. The growth rate is a factor in determining how great a burden would be imposed on a country by the changing

[62] USD/BRL—https://fxtop.com/en/historical-exchange-rates.php?A=1&C1=USD&C2=BRL&MA=1&DD1=01&MM1=01&YYYY1=1970&B=1&P=&I=1&DD2=25&MM2=02&YYYY2=2022&btnOK=Go%21

[63] Population Growth—https://www.indexmundi.com/world/population_growth_rate.html#:~:text=Population%20growth%20rate%3A,1.03%25%20(2020%20est.)&text=Definition%3A%20The%20average%20annual%20percent,entering%20and%20leaving%20a%20country

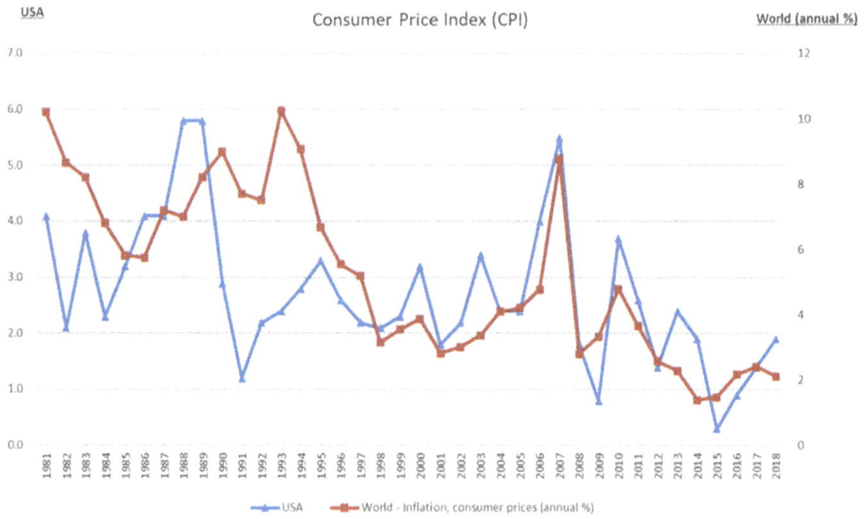

Fig. 9.18 CPI—the United States and the world

needs of its people for infrastructure (e.g., schools, hospitals, housing, roads), resources (e.g., food, water, electricity), and jobs. Rapid population growth can be seen as threatening by neighboring countries (Fig. 9.18).

Data Source: The World Bank – Population growth (annual %)[64]

9.3.5 *Inflation, Consumer Prices (Annual %) (CPI)*

Definition[65]: Inflation, as measured by the Consumer Price Index, reflects the annual percentage change in the cost to the average consumer of acquiring a basket of goods and services that may be fixed or changed at specified intervals, such as yearly. The economic theory states that there are two sources of inflation, cost-push and demand-pull inflation. In a country where there is demand-pull inflation, due to increasing demand for food, producers are expected to invest more in the agricultural sector, resulting in increased production, and therefore, increasing demand-pull inflation should lead to increasing percentage contribution of agriculture to GDP. CPI's coefficient in this case is hypothesized to be positive. However, where cost-push inflation results, because of a decrease in aggregate agricultural supply, which may be caused by either an increase in wages or an increase in the prices of raw materials, the greater costs of agricultural production result in the amount of agricultural

[64]Population growth (annual %)—https://data.worldbank.org/indicator/SP.POP.GROW?end=2017&locations=1W&start=1981&view=chart

[65]Inflation—https://www.indexmundi.com/facts/indicators/FP.CPI.TOTL.ZG

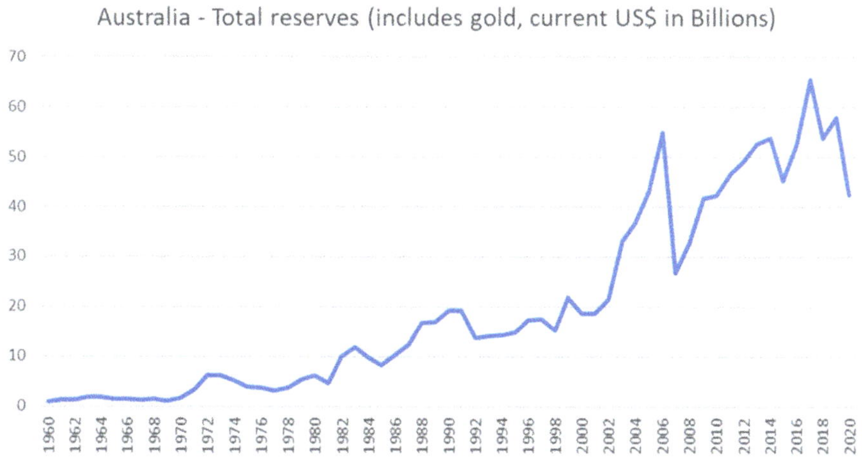

Fig. 9.19 Total reserves.

production falling, thus reducing the percentage contribution of agriculture to GDP. The coefficient of CPI in this case is expected to be negative. Thus, the coefficient of CPI is hypothesized to have either a negative or a positive sign depending on the source of inflation in the countries (please see Fig. 9.19).

Data Source: Inflation, Consumer Prices[66]

The CPI in the United States reached[67] greater than 5% in 1989, 1990, and 2008. The world had a double-digit (10.22%) inflation in 1982 [36].

9.3.6 Ratio of Export to Import Agricultural Products (EXIM)

The ratio of agricultural trade is the ratio of exports and imports of total agricultural merchandise for a country.

$$\text{EXIM} = \frac{\text{Exports}}{\text{Imports}}$$

Data Source: Agricultural raw material exports (% of merchandise exports)[68]

---

[66] Inflation—https://data.worldbank.org/indicator/FP.CPI.TOTL.ZG?end=2017&locations=1W&start=1981&view=chart

[67] Food Price Outlook—https://www.ers.usda.gov/data-products/food-price-outlook/food-price-outlook/#Consumer%20Price%20Index

[68] Agricultural raw materials exports (% of merchandise exports)—https://data.worldbank.org/indicator/TX.VAL.AGRI.ZS.UN?end=2017&locations=1W&start=1981&view=chart

9.3.7 Agriculture, Forestry, and Fishing, Value Added (% of GDP)

Agriculture corresponds to ISIC divisions 1–5 and includes forestry, hunting, and fishing, as well as cultivation of crops and livestock production. Value added is the net output of a sector after adding up all outputs and subtracting intermediate inputs. It is calculated without making deductions for depreciation of fabricated assets or depletion and degradation of natural resources. The country with the highest value in the world is Somalia, at 62.74. The country with the lowest value in the world is Singapore, at 0.02.

Data Source: Agriculture, forestry, and fishing, value added (% of GDP) – World[69]

9.3.8 Total Reserves (Includes Gold, Current US$)

The balance of payments records an economy's transactions with the rest of the world. Balance of payments accounts are divided into two groups: the current account, which records transactions in goods, services, primary income, and secondary income, and the capital and financial account, which records capital transfers; acquisition or disposal of non-produced, nonfinancial assets; and transactions in financial assets and liabilities. The current account balance is one of the most analytically useful indicators of an external imbalance.[70] Total reserves (includes gold, current US$) in Australia was reported at 42.54 billion USD in 2020, according to the World Bank collection of development indicators, compiled from officially recognized sources (please see Fig. 9.20). Australia's—total reserves (includes gold, current US$)—actual values, historical data, forecasts, and projections were sourced from the World Bank on March of 2022. I have added total reserves for predicting sugarcane $ production for the Australian economy and did not influence the model.

Total reserves play an important role in establishing food security as reserves play a crucial role in stabilizing the exchange rate and bolster liquidity within the economy. For instance, in some countries, such as Yemen, famine was prevented 2 years ago, 2020, when member states and donors provided urgently needed resources.[71]

---

[69] Agriculture, forestry, and fishing, value added (% of GDP)—World—https://data.worldbank.org/indicator/NV.AGR.TOTL.KD.ZG?end=2017&locations=1W&start=1981&view=chart

[70] Total Reserves—https://data.worldbank.org/indicator/FI.RES.TOTL.CD

[71] Yemen: Acute Food Insecurity Situation October–December 2020 and Projection for January–June 2021—https://www.ipcinfo.org/ipc-country-analysis/details-map/en/c/1152947/

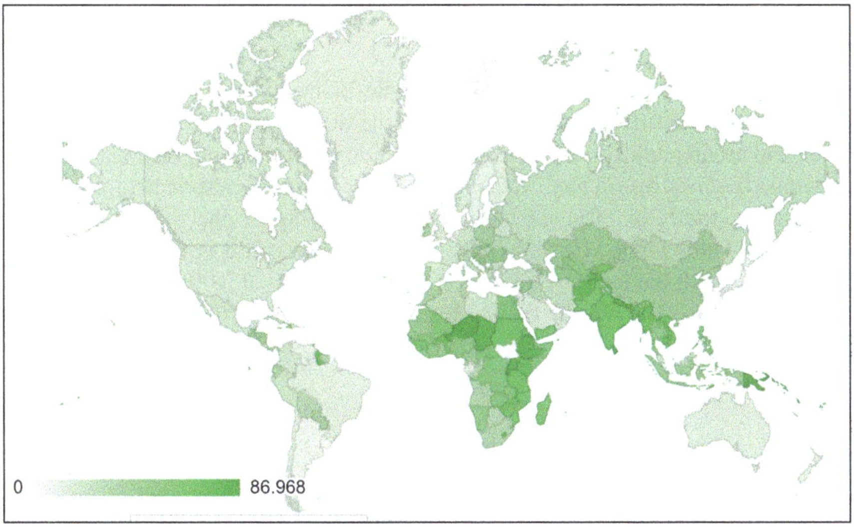

Fig. 9.20 Rural population

9.4 Farm Inputs

Real economic activity variables represent mostly demand of the agriculture, some of which are critical.

9.4.1 Employment in Agriculture (% of Total Employment) (Modeled ILO Estimate)

Employment is defined as persons of working age who are engaged in any activity to produce goods or provide services for pay or profit, whether at work during the reference period or not at work due to temporary absence from a job, or to working-time arrangement. The agriculture[72] sector consists of activities in agriculture, hunting, forestry, and fishing, in accordance with division 1 (ISIC 2) or categories A-B (ISIC 3) or category A (ISIC 4).

Source: Agriculture, forestry, and fishing, value added (annual % growth)[73]

[72]Employment in agriculture—https://www.indexmundi.com/facts/indicators/SL.AGR.EMPL.ZS

[73]Employment in Agriculture—https://data.worldbank.org/indicator/SL.AGR.EMPL.ZS?end=2017&locations=1W&start=1981&view=chart

9.4.2 *Rural Population (% of Total Population)*

Definition[74]: Rural population refers to people living in rural areas as defined by national statistical offices. It is calculated as the difference between the total population and urban population. The map below shows how rural population (% of total population) varies by country. The shade of the country corresponds to the magnitude of the indicator. The darker the shade, the higher the value (please see Fig. 9.20). The country with the highest value in the world is Burundi, at 86.97. The country with the lowest value in the world is Nauru, at 0.00 [37].

Data Source: The World Bank—Rural population (% of total population) – World[75]

9.5 Food Security Indicators

The following food security indicators are analyzed to see the causation of economic, farm levels, and other drivers on the food security.

9.5.1 *Shock Indicators: Per Capita Food Production Variability*

Per capita food production variability (please see Fig. 9.21) corresponds to the variability of the "food net per capita production value in constant 2004-2006 international $" as disseminated in FAOSTAT.[76]

9.5.2 *Domestic Food Price Volatility*

Market-level analyses are an important method of measuring food security and can serve many purposes, including estimating domestic supply against population need, evaluating market response to changes in supply or demand, and providing insight on the consumer prices of food versus those of other goods.[77] The domestic food price volatility index measures the variability in the relative price of food in a

[74]Rural Population—https://www.indexmundi.com/facts/indicators/SP.RUR.TOTL.ZS

[75]The World Bank—Rural population (% of total population)—https://data.worldbank.org/indicator/SP.RUR.TOTL.ZS?end=2017&locations=1W&start=1981&view=chart

[76]Per capita food production variability —http://fsis.sd/SD/EN/FoodSecurity/Indicators/SSI/20/

[77]Volatility of food prices standard deviations of prices over time—https://inddex.nutrition.tufts.edu/data4diets/indicator/volatility-food-prices

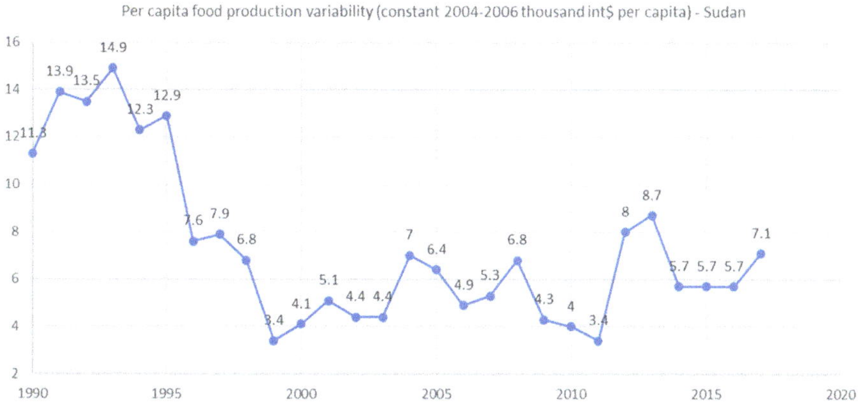

Fig. 9.21 Food production variability

country. The indicator is calculated from the monthly domestic food price level index using monthly consumer and general food price indices and purchasing power parity data from the International Comparison Program conducted by the World Bank (see the Relative Price of Food Indicator for more information). Month-to-month growth rates are calculated, and the standard deviation of these growth rates are calculated over the previous 8 months (8 months rolling standard deviation). The average of these standard deviations is then computed to obtain an annual volatility indicator. The indicator is available for 130 countries for which monthly general and food consumer price indices and purchasing power parity data are available. Aggregates are available for all microregions according to the M-49 classification and those microregions with adequate representability of countries with monthly consumer price indices in terms of GDP. The domestic food price volatility compares the variations of the domestic food price index across countries and time.[78]

The method of construction for this indicator can be calculated on both an annual and monthly basis and is reported as the standard deviation around the mean of the price index over the reference period. It is based on a monthly domestic consumer food price index.

Australia sugarcane	

[78]Domestic food price volatility—http://fsis.sd/SD/EN/FoodSecurity/Indicators/SSI/19/

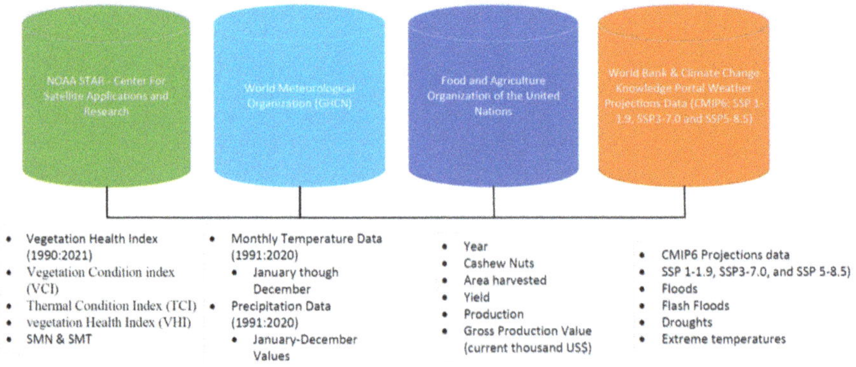

Fig. 9.22 Data sources

9.6 Machine Learning Model: Australia Macroeconomic Drivers and Sugarcane Production Predictive Model

Let us build a Queensland, Australia, sugarcane production model:

Software code for this model: FinalSugarCaneAustraliaMacroEconomicLinkageModel (Jupyter Notebook Code)

9.6.1 Data Sources

The data sources for the ML model is from FAO Statistics Data,[79] World Bank Climate Change Data specifically Australia–Queensland,[80] World Metrological Organization Data,[81] FAO Earth Observation,[82] and NOAA STAR–STAR–Global Vegetation Health Products: Province-Averaged VH[83] (please see Fig. 9.22). Please refer to Appendix C for world data sources, Appendix D for US data sources, and Appendix E for economics data sources.

[79]FAO Statistics data—https://www.fao.org/faostat/en/#data/QCL

[80]Climate Change Data—https://climateknowledgeportal.worldbank.org/download-data

[81]World Metrological Organization Data—https://climatedata-catalogue.wmo.int/explore

[82]Earth Observation Mozambique—https://www.fao.org/giews/earthobservation/country/index.jsp?lang=en&code=AUS

[83]NOAA STAR—VHI—https://www.star.nesdis.noaa.gov/smcd/emb/vci/VH/vh_adminMeanByCrop.php?type=Province_Weekly_MeanPlot

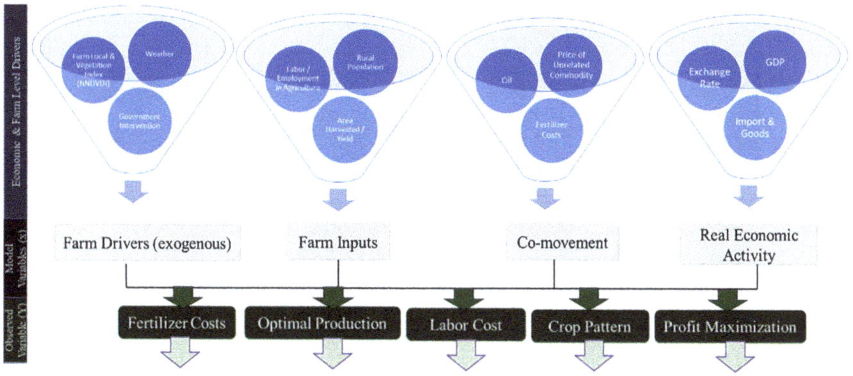

Fig. 9.23 Economic–agri linkage

9.6.2 *Sugarcane Production Model*

Agriculture has been one of the pillars supporting the country's economy, accounting for 12% of the nation's GDP.[84] Due to the expansive arable land in the country, large-scale farming activities thrive, consisting of some of the largest beef ranches in the world as well as huge wheat and barley fields. Most of the agricultural commodities produced in Australia are sold to international markets with 80% of all wheat, 90% of all wool, and over 50% of barley being exported to various international destinations. The top consumers of these commodities include China, the United States, Japan, and Indonesia. The Australian landscape is predominantly arid and hence only suitable for livestock rearing and specifically beef production with estimates putting the total area under beef farming being 60% of the country. As part of the modeling effort, this study identified four key drivers of wheat futures prices in each futures market: real economic activity, co-movement, precautionary demand, and net supply (see Figs. 9.3 and 9.23). Each of these drivers encapsulates numerous economic phenomena, as discussed earlier. Structural vector autoregression, or SVAR, is used to identify the influence of each driver on observed wheat futures prices, with the econometric estimation procedure[85] run separately for each futures market and results compared across markets [19].

[84] Top 10 Agricultural Exports of Australia—https://www.worldatlas.com/articles/top-10-agricultural-exports-of-australia.html

[85] Deconstructing Wheat Price Spikes: A Model of Supply and Demand, Financial Speculation, and Commodity Price Comovement—https://www.ers.usda.gov/webdocs/publications/45199/46439_err165.pdf?v=0

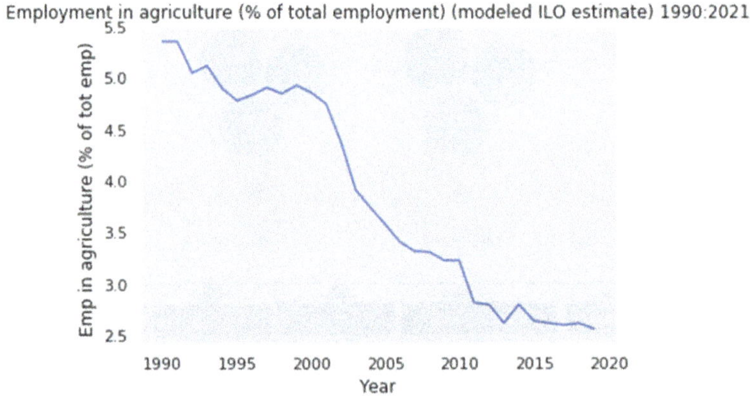

Fig. 9.24 Employment in agriculture

9.6.2.1 Employment in Agriculture (% of Total Employment) (Modeled ILO Estimate)

Employment[86] is defined as persons of working age who were engaged in any activity to produce goods or provide services for pay or profit, whether at work during the reference period or not at work due to temporary absence from a job, or to working-time arrangement. The median[87] age for workers in this industry is 51 years and median weekly earnings are around $1,053 per week [38]. Employment in agriculture (% of total employment) (please see Fig. 9.24) is decreasing at the rate of 2.47% (CAGR). Farming in Australia,[88] as in many countries, is on the verge of a workforce crisis [39].

9.6.2.2 Agriculture, Forestry, and Fishing, Value Added (Constant 2015 US$)

The gross value of Australia's agricultural production is forecast to reach a record $78 billion in 2021–2022[89] (please see figure—source: ABARES; Australian Bureau of Statistics[90]) [40]. This $5.4 billion upward revision from the outlook

[86] Employment in agriculture (% of total employment) (modeled ILO estimate)—https://databank.worldbank.org/metadataglossary/jobs/series/SL.AGR.EMPL.ZS

[87] Snapshot of Australia's Agricultural Workforce—https://www.awe.gov.au/abares/products/insights/snapshot-of-australias-agricultural-workforce

[88] Farming on the verge of a workforce crisis—https://www2.deloitte.com/au/en/pages/consumer-business/articles/farming-verge-workforce-crisis.html

[89] Agricultural overview: December quarter 2021—https://www.awe.gov.au/abares/research-topics/agricultural-outlook/agriculture-overview

[90] Australian Agriculture Overview—https://www.awe.gov.au/abares/research-topics/agricultural-outlook/agriculture-overview

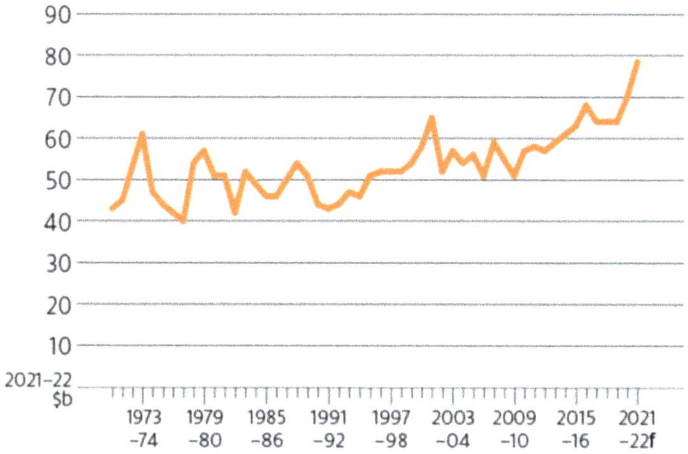

Fig. 9.25 Agriculture value added

issued in September is the result of further improvements in domestic growing conditions, downgrades for key overseas competitors driving prices higher, and steep increases in logistics and fertilizer costs worldwide. Prices are at multiyear highs for many agricultural commodities. Many of the factors driving international prices are driven by poor seasonal conditions, so there is considerable uncertainty about how long prices will remain at these levels—posing a risk to the forecast values being realized. Finally, the value of agricultural exports is forecast to be a record $61 billion (please see Fig. 9.25).

9.6.2.3 Fertilizer Consumption (Kilograms Per Hectare of Arable Land)

Fertilizer consumption[91] measures the quantity of plant nutrients used per unit of arable land. Fertilizer consumption (kilograms per hectare of arable land) in the world was reported at 137 kg in 2018.[92] Fertilizer products cover nitrogenous, potash, and phosphate fertilizers (including ground rock phosphate). Traditional nutrients—animal and plant manures—are not included. For data dissemination, FAO has adopted the concept of a calendar year (January to December). Some countries compile fertilizer data on a calendar year basis, while others are on a split-year basis. Australia's fertilizer consumption per hectare of arable land was 85.87 kg in 2018 and is lower than the world consumption. Though Australia fertilizer consumption fluctuated substantially in recent years, it tended to increase through the 1969–2018 period, ending at 85.9 kg per hectare in 2018 (please see Fig. 9.26).

[91] Fertilizer consumption (% of fertilizer production)—https://databank.worldbank.org/metadataglossary/all/series?search=Fertilizer%20consumption%20(kilograms%20per%20he

[92] World—Fertilizer Consumption (kilograms Per Hectare Of Arable Land)—https://tradingeconomics.com/world/fertilizer-consumption-kilograms-per-hectare-of-arable-land-wb-data.html

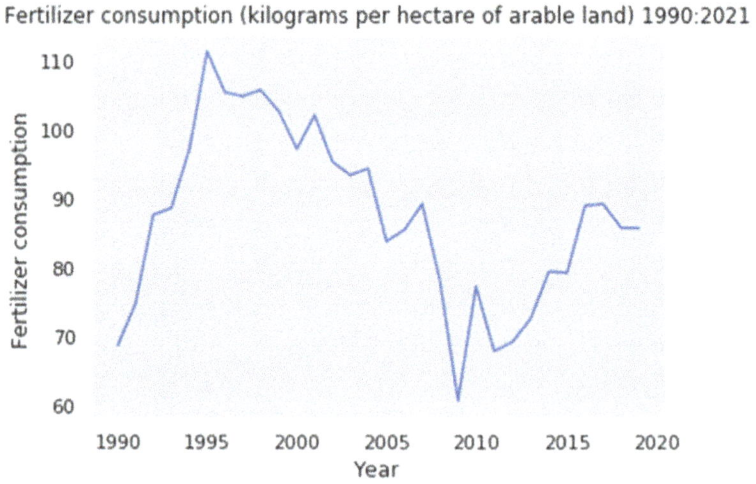

Fig. 9.26 Fertilizer consumption

Data Signal
Data signal from Australian fertilizer consumption data: higher gas prices and the great recession, 2008, pushed the fertilizer prices up, making it more expensive for farmers to buy the necessary fertilizers!

In 2009, Australia fertilizer consumption fell to the lowest 60.927 kg due to peak prices[93] (higher gas prices and 2008 the great recession) [41]. Many commodity prices, including fertilizers, increased dramatically from mid-2006 until late 2008. Both fertilizer indices hit all-time highs in 2008 (see[94] Fig. 9.1). Rising demand in emerging markets and long-term supply concerns contributed to the record-setting commodity prices. Later in the year, the 2008 financial crisis disrupted all markets, leading to falling commodity prices, including fertilizers [41].

The Agricultural Marketing Service (AMS), an agency of the USDA, last reported Illinois fertilizer prices on July 29, 2021, when prices were $746 per ton for anhydrous ammonia, $717 per ton for DAP, and $600 ton for potash. July 29, 2021, prices were much higher than earlier-year levels on July 30, 2020 (please Fig. 9.27):

- Anhydrous ammonia increased from $487 per ton in 2020 to $746 per ton in 2021, increasing $259 per ton, or a 53% increase. The last time the anhydrous ammonia price was above $746 per ton was in June 2014.
- DAP increased from $390 per ton to $717 per ton, an increase of $327 per ton, or 83%. The last time prices the DAP prices were above $717 per ton was in December 2008.

[93]Weekly Farm Economics 2021 Fertilizer Price Increases in Perspective, with Implications for 2022 Costs—https://farmdocdaily.illinois.edu/2021/08/2021-fertilizer-price-increases-in-perspective-with-implications-for-2022-costs.html

[94]World Bank—Pink Sheet—https://thedocs.worldbank.org/en/doc/5d903e848db1d1b83e0ec8f744e55570-0350012021/related/CMO-Historical-Data-Monthly.xlsx

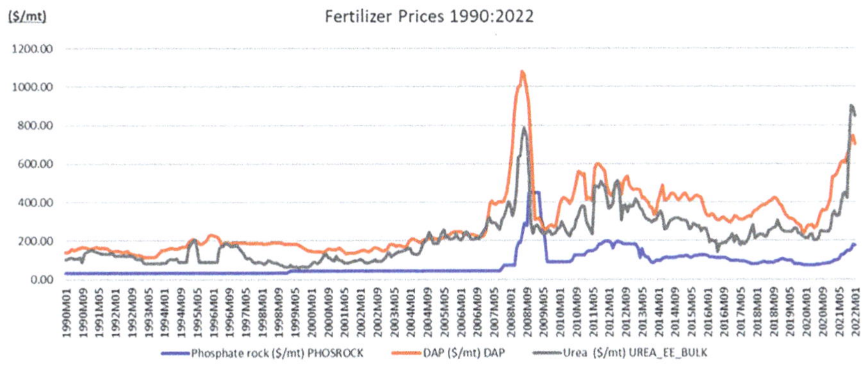

Fig. 9.27 Fertilizer prices 1990:2022

- Potash increased from $350 per ton in 2020 to $600 in 2021, increasing $250 per ton, or 71%. The last time the potash price was above $600 per ton was in November 2012.

9.6.2.4 GDP Per Capita Growth (Annual %): NY.GDP.PCAP.KD.ZG

The annual percentage growth rate of GDP per capita[95] based on constant local currency (Australian dollar). Aggregates are based on constant 2010 US dollars. GDP per capita is gross domestic product divided by the midyear population.[96] The value for GDP per capita growth (annual %) in Australia was 0.571 as of 2019. As the graph below shows, over the past 30 years, this indicator reached a maximum value of 3.74 in 1999 and a minimum value of –1.656 in 1991 (please see Fig. 9.28).

In terms of dollars:

- Australia GDP per capita for 2019 was $55,057, a 4.01% decline from 2018.
- Australia GDP per capita for 2018 was $57,355, a 6.16% increase from 2017.
- Australia GDP per capita for 2017 was $54,028, an 8.12% increase from 2016.

9.6.2.5 Rural Population (% of Total Population)

Rural population refers to people living in rural areas as defined by national statistical offices.[97] It is calculated as the difference between total population and urban population. The rural population (% of total population) in Australia was

[95] GDP Per Capita—https://databank.worldbank.org/metadataglossary/ida-results-measurement-system,-tier-i-database-%E2%80%93-wdi/series/NY.GDP.PCAP.KD.ZG

[96] Meta Data Glossary—https://databank.worldbank.org/metadataglossary/all/series

[97] Rural Population (% of total population)—https://data.worldbank.org/indicator/SP.RUR.TOTL.ZS?locations=AU

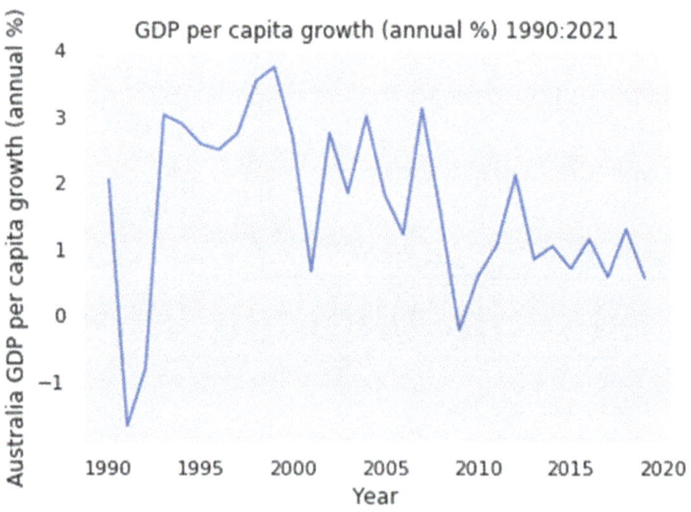

Fig. 9.28 GDP per capita growth

reported at 13.76% in 2020, according to the World Bank collection of development indicators, compiled from officially recognized sources. Australia's—rural population—actual values, historical data, forecasts, and projections were sourced from the World Bank in February of 2022.

Australia's major cities continue to grow while rural areas face population decline (please see Fig. 9.29). City infrastructure and housing are facing unprecedented pressure. Meanwhile, some rural areas are struggling to attract workers to fill well-paying positions and infrastructure is underutilized. These issues[98] are not new [42].

The latest population data from the Australian Bureau of Statistics, released on 27 March, reveals that these trends are only intensifying.

"Many rural areas are experiencing population decline. How long can this continue?"

The reasons why people leave regional and rural areas are not complex: lack of appropriate medical care, aged care, education, and training. These are the same reasons many people choose not to leave major cities. These reasons can be addressed with well-targeted government investment.

On average, people living in remote and very remote areas are younger than those in major cities. In 2017[99]:

- Remote and very remote areas had a higher proportion of people aged 0–14 years (22%) than major cities (19%) and inner regional and outer regional (19%) areas.

[98] City congestion and rural population decline— https://www.canberratimes.com.au/story/6016548/city-congestion-and-rural-population-decline/

[99] Remote and Rural Population—https://www.aihw.gov.au/reports/australias-health/rural-and-remote-health

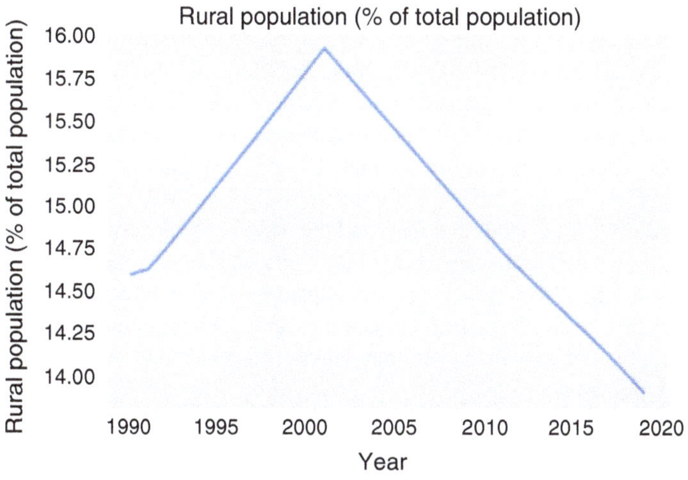

Fig. 9.29 Rural population

- Remote and very remote areas also had a higher proportion of people aged 25–44 years (30%) than inner regional and outer regional areas (23%) but was like major cities (30%).
- Remote and very remote areas had a lower proportion of people aged 65 years and over (11%) than inner regional and outer regional areas (19%) and major cities (14%).

In general, people aged 15 years and over living in metropolitan areas are more likely to be employed than people living outside these areas. This may be due to lower levels of access to work outside metropolitan areas and the smaller range of employment and career opportunities in these areas (ABS 2019c[100]).

People living in rural and remote areas also generally have lower incomes but pay higher prices for goods and services (NRHA 2014). In 2017–2018, Australians living outside capital cities had, on average, 19% less household income per week compared with those living in capital cities and 30% less mean household net worth (ABS 2019b[101]).

9.6.2.6 Exchange Rate (USD/AUD)

The portfolio-balance model states that a commodity-exporting country's exchange rate is heavily dependent on foreign-determined asset supply and demand

[100] Labour Force, Australia, Detailed—Electronic Delivery, Dec 2018—https://www.abs.gov.au/Ausstats/abs@.nsf/0/1EB7192B0FC044A7CA2583AE0012ED32?OpenDocument

[101] Household Income and Wealth, Australia—https://www.abs.gov.au/statistics/economy/finance/household-income-and-wealth-australia/latest-release

fluctuations. Thus, commodity price increases lead to a balance of payment surplus[102] and an increase in foreign holdings of the country's currency [43].

Currencies are moved by many factors, including supply and demand, politics, interest rates, speculation, and economic growth. More specifically, since economic growth and exports are directly related to a country's domestic industry, it is natural for some currencies to be heavily correlated[103] with commodity prices [44].

Three currencies that have the tightest correlations with commodities are the Australian dollar, the Canadian dollar, and the New Zealand dollar. Other currencies are also impacted by commodity prices but have a weaker correlation than the above three, such as the Swiss franc and the Japanese yen—which tend to rise when commodity prices fall. The truth is countries that rely on commodity exports for their economy can see their national currency's exchange rates fluctuate with commodity prices.

The exchange rate plays an important role for Australian exports and imports.[104] Domestically, a depreciation of the Australian dollar encourages substitution from imports to domestically produced goods and services, as imported products become relatively more expensive. A lower exchange rate also makes Australian exports more competitive in world markets, as exported goods and services become relatively cheaper in foreign currency terms [45].

Movements in the exchange rate assist the economy in adjusting to structural change, such as that experienced in Australia during the recent terms of trade boom. Strong demand for resources from the rapidly growing Chinese economy caused commodity prices—and thus Australia's terms of trade—to rise substantially between 2001 and 2011. Associated with this, the Australian dollar appreciated by around 80% in real trade-weighted terms over that period (please see Fig. 9.30 [29]).

Generally, trade in many agricultural commodities is denominated in USD. It is imperative to add the exchange rate of the local currency, in this case Australian dollar (AUD), to model the production value in terms of overall value. A depreciating USD (that is, increased in Australian currency during years July 1974, USD = $0.6 \times$ AUD, and January 2012, USD = $0.9 \times$ AUD) causes dollar-denominated international commodity prices to rise, although not to the full extent of the depreciation (please see Fig. 9.31). The opposite occurs when the dollar appreciates (July 2001, USD = $2.1 \times$ AUD, and July 2021, USD = $1.7 \times$ AUD).

For example, consider below sugarcane exports from Australia (US Sugar #11 Futures[105]). The Sugar ICE No. 11 contract is the world benchmark contract for raw sugar trading and is available on the Intercontinental Exchange[106] (ICE). One of the

[102]The Relationship between Commodity Prices and Currency Exchange Rates: Evidence from the Futures Markets—https://www.nber.org/system/files/chapters/c11859/revisions/c11859.rev1.pdf

[103]Commodity prices and currency movements—https://www.investopedia.com/articles/forex/06/commoditycurrencies.asp#toc-trading-currencies-as-a-supplement-to-trading-oil-or-gold

[104]Sensitivity of Australian Trade to the Exchange Rate—https://www.rba.gov.au/publications/bulletin/2016/sep/2.html

[105]US Sugar #11 Futures—https://www.investing.com/commodities/us-sugar-no11-historical-data

[106]The ICE Sugar Contract 11—https://tradingeconomics.com/commodity/sugar

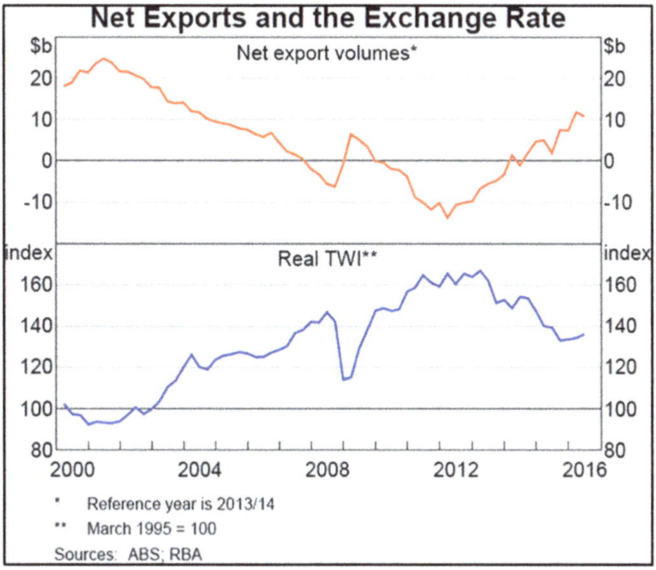

Fig. 9.30 Net exports and the exchange rate

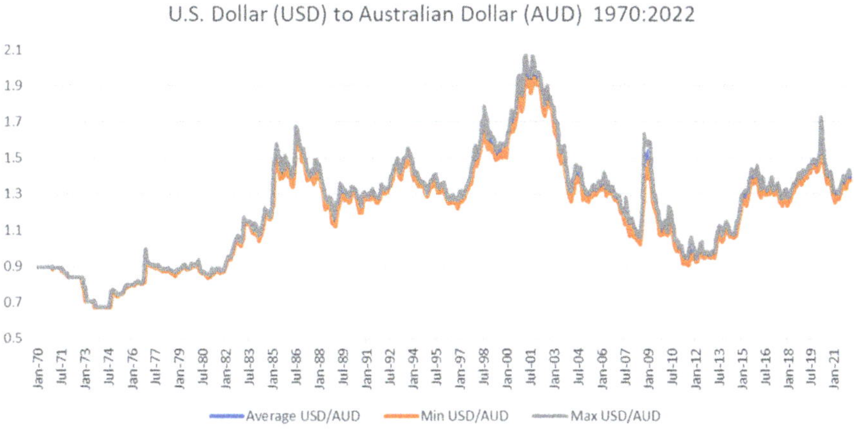

Fig. 9.31 USD to AUD exchange rate

biggest producer and exporter of sugar in the world is Australia (#10 largest producer[107] with 4,364,000 tons sugar production in 2022). Now, let us analyze

[107] World Largest Sugar Producers 2022—https://worldpopulationreview.com/country-rankings/sugar-producing-countries

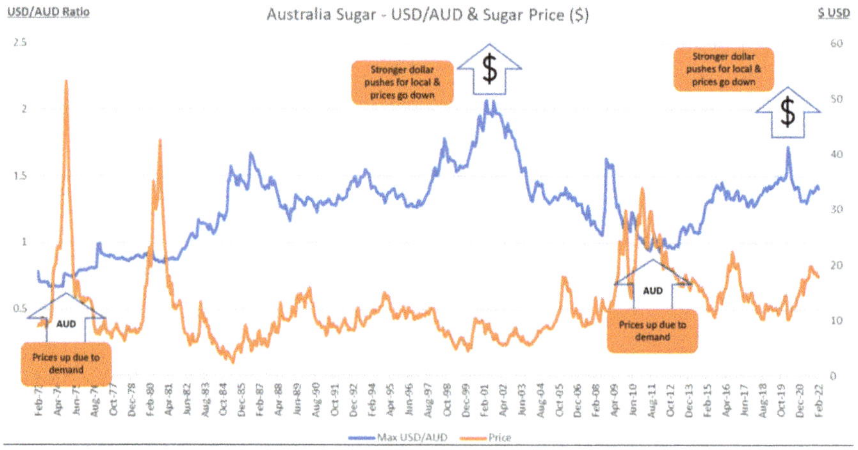

Fig. 9.32 Australia sugar

the signal between USD/Australia dollar exchange rate[108] 1960–2022 and ICE 11 futures. As indicated in the graph, higher USD/AUD values (i.e., $ appreciation in USD) coincided with a drop in ICE 11 futures price, and vice versa. That is, a drop in USD/AUD (depreciation in AUD) shows an increase in sugar futures due to increased demand (please see Fig. 9.32).

A similar behavior can be observed in staple cereals also. Wheat is one of the most important crops to humankind as it is a staple of many diets around the world. According to FAOSTAT, China produces more wheat than any other country, followed by India, Russia, and the United States. Here is an overview of the world's top ten wheat-producing countries. Australia is the sixth largest producer of wheat with 31,818,744 tons in 2022 and is one of the major winter crops grown in Australia. Western Australia, Victoria, New South Wales, and Queensland are the top wheat-producing states on the continent.[109] The crop is sown in the autumn months and harvested in the spring or winter, depending on the environmental conditions. Western Australia is the largest exporter of wheat, especially to those countries in Asia and the Middle East, generating earnings of around $2 billion annually from such sales. For the past 30 years, there has been a constant 1% annual rise in Western Australia's wheat production. The east coast of Australia, on the other hand, produces wheat that is meant for domestic consumption and feedstock.

[108] The USA/AUD Exchange Rate—https://fxtop.com/en/historical-exchange-rates.php?A=1&C1=USD&C2=AUD&MA=1&DD1=01&MM1=01&YYYY1=1970&B=1&P=&I=1&DD2=25&MM2=02&YYYY2=2022&btnOK=Go%21

[109] Top Wheat Producing Countries, 2022—https://www.worldatlas.com/articles/top-wheat-producing-countries.html

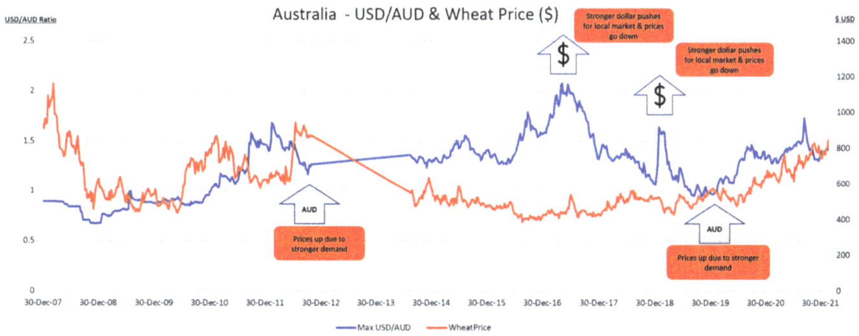

Fig. 9.33 Exchange rate and wheat price

Consider the above wheat price movements (US wheat futures[110]) versus USD/AUD exchange rates. Given Australia is one top producer (#6 producer[111] with 31,818,744 tons wheat production in 2022), the impact of AUD exchange rates can be seen on the wheat prices (please see Fig. 9.33). A signal between USD/-Australia dollar exchange rate[112] in 1960–2022 and wheat futures can be observed. As indicated in the graph, higher USD/AUD values (i.e., appreciation in USD) coincided with a drop in wheat future price, and vice versa. That is, a drop in USD/AUD (depreciation in AUD ) shows an increase in wheat futures due to increased demand.

Data Signal (Commodity Prices vs. Currency Exchange Rates)
Countries that rely on commodity exports for their economy can see their national currency's exchange rates fluctuate with commodity prices.

Here are the time series events of wheat price increases:

(1) 2007 and 2008 wheat prices[113]

Adverse weather conditions: For wheat—a major food grain—*adverse weather* in several primary growing areas in 2006/2007 and 2007/2008 resulted in below normal global production. The lower production levels contributed to the sharp increase in global wheat prices in 2007 and early 2008. Macroeconomic factors,

(continued)

[110]US Wheat Futures—https://www.investing.com/commodities/us-wheat-historical-data

[111]World Largest Sugar Producers 2022—https://worldpopulationreview.com/country-rankings/sugar-producing-countries

[112]The USA/AUD Exchange Rate—https://fxtop.com/en/historical-exchange-rates.php?A=1&C1=USD&C2=AUD&MA=1&DD1=01&MM1=01&YYYY1=1970&B=1&P=&I=1&DD2=25&MM2=02&YYYY2=2022&btnOK=Go%21

[113]Factors Behind the Rise in May 2009 Global Rice Prices in 2008— https://www.ers.usda.gov/webdocs/outlooks/38489/13518_rcs09d01_1_.pdf?v=9.9

	such as the declining dollar and a shift of funds from equities and real estate into commodities, also contributed to rising global commodity prices. For corn, wheat, and soybeans, the primary factors behind the price increases were increased use of biofuels, changing diets in China and India, major weather problems, and tight stocks-to-use ratios. Increased participation in futures markets by nontraditional investors likely contributed to greater price volatility.
	Commodity speculation, mostly by Middle Eastern countries, faced with a need to invest the revenues resulting from their *quadrupled oil* prices has also driven up prices. Algeria, for example, has bought about 3 million tons R sugar and Kuwait about 2 million within the last month-equivalent to about one-third of the American consumption and considerably more than their own domestic needs.
(2) 2006 and 2007 wheat prices increase[114]	
	*Wheat crop shortfall*: In India, the restrictions and ban on non-basmati rice exports were also a response to large imports of wheat in 2006/2007 and 2007/2008. India was an exporter of wheat from 2000/2001 to 2005/2006, importing very little wheat. However, with wheat crop shortfalls in 2006/2007 and 2007/2008, the Government of India curtailed wheat exports and imported several million tons of wheat to meet domestic demand [46].
(3) 1985 sugar prices—fell [46, 47]	
	The price of sugar futures fell to new lows in April 21, 1985, as the worldwide *surplus* weighed on the market [46]. Sugar prices fell again, bottoming in May 1985 at less than US$0.03 per pound and averaging US$0.04 per pound for the year. In early 1990, prices had recovered to US$0.14 cents per pound and then declined[115] to approximately US$0.08 to US$0.09 per pound [47].
(4) 1984 spike in sugar prices	
	Brazil is the largest sugar-producing country in the world. The production of sugar has increased 352% since 1990. About 34% of Brazilian sugar consumed domestically is converted into ethanol for fuel. Exports have risen from 1.2 million metric tons in 1990 to 26.9 million metric tons in 2011. Sugar that is converted into ethanol is subsidized at prices higher than the world price. Recent increases in the world oil price have increased the price of ethanol[116] which in turn increased Brazil's conversion of sugar into ethanol, reducing potential sugar exports from Brazil [34]. That reduction in the growth of sugar exports could be one of the main forces for world sugar price increases. Brazil decreased its exports by 7.8% in 2011 which provided strength for sugar prices in 2011.

9.6.2.7 Vegetation Index

Crop production is directly affected by the weather and climatic conditions. In many parts of the world, for example, Mozambique, Myanmar, Thailand, and Mediterranean basins, natural vegetation in general and crop production has always been

[114] Factors Behind the Rise in May 2009 Global Rice Prices in 2008— https://www.ers.usda.gov/webdocs/outlooks/38489/13518_rcs09d01_1_.pdf?v=9.9

[115] Philippine—Sugar—http://countrystudies.us/philippines/64.htm

[116] 2012 Outlook of the U.S. and World Sugar Markets, 2011-2021—https://ageconsearch.umn.edu/record/128037/files/AAE692.pdf

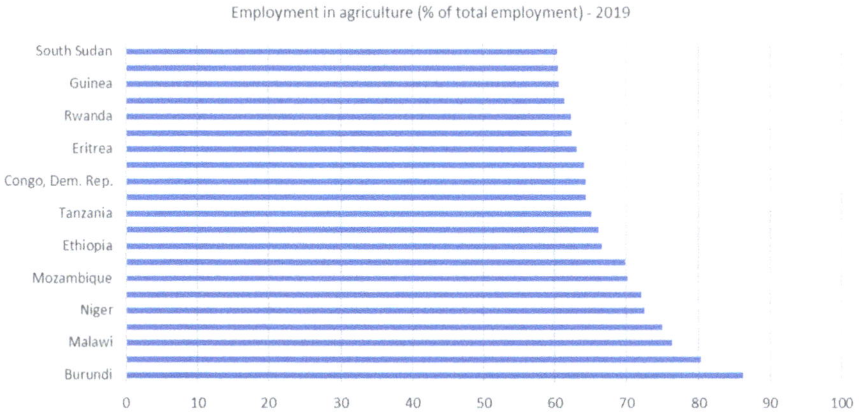

Fig. 9.34 Employment in agriculture

affected by large natural climate variability and is expected to continue to be affected in the future. Particularly, seasonal changes in precipitation and temperature and their seasonal variability affect crop production, especially in regions where crops are highly dependent on precipitation. Because of the long-term influence of precipitation and temperature on crop production, drought is a major cause of unexpected crops. The Vegetation Index captures such drought signals. The Vegetation Condition Index[117] (VCI) evaluates the current vegetation health in comparison to the historical trends. The VCI relates current dekadal Normalized Difference Vegetation Index (NDVI) to its long-term minimum and maximum, normalized by the historical range of NDVI values for the same dekad. The VCI was designed to separate the weather-related component of the NDVI from the ecological element. Weather has both direct and indirect effects in the Australian GDP. In general, this is true for other countries.

Weather Impact on GDP: Direct Effect

Agriculture is the major employer in the world with many small and marginalized farmers' livelihood depending on it. The following is the list of countries with employment in agriculture as percentage of total employment (FAOSTAT 2019)[118] with Burundi, Malawi, and Niger leading the pack. It is clear that the more a country's economy depends on agriculture for employment, the more impact the weather and climate change would have on the country's economy (please see Fig. 9.34).

[117] Vegetation Index—https://www.fao.org/giews/earthobservation/country/index.jsp?lang=en&code=AUS

[118] Employment in agriculture (% of total employment) (modeled ILO estimate)—https://data.worldbank.org/indicator/SL.AGR.EMPL.ZS

The same is applied to Australia. Australia has 3,588,950 sq. km agricultural land[119] (The World Bank 2019) with 46.65% agricultural land as the total percent of land area[120] (The World Bank 2019). The gross value for Australian agriculture increased slightly to $61 billion in 2019–2020 despite drought conditions. In 2020, agriculture contributed around 2.01% to the GDP of Australia, 25.46% came from industry, and 66.28% from the services sector.[121] The total gross value of crops decreased by 5% to $28 billion due to a large fall in the value of cereal and broadacre crops, particularly wheat. Drought conditions impacted broadacre crops in New South Wales, Queensland, and Western Australia. Results were mixed for horticultural crops.[122] The following are key crop results for 2019–2020 [48]:

- $5.4 billion for fruit and nuts (up to 9% from 2018 to 2019)
- $4.9 billion for wheat (down to 20%)
- $4.2 billion for vegetables (down to 4%)
- $3.0 billion for barley (steady)
- $2.5 billion for hay (up to 24%)
- $1.4 billion for canola (up to 1%)
- $252 million for cotton (down to 78%)

Impacts of drought are evident in agricultural activity estimates for the 2019–2020 reference year across several ABS' agricultural collections, including the Value of Agricultural Commodities Produced. Many farming areas across Australia experienced drought throughout the reference period, with New South Wales and Queensland particularly impacted. In addition, the impact on estimates from the bushfire activity in New South Wales, Victoria, and Queensland in early 2020 and flooding experienced in parts of New South Wales and Queensland are varied, with some respondents reporting impacts on their agricultural activity from these events. In recent months Australia has experienced a number of natural disasters: large parts of Australia have been experiencing ongoing droughts, bushfires, damaging hailstorms, floods, and the global COVID-19 outbreak. These events will have had an impact on all aspects of the Australian economy to varying degrees.[123] In 2017–2018, the Value of Agricultural Commodities Produced fell 3% to $59 billion. Detailed information about where and which commodities were most impacted for these early stages are published in the 2017–2018 Value of Agricultural Commodities Produced publication [49].

[119]The World Bank (AG.LND.AGRI.K2)—https://data.worldbank.org/indicator/AG.LND.AGRI.K2

[120]The World Bank (AG.LND.AGRI.ZS)—https://data.worldbank.org/indicator/AG.LND.AGRI.ZS

[121]Distribution of gross domestic product (GDP) across economic sectors Australia 2020—https://www.statista.com/statistics/375558/australia-gdp-distribution-across-economic-sectors/

[122]Value of Agricultural Commodities Produced, Australia—https://www.abs.gov.au/statistics/industry/agriculture/value-agricultural-commodities-produced-australia/latest-release

[123]Measuring natural disasters in the Australian economy—https://www.abs.gov.au/statistics/research/measuring-natural-disasters-australian-economy

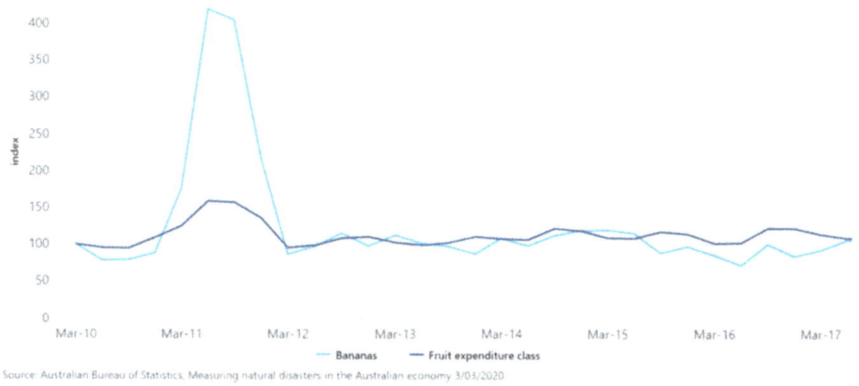

Fig. 9.35 Cyclone and crops

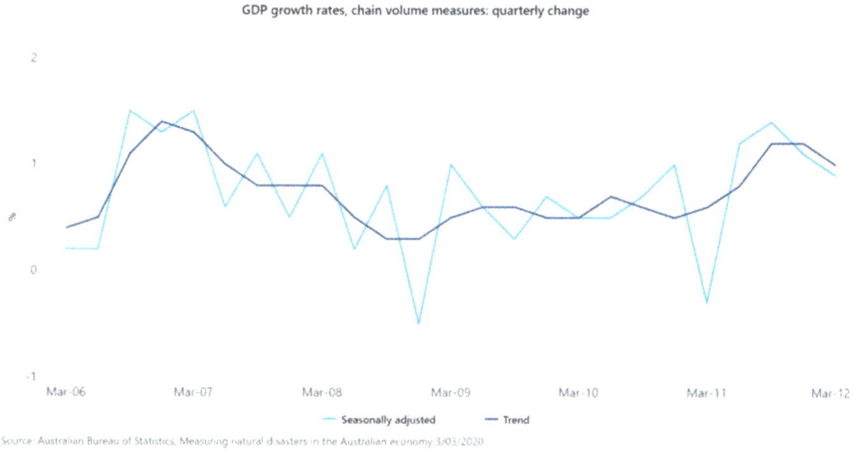

Fig. 9.36 GDP growth rates

Cyclones and Floods

Tropical cyclone Yasi in 2011 in North Queensland caused an estimated $300 million loss in crops. About 75% of banana crops were damaged, resulting in a rise of over 400% in the price of bananas over the 12 months to June 2011 (please see Fig. 9.35). This saw an increase of fruit prices in the CPI of over 60% compared to the previous year [49].

Floods

In December 2010, Queensland experienced widespread flooding. The flooding led to lower production as measured in Australian National Accounts: National Income, Expenditure and Product December 2010 and March 2011 quarters (please see Fig. 9.36).

Weather Impact on GDP: Indirect Effect

Weather not only has a direct impact on the GDP, but it also has indirect impact. For instance, consider price movements of essential commodities due to weather issues that are in the other parts of agricultural economies. The network effect is very evident in the following commodity movement.

To meet demands of food, the world must, by 2050, produce 49% more food than in 2012 as populations grow and diets change. At the same time, almost 80% of the poor live in rural areas where people depend on agriculture, fisheries, or forestry as their main source of income and food. If temperatures continue to rise, then progress towards eradicating hunger and ensuring the sustainability of our natural resource base to achieve the 2030 Agenda for Sustainable Development will be at risk. Climate change will bring further extreme weather events, land degradation and desertification, water scarcity, rising sea levels, and shifting climates—all of which will make the rural poor the first victims, hampering efforts to feed the whole planet. Agricultural economic opportunities shrink due to climate change—employment days would cut, for instance, agricultural farmers due to shrinkage of crop—cut harvest months from 3 months, normally, to a month, due to cold snap and drought[124,125] [50, 51].

9.6.3 *Model Development*

The following are the steps to build the model.

Step 1: Load into Python DataFrame—Load ML-specific libraries:

```
import numpy as np
import pandas as pd
import functools
from datetime import date
import plotly.graph_objects as go
import seaborn as sns; sns.set()
import statsmodels.api as sm
from statsmodels.tsa.arima_process import ArmaProcess
from scipy import stats
from scipy.stats import pearsonr
import matplotlib.pyplot as plt
```

(continued)

[124] Coffee Prices Soar After Bad Harvests and Insatiable Demand—https://www.wsj.com/articles/coffee-prices-soar-after-bad-harvests-and-insatiable-demand-11626093703?mod=article_inline

[125] Coffee Prices Jump to Six-Year High as Brazilian Frost Threatens Crop—https://www.wsj.com/articles/coffee-prices-jump-to-six-year-high-as-brazilian-frost-threatens-crop-11627380128?mod=article_inline

```
from sklearn.preprocessing import MinMaxScaler
from sklearn.model_selection import train_test_split
from sklearn.linear_model import LinearRegression
from sklearn.metrics import r2_score
from sklearn import metrics
from sklearn.impute import SimpleImputer
import eli5
```

Load all the required libraries to build the machine learning model.

Step 2: Load Australia data sources:

```
AusKeyEconDf = pd.read_csv("Australia.csv")
AusKeyEconDf.head(3)
```

Series Name index	Agriculture, forestry, and fishing, value added (constant 2015 US$)	Employment in agriculture (% of total employment) (modeled ILO estimate)	Fertilizer consumption (% of fertilizer production)	Fertilizer consumption (kilograms per hectare of arable land)	GDP per capita growth (annual %)	Imports of goods and services (% of GDP)	Labor force participation rate, total (% of total population ages 15-64) (modeled ILO estimate)	Rural population (% of total population)
1971	..	..	107.953502383716	63.3498843210536	0.542875961259725	12.9126526164591	..	15.84
1972	..	..	103.480938546354	76.723646389394	2.01650232004111	11.951252356585	..	15.68
1973	..	..	112.260815460069	102.760318949343	1.06226509266496	10.9991772190002	..	15.522
1974	..	..	93.6091057823662	58.6564487752064	1.50339189733786	13.1911087554245	..	15.364
1975	12608515574.8426	..	113.64632208264	55.3202129227142	0.0956650322482062	14.7074091221572	..	15.208
1976	13651571419.8604	..	115.299553914782	72.5115644002211	1.56782593896072	13.3574007220217	..	15.053
1977	14016384055.31	..	111.397203132159	72.0753338273934	2.43226026838195	14.6255039694085	..	14.9
1978	13557156882.1648	..	120.162082886958	72.8210393748122	-0.272382968871057	14.5660075741717	..	14.748
1979	16500706720.6603	..	114.370268279163	74.7907382744644	2.93241749918668	15.3305440739462	..	14.597
1980	14471757573.8553	..	118.475427448725	69.5025643764759	1.78708039101825	15.8814442942653	..	14.448
1981	12687515494.1389	..	122.801919350932	66.7682376824317	1.71191514327921	16.7108457450299	..	14.3
1982	14762708496.1417	..	138.642439903846	59.3275266162629	1.61420796983991	16.8005366166999	..	14.33
1983	11535272768.7929	..	149.954019090681	62.0257623772664	-3.43711705686596	15.5968144563307	..	14.359
1984	16529609130.1589	..	156.366052303861	57.7588389530337	3.41834533995349	14.9780601218154	..	14.389
1985	16562365194.2574	..	157.134831460674	55.3716071173019	3.82274903959301	17.2729744195332	..	14.418
1986	15405626538.5449	..	143.321280888146	58.0646307491415	2.33980885428716	18.0473622436846	..	14.448
1987	15570049134.8039	..	139.338982252081	70.8993709183674	0.999993493461133	17.0615815695509	..	14.478
1988	15234138908.853	..	146.204466444352	77.3587373180415	4.0275705632889	16.6189447339952	..	14.507
1989	15906601636.5214	..	188.78628236246	79.8789907000628	2.12125930120597	16.9536459818821	..	14.537
1990	17209779167.0271	..	236.502153410241	68.9170287803727	2.05192904909737	17.0435698236889	73.62	14.567

Step 3: Load crude oil data sources:

```
CrudeOilDf = pd.read_excel("CrudeOil.xlsx")
CrudeOilDf.set_index("Year",inplace = True)
CrudeOilDf
```

Output:

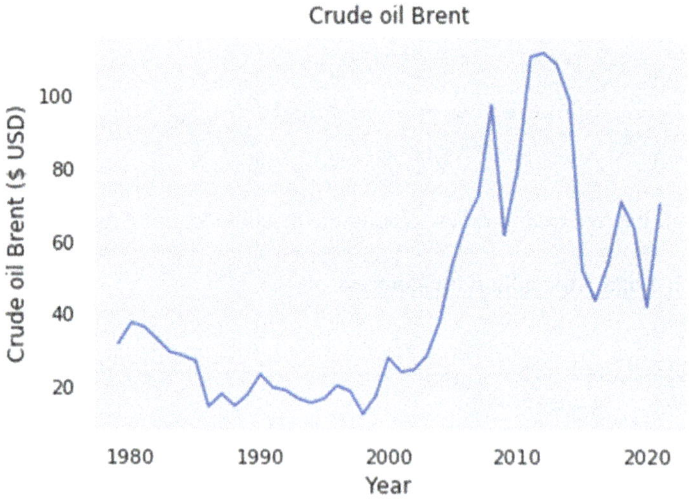

Step 4: Load sugar production data sources:

```
AustraliaSugarProductionDf = pd.read_csv
("SugarCaneProductionValue.csv")
AustraliaSugarProductionDf

AustraliaSugarProductionDf = AustraliaSugarProductionDf
[["Year","Value"]]
AustraliaSugarProductionDf.set_index("Year", inplace = True)
AustraliaSugarProductionDf
```

Plot sugar production:

```
import matplotlib.pyplot as plt
import numpy as np

# Data for plotting
t = AustraliaSugarProductionDf.index
s = AustraliaSugarProductionDf['production Value 1000 US$']

fig, ax = plt.subplots()
ax.plot(t, s)

ax.set(xlabel='Year', ylabel='production Value 1000 US$',
   title='production Value 1000 US$')
ax.grid()

plt.show()
```

Output:

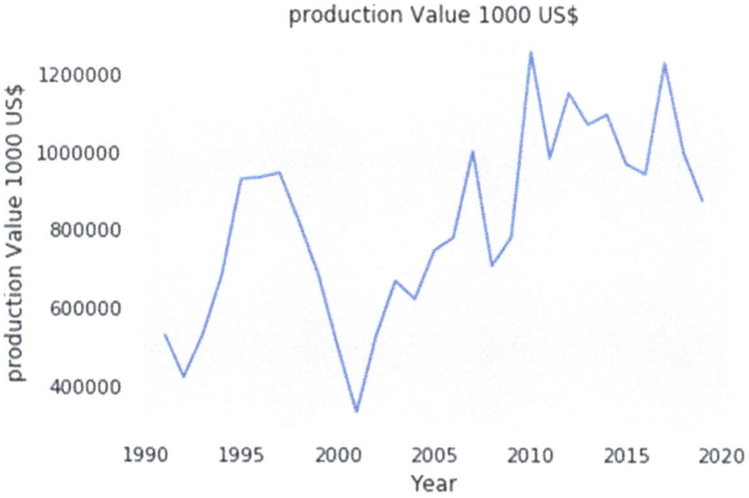

Step 5: Load US dollar to Australian dollar exchange rate

```
AustraliaExhangeRateDf= pd.read_csv("ExchangeRate-Australia.
csv")
AustraliaExhangeRateDf
```

Plot USD/AUD:

```
import matplotlib.pyplot as plt
import numpy as np

# Data for plotting
t = AustraliaExhangeRateDf.index
s = AustraliaExhangeRateDf['Standard local currency units per
USD']

fig, ax = plt.subplots()
ax.plot(t, s)

ax.set(xlabel='Year', ylabel='Standard local currency units per
USD',
    title='USD/AUD Exchange Rate')
ax.grid()

plt.show()
```

Output:

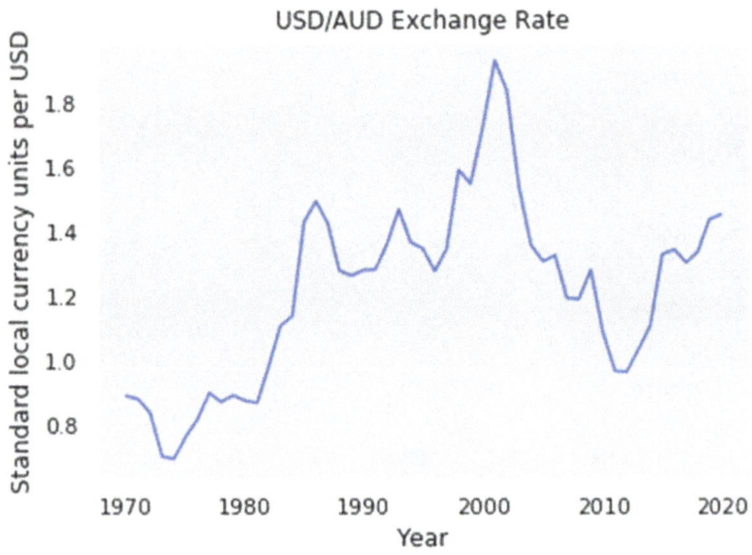

Please see the observed USD/AUD versus sugarcane:

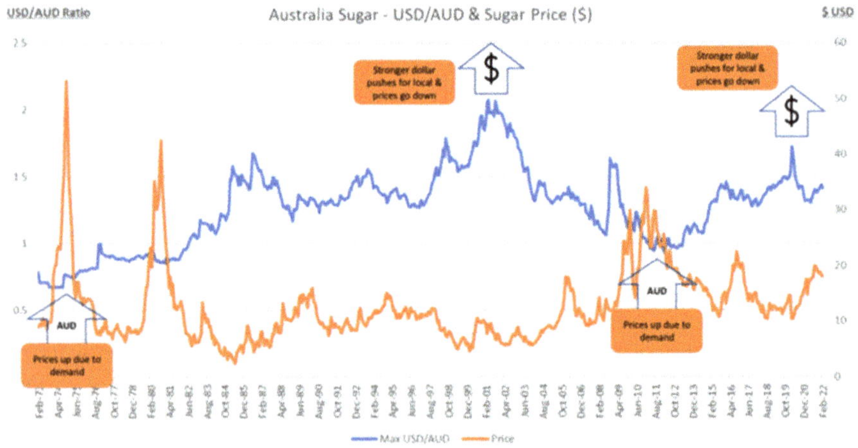

Step 6: Load carbon emissions

```
AustraliaEmissionCODf = pd.read_csv("CO2-Australia.csv")
AustraliaEmissionCODf
```

I have been using the same code to draw the graph.
Output: as you can see, the emissions are growing at a constant rate.

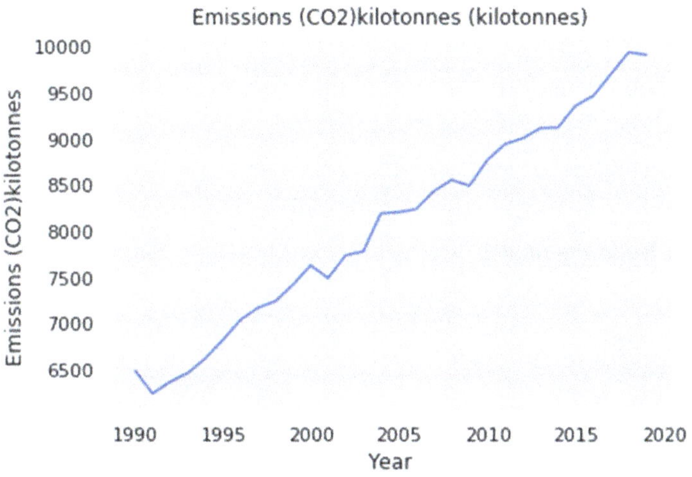

Step 7: Load sugar harvest and production

```
AustraliaSugarCaneAreaHarvested_df = pd.read_csv
("AreaHarvest_SugarCane_Australia.csv")
AustraliaSugarCaneAreaHarvested_df
```

Output:

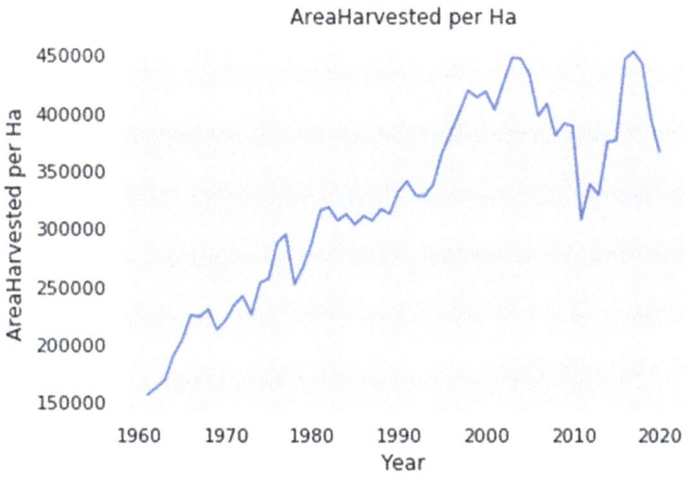

Step 8: Merge all related Australian sugarcane DataFrame

```
AusKeyEconSugarProdDF = pd.merge
(AusKeyEconomicDf,AustraliaSugarProductionDf,left_on
=AusKeyEconomicDf.index,right_on=AustraliaSugarProductionDf.
index,how = "outer")
AusKeyEconSugarProdDF.rename(columns = {"key_0":"Year"},
inplace = True)
AusKeyEconSugarProdDF.set_index(["Year"],inplace = True)
```

Step 9: Build the model

```
X = final_df[[
    'Employment in agriculture (% of total employment) (modeled ILO
estimate)',
    'Fertilizer consumption (kilograms per hectare of arable
land)',
    'GDP per capita growth (annual %)',
    'Imports of goods and services (% of GDP)',
   'Labor force participation rate, total (% of total population
ages 15-64) (modeled ILO estimate)',
    'Rural population (% of total population)',
    'Standard local currency units per USD',
    'Crude oil Brent', 'AreaHarvested per Ha'
    ]]


y = final_df['production Value 1000 US$']
```

Train the model:

```
X_train,X_test, y_train, y_test = train_test_split(X,y,
test_size = 0.2,random_state = 0)


regressor = LinearRegression()
regressor.fit(X_train,y_train)
```

Predict the test values:

```
# predicting the test
y_pred = regressor.predict(X_test)
y_pred
```

Output:

```
array([ 632194.49531877, 1105396.87173658, 701021.81233884,
    490423.93887014, 1122919.03907904, 1098833.86293743])
```

Step 10: Plot the model

```
plt.scatter(y_test, y_pred)
plt.xlabel('Actual Labels')
plt.ylabel('Predicted Labels')
plt.title('Predictions')
# overlay the regression line
z = np.polyfit(y_test, y_pred, 1)
p = np.poly1d(z)
plt.plot(y_test,p(y_test), color='magenta')
plt.show()
```

Output:

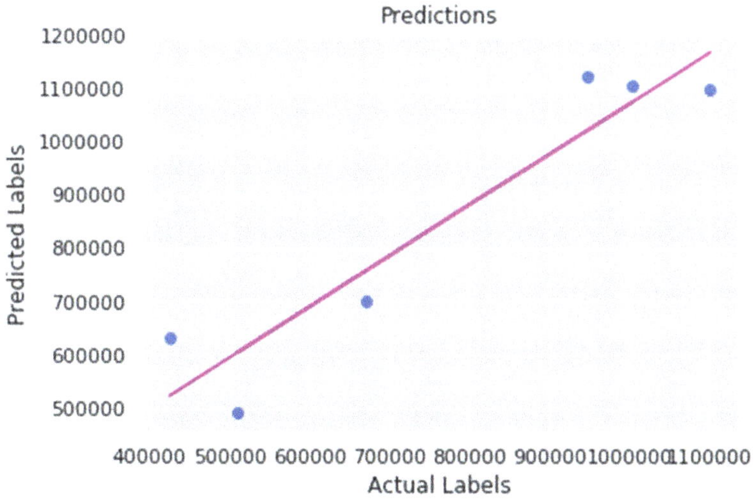

Step 11: Model equation

Develop the model equation: Y = mx + c form

Use the regressor coefficients: develop mx part.

```
mx=""
for ifeature in range(len(X.columns)):
  if regressor.coef_[ifeature] <0:
```

(continued)

```
    # format & beautify the equation
   mx += " - " + "{:.2f}".format(abs(regressor.coef_[ifeature])) +
" * " + X.columns[ifeature]
  else:
    if ifeature == 0:
     mx += "{:.2f}".format(regressor.coef_[ifeature]) + " * " + X.
columns[ifeature]
    else:
     mx += " + " + "{:.2f}".format(regressor.coef_[ifeature]) + " *
" + X.columns[ifeature]

print(mx)
```

Output:

- 60901.20 * Employment in agriculture (% of total employment) (modeled ILO estimate) + 9301.76 * Fertilizer consumption (kilograms per hectare of arable land) - 2406.18 * GDP per capita growth (annual %) + 32178.90 * Imports of goods and services (% of GDP) - 4751.42 * Labor force participation rate, total (% of total population ages 15-64) (modeled ILO estimate) - 151240.57 * Rural population (% of total population) - 869682.43 * Standard local currency units per USD - 558.99 * Crude oil Brent + 0.94 * AreaHarvested per Ha

Combine with intercept:

```
# y=mx+c
if(regressor.intercept_ <0):
  print("The formula for the " + y.name + " linear regression line
(y=mx+c) is = " + " - {:.2f}".format(abs(regressor.intercept_)) +
" + " + mx)
else:
  print("The formula for the " + y.name + " linear regression line
(y=mx+c) is = " + "{:.2f}".format(regressor.intercept_) + " + " +
mx)
```

Output:

```
The formula for the production Value 1000 US$ linear regression line (y=mx+c) is = 3052507.92 + - 60901.20 * Employment in agriculture (% of total employment) (modeled ILO estimate) + 9301.76 * Fertilizer consumption (kilograms per hectare of arable land) - 2406.18 * GDP per capita growth (annual %) + 32178.90 * Imports of goods and services (% of GDP) - 4751.42 * Labor force participation rate, total (% of total population ages 15-64) (modeled ILO estimate) - 151240.57 * Rural population (% of total population) - 869682.43 * Standard local currency units per USD - 558.99 * Crude oil Brent + 0.94 * AreaHarvested per Ha
```

Model score:

```
r2_score(y_test,y_pred)
```

Output:
$R^2 = 0.7789$ or 77.89%

Myanmar Rice Production

9.7 Machine Learning Model: Myanmar Rice Production Macroeconomic Linkage Predictive Model

Let us build the Queensland, Australia, sugarcane production model:

Software code for this model:
FinalMyanmar_Rice_Production_MacroEconomic_LinkageModel.ipynb(Jupyter Notebook Code)

9.7.1 Data Sources

The data sources for the ML model is from FAO Statistics Data,[126] World Bank Climate Change Data specifically Myanmar,[127] World Metrological Organization Data,[128] FAO Earth Observation,[129] and NOAA STAR – STAR – Global Vegetation Health Products: Province-Averaged VH.[130]

[126] FAO Statistics data—https://www.fao.org/faostat/en/#data/QCL

[127] Climate Change Data—https://climateknowledgeportal.worldbank.org/download-data

[128] World Metrological Organization Data—https://climatedata-catalogue.wmo.int/explore

[129] Earth Observation Mozambique—https://www.fao.org/giews/earthobservation/country/index.jsp?lang=en&code=AUS

[130] NOAA STAR—VHI—https://www.star.nesdis.noaa.gov/smcd/emb/vci/VH/vh_adminMeanByCrop.php?type=Province_Weekly_MeanPlot

9.7.2 *Model Development*

The following are the steps to build the model.

Step 1: Load into Python DataFrame: load ML-specific libraries:

```
import numpy as np
import pandas as pd
import functools
from datetime import date
import plotly.graph_objects as go
import seaborn as sns; sns.set()
import statsmodels.api as sm
from statsmodels.tsa.arima_process import ArmaProcess
from scipy import stats
from scipy.stats import pearsonr
import matplotlib.pyplot as plt
from sklearn.preprocessing import MinMaxScaler
from sklearn.model_selection import train_test_split
from sklearn.linear_model import LinearRegression
from sklearn.metrics import r2_score
from sklearn import metrics
from sklearn.impute import SimpleImputer
import eli5
```

Load all the required libraries to build the machine learning model.

Step 2: Load Myanmar exchange data:

```
exchange_df = pd.read_csv('Myanmar_Exchange.csv')
exchange_df.head(5)
```

	Domain Code	Domain	Area Code (FAO)	Area	ISO Currency Code (FAO)	Currency	Item Code	Item	Year Code	Year	Unit	Value	Flag	Flag Description	Note
0	PE	Exchange rates - Annual	28	Myanmar	MMK	Kyat	5540	Standard local currency units per USD	1970	1970	NaN	4.761900	X	International reliable sources	Data from UNSD AMA
1	PE	Exchange rates - Annual	28	Myanmar	MMK	Kyat	5540	Standard local currency units per USD	1971	1971	NaN	4.764843	X	International reliable sources	Data from UNSD AMA
2	PE	Exchange rates - Annual	28	Myanmar	MMK	Kyat	5540	Standard local currency units per USD	1972	1972	NaN	5.459485	X	International reliable sources	Data from UNSD AMA
3	PE	Exchange rates - Annual	28	Myanmar	MMK	Kyat	5540	Standard local currency units per USD	1973	1973	NaN	4.931048	X	International reliable sources	Data from UNSD AMA
4	PE	Exchange rates - Annual	28	Myanmar	MMK	Kyat	5540	Standard local currency units per USD	1974	1974	NaN	4.862517	X	International reliable sources	Data from UNSD AMA

Step 3: Load Myanmar rice production data:

```
production_df = pd.read_csv("Myanmar.csv")
production_df.head(2)
```

Output:

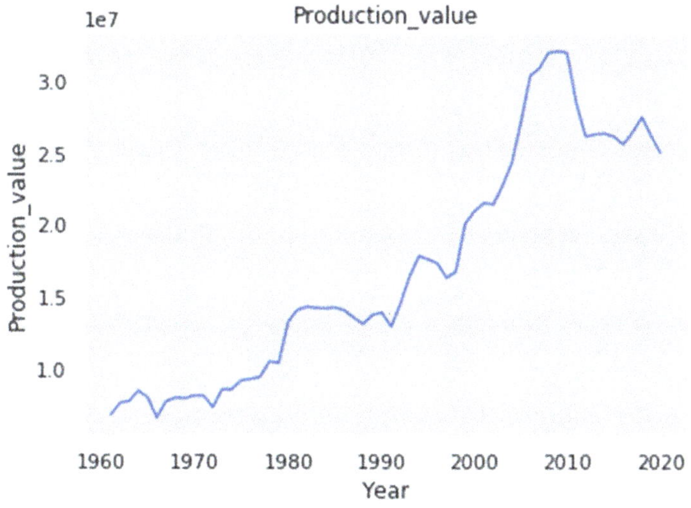

Exchange rate:

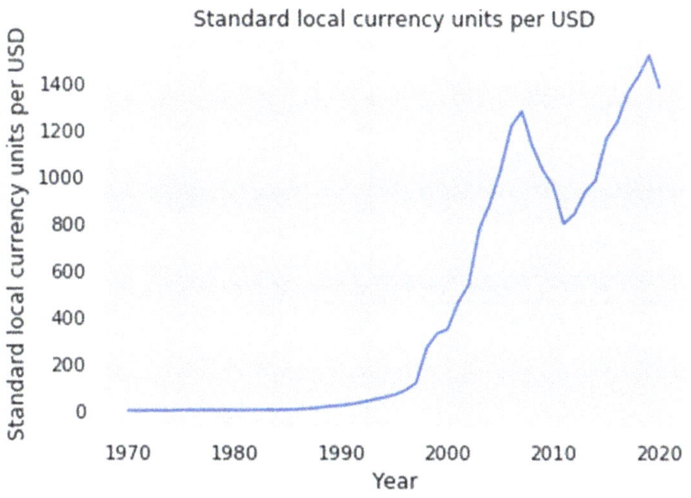

Step 4: Load Myanmar rice input data—fertilizer cost:

```
fertilizer_df = pd.read_csv('Myanmar_fertilizer.csv')
fertilizer_df.set_index("Year Code",inplace = True)
```

Plot fertilizer usage:

```
import matplotlib.pyplot as plt
import numpy as np

# Data for plotting
t = fertilizer_df.index
s = fertilizer_df['Nutrient_nitrogen_Value']

fig, ax = plt.subplots()
ax.plot(t, s)

ax.set(xlabel='Year', ylabel='Nutrient_nitrogen_Value',
   title='Nutrient_nitrogen_Value')
ax.grid()

plt.show()
```

Output:

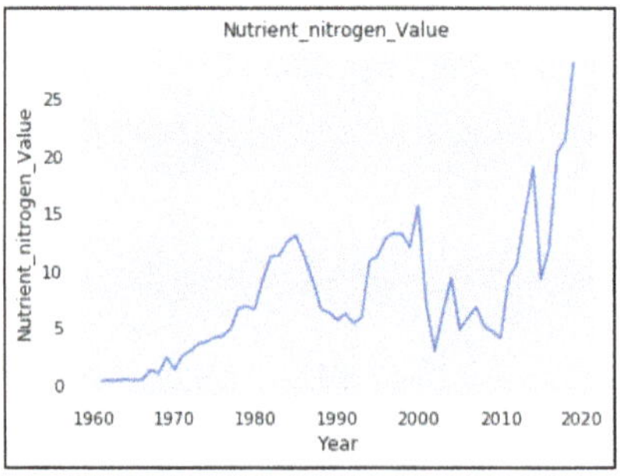

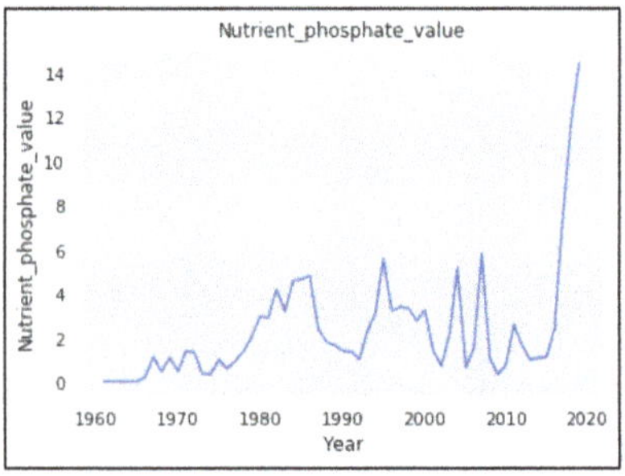

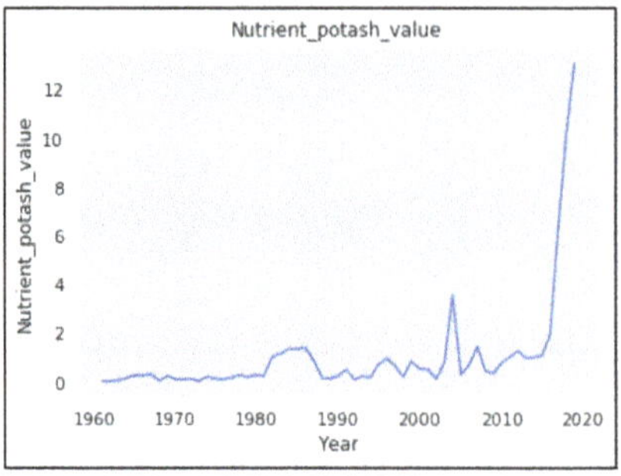

Step 5: Load Myanmar macroeconomic indicator data

```
indicatorsdf = pd.read_csv("Myanmar_Indicators.csv")
indicatorsdf
```

Output:

	Series Name	GDP growth (annual %)	Agriculture, forestry, and fishing, value added (% of GDP)	Inflation, consumer prices (annual %)	Rural population (% of total population)	Fertilizer consumption (kilograms per hectare of arable land)	Labor force participation rate, total (% of total population ages 15-64) (modeled ILO estimate	Employment in agriculture (% of total employment) (modeled ILO estimate)
0	1971	4.544330	NaN	2.087912	76.785	4.415547	NaN	NaN
1	1972	3.265764	NaN	7.642626	76.394	4.817352	NaN	NaN
2	1973	0.713379	NaN	25.200000	76.097	4.370901	NaN	NaN
3	1974	2.171778	NaN	25.211661	76.087	4.706762	NaN	NaN
4	1975	4.732067	NaN	31.655663	76.078	5.757391	NaN	NaN
5	1976	5.136158	NaN	22.384402	76.067	5.375604	NaN	NaN
6	1977	6.014451	NaN	-1.156687	76.058	6.488377	NaN	NaN
7	1978	6.243224	NaN	-6.044706	76.048	8.947897	NaN	8531.000000
8	1979	5.838772	NaN	5.672295	76.038	9.731368	NaN	8697.000000
9	1980	6.604599	NaN	0.608153	76.027	10.466938	NaN	8864.000000
10	1981	7.116933	NaN	0.318445	76.018	13.076523	NaN	9034.000000
11	1982	5.968237	NaN	5.304740	76.007	17.505198	NaN	9205.000000
12	1983	4.980957	NaN	5.650083	75.958	16.612920	NaN	9400.000000
13	1984	4.667729	NaN	4.846312	75.788	19.622563	NaN	9590.000000
14	1985	3.866811	NaN	6.807567	75.619	20.232148	NaN	9772.000000

Plot GDP growth:

```
import matplotlib.pyplot as plt
import numpy as np

# Data for plotting
t = indicatorsdf1.index
s = indicatorsdf1['GDP growth (annual %)']

fig, ax = plt.subplots()
ax.plot(t, s)

ax.set(xlabel='Year', ylabel='GDP growth (annual %)',
    title='GDP growth (annual %)')
ax.grid()

plt.show()
```

Output:

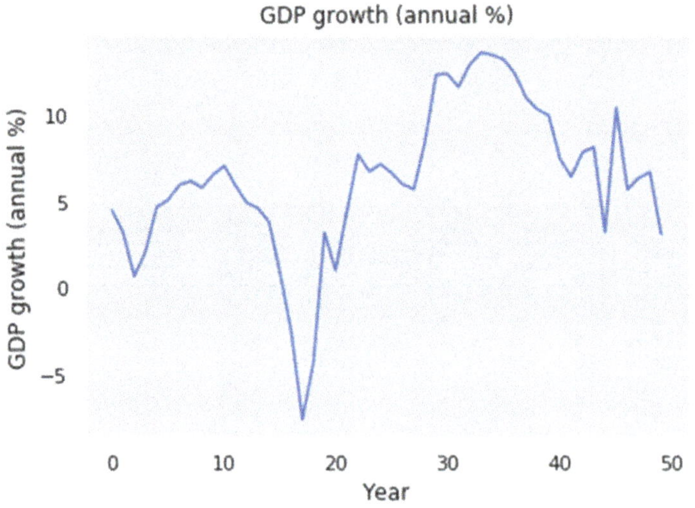

Step 6: Merge all related Australian sugarcane DataFrame

```
df4 = pd.merge(production_df,indicatorsdf,on = "Year",how =
"outer")
```

Output:

```
Index(['Year', 'Area_harvested_Value', 'Yield_value',
'Production_value',
    'GDP growth (annual %)',
    'Agriculture, forestry, and fishing, value added (% of GDP)',
    'Inflation, consumer prices (annual %)',
    'Rural population (% of total population)',
    'Fertilizer consumption (kilograms per hectare of arable
land)',
    'Labor force participation rate, total (% of total population
ages 15-64) (modeled ILO estimate',
    'Employment in agriculture, forestry and fishing by age, total
(15+)'],
   dtype='object')
```

Step 7: Build the model

```
X = final_df1[['Standard local currency units per USD',
    'Employment in agriculture, forestry and fishing by age, total
(15+)',

    'Producer Price Index (2014-2016 = 100)',
    'Crude oil, Brent',
     'GDP growth (annual %)',
     'Apr', 'May', 'Jun','July'
    ]].values


y = final_df1['ProductionValue'].values
```

Train the model:

```
X_train,X_test, y_train, y_test = train_test_split(X,y,
test_size = 0.2,random_state = 0)


regressor = LinearRegression()
regressor.fit(X_train,y_train)
```

Predict the test values:

```
# predicting the test
y_pred = regressor.predict(X_test)
y_pred
```

Output:

```
array([16266156.48612512, 24287971.53330033, 24236706.46068561,
    22113583.1336503, 25079070.61625148, 23264371.48018068,
    27498150.07224797])
```

Step 8: Plot the model

```
plt.scatter(y_test, y_pred)
    plt.xlabel('Actual Labels')
    plt.ylabel('Predicted Labels')
```

(continued)

```
plt.title('Predictions')
# overlay the regression line
z = np.polyfit(y_test, y_pred, 1)
p = np.poly1d(z)
plt.plot(y_test,p(y_test), color='magenta')
plt.show()
```

Output:

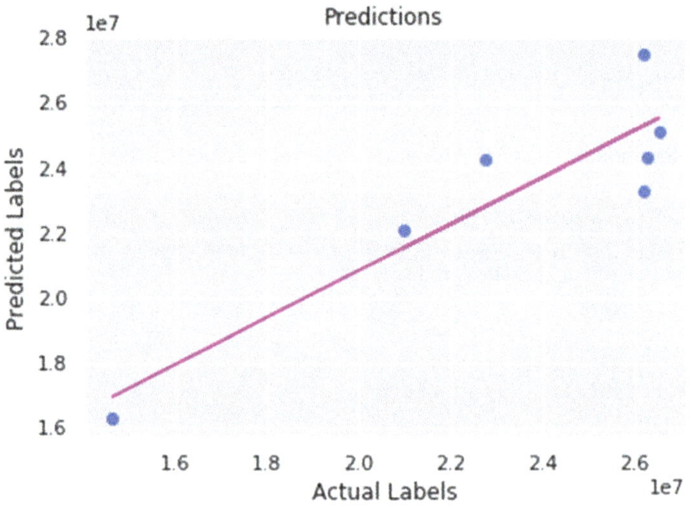

Step 9: Model equation

Develop the model equation: Y = mx + c form

Use the regressor coefficients: develop mx part.

```
mx=""
for ifeature in range(len(X.columns)):
  if regressor.coef_[ifeature] <0:
    # format & beautify the equation
   mx += " - " + "{:.2f}".format(abs(regressor.coef_[ifeature])) +
" * " + X.columns[ifeature]
  else:
    if ifeature == 0:
     mx += "{:.2f}".format(regressor.coef_[ifeature]) + " * " + X.
columns[ifeature]
    else:
```

(continued)

```
    mx += " + " + "{:.2f}".format(regressor.coef_[ifeature]) + " *
" + X.columns[ifeature]

print(mx)
```

Output:
6674.47 * Standard local currency units per USD - 102.72 * Employment in agriculture, forestry, and fishing by age, total (15+) - 20051.86 * Producer Price Index (2014-2016 = 100) + 82899.89 * Crude oil, Brent + 392694.35 * GDP growth (annual %) + 13295.39 * Apr - 27006.83 * May + 8550.51 * Jun + 4027.57 * July
Combine with intercept:

```
# y=mx+c
if(regressor.intercept_ <0):
  print("The formula for the " + y.name + " linear regression line
(y=mx+c) is = " + " - {:.2f}".format(abs(regressor.intercept_)) +
" + " + mx)
else:
  print("The formula for the " + y.name + " linear regression line
(y=mx+c) is = " + "{:.2f}".format(regressor.intercept_) + " + " +
mx)
```

Output:

```
The formula for the ProductionValue linear regression line (y=mx+c) is =
13704948.58 + 6674.47 * Standard local currency units per USD - 102.72 *
Employment in agriculture, forestry and fishing by age, total (15+) - 20051.86
* Producer Price Index (2014-2016 = 100) + 82899.89 * Crude oil, Brent +
392694.35 * GDP growth (annual %) + 13295.39 * Apr - 27006.83 * May +
8550.51 * Jun + 4027.57 * July
```

Model score:

```
r2_score(y_test,y_pred)
```

Output:
$R^2 = 0.8078$ or 80.78%

9.8 Linkage Models and Data Flow

In the following section, application of the data flow is applied on various price spike, agricultural production, and climate change models to understand the lineage of behavior of economic models, farm inputs, real economic activities, and co-movement variables.

9.8.1 Brazil, 1984, Sugar Price Spike: Increase in Oil Prices Has Reduced Sugar in the World Market

Brazil is the largest sugar-producing country in the world. The production of sugar has increased to 352% since 1990. About 34% of Brazilian sugar consumed domestically is converted into ethanol for fuel. Exports have risen from 1.2 million metric tons in 1990 to 26.9 million metric tons in 2011. Sugar that is converted into ethanol is subsidized at prices higher than the world price. Recent increases in the world oil price have increased the price of ethanol[131] which in turn increased Brazil's conversion of sugar into ethanol, reducing potential sugar exports from Brazil [34]. That reduction in the growth of sugar exports could be one of the main forces for world sugar price increases. Brazil decreased its exports by 7.8% in 2011 which provided strength for sugar prices in 2011 (please see Fig. 9.37).

9.8.2 2008 Trade Restrictions by Major Rice Suppliers

Global rice[132] prices rose to record highs in the spring of 2008, with trading prices tripling from November 2007 to late April 2008. The price increase was not due to crop failure or a particularly tight global rice supply situation [36]. Instead, *trade restrictions by major suppliers, panic buying by several large importers, a weak dollar, and record oil prices* were the immediate cause of the rise in rice prices. Some of the main rice-producing countries have imposed export curbs and this has combined with low global stocks to drive rice higher.[133] Because rice is critical to the diet of about half the world's population, the rapid increase in global rice prices in late 2007 and early 2008 had a detrimental impact on those rice consumers' well-being. Although rice prices have dropped more than 40% from their April 2008 highs, they remain well above pre-2007 levels (please see Fig. 9.38).

[131] 2012 Outlook of the U.S. and World Sugar Markets, 2011-2021—https://ageconsearch.umn.edu/record/128037/files/AAE692.pdf

[132] Factors Behind the Rise in Global Rice Prices in 2008— https://www.ers.usda.gov/publications/pub-details/?pubid=38490

[133] US Rice Jumps to Record High on Supply Fears—https://www.cnbc.com/2008/04/23/us-rice-jumps-to-record-high-on-supply-fears.html

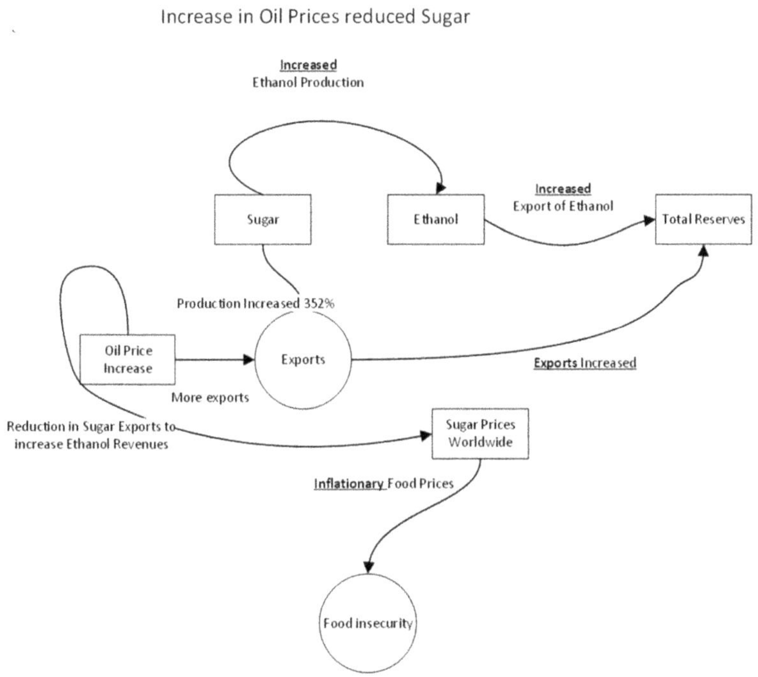

Fig. 9.37 Brazil, 1984, Sugar Prices Spike—Increase in Oil Prices has reduced Sugar in the World Market

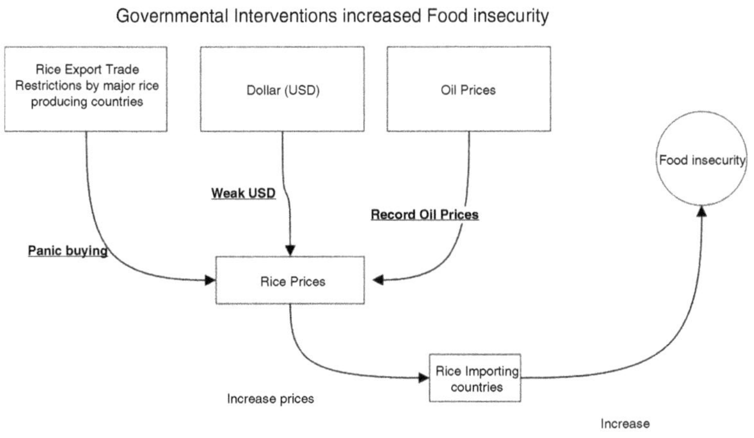

Fig. 9.38 Trade restrictions by major rice suppliers, 2008

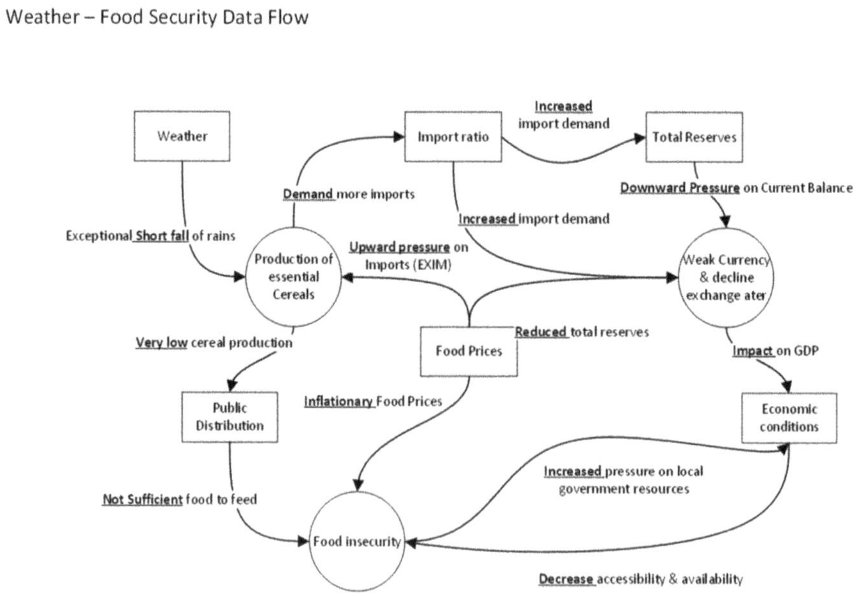

Fig. 9.39 Weather vs. food security

9.8.3 *Weather: Food Security Data Flow*

Adverse weather conditions: For wheat—a major food grain—adverse weather in several primary growing areas in 2006/2007 and 2007/2008 resulted in below normal global production. The lower production levels contributed to the sharp increase in global wheat prices in 2007 and early 2008. Macroeconomic factors, such as the declining dollar and a shift of funds from equities and real estate into commodities, also contributed to rising global commodity prices. For corn, wheat, and soybeans, the primary factors behind the price increases were increased use of biofuels, changing diets in China and India, major weather problems, and tight stocks-to-use ratios. Increased participation in futures markets by nontraditional investors likely contributed to greater price volatility (please see Fig. 9.39). Please check 2007 and 2008 wheat prices.[134]

[134]Factors Behind the Rise in May 2009 Global Rice Prices in 2008—https://www.ers.usda.gov/webdocs/outlooks/38489/13518_rcs09d01_1_.pdf?v=9.9

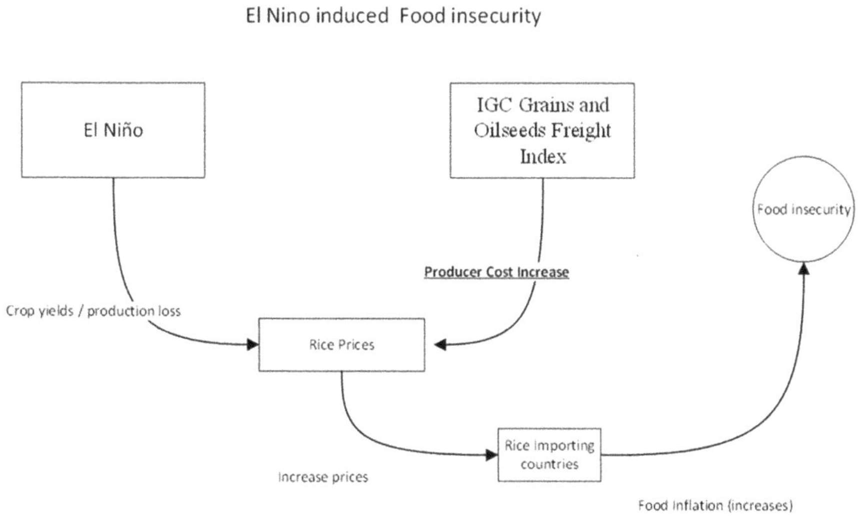

Fig. 9.40 El Niño-induced food security

9.8.4 El Niño Triggers Macroeconomic Waves (Albeit, Not for Faint Surfers)

Global rice production in 2009 was impaired by the return of El Niño conditions in Asia—resulting in increase in rice prices during the end of 2009 (please see Fig. 9.40).

El Niño and La Niña are the warm and cool phases[135] of a recurring climate pattern across the tropical Pacific—the El Niño–Southern Oscillation, or "ENSO" for short.[136] The pattern shifts back and forth irregularly every 2 to 7 years, and each phase triggers predictable disruptions of temperature and precipitation. These changes disrupt the large-scale air movements in the tropics, triggering a cascade of global side effects.[137] El Niño is a climatic event that recur every 2 to 7 years. Please see below figure to access years ELNO occurred, during 1997–1998, when it had strong adverse effects on agricultural production in South America and Southeast Asia. In South America, it was accompanied with floods and storms in Peru, Ecuador, and Chile. In Southeast Asia, in Papua New Guinea, Indonesia, and the Philippines, it was associated with severe drought (La Niña). An El Niño episode is

[135] Cold & Warm Episodes by Season—https://origin.cpc.ncep.noaa.gov/products/analysis_monitoring/ensostuff/ONI_v5.php

[136] ONI—https://origin.cpc.ncep.noaa.gov/products/analysis_monitoring/ensostuff/ONI_change.shtml

[137] El Niño & La Niña (El Niño-Southern Oscillation)—https://www.climate.gov/enso

normally preceded in December to January by an unusual warming of the Pacific waters along the Peruvian coast.[138]

January 2020, COVID-19, and Rice Prices
Despite projections for near-record global rice supplies in 2019/2020,[139] trading prices for rice in mid-April 2020 rose to their highest points in more than 7 years after several major exporting countries imposed export bans and restrictions in March 2020 as a response to heightened concerns over domestic food security following the outbreak of COVID-19 [52].

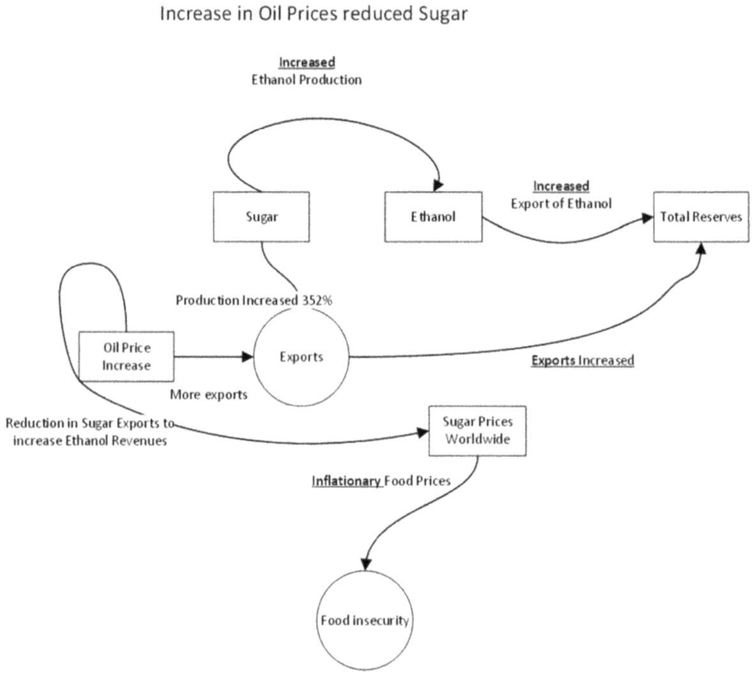

9.9 Modeling Exogenous Varialbes and Linkage Models

Weather has a profound impact on the agriculture, food security, commerce, rural economies, and overall connectedness of the world. In essence, counting weather is the most important ML variable as part of the linkage models. A favorable weather

[138] El Nino—https://en.wikipedia.org/wiki/El_Ni%C3%B1o

[139] Global trading prices for rice rose to highest level in 7 years due to export restrictions—https://www.ers.usda.gov/data-products/chart-gallery/gallery/chart-detail/?chartId=98504

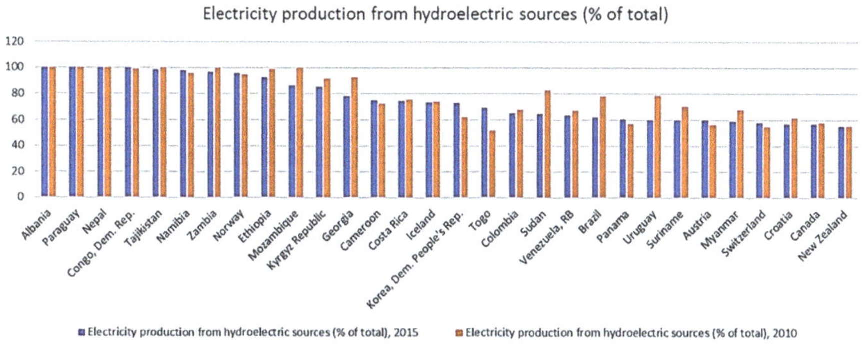

Fig. 9.41 Hydroelectric sources (% of total)

improves the economies and improves the account balance of countries. Converse is true too. That is, weather that does not provide enough water will profoundly affects agricultural crop and increases the overall energy bill and in many cases increases the account balance, thereby decreasing the exchange rate, which results in more severe food insecurity.

Let us model the impact of the short rainfall/missed mansoon or drought conditions on the countries' economies. As we know drought causes decreased agriculture productivity or in most cases crop failuers due to drought. The impact is multifold, one on agriculture and others on the energy generation of a country, if the country depends on electricity production from hydroelectric sources. The folowing figure shows the top 30 countries that are very reliant on hydro sources.

The map below shows how Eelectricity production from hydroelectric sources (% of total) varies by country. The shade of the country corresponds to the magnitude of the indicator. The darker the shade, the higher the value. The country with the highest value in the world is Albania, with a value of 100.00 (please see Fig. 9.41). The country with the lowest value in the world is Kuwait, with a value of 0.00.[140]

Electrical energy from hydropower is derived from turbines being driven by flowing water in rivers, with or without man-made dams forming reservoirs. Presently, hydropower is the world's largest source of renewable electricity. Hydropower represents the largest share of renewable electricity production. It was second only to wind power for newly-built capacities between 2005 and 2010. The International Energy Agency (IEA) estimates that hydropower could produce up to 6,000 terawatt-hours in 2050, roughly twice as much as today. Hydropower's storage capacity and fast response characteristics are especially valuable to meet sudden fluctuations in electricity demand and to match supply from less flexible electricity sources and variable renewable sources, such as solar photovoltaic (PV) and wind power. Use of energy is important in improving people's standard of living. But electricity generation also can damage the environment. Whether such damage

[140]Electricity production from hydroelectric sources (% of total)—https://www.indexmundi.com/facts/indicators/EG.ELC.HYRO.ZS

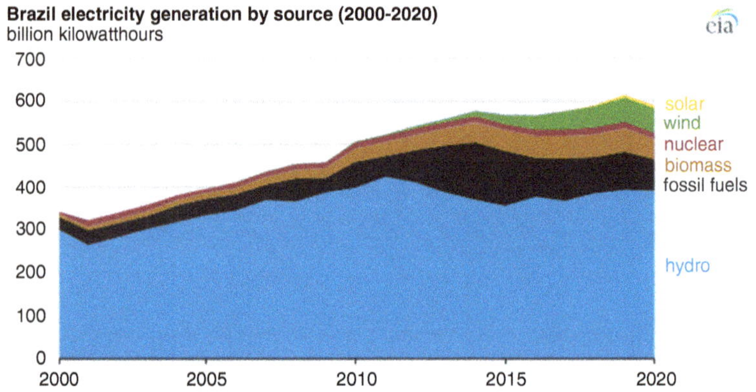

Fig. 9.42 Brazil electricity

occurs depends largely on how electricity is generated. For example, burning coal releases twice as much carbon dioxide—a major contributor to global warming—as does burning an equivalent amount of natural gas. Anthropogenic carbon dioxide emissions result primarily from fossil fuel combustion and cement manufacturing. In combustion different fossil fuels release different amounts of carbon dioxide for the same level of energy use: oil releases about 50% more carbon dioxide than natural gas, and coal releases about twice as much. Nuclear energy does not generate carbon dioxide emissions, but it produces other dangerous waste products.

Brazil[141] largely relies on hydropower for electricity generation; in 2020, hydropower supplied 66% of its electricity demand. Wind and solar generation have grown quickly in recent years and had a combined 11% share of the country's electricity generation in 2020. Biomass accounted for an 8% share. Fossil fuel-fired plants made up another 12% of electricity generation, while nuclear power accounted for 2%.

Most of Brazil's hydropower capacity is located north in the Amazon River Basin, but electricity demand centers are mainly along the eastern coast, particularly in the south (please see Fig. 9.42).

9.9.1 Drought in Brazil[142]

In Brazil, where hydroelectric power is the top source of electricity at 61%, drought recently cut water flows into hydro dams to a 91-year low, the country's mines and energy minister said.

---

[141] Hydropower—https://www.eia.gov/todayinenergy/detail.php?id=49436#:~:text=Brazil%20largely%20relies%20on%20hydropower,accounted%20for%20an%208%25%20share

[142] Droughts shrink hydropower, pose risk to global push to clean energy—https://www.reuters.com/business/sustainable-business/inconvenient-truth-droughts-shrink-hydropower-pose-risk-global-push-clean-energy-2021-08-13/

Fig. 9.43 Brazil battered by drought (Brazilian drought—https://earthobservatory.nasa.gov/images/148468/brazil-battered-by-drought)

To offset the drop in hydropower, the country is seeking to activate thermoelectric plants, mainly powered by natural gas, threatening to drive up greenhouse gas emissions. In July, sector regulator Aneel raised the most expensive electricity rate by 52%, due to the drought crisis. Severe weather events like the current drought will become increasingly frequent with climate change, and Brazilians will need to change their attitudes about water (Fig. 9.43).

"People always thought that water is unlimited, but it really isn't,"
José Marengo,
a climatologist at the government's disaster monitoring center

Images: credits to NOAA Earth Observatory[143]

Low water levels are noticeable around several lakes in the Paraná River basin, home to several hydroelectric dams and reservoirs that help power the region. Seven of the 14 main reservoirs nearby stood at their lowest levels since 1999. Brazil's Electric Sector Monitoring Committee has eased restrictions on some hydroelectric dams in order to meet electricity needs and prevent power outages.

Prolonged dry conditions have caused the worst drought in central and southern Brazil in almost a century, according to Brazilian government agencies. The drought is expected to cause crop losses, water scarcity, and increased fire activity in the Amazon rainforest and Pantanal wetlands.

Brazil's National Water and Basic Sanitation Agency (ANA) has declared a "critical situation" of water resources in the Paraná River basin from June to November 2021. The country's national meteorological system also warned of water shortages for Minas Gerais, Goiás, Mato Grosso do Sul, Paraná, and São Paulo through September 2021 (Fig. 9.44).

[143] Brazilian Drought—https://earthobservatory.nasa.gov/images/148468/brazil-battered-by-drought

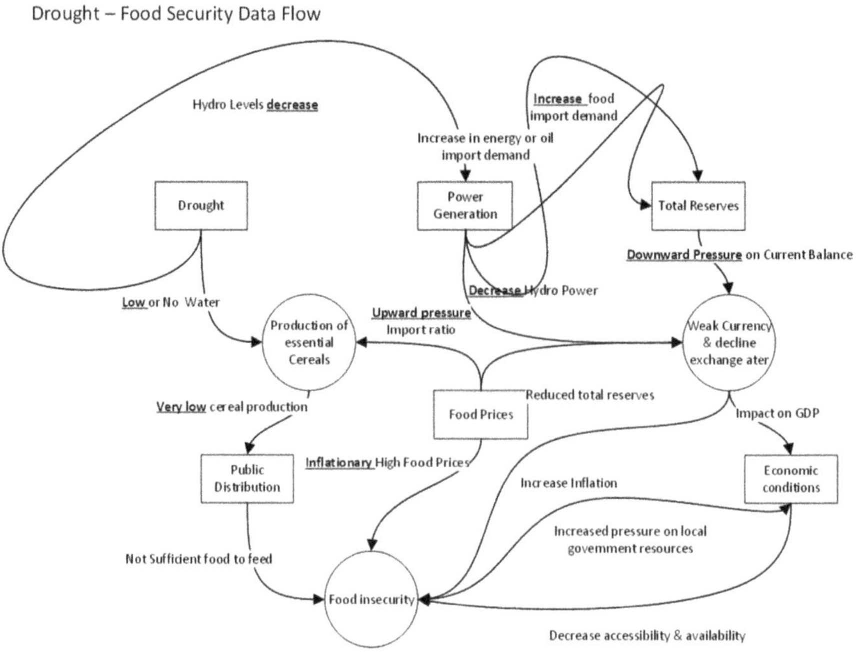

Fig. 9.44 Drought and food security data flow

After reading this chapter, you should be able to answer linkage models that provide a machine learning framework to model the variables from multidisciplinary field datasets that include agriculture, meteorological, macroeconomics, food security, climate change risk models, resolution imaging spectroradiometer, co-movement variables, and farm inputs such as labor and fertilizer costs. You should be able to build machine learning models for use cases such as sugarcane production and rice production predictive model.

References

1. Tassilo Hummel and Gus Trompiz, Macron tells French farmers: Ukraine war will weigh on you, and it will last, February 25, 2022, https://www.yahoo.com/news/macron-tells-french-farmers-ukraine-075753530.html, March 01, 2022
2. Sutanuka Ghosal, Prices of basmati rice, edible oils and pulses ride high going into festive season, October 05, 2021 05:51 pm IST, https://economictimes.indiatimes.com/news/economy/agriculture/prices-of-basmati-rice-edible-oils-and-pulses-ride-high-going-into-festive-season/articleshow/86782579.cms?from=mdr, Access Date: March 01, 2022

3. The United Nations, Low monsoon rains in India means less rice for the world in 2012 – UN agency, 6 August 2012, https://news.un.org/en/story/2012/08/417072, March 01, 2022
4. Gus Trompiz in Paris and Colin Packham, Paris wheat sets new record highs as traders fret on world supplies, 11/23/2021, https://www.agriculture.com/markets/newswire/grains-paris-wheat-sets-new-record-highs-as-traders-fret-on-world-supplies, Access Date: March 01, 2022
5. FAO, IFAD, UNICEF, WFP and WHO. 2021. The State of Food Security and Nutrition in the World 2021. Transforming food systems for food security, improved nutrition and affordable healthy diets for all. Rome, FAO. https://doi.org/10.4060/cb4474en
6. Kibrom T. Sibhatu, Vijesh V. Krishna, and Matin Qaim, Production diversity and dietary diversity in smallholder farm households, August 10, 2015, https://www.pnas.org/doi/10.1073/pnas.1510982112, Access Date: March 01, 2022
7. Samuel Benin, James Thurlow, Xinshen Diao, Christen Mccool and Franklin, Agricultural growth and investment options for poverty reduction in Malawi, 2008, https://www.ifpri.org/publication/agricultural-growth-and-investment-options-poverty-reduction-malawi-0, Access Date: January 29, 2022
8. Karl Pauw, James Thurlow and Dirk Van Seventer, Droughts and floods in Malawi Assessing the economywide effects, 2010, https://www.ifpri.org/publication/droughts-and-floods-malawi, Access Date: January 29, 2022
9. Jacques Bughin, Jeongmin Seong, James Manyika, Michael Chui, and Raoul Joshi, Notes from the AI Frontier Modeling the Impact of AI on the World Economy, September 2018, https://www.mckinsey.com/~/media/McKinsey/Featured%20Insights/Artificial%20Intelligence/Notes%20from%20the%20frontier%20Modeling%20the%20impact%20of%20AI%20on%20the%20world%20economy/MGI-Notes-from-the-AI-frontier-Modeling-the-impact-of-AI-on-the-world-economy-September-2018.ashx, Access Date: November 22, 2021
10. GIEWS, Crop Prospects and Food Situation, 4 March, 7 July, 29 September, 8 December, https://www.fao.org/giews/reports/crop-prospects/en/, Access Date: November 22, 2021
11. FAO. 2021. Crop Prospects and Food Situation - Quarterly Global Report No. 4, December 2021. Rome., https://doi.org/10.4060/cb7877en, Access Date: November 22, 2021
12. FAO. 2018. Crop Prospects and Food Situation - Quarterly Global Report No. 2, June 2018. Rome., https://www.fao.org/3/I9666EN/i9666en.pdf, Access Date: November 22, 2021
13. FAO. 2021. Crop Prospects and Food Situation - Quarterly Global Report No. 1, March 2021. Rome, https://doi.org/10.4060/cb3672en, Access Date: November 22, 2021
14. FAO. 2016. Crop Prospects and Food Situation - Quarterly Global Report No. 4, December 2016. Rome., https://www.fao.org/3/i6558e/i6558e.pdf, Access Date: March 02, 2022
15. Eliana Favari, Michael Geiger, Sharad Tandon, Siddharth Krishnaswamy, The critical role of escalating food prices in Yemen's food security crisis, November 02, 2021, https://blogs.worldbank.org/arabvoices/critical-role-escalating-food-prices-yemens-food-security-crisis, Access Date: March 02, 2022
16. FAO. 2008. Crop Prospects and Food Situation - Quarterly Global Report No. 1, February 2008. Rome., https://www.fao.org/3/ah881e/ah881e00.pdf, Access Date: March 02, 2022
17. Food and Agriculture Organization of the United Nations, Handbook for Defining and Setting up a Food Security Information and Early Warning System (FSIEWS), 2000, https://www.fao.org/3/X8622E/x8622e00.htm#TopOfPage, Access Date: January 22, 2022
18. FAO, Food Security Information and Knowledge Sharing System - Sudan, 2016, https://www.fao.org/family-farming/detail/en/c/450039/, Access Date: January 22, 2022
19. Joseph P. Janzen, Colin A. Carter, Aaron D. Smith, and Michael K. Adjemian, Deconstructing Wheat Price Spikes: A Model of Supply and Demand, Financial Speculation, and Commodity Price Comovement, April 2014, https://www.ers.usda.gov/webdocs/publications/45199/46439_err165.pdf?v=0, Access Date: February 27, 2022
20. Raju, Totakura & Bavise, Aayush & Chauhan, Pradeep. (2020). Analysing Volatility spillovers between grain and freight markets. Pomorstvo. 34. 10.31217/p.34.2.23, Access Date: February 28, 2022

21. International Grains Council (IGC), Covid 19 crisis and the Grains/oilseeds and rice trade Agriculture Committee of WTO, July 2020, https://www.wto.org/english/tratop_e/agric_e/petit_28jul20_e.pdf, Access Date: February 27, 2022
22. The International Grains Council (IGC), Statement by the International Grains Council 99th Meeting of the WTO Committee on Agriculture 23-24 September 2021 "COVID-19 And Agriculture", 14 September 2021, https://docs.wto.org/dol2fe/Pages/SS/directdoc.aspx?filename=q:/G/AG/GEN191.pdf&Open=True, Access Date: February 27, 2022
23. Duygu Akkaya, Kostas Bimpikis and Hau Lee, Government Interventions to Promote Agricultural Innovation, 2018, https://web.stanford.edu/~kostasb/publications/agricultural_innovation.pdf, Access Date: February 27, 2022
24. Fred Gale, Growth and Evolution in China's Agricultural Support Policies, August 2013, https://www.ers.usda.gov/webdocs/publications/45115/39368_err153.pdf?v=41491, Access Date: February 27, 2022
25. Pratik Parija, India to Boost Farm Credit and Infrastructure to Spur Output, 07:55 AM IST, 01 Feb 2021, https://www.bloombergquint.com/business/india-plans-to-enhance-farm-credit-to-boost-agricultural-output, Access Date: February 27, 2022
26. Tyler Smith, as farmers go, so goes the nation, March 25, 2019, https://www.aeaweb.org/research/farmer-recovery-great-depression, Access Date: February 27, 2021
27. The EU, Financial needs in the agriculture and agri-food sectors in France, June 2020, https://www.fi-compass.eu/sites/default/files/publications/financial_needs_agriculture_agrifood_sectors_France_0.pdf, Access Date: February 27, 2021
28. Mike Dorning, USDA Warns Fertilizer Firms on Price Gouging Over Ukraine, February 24, 2022, 9:17 AM PST, https://www.bloomberg.com/news/articles/2022-02-24/usda-chief-warns-fertilizer-firms-on-price-gouging-over-ukraine, Access Date: March 4, 2022
29. Isadore Barmash, Soaring Sugar Cost Arouses Consumers And U.S. Inquiries, Nov. 15, 1974, https://www.nytimes.com/1974/11/15/archives/soaring-sugar-cost-arouses-consumers-and-u-s-inquiries-what-sent.html, Access Date: February 21, 2022
30. Mr. Helmut Ahlfeld,FAO - Cuba Conference, Cuba - SUGAR MARKETS IN DISARRAY,7-9 December 1999, https://www.fao.org/3/X4988E/x4988e05.htm, Access Date: February 21 2022
31. Elizabeth M. Fowler, Commodities; Price Rises For Sugar Continue, June 6, 1983, https://www.nytimes.com/1983/06/06/business/commodities-price-rises-for-sugar-continue.html, Access Date: February 21 2022
32. The Associated Press, FUTURES/OPTIONS ; Sugar Surplus Sends Prices Plummeting, April 24, 1985, https://www.nytimes.com/1985/04/24/business/futures-options-sugar-surplus-sends-prices-plummeting.html, Access Date: February 21, 2022
33. Ronald E. Dolan, ed. Philippines: A Country Study. Washington: GPO for the Library of Congress, 1991, Access Date: February 21, 2022
34. Won W. Koo and Richard D. Taylor, 2012 Outlook of the U.S. and World Sugar Markets, 2011-2021, April 2012, https://ageconsearch.umn.edu/record/128037/files/AAE692.pdf, Access Date: February 21 2022
35. Matt Egan, Don't expect grocery store prices to come down anytime soon, February 8, 2022, https://www.cnn.com/2022/02/08/business/grocery-store-prices-goldman/index.html, Access Date: February 21 2022
36. Matthew MacLachlan, Food Price Outlook, February 25, 2022, https://www.ers.usda.gov/data-products/food-price-outlook/food-price-outlook/#Consumer%20Price%20Index, Access Date: March 13, 2022
37. W. Arthur Lewis, Economic Development with Unlimited Supplies of Labour, First published: May 1954, https://doi.org/10.1111/j.1467-9957.1954.tb00021.x, Access Date: September 18, 2019
38. Bill Binks, Nyree Stenekes, Heleen Kruger and Robert Kancans, Snapshot of Australia's Agricultural Workforce, 2016, https://www.awe.gov.au/abares/products/insights/snapshot-of-australias-agricultural-workforce, Access Date: January 08, 2022

39. Rob McConnel, Farming on the verge of a workforce crisis, 2016, https://www2.deloitte.com/au/en/pages/consumer-business/articles/farming-verge-workforce-crisis.html, Access Date: January 08, 2022
40. Andrew Cameron, Agricultural overview: March quarter 2022, 2022, https://www.awe.gov.au/abares/research-topics/agricultural-outlook/agriculture-overview, Access Date: March 13, 2022
41. Schnitkey, G., N. Paulson, C. Zulauf and K. Swanson. "2021 Fertilizer Price Increases in Perspective, with Implications for 2022 Costs." farmdoc daily (11):114, Department of Agricultural and Consumer Economics, University of Illinois at Urbana-Champaign, August 3, 2021.
42. Amanda Davies, City congestion and rural population decline, April 11 2019 - 7:00AM, https://www.canberratimes.com.au/story/6016548/city-congestion-and-rural-population-decline/, Access Date: February 26, 2022
43. Kalok Chan, Yiuman Tse, and Michael Williams, The Relationship between Commodity Prices and Currency Exchange Rates: Evidence from the Futures Markets, May 2009, https://www.nber.org/system/files/chapters/c11859/revisions/c11859.rev1.pdf, Access Date: February 27, 2022
44. Cory Mitchell, Commodity Prices and Currency Movements, Updated October 31, 2021, https://www.investopedia.com/articles/forex/06/commoditycurrencies.asp#toc-trading-currencies-as-a-supplement-to-trading-oil-or-gold, Access Date: February 27, 2022
45. Reserve Bank of Australia, Sensitivity of Australian Trade to the Exchange Rate, Bulletin – September Quarter 2016, https://www.rba.gov.au/publications/bulletin/2016/sep/2.html, Access Date: February 27, 2022
46. Nathan Childs and James Kiawu, Factors Behind the Rise in May 2009 Global Rice Prices in 2008, 2008 RCS-09D-01, https://www.ers.usda.gov/webdocs/outlooks/38489/13518_rcs09d01_1_.pdf?v=9.9, Access Date: February 27, 2022
47. Ronald E. Dolan, ed. Philippines: A Country Study. Washington: GPO for the Library of Congress, 1991, http://countrystudies.us/philippines/, Access Date: March 01. 2022
48. Australian Bureau of Statistics, Value of Agricultural Commodities Produced, Australia, 14/05/2021, https://www.abs.gov.au/statistics/industry/agriculture/value-agricultural-commodities-produced-australia/latest-release, Access Date: March 01. 2022
49. Australian Bureau of Statistics, Measuring natural disasters in the Australian economy, 3/03/2020, https://www.abs.gov.au/statistics/research/measuring-natural-disasters-australian-economy, Access Date: March 01. 2022
50. Will Horner and Jeffrey T. Lewis, Coffee Prices Soar After Bad Harvests and Insatiable Demand, July 12, 2021, 8:41 am ET, https://www.wsj.com/articles/coffee-prices-soar-after-bad-harvests-and-insatiable-demand-11626093703?mod=article_inline, Access Date: February 03, 2022
51. Jeffrey T. Lewis and Joe Wallace, Coffee Prices Jump to Six-Year High as Brazilian Frost Threatens Crop, Updated July 27, 2021, 4:50 pm ET, https://www.wsj.com/articles/coffee-prices-jump-to-six-year-high-as-brazilian-frost-threatens-crop-11627380128?mod=article_inline, Access Date: February 03, 2022
52. Nathan Childs, Global trading prices for rice rose to highest level in 7 years due to export restrictions, May 27, 2020, https://www.ers.usda.gov/data-products/chart-gallery/gallery/chart-detail/?chartId=98504, Access Date: March 04, 2022

Chapter 10
Heuristics and Agricultural Production Modeling

This chapter covers the following:

- Agricultural linkage models and production models using heuristic techniques
- Development of linkage models
- CMIP6, SSP projections
- Observe impact of climate changes on the agricultural resources

This chapter introduces agricultural linkage models, production models using heuristic techniques. The heuristic models developed are based on the economical linkage models built in prior chapters. Next, the chapter introduces linkage and projection models and investigates the food demand and agricultural harvest need for the future, at the end of the century, 2100. Finally, the chapter concludes with the Thailand Nakhon Sawan Rice Production linear programming (LP) linkage model (SSP5_8.5 – 2080–2099) and LP linkage model (SSP3_7.0 – 2080-2099).

Food security is an economic argument. The key drivers of economics are, in a nutshell, the enablers of food security. Conversely, the drivers that adversely impact economics are drivers of food insecurity. The key drivers of food insecurity are food prices, food price instability, oil prices, deficient grain stock levels, higher exchange rate volatility, rising demand for food due to population growth, commodity futures speculations, climate change, political instability/conflicts, and abrupt change in consumption patterns and pressure on food production rate from climate change, scarcity of natural resource availability due to land degradation and water scarcity, food tradeoff biofuel production, and a lack public and private investment in infrastructure. The degree of importance of each key driver varies between countries and regions according to their unique set of physical, economic, and social circumstances. Given climate change, increasing demand, and growing population, we need to generate more food in the future.

C. Vuppalapati, *Artificial Intelligence and Heuristics for Enhanced Food Security*, International Series in Operations Research & Management Science 331,
https://doi.org/10.1007/978-3-031-08743-1_10

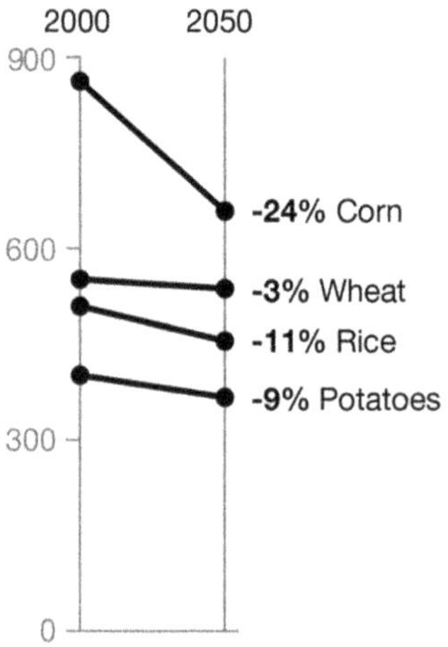

Fig. 10.1 Global production change (millions of tons)

To meet the demands of food, the world must, by 2050, produce 49% more food than in 2012 as populations grow and diets change.[1,2] [1, 2]

One way is to investigate the Food Security Monitoring System (FSMS) framework. The Food Security Monitoring System[3] (FSMS), generally, covers key indicators, 35 indicators, that are classified[4] along the four dimensions of food security [3]: availability (5 indicators), access (10 indicators), utilization (13 indicators), and stability (7 indicators). The indicators are classified along the four dimensions of food security: availability, access, utilization, and stability.[5] Please see the Food Security Monitoring System (FSMS)/Appendix—Food Security and Nutrition for all lists of food security indicators. To enhance food security, it is important to look through the framework of economic-and-food security linkage.

Agriculture is one of the largest worldwide employers and one of the biggest industries; in 2020, agriculture contributed to 4.346% (see Fig. 10.1) of the global gross domestic product (GDP), and in some developing countries, it can account for

[1]Coffee Prices Soar After Bad Harvests and Insatiable Demand—https://www.wsj.com/articles/coffee-prices-soar-after-bad-harvests-and-insatiable-demand-11626093703?mod=article_inline

[2]Coffee Prices Jump to Six-Year High as Brazilian Frost Threatens Crop—https://www.wsj.com/articles/coffee-prices-jump-to-six-year-high-as-brazilian-frost-threatens-crop-11627380128?mod=article_inline

[3]Sudan—Food Security Monitoring System, http://fsis.sd/SD/EN/FoodSecurity/Monitoring/

[4]Handbook for Defining and Setting up a Food Security Information and Early Warning System (FSIEWS)—https://www.fao.org/3/X8622E/x8622e00.htm#TopOfPage

[5]Food Security Information and Knowledge Sharing System—Sudan—https://www.fao.org/family-farming/detail/en/c/450039/

almost one-third of the GDP, occupying nearly 80% of the workforce[6] of total export earnings and employment[7] [4]. Climate shocks[8] therefore have a potentially profound direct effect on the agricultural sector and farm households while also indirectly affecting other economic sectors and nonfarm households through price and production linkages [5].

10.1 Linkage and Projection Models

Climate change has a major impact on soil, and changes in land use and soil can either accelerate or slow down climate change. Without healthier soil and sustainable land and soil management, we cannot tackle the climate crisis, produce enough food, and adapt to a changing climate. The answer might lie in preserving and restoring key ecosystems and letting nature capture carbon from the atmosphere.[9] Soil management is integral to managing climate change.

Climate change may benefit some plants by lengthening growing seasons and increasing carbon dioxide. Yet other effects of a warmer world, such as more pests, droughts, and flooding, will be less benign. What are the adaptation plans? Using an aggressive climate model (known as HadGEM2), researchers at the International Food Policy Research Institute (IFPRI) project claimed that by 2050, suitable croplands for four top commodities—corn, potatoes, rice, and wheat—will shift, in some cases pushing farmers to plant new crops. The same can be extrapolated for 2100, at the end of the century. Some of the crop's (corn, wheat, rice, and potatoes) yield could be drastically reduced due to much more increased growing season (please see Figs. 10.1, 10.2, and 10.3).

Climate change is likely to be most forgiving of wheat,[10] but not enough to offset losses from other major crops [6].

[6] Malawi—https://ricepedia.org/malawi & http://irri.org/resources/publications/books/rice-almanac-4th-edition

[7] Agricultural growth and investment options for poverty reduction in Malawi—https://www.ifpri.org/publication/agricultural-growth-and-investment-options-poverty-reduction-malawi-0

[8] Droughts and floods in Malawi Assessing the economy wide effects—https://www.ifpri.org/publication/droughts-and-floods-malawi

[9] Soil, land and climate change—https://www.eea.europa.eu/signals/signals-2019-content-list/articles/soil-land-and-climate-change

[10] Crop Changes—https://www.nationalgeographic.com/climate-change/how-to-live-with-it/crops.html

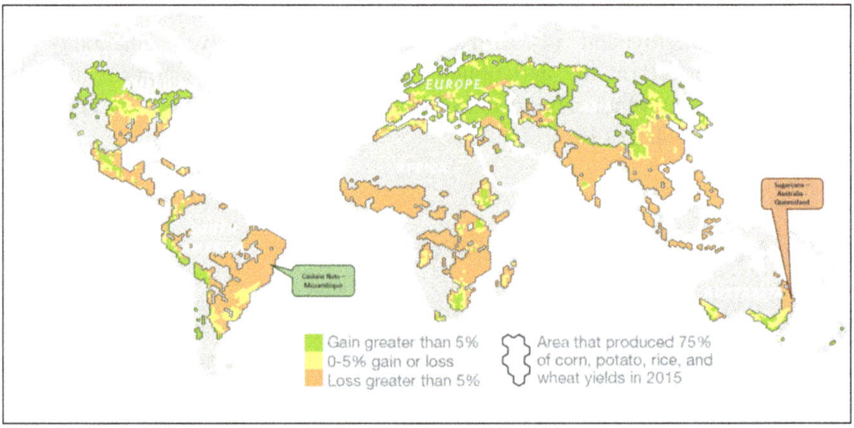

Fig. 10.2 Change in potential average yields for corn, potatoes, rice, and wheat in 2050 [6]

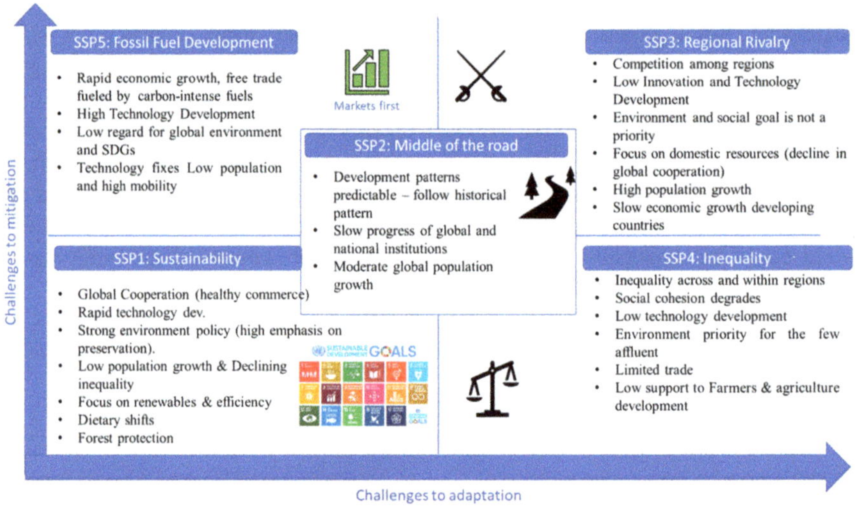

Fig. 10.3 Shared socioeconomic pathways [7]

Reference: shared socioeconomic pathways.[11] To apply linkage models [8] developed in the previous chapters for CMIP6 simulations for the years 2040–2059 and 2080–2099, we need to derive the following values of the following model variables (please see Fig. 10.6):

- Available harvest land for cultivation (supply side)
- Temperature and humidity values as per climate projections

[11] Climate Model—Sea Surface Temperature Change: SSP2 (Middle of the Road)—2015–2100—https://sos.noaa.gov/catalog/datasets/724/

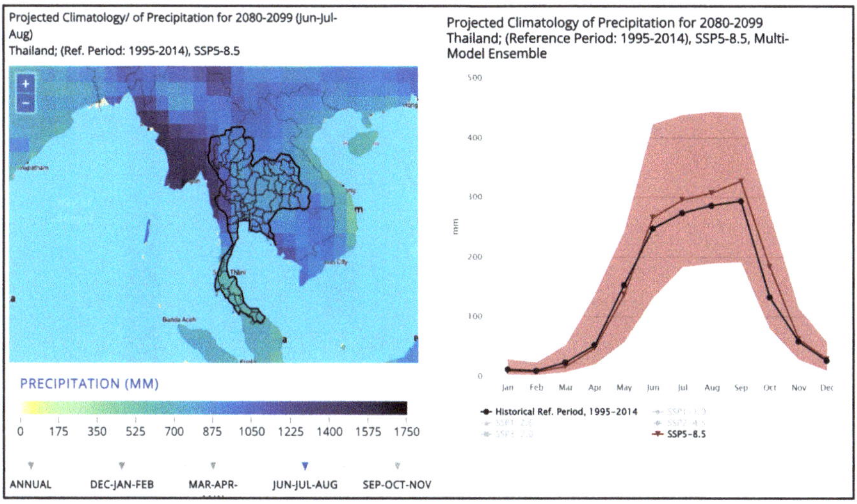

Fig. 10.4 Thailand, Nakhon Sawan precipitation

- Required production of commodity (for human consumption)
- Animal husbandry (for livestock feed)
- Population growth (demand side)

Temperature and humidity values are derived from the Climate Change Portal for CMIP6 and various SSP scenarios. It is important to note that SSPs are meant to provide insight into future climates based on defined emissions, mitigation efforts, and development paths. For instance, consider Thailand, Nakhon Sawan.[12] The projected climate change values for 2089–2099, SSP5-8.5, results in a drop of 15% production (please see Fig. 10.4).

SSP5-85 drops are the steepest due to climate change and other impacts of climate change (please see Figs. 10.5 and 10.6).

10.1.1 Available Harvest Land for Cultivation

Agricultural land covers more than one-third of the world's land area. Agricultural land constitutes only a part of any country's total area, which can include areas not suitable for agriculture, such as forests, mountains, and inland water bodies. Arable land includes land defined by the FAO as land under temporary crops (double-cropped areas are counted once), temporary meadows for mowing or for pasture, land under market or kitchen gardens, and temporarily fallow land. Land abandoned

[12] Thailand—https://climateknowledgeportal.worldbank.org/country/thailand/climate-data-projections

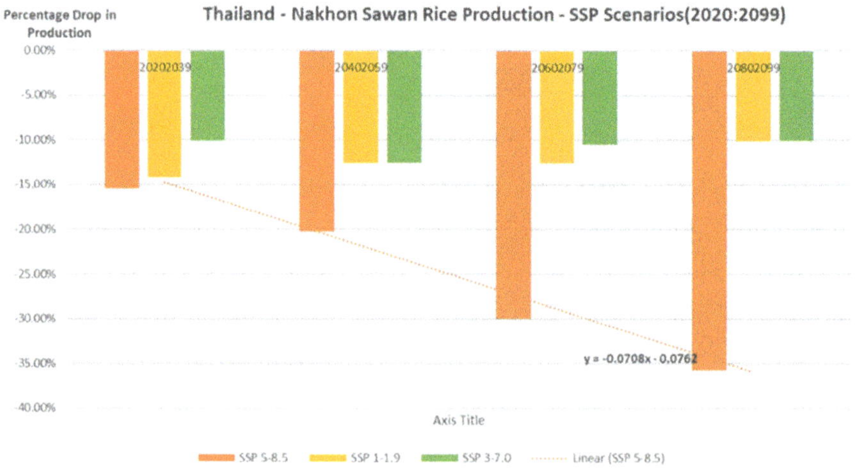

Fig. 10.5 Thailand—rice reduction

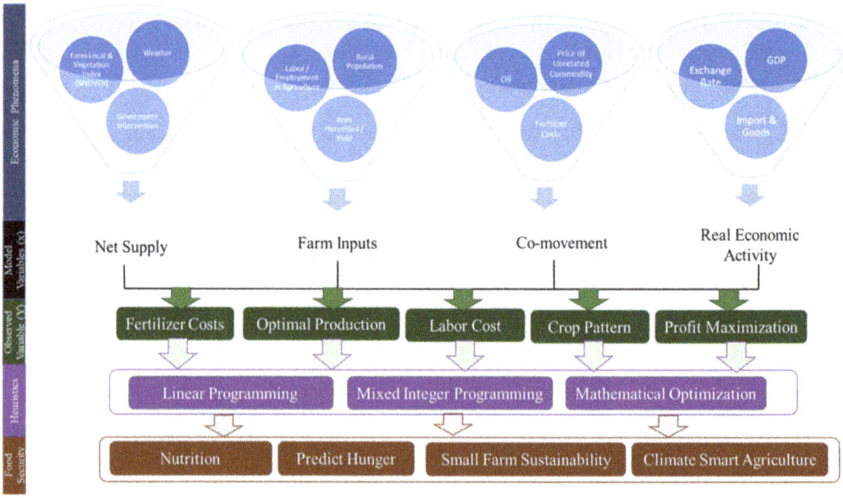

Fig. 10.6 Linkage model [8]

because of shifting cultivation is excluded. Agriculture is still a major sector in many economies, and agricultural activities provide developing countries with food and revenue. But agricultural activities also can degrade natural resources. Poor farming practices can cause soil erosion and loss of soil fertility. Efforts to increase productivity by using chemical fertilizers, pesticides, and intensive irrigation have environmental costs and health impacts. Excessive use of chemical fertilizers can alter the chemistry of soil. Pesticide poisoning is common in developing countries. And salinization of irrigated land diminishes soil fertility. Thus, inappropriate use of

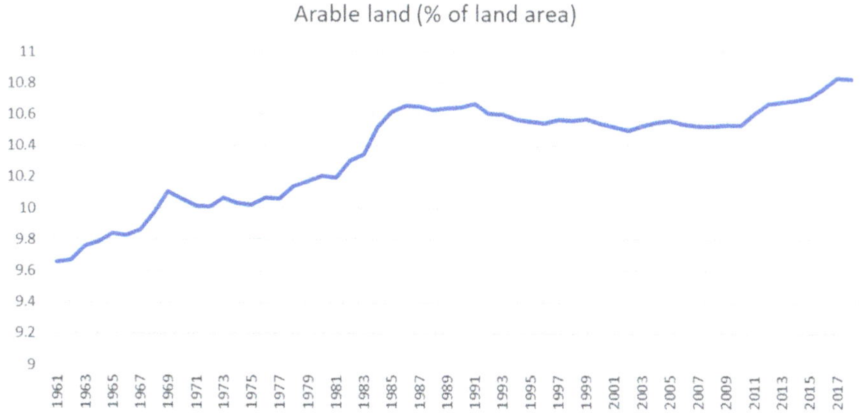

Fig. 10.7 Arable land

inputs for agricultural production has far-reaching effects. There is significant geographic variation in the availability of land considered suitable for agriculture. Increasing population and demand from other sectors place growing pressure on available resources. Arable land (% of land area) in the world was reported at 10.83% in 2018, according to the World Bank collection of development indicators, compiled from officially recognized sources[13] (please see Fig. 10.7).

According to the FAO, the world's cultivated area has grown by 12% over the last 50 years. The global irrigated area has doubled over the same period, accounting for most of the net increase in cultivated land. Agriculture already uses 11% of the world's land surface for crop production. It also makes use of 70% of all water withdrawn from aquifers, streams, and lakes. Agricultural policies have primarily benefitted farmers with productive land and access to water, bypassing most small-scale producers who are still locked in a poverty trap of high vulnerability, land degradation, and climatic uncertainty. Land resources are central to agriculture and rural development and are intrinsically linked to global challenges of food insecurity and poverty, climate change adaptation and mitigation, and degradation and depletion of natural resources that affect the livelihoods of millions of rural people across the world. In many industrialized countries, agricultural land is subject to zoning regulations. In the context of zoning, agricultural land (or more properly agriculturally zoned land) refers to plots that may be used for agricultural activities, regardless of the physical type or quality of land. FAO's agricultural land data contains a wide range of information on variables that are significant for understanding the structure of a country's agricultural sector, making economic plans and policies for food security, and deriving environmental indicators, including those related to investment in agriculture and data on gross crop area and net crop area which are useful for policy formulation and monitoring.

[13]Arable Land—https://data.worldbank.org/indicator/AG.LND.ARBL.ZS

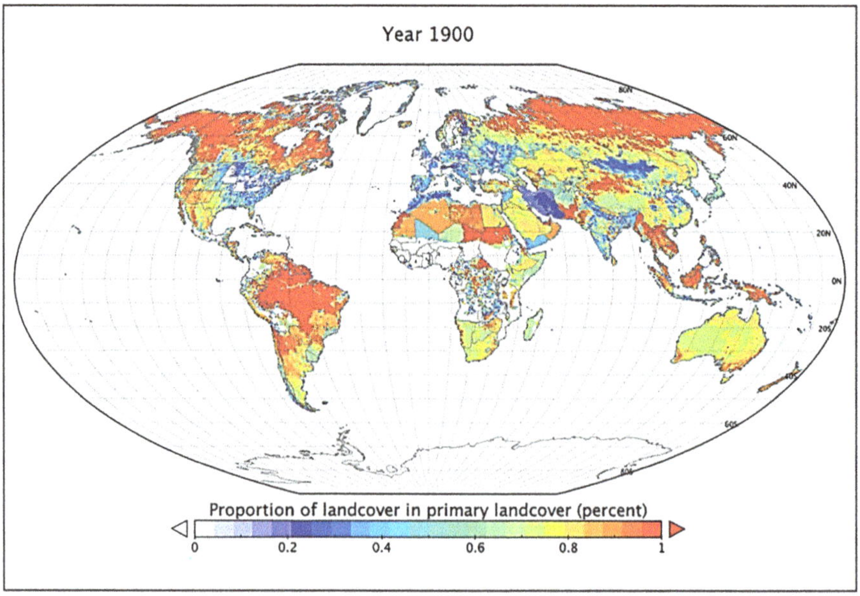

Fig. 10.8 Proportion of landcover in primary landcover in year 1900 and projected to year 2100 under a medium-high emission pathway by the AIM model (from LUHa_u2.v1 and LUHa_u2.v1_aim.v1.1) [10]

Nevertheless, there are stressors related to arable land: the world has lost a third of its arable land due to erosion or pollution in the past 40 years,[14] with potentially disastrous consequences as global demand for food soars, as scientists have warned [9]. New research has calculated that nearly 33% of the world's adequate or high-quality food-producing land has been lost at a rate that far outstrips the pace of natural processes to replace diminished soil. The steep decline in soil has occurred at a time when the world's demand for food is rapidly increasing. It is estimated that the world will need to grow 50% more food by 2050 to feed an anticipated population of 9 billion people. According to the UN's Food and Agriculture Organization, the increase in food production will be most needed in developing countries.

To assess and run the LP models to assess the changes required to meet 2100 demand, we need two important demand and supply aspects of the linkage model (please see Fig. 10.8):

1. Harvest land
2. Production needs due to increased food demand

[14]Earth has lost a third of arable land in past 40 years, scientists say—https://www.theguardian.com/environment/2015/dec/02/arable-land-soil-food-security-shortage

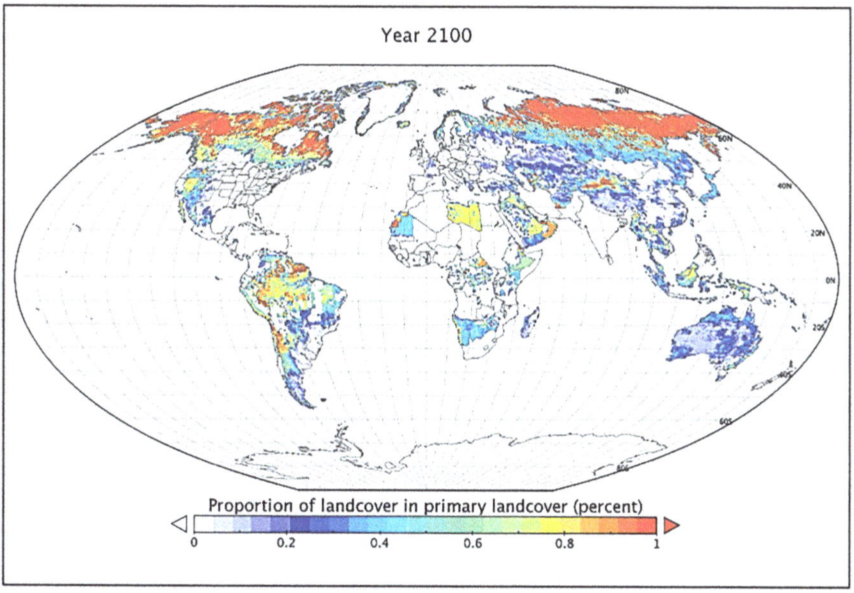

Fig. 10.9 Proportion of land cover in primary land cover in year 1900 and projected to year 2100 under a medium–high emission pathway by the AIM model (from LUHa_u2.v1 and LUHa_u2.v1_aim.v1.1) [10]

Harvest Land: LUH1—Harmonized Global Land Use for Years 1500–2100
These data represent fractional land use and land cover patterns annually for the years 1500–2100 for the globe at 0.5° (~50-km) spatial resolution. Land use categories of cropland, pasture, primary land, secondary (recovering) land, and urban land and underlying annual land-use transitions are included (LUH1[15]). Annual data on age and biomass density of secondary land, as well as annual wood harvest, are included for each grid cell [10, 11]. Please see Fig. 10.9.

A key feature of these data is that historical reconstructions of land use were harmonized (computationally adjusted to minimize differences at the transition period) with modeled future scenarios, allowing for a seamless examination of historical and future land use. The output data present a single consistent, spatially gridded set of land-use change scenarios for studies of human impacts on the past, present, and future earth system.[16] For additional information about the algorithms, inputs, and options used in creating the land use transitions data, see [10, 12, 13].

[15]LUH1—https://daac.ornl.gov/VEGETATION/guides/Land_Use_Harmonization_V1.html

[16]The underpinnings of land-use history: three centuries of global gridded land-use transitions, wood-harvest activity, and resulting secondary lands—https://onlinelibrary.wiley.com/doi/abs/10.1111/j.1365-2486.2006.01150.x

10.1.2 CMIP6: SSP8.5 Harvest Land

An important feature of RCP8.5[17] was the increase in cultivated land of about 185 million ha from 2000 to 2050 and another 120 million ha from 2050 to 2100 [14]. While aggregate arable land use in developed countries slightly decreased, all the net increases occurred in developing countries. This strong increase in agricultural resource use was driven by the increasing population, which rose to 12 billion people by 2100. Moreover, faster phenological development of crops due to climate warming is one of the main drivers for potential future yield reductions.[18] Yield improvements and intensification were assumed to account for most of the needed production increases: while *global agricultural output in the scenario increased by 135% by 2080, cultivated land expanded by only 16% above 2005 levels.* While agricultural land was expanding, forest cover declined over the century by 300 million ha from 2000 to 2050 and another 150 million ha from 2050 to 2100 [15].

In total over all crops, the global elevated-, high-, and serious-risk areas increase with warming climate, while the medium- and low-risk areas decrease (below figure). Accordingly, 39% of global cropland areas are exposed to elevated, high, or serious risk for nonexisting adapted varieties under SSP5-8.5 (16% in SSP1-2.6, 23% in SSP2-4.5, and 30% in SSP3-7.0) (Fig. 10.10 below).

The figure shows the share of areas with different risks for nonexisting adapted varieties on global cropland. The bars show the model median over five climate models (summed for the harvested area of maize, wheat, rice, and soybean) for low, medium, elevated, high, and serious risk, respectively. The scenarios (SSP1-2.6, SSP2-4.5, SSP3-7.0, and SSP5-8.5) refer to 2070–2100. *Harvested areas are kept constant for all scenarios* [14].

For linkage modeling, we apply 16% harvest land above 2005 levels and production output by 135% to meet future population needs. Our assumption found that 39% (SSP5-8.5) of global cropland could require new crop varieties to avoid yield loss from climate change by the end of the century.

10.1.3 CMIP6: SSP3-7.0 Harvest Land

The land productivity was assumed to be extended from historical trends but does not exceed the regional potential. Urban land use increased due to population and

[17] Harmonization of Land-Use Scenarios for the Period 1500-2100: 600 Years of Global Gridded Annual Land-Use Transitions, Wood Harvest, and Resulting Secondary Lands—https://www.researchgate.net/publication/227300487_Harmonization_of_Land-Use_Scenarios_for_the_Period_1500-2100_600_Years_of_Global_Gridded_Annual_Land-Use_Transitions_Wood_Harvest_and_Resulting_Secondary_Lands

[18] Large potential for crop production adaptation depends on available future varieties-https://onlinelibrary.wiley.com/doi/full/10.1111/gcb.15649

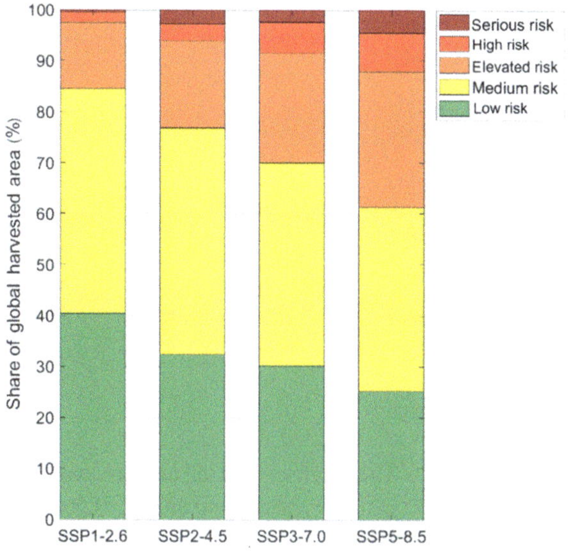

Fig. 10.10 Share of areas with different risks for non-existing adapted varieties on global cropland [14]

economic growth, while cropland areas expanded due to increasing food demand. Grassland area declined while total forested area extent remained constant throughout the century. Harvested areas are kept constant for all scenarios as mentioned above—6% harvest land above 2005 levels and production output by 135% to meet future population needs.

10.1.4 CMIP6: SSP2-4.5 Harvest Land

Agricultural land declined slightly due to afforestation, yet food demand was still met through crop yield improvements, dietary shifts, production efficiency, and international trade. For linkage modeling, we apply -5% harvest land above 2005 levels and production output by 110%. That is, 95% harvest land above 2005 levels and production output by 110% to meet future population needs.

10.1.5 CMIP6: SSP1-2.6 and SSP1-1.9 Harvest Land

SSP1-1.9 and SSP1-2.6 (RCP 2.6) represented a very low emission scenario and explores the feasibility of limiting climate change to less than 2 °C by limiting radiative forcing to a peak of 3 Wm−2 in the mid-century, declining to 2.6 Wm−2 in 2100. The overall trends in land use and land cover were mainly determined by demand, trade, and production of agricultural products and bioenergy modeled at the level of 24 world regions. Land use was downscaled to the grid level by a set of

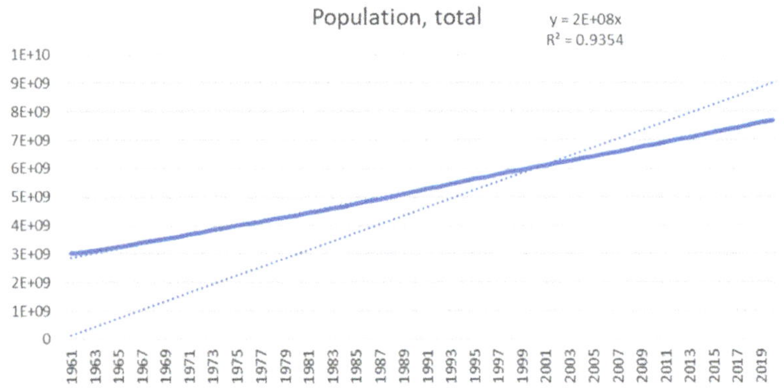

Fig. 10.11 Population total

allocation rules. These rules allocated land use based on (1) agricultural productivity, (2) proximity to existing agricultural areas, (3) proximity to current water bodies and cities, and (4) a random factor (MNP 2006). Land-use change in the RCP2.6 followed a trend of agricultural land relocating from high-income to low-income regions, and a clear increase in bioenergy production was evident with a new area for bioenergy crops occurring near current agricultural areas.

For linkage modeling, we apply the same harvest land percentage as of 2005 levels and production output same as 2005. Other factors that model do not consider rural population and employment of agriculture. Harvested areas are kept constant for all scenarios.

10.1.6 World Population, Total

Total population[19] is based on the de facto definition of population, which counts all residents regardless of legal status or citizenship. The values shown are midyear estimates. Increases in human population, whether because of immigration or more births than deaths, can impact natural resources and social infrastructure. This can place pressure on a country's sustainability. A significant growth in population will negatively impact the *availability of land for agricultural production* and will aggravate demand for food, energy, water, social services, and infrastructure. On the other hand, decreasing population size – a result of fewer births than deaths and people moving out of a country – can impact a government's commitment to maintain services and infrastructure (please see Fig. 10.11).

[19]World Bank—SP.POP.TOTL—https://data.worldbank.org/indicator/SP.POP.TOTL

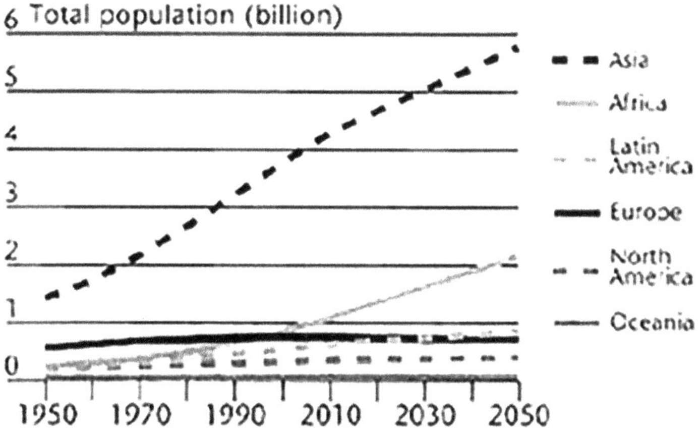

Fig. 10.12 Population growth 1950-2020 [16]

Projecting World Population Growth

During the period 1995 (5622085293 people or 5622 million people) to 2050 (projected 9800 million), the world's population is projected to increase by some 72%, from 5700 (approximately) million to 9 800 million people, and then begin to stabilize. Calculating how much food this future population will require depends on the following (please see Fig. 10.12):

- Demographic issues, such as the number of people to be fed, their age, size, and level of activity
- The extent to which undernutrition can be overcome by 2050
- Changes in diet (see over)

Although it took 123 years for the world population to grow from one billion to two billion, it took only 33 years for the third billion, 14 years for the fourth billion, and 13 years for the fifth billion. In 2000 (6034484369 people), the world will pass the six billion mark (please see Fig. 10.13).

10.1.7 More Food Required in the Future: Changing Demographics

When calculating the energy requirement of a future population,[20] factors, such as the expected physical size of people and, to a small degree, changes in the population's age structure, must be taken into consideration [15, 18]. The aging of the population and its increase in height because of better nutrition will increase

[20]FOOD NEEDS AND POPULATION—https://www.fao.org/3/x0262e/x0262e23.htm

Fig. 10.13 World population growth [15]

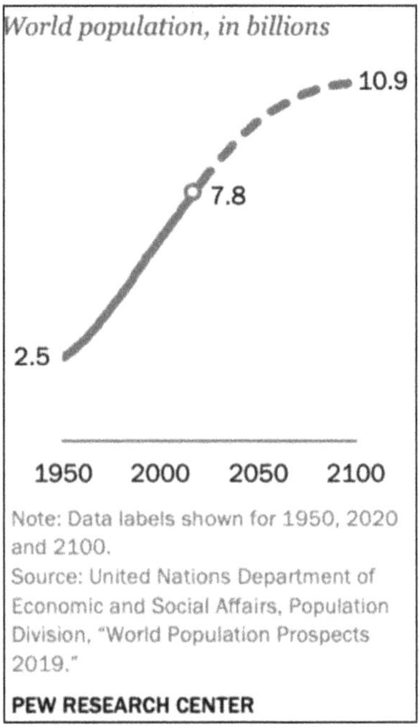

energy requirements, whereas declining fertility and increasing urbanization (people are less physically active in cities than when working the land) will decrease energy requirements. Overall, these demographic changes mean that food requirements of developing countries may have to double in terms of plant energy (see over for note on measurements) by 2050. Sub-Saharan Africa may have to more than triple plant energy production (Contributed by United Nations Population Fund[21] [UNFPA], excerpted from The State of World Population 1996).

The world's population is projected to nearly[22] stop growing by the end of the century [17].

Appropriate actions in agriculture, forestry, and fisheries can mitigate greenhouse gas emission and promote climate adaptation—with efforts to reduce emissions and adjust practices to the new reality often enhancing and supporting one another. For millions of people, especially rural family farmers in developing countries, our actions can make the difference between poverty and prosperity, between hunger and food security and nutrition. Agriculture—where the fight against hunger and climate change comes together—can unlock solutions[23] [17].

[21] United Nations Population Fund—https://www.unfpa.org/

[22] World's population is projected to nearly stop growing by the end of the century—https://www.pewresearch.org/fact-tank/2019/06/17/worlds-population-is-projected-to-nearly-stop-growing-by-the-end-of-the-century/

[23] FAO'S WORK ON CLIMATE CHANGE—https://www.fao.org/3/i8037e/i8037e.pdf

Thailand Rice Yield Production

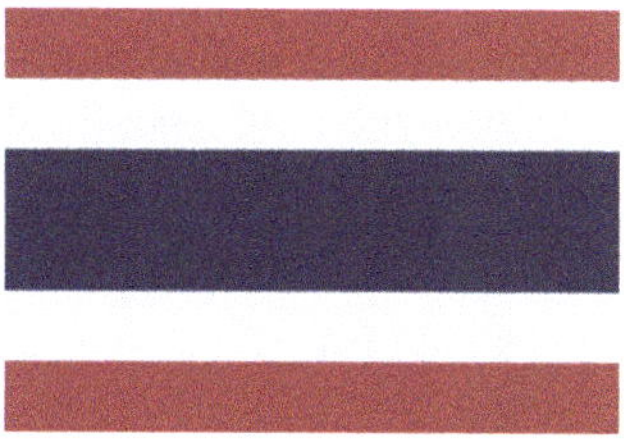

10.2 Thailand Nakhon Sawan Rice Production LP Linkage Model (SSP5_8.5: 2080–2099)

As developed in Chapter 5, Thailand Nakhon Sawan Reice Land Production equation, we will develop a linear programming to run the optimal solution.

The formula for the ProductionRicePaddy linear regression line is

$$\begin{aligned} &= -39630641.34 + 7005.45^{*}\ \text{Aug} - \text{prcp} - 2268.66^{*}\ \text{Sep} - \text{prcp} \\ &\quad - 4687.05^{*}\ \text{Oct} - \text{prcp} - 1877460.25^{*}\ \text{Aug} - \text{minTemp} \\ &\quad + 1082500.91^{*}\ \text{Sep} - \text{minTemp} - 137420.29^{*}\ \text{Oct} - \text{minTemp} \\ &\quad + 1210260.26^{*}\ \text{Aug} - \text{maxTemp} - 1110296.36^{*}\ \text{Sep} - \text{maxTemp} \\ &\quad + 749749.23^{*}\ \text{Oct} - \text{maxTemp} + 6.20^{*}\ \text{area harvested RicePaddy} \end{aligned} \tag{10.1}$$

The optimal solution needs two aspects:

- Area harvested
  - Area harvested is baselined with 2015 area harvest – 9717975 ha.
  - By 2100, area harvested could be increased[24] by 16%. That is, the increase in harvested areas to fulfill agricultural demand to meet population demand is **11272851 ha** [14].
  - A lower range of harvested area to develop simulation includes 5% of 11272851 ha–10709208.45 ha. We have tried iteratively for no optimal solution to solve the above equation.
- Rice production

[24] Harmonization of Land-Use Scenarios for the Period 1500-2100: 600 Years of Global Gridded Annual Land-Use Transitions, Wood Harvest, and Resulting Secondary Lands—https://www.researchgate.net/publication/227300487_Harmonization_of_Land-Use_Scenarios_for_the_Period_1500-2100_600_Years_of_Global_Gridded_Annual_Land-Use_Transitions_Wood_Harvest_and_Resulting_Secondary_Lands

- In 2015, Thailand rice production was 27702191 tons.
- In 2100, Thailand rice production needs to be 135% higher than the year 2015 production to meet the 2100 population demand: 65100148.85 tons.
- A lower range of rice production to develop simulation includes 5% of 65100148.85 tons – 61845141.41 tons.

Thailand Nakhon Sawan Rice Production: Harvest Pattern for SSP5-8.5

The software code for this model can be found on GitHub Link: Thailand_NakhonSawan_Rice_Production_CMIP6_SSP585_2020_2099.ipynb (Jupyter notebook Code)

10.2.1 Step 1: Define Thailand Nakhon Sawan linear programming (LP) variables

Define LP variables as per Eq. (10.1).

```
# Instantiate a Glop solver, naming it LinearExample.

  ### Run
cmip6_20802099_climatologymean_90th_ssp5_8p5_multimodelensembel
  solver = pywraplp.Solver.CreateSolver('GLOP')

  # Create the two variables and let them take on any non-negative
value.
  Area_harvested_RicePaddy = solver.IntVar(0, solver.infinity(),
'Area_harvested_RicePaddy')
  Aug_Precp = solver.IntVar(0, solver.infinity(), 'Aug_Precp')
  Sep_Precp = solver.IntVar(0, solver.infinity(), 'Sep_Precp')
  Oct_Precp = solver.IntVar(0, solver.infinity(), 'Oct_Precp')

  Aug_avgtemp = solver.IntVar(0, solver.infinity(), 'Aug_avgtemp')
  Sep_avgtemp = solver.IntVar(0, solver.infinity(), 'Sep_avgtemp')
  Oct_avgtemp = solver.IntVar(0, solver.infinity(), 'Oct_avgtemp')

  Aug_mintemp = solver.IntVar(0, solver.infinity(), 'Aug_mintemp')
  Sep_mintemp = solver.IntVar(0, solver.infinity(), 'Sep_mintemp')
  Oct_mintemp = solver.IntVar(0, solver.infinity(), 'Oct_mintemp')

  Aug_maxtemp = solver.IntVar(0, solver.infinity(), 'Aug_maxtemp')
  Sep_maxtemp = solver.IntVar(0, solver.infinity(), 'Sep_maxtemp')
  Oct_maxtemp = solver.IntVar(0, solver.infinity(), 'Oct_maxtemp')
  print('Number of variables =', solver.NumVariables())
```

10.2.2 Step 2: Define Variable Constraints

For each variable, define the upper and lower limits as per CMIP6 SSP5.85. The mean temperature anomaly for 2080–2099, Thailand, Nakha Sawan:

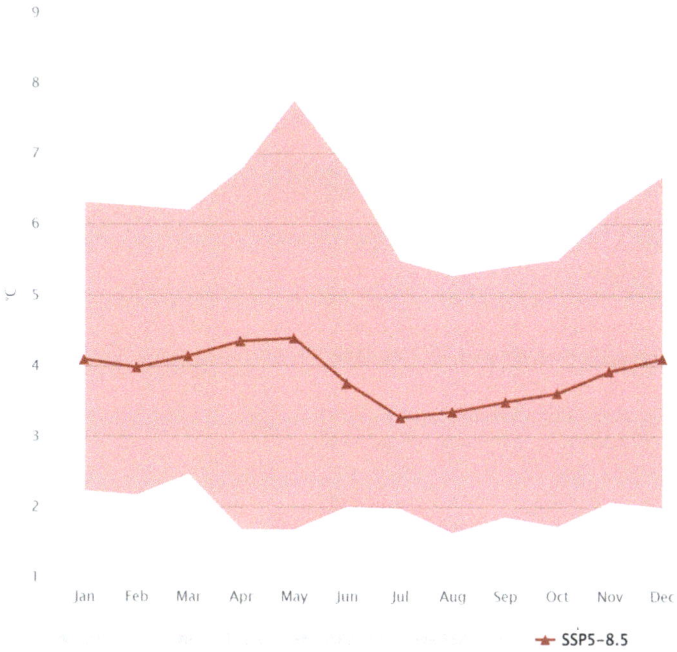

<table>
<tr><th>Projected mean temperature anomaly includes</th><th>The projected mean temperatures[25]:</th></tr>
<tr><td>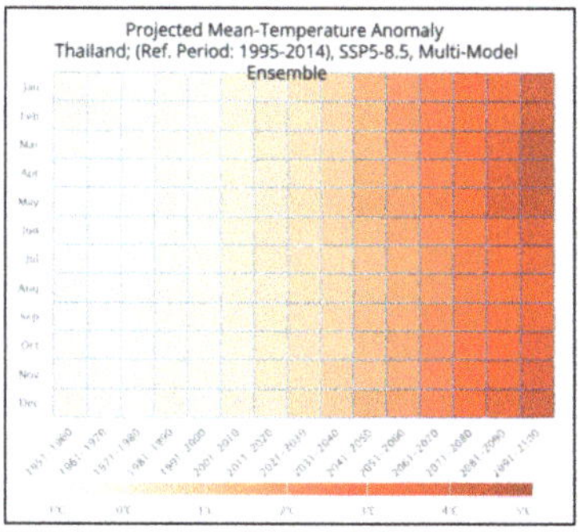
</td><td>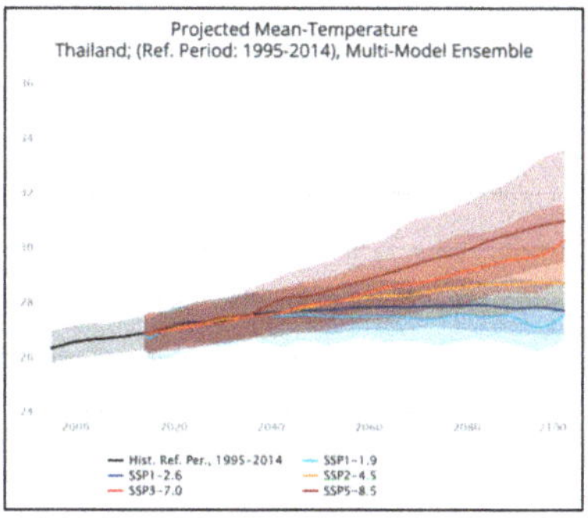
</td></tr>
</table>

[25] Climate Projections—https://climateknowledgeportal.worldbank.org/country/thailand/climate-data-projections-expert

```
# Thailand Nakhon Sawan area harvested in 2015 - 9717975 ha
  # Production 2015 - 27702191.0 tonnes
  # while global agricultural output in the scenario increased by
135% by 2080,
  # cultivated land expanded by only 16% above 2005 levels

  solver.Add(Area_harvested_RicePaddy <=11272851) ##7385661)
#7385661.36) ### 2020 Harvest Limits
  solver.Add(Area_harvested_RicePaddy >= 10709208.45)

  solver.Add(Aug_Precp <= 503.56)
  solver.Add(Aug_Precp >= 455.61)

  solver.Add(Sep_Precp <= 505.93)
  solver.Add(Sep_Precp >= 457.74)

  solver.Add(Oct_Precp <= 292.86)
  solver.Add(Oct_Precp >= 264.97)

  solver.Add(Aug_avgtemp <= 34.17)
  solver.Add(Aug_avgtemp >= 30.92)

  solver.Add(Sep_avgtemp <= 32.56)
  solver.Add(Sep_avgtemp >= 30.56)

  solver.Add(Oct_avgtemp <= 34.18)
  solver.Add(Oct_avgtemp >= 30.93)

  solver.Add(Aug_mintemp <= 30.35)
  solver.Add(Aug_mintemp >= 27.64)

  solver.Add(Sep_mintemp <= 30.35)
  solver.Add(Sep_mintemp >= 27.46)

  solver.Add(Oct_mintemp <= 29.91)
  solver.Add(Oct_mintemp >= 27.06)

  solver.Add(Aug_maxtemp <= 38.39)
  solver.Add(Aug_maxtemp >= 34.74)

  solver.Add(Sep_maxtemp <= 38.39)
  solver.Add(Sep_maxtemp >= 34.74)

  solver.Add(Oct_maxtemp <= 38.88)
  solver.Add(Oct_maxtemp >= 35.17)
```

10.2.3 Step 3: Define Model Constraints

The following code defines LP model constraints:

```
solver.Add( - 945022.83 + 7154.86 * Aug_Precp - 387.06 * Sep_Precp -
      9361.49 * Oct_Precp - 2246498.55 * Aug_avgtemp + 4974841.27 *
Sep_avgtemp - 2714371.41 * Oct_avgtemp -
       3373018.11 * Aug_mintemp - 1375706.10 * Sep_mintemp +
1361518.48 * Oct_mintemp + 2248087.38 * Aug_maxtemp -
       2399233.33 * Sep_maxtemp + 1476058.59 * Oct_maxtemp + 6.92 *
Area_harvested_RicePaddy >= 61845141.41)

  # Optimal Land

  print('Number of constraints =', solver.NumConstraints())
```

Here constraint is based on production needs for 2100.

10.2.4 Step 4: Define Objective Function and Solve It

```
# Objective function: The product value per tone
  solver.Maximize(- 945022.83 + 7154.86 * Aug_Precp - 387.06 *
Sep_Precp -
      9361.49 * Oct_Precp - 2246498.55 * Aug_avgtemp + 4974841.27 *
Sep_avgtemp - 2714371.41 * Oct_avgtemp -
       3373018.11 * Aug_mintemp - 1375706.10 * Sep_mintemp +
1361518.48 * Oct_mintemp + 2248087.38 * Aug_maxtemp -
       2399233.33 * Sep_maxtemp + 1476058.59 * Oct_maxtemp + 6.92 *
Area_harvested_RicePaddy)

  # Solve the system.
  status = solver.Solve()
```

10.2.5 Step 5: Output the Model

```
if status == pywraplp.Solver.OPTIMAL:
    print('Solution:')
```

(continued)

```
    print('Objective value =', solver.Objective().Value())
    print('Area_harvested_RicePaddy =',
Area_harvested_RicePaddy.solution_value())
    print('Nov_Precp =', Aug_Precp.solution_value())
    print('Sep_prcp =', Sep_Precp.solution_value())
    print('Oct_prcp =', Oct_Precp.solution_value())

    print('Aug_minTemp =', Aug_mintemp.solution_value())
    print('Sep_minTemp =', Sep_mintemp.solution_value())
    print('Oct_minTemp =', Oct_mintemp.solution_value())

    print('Aug_avgtemp =', Aug_avgtemp.solution_value())
    print('Sep_avgtemp =', Sep_avgtemp.solution_value())
    print('Oct_avgtemp =', Oct_avgtemp.solution_value())

    print('Aug_maxTemp =', Aug_maxtemp.solution_value())
    print('Sep_maxTemp =', Sep_maxtemp.solution_value())
    print('Oct_maxTemp =', Oct_maxtemp.solution_value())

    print('Area_harvested_RicePaddy =',
Area_harvested_RicePaddy.solution_value())

  else:
    print('The problem does not have an optimal solution.')

  print('\nAdvanced usage:')
  print('Problem solved in %f milliseconds' % solver.wall_time())
  print('Problem solved in %d iterations' % solver.iterations())

solveThailandNakhonSawanRiceProd_SSP5_85_20802099()
```

Output:

```
Number of variables = 13
   Number of constraints = 27
   The problem does not have an optimal solution.
   Advanced usage:
   Problem solved in 1.000000 milliseconds
   Problem solved in 0 iterations
```

Despite a 5% reduction (-3255007.443 tons of rice) over the required rice production demand for 2100, the LP linkage model could not be solved.

% change (model tuning)	Actual rice production (tons), year 2015	Required production (tons), year 2100	Actual area harvested (ha), year 2015	Required area harvested (ha), year 2100	Feasible area harvested (ha), year 2100	Feasible production (tons), year 2100
5%	27702191	65100148.85	9717975	11272851	10709208.45	61845141.41

10.2.6 Step 6: Redefine Objective Function (Seven Percentage Reduction in Harvested Land and Production)

```
solver.Add( - 945022.83 + 7154.86 * Aug_Precp - 387.06 * Sep_Precp -
      9361.49 * Oct_Precp - 2246498.55 * Aug_avgtemp + 4974841.27 *
Sep_avgtemp - 2714371.41 * Oct_avgtemp -
       3373018.11 * Aug_mintemp - 1375706.10 * Sep_mintemp +
1361518.48 * Oct_mintemp + 2248087.38 * Aug_maxtemp -
       2399233.33 * Sep_maxtemp + 1476058.59 * Oct_maxtemp + 6.92 *
Area_harvested_RicePaddy >= 10145565.9)

  # Optimal Land

  print('Number of constraints =', solver.NumConstraints())

  # Objective function: The product value per tone
  solver.Maximize(- 945022.83 + 7154.86 * Aug_Precp - 387.06 *
Sep_Precp -
      9361.49 * Oct_Precp - 2246498.55 * Aug_avgtemp + 4974841.27 *
Sep_avgtemp - 2714371.41 * Oct_avgtemp -
       3373018.11 * Aug_mintemp - 1375706.10 * Sep_mintemp +
1361518.48 * Oct_mintemp + 2248087.38 * Aug_maxtemp -
       2399233.33 * Sep_maxtemp + 1476058.59 * Oct_maxtemp + 6.92 *
Area_harvested_RicePaddy)

  # Solve the system.
  status = solver.Solve()
```

Output:

```
Number of variables = 13
  Number of constraints = 27
  The problem does not have an optimal solution.
  Advanced usage:
  Problem solved in 1.000000 milliseconds
  Problem solved in 0 iterations
```

Despite a 7% reduction (-4557010.42 tons of rice) over the required rice production demand for 2100, the LP linkage model could not be solved.

% change (model tuning)	Actual rice production (tons), year 2015	Required production (tons), year 2100	Actual area harvested (ha), year 2015	Required area harvested (ha), year 2100	Feasible area harvested (ha), year 2100	Feasible Production (tons), year 2100
5%	27702191	65100148.85	9717975	11272851	10709208.45	61845141.41
7%	27702191	65100148.85	9717975	11272851	10483751.43	60543138.43

10.2.7 Step 7: Define Objective Function and Solve It (with Ten Percentage Drop in Rice Production)

Rerun the LP with 10% drop in expected production, 2100.

```
solver.Add(Area_harvested_RicePaddy <=11272851) ##7385661)
#7385661.36) ### 2020 Harvest Limits
  solver.Add(Area_harvested_RicePaddy >= 10145565.9)

solver.Add( - 945022.83 + 7154.86 * Aug_Precp - 387.06 * Sep_Precp -
      9361.49 * Oct_Precp - 2246498.55 * Aug_avgtemp + 4974841.27 *
Sep_avgtemp - 2714371.41 * Oct_avgtemp -
       3373018.11 * Aug_mintemp - 1375706.10 * Sep_mintemp +
1361518.48 * Oct_mintemp + 2248087.38 * Aug_maxtemp -
       2399233.33 * Sep_maxtemp + 1476058.59 * Oct_maxtemp + 6.92 *
Area_harvested_RicePaddy >= 58590133.97)
```

Despite a 10% reduction (-6510014.885 tons of rice) over the required rice production demand for 2100, the LP linkage model could not be solved.

10.2.8 Step 8: Define Objective Function and Solve It (with Fifteen Percentage Drop in Rice Production)

Rerun the LP with 15% drop in expected production, 2100.

```
solver.Add(Area_harvested_RicePaddy <=11272851) ##7385661)
#7385661.36) ### 2020 Harvest Limits
  solver.Add(Area_harvested_RicePaddy >= 9581923.35)
```

(continued)

```
solver.Add( - 945022.83 + 7154.86 * Aug_Precp - 387.06 * Sep_Precp -
      9361.49 * Oct_Precp - 2246498.55 * Aug_avgtemp + 4974841.27 *
Sep_avgtemp - 2714371.41 * Oct_avgtemp -
      3373018.11 * Aug_mintemp - 1375706.10 * Sep_mintemp +
1361518.48 * Oct_mintemp + 2248087.38 * Aug_maxtemp -
      2399233.33 * Sep_maxtemp + 1476058.59 * Oct_maxtemp + 6.92 *
Area_harvested_RicePaddy >= 55335126.52)
```

We could solve the solution with 15% reduction from the expected production for 2100.

Output:

```
Number of variables = 13
Number of constraints = 27
Solution:
Objective value =
56631683.69940001
Area_harvested_RicePaddy =
11272851.0
Nov_Precp = 503.56
Sep_prcp = 457.74
Oct_prcp = 264.97
Aug_minTemp = 27.64
Sep_minTemp = 27.46
Oct_minTemp = 29.91
Aug_avgtemp = 30.92
Sep_avgtemp = 32.56
Oct_avgtemp = 30.93
Aug_maxTemp = 38.39
Sep_maxTemp = 34.74
Oct_maxTemp = 38.88
Area_harvested_RicePaddy =
11272851.0
```

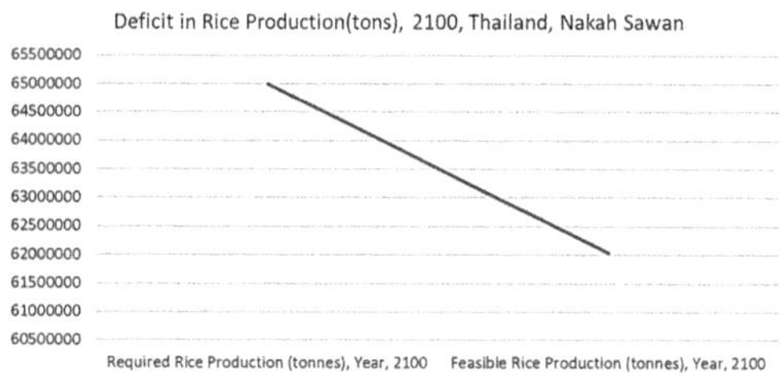

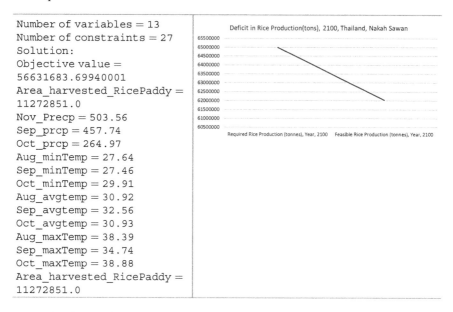

As clearly seen from the above LPs, the rice production would drop by 15% than required. Climate change would have similar production losses for rice cultivation across the world. It would be prudent to have climate-smart technologies and adaptation policies. Similar results could be found as stated by other studies.[26] Climate change affects global agricultural production and threatens food security. To counter the effect of faster maturity, adapted varieties would require more heat units to regain the previous growing period length (please see Table 10.1).

[26]Large potential for crop production adaptation depends on available future varieties-https://onlinelibrary.wiley.com/doi/full/10.1111/gcb.15649

Table 10.1 Summary table

% change (model tuning)	Actual rice production (tons), year 2015	Required production (tons), year 2100	Actual area harvested (ha), year 2015	Required area harvested (ha), year 2100	Feasible area harvested (ha), year 2100	Feasible production (tons), year 2100	Solved
5%	27702191	65100148.85	9717975	11272851	10709208.45	61845141.41	✗
7%	27702191	65100148.85	9717975	11272851	10483751.43	60543138.43	✗
10%	27702191	65100148.85	9717975	11272851	10145565.9	58590133.97	✗
15%	27702191	65100148.85	9717975	11272851	9581923.35	55335126.52	✓

10.3 Thailand Nakhon Sawan Rice Production LP Linkage Model (SSP3_7.0: 2080–2099)

As developed in Chapter 5, Thailand Nakhon Sawan Reice Land Production equation, we will develop linear programming to run the optimal solution.

The formula for the ProductionRicePaddy linear regression line is

$$\begin{aligned} &= -39630641.34 + 7005.45^{*}\ \text{Aug} - \text{prcp} - 2268.66^{*}\ \text{Sep} - \text{prcp} \\ &\quad - 4687.05^{*}\ \text{Oct} - \text{prcp} - 1877460.25^{*}\ \text{Aug} - \text{minTemp} \\ &\quad + 1082500.91^{*}\ \text{Sep} - \text{minTemp} - 137420.29^{*}\ \text{Oct} - \text{minTemp} \\ &\quad + 1210260.26^{*}\ \text{Aug} - \text{maxTemp} - 1110296.36^{*}\ \text{Sep} - \text{maxTemp} \\ &\quad + 749749.23^{*}\ \text{Oct} - \text{maxTemp} + 6.20^{*}\ \text{Area harvested RicePaddy} \end{aligned} \tag{10.2}$$

The optimal solution needs two aspects: please find climatology details as listed below.[27]

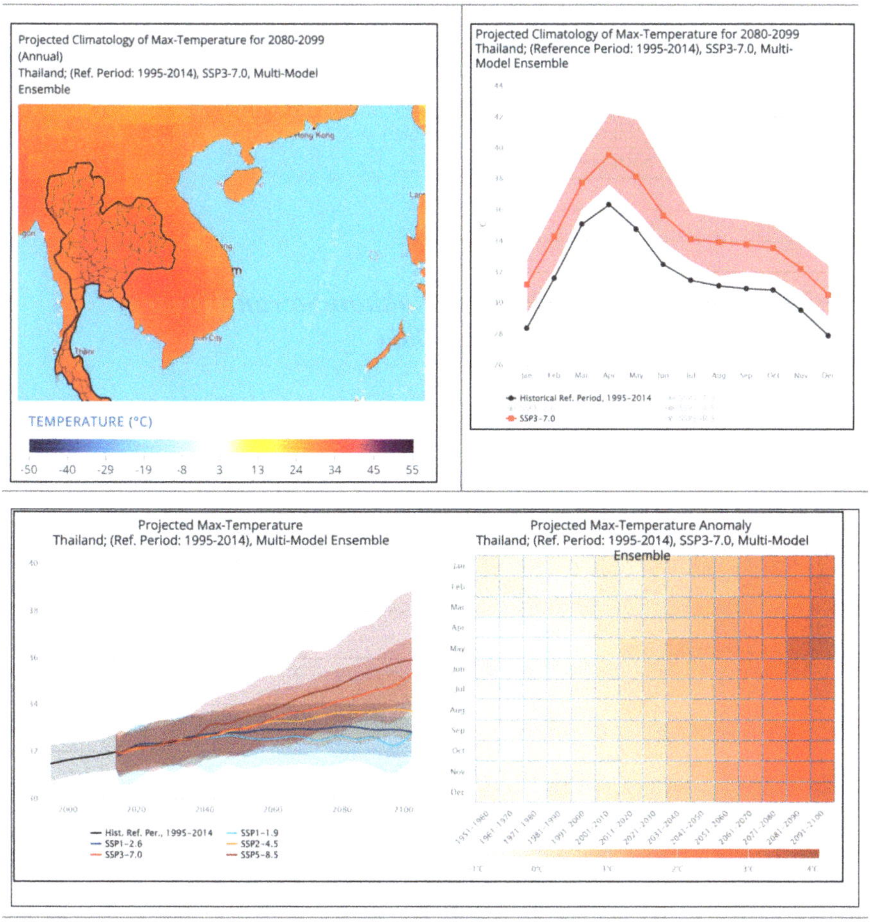

[27] Thailand—https://climateknowledgeportal.worldbank.org/country/thailand/climate-data-projections-expert

- Area harvested
  - Area harvested is baselined with the 2015 area harvest – 9717975 ha.
  - By 2100, the area harvested could be increased[28] by 16%. That is, the increase in harvested areas to fulfill agricultural demand to meet population demand is **11272851 ha**.
  - A lower range of harvested area to develop simulation includes 5% of 11272851 ha – 10709208.45 ha. We have tried iteratively for no optimal solution to solve the above equation.
- Rice production
  - In 2015, Thailand rice production was 27702191 tons.
  - In 2100, Thailand rice production needs to be 135% higher than the year 2015 production to meet the 2100 population demand: 65100148.85 tons.
  - A lower range of rice production to develop simulation includes 5% of 65100148.85 tons – 61845141.41 tons.

Thailand Nakhon Sawan Rice Production: Harvest Pattern for SSP3-7.0

The software code for this model can be found on GitHub Link: Thailand_NakhonSawan_Rice_Production_CMIP6_SSP370_2020_2099.ipynb (Jupyter notebook Code)

10.3.1 *Step 1: Define Thailand Nakhon Sawan Linear Programming (LP) Variables*

Define LP variables as per Eq. (10.1).

```
def solveThailandNakhonSawanRiceProd_SSP3_70_20802099():
  """Linear programming sample."""
  # Instantiate a Glop solver, naming it LinearExample.

      ### Run cmip6_20802099_climatologymean_90th_ssp5_8p5_
multimodelensembel
  solver = pywraplp.Solver.CreateSolver('GLOP')
```

(continued)

[28] Harmonization of Land-Use Scenarios for the Period 1500-2100: 600 Years of Global Gridded Annual Land-Use Transitions, Wood Harvest, and Resulting Secondary Lands—https://www.researchgate.net/publication/227300487_Harmonization_of_Land-Use_Scenarios_for_the_Period_1500-2100_600_Years_of_Global_Gridded_Annual_Land-Use_Transitions_Wood_Harvest_and_Resulting_Secondary_Lands

```
  # Create the two variables and let them take on any non-negative
value.
  Area_harvested_RicePaddy = solver.IntVar(0, solver.infinity(),
'Area_harvested_RicePaddy')
  Aug_Precp = solver.IntVar(0, solver.infinity(), 'Aug_Precp')
  Sep_Precp = solver.IntVar(0, solver.infinity(), 'Sep_Precp')
  Oct_Precp = solver.IntVar(0, solver.infinity(), 'Oct_Precp')

  Aug_avgtemp = solver.IntVar(0, solver.infinity(),
'Aug_avgtemp')
  Sep_avgtemp = solver.IntVar(0, solver.infinity(),
'Sep_avgtemp')
  Oct_avgtemp = solver.IntVar(0, solver.infinity(),
'Oct_avgtemp')

  Aug_mintemp = solver.IntVar(0, solver.infinity(),
'Aug_mintemp')
  Sep_mintemp = solver.IntVar(0, solver.infinity(),
'Sep_mintemp')
  Oct_mintemp = solver.IntVar(0, solver.infinity(),
'Oct_mintemp')

  Aug_maxtemp = solver.IntVar(0, solver.infinity(),
'Aug_maxtemp')
  Sep_maxtemp = solver.IntVar(0, solver.infinity(),
'Sep_maxtemp')
  Oct_maxtemp = solver.IntVar(0, solver.infinity(),
'Oct_maxtemp')

  print('Number of variables =', solver.NumVariables())
```

10.3.2 Step 2: Define Variable Constraints

Please add variable thresholds as per SSP3_7.0 – year 2080–2099 values.

```
 solver.Add(Area_harvested_RicePaddy <=11272851) ##7385661)
#7385661.36) ### 2020 Harvest Limits
  solver.Add(Area_harvested_RicePaddy >= 9018280.8)

  solver.Add(Aug_Precp <= 447.55)
  solver.Add(Aug_Precp >= 425.17)
```

(continued)

```
solver.Add(Sep_Precp <= 443.2)
solver.Add(Sep_Precp >= 421.04)

solver.Add(Oct_Precp <= 174.88)
solver.Add(Oct_Precp >= 166.13)

solver.Add(Aug_avgtemp <= 29.02)
solver.Add(Aug_avgtemp >= 27.56)

solver.Add(Sep_avgtemp <= 28.79)
solver.Add(Sep_avgtemp >= 27.35)

solver.Add(Oct_avgtemp <= 28.25)
solver.Add(Oct_avgtemp >= 26.83)

solver.Add(Aug_mintemp <= 25.58)
solver.Add(Aug_mintemp >= 24.30)

solver.Add(Sep_mintemp <= 25.05)
solver.Add(Sep_mintemp >= 23.79)

solver.Add(Oct_mintemp <= 23.73)
solver.Add(Oct_mintemp >= 22.54)

solver.Add(Aug_maxtemp <= 32.58)
solver.Add(Aug_maxtemp >= 30.95)

solver.Add(Sep_maxtemp <= 32.47)
solver.Add(Sep_maxtemp >= 30.84)

solver.Add(Oct_maxtemp <= 33.19)
solver.Add(Oct_maxtemp >= 31.53)
```

10.3.3 Step 3: Define Model Constraints

The following code defines LP model constraints:

```
solver.Add( - 945022.83 + 7154.86 * Aug_Precp - 387.06 * Sep_Precp -
     9361.49 * Oct_Precp - 2246498.55 * Aug_avgtemp + 4974841.27 *
Sep_avgtemp - 2714371.41 * Oct_avgtemp -
      3373018.11 * Aug_mintemp - 1375706.10 * Sep_mintemp +
1361518.48 * Oct_mintemp + 2248087.38 * Aug_maxtemp -
```

(continued)

```
      2399233.33 * Sep_maxtemp + 1476058.59 * Oct_maxtemp + 6.92 *
Area_harvested_RicePaddy >= 61845141.41)

  # Optimal Land

  print('Number of constraints =', solver.NumConstraints())
```

Here constraint is based on production needs for 2100.

10.3.4 Step 4: Define Objective Function and Solve It

```
# Objective function: The product value per tone
  solver.Maximize(- 945022.83 + 7154.86 * Aug_Precp - 387.06 *
Sep_Precp -
      9361.49 * Oct_Precp - 2246498.55 * Aug_avgtemp + 4974841.27 *
Sep_avgtemp - 2714371.41 * Oct_avgtemp -
       3373018.11 * Aug_mintemp - 1375706.10 * Sep_mintemp +
1361518.48 * Oct_mintemp + 2248087.38 * Aug_maxtemp -
      2399233.33 * Sep_maxtemp + 1476058.59 * Oct_maxtemp + 6.92 *
Area_harvested_RicePaddy)

  # Solve the system.
  status = solver.Solve()
```

10.3.5 Step 5: Output the Model

```
if status == pywraplp.Solver.OPTIMAL:
    print('Solution:')
    print('Objective value =', solver.Objective().Value())
    print('Area_harvested_RicePaddy =',
Area_harvested_RicePaddy.solution_value())
    print('Nov_Precp =', Aug_Precp.solution_value())
    print('Sep_prcp =', Sep_Precp.solution_value())
    print('Oct_prcp =', Oct_Precp.solution_value())

    print('Aug_minTemp =', Aug_mintemp.solution_value())
    print('Sep_minTemp =', Sep_mintemp.solution_value())
    print('Oct_minTemp =', Oct_mintemp.solution_value())
```

(continued)

```
        print('Aug_avgtemp =', Aug_avgtemp.solution_value())
        print('Sep_avgtemp =', Sep_avgtemp.solution_value())
        print('Oct_avgtemp =', Oct_avgtemp.solution_value())

        print('Aug_maxTemp =', Aug_maxtemp.solution_value())
        print('Sep_maxTemp =', Sep_maxtemp.solution_value())
        print('Oct_maxTemp =', Oct_maxtemp.solution_value())

        print('Area_harvested_RicePaddy =',
  Area_harvested_RicePaddy.solution_value())

    else:
        print('The problem does not have an optimal solution.')

    print('\nAdvanced usage:')
    print('Problem solved in %f milliseconds' % solver.wall_time())
    print('Problem solved in %d iterations' % solver.iterations())

solveThailandNakhonSawanRiceProd_SSP5_85_20802099()
```

We could solve the solution with 15% reduction from the expected production for 2100.

Output:

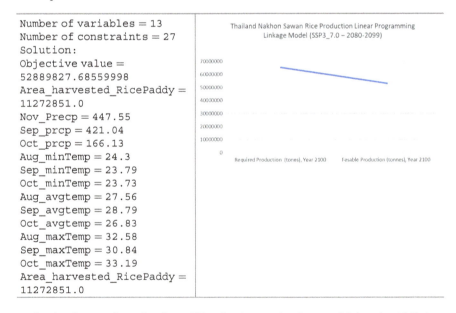

```
Number of variables = 13
Number of constraints = 27
Solution:
Objective value =
52889827.68559998
Area_harvested_RicePaddy =
11272851.0
Nov_Precp = 447.55
Sep_prcp = 421.04
Oct_prcp = 166.13
Aug_minTemp = 24.3
Sep_minTemp = 23.79
Oct_minTemp = 23.73
Aug_avgtemp = 27.56
Sep_avgtemp = 28.79
Oct_avgtemp = 26.83
Aug_maxTemp = 32.58
Sep_maxTemp = 30.84
Oct_maxTemp = 33.19
Area_harvested_RicePaddy =
11272851.0
```

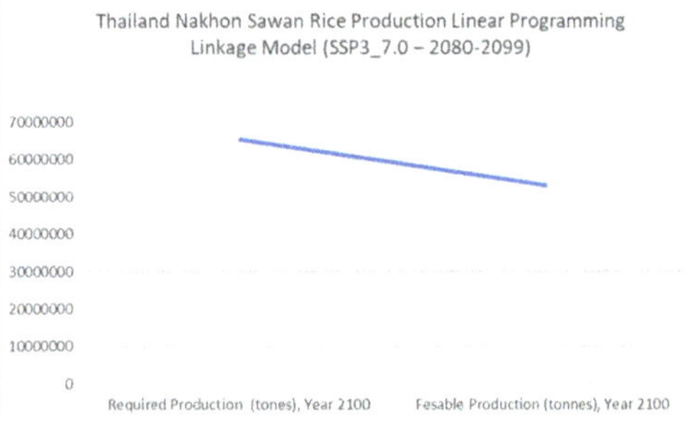

As clearly seen from the above LPs, the rice production would drop by 15% than required. Climate change would have similar production losses for rice cultivation across the world. It would be prudent to have climate-smart technologies and adaptation policies. Similar results could be found as stated by other studies.[29] Climate change affects global agricultural production and threatens food security. To counter the effect of faster maturity, adapted varieties would require more heat units to regain the previous growing period length (please see Table 10.2).

The linkage[30] between macroeconomic indicators and real economy is analyzed as part of the book and machine learning applied to assess the relationship between economic cycles and macroeconomic/financial indicators. As one would expect, most macroeconomic and financial variables exhibit procyclical behavior during recessions. In addition, recessions are characterized by sharp declines in (residential) investment, industrial production, imports, and housing and equity prices, modest declines in consumption and exports, and some decrease in employment rates. Two key policy-related variables – short-term interest rates and fiscal expenditures – often behave countercyclical during recessions [19].

The purpose of ML is to study the interrelations between recessions and the effects of its spread across the globe. Recessions in many advanced countries [19] were bunched in four periods over the past 40 years –the mid-1970s, the early 1980s, the early 1990s, and the early 2000s – and often coincided with global shocks. Just as many countries experience synchronized recessions, countries also go through simultaneous episodes of credit contractions. Moreover, declines in house and equity prices tend to occur at the same time. The goal is to uncover macroeconomic signals to predict the future prognostics.

After reading this chapter, you should be able to answer agricultural linkage models and production models using heuristic techniques. You should be able to develop linkage models of machine learning and heuristics. You should be able to overlay linkage models with CMIP6, SSP projections on top of linkage models to observe the impact of climate change on agricultural resources.

[29]Large potential for crop production adaptation depends on available future varieties-https://onlinelibrary.wiley.com/doi/full/10.1111/gcb.15649

[30]What Happens During Recessions, Crunches and Busts?—https://www.imf.org/external/pubs/ft/wp/2008/wp08274.pdf

Table 10.2 Summary table

% change (model tuning)	Actual rice production (tons), year 2015	Required production (tons), year 2100	Actual area harvested (ha), year 2015	Required area harvested (ha), year 2100	Feasible area harvested (ha), year 2100	Feasible production (tons), year 2100	Solved
5%	27702191	65100148.85	9717975	11272851	10709208.45	61845141.41	✗
7%	27702191	65100148.85	9717975	11272851	10483751.43	60543138.43	✗
10%	27702191	65100148.85	9717975	11272851	10145565.9	58590133.97	✗
15%	27702191	65100148.85	9717975	11272851	9581923.35	52889827.68	✓

References

1. Will Horner and Jeffrey T. Lewis, Coffee Prices Soar After Bad Harvests and Insatiable Demand, July 12, 2021, 8:41 am ET, https://www.wsj.com/articles/coffee-prices-soar-after-bad-harvests-and-insatiable-demand-11626093703?mod=article_inline, Access Date: February 03, 2022
2. Jeffrey T. Lewis and Joe Wallace, Coffee Prices Jump to Six-Year High as Brazilian Frost Threatens Crop, Updated July 27, 2021, 4:50 pm ET, https://www.wsj.com/articles/coffee-prices-jump-to-six-year-high-as-brazilian-frost-threatens-crop-11627380128?mod=article_inline, Access Date: February 03, 2022
3. Food and Agriculture Organization of the United Nations, Handbook for Defining and Setting up a Food Security Information and Early Warning System (FSIEWS), 2000, https://www.fao.org/3/X8622E/x8622e00.htm#TopOfPage, Access Date: January 22, 2022
4. Samuel Benin, James Thurlow, Xinshen Diao, Christen Mccool and Franklin, Agricultural growth and investment options for poverty reduction in Malawi, 2008, https://www.ifpri.org/publication/agricultural-growth-and-investment-options-poverty-reduction-malawi-0, Access Date: January 29, 2022
5. Karl Pauw, James Thurlow and Dirk Van Seventer, Droughts and floods in Malawi Assessing the economywide effects, 2010, https://www.ifpri.org/publication/droughts-and-floods-malawi, Access Date: January 29, 2022
6. National Geographic, Climate Change 5 Ways, It Will Affect You, 2015, https://www.nationalgeographic.com/climate-change/how-to-live-with-it/health.html, Access Date: February 03, 2022
7. Zeke Hausfather, Explainer: How "Shared Socioeconomic Pathways" explore future climate change, 19 April 2018, https://www.carbonbrief.org/explainer-how-shared-socioeconomic-pathways-explore-future-climate-change, Access Date: February 03 2022
8. Joseph P. Janzen, Colin A. Carter, Aaron D. Smith, and Michael K. Adjemian, Deconstructing Wheat Price Spikes: A Model of Supply and Demand, Financial Speculation, and Commodity Price Comovement, April 2014, https://www.ers.usda.gov/webdocs/publications/45199/46439_err165.pdf?v=0, Access Date: February 27, 2022
9. Oliver Milman, Earth has lost a third of arable land in past 40 years, scientists say, Wed 2 Dec 2015 05.00 EST, https://www.theguardian.com/environment/2015/dec/02/arable-land-soil-food-security-shortage, Access Date: March 06 2022
10. Oak Ridge National Laboratory Distributed Active Archive Center (ORNL DAAC), LUH1: Harmonized Global Land Use for Years 1500–2100, V1, Revision Date: September 23, 2014, https://daac.ornl.gov/VEGETATION/guides/Land_Use_Harmonization_V1.html, Access Date: March 14, 2022
11. Chini, L.P., G.C. Hurtt, and S. Frolking. 2014. LUH1: Harmonized Global Land Use for Years 1500–2100, V1. ORNL DAAC, Oak Ridge, Tennessee, USA. https://doi.org/10.3334/ORNLDAAC/1248
12. G. C. Hurtt, S. Frolking, M. G. Fearon, B. Moore, E. Shevliakova, S. Malyshev, S. W. Pacala, R. A. Houghton, The underpinnings of land-use history: three centuries of global gridded land-use transitions, wood-harvest activity, and resulting secondary lands, First published: 22 May 2006, https://doi.org/10.1111/j.1365-2486.2006.01150.x, Access Date: March 06, 2022
13. Hurtt, G. Chini, L. Frolking, Steve Betts, Richard Feddema, Johannes Fischer, Günther Fisk, J. Hibbard, Kathy Houghton, Richard Janetos, Anthony Jones, Chris Kindermann, Georg Kinoshita, T. Klein Goldewijk, Kees Riahi, Keywan Shevliakova, Elena Smith, Shanae Stehfest, E. Thomson, A. & Wang, Yingping. (2011). Harmonization of Land-Use Scenarios for the Period 1500-2100: 600 Years of Global Gridded Annual Land-Use Transitions, Wood Harvest, and Resulting Secondary Lands. Climatic Change. 109. 117-161. https://doi.org/10.1007/s10584-011-0153-2.
14. Hurtt, G.C., Chini, L.P., Frolking, S. et al. Harmonization of land-use scenarios for the period 1500–2100: 600 years of global gridded annual land-use transitions, wood harvest, and

resulting secondary lands. Climatic Change 109, 117 (2011). https://doi.org/10.1007/s10584-011-0153-2
15. FAO World Food Summit, Food for All, Rome 13-17 November 1996, https://www.fao.org/3/x0262e/x0262e00.htm#TopOfPage, Access Date: March 14, 2022
16. Florian Zabel, Christoph Müller, Joshua Elliott, Sara Minoli, Jonas Jägermeyr, Julia M. Schneider, James A. Franke, Elisabeth Moyer, Marie Dury, Louis Francois, Christian Folberth, Wenfeng Liu, Thomas A.M. Pugh, Stefan Olin, Sam S. Rabin, Wolfram Mauser, Tobias Hank, Alex C. Ruane, Senthold Asseng, Large potential for crop production adaptation depends on available future varieties, 17 May 2021, https://onlinelibrary.wiley.com/doi/full/10.1111/gcb.15649, Access Date: March 07, 2022
17. FAO, FAO'S work on climate change United Nations Climate Change Conference 2017, 2017, https://www.fao.org/3/i8037e/i8037e.pdf, Access Date: January 30, 2022
18. Anthony Cilluffo and Neil G. Ruiz, World's population is projected to nearly stop growing by the end of the century, June 17, 2019, https://www.pewresearch.org/fact-tank/2019/06/17/worlds-population-is-projected-to-nearly-stop-growing-by-the-end-of-the-century/, Access Date: March 06, 2022
19. Stijn Claessens, M. Ayhan Kose, and Marco E. Terrones, What Happens During Recessions, Crunches and Busts? 2008, https://www.imf.org/external/pubs/ft/wp/2008/wp08274.pdf, Access Date: March 14, 2022

Part IV
Conclusion

Chapter 11
Future

The world is at a critical juncture. In 2020, nearly one in three people did not have access to adequate food—between 720 and 811 million people faced hunger. Compared with 2019, 46 million more people in Africa, almost 57 million more in Asia, and about 14 million more in Latin America and the Caribbean were affected by hunger. Based on forecasts of global population growth, the current deficit to feed people around the world, and increased demand for greener fuel and biodiesel, food security will remain an important economic development issue over the next several decades. As food-versus-fuel tension becomes more intense, the day will come when more agricultural products will be used for energy than food. Adding to the conundrum, the COVID-19 pandemic has changed the face of the earth in terms of supply chain, resource availability, and human labor and has exposed our vulnerabilities in food security to an even greater extent. In essence, humanity is at a critical juncture, and what this unprecedented movement in our lives has thrusted upon us—the practitioners of the agriculture and technologists of the world—is to innovate and become more productive to address the multipronged food security challenges.

Agricultural innovation is key to overcoming concerns of food security; the infusion of data science, artificial intelligence, sensor technologies, and heuristic modeling with traditional agricultural practices such as soil engineering, fertilizers, and agronomy is one of the best ways to achieve this. Data science helps farmers to unravel patterns in equipment usage, transportation and storage costs, yield per hectare, and weather trends to better plan and spend resources. Artificial intelligence (AI) enables farmers to learn from fellow worldwide farmers to apply the best techniques that are transferred learning from the AI. Heuristic modeling is an essential software technique that codifies farmers' tacit knowledge such as better seed per soil, better feed for dairy cattle breed, or production practices to match weather pattern that was acquired over years of their hard work to share with worldwide farmers to improve overall production efficiencies, the best antidote to the food security issue. In addition to the paradigm shift, economic sustainability of small farms is a major enabler of food security.

C. Vuppalapati, *Artificial Intelligence and Heuristics for Enhanced Food Security*, International Series in Operations Research & Management Science 331,
https://doi.org/10.1007/978-3-031-08743-1_11

Appendix 1: Food Security and Nutrition

The 17 Sustainable Development Goals (SDGs)

The 2030 Agenda for Sustainable Development, adopted by all United Nations Member States in 2015, provides a shared blueprint for peace and prosperity for people and the planet, now and into the future. At its heart are the 17 Sustainable Development Goals[1] (SDGs), which are an urgent call for action by all countries—developed and developing—in a global partnership. They recognize that ending poverty and other deprivations must go hand in hand with strategies that improve health and education, reduce inequality, and spur economic growth—all while tackling climate change and working to preserve our oceans and forests.

[1]The 17 Goals of Sustainable Development—https://sdgs.un.org/goals

C. Vuppalapati, *Artificial Intelligence and Heuristics for Enhanced Food Security*, International Series in Operations Research & Management Science 331,
https://doi.org/10.1007/978-3-031-08743-1

Food Security Monitoring System (FSMS)

The Food Security Monitoring System[2] (FSMS), generally, covers key indicators, 35 indicators, that are classified along the four dimensions of food security[3]: availability (5 indicators), access (10 indicators), utilization (13 indicators), and stability (7 indicators).

Number	Type of indicator	Source
	Availability indicators	
1	Average dietary energy supply adequacy	FAO
2	Average value of food production	FAO
3	Share of dietary energy supply derived from cereals, roots, and tubers	FAO
4	Average vegetarian protein supply	FAO
5	Average supply of protein of animal origin	FAO
	Accessibility indicators	

(continued)

[2]Sudan—Food Security Monitoring System, http://fsis.sd/SD/EN/FoodSecurity/Monitoring/

[3]Handbook for Defining and Setting up a Food Security Information and Early Warning System (FSIEWS)—https://www.fao.org/3/X8622E/x8622e00.htm#TopOfPage

Number	Type of indicator	Source
6	Gross domestic product per capita (in purchasing power equivalent)	WB
7	Domestic food price level index	WB/ILO/FAO
8	Percent of paved roads over total roads	WB
9	Road density	World Road Statistics
10	Rail lines density	WB
11	Prevalence of undernourishment (PoU)	FAO
12	Number of people undernourished	FAO
13	Share of food expenditure of the poor	FAO
14	Depth of the food deficit	FAO
15	Prevalence of food inadequacy	FAO
	Utilization indicators	
16	Access to improved water sources	WHO/UNICEF
17	Access to improved sanitation facilities	WHO/UNICEF
18	Percentage of children under 5 years of age affected by wasting	WHO/UNICEF
19	Percentage of children under 5 years of age who are stunted	WHO/UNICEF
20	Percentage of children under 5 years of age who are underweight	WHO/UNICEF
21	Percentage of adults who are underweight	WHO
22	Prevalence of anemia among pregnant women	WHO/WB
23	Prevalence of anemia among children under 5 years of age	WHO/WB
24	Prevalence of vitamin A deficiency	WHO
25	Prevalence of iodine deficiency	WHO
26	Prevalence of undernourishment (PoU)	FAO
27	Mortality rate	WHO/WB/UNICEF
28	Food consumption	
	Stability indicators	
29	Cereal import dependency ratio	FAO
30	Percent of arable land equipped for irrigation	FAO
31	Value of food imports over total merchandise exports	FAO
32	Political stability and absence of violence/terrorism	WB/WWGI
33	Domestic food price volatility	WB/ILO/FAO
34	Per capita food production variability	FAO
35	Per capita food supply variability	FAO

NHANES 2019–2020 Questionnaire Instruments: Food Security

Questionnaires are administered to NHANES participants both at home and in the MECs. The questionnaires and a brief description of each section follows[4].

- FSQ.041 In the last 12 months, since last {DISPLAY CURRENT MONTH AND LAST YEAR}, did {you/you or other adults in your household} ever cut the size of your meals or skip meals because there wasn't enough money for food?
- FSQ.061 In the last 12 months, did you ever eat less than you felt you should because there wasn't enough money for food?
- FSQ.071 [In the last 12 months], were you ever hungry but didn't eat because there wasn't enough money for food?
- FSQ.081 [In the last 12 months], did you lose weight because there wasn't enough money for food?
- FSQ.092 [In the last 12 months], did {you/you or other adults in your household} ever not eat for a whole day because there wasn't enough money for food?
- FSQ.111 In the last 12 months, since {DISPLAY CURRENT MONTH AND LAST YEAR} did you ever cut the size of {CHILD'S NAME/any of the children's} meals because there wasn't enough money for food?
- [In the last 12 months], did {CHILD'S NAME/any of the children} ever skip meals because there wasn't enough money for food?

Macroeconomic Signals That Could Help Predict Economic Cycles

Signal #	Type	Description
Oil price swings	Leading indicator	Oil price swings[5] appear to be consistent and frequent historical precursors to US recessions[6]. A spike in oil prices has preceded or coincided with 10 out of 12 post-WWII recessions.
Monetary trends	Lagging indicator	A period of expansionary monetary policy in the years prior to the recession, sometimes to help fund government war spending or to reinflate the economy after the previous round of recession. Once the resulting debt bubbles pop or the end of a war leads to cutbacks in monetary expansion, several years' worth of overextended, debt-based

(continued)

[4]Food Security (FSQ)—https://wwwn.cdc.gov/nchs/data/nhanes/2019-2020/questionnaires/FSQ_Family_K.pdf

[5]Oil Price Swings—https://www.macrotrends.net/1369/crude-oil-price-history-chart

[6]Oil futures volatility and the economy—https://www.eurekalert.org/news-releases/843803

Signal #	Type	Description
		investments, and malinvestments tend to be wiped out in a process of debt deflation[7,8] in a relatively short period. This spikes unemployment and drags down GDP.
During recession, GDP and credit contraction occur	Lagging indicator	Country GDP/credit contraction, in percent[9].

List of World Development Indicators (WDI)

Indicator	Code
Growth and economic structure	
GDP (current US$)	NY.GDP.MKTP.CD
GDP growth (annual %)	NY.GDP.MKTP.KD.ZG
Agriculture, value added (annual % growth)	NV.AGR.TOTL.KD.ZG
Industry, value added (annual % growth)	NV.IND.TOTL.KD.ZG
Manufacturing, value added (annual % growth)	NV.IND.MANF.KD.ZG
Services, value added (annual % growth)	NV.SRV.TOTL.KD.ZG
Final consumption expenditure (annual % growth)	NE.CON.TOTL.KD.ZG
Gross capital formation (annual % growth)	NE.GDI.TOTL.KD.ZG
Exports of goods and services (annual % growth)	NE.EXP.GNFS.KD.ZG
Imports of goods and services (annual % growth)	NE.IMP.GNFS.KD.ZG
Agriculture, value added (% of GDP)	NV.AGR.TOTL.ZS
Industry, value added (% of GDP)	NV.IND.TOTL.ZS
Services, value added (% of GDP)	NV.SRV.TOTL.ZS
Final consumption expenditure (% of GDP)	NE.CON.TOTL.ZS
Gross capital formation (% of GDP)	NE.GDI.TOTL.ZS
Exports of goods and services (% of GDP)	NE.EXP.GNFS.ZS
Imports of goods and services (% of GDP)	NE.IMP.GNFS.ZS
Income and savings	
GNI per capita, Atlas method (current US$)	NY.GNP.PCAP.CD
GNI per capita, PPP (current international $)	NY.GNP.PCAP.PP.CD
Population, total	SP.POP.TOTL
Gross savings (% of GDP)	NY.GNS.ICTR.ZS

(continued)

[7]Michael Assous (2013) Irving Fisher's debt deflation analysis: From the Purchasing Power of Money (1911) to the Debt-deflation Theory of the Great Depression (1933), The European Journal of the History of Economic Thought, 20:2, 305–322, DOI: 10.1080/09672567.2012.762936

[8]Debt Deflation—https://www.investopedia.com/terms/d/debtdeflation.asp

[9]What Happens During Recessions, Crunches and Busts?—https://www.imf.org/external/pubs/ft/wp/2008/wp08274.pdf

Indicator	Code
Adjusted net savings, including particulate emission damage (% of GNI)	NY.ADJ.SVNG.GN.ZS
Balance of payments	
Export value index (2000 = 100)	TX.VAL.MRCH.XD.WD
Import value index (2000 = 100)	TM.VAL.MRCH.XD.WD
Personal remittances, received (% of GDP)	BX.TRF.PWKR.DT.GD.ZS
Current account balance (% of GDP)	BN.CAB.XOKA.GD.ZS
Foreign direct investment, net inflows (% of GDP)	BX.KLT.DINV.WD.GD.ZS

Wheat Data Disappearance and End Stocks

This data product contains statistics on wheat—including the five classes of wheat: hard red winter, hard red spring, soft red winter, white, and durum and rye. It includes data published in the monthly Wheat Outlook and previously annual Wheat Yearbook. Data are monthly, quarterly, and/or annual, depending upon the data series. Most data are on a marketing-year basis, but some are calendar year[10].

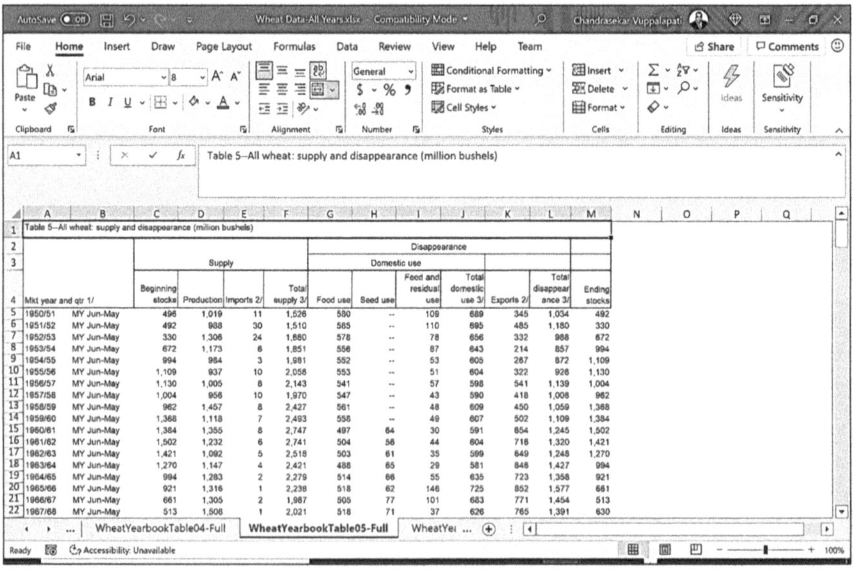
Table 5--All wheat: supply and disappearance (million bushels)

Mkt year and qtr 1/		Supply: Beginning stocks	Production	Imports 2/	Total supply 3/	Disappearance: Domestic use: Food use	Seed use	Feed and residual use	Total domestic use 3/	Exports 2/	Total disappearance 3/	Ending stocks
1950/51	MY Jun-May	496	1,019	11	1,526	580	--	109	689	345	1,034	492
1951/52	MY Jun-May	492	988	30	1,510	585	--	110	695	485	1,180	330
1952/53	MY Jun-May	330	1,306	24	1,660	578	--	78	656	332	988	672
1953/54	MY Jun-May	672	1,173	6	1,851	556	--	87	643	214	857	994
1954/55	MY Jun-May	994	984	3	1,981	552	--	53	605	267	872	1,109
1955/56	MY Jun-May	1,109	937	10	2,056	553	--	51	604	322	926	1,130
1956/57	MY Jun-May	1,130	1,005	8	2,143	541	--	57	598	541	1,139	1,004
1957/58	MY Jun-May	1,004	956	10	1,970	547	--	43	590	418	1,008	962
1958/59	MY Jun-May	962	1,457	8	2,427	561	--	48	609	450	1,059	1,368
1959/60	MY Jun-May	1,368	1,118	7	2,493	558	--	49	607	502	1,109	1,384
1960/61	MY Jun-May	1,384	1,355	8	2,747	497	64	30	591	654	1,245	1,502
1961/62	MY Jun-May	1,502	1,232	6	2,741	504	56	44	604	718	1,320	1,421
1962/63	MY Jun-May	1,421	1,092	5	2,518	503	61	35	599	649	1,248	1,270
1963/64	MY Jun-May	1,270	1,147	4	2,421	488	65	29	581	846	1,427	994
1964/65	MY Jun-May	994	1,283	2	2,279	514	66	55	635	723	1,358	921
1965/66	MY Jun-May	921	1,316	1	2,238	518	62	146	725	852	1,577	661
1966/67	MY Jun-May	661	1,305	2	1,967	505	77	101	683	771	1,454	513
1967/68	MY Jun-May	513	1,508	1	2,021	518	71	37	626	765	1,391	630

Hard red wheat contracts—https://www.barchart.com/futures/quotes/KEN20/interactive-chart

[10]Wheat Data—https://www.ers.usda.gov/data-products/wheat-data/

Food Aids

PL-480 or Food for Peace

The Government of India has received generous PL-480 from the United States during 1956–1968. USAID and PL-480, 1961–1969[11]. The administrations of John F. Kennedy and Lyndon B. Johnson marked a revitalization of the US foreign assistance program, signified a growing awareness of the importance of humanitarian aid as a form of diplomacy, and reinforced the belief that American security was linked to the economic progress and stability of other nations.

Johnson with Gandhi, March 28, 1966 (White House Photo Office)

The United Nations: 17 Sustainable Development Goals (SDGs)

The 17 Sustainable Development Goals (SDGs) are an urgent call for action by all countries—developed and developing—in a global partnership. They recognize that ending poverty and other deprivations must go together with strategies that improve health and education, reduce inequality, and spur economic growth—all while tackling climate change and working to preserve our oceans and forests[12].

[11] USAID – PL480—https://history.state.gov/milestones/1961-1968/pl-480

[12] The 17 Goals—https://sdgs.un.org/goals

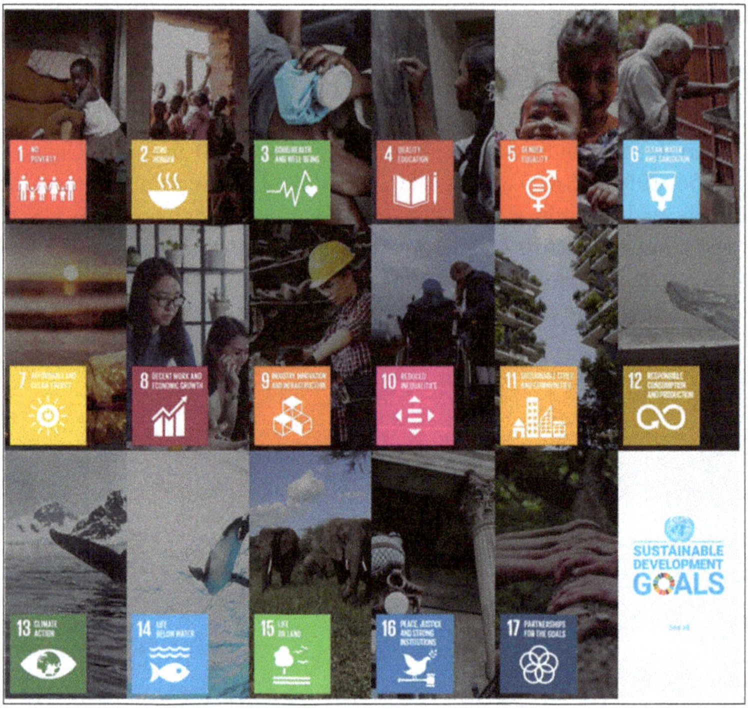

The Statistical Distributions of Commodity Prices in Both Real and Nominal Terms

Statistical properties of commodities[13]:

Decade	Variable	Mean	Std dev	Coeff. of variation	Skewness	Kurtosis
1970/ 1980	Banana, n	216.8	64.9	29.9	0.6	−0.6
	Banana, r	543.5	89.0	16.4	0.5	−0.0
	Cocoa, n	1777.9	1202.5	67.6	0.7	−1.0
	Cocoa, r	3970.1	1765.4	44.5	0.9	−0.2
	Coffee, n	2219.8	1412.4	63.6	1.2	0.8
	Coffee, r	5159.0	2172.5	42.1	1.9	4.1
	Cotton, n	1243.3	438.2	35.2	−0.0	−1.4
	Cotton, r	3059.3	666.7	21.8	1.7	2.9

(continued)

[13]The statistical distributions of commodity prices in both real and nominal terms—http://www.fao.org/3/y4344e/y4344e0i.htm

Decade	Variable	Mean	Std dev	Coeff. of variation	Skewness	Kurtosis
	Jute, n	322.7	57.9	17.9	1.0	−0.2
	Jute, r	845.0	201.4	23.8	0.3	−1.4
	Maize, n	98.6	26.7	27.1	−0.4	−0.8
	Maize, r	226.2	51.8	22.9	0.4	−0.7
	Palm oil, n	466.0	164.8	35.4	−0.1	−1.2
	Palm oil, r	1028.6	247.7	24.1	1.3	1.8
	Rape oil, n	496.7	170.7	34.4	0.0	−0.5
	Rape oil, r	1108.4	299.0	27.0	1.3	1.7
	Rapeseed, n	263.6	81.3	30.8	−0.3	−0.6
	Rapeseed, r	592.2	148.5	25.1	1.2	0.9
	Rice, n	285.4	127.2	44.6	0.6	−0.0
	Rice, r	682.2	254.6	37.3	2.0	3.3
	Rubber, n	676.0	271.5	40.2	0.5	−0.5
	Rubber, r	1641.9	344.7	21.0	1.1	2.1
	Sisal, n	476.6	271.9	57.0	0.9	0.0
	Sisal, r	1141.2	553.3	48.5	1.6	1.5
	Soybeans, n	237.9	69.5	29.2	0.0	−0.1
	Soybeans, r	540.2	149.1	27.6	2.4	9.0
	Soymeal, n	196.3	68.9	35.1	1.5	5.4
	Soymeal, r	451.0	181.5	40.3	3.4	14.5
	Sugar, n	243.4	198.8	81.7	2.5	7.4
	Sugar, r	595.4	424.3	71.3	2.5	7.1
	Sunflmeal, n	153.3	44.8	29.2	0.8	2.4
	Sunflmeal, r	353.6	118.6	33.5	2.6	8.4
	Tea, n	1532.3	607.3	39.6	1.7	3.7
	Tea, r	3802.9	892.6	23.5	2.4	8.4
	Wheat, n	123.5	43.8	35.5	−0.1	−1.0
	Wheat, r	284.0	87.5	30.8	1.7	2.5
1980/ 1990	Banana, n	406.3	83.4	20.5	1.2	2.2
	Banana, r	557.4	104.4	18.7	0.7	−0.1
	Cocoa, n	2060.1	428.7	20.8	0.3	0.6
	Cocoa, r	2882.9	769.7	26.7	−0.1	−0.5
	Coffee, n	2892.9	569.4	19.7	0.9	1.2
	Coffee, r	4031.8	990.4	24.6	0.2	−0.3
	Cotton, n	1625.3	319.7	19.7	−0.4	−0.3
	Cotton, r	2254.3	524.6	23.3	−0.3	−1.1
	Jute, n	361.1	138.7	38.4	2.5	5.8
	Jute, r	507.7	247.0	48.6	2.7	6.4
	Maize, n	112.8	22.6	20.0	−0.3	−0.5
	Maize, r	154.5	42.2	27.3	0.1	−1.1
	Palm oil, n	448.4	157.0	35.0	0.9	0.5
	Palm oil, r	622.5	273.7	44.0	0.9	0.2
	Rape oil, n	454.1	124.7	27.5	0.9	0.6

(continued)

Decade	Variable	Mean	Std dev	Coeff. of variation	Skewness	Kurtosis
	Rape oil, r	626.1	232.4	37.1	1.0	0.1
	Rapeseed, n	267.5	58.7	21.9	0.1	−0.0
	Rapeseed, r	369.1	115.3	31.2	0.1	−0.7
	Rice, n	294.4	92.7	31.5	1.0	0.2
	Rice, r	404.1	132.7	32.8	1.0	0.4
	Rubber, n	997.3	203.3	20.4	1.2	0.8
	Rubber, r	1379.5	313.5	22.7	0.8	−0.5
1980/1990	Sisal, n	594.5	85.8	14.4	1.3	1.1
	Sisal, r	822.7	145.6	17.7	0.4	−0.1
	Soybeans, n	259.6	43.1	16.6	0.6	−0.6
	Soybeans, r	351.0	71.6	20.4	0.5	−0.5
	Soymeal, n	219.3	41.8	19.1	0.4	−0.4
	Soymeal, r	294.4	56.1	19.0	0.5	−0.3
	Sugar, n	236.6	168.3	71.1	1.9	3.7
	Sugar, r	320.9	224.0	69.8	2.0	3.6
	Sunflmeal, n	143.2	38.5	26.9	0.3	−0.5
	Sunflmeal, r	193.7	58.6	30.3	0.6	−0.9
	Tea, n	2131.0	564.5	26.5	1.8	2.8
	Tea, r	2991.0	1035.2	34.6	1.6	2.0
	Wheat, n	149.2	23.1	15.5	−0.5	−0.7
	Wheat, r	203.2	39.8	19.6	−0.4	−1.2
1990/2000	Banana, n	480.5	104.3	21.7	0.7	0.5
	Banana, r	514.0	115.3	22.4	0.7	0.3
	Cocoa, n	1340.0	226.5	16.9	0.0	−0.9
	Cocoa, r	1434.4266.9	18.6	0.5	−0.5	
	Coffee, n	2095.6	805.6	38.4	0.9	0.1
	Coffee, r	2238.4	860.0	38.4	0.9	0.2
	Cotton, n	1632.6	307.1	18.8	0.3	0.0
	Cotton, r	1741.3	306.7	17.6	−0.1	−0.5
	Jute, n	329.7	80.7	24.5	0.6	−0.6
	Jute, r	351.1	80.2	22.8	0.5	−0.8
	Maize, n	111.2	23.9	21.5	2.0	4.9
	Maize, r	118.1	24.1	20.4	1.6	3.7
	Palm oil, n	477.6	125.7	26.3	0.3	−1.3
	Palm oil, r	509.1	139.1	27.3	0.4	−0.9
	Rape oil, n	507.2	100.2	19.8	0.1	−1.3
	Rape oil, r	541.5	117.1	21.6	0.2	−1.2
	Rapeseed, n	248.9	47.4	19.0	−0.0	−1.5
	Rapeseed, r	265.7	55.1	20.7	0.0	−1.3
	Rice, n	285.5	41.6	14.6	0.1	−0.3
	Rice, r	304.1	44.4	14.6	0.2	−0.5
	Rubber, n	993.4	66.9	6.7	0.2	−1.0
	Rubber, r	1061.2	71.4	6.7	−0.2	−0.5

(continued)

Decade	Variable	Mean	Std dev	Coeff. of variation	Skewness	Kurtosis
	Sisal, n	700.4	109.6	15.6	−0.2	−0.6
	Sisal, r	749.0	121.5	16.2	−0.3	−0.3
	Soybeans, n	250.6	33.7	13.5	0.5	−0.2
	Soybeans, r	267.1	39.4	14.8	0.2	−0.5
	Soymeal, n	205.5	40.0	19.5	0.6	−0.2
	Soymeal, r	218.8	43.8	20.0	0.6	−0.2
	Sugar, n	234.7	51.3	21.9	−0.1	−0.4
	Sugar, r	251.0	55.6	22.2	−0.1	−0.1
	Sunflmeal, n	116.8	24.5	21.0	−0.2	−0.7
	Sunflmeal, r	124.3	26.3	21.2	−0.3	−0.7
	Tea, n	2002.6	334.3	16.7	1.1	1.6
	Tea, r	2147.0	409.0	19.0	1.2	2.1
	Wheat, n	146.4	32.2	22.0	1.1	1.3
	Wheat, r	155.5	32.3	20.8	0.7	0.8

(*r* real, *n* nominal)

The kurtosis for a normal distribution is 3.

Poverty Thresholds for 2019 by Size of Family and Number of Related Children Under 18 Years

Size of family unit	Weighted average thresholds	Related children under 18 years								
		None	One	Two	Three	Four	Five	Six	Seven	Eight or more
One person (unrelated individual):	13,011									
Under age 65 years	13,300	13,300								
Aged 65 years and older	12,261	12,261								
Two people:	16,521									
House-holder under age 65 years	17,196	17,120	17,622							
House-holder aged 65 years and older	15,468	15,453	17,555							
Three people	20,335	19,998	20,578	20,598						
Four people	26,172	26,370	26,801	25,926	26,017					
Five people	31,021	31,800	32,263	31,275	30,510	30,044				
Six people	35,129	36,576	36,721	35,965	35,239	34,161	33,522			
Seven people	40,016	42,085	42,348	41,442	40,811	39,635	38,262	36,757		

(continued)

Size of family unit	Weighted average thresholds	Related children under 18 years								
		None	One	Two	Three	Four	Five	Six	Seven	Eight or more
Eight people	44,461	47,069	47,485	46,630	45,881	44,818	43,470	42,066	41,709	
Nine people or more	52,875	56,621	56,895	56,139	55,503	54,460	53,025	51,727	51,406	49,426
Source: US Census Bureau[14]										

[14]Source: https://www.census.gov/data/tables/time-series/demo/income-poverty/historical-poverty-thresholds.html

Appendix 2: Agriculture

Agricultural Data Surveys

USDA NASS

The US Department of Agriculture—National Analytics Statistics Services: Online Survey Form[15]

[15]NASS Online Survey—https://www.agcounts.usda.gov/static/cawi/layouts/cawi/breeze/index.html

C. Vuppalapati, *Artificial Intelligence and Heuristics for Enhanced Food Security*, International Series in Operations Research & Management Science 331,
https://doi.org/10.1007/978-3-031-08743-1

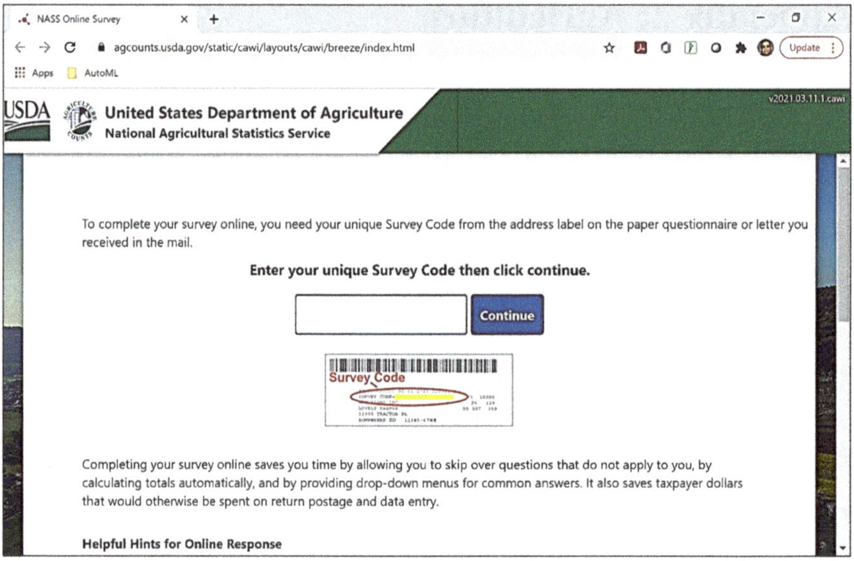

USDA: The Foreign Agricultural Service (FAS) Reports and Databases

World Agricultural Production

Current: https://www.fas.usda.gov/data/world-agricultural-production

Archive: https://usda.library.cornell.edu/concern/publications/5q47rn72z?locale=en

USDA's Foreign Agricultural Service (FAS) publishes a monthly report on crop acreage, yield, and production in major countries worldwide. Sources include reporting from FAS's worldwide offices, official statistics of foreign governments, and analysis of economic data and satellite imagery. The reports reflect official USDA estimates released in the monthly World Agricultural Supply and Demand Estimates (WASDE).

World Markets and Trade

Current: https://www.fas.usda.gov/data

Archive: https://usda.library.cornell.edu/catalog?f%5Bmember_of_collections_ssim%5D%5B%5D=Foreign+Agricultural+Service&locale=en

USDA's Foreign Agricultural Service (FAS) publishes monthly and quarterly reports which include data on US and global trade, production, consumption, and stocks, as well as analysis of developments affecting world trade in oilseeds, grains, cotton, livestock, and poultry. The reports reflect official USDA estimates released in the monthly World Agricultural Supply and Demand Estimates (WASDE).

Global Agricultural Information Network (GAIN)

https://www.fas.usda.gov/databases/global-agricultural-information-network-gain

USDA's Foreign Agricultural Service (FAS) provides timely reports on foreign markets through the Global Agriculture Information Network (GAIN) database. An average of 2,000 reports are added each year, with reports going back to 1995. GAIN reports are compiled by FAS' global market intelligence network, which includes FAS foreign service officers and locally engaged staff in over 90 overseas offices worldwide. They provide on-the-ground intelligence, insight, and analysis on nearly 200 countries, delivering information on foreign agricultural markets, crop conditions, and agro-political dynamics of interest to US agriculture. GAIN reports contain assessments of commodity and trade issues made by USDA staff and are not necessarily statements of official US government policy.

Production, Supply, and Distribution (PS&D) Online

https://apps.fas.usda.gov/psdonline/app/index.html#/app/home

PSD Online is the public repository for USDA's Official Production, Supply, and Distribution forecast data and reports for key agricultural commodities. PSD Online data are reviewed and updated monthly by an interagency committee chaired by USDA's World Agricultural Outlook Board (WAOB). The committee consists of representatives from Foreign Agricultural Service (FAS), the Economic Research Service (ERS), the Farm Service Agency (FSA), and the Agricultural Marketing Service (AMS).

USDA and NASA Global Agricultural Monitoring (GLAM)

https://glam1.gsfc.nasa.gov/

The USDA and NASA Global Agricultural Monitoring (GLAM) system provides near real-time and science-quality moderate-resolution imaging spectroradiometer (MODIS) Normalized Difference Vegetation Index (NDVI) from the satellites Terra and Aqua. The public can view and retrieve MODIS 8-day composited, global NDVI satellite imagery and time series data. GLAM was developed by NASA's Global Inventory Modeling and Mapping Studies (GIMMS) group for USDA's Foreign Agricultural Service.

Global Agricultural and Disaster Assessment System (GADAS)

https://geo.fas.usda.gov/GADAS/index.html

USDA's Foreign Agricultural Service (FAS) provides the Global Agricultural and Disaster Assessment System (GADAS), a web-based Geographic Information System (GIS) tool which integrates a vast array of highly detailed earth observation data streams, particularly targeted towards agricultural and disaster assessment analysis. GADAS is an interactive website which provides analysts with a wide variety of routine geospatial products (maps, charts, tables) they require for comprehensive situational investigations and recurring assessments. GADAS is an interactive global web analysis system, capable of displaying, comparing, analyzing, and sharing geospatial data[16].

Export Sales Reporting

https://apps.fas.usda.gov/esrquery/

USDA's Export Sales Reporting Program monitors US agricultural export sales on a daily and weekly basis. Export sales reporting provides a constant stream of up-to-date market information for 40 US agricultural commodities sold abroad. The weekly US Export Sales report is the most currently available source of US export sales data. The data is used to analyze overall levels of export demand, determine where markets exit, and assess the relative position of US commodities in foreign markets

Global Agricultural Trade System (GATS)

https://apps.fas.usda.gov/gats/default.aspx

The Global Agricultural Trade System (GATS) is a searchable database containing monthly US Census Bureau trade data organized by agricultural commodity and agricultural-related product groups. Trade data is searchable by partner countries and partner groups. Historical US agricultural trade data is available back to 1967. In addition, UN trade statistics (UN Comtrade) may be queried through GATS. UN trade data is available for nearly 200 countries or areas, dating from the inception of the Harmonized System (HS) of trade codes in 1989 to present. The database is continuously updated. US trade data is updated monthly according to the US Census Bureau's reporting system. UN Comtrade data are updated in GATS after nationally submitted data to the UN are standardized by the UN Statistical Division and added to the UN Comtrade database.

[16]World Agricultural Production—https://apps.fas.usda.gov/psdonline/circulars/production.pdf

Data Sources

UN DATA MARTS

Source: http://data.un.org/Explorer.aspx?d=FAO

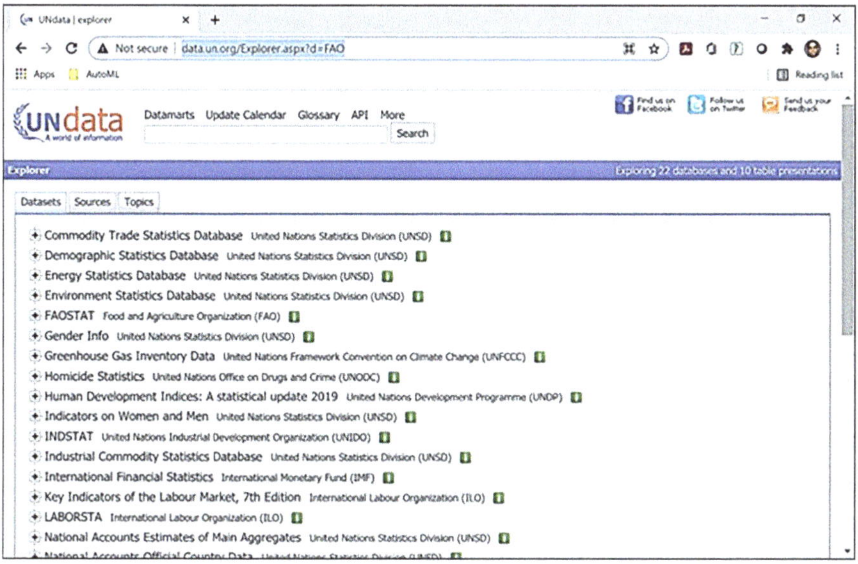

World Development Indicator tool[17]:

[17]The World Bank Development Indicator—https://databank.worldbank.org/reports.aspx?source=2&type=metadata&series=SI.POV.DDAY

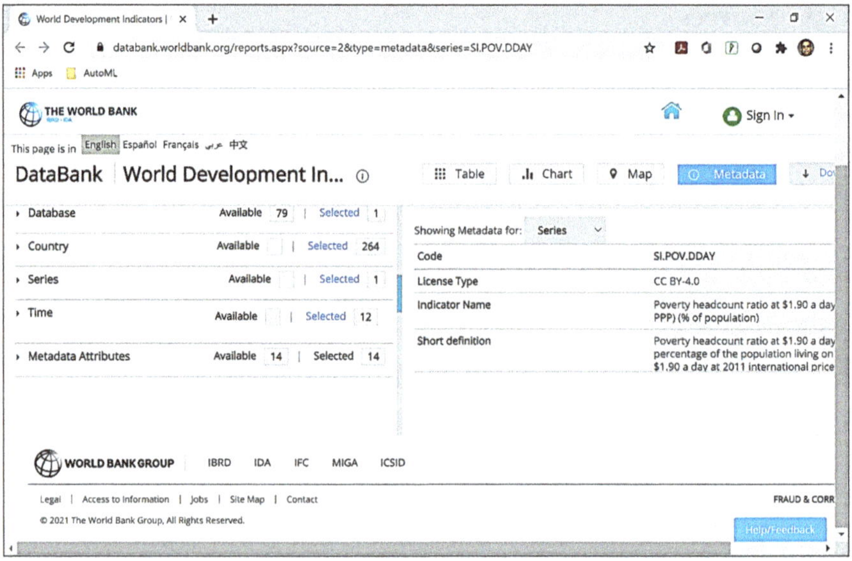

Food and Agriculture Organization of the United Nations

FAOSTAT Data[18]: FAO provides the following data domains: Production, Prices, Emissions-Agriculture, Inputs, Trade, Population, Emissions-Land Use, Investment, Macro Economic Data, and others.

[18]FAPSTAT—http://www.fao.org/faostat/en/#data

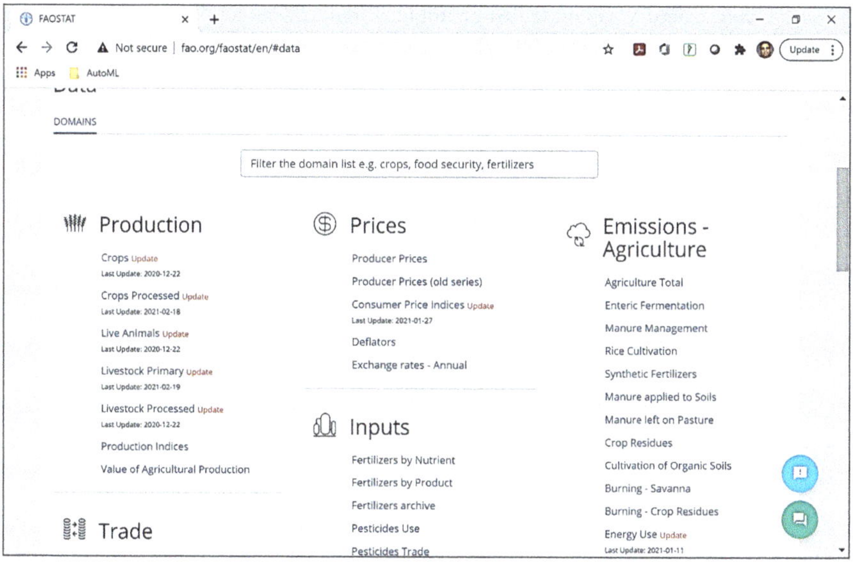

Conversion Factors

Here are a few useful conversion factors[19]:

- 1 bushel of wheat = 60 pounds or 77.2 kg/hectoliter
- 1 metric ton = 2,204.622 pounds
- Bushels × 0.0272155 = metric tons
- Metric tons × 36.7437 = bushels
- Price per bushel × 36.7437 = price per metric ton
- Price per metric ton × 0.0272155 = price per bushel
- 1 acre = 0.4047 hectares
- 1 hectare = 2.4710 acres
- Bushels/acre × 0.06725 = metric tons/hectare
- Metric tons/hectare × 14.87 = bushels/acre

National Dairy Development Board (NDDB), India

The National Dairy Development Board's (NDDB)[20] creation is rooted in the conviction that our nation's socioeconomic progress lies largely on the development of rural India.

[19] The USDA Conversion Factors—https://www.ers.usda.gov/data-products/wheat-data/documentation/

[20] About NDDB—https://www.nddb.coop/

The Dairy Board was created to promote, finance, and support producer-owned and controlled organizations. NDDB's programs and activities seek to strengthen farmer-owned institutions and support national policies that are favorable to the growth of such institutions. Fundamental to NDDB's efforts are cooperative strategies and principles.

NDDB's efforts transformed India's rural economy by making dairying a viable and profitable economic activity for millions of milk producers while addressing the country's need for self-sufficiency in milk production.

NDDB has been reaching out to dairy farmers by implementing other income-generating innovative activities and offering them sustainable livelihood.

Milk production[21] and per capita availability of milk in India[22]

Year	Production (million tons)	Per capita availability (g/day)
1991–1992	55.6	178
1992–1993	58.0	182
1993–1994	60.6	186
1994–1995	63.8	192
1995–1996	66.2	195
1996–1997	69.1	200
1997–1998	72.1	205
1998–1999	75.4	210
1999–2000	78.3	214
2000–2001	80.6	217
2001–2002	84.4	222
2002–2003	86.2	224
2003–2004	88.1	225
2004–2005	92.5	233
2005–2006	97.1	241
2006–2007	102.6	251
2007–2008	107.9	260
2008–2009	112.2	266
2009–2010	116.4	273
2010–2011	121.8	281
2011–2012	127.9	290
2012–2013	132.4	299
2013–2014	137.7	307
2014–2015	146.3	322
2015–2016	155.5	337
2016–2017	165.4	355
2017–2018	176.3	375
2018–2019	187.7	394

[21] Milk Production in India—https://www.nddb.coop/information/stats/milkprodindia

[22] NDDB Daily Milk Data—https://www.nddb.coop/sites/default/files/statistics/Mp%20India-ENG-2019.pdf

Worldwide: Artificial Intelligence (AI) Readiness

In terms of readiness for AI, countries appear to fall into four groups [23]—varying conditions among countries imply different degrees of AI adoption and absorption and therefore economic impact.

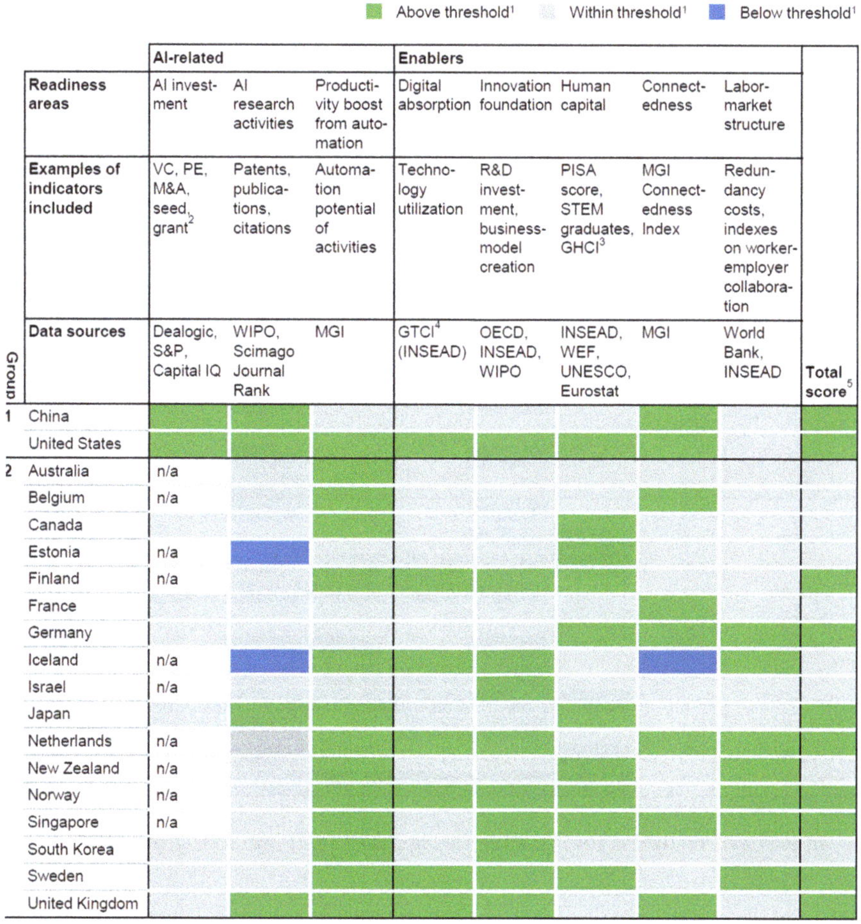
Above threshold¹
Within threshold¹
Below threshold¹
AI-related
Enablers
Readiness areas
AI investment
AI research activities
Productivity boost from automation
Digital absorption
Innovation foundation
Human capital
Connectedness
Labor-market structure
Examples of indicators included
VC, PE, M&A, seed, grant²
Patents, publications, citations
Automation potential of activities
Technology utilization
R&D investment, business-model creation
PISA score, STEM graduates, GHCI³
MGI Connectedness Index
Redundancy costs, indexes on worker-employer collaboration
Data sources
Dealogic, S&P, Capital IQ
WIPO, Scimago Journal Rank
MGI
GTCI⁴ (INSEAD)
OECD, INSEAD, WIPO
INSEAD, WEF, UNESCO, Eurostat
MGI
World Bank, INSEAD
Total score⁵
Group
1 China
United States
2 Australia
Belgium
Canada
Estonia
Finland
France
Germany
Iceland
Israel
Japan
Netherlands
New Zealand
Norway
Singapore
South Korea
Sweden
United Kingdom
n/a

[23] AI Readiness—https://www.mckinsey.com/~/media/McKinsey/Featured%20Insights/Artificial%20Intelligence/Notes%20from%20the%20frontier%20Modeling%20the%20impact%20of%20AI%20on%20the%20world%20economy/MGI-Notes-from-the-AI-frontier-Modeling-the-impact-of-AI-on-the-world-economy-September-2018.ashx

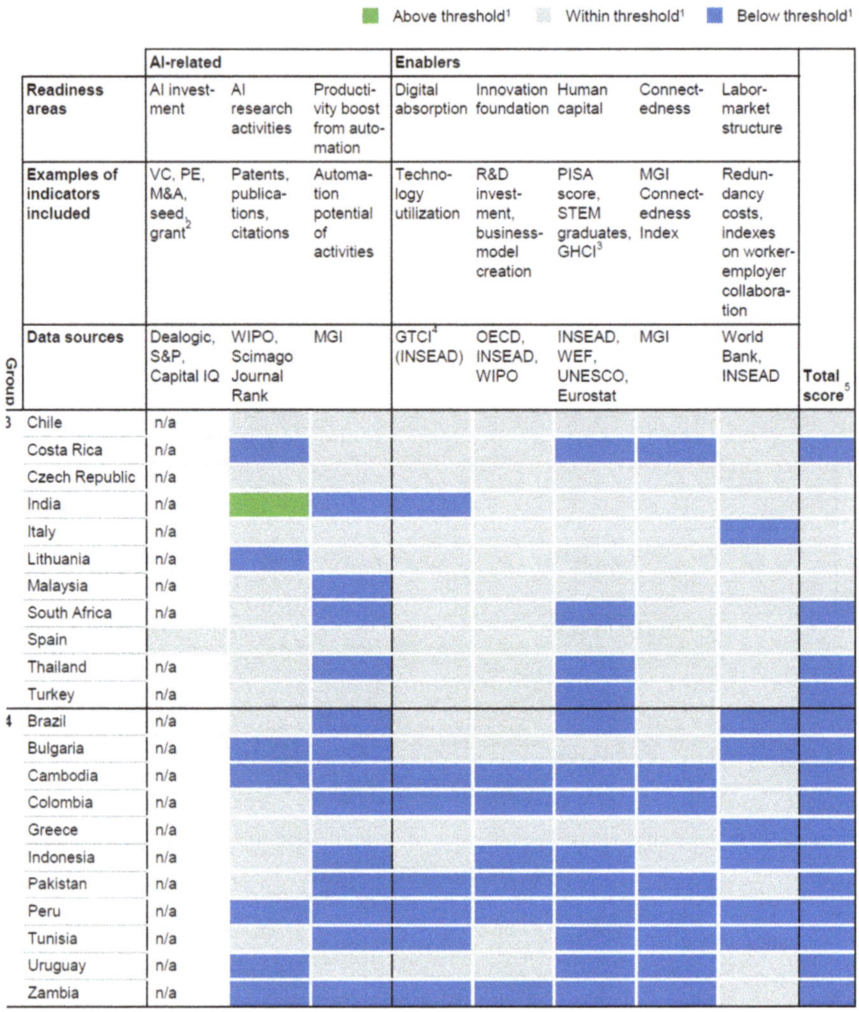

In terms of readiness for AI, countries appear to fall into four groups[24]

[24] AI Readiness—https://www.mckinsey.com/~/media/McKinsey/Featured%20Insights/Artificial%20Intelligence/Notes%20from%20the%20frontier%20Modeling%20the%20impact%20of%20AI%20on%20the%20world%20economy/MGI-Notes-from-the-AI-frontier-Modeling-the-impact-of-AI-on-the-world-economy-September-2018.ashx

Appendix 3: Data World

Global Historical Climatology Network Monthly (GHCNm)

The Global Historical Climatology Network monthly (GHCNm[25]) dataset provides monthly climate summaries from thousands of weather stations around the world. The initial version was developed in the early 1990s, and subsequent iterations were released in 1997, 2011, and most recently 2018. The period of record for each summary varies by station, with the earliest observations dating to the eighteenth century. Some station records are purely historical and are no longer updated, but many others are still operational and provide short time delay updates that are useful for climate monitoring. The current version (GHCNm version 4) consists of mean monthly temperature data, as well as a beta release of monthly precipitation data.

NCEI uses GHCN monthly to monitor long-term trends in temperature and precipitation. It has also been employed in several international climate assessments, including the Intergovernmental Panel on Climate Change 4th Assessment Report, the Arctic Climate Impact Assessment, and the "State of the Climate" report published annually by the Bulletin of the American Meteorological Society.

Select a monthly time series: historical observations

To retrieve historical data of a weather station, please consider the following[26]:

- Step 1: Select database for historical observations.

In the selection below, the minimum temperature was selected GHCN-M (all).

[25] GHCNm—https://www.ncei.noaa.gov/products/land-based-station/global-historical-climatology-network-monthly

[26] GHCNm—https://climatedata-catalogue.wmo.int/explore

C. Vuppalapati, *Artificial Intelligence and Heuristics for Enhanced Food Security*, International Series in Operations Research & Management Science 331,
https://doi.org/10.1007/978-3-031-08743-1

GHCN-M (adjusted)	GHCN-M (all)	other
○precipitation	○precipitation	○GLOSS sealevel
○mean temperature	○mean temperature	○world river discharge (RivDis)
○minimum temperature	◉minimum temperature	○USA river discharge (HCDN)
○maximum temperature	○maximum temperature	○N-America snowcourses (NRCS)
(full lists)	○sealevel pressure	○european SLP (ADVICE)

- Step 2: Select stations.

Select a location on the world map to obtain the data for the nearest weather station.

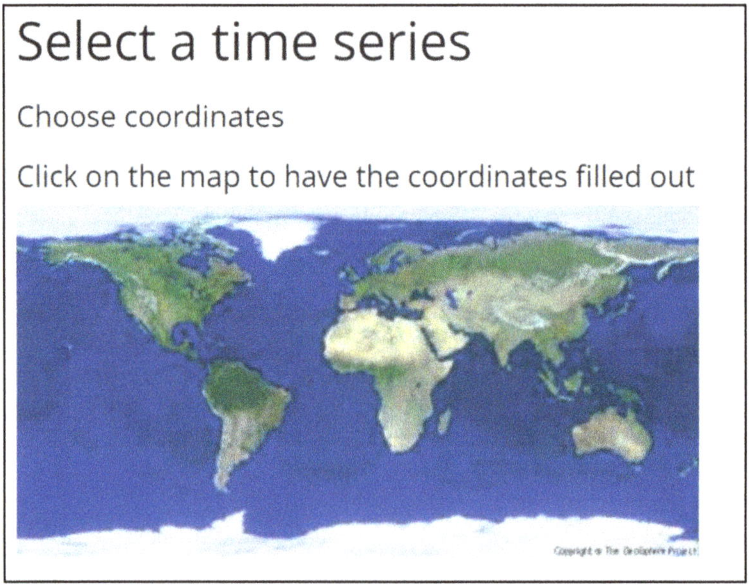

In the coordinates screen, the location of Brazil was selected and the coordinates are populated in stations near fields.

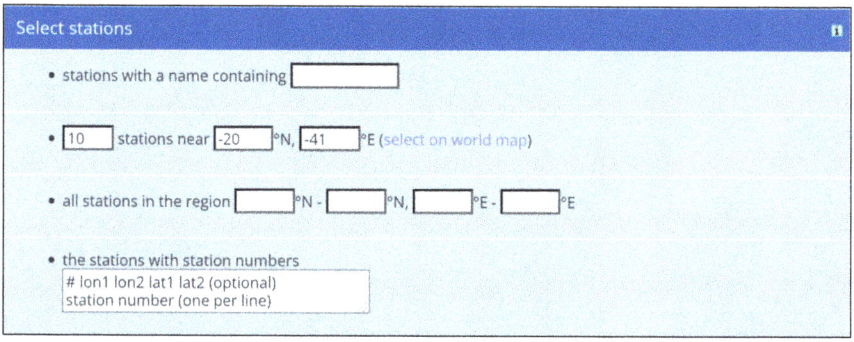

- Step 3: Choose the time and distance.

Provide stations to be retrieved with the time series data keyed in. The following selection can be interpreted as: "*At least 10 years of data in the monthly season starting in all months in year 2000–2021.*"

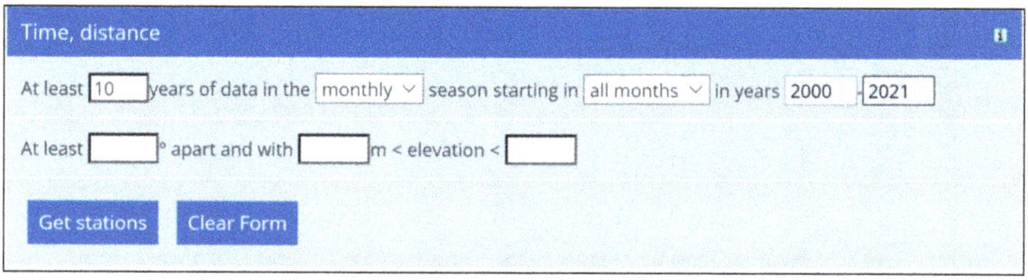

Select the station of the nearest region for which data is to be analyzed. In the above case, I have selected

Montes Claros (Brazil)
Coordinates: −16.72N, −43.87E, 646.0m (prob: 849m)
WMO station code: 83437 (get data)
Associated with urban area (pop. 152000)
Terrain: hilly WARM GRASS/SHRUB
Found 18 years with data in 2002–2019

The selected data is as described in the figure below:

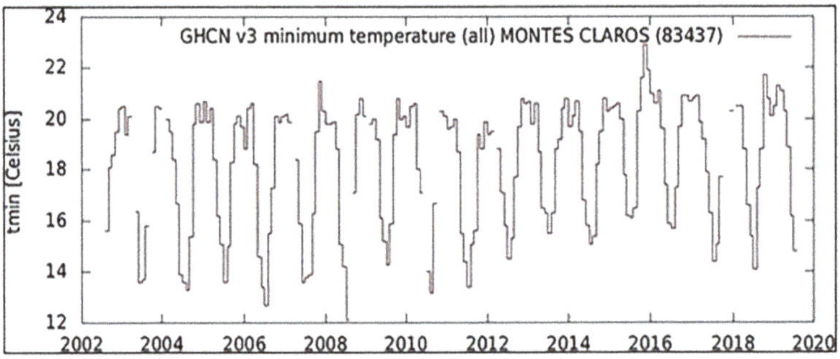

Montes Claros (Brazil), coordinates, −16.72N, −43.87E, 646.0m (prob: 849m); WMO station code, 83437 Montes Claros, tmin [Celsius] daily minimum temperature (unadjusted) from GHCN-M version 3.3.0.20190817 (eps, pdf, metadata, raw data, netcdf)

The raw data looks as follows: https://climexp.knmi.nl/data/ma83437.dat

```
# file :: ./data/ma83781.dat
# history ::  2022-08-01  0:59:50 ./bin/climatology ./data/ma83781.dat\n 2022-08-01  0:59:49 dat2nc
./data/ma83781.dat n 83781 ./data/ma83781.nc.246901\n
# institution :: KNMI Climate Explorer and NOAA/NCEI
# latitude ::  -23.50 degrees_north
# longitude ::  -46.62 degrees_east
# references :: Durre, I., M.J. Menne, and R.S. Vose, 2008: Strategies for evaluating quality assurance
procedures. Journal of Applied Meteorology and Climatology, 47(6), 1785-1791.\\nLawrimore, J. H., M. J. Menne, B.
E. Gleason, C. N. Williams, D. B. Wuertz, R. S. Vose, and J. Rennie (2011), An overview of the Global Historical
Climatology Network monthly mean temperature data set, version 3, J. Geophys. Res., 116, D19121,
doi:10.1029/2011JD016187.
# scripturl01 :: http://climexp.knmi.nl/getminall.cgi?STATION=SAO_PAULO&WMO=83781&id=someone@somewhere
# source_doi :: 10.7289/V5X34VDR accessed Tue May 10 08:42:28 UTC 2022
# source_url :: https://www.ncdc.noaa.gov/ghcnm/v3.php
# station_code :: 83781
# station_country :: BRAZIL
# station_name :: SAO_PAULO
#2000 + date        mean          2.5%          17%          50%          83%
97.5%
20000101        19.6235       18.0850       18.6560       19.5998       20.4220
21.4150
20000201        19.8941       17.8700       19.0000       19.9998       20.8610
21.4150
20000301        19.1533       17.6625       18.6050       19.0998       19.8850
```

Datasets for Brazil

Dataset access: INMET :: BDMEP (https://bdmep.inmet.gov.br/)

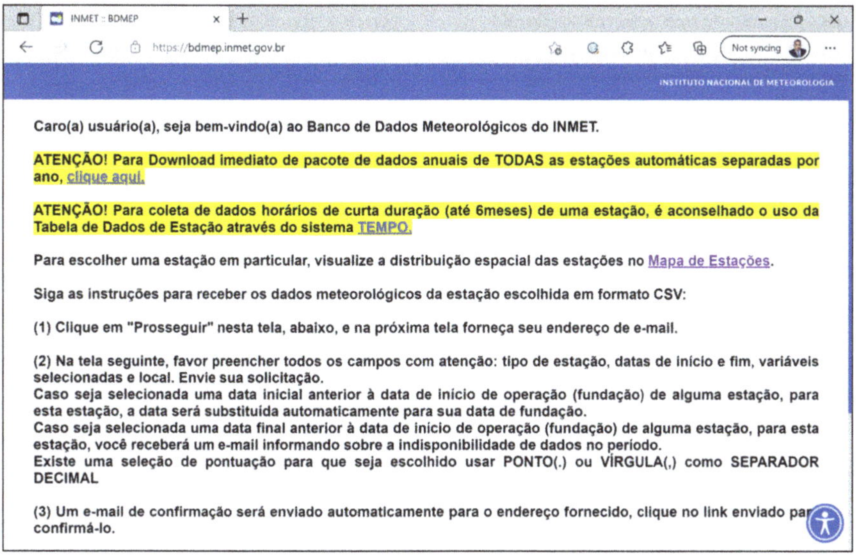

Portuguese	English
Caro(a) usuário(a), seja bem-vindo(a) ao Banco de Dados Meteorológicos do INMET.	Dear user, welcome to the INMET Meteorological Database.
ATENÇÃO! Para Download imediato de pacote de dados anuais de TODAS as estações automáticas separadas por ano, clique aqui.	ATTENTION! For immediate download of annual data package of ALL automatic stations separated by year, click here.
ATENÇÃO! Para coleta de dados horários de curta duração (até 6meses) de uma estação, é aconselhado o uso da Tabela de Dados de Estação através do sistema TEMPO.	ATTENTION! For short-term (up to 6 months) hourly data collection from a station, it is advisable to use the Station Data Table through the TEMPO system.
Para escolher uma estação em particular, visualize a distribuição espacial das estações no Mapa de Estações.	To choose a particular station, view the spatial distribution of stations on the Station Map.
Siga as instruções para receber os dados meteorológicos da estação escolhida em formato CSV:	Follow the instructions to receive the weather data of the chosen station in CSV format:
(1) Clique em "Prosseguir" nesta tela, abaixo, e na próxima tela forneça seu endereço de e-mail.	(1) Click "Continue" on this screen below, and on the next screen provide your email address.
(2) Na tela seguinte, favor preencher todos os campos com atenção: tipo de estação, datas de início e fim, variáveis selecionadas e local. Envie sua solicitação.	(2) On the next screen, please fill in all the fields carefully: type of station, start and end dates, selected variables and location. Submit your request.
Caso seja selecionada uma data inicial anterior à data de início de operação (fundação) de alguma estação, para esta estação, a data será substituída automaticamente para sua data de fundação.	If an initial date is selected before the start of operation (foundation) date of any station, for this station, the date will be automatically substituted for its foundation date.
Caso seja selecionada uma data final anterior à data de início de operação (fundação) de alguma estação, para esta estação, você receberá um e-mail informando sobre a indisponibilidade de dados no período.	If an end date is selected before the start of operation (foundation) date of any station, for this station, you will receive an email informing you about the unavailability of data in the period.
Existe uma seleção de pontuação para que seja escolhido usar PONTO(.) ou VÍRGULA(,) como SEPARADOR DECIMAL	There is a punctuation selection so that you can choose to use DOT(.) or COMMA(,) as DECIMAL SEPARATOR
(3) Um e-mail de confirmação será enviado automaticamente para o endereço fornecido, clique no link enviado para confirmá-lo.	(3) A confirmation email will be sent automatically to the provided address, click on the sent link to confirm it.
Para efeito de controle, existe uma fila de processamento e quando sua requisição for iniciada você receberá um e-mail alertando.	For control purposes, there is a processing queue and when your request is started you will receive an email alerting you.
Ao término do processo, um e-mail de conclusão será enviado contendo o link de acesso aos dados selecionados; após 48h esses dados serão apagados.	At the end of the process, a completion email will be sent containing the access link to the selected data; after 48 h these data will be deleted.
Para mais de uma requisição, favor aguardar o término do primeiro processamento.	For more than one request, please wait for the first processing to finish.

The National Meteorological Service of Brazil[27], hereafter INMET, has provided the Brazilian Climate data (Brazilian-CD) online and freely available as text CSV to the public. The Brazilian-CD comprises data from all observing weather stations operated and maintained by INMET, either automatic or conventional stations, with the total number varying according to year, from about 400 in 2000 to 834 stations in 2020. The WMO SMM-CD_NRP was used to assess the temperature, humidity, and precipitation data of the Brazilian-CD. The averaged stewardship maturity rating levels for these categories are 2.5 and 1.7, respectively. The assessment was originally made by an expert from INMET, which was moderated by one member of the WMO ET-DDS team. The online publication of the Brazilian CD meets important criteria for publicity, such as data access, portability, and preservation, in the operational data management category, but also in this category there is room for improving the documentation and data integrity, due to lack of information, or even application, of such aspects in the website. Regarding the data stewardship category, it was noticed that no quality assurance or quality control procedures are informed in the online documentation of this CD, but a further communication to the assessment point-of-contact (POC) had clarified that the dataset is under a routine procedure of quality control. Furthermore, despite the CD being provided by INMET, no explicit information on governance or POC is given, which lowers the score for this aspect. The Brazilian CD is available online with minimal metadata information, regarding the observing station location, altitude, and period of operation. Besides, climate normal parameters can be found as figures and table, by simply consulting the map of the station, though such information is not integrated in the provided CSV downloaded file. Nonetheless, substantial progress has been made recently by INMET for improving the informational content of the Brazilian Climate data to users, as nationally reported and also referred by relevant centers as NOAA CPC and IRI.

[27] World Metrological Data—https://climatedata-catalogue.wmo.int/beta/datasets-brazil

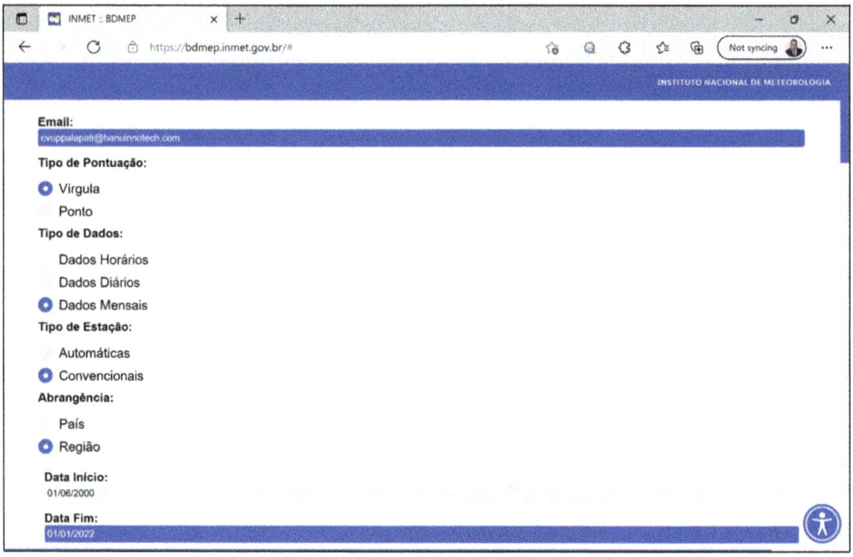
INMET :: BDMEP
https://bdmep.inmet.gov.br/#
Not syncing
INSTITUTO NACIONAL DE METEOROLOGIA
Email:
cvuppalapati@hanuinnotech.com
Tipo de Pontuação:
Vírgula
Ponto
Tipo de Dados:
Dados Horários
Dados Diários
Dados Mensais
Tipo de Estação:
Automáticas
Convencionais
Abrangência:
País
Região
Data Início:
01/06/2000
Data Fim:
01/01/2022

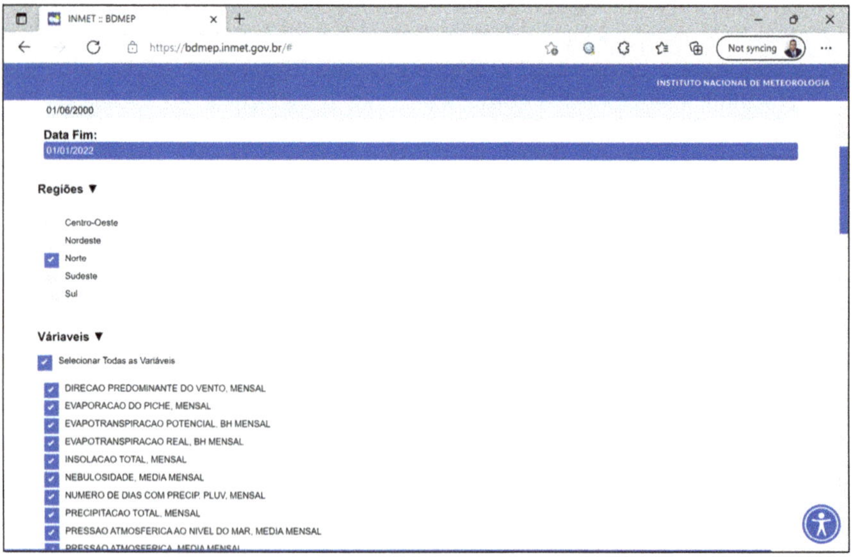
INMET :: BDMEP
https://bdmep.inmet.gov.br/#
Not syncing
INSTITUTO NACIONAL DE METEOROLOGIA
01/06/2000
Data Fim:
01/01/2022
Regiões ▼
Centro-Oeste
Nordeste
Norte
Sudeste
Sul
Váriaveis ▼
Selecionar Todas as Variáveis
DIRECAO PREDOMINANTE DO VENTO, MENSAL
EVAPORACAO DO PICHE, MENSAL
EVAPOTRANSPIRACAO POTENCIAL, BH MENSAL
EVAPOTRANSPIRACAO REAL, BH MENSAL
INSOLACAO TOTAL, MENSAL
NEBULOSIDADE, MEDIA MENSAL
NUMERO DE DIAS COM PRECIP. PLUV, MENSAL
PRECIPITACAO TOTAL, MENSAL
PRESSAO ATMOSFERICA AO NIVEL DO MAR, MEDIA MENSAL

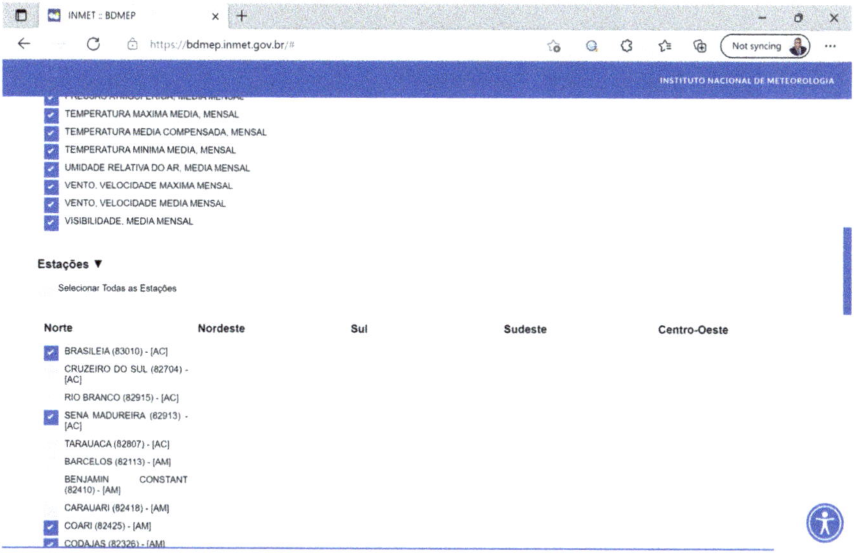
INMET :: BDMEP
https://bdmep.inmet.gov.br/#
Not syncing
INSTITUTO NACIONAL DE METEOROLOGIA
TEMPERATURA MAXIMA MEDIA, MENSAL
TEMPERATURA MEDIA COMPENSADA, MENSAL
TEMPERATURA MINIMA MEDIA, MENSAL
UMIDADE RELATIVA DO AR, MEDIA MENSAL
VENTO, VELOCIDADE MAXIMA MENSAL
VENTO, VELOCIDADE MEDIA MENSAL
VISIBILIDADE, MEDIA MENSAL
Estações ▼
Selecionar Todas as Estações
Norte
Nordeste
Sul
Sudeste
Centro-Oeste
BRASILEIA (83010) - [AC]
CRUZEIRO DO SUL (82704) - [AC]
RIO BRANCO (82915) - [AC]
SENA MADUREIRA (82913) - [AC]
TARAUACA (82807) - [AC]
BARCELOS (82113) - [AM]
BENJAMIN CONSTANT (82410) - [AM]
CARAUARI (82418) - [AM]
COARI (82425) - [AM]
CODAJAS (82326) - [AM]

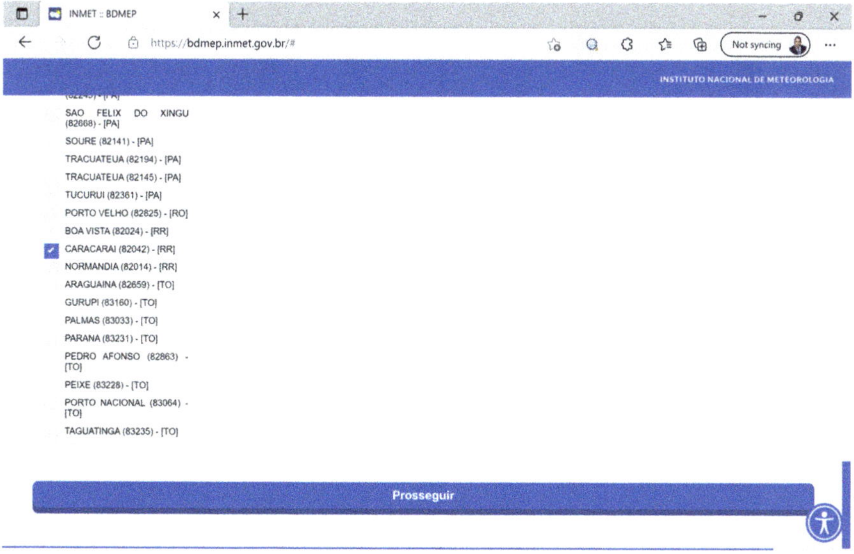
INMET :: BDMEP
https://bdmep.inmet.gov.br/#
Not syncing
INSTITUTO NACIONAL DE METEOROLOGIA
SAO FELIX DO XINGU (82668) - [PA]
SOURE (82141) - [PA]
TRACUATEUA (82194) - [PA]
TRACUATEUA (82145) - [PA]
TUCURUI (82361) - [PA]
PORTO VELHO (82825) - [RO]
BOA VISTA (82024) - [RR]
CARACARAI (82042) - [RR]
NORMANDIA (82014) - [RR]
ARAGUAINA (82659) - [TO]
GURUPI (83160) - [TO]
PALMAS (83033) - [TO]
PARANA (83231) - [TO]
PEDRO AFONSO (82863) - [TO]
PEIXE (83228) - [TO]
PORTO NACIONAL (83064) - [TO]
TAGUATINGA (83235) - [TO]
Prosseguir

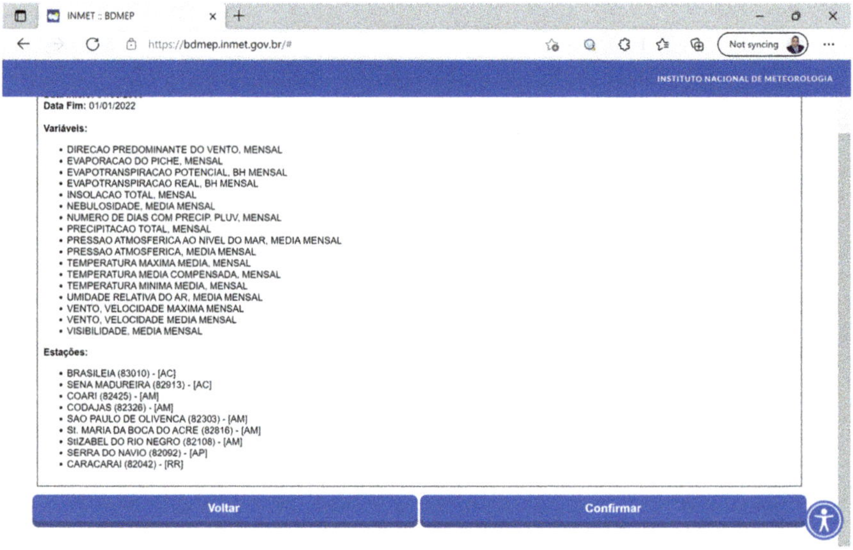
INMET :: BDMEP
https://bdmep.inmet.gov.br/#
Not syncing
INSTITUTO NACIONAL DE METEOROLOGIA
Data Fim: 01/01/2022
Variáveis:
• DIRECAO PREDOMINANTE DO VENTO, MENSAL
• EVAPORACAO DO PICHE, MENSAL
• EVAPOTRANSPIRACAO POTENCIAL, BH MENSAL
• EVAPOTRANSPIRACAO REAL, BH MENSAL
• INSOLACAO TOTAL, MENSAL
• NEBULOSIDADE, MEDIA MENSAL
• NUMERO DE DIAS COM PRECIP. PLUV, MENSAL
• PRECIPITACAO TOTAL, MENSAL
• PRESSAO ATMOSFERICA AO NIVEL DO MAR, MEDIA MENSAL
• PRESSAO ATMOSFERICA, MEDIA MENSAL
• TEMPERATURA MAXIMA MEDIA, MENSAL
• TEMPERATURA MEDIA COMPENSADA, MENSAL
• TEMPERATURA MINIMA MEDIA, MENSAL
• UMIDADE RELATIVA DO AR, MEDIA MENSAL
• VENTO, VELOCIDADE MAXIMA MENSAL
• VENTO, VELOCIDADE MEDIA MENSAL
• VISIBILIDADE, MEDIA MENSAL
Estações:
• BRASILEIA (83010) - [AC]
• SENA MADUREIRA (82913) - [AC]
• COARI (82425) - [AM]
• CODAJAS (82326) - [AM]
• SAO PAULO DE OLIVENCA (82303) - [AM]
• St. MARIA DA BOCA DO ACRE (82816) - [AM]
• StIZABEL DO RIO NEGRO (82108) - [AM]
• SERRA DO NAVIO (82092) - [AP]
• CARACARAI (82042) - [RR]
Voltar
Confirmar

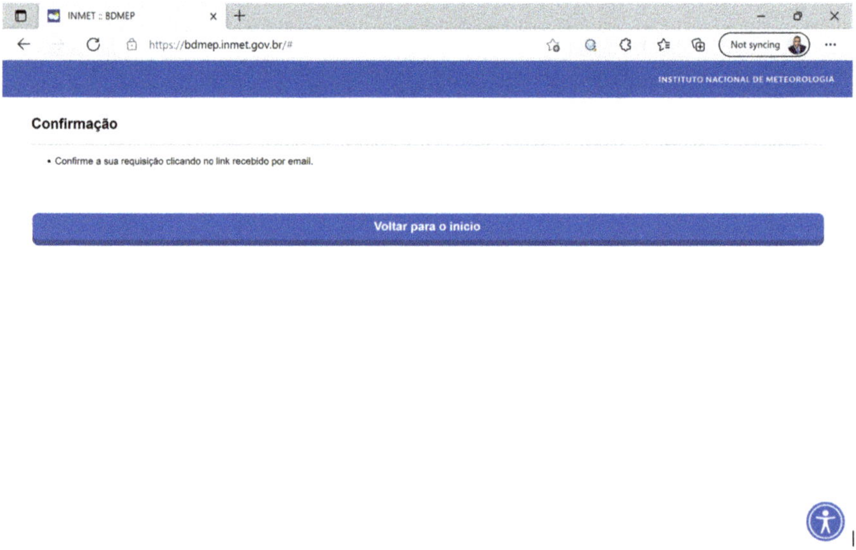
INMET :: BDMEP
https://bdmep.inmet.gov.br/#
Not syncing
INSTITUTO NACIONAL DE METEOROLOGIA
Confirmação
• Confirme a sua requisição clicando no link recebido por email.
Voltar para o inicio

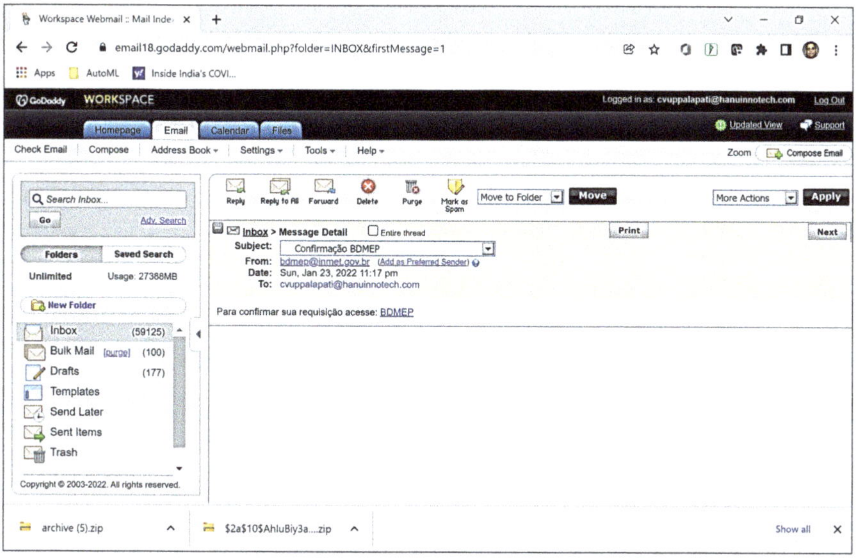

Confirm by clicking the link.

Portages	English
Sua requisição foi confirmada e está na fila para ser processada!	Your request has been confirmed and is queued to be processed!

Portages	English
Sua requisição começou a ser processada. Aguarde novos emails para acessar os dados selecionados. Caso tenha selecionado muitas estações e/ou um período de tempo muito longo, este processamento poderá levar até 24h para terminar. Este é um e-mail automático, não responda.	Your request has started to be processed. Wait for new emails to access the selected data. If you have selected many stations and/or a very long period of time, this processing may take up to 24 h to complete. This is an automated email, do not reply.

Email confirmation to download the data:

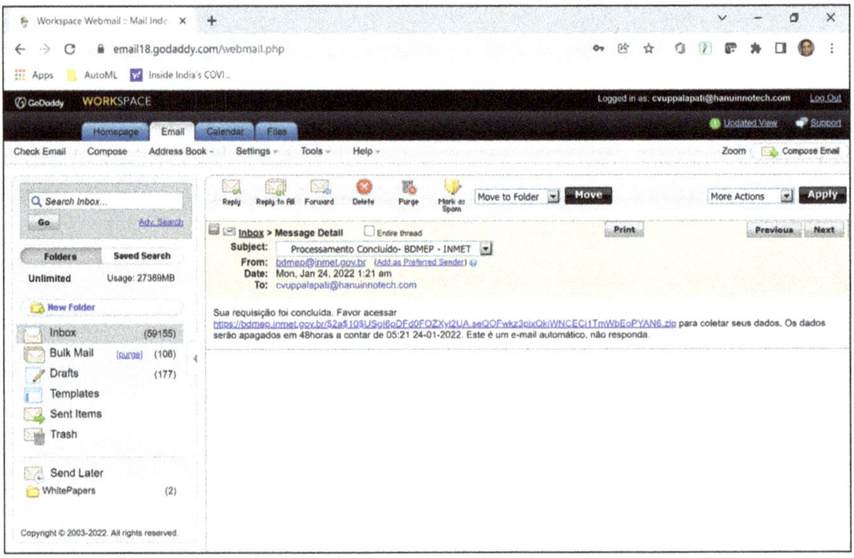

Sua requisição foi concluída. Favor acessar https://bdmep.inmet.gov.br/$2a$10 $USgI6pDFd0FOZXyl2UA.seQOFwkz3 pixQkiWNCECi1TmWbEoPYAN6.zip para coletar seus dados. Os dados serão apagados em 48horas a contar de 05:21 24-01-2022. Este é um e-mail automático, não responda.	Your request has been completed. Please access https://bdmep.inmet.gov.br/$2a$10 $USgI6pDFd0FOZXyl2UA.seQOFwkz3 pixQkiWNCECi1TmWbEoPYAN6.zip to collect your data. Data will be deleted within 48 h from 05:21 2022-01-24. This is an automated email, do not reply.

Source: https://climatedata-catalogue.wmo.int/explore

Labor Force Statistics from the Current Population Survey

IMF Country Index Weights

Country Indexes and Weights provide macro-level consumption and social behavior of people in terms of purchase preferences, food and nonalcoholic beverages, alcoholic beverages, clothing, and others.

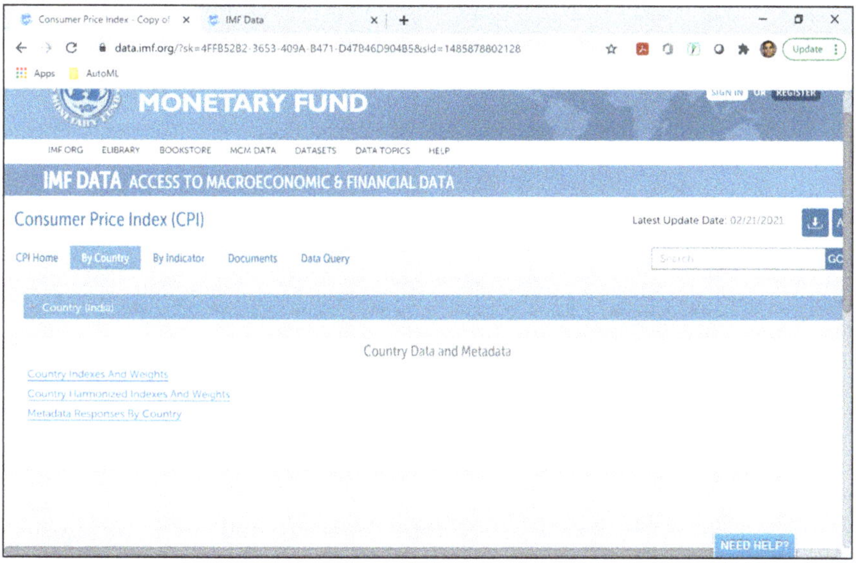

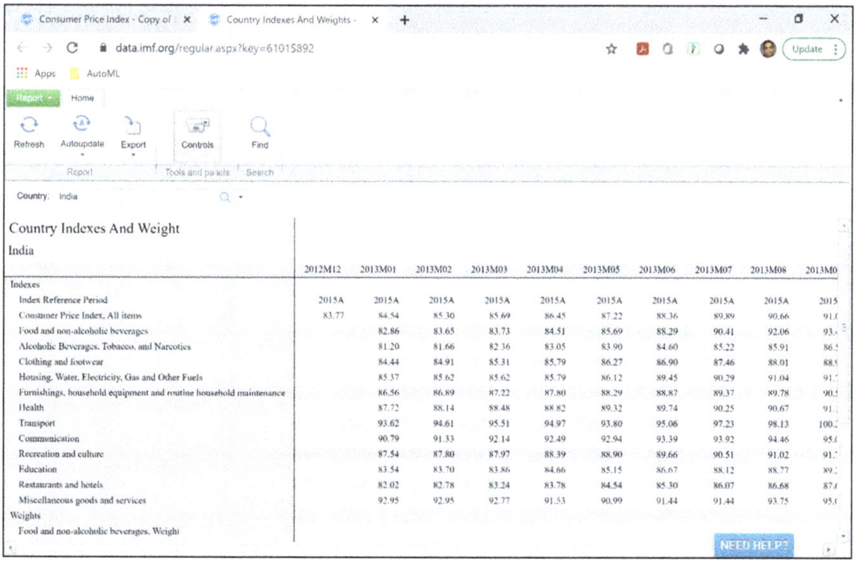

https://data.imf.org/regular.aspx?key=61015892

https://data.imf.org/?sk=388dfa60-1d26-4ade-b505-a05a558d9a42&sId=1479329328660

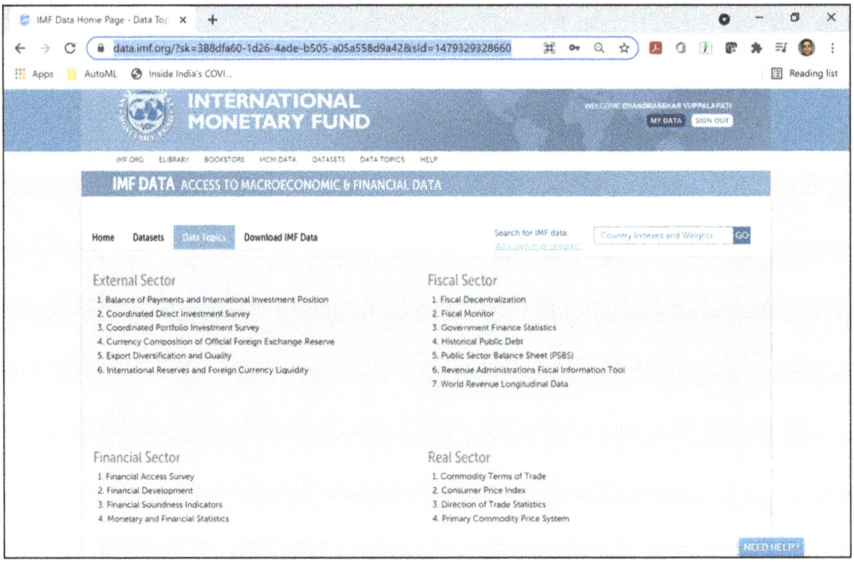

IMF Data Bulk Download

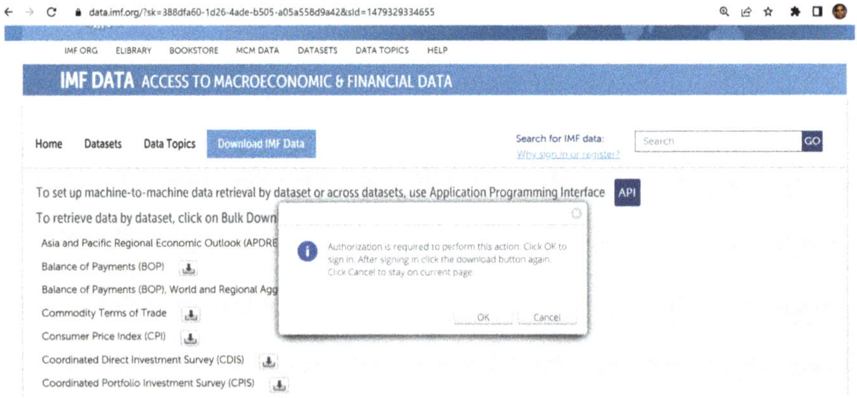

Download macroeconomic bulk data—https://data.imf.org/?sk=388DFA60-1D26-4ADE-B505-A05A558D9A42&sId=1479329132316

https://data.imf.org/?sk=388DFA60-1D26-4ADE-B505-A05A558D9A42&sId=1479329334655

World Bank Data

Macroeconomic world bank data[28]:

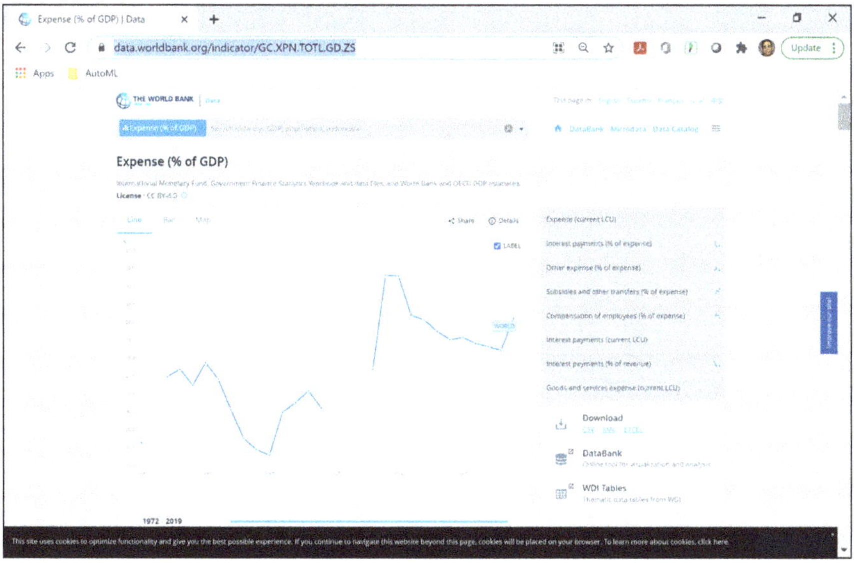

Source: https://data.worldbank.org/indicator/GC.XPN.TOTL.GD.ZS

The World Bank Development Indicators

The World Development Indicator tool[29]:

[28]The World Bank Data—https://www.worldbank.org/en/topic/macroeconomics

[29]The World Bank Development Indicator—https://databank.worldbank.org/reports.aspx?source=2&type=metadata&series=SI.POV.DDAY

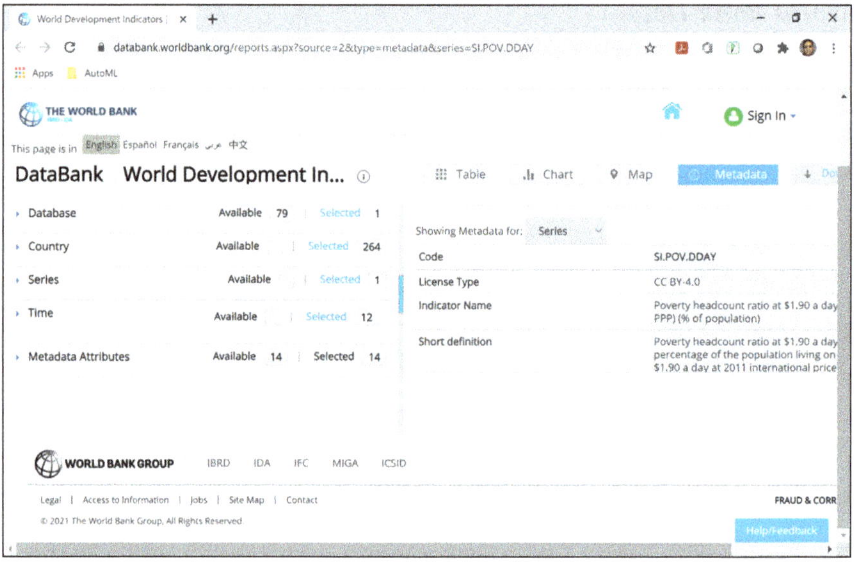

Food and Agriculture Organization of the United Nations

FAOSTAT Data[30]: FAO provides the following data domains: Production, Prices, Emissions-Agriculture, Inputs, Trade, Population, Emissions-Land Use, Investment, Macro Economic Data, and others.

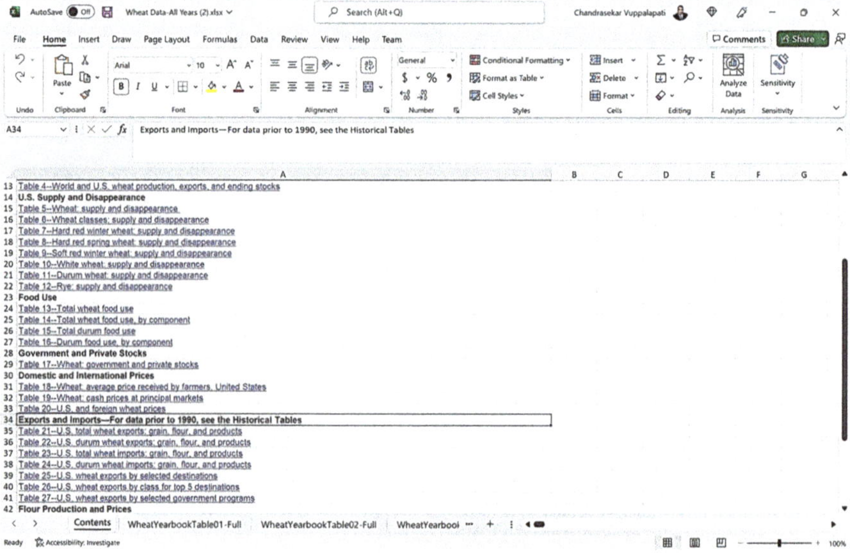

[30]FAPSTAT—http://www.fao.org/faostat/en/#data

Appendix 4: US Data

USDA Datasets

USDA datasets[31]:

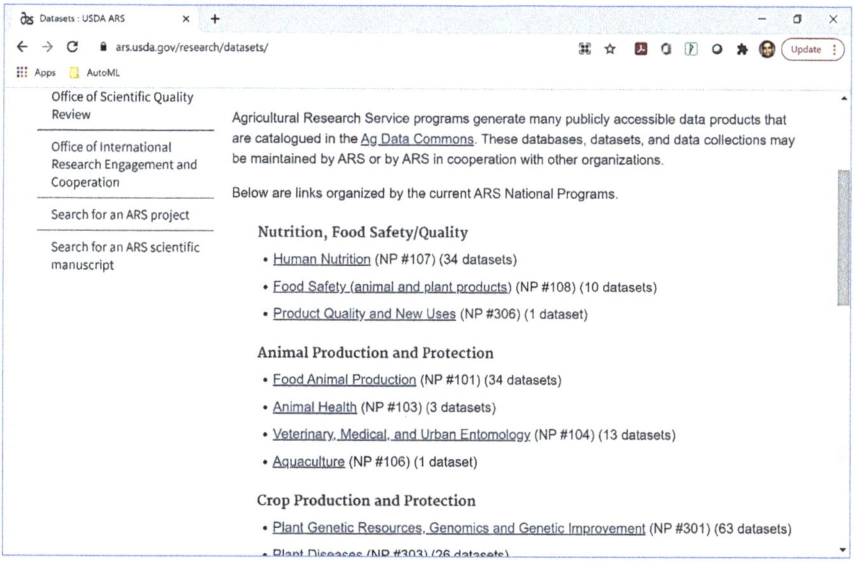

[31] USDA Datasets—https://www.ars.usda.gov/research/datasets/

C. Vuppalapati, *Artificial Intelligence and Heuristics for Enhanced Food Security*, International Series in Operations Research & Management Science 331,
https://doi.org/10.1007/978-3-031-08743-1

DATA.GOV

Explore data[32] that can help inform agriculture investment, innovation, and policy strategy. If you are interested in agricultural production, food security, rural development, nutrition, natural resources, and regional food systems, this page is for you.

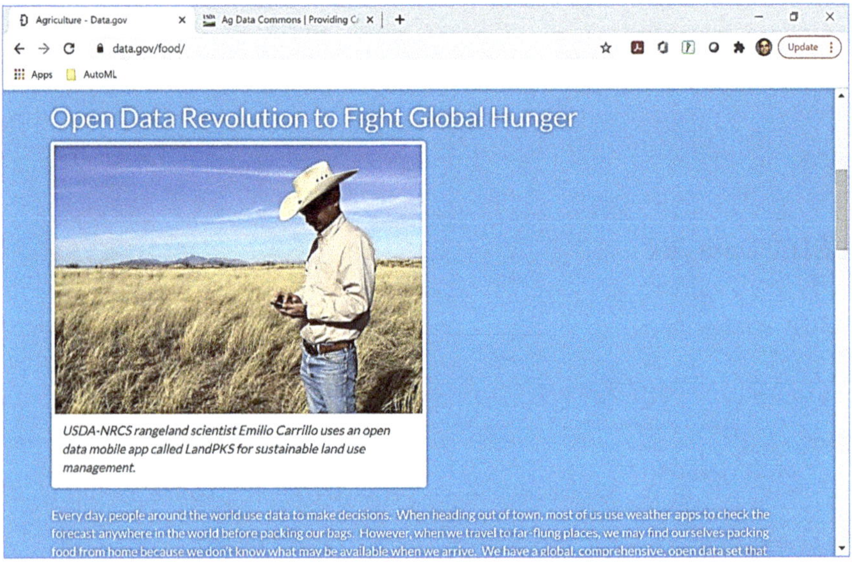

Wheat Data—https://www.ers.usda.gov/data-products/wheat-data.aspx

Wheat Data Disappearance and End Stocks

This data product contains statistics on wheat—including the five classes of wheat: hard red winter, hard red spring, soft red winter, white, and durum and rye. It includes data published in the monthly Wheat Outlook and previously annual Wheat Yearbook. Data are monthly, quarterly, and/or annual, depending upon the data series. Most data are on a marketing-year basis, but some are calendar year[33].

[32] Data.gov—https://www.data.gov/food/

[33] Wheat Data—https://www.ers.usda.gov/data-products/wheat-data/

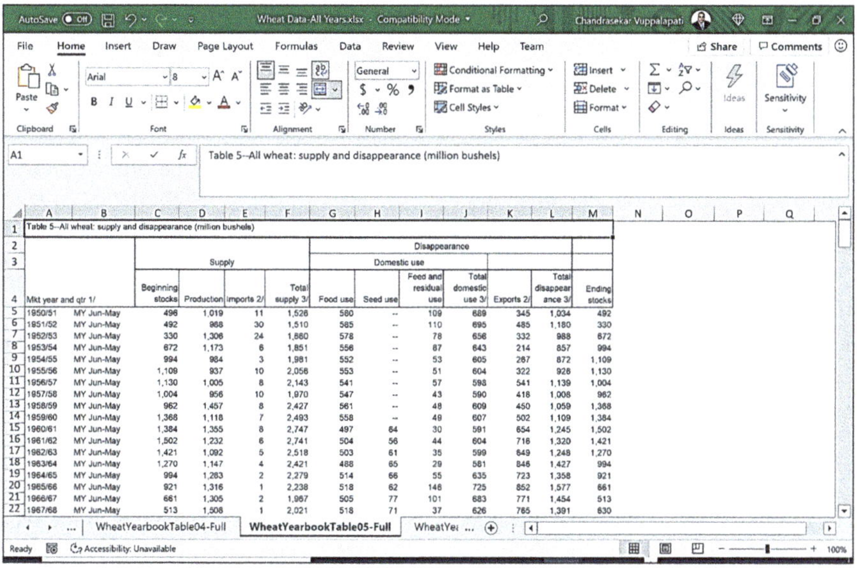

Table 5--All wheat: supply and disappearance (million bushels)

Mkt year and qtr 1/		Supply: Beginning stocks	Production	Imports 2/	Total supply 3/	Disappearance: Domestic use: Food use	Seed use	Feed and residual use	Total domestic use 3/	Exports 2/	Total disappearance 3/	Ending stocks
1950/51	MY Jun-May	496	1,019	11	1,526	580	--	109	689	345	1,034	492
1951/52	MY Jun-May	492	988	30	1,510	585	--	110	695	485	1,180	330
1952/53	MY Jun-May	330	1,306	24	1,660	578	--	78	656	332	988	672
1953/54	MY Jun-May	672	1,173	6	1,851	556	--	87	643	214	857	994
1954/55	MY Jun-May	994	984	3	1,981	552	--	53	605	267	872	1,109
1955/56	MY Jun-May	1,109	937	10	2,056	553	--	51	604	322	928	1,130
1956/57	MY Jun-May	1,130	1,005	8	2,143	541	--	57	598	541	1,139	1,004
1957/58	MY Jun-May	1,004	956	10	1,970	547	--	43	590	418	1,008	962
1958/59	MY Jun-May	962	1,457	8	2,427	561	--	48	609	450	1,059	1,368
1959/60	MY Jun-May	1,368	1,118	7	2,493	558	--	49	607	502	1,109	1,384
1960/61	MY Jun-May	1,384	1,355	8	2,747	497	64	30	591	654	1,245	1,502
1961/62	MY Jun-May	1,502	1,232	6	2,741	504	56	44	604	716	1,320	1,421
1962/63	MY Jun-May	1,421	1,092	5	2,518	503	61	35	599	649	1,248	1,270
1963/64	MY Jun-May	1,270	1,147	4	2,421	488	65	29	581	846	1,427	994
1964/65	MY Jun-May	994	1,283	2	2,279	514	66	55	635	723	1,358	921
1965/66	MY Jun-May	921	1,316	1	2,238	518	62	146	725	852	1,577	661
1966/67	MY Jun-May	661	1,305	2	1,967	505	77	101	683	771	1,454	513
1967/68	MY Jun-May	513	1,508	1	2,021	518	71	37	626	765	1,391	630

Hard red wheat contracts—https://www.barchart.com/futures/quotes/KEN20/interactive-chart

Dollars/Bushel: Dollars/Ton Converter

Dollars/Bushel converter[34]:

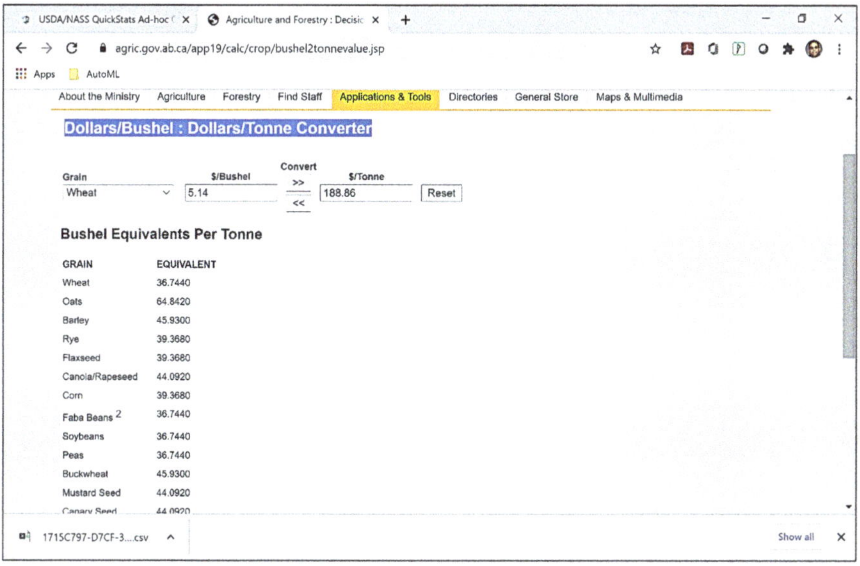

[34] Dollar Bushel Converter—https://www.agric.gov.ab.ca/app19/calc/crop/bushel2tonnevalue.jsp

NOAA: Storm Events Database

Storm Events Database

The Storm Events Database contains the records used to create the official NOAA Storm Data publication, documenting[35]:

- The occurrence of storms and other significant weather phenomena having sufficient intensity to cause loss of life, injuries, significant property damage, and/or disruption to commerce;
- Rare, unusual, weather phenomena that generate media attention, such as snow flurries in South Florida or the San Diego coastal area
- Other significant meteorological events, such as record maximum or minimum temperatures or precipitation that occur in connection with another event.

The database currently contains data from January 1950 to November 2020, as entered by NOAA's National Weather Service (NWS). Due to changes in the data collection and processing procedures over time, there are unique periods of record available depending on the event type. NCEI has performed data reformatting and standardization of event types but has not changed any data values for locations, fatalities, injuries, damage, narratives, and any other event-specific information. Please refer to the Database Details page for more information.

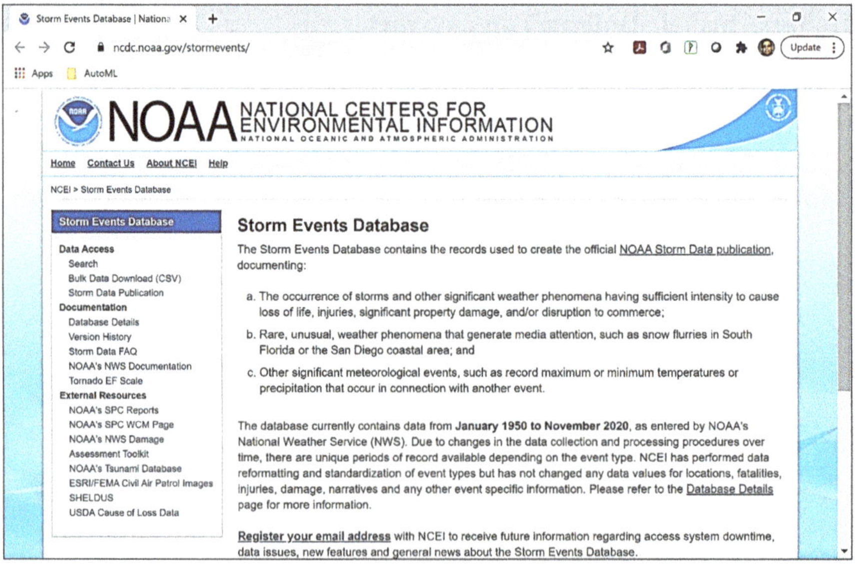

[35] Storm Events Database—https://www.ncdc.noaa.gov/stormevents/

Storm Data Export Format: describe the storm database metadata[36].

Storm Data Event Table

The only events permitted in Storm Data are listed in the table below. The chosen event name should be the one that most accurately describes the meteorological event leading to fatalities, injuries, damage, etc. However, significant events, such as tornadoes, having no impact or causing no damage, should also be included in Storm Data[37].

Astronomical Low Tide	Z	Lightning	C
Avalanche	Z	Marine Dense Fog	M
Blizzard	Z	Marine Hail	M
Coastal Flood	Z	Marine Heavy Freezing Spray	M
Cold/Wind Chill	Z	Marine High Wind	M
Debris Flow	C	Marine Hurricane/Typhoon	M
Dense Fog	Z	Marine Lightning	M
Dense Smoke	Z	Marine Strong Wind	M
Drought	Z	Marine Thunderstorm Wind	M
Dust Devil	C	Marine Tropical Depression	M
Dust Storm	Z	Marine Tropical Storm	M
Excessive Heat	Z	Rip Current	Z
Extreme Cold/Wind Chill	Z	Seiche	Z
Flash Flood	C	Sleet	Z
Flood	C	Sneaker Wave	Z
Frost/Freeze	Z	Storm Surge/Tide	Z
Funnel Cloud	C	Strong Wind	Z
Freezing Fog	Z	Thunderstorm Wind	C
Hail	C	Tornado	C
Heat	Z	Tropical Depression	Z
Heavy Rain	C	Tropical Storm	Z
Heavy Snow	Z	Tsunami	Z
High Surf	Z	Volcanic Ash	Z
High Wind	Z	Waterspout	M
Hurricane (Typhoon)	Z	Wildfire	Z
Ice Storm	Z	Winter Storm	Z
Lake-Effect Snow	Z	Winter Weather	Z
Lakeshore Flood	Z		

Legend: There are three designators: C - County/Parish; Z - Zone; and M – Marine Zone.

[36]Storm Data Export Format—https://www1.ncdc.noaa.gov/pub/data/swdi/stormevents/csvfiles/Storm-Data-Export-Format.pdf

[37]NATIONAL WEATHER SERVICE INSTRUCTION 10-1605 STORM DATA PREPARATION https://www.nws.noaa.gov/directives/sym/pd01016005curr.pdf

Appendix 5: Economic Frameworks and Macroeconomics

Macroeconomic Signals That Could Help Predict Economic Cycles

Signal #	Type	Description
Oil price swings	Leading indicator	Oil price swings[38] appear to be consistent and frequent historical precursors to US recessions[39]. A spike in oil prices has preceded or coincided with 10 out of 12 post-WWII recessions.
Monetary trends	Lagging indicator	A period of expansionary monetary policy in the years prior to the recession sometimes to help fund government war spending or to reinflate the economy after the previous round of recession. Once the resulting debt bubbles pop or the end of a war leads to cutbacks in monetary expansion, several years' worth of overextended, debt-based investments and malinvestments tend to be wiped out in a process of debt deflation[40,41] in a relatively short period. This spikes unemployment and drags down GDP.

(continued)

[38] Oil Price Swings—https://www.macrotrends.net/1369/crude-oil-price-history-chart

[39] Oil futures volatility and the economy—https://www.eurekalert.org/news-releases/843803

[40] Michael Assous (2013) Irving Fisher's debt deflation analysis: From the Purchasing Power of Money (1911) to the Debt-deflation Theory of the Great Depression (1933), The European Journal of the History of Economic Thought, 20:2, 305–322, DOI: 10.1080/09672567.2012.762936

[41] Debt Deflation—https://www.investopedia.com/terms/d/debtdeflation.asp

C. Vuppalapati, *Artificial Intelligence and Heuristics for Enhanced Food Security*, International Series in Operations Research & Management Science 331,
https://doi.org/10.1007/978-3-031-08743-1

Signal #	Type	Description
During recession, GDP and credit contraction occur	Lagging indicator	Country GDP/credit contraction, in percent[42].

The US Recessions

The bottom line: So, what do all these very different recessions have in common? In many cases, the most important single factor is a period of expansionary monetary policy in the years prior to the recession, sometimes to help fund government war spending or to reinflate the economy after the previous round of recession.

Once the resulting debt bubbles pop or the end of a war leads to cutbacks in monetary expansion, several years' worth of overextended, debt-based investments and malinvestments tend to be wiped out in a process of debt deflation in a relatively short period. This spikes unemployment and drags down GDP.

Beyond the underlying monetary trends, real economic shocks often help to trigger the turning point into recession. For one, oil price swings appear to be consistent and frequent historical precursors to US recessions[43].

A spike in oil prices has preceded or coincided with 10 out of 12 post-WWII recessions[44]. This highlights that while global integration of economies allowing for more effective cooperative efforts between governments has increased over time, the integration itself ties the world economies more closely together, making them more susceptible to problems outside their borders.

Recession	Details
The Great Recession (December 2007 to June 2009)[45]	The collapse of the housing bubble of the 2000s and resulting record foreclosures and a financial crisis that threw markets worldwide into a tailspin. Summary: Duration: 18 months[46] GDP decline: 4.3%[47] Peak unemployment rate: 10.0%[48]

(continued)

[42] What Happens During Recessions, Crunches and Busts?—https://www.imf.org/external/pubs/ft/wp/2008/wp08274.pdf

[43] Oil futures volatility and the economy—https://www.eurekalert.org/news-releases/843803

[44] Crude Oil Prices—70 Year Historical Chart—https://www.macrotrends.net/1369/crude-oil-price-history-chart

[45] The Great Recession—https://www.federalreservehistory.org/essays/great-recession-of-200709

[46] Business Cycle Dating—https://www.cepr.net/clearing-up-some-facts-about-the-depression-of-1946/

[47] The Great Recession—https://www.federalreservehistory.org/essays/great-recession-of-200709

[48] The Great Recession—https://www.federalreservehistory.org/essays/great-recession-of-200709

Recession	Details
Dot-com recession (March 2001 to November 2001)	The dot-com bubble burst in 2000, when an overinflated Nasdaq lost more than 75% of its value and wiped out a generation of tech investors. Those losses left the stock market in a vulnerable place that got worse in the fall of 2001, when the devastation of the September 11, 2001[49], terrorist attacks and a series of major accounting scandals at corporations like Enron and Swissair spurred a stock market crash. The resulting recession was relatively short, at just 8 months, and also shallow, as GDP dipped only 0.6% and unemployment reached 5.5%. Summary: Duration: 8 months[50] GDP decline: 0.3% Peak unemployment rate: 5.5%[51]
Gulf War recession (July 1990 to March 1991)	A mild recession kicked off in 1990, as the Federal Reserve had been slowly raising interest rates for over 2 years to keep inflation in check. Those moves slowed down the economy, which then took a hit when Iraq invaded Kuwait in the summer of 1990 (followed by US involvement and the Gulf War) and caused global oil prices to more than double. The recession lasted just 8 months, with GDP declining[52] 1.1% during that period and unemployment reaching roughly 7%. Summary: Duration: 8 months[53] GDP decline: 1.5% Peak unemployment rate: 6.8%
Energy crisis recession (July 1981 to November 1982)	In 1981, having emerged from a recession just a year before (the early 1980s are described as having a "double-dip recession because they were too close together), the Federal Reserve tried to tame rising inflation with stricter monetary policy that raised interest rates and slowed the economy. Those policies managed to reduce inflation to around 4% by 1983, but the cost was a 16-month recession that saw GDP drop by around 3% and unemployment spiked to 10.8%. Summary:

(continued)

[49] 2001 Recession—https://files.stlouisfed.org/files/htdocs/publications/review/03/09/Kliesen.pdf

[50] Business Cycle Dating—https://www.cepr.net/clearing-up-some-facts-about-the-depression-of-1946/

[51] The Current Economic Recession: How Long, How Deep, and How Different From the Past?—https://web.archive.org/web/20091010043009/http://fpc.state.gov/documents/organization/7962.pdf

[52] Deep Recessions, Fast Recoveries, and Financial Crises: Evidence from the American Record—https://www.nber.org/system/files/working_papers/w18194/w18194.pdf

[53] Business Cycle Dating—https://www.cepr.net/clearing-up-some-facts-about-the-depression-of-1946/

Recession	Details
	Duration: 16 months GDP decline: 2.9% Peak unemployment rate: 10.8%
1980 Recession (January 1980 to July 1980)	Inflation rates rose throughout the late 1970s, reaching double-digit levels in 1979 and peaking at 22%[54] in 1980. As a result, the Federal Reserve raised interest rates to stop the rising inflation, which slowed down the economy (GDP dropped over 2%[55]) and caused unemployment to spike to 7.8%. The Fed lowered interest rates again in mid-1980, giving the economy a chance to rebound and ending a brief, 6-month recession. Summary: Duration: 6 months GDP decline: 2.2% Peak unemployment rate: 7.8%
Oil embargo recession (November 1973 to March 1975)	In the fall of 1973, the Organization of the Petroleum Exporting Countries, or OPEC, put an embargo on oil imports from multiple countries, including the United States, over their support of Israel's military. Oil prices roughly quadrupled[56] as a result, putting a major crunch on the economy as gas prices soared for consumers, reducing their spending on other items. The resulting recession lasted for 16 months (it even outlasted the oil embargo itself, which OPEC lifted in 1974) and saw GDP decline by 3.4%[57] while unemployment climbed from 4.8% to nearly 9%. Summary: Duration: 16 months GDP decline: 3% Peak unemployment rate: 8.6%
Recession of 1969–1970 (December 1969 to November 1970)	The "mild recession[58]" that ensued caused unemployment to peak at around 6% while the GDP dropped <1% before the Fed eased its monetary policies to restart economic growth in 1970. Summary: Duration: 11 months GDP decline: 0.6% Peak unemployment rate: 5.9%
Recession of 1960–1961 (April 1960 to February 1961)	Even though two previous recessions in the 1950s stemmed from tighter monetary policies giving rise to

(continued)

[54] Mysteries of monetary policy—https://www.aei.org/articles/mysteries-monetary-policy/

[55] The 2001 Recession: How Was It Different and What Developments May Have Caused It?—https://files.stlouisfed.org/files/htdocs/publications/review/03/09/Kliesen.pdf

[56] Oil Shock of 1973–74—https://www.federalreservehistory.org/essays/oil-shock-of-1973-74

[57] Deep Recessions, Fast Recoveries, and Financial Crises: Evidence from the American Record—https://www.nber.org/system/files/working_papers/w18194/w18194.pdf

[58] Deep Recessions, Fast Recoveries, and Financial Crises: Evidence from the American Record—https://www.nber.org/system/files/working_papers/w18194/w18194.pdf

Recession	Details
	interest rates, the Federal Reserve began slowly raising interest rates following the end of the previous recession in 1958, leading to another short-lived recession at the start of the 1960s. The 10-month recession saw the GDP drop by nearly 2% and unemployment peaked at 6.9%, while President John F. Kennedy spurred a rebound in 1961 with stimulus spending that included tax cuts and expanded unemployment and Social Security benefits. Summary: Duration: 10 months GDP decline: 1.6% Peak unemployment rate: 6.9%
Recession of 1957–1958 (August 1957 to April 1958)	This recession in the late 1950s lasted 8 months. GDP fell by 3.7% and unemployment peaked at 7.4% as the government's tighter monetary policy in the mid-1950s raised interest rates in an effort to curb inflation. As a result, consumer prices also continued to rise, which led to a decline in spending. Meanwhile, a global recession (which also happened to coincide with the 1957 Asian flu pandemic that killed 1.1 million people[59] worldwide) further hurt the US economy as the country's exports declined by more than $4 billion[60]. Summary: Duration: 8 months13 GDP decline: 3.7%18 Peak unemployment rate: 7.4%24
Post-Korean War recession (July 1953 to May 1954)	As with previous postwar recessions, this downturn was spurred by a shift in government spending after the end of the Korean War (which lasted from 1950 to 1953). The country's GDP dropped by 2.2% and unemployment peaked at roughly 6%[61], as the government wound down security spending following the war and the US Federal Reserve tightened monetary policy[62] to curb inflation (which includes increasing interest rates). However, spiking interest rates hurt consumer confidence in the economy and decreased consumer demand. The Fed eased its policies in 1954, allowing the economy to

(continued)

[59] 1957–1958 Pandemic (H2N2 virus)—https://www.cdc.gov/flu/pandemic-resources/1957-1958-pandemic.html

[60] The 1957–58 Recession in World Trade—https://fraser.stlouisfed.org/files/docs/publications/FRB/pages/1955-1959/14330_1955-1959.pdf

[61] Unemployment rate and long-term unemployment rate, January 1948–December 2011, seasonally adjusted (in percent)—https://www.bls.gov/spotlight/2012/recession/data_cps_unemp_1948.htm

[62] Deep Recessions, Fast Recoveries, and Financial Crises: Evidence from the American Record—https://www.nber.org/papers/w18194

Recession	Details
	rebound after 10 months. Summary: Duration: 10 months[63] GDP decline: 2.7%[64] Peak unemployment rate: 5.9%[65]
Post-WWII slump (November 1948 to October 1949)	After the war there was an 8-month recession (see below), but the economic challenges stemming from the end of World War II again caught up with the US economy during the last stretch of the 1940s. But this 11-month recession—in which the country's GDP dropped by <2%—was considered "very mild[66]" by economists, who attribute the downturn in part to consumer demand leveling off after previously spiking when wartime rationing efforts ceased. Economists also point to a decline in fixed investments, while the influx of veterans returning from war and competing[67] for limited civilian jobs helped the unemployment rate climb as high as 7.9%[68], according to the US Bureau of Labor Statistics. Summary: Duration: 11 months[69] GDP decline: 1.7%1 Peak unemployment rate: 5.7%

(continued)

[63] Business Cycle Dating—https://www.cepr.net/clearing-up-some-facts-about-the-depression-of-1946/

[64] The Current Economic Recession: How Long, How Deep, and How Different From the Past?—https://web.archive.org/web/20091010043009/http://fpc.state.gov/documents/organization/7962.pdf

[65] Labor Force Statistics from the Current Population Survey 1948–2021—https://data.bls.gov/timeseries/LNS14000000?years_option=all_years

[66] A Case Study: The 1948–1949 Recessions—https://www.nber.org/system/files/chapters/c2798/c2798.pdf

[67] A Review of Past Recessions—https://www.investopedia.com/articles/economics/08/past-recessions.asp

[68] Unemployment rate and long-term unemployment rate, January 1948–December 2011, seasonally adjusted (in percent)—https://www.bls.gov/spotlight/2012/recession/data_cps_unemp_1948.htm

[69] The Current Economic Recession: How Long, How Deep, and How Different From the Past?—https://web.archive.org/web/20091010043009/http://fpc.state.gov/documents/organization/7962.pdf

Recession	Details
Post-World War II recession (February 1945 to October 1945)	This downturn was caused primarily by a significant drop in government spending and GDP (which fell 11%) as the United States pivoted from a wartime economy built around manufacturing supplies for the World War II effort to a peacetime economy focused on creating civilian jobs for returning veterans. Summary[70]: Duration: 8 months[71] GDP decline: 10.9%[72] Peak unemployment rate: 5.2%[73]
The Roosevelt recession (May 1937 to June 1938)[74]	This recession was essentially a 13-month pause in the nation's recovery from the Great Depression and modern economists have called the episode a "cautionary tale." In 1937, President Franklin D. Roosevelt cut government spending at a time when the country's economic recovery was still fragile enough to be derailed. As a result, unemployment jumped from roughly 14% to nearly 20% and the real GDP fell by 10%. The following year, Roosevelt signed a $3.75 billion spending bill that restarted the economic recovery. Summary[75]: Duration: 13 months GDP decline: 10% Peak unemployment rate: 20%[76]

Labor Force Statistics from the Current Population Survey

Series Id: LNS14000000[77] Seasonally adjusted
Series title: (Seas) Unemployment rate

[70] A Review of Past Recessions—https://www.investopedia.com/articles/economics/08/past-recessions.asp

[71] Business Cycle Dating—https://www.cepr.net/clearing-up-some-facts-about-the-depression-of-1946/

[72] Clearing Up Some Facts About the Depression of 1946—https://www.cepr.net/clearing-up-some-facts-about-the-depression-of-1946/

[73] ANNUAL ESTIMATES OF UNEMPLOYMENT IN THE UNITED STATES, 1900–1954—https://www.nber.org/system/files/chapters/c2644/c2644.pdf

[74] How many recessions you've actually lived through and what happened in every one—https://www.cnbc.com/2020/04/09/what-happened-in-every-us-recession-since-the-great-depression.html

[75] A Review of Past Recessions—https://www.investopedia.com/articles/economics/08/past-recessions.asp

[76] Recession of 1937–38—https://www.federalreservehistory.org/essays/recession-of-1937-38

[77] U.S. Unemployment data—https://data.bls.gov/timeseries/LNS14000000?years_option=all_years

Labor force status: Unemployment rate
Type of data: Percent or rate
Age: 16 years and over

Year	Jan	Feb	Mar	Apr	May	Jun	Jul	Aug	Sep	Oct	Nov	Dec
1948	3.4	3.8	4.0	3.9	3.5	3.6	3.6	3.9	3.8	3.7	3.8	4.0
1949	4.3	4.7	5.0	5.3	6.1	6.2	6.7	6.8	6.6	7.9	6.4	6.6
1950	6.5	6.4	6.3	5.8	5.5	5.4	5.0	4.5	4.4	4.2	4.2	4.3
1951	3.7	3.4	3.4	3.1	3.0	3.2	3.1	3.1	3.3	3.5	3.5	3.1
1952	3.2	3.1	2.9	2.9	3.0	3.0	3.2	3.4	3.1	3.0	2.8	2.7
1953	2.9	2.6	2.6	2.7	2.5	2.5	2.6	2.7	2.9	3.1	3.5	4.5
1954	4.9	5.2	5.7	5.9	5.9	5.6	5.8	6.0	6.1	5.7	5.3	5.0
1955	4.9	4.7	4.6	4.7	4.3	4.2	4.0	4.2	4.1	4.3	4.2	4.2
1956	4.0	3.9	4.2	4.0	4.3	4.3	4.4	4.1	3.9	3.9	4.3	4.2
1957	4.2	3.9	3.7	3.9	4.1	4.3	4.2	4.1	4.4	4.5	5.1	5.2
1958	5.8	6.4	6.7	7.4	7.4	7.3	7.5	7.4	7.1	6.7	6.2	6.2
1959	6.0	5.9	5.6	5.2	5.1	5.0	5.1	5.2	5.5	5.7	5.8	5.3
1960	5.2	4.8	5.4	5.2	5.1	5.4	5.5	5.6	5.5	6.1	6.1	6.6
1961	6.6	6.9	6.9	7.0	7.1	6.9	7.0	6.6	6.7	6.5	6.1	6.0
1962	5.8	5.5	5.6	5.6	5.5	5.5	5.4	5.7	5.6	5.4	5.7	5.5
1963	5.7	5.9	5.7	5.7	5.9	5.6	5.6	5.4	5.5	5.5	5.7	5.5
1964	5.6	5.4	5.4	5.3	5.1	5.2	4.9	5.0	5.1	5.1	4.8	5.0
1965	4.9	5.1	4.7	4.8	4.6	4.6	4.4	4.4	4.3	4.2	4.1	4.0
1966	4.0	3.8	3.8	3.8	3.9	3.8	3.8	3.8	3.7	3.7	3.6	3.8
1967	3.9	3.8	3.8	3.8	3.8	3.9	3.8	3.8	3.8	4.0	3.9	3.8
1968	3.7	3.8	3.7	3.5	3.5	3.7	3.7	3.5	3.4	3.4	3.4	3.4
1969	3.4	3.4	3.4	3.4	3.4	3.5	3.5	3.5	3.7	3.7	3.5	3.5
1970	3.9	4.2	4.4	4.6	4.8	4.9	5.0	5.1	5.4	5.5	5.9	6.1
1971	5.9	5.9	6.0	5.9	5.9	5.9	6.0	6.1	6.0	5.8	6.0	6.0
1972	5.8	5.7	5.8	5.7	5.7	5.7	5.6	5.6	5.5	5.6	5.3	5.2
1973	4.9	5.0	4.9	5.0	4.9	4.9	4.8	4.8	4.8	4.6	4.8	4.9
1974	5.1	5.2	5.1	5.1	5.1	5.4	5.5	5.5	5.9	6.0	6.6	7.2
1975	8.1	8.1	8.6	8.8	9.0	8.8	8.6	8.4	8.4	8.4	8.3	8.2
1976	7.9	7.7	7.6	7.7	7.4	7.6	7.8	7.8	7.6	7.7	7.8	7.8
1977	7.5	7.6	7.4	7.2	7.0	7.2	6.9	7.0	6.8	6.8	6.8	6.4
1978	6.4	6.3	6.3	6.1	6.0	5.9	6.2	5.9	6.0	5.8	5.9	6.0
1979	5.9	5.9	5.8	5.8	5.6	5.7	5.7	6.0	5.9	6.0	5.9	6.0
1980	6.3	6.3	6.3	6.9	7.5	7.6	7.8	7.7	7.5	7.5	7.5	7.2
1981	7.5	7.4	7.4	7.2	7.5	7.5	7.2	7.4	7.6	7.9	8.3	8.5
1982	8.6	8.9	9.0	9.3	9.4	9.6	9.8	9.8	10.1	10.4	10.8	10.8
1983	10.4	10.4	10.3	10.2	10.1	10.1	9.4	9.5	9.2	8.8	8.5	8.3
1984	8.0	7.8	7.8	7.7	7.4	7.2	7.5	7.5	7.3	7.4	7.2	7.3
1985	7.3	7.2	7.2	7.3	7.2	7.4	7.4	7.1	7.1	7.1	7.0	7.0
1986	6.7	7.2	7.2	7.1	7.2	7.2	7.0	6.9	7.0	7.0	6.9	6.6
1987	6.6	6.6	6.6	6.3	6.3	6.2	6.1	6.0	5.9	6.0	5.8	5.7

(continued)

Year	Jan	Feb	Mar	Apr	May	Jun	Jul	Aug	Sep	Oct	Nov	Dec
1988	5.7	5.7	5.7	5.4	5.6	5.4	5.4	5.6	5.4	5.4	5.3	5.3
1989	5.4	5.2	5.0	5.2	5.2	5.3	5.2	5.2	5.3	5.3	5.4	5.4
1990	5.4	5.3	5.2	5.4	5.4	5.2	5.5	5.7	5.9	5.9	6.2	6.3
1991	6.4	6.6	6.8	6.7	6.9	6.9	6.8	6.9	6.9	7.0	7.0	7.3
1992	7.3	7.4	7.4	7.4	7.6	7.8	7.7	7.6	7.6	7.3	7.4	7.4
1993	7.3	7.1	7.0	7.1	7.1	7.0	6.9	6.8	6.7	6.8	6.6	6.5
1994	6.6	6.6	6.5	6.4	6.1	6.1	6.1	6.0	5.9	5.8	5.6	5.5
1995	5.6	5.4	5.4	5.8	5.6	5.6	5.7	5.7	5.6	5.5	5.6	5.6
1996	5.6	5.5	5.5	5.6	5.6	5.3	5.5	5.1	5.2	5.2	5.4	5.4
1997	5.3	5.2	5.2	5.1	4.9	5.0	4.9	4.8	4.9	4.7	4.6	4.7
1998	4.6	4.6	4.7	4.3	4.4	4.5	4.5	4.5	4.6	4.5	4.4	4.4
1999	4.3	4.4	4.2	4.3	4.2	4.3	4.3	4.2	4.2	4.1	4.1	4.0
2000	4.0	4.1	4.0	3.8	4.0	4.0	4.0	4.1	3.9	3.9	3.9	3.9
2001	4.2	4.2	4.3	4.4	4.3	4.5	4.6	4.9	5.0	5.3	5.5	5.7
2002	5.7	5.7	5.7	5.9	5.8	5.8	5.8	5.7	5.7	5.7	5.9	6.0
2003	5.8	5.9	5.9	6.0	6.1	6.3	6.2	6.1	6.1	6.0	5.8	5.7
2004	5.7	5.6	5.8	5.6	5.6	5.6	5.5	5.4	5.4	5.5	5.4	5.4
2005	5.3	5.4	5.2	5.2	5.1	5.0	5.0	4.9	5.0	5.0	5.0	4.9
2006	4.7	4.8	4.7	4.7	4.6	4.6	4.7	4.7	4.5	4.4	4.5	4.4
2007	4.6	4.5	4.4	4.5	4.4	4.6	4.7	4.6	4.7	4.7	4.7	5.0
2008	5.0	4.9	5.1	5.0	5.4	5.6	5.8	6.1	6.1	6.5	6.8	7.3
2009	7.8	8.3	8.7	9.0	9.4	9.5	9.5	9.6	9.8	10.0	9.9	9.9
2010	9.8	9.8	9.9	9.9	9.6	9.4	9.4	9.5	9.5	9.4	9.8	9.3
2011	9.1	9.0	9.0	9.1	9.0	9.1	9.0	9.0	9.0	8.8	8.6	8.5
2012	8.3	8.3	8.2	8.2	8.2	8.2	8.2	8.1	7.8	7.8	7.7	7.9
2013	8.0	7.7	7.5	7.6	7.5	7.5	7.3	7.2	7.2	7.2	6.9	6.7
2014	6.6	6.7	6.7	6.2	6.3	6.1	6.2	6.1	5.9	5.7	5.8	5.6
2015	5.7	5.5	5.4	5.4	5.6	5.3	5.2	5.1	5.0	5.0	5.1	5.0
2016	4.8	4.9	5.0	5.1	4.8	4.9	4.8	4.9	5.0	4.9	4.7	4.7
2017	4.7	4.6	4.4	4.5	4.4	4.3	4.3	4.4	4.2	4.1	4.2	4.1
2018	4.0	4.1	4.0	4.0	3.8	4.0	3.8	3.8	3.7	3.8	3.8	3.9
2019	4.0	3.8	3.8	3.7	3.7	3.6	3.6	3.7	3.5	3.6	3.6	3.6
2020	3.5	3.5	4.4	14.8	13.3	11.1	10.2	8.4	7.8	6.9	6.7	6.7
2021	6.3	6.2	6.0	6.1	5.8	5.9	5.4	5.2	4.8			

United Nations Statistics Department (UNSD)

The following table provides the main economic indicators. MEI and food security show correlation, especially during economic downturns.

The OCED: Main Economic Indicators (MEI)

National accounts 1. GDP (value) 2. GDP (volume) 3. Implicit price value	*OCED composite leading indicator* 42. Trend restored 43. Six-month rate of change (annual rate)	*Labor* 78. Employment: total 79. – Employment: agriculture 80. – Employment: industry 81 – Employment: services 82. Total employees 83. – Part-time employees 84. – Temporary employees 85. Total unemployment (level) 86. Total unemployment (rate) 87. Unemployment: short-term index 88. Worked hours 89. Job vacancies
Production 4. Industry excluding construction 5. Manufacturing 6. – Consumer goods: total 7. – Consumer nondurable goods 8. – Consumer durable goods 9. – Investment goods 10. Intermediate goods including energy 11. Intermediate goods excluding energy 12. Energy 13. Construction 14. Services 15. Rate of capacity utilization	*Construction* 44. Orders/permits: total construction 45. Orders/permits: residuals 46. Work put in place: total construction 47. Work put in place: residential	*Wages* 90. Hourly earnings: all activities 91. Hourly earnings: manufacturing 92. United labor costs: manufacturing
Commodity output 16. Cement 17. Crude steel 18. Crude petroleum 19. Natural gas 20. Commercial vehicles 21. Passenger cars	*Business tendency surveys* 48. Industrial business climate 49. Industry production: future tendency 50. Industrial order inflow: tendency 51. Industrial order books: level 52. Industrial finished goods stocks: level 53. Industrial export order books or demand: level 54. Industrial rate of capacity utilization 55. Industrial employment: future tendency 56. Industrial selling prices: future tendency	*Producer prices* 93. Total 94. Manufacturing 95. – Consumer goods 96. – Investment goods 97. – Intermediate goods including energy 98. – Intermediate goods excluding energy 99. – Energy 100. Food 101. Services

(continued)

	57. Construction order inflows: future tendency 58. Construction employment: future tendency 59. Retail/wholesale: present business situation 60. Retail/wholesale business situation: future tendency 61. Retail/wholesale stocks: level 62. Other services: present business situation 63. Other services' business situation: future tendency 64. Other services' employment: future tendency	
Manufacturing—sales (volume) 22. Total 23. – Domestic 24. – Export 25. Consumer goods 26. – Consumer nondurable goods 27. – Consumer durable goods 28. Investment goods 29. Intermediate goods including energy	*Consumer tendency surveys* 65. Consumer confidence indicator 66. Consumer expected inflation 67. Consumer expected economic situation	*Consumer prices* 102. Total 103 Food 104. All items less food and energy 105. Energy 106. All services less rent 107. Rent 108. National core inflation
Manufacturing—new orders (volume) 30. Total 31. – Domestic 32. – Export 33. Consumer goods 34. – Consumer nondurable goods 35. – Consumer durable goods 36. Investment goods 37. Intermediate goods including energy	*Retail sales* 68. Total retail sales (value) 69. – Total retail sales (volume) 70. New passenger car registration (level)	*Domestic finance* 109. Narrow money 110. Broad money 111. Domestic credit to total economy 112. New capital issues 113. Fiscal balance 114. Public debt
Manufacturing—stocks (volume) 38. Total 39. Finished goods 40. Work in progress 41. Intermediate goods	*International trade* 71. Imports c.i.f or FOB (value) 72. Exports c.i.f or f FOB (value) 73. Net trade (value) 74. Imports c.i.f or FOB (volume) 75. Exports c.i.f or FOB (volume) 76. Import prices 77. Export prices	*Balance of payments* 115. Current account balance 116. – Balance on goods 117. – Balance on services 118. – Balance on income 119. – Balance on current transfer 120. Capital and financial account balance 121. – Reserve assets 122. Net errors and omission

(continued)

	Interest rates—share price 123. Three-month interest rate 124. Prime interest rate 125. Long-term interest rate 126. All shares price index
	Foreign finances 127. US dollar exchange rate: spot 128. Euro exchange rate: spot 129. Reserve assets excluding gold

Reserve Bank of India: Handbook of Statistics on Indian Economy

Macroeconomic aggregates (at current prices)[78] and component of GDP[79]:

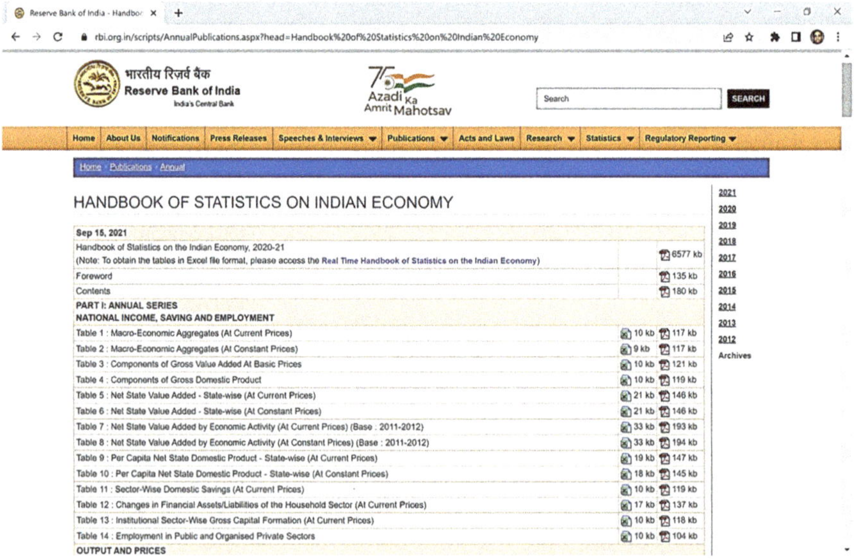

[78] Macro-economic aggregates—https://www.rbi.org.in/scripts/PublicationsView.aspx?id=20406
[79] Components of GDP—https://www.rbi.org.in/scripts/PublicationsView.aspx?id=20409

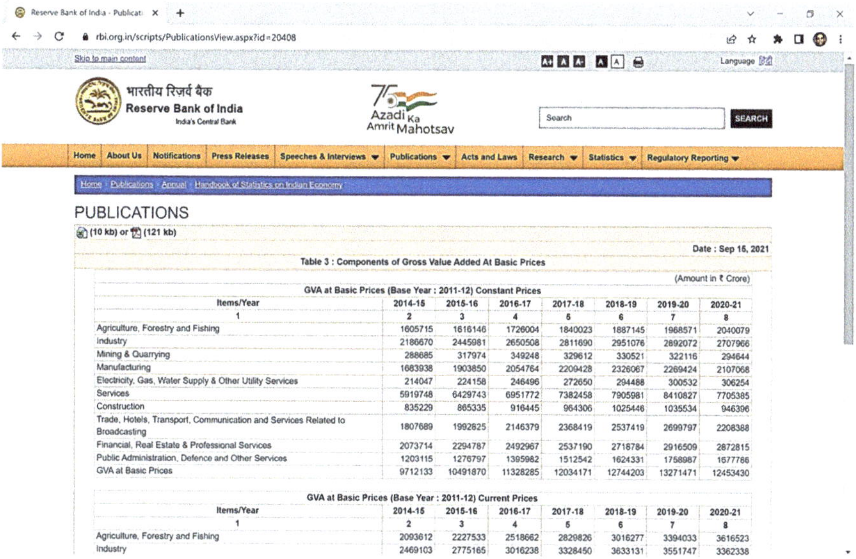

Table 3 : Components of Gross Value Added At Basic Prices

(Amount in ₹ Crore)

GVA at Basic Prices (Base Year : 2011-12) Constant Prices

Items/Year	2014-15	2015-16	2016-17	2017-18	2018-19	2019-20	2020-21
1	2	3	4	5	6	7	8
Agriculture, Forestry and Fishing	1605715	1616146	1726004	1840023	1887145	1968571	2040079
Industry	2186670	2445981	2650508	2811690	2951076	2892072	2707966
Mining & Quarrying	288685	317974	349248	329612	330521	322116	294644
Manufacturing	1683938	1903850	2054764	2209428	2326067	2269424	2107068
Electricity, Gas, Water Supply & Other Utility Services	214047	224158	246496	272650	294488	300532	306254
Services	5919748	6429743	6951772	7382458	7905981	8410827	7705385
Construction	835229	865335	916445	964306	1025446	1035534	946396
Trade, Hotels, Transport, Communication and Services Related to Broadcasting	1807689	1992825	2146379	2368419	2537419	2699797	2208388
Financial, Real Estate & Professional Services	2073714	2294787	2492967	2537190	2718784	2916509	2872815
Public Administration, Defence and Other Services	1203115	1276797	1395982	1512542	1624331	1758987	1677786
GVA at Basic Prices	9712133	10491870	11328285	12034171	12744203	13271471	12453430

GVA at Basic Prices (Base Year : 2011-12) Current Prices

Items/Year	2014-15	2015-16	2016-17	2017-18	2018-19	2019-20	2020-21
1	2	3	4	5	6	7	8
Agriculture, Forestry and Fishing	2093612	2227533	2518662	2829826	3016277	3394033	3616523
Industry	2469103	2775165	3016238	3328450	3633131	3551747	3362338

Components of gross value added at basic prices[80]
https://www.rbi.org.in/scripts/PublicationsView.aspx?id=19736

Department of Commerce

Bureau of Economic Analysis: The United States Department of Commerce

- NIPA Handbook: Concepts and Methods of the US National Income and Product Accounts
  NIPA Handbook: Concepts and Methods of the U.S. National Income and Product Accounts | U.S. Bureau of Economic Analysis (BEA)

United Nations Data Sources

UN DATA MARTS

Source: http://data.un.org/Explorer.aspx?d=FAO

[80]Components of Gross Value—https://www.rbi.org.in/scripts/PublicationsView.aspx?id=20408

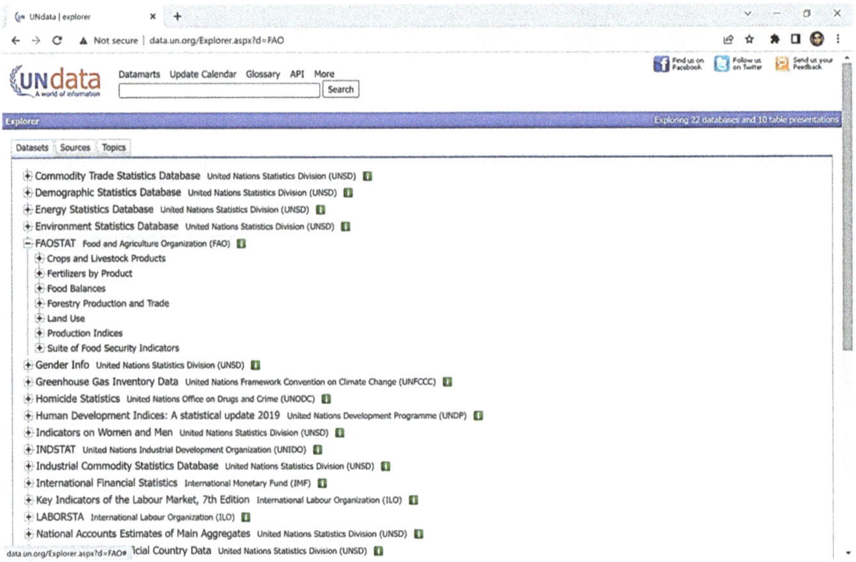

IHS Global Economy Data

IHS[81] provides a global economic database covering 200+ countries. The data includes balance of payments, cyclical indicators, finance and financial markets, government finance, housing and construction, output, capacity and capacity utilization, merchandise trade, national accounts, labor market, population, prices, and wholesale and retail trade.

US Commodities Futures Data

Investing.com futures prices[82]:

[81] IHS Global Economy Data—https://ihsmarkit.com/products/global-economic-data.html

[82] Investing.com—https://www.investing.com/commodities/us-soybeans-historical-data

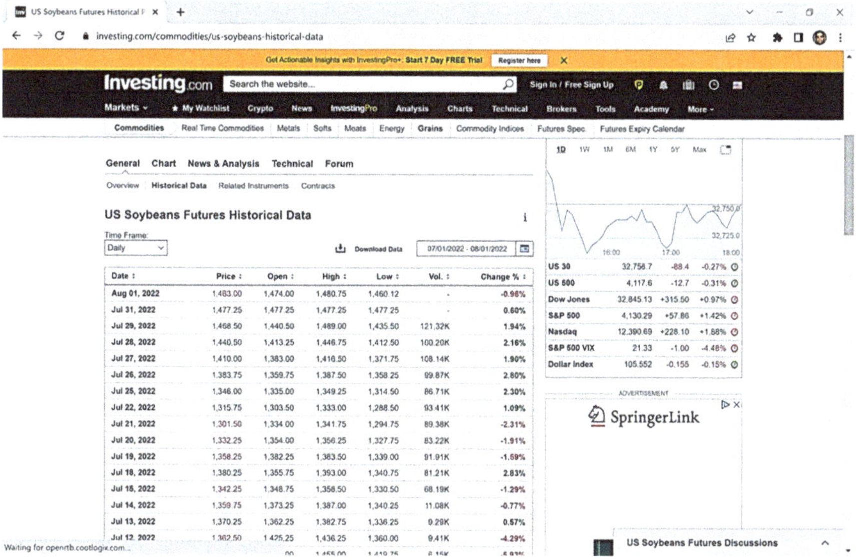

Date	Price	Open	High	Low	Vol.	Change %
Aug 01, 2022	1,463.00	1,474.00	1,480.75	1,460.12	-	-0.96%
Jul 31, 2022	1,477.25	1,477.25	1,477.25	1,477.25	-	0.60%
Jul 29, 2022	1,468.50	1,440.50	1,489.00	1,435.50	121.32K	1.94%
Jul 28, 2022	1,440.50	1,413.25	1,446.75	1,412.50	100.20K	2.16%
Jul 27, 2022	1,410.00	1,383.00	1,416.50	1,371.75	108.14K	1.90%
Jul 26, 2022	1,383.75	1,359.75	1,387.50	1,358.25	99.87K	2.80%
Jul 25, 2022	1,346.00	1,335.00	1,349.25	1,314.50	86.71K	2.30%
Jul 22, 2022	1,315.75	1,303.50	1,333.00	1,288.50	93.41K	1.09%
Jul 21, 2022	1,301.50	1,334.00	1,341.75	1,294.75	89.38K	-2.31%
Jul 20, 2022	1,332.25	1,354.00	1,356.25	1,327.75	83.22K	-1.91%
Jul 19, 2022	1,358.25	1,382.25	1,383.50	1,339.00	91.91K	-1.59%
Jul 18, 2022	1,380.25	1,355.75	1,393.00	1,340.75	81.21K	2.83%
Jul 15, 2022	1,342.25	1,348.75	1,358.50	1,330.50	68.19K	-1.29%
Jul 14, 2022	1,359.75	1,373.25	1,387.00	1,340.25	11.08K	-0.77%
Jul 13, 2022	1,370.25	1,362.25	1,382.75	1,336.25	9.29K	0.57%
Jul 12, 2022	1,362.50	1,425.25	1,436.25	1,360.00	9.41K	-4.29%

IMF Data Access to Macroeconomic and Financial Data

The International Monetary Fund (IMF[83]) covers the following topics:

- External sector
- Fiscal sector
- Financial sector
- Real sector
- Cross domain

[83] IMF—https://data.imf.org/?sk=388DFA60-1D26-4ADE-B505-A05A558D9A42&sId=1479329328660

Some of the macro-level indicators are as follows: real GDP, percentage change from previous year, CPI, percentage change from previous year, external trade, and real effective exchange rate.

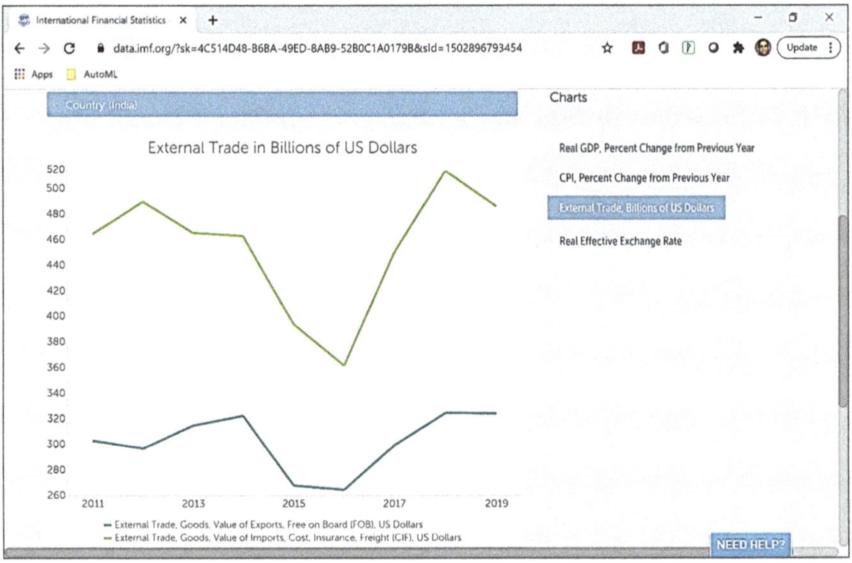

https://data.imf.org/?sk=4C514D48-B6BA-49ED-8AB9-52B0C1A0179B&sId=1502896793454

IMF Country Index Weights

Country Indexes and Weights provide macro-level consumption and social behavior of people in terms of purchase preferences, food and non-alcoholic beverages, alcoholic beverages, clothing, and others.

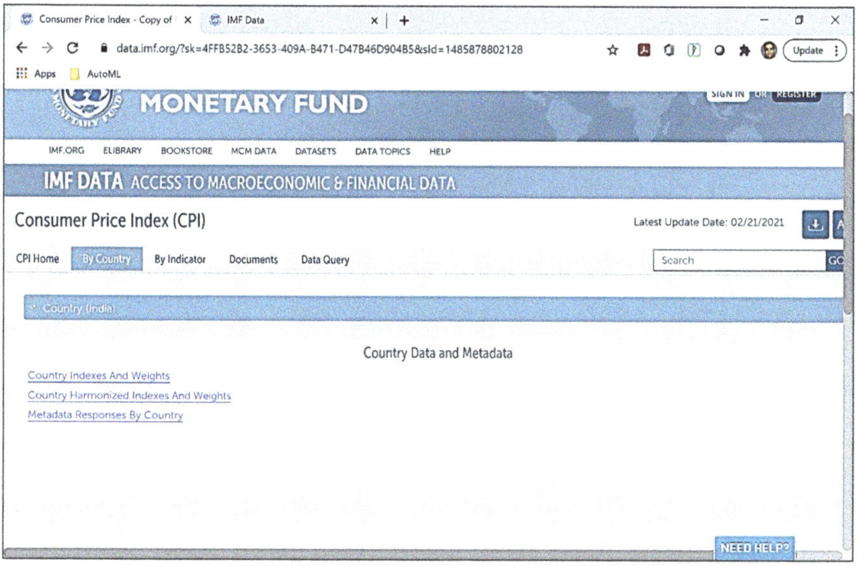

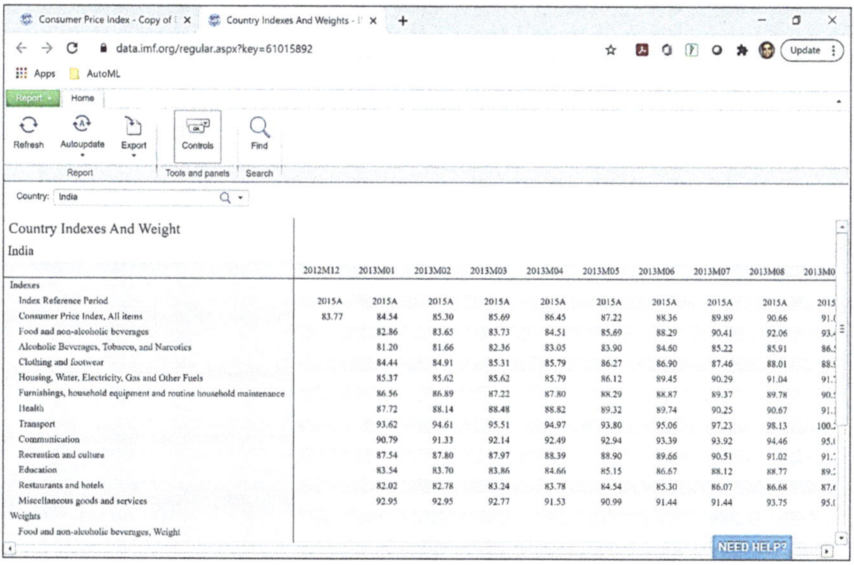

https://data.imf.org/regular.aspx?key=61015892

https://data.imf.org/?sk=388dfa60-1d26-4ade-b505-a05a558d9a42&sId=1479329328660

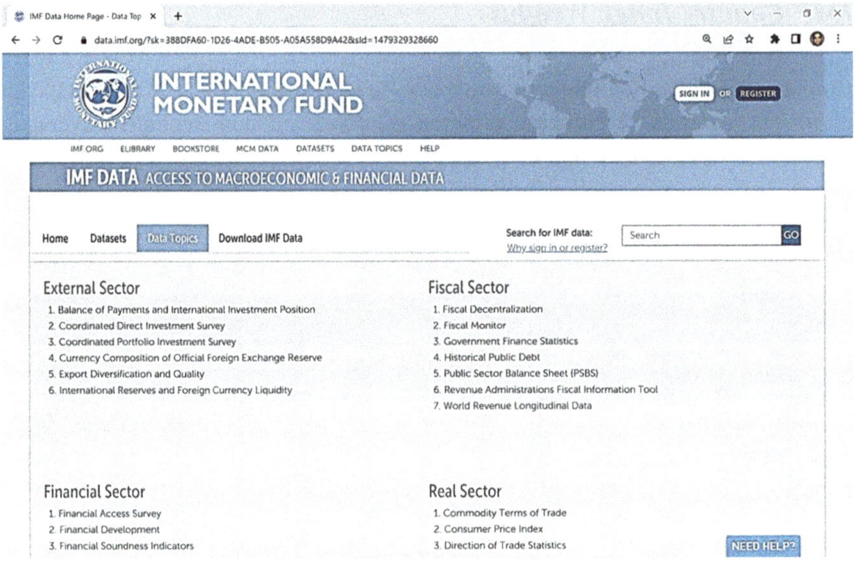

IMF Data Bulk Download

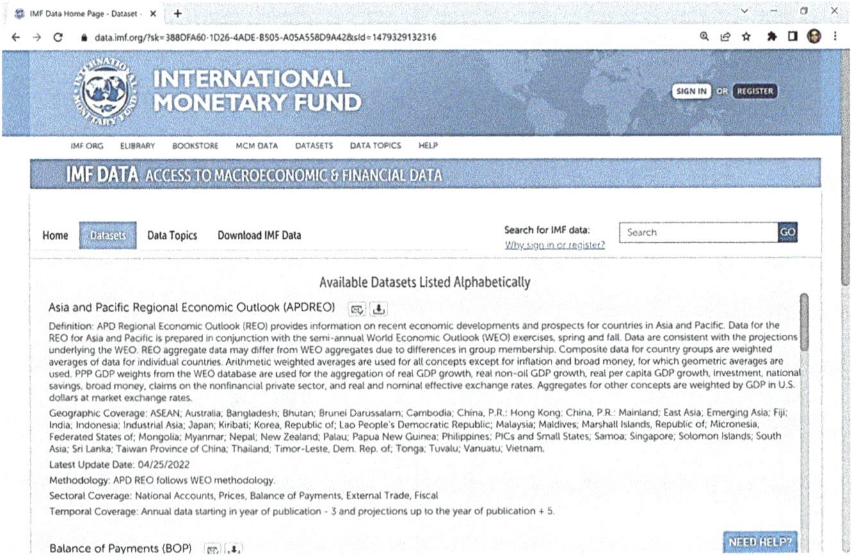

Download macroeconomic bulk data—https://data.imf.org/?sk=388DFA60-1D26-4ADE-B505-A05A558D9A42&sId=1479329132316

https://data.imf.org/?sk=388DFA60-1D26-4ADE-B505-A05A558D9A42&sId=1479329334655

World Bank Data

Macroeconomic world bank data[84]:

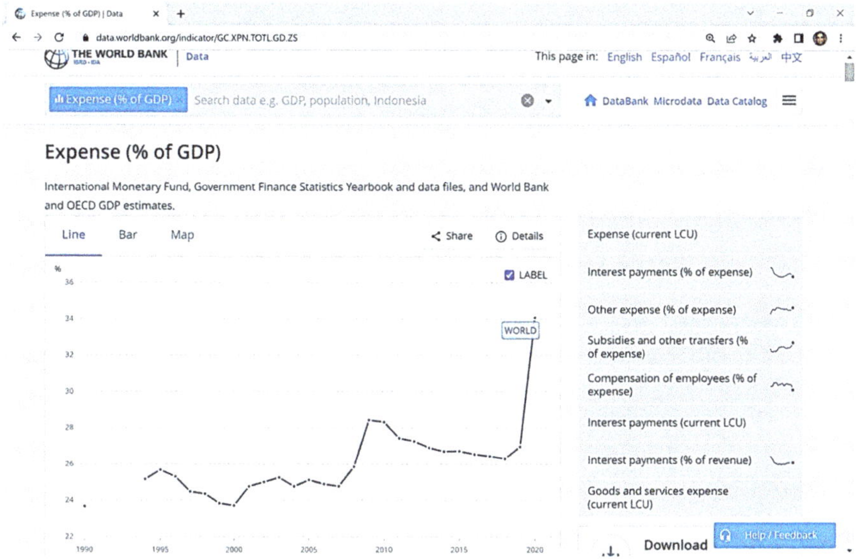

Source: https://data.worldbank.org/indicator/GC.XPN.TOTL.GD.ZS

The World Bank Development Indicators

World Development Indicator tool[85]:

[84]The World Bank Data—https://www.worldbank.org/en/topic/macroeconomics

[85]The World Bank Development Indicator—https://databank.worldbank.org/reports.aspx?source=2&type=metadata&series=SI.POV.DDAY

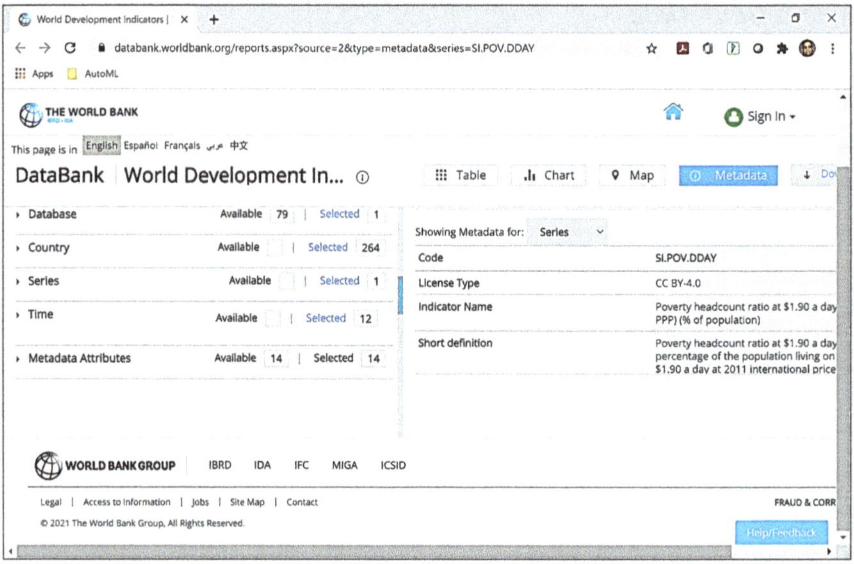

Food and Agriculture Organization of the United Nations

FAOSTAT Data[86]: FAO provides the following data domains: Production, Prices, Emissions-Agriculture, Inputs, Trade, Population, Emissions-Land Use, Investment, Macro Economic Data, and others.

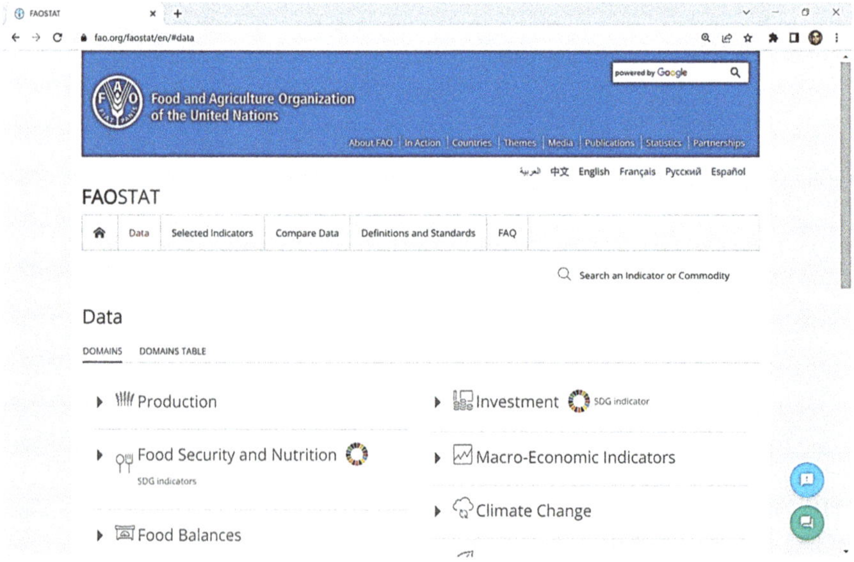

[86]FAPSTAT—http://www.fao.org/faostat/en/#data

Wheat Data Disappearance and End Stocks

This data product contains statistics on wheat—including the five classes of wheat: hard red winter, hard red spring, soft red winter, white, and durum and rye. It includes data published in the monthly Wheat Outlook and previously annual Wheat Yearbook. Data are monthly, quarterly, and/or annual, depending upon the data series. Most data are on a marketing-year basis, but some are calendar year[87].

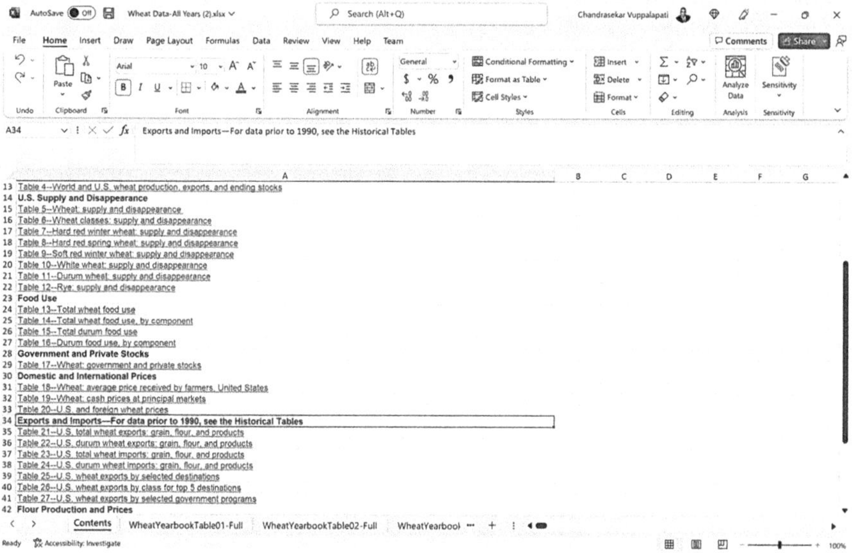

Hard red wheat contracts—https://www.barchart.com/futures/quotes/KEN20/interactive-chart

[87] Wheat Data—https://www.ers.usda.gov/data-products/wheat-data/

Poverty Thresholds for 2019 by Size of Family and Number of Related Children Under 18 Years

Size of family unit	Weighted average thresholds	Related children under 18 years								
		None	One	Two	Three	Four	Five	Six	Seven	Eight or more
One person (unrelated individual):	13,011									
Under age 65 years	13,300	13,300								
Aged 65 years and older	12,261	12,261								
Two people:	16,521									
Householder under age 65	17,196	17,120	17,622							
Householder aged 65 and older	15,468	15,453	17,555							
Three people	20,335	19,998	20,578	20,598						
Four people	26,172	26,370	26,801	25,926	26,017					
Five people	31,021	31,800	32,263	31,275	30,510	30,044				
Six people	35,129	36,576	36,721	35,965	35,239	34,161	33,522			
Seven people	40,016	42,085	42,348	41,442	40,811	39,635	38,262	36,757		
Eight people	44,461	47,069	47,485	46,630	45,881	44,818	43,470	42,066	41,709	
Nine people or more	52,875	56,621	56,895	56,139	55,503	54,460	53,025	51,727	51,406	49,426
Source: US Census Bureau										

Rice Production Manual

The Rice Production Manual provides a comprehensive view of rice and agriculture[88].

[88] Rice Production Manual—http://rice.ucanr.edu/files/288581.pdf

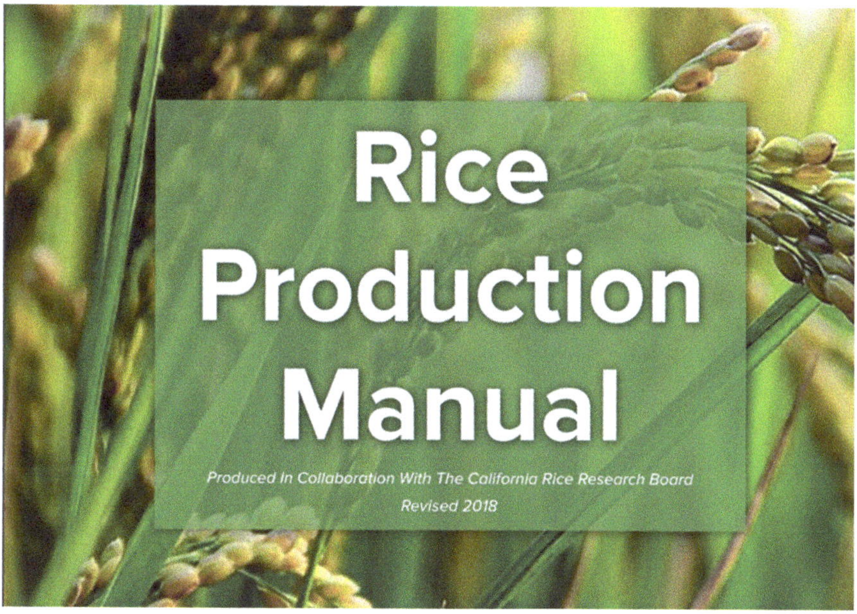
Rice
Production
Manual
Produced In Collaboration With The California Rice Research Board
Revised 2018

Lightning Source UK Ltd.
Milton Keynes UK
UKHW021200210922
409191UK00001B/1

9 783031 087424